AN INTRODUCTION T
DYNAMICS AND (

Second Edition

Symbolic dynamics is a mature yet rapidly developing area of dynamical systems. It has established strong connections with many areas, including linear algebra, graph theory, probability, group theory, and the theory of computation, as well as data storage, statistical mechanics, and C^*-algebras.

This Second Edition maintains the introductory character of the original 1995 edition as a general textbook on symbolic dynamics and its applications to coding. It is written at an elementary level and aimed at students, well-established researchers, and experts in mathematics, electrical engineering, and computer science. Topics are carefully developed and motivated with many illustrative examples. There are more than 500 exercises to test the reader's understanding. In addition to a chapter in the First Edition on advanced topics and a comprehensive bibliography, the Second Edition includes a detailed Addendum, with companion bibliography, describing major developments and new research directions since publication of the First Edition.

Douglas Lind is Professor Emeritus of Mathematics at the University of Washington. He was department chair, is an Inaugural Fellow of the American Mathematical Society, and served in many governance roles for the Mathematical Sciences Research Institute, including chairing the committee that designed the 2006 addition.

Brian Marcus is Professor of Mathematics at the University of British Columbia. He shared the 1993 Leonard Abraham Prize Paper award of the IEEE Communications Society. He is currently the UBC Site Director of the Pacific Institute for the Mathematical Sciences (PIMS) and is a Fellow of the AMS and IEEE.

CAMBRIDGE
MATHEMATICAL LIBRARY

Cambridge University Press has a long and honourable history of publishing in mathematics and counts many classics of the mathematical literature within its list. Some of these titles have been out of print for many years now and yet the methods which they espouse are still of considerable relevance today.

The *Cambridge Mathematical Library* provides an inexpensive edition of these titles in a durable paperback format and at a price that will make the books attractive to individuals wishing to add them to their own personal libraries. Certain volumes in the series have a foreword, written by a leading expert in the subject, which places the title in its historical and mathematical context.

A complete list of books in the series can be found at www.cambridge.org/mathematics. Recent titles include the following:

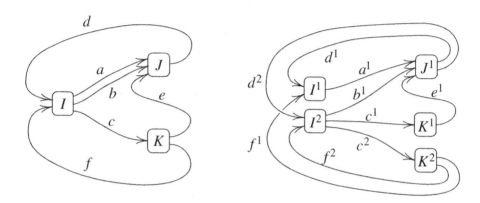

An example of state splitting, a fundamental operation in symbolic dynamics. By partitioning outgoing edges, the states of the graph on the left are split to form the graph on the right.

AN INTRODUCTION TO SYMBOLIC DYNAMICS AND CODING

Second Edition

DOUGLAS LIND

University of Washington

BRIAN MARCUS

University of British Columbia

CAMBRIDGE
UNIVERSITY PRESS

CAMBRIDGE
UNIVERSITY PRESS

University Printing House, Cambridge CB2 8BS, United Kingdom

One Liberty Plaza, 20th Floor, New York, NY 10006, USA

477 Williamstown Road, Port Melbourne, VIC 3207, Australia

314–321, 3rd Floor, Plot 3, Splendor Forum, Jasola District Centre, New Delhi – 110025, India

79 Anson Road, #06–04/06, Singapore 079906

Cambridge University Press is part of the University of Cambridge.

It furthers the University's mission by disseminating knowledge in the pursuit of
education, learning, and research at the highest international levels of excellence.

www.cambridge.org
Information on this title:www.cambridge.org/9781108820288
DOI: 10.1017/9781108899727

© Cambridge University Press 1995, 2021

This publication is in copyright. Subject to statutory exception
and to the provisions of relevant collective licensing agreements,
no reproduction of any part may take place without the written
permission of Cambridge University Press.

First published 1995
Second edition 2021

Printed in the United Kingdom by TJ Books Ltd, Padstow Cornwall

A catalogue record for this publication is available from the British Library.

ISBN 978-1-108-82028-8 Paperback

Cambridge University Press has no responsibility for the persistence or accuracy
of URLs for external or third-party internet websites referred to in this publication
and does not guarantee that any content on such websites is, or will remain,
accurate or appropriate.

Dedicated to the memory of Rufus Bowen,
our friend and teacher
1947–1978

CONTENTS

PREFACE TO FIRST EDITION

Symbolic dynamics is a rapidly growing part of dynamical systems. Although it originated as a method to study general dynamical systems, the techniques and ideas have found significant applications in data storage and transmission as well as linear algebra. This is the first general textbook on symbolic dynamics and its applications to coding, and we hope that it will stimulate both engineers and mathematicians to learn and appreciate the subject.

Dynamical systems originally arose in the study of systems of differential equations used to model physical phenomena. The motions of the planets, or of mechanical systems, or of molecules in a gas can be modeled by such systems. One simplification in this study is to discretize time, so that the state of the system is observed only at discrete ticks of a clock, like a motion picture. This leads to the study of the iterates of a single transformation. One is interested in both quantitative behavior, such as the average time spent in a certain region, and also qualitative behavior, such as whether a state eventually becomes periodic or tends to infinity. Symbolic dynamics arose as an attempt to study such systems by means of discretizing space as well as time. The basic idea is to divide up the set of possible states into a finite number of pieces, and keep track of which piece the state of the system lies in at every tick of the clock. Each piece is associated with a "symbol," and in this way the evolution of the system is described by an infinite sequence of symbols. This leads to a symbolic dynamical system that mirrors and helps us to understand the dynamical behavior of the original system.

In their fundamental paper [MorH1] written over fifty years ago, Morse and Hedlund named the subject of symbolic dynamics and described its philosophy as follows.

> *The methods used in the study of recurrence and transitivity frequently combine classical differential analysis with a more abstract symbolic analysis. This involves a characterization of the ordinary dynamical trajectory by an unending sequence of symbols termed a symbolic trajectory such that the properties of recurrence and transitivity of the dynamical trajectory are reflected in analogous properties of its symbolic trajectory.*

One example of this idea that you are very familiar with is the decimal expansion of real numbers in the unit interval $[0, 1)$. Here the transformation is given by multiplying a number $x \in [0, 1)$ by 10 and keeping the its fractional part $\{10x\}$. We partition $[0, 1)$ into ten equal subintervals $[0, 1/10), [1/10, 2/10), \ldots, [9/10, 1)$, and we use the "symbol" j for the interval $[j/10, (j+1)/10)$. Let x be a number in $[0, 1)$. Using the transformation $x \mapsto \{10x\}$ and this partition, we write down an infinite sequence of symbols corresponding to any given $x = .x_1x_2 \ldots$ as follows. Since $x \in [x_1/10, (x_1 + 1)/10)$, the first symbol is just the first digit x_1 of its decimal expansion. Since $\{10x\} \in [x_2/10, (x_2 + 1)/10)$, the next symbol is the first digit of the decimal expansion of $\{10x\}$, which is simply the second digit x_2 of the decimal expansion of x. Continuing in this way, we see that the symbolic sequence derived from x is none other than the sequence of digits from its decimal expansion.

Symbolic dynamics began when Jacques Hadamard [Had] applied this idea in 1898 to more complicated systems called geodesic flows on surfaces of negative curvature. The main point of his work is that there is a simple description of the possible sequences that can arise this way. He showed that there is a finite set of forbidden pairs of symbols, and that the possible sequences are exactly those that do not contain any forbidden pair. This is an example of one of the fundamental objects of study in symbolic dynamics called a shift of finite type. Later discoveries of Morse, Hedlund, and others in the 1920's, 1930's, and 1940's showed that in many circumstances such a finite description of the dynamics is possible. These ideas led in the 1960's and 1970's to the development of powerful mathematical tools to investigate a class of extremely interesting mappings called hyperbolic diffeomorphisms. We will see in §6.5 some examples of this.

Another source of inspiration and questions in symbolic dynamics comes from data storage and transmission. As we discuss in §2.5, when information is stored as bits on a magnetic or optical disk, there are physical reasons why it is not stored verbatim. For example, the bits on the surface of a compact audio disk are written in a long sequence obeying the constraint that between successive 1's there are at least two 0's but no more than ten 0's. How can one efficiently transform arbitrary data (such as a computer program or a Beethoven symphony) into sequences that satisfy such kinds of constraints? What are the theoretical limits of this kind of transformation? We are again confronted with a space of sequences having a finite description, and we are asking questions about such spaces and ways to encode and decode data from one space (the space of arbitrary sequences) to another (the space of constrained sequences). One of the main results in this book, the Finite-State Coding Theorem of Chapter 5, tells us when such codes are possible, and gives us an algorithm for finding them. This has led to new codes and a deeper understanding of code constructions. In particular, it has yielded a new and useful technique for code construction called the state-splitting algorithm.

While symbolic dynamics has drawn heavily on other mathematical disciplines such as linear algebra for its tools, it has also contributed new tools for other areas. For example, some deep work of Boyle and Handelman in symbolic dynamics described in Chapter 11 has led to the complete solution of the problem of characterizing the possible sets of nonzero eigenvalues of real nonnegative matrices. This solved a variant of a problem which has been perplexing linear algebraists for decades.

This book is intended to serve as an introduction to symbolic dynamics, its basic ideas, techniques, problems, and spirit. We will focus on symbolic dynamical systems that use just a finite number of symbols, on sequences of such symbols that are infinite in both directions (unlike the decimal expansion above), and spaces of sequences that have finite memory properties. Our aim is to study coding problems for such systems. In particular, we concentrate on the following:

- Find complete sets of necessary and sufficient conditions for the existence of various types of codes from one symbolic dynamical system to another.
- Completely determine the values of certain properties, invariant under a natural notion of conjugacy, such as entropy, numbers of periodic sequences, and so on.
- Explore interactions with information and coding theory, linear algebra, and general dynamical systems.

On the other hand, there are many important topics within symbolic dynamics that we do not treat in any depth in this book. Many of these topics are briefly discussed in Chapter 13.

The book is organized as follows. We start in Chapter 1 with the basic notions in symbolic dynamical systems, such as full shifts, shift spaces, irreducibility, and sliding block codes. In Chapter 2 we focus on a special class of shift spaces called shifts of finite type, and in Chapter 3 we study a generalization of these called sofic shifts. Entropy, an important numerical invariant, is treated in Chapter 4, which also includes an introduction to the fundamental Perron-Frobenius theory of nonnegative matrices. Building on the material in the first four chapters, we develop in Chapter 5 the state-splitting algorithm for code construction and prove the Finite-State Coding Theorem. Taken together, the first five chapters form a self-contained introduction to symbolic dynamics and its applications to practical coding problems.

Starting with Chapter 6 we switch gears. This chapter provides some background for the general theory of dynamical systems and the place occupied by symbolic dynamics. We show how certain combinatorial ideas like shift spaces and sliding block codes studied in earlier chapters are natural expressions of fundamental mathematical notions such as compactness and continuity. There is a natural notion of two symbolic systems being "the same," or conjugate, and Chapter 7 deals with the question of when two shifts of finite type are conjugate. This is followed in Chapters 8 and 9 with the classification of shifts of finite type and sofic shifts using weaker notions, namely finite equivalence and almost conjugacy. Chapters 7, 8, and 9 treat problems of coding between systems with equal entropy. In Chapter 10 we treat the case of unequal entropy and, in particular, determine when one shift of finite type can be embedded in another. The set of numbers which can occur as entropies of shifts of finite type is determined in Chapter 11. Also in that chapter we draw on the results of Chapter 10 to prove a partial result towards characterizing the zeta functions of shifts of finite type (the zeta function is a function which determines the numbers of periodic sequences of all periods). Some of these results in turn are used in Chapter 12 to classify shifts of finite type and sofic shifts up to a notion that sits in between conjugacy and almost conjugacy. The main result of Chapter 12 may be regarded as a generalization of the Finite-State Coding Theorem, and tells us when one sofic shift can be encoded into another in a natural way. Chapter 13 contains a survey of more advanced results and recent literature on the subject. This is a starting point for students wishing to become involved in areas of current research.

The mathematical prerequisites for this book are relatively modest. A reader who has mastered undergraduate courses in linear algebra, abstract algebra, and calculus should be well prepared. This includes upper-division undergraduate mathematics majors, mathematics graduate students, and graduate students studying information and storage systems in computer science and electrical engineering. We have tried to start from scratch with basic material that requires little background. Thus in the first five chapters no more than basic linear algebra and one-variable calculus is needed: matrices, linear transformations, similarity, eigenvalues, eigenvectors, and notions of convergence and continuity (there are also a few references to basic abstract algebra, but these are not crucial). Starting with Chapter 6, the required level of mathematical sophistication increases. We hope that the foundation laid in the first five chapters will inspire students to continue reading. Here the reader should know something more of linear algebra, including the Jordan canonical form, as well as some abstract algebra including groups, rings, fields, and ideals, and some basic notions from the topology of Euclidean space such as open set, closed set, and compact set, which are quickly reviewed.

The book contains over 500 exercises, ranging from simple checks of understanding to much more difficult problems and projects. Those that seem harder have been marked with a *. There is a detailed index as well as a notation index.

Theorems, propositions, lemmas, and other items are numbered consecutively within sections using three parts: $k.m.n$ refers to item n in section m of Chapter k. For instance, Proposition 5.2.3 refers to the third item in Section 2 of Chapter 5. Equations are numbered separately with the same three-part system, but with a different style, so that (3-2-4) refers to the fourth equation in Section 2 of Chapter 3.

We use the standard notations of $\mathbb{Z}$, $\mathbb{Q}$, $\mathbb{R}$, and $\mathbb{C}$ for the sets of integers, rationals, reals, and complexes, and use $\mathbb{Z}^+$, $\mathbb{Q}^+$, and $\mathbb{R}^+$ to denote the nonnegative elements of the corresponding sets. All matrices are assumed to be square unless stated (or context demands) otherwise.

The authors have taught courses based on preliminary versions of this material at the University of Washington, Stanford University and the University of California at Berkeley. They have also given lecture series at the Mathematical Sciences Research Center in Berkeley and the IBM Almaden Research Center. Based on these experiences, here are some suggestions on how to organize the material in this book into courses of different lengths. Chapters 1–5 would constitute a solid one-quarter course introducing symbolic dynamics from scratch and concluding with applications to coding. Chapters 1–7 could form a one-semester course that also includes more theoretical material and finishes with a detailed discussion of the conjugacy problem. The entire book can be covered in a one-year course that would bring students to the frontiers of current research. Of course, the material can be covered more quickly for students with more than the minimum required background.

This book has benefited enormously from suggestions, comments, and corrections made by many people (students, faculty and industrial researchers) on preliminary versions of this manuscript, in lectures on these versions, and in answers to our queries. These include Paul Algoet. Jonathan Ashley, Leslie Badoian, Mignon Belongie, Mike Boyle, Karen Brucks, Elise Cawley, Phil Chou, Wu Chou, Ethan Coven, Jim Fitzpatrick, Dave Forney, Lisa Goldberg, Michael Gormish, Michael Hollander, Danrun Huang, Natasha Jonoska, Sampath Kannan, Bruce Kitchens, Wolfgang Krieger, David Larsen, Nelson Markley, Lee Neuwirth, Kyewon Koh Park, Karl Petersen, Ronny Roth, Paul Siegel, Sylvia Silberger, N. T. Sindhushayana, Serge Troubetzkoy, Paul Trow, Selim Tuncel, Jeff Von Limbach, Jack Wagoner, Peter Walters, Barak Weiss, Susan Williams, and Amy Wilkinson. We are happy to thank Roy Adler, who first suggested the possibility of a textbook on symbolic dynamics, especially one that requires only a minimum amount of mathematical background. Adler's 1984 set of unpublished lecture notes from Brown University and his notes on Markov partitions were valuable sources. We are also grateful to Elza Erkip, Amos Lapidoth, and Erik Ordentlich, who earned their just desserts by working so many of the exercises in the first five chapters. Special thanks are due to Zhe-xian Wan for very detailed and helpful comments on our manuscript and to Brian Hopkins for clarifying a multitude of murky passages and for eliminating an uncountable number of needless commas. This book is typeset using the $\mathcal{A}_{\mathcal{M}}\mathcal{S}$-TEX extensions to TEX, and we thank Michael Spivak for his help with this. Finally, we wish to express our gratitude to the National Science Foundation (grants DMS-9004252 and DMS-9303240), the University of Washington, and the IBM Corporation for their support of this work.

Current information regarding this book, including selected corrections and suggestions from readers, may be found using the World Wide Web at the address https://sites.math.washington.edu/SymbolicDynamics

This address can also be used to submit items for inclusion.

The first author is profoundly grateful to John Whittier Treat for many things, both spoken and unspoken, over many years.

The second author would like to thank Yvonne, Nathaniel, Mom and Dad for a life filled with love and support, without which this book could not have been written.

Notes on the Second Printing

The Second Printing (1999) has provided us the opportunity to fix some minor errors. In addition, we have made a few changes to reflect the recent solution by Kim and Roush [KimR12] of the Shift Equivalence Problem described in §7.3.

A complete list of changes made for the Second Printing is available at the Web site displayed above. We will continue to list further corrections at this Web site.

PREFACE TO SECOND EDITION

The twenty five years since our book was written have seen enormous advances in areas related to symbolic dynamics, with long-standing problems solved and completely new research directions undergoing vigorous development. We wanted to preserve the introductory character of the original book, while providing an extended account of many of these new topics in a separate Addendum. Therefore we restricted changes in our original book to clarifications and correcting minor errors, together with noting the solutions to the Spectral Conjecture and the Road Problem.

The Addendum surveys topics that appear to us to be important and interesting. These include the characterization of the entropies of higher dimensional shifts of finite type, the pivotal use of ideas from computation theory and logic to attack dynamical problems in higher dimensional shifts, and the extension of entropy theory to actions of sofic groups. So many new and interesting developments are happening that we could not possibly be comprehensive. But we have provided an extensive separate Addendum Bibliography that we hope will guide the reader to find out more about something that catches their eye.

We are very grateful for the help we have received from Sebastián Barbieri, Lewis Bowen, Raimundo Briceño, Justin Cai, Tullio Ceccherini-Silberstein, Nishant Chandgotia, Michel Coornaert, Madeline Doering, Søren Eilers, Rafael Frongillo, Ricardo Gomez, Ben Hayes, David Kerr, Bruce Kitchens, Bryna Kra, Hanfeng Li, Tom Meyerovitch, Ronnie Pavlov, Ian Putnam, Brendan Seward, and Marcelo Sobottka.

We are especially indebted to Mike Boyle, whose list of open problems in symbolic dynamics played an essential part in structuring the Addendum, and whose encyclopedic knowledge guided us at many points to provide perspectives we hope readers will find valuable.

We are also extremely grateful to Luisa Borsato and Sophie MacDonald for their careful reading of our original text and making many excellent suggestions and corrections.

Reconstructing this book from the original TeX files (from an incomplete backup) using the L^A_MS-TeX package was only possible with the help of Ian Allison and Will Kranz.

Finally, it is a great pleasure to thank Roger Astley for his initiative in proposing this second edition, Clare Dennison for overseeing production of the book, and Irene Pizzie for her meticulous copyediting.

Douglas Lind, *Seattle*
Brian Marcus, *Vancouver*
September, 2020

CHAPTER 1

SHIFT SPACES

Shift spaces are to symbolic dynamics what shapes like polygons and curves are to geometry. We begin by introducing these spaces, and describing a variety of examples to guide the reader's intuition. Later chapters will concentrate on special classes of shift spaces, much as geometry concentrates on triangles and circles. As the name might suggest, on each shift space there is a shift map from the space to itself. Together these form a "shift dynamical system." Our main focus will be on such dynamical systems, their interactions, and their applications.

In addition to discussing shift spaces, this chapter also connects them with formal languages, gives several methods to construct new shift spaces from old, and introduces a type of mapping from one shift space to another called a sliding block code. In the last section, we introduce a special class of shift spaces and sliding block codes which are of interest in coding theory.

§1.1. Full Shifts

Information is often represented as a sequence of discrete symbols drawn from a fixed finite set. This book, for example, is really a very long sequence of letters, punctuation, and other symbols from the typographer's usual stock. A real number is described by the infinite sequence of symbols in its decimal expansion. Computers store data as sequences of 0's and 1's. Compact audio disks use blocks of 0's and 1's, representing signal samples, to digitally record Beethoven symphonies.

In each of these examples, there is a finite set $\mathcal{A}$ of *symbols* which we will call the *alphabet*. Elements of $\mathcal{A}$ are also called *letters*, and they will typically be denoted by a, b, c, ..., or sometimes by digits like 0, 1, 2, ..., when this is more meaningful. Decimal expansions, for example, use the alphabet $\mathcal{A} = \{0, 1, \ldots, 9\}$.

Although in real life sequences of symbols are finite, it is often extremely useful to treat long sequences as infinite in both directions (or *bi-infinite*).

1

This is analogous to using real numbers, continuity, and other ideas from analysis to describe physical quantities which, in reality, can be measured only with finite accuracy.

Our principal objects of study will therefore be collections of bi-infinite sequences of symbols from a finite alphabet $\mathcal{A}$. Such a sequence is denoted by $x = (x_i)_{i \in \mathbb{Z}}$, or by

$$x = \ldots x_{-2}x_{-1}x_0x_1x_2 \ldots,$$

where each $x_i \in \mathcal{A}$. The symbol x_i is the ith *coordinate* of x, and x can be thought of as being given by its coordinates, or as a sort of infinite "vector." When writing a specific sequence, you need to specify which is the 0th coordinate. This is conveniently done with a "decimal point" to separate the x_i with $i \geqslant 0$ from those with $i < 0$. For example,

$$x = \ldots 010.1101 \ldots$$

means that $x_{-3} = 0$, $x_{-2} = 1$, $x_{-1} = 0$, $x_0 = 1$, $x_1 = 1$, $x_2 = 0$, $x_3 = 1$, and so on.

Definition 1.1.1. If $\mathcal{A}$ is a finite alphabet, then the *full $\mathcal{A}$-shift* is the collection of all bi-infinite sequences of symbols from $\mathcal{A}$. The *full r-shift* (or simply *r-shift*) is the full shift over the alphabet $\{0, 1, \ldots, r-1\}$.

The full $\mathcal{A}$-shift is denoted by

$$\mathcal{A}^{\mathbb{Z}} = \{x = (x_i)_{i \in \mathbb{Z}} : x_i \in \mathcal{A} \text{ for all } i \in \mathbb{Z}\}.$$

Here $\mathcal{A}^{\mathbb{Z}}$ is the standard mathematical notation for the set of all functions from $\mathbb{Z}$ to $\mathcal{A}$, and such functions are just the bi-infinite sequences of elements from $\mathcal{A}$. Each sequence $x \in \mathcal{A}^{\mathbb{Z}}$ is called a *point* of the full shift. Points from the full 2-shift are also called *binary sequences*. If $\mathcal{A}$ has size $|\mathcal{A}| = r$, then there is a natural correspondence between the full $\mathcal{A}$-shift and the full r-shift, and sometimes the distinction between them is blurred. For example, it can be convenient to refer to the full shift on $\{+1, -1\}$ as the full 2-shift.

Blocks of consecutive symbols will play a central role. A *block* (or *word*) over $\mathcal{A}$ is a finite sequence of symbols from $\mathcal{A}$. We will write blocks without separating their symbols by commas or other punctuation, so that a typical block over $\mathcal{A} = \{a, b\}$ looks like $aababbabbb$. It is convenient to include the sequence of *no* symbols, called the *empty block* (or *empty word*) and denoted by ε. The *length* of a block u is the number of symbols it contains, and is denoted by $|u|$. Thus if $u = a_1 a_2 \ldots a_k$ is a nonempty block, then $|u| = k$, while $|\varepsilon| = 0$. A *k-block* is simply a block of length k. The set of all k-blocks over $\mathcal{A}$ is denoted $\mathcal{A}^k$. A *subblock* or *subword* of $u = a_1 a_2 \ldots a_k$ is a block of the form $a_i a_{i+1} \ldots a_j$, where $1 \leqslant i \leqslant j \leqslant k$. By convention, the empty block ε is a subblock of every block.

If x is a point in $\mathcal{A}^{\mathbb{Z}}$ and $i \leqslant j$, then we will denote the block of coordinates in x from position i to position j by

$$x_{[i,j]} = x_i x_{i+1} \ldots x_j .$$

If $i > j$, define $x_{[i,j]}$ to be ε. It is also convenient to define

$$x_{[i,j)} = x_i x_{i+1} \ldots x_{j-1} .$$

By extension, we will use the notation $x_{[i,\infty)}$ for the *right-infinite sequence* $x_i x_{i+1} x_{i+2} \ldots$, although this is not really a block since it has infinite length. Similarly, $x_{(-\infty,i]} = \ldots x_{i-2} x_{i-1} x_i$. The *central $(2k+1)$-block of x* is $x_{[-k,k]} = x_{-k} x_{-k+1} \ldots x_k$. We sometimes will write $x_{[i]}$ for x_i, especially when we want to emphasize the index i.

Two blocks u and v can be put together, or *concatenated*, by writing u first and then v, forming a new block uv having length $|uv| = |u| + |v|$. Note that uv is in general not the same as vu, although they have the same length. By convention, $\varepsilon u = u \varepsilon = u$ for all blocks u. If $n \geqslant 1$, then u^n denotes the concatenation of n copies of u, and we put $u^0 = \varepsilon$. The law of exponents $u^m u^n = u^{m+n}$ then holds for all integers $m, n \geqslant 0$. The point $\ldots uuu.uuu \ldots$ is denoted by u^∞.

The index i in a point $x = (x_i)_{i \in \mathbb{Z}}$ can be thought of as indicating time, so that, for example, the time-0 coordinate of x is x_0. The passage of time corresponds to shifting the sequence one place to the left, and this gives a map or transformation from a full shift to itself.

Definition 1.1.2. The *shift map* σ on the full shift $\mathcal{A}^{\mathbb{Z}}$ maps a point x to the point $y = \sigma(x)$ whose ith coordinate is $y_i = x_{i+1}$.

The operation σ, pictured below, maps the full shift $\mathcal{A}^{\mathbb{Z}}$ onto itself. There

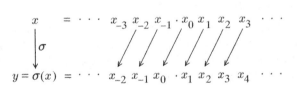

is also the inverse operation σ^{-1} of shifting one place to the right, so that σ is both one-to-one and onto. The composition of σ with itself $k > 0$ times $\sigma^k = \sigma \circ \ldots \circ \sigma$ shifts sequences k places to the left, while $\sigma^{-k} = (\sigma^{-1})^k$ shifts the same amount to the right. This shifting operation is the reason $\mathcal{A}^{\mathbb{Z}}$ is called a full shift ("full" since all sequences of symbols are allowed).

The shift map is useful for expressing many of the concepts in symbolic dynamics. For example, one basic idea is that of codes, or rules, which transform one sequence into another. For us, the most important codes are those that do not change with time. Consider the map $\phi\colon \{0,1\}^{\mathbb{Z}} \to \{0,1\}^{\mathbb{Z}}$ defined by the rule $\phi(x) = y$, where $y_i = x_i + x_{i+1}$ (mod 2). Then ϕ is a coding rule that replaces the symbol at index i with the sum modulo 2 of itself and its right neighbor. The coding operation ϕ acts the same at each coordinate, or is *stationary*, i.e., independent of time.

Another way to say this is that applying the rule ϕ and then shifting gives exactly the same result as shifting and then applying ϕ. Going through the following diagram to the right and then down gives the same result as going down and then to the right.

$$
\begin{array}{ccc}
x & \xrightarrow{\ \ \sigma\ \ } & \sigma(x) \\
\phi\downarrow & & \downarrow\phi \\
\phi(x) & \xrightarrow{\ \ \sigma\ \ } & \sigma(\phi(x)) = \phi(\sigma(x))
\end{array}
$$

We can express this as $\sigma \circ \phi = \phi \circ \sigma$, or in terms of the coordinates by $\sigma(\phi(x))_{[i]} = \phi(\sigma(x))_{[i]}$, since both equal $x_{i+1} + x_{i+2}$ (mod 2). Recall that when two mappings f and g satisfy $f \circ g = g \circ f$, they are said to *commute*. Not all pairs of mappings commute (try: $f =$ "put on socks" and $g =$ "put on shoes"). Using this terminology, a code ϕ on the full 2-shift is stationary if it commutes with the shift map σ, which we can also express by saying that the following diagram commutes.

$$
\begin{array}{ccc}
\{0,1\}^{\mathbb{Z}} & \xrightarrow{\ \ \sigma\ \ } & \{0,1\}^{\mathbb{Z}} \\
\phi\downarrow & & \downarrow\phi \\
\{0,1\}^{\mathbb{Z}} & \xrightarrow{\ \ \sigma\ \ } & \{0,1\}^{\mathbb{Z}}
\end{array}
$$

We will discuss codes in more detail in §1.5.

Points in a full shift which return to themselves after a finite number of shifts are particularly simple to describe.

Definition 1.1.3. A point x is *periodic for* σ if $\sigma^n(x) = x$ for some $n \geqslant 1$, and we say that x has *period* n under σ. If x is periodic, the smallest positive integer n for which $\sigma^n(x) = x$ is the *least period* of x. If $\sigma(x) = x$, then x is called a *fixed point* for σ.

If x has least period k, then it has period $2k, 3k, \ldots$, and every period of x is a multiple of k (see Exercise 1.1.5). A fixed point for σ must have the form a^{∞} for some symbol a, and a point of period n has the form u^{∞} for some n-block u.

Iteration of the shift map provides the "dynamics" in symbolic dynamics (see Chapter 6). Naturally, the "symbolic" part refers to the symbols used to form sequences in the spaces we will study.

EXERCISES

1.1.1. How many points $x \in \mathcal{A}^{\mathbb{Z}}$ are fixed points? How many have period n? How many have least period 12?

1.1.2. For the full $\{+1, -1\}$-shift and $k \geqslant 1$, determine the number of k-blocks having the property that the sum of the symbols is 0.

1.1.3. Let ϕ be the coding rule from this section.
  (a) Prove that ϕ maps the full 2-shift *onto* itself, i.e., that given a point y in the 2-shift, there is an x with $\phi(x) = y$.
  (b) Find the number of points x in the full 2-shift with $\phi^n(x) = 0^\infty$ for $n = 1, 2,$ or 3. Can you find this number for every n?
  *(c) Find the number of points x with $\phi^n(x) = x$ for $n = 1, 2,$ or 3. Can you find this number for every n?

1.1.4. For each k with $1 \leqslant k \leqslant 6$ find the number of k-blocks over $\mathcal{A} = \{0, 1\}$ having no two consecutive 1's appearing. Based on your result, can you guess, and then prove, what this number is for every k?

1.1.5. Determine the least period of u^∞ in terms of properties of the block u. Use your solution to show that if x has period n, then the least period of x divides n.

1.1.6. (a) Describe those pairs of blocks u and v over an alphabet $\mathcal{A}$ such that $uv = vu$.
  *(b) Describe those sequences $u_1, u_2, \ldots, u_n$ of n blocks for which all n concatenations $u_1 u_2 \ldots u_n,\ u_2 \ldots u_n u_1,\ \ldots,\ u_n u_1 u_2 \ldots u_{n-1}$ of the cyclic permutations are equal.

§1.2. Shift Spaces

The symbol sequences we will be studying are often subject to constraints. For example, Morse code uses the symbols "dot," "dash," and "pause." The ordinary alphabet is transmitted using blocks of dots and dashes with length at most six separated by a pause, so that any block of length at least seven which contains no pause is forbidden to occur (the only exception is the SOS signal). In the programming language Pascal, a program line such as `sin(x)***2 := y` is not allowed, nor are lines with unbalanced parentheses, since they violate Pascal's syntax rules. The remarkable error correction in compact audio disks results from the use of special kinds of binary sequences specified by a finite number of conditions. In this section we introduce the fundamental notion of shift space, which will be the subset of points in a full shift satisfying a fixed set of constraints.

If $x \in \mathcal{A}^{\mathbb{Z}}$ and w is a block over $\mathcal{A}$, we will say that w *occurs in* x if there are indices i and j so that $w = x_{[i,j]}$. Note that the empty block ε occurs in every x, since $\varepsilon = x_{[1,0]}$. Let $\mathcal{F}$ be a collection of blocks over $\mathcal{A}$, which we will think of as being the *forbidden blocks*. For any such $\mathcal{F}$, define $X_{\mathcal{F}}$ to be the subset of sequences in $\mathcal{A}^{\mathbb{Z}}$ which do *not* contain any block in $\mathcal{F}$.

Definition 1.2.1. A *shift space* (or simply *shift*) is a subset X of a full shift $\mathcal{A}^{\mathbb{Z}}$ such that $X = X_{\mathcal{F}}$ for some collection $\mathcal{F}$ of forbidden blocks over $\mathcal{A}$.

The collection $\mathcal{F}$ may be finite or infinite. In any case it is at most countable since its elements can be arranged in a list (just write down its blocks of length 1 first, then those of length 2, and so on). For a given shift space there may be many collections $\mathcal{F}$ describing it (see Exercise 1.2.4). Note that the empty set $\varnothing$ is a shift space, since putting $\mathcal{F} = \mathcal{A}$ rules out every point. When a shift space X is contained in a shift space Y, we say that X is a *subshift* of Y.

In the equation $X = \mathsf{X}_\mathcal{F}$, the notation X refers to the operation of forming a shift space, while X denotes the resulting set. We will sometimes use similar typographical distinctions between an operation and its result, for example in §2.2 when forming an adjacency matrix from a graph. By use of such distinctions, we hope to avoid the type of nonsensical equations such as "$y = y(x)$" you may have seen in calculus classes.

Example 1.2.2. X is $\mathcal{A}^{\mathbb{Z}}$, where we can take $\mathcal{F} = \varnothing$, reflecting the fact that there are no constraints. □

Example 1.2.3. X is the set of all binary sequences with no two 1's next to each other. Here $X = \mathsf{X}_\mathcal{F}$, where $\mathcal{F} = \{11\}$. This shift is called the *golden mean shift* for reasons which will surface in Chapter 4. □

Example 1.2.4. X is the set of all binary sequences so that between any two 1's there are an even number of 0's. We can take for $\mathcal{F}$ the collection

$$\{10^{2n+1}1 : n \geqslant 0\}.$$

This example is naturally called the *even shift*. □

In the following examples, the reader will find it instructive to list an appropriate collection $\mathcal{F}$ of forbidden blocks for which $X = \mathsf{X}_\mathcal{F}$.

Example 1.2.5. X is the set of all binary sequences for which 1's occur infinitely often in each direction, and such that the number of 0's between successive occurrences of a 1 is either 1, 2, or 3. This shift is used in a common data storage method for hard disk drives (see §2.5). For each pair (d, k) of nonnegative integers with $d \leqslant k$, there is an analogous (d, k) *run-length limited shift*, denoted by $X(d, k)$, and defined by the constraints that 1's occur infinitely often in each direction, and there are at least d 0's, but no more than k 0's, between successive 1's. Using this notation, our example is $X(1, 3)$. □

Example 1.2.6. To generalize the previous examples, fix a nonempty subset S of $\{0, 1, 2, \dots\}$. If S is finite, define $X = X(S)$ to be the set of all binary sequences for which 1's occur infinitely often in each direction, and such that the number of 0's between successive occurrences of a 1 is an integer in S. Thus a typical point in $X(S)$ has the form

$$x = \dots 1\, 0^{n_{-1}}\, 1\, 0^{n_0}\, 1\, 0^{n_1}\, 1 \dots,$$

where each $n_j \in S$. For example, the (d, k) run-length limited shift corresponds to $S = \{d, d+1, \ldots, k\}$.

When S is infinite, it turns out that to obtain a shift space we need to allow points that begin or end with an infinite string of 0's (see Exercise 1.2.8). In this case, we define $X(S)$ the same way as when S is finite, except that we do *not* require that 1's occur infinitely often in each direction. In either case, we refer to $X(S)$ as the *S-gap shift*.

Observe that the full 2-shift is the S-gap shift with $S = \{0, 1, 2, \ldots\}$, the golden mean shift corresponds to $S = \{1, 2, 3, \ldots\}$, and the even shift to $S = \{0, 2, 4, \ldots\}$. As another example, for $S = \{2, 3, 5, 7, 11, \ldots\}$ the set of primes, we call $X(S)$ the *prime gap shift*. $\square$

Example 1.2.7. For each positive integer c, the *charge constrained shift*, is defined as the set of all points in $\{+1, -1\}^{\mathbb{Z}}$ so that for every block occurring in the point, the algebraic sum s of the $+1$'s and -1's satisfies $-c \leqslant s \leqslant c$. These shifts arise in engineering applications and often go by the name "DC-free sequences." See Immink [Imm2, Chapter 6]. $\square$

Example 1.2.8. Let $\mathcal{A} = \{e, f, g\}$, and X be the set of points in the full $\mathcal{A}$-shift for which e can be followed only by e or f, f can be followed only by g, and g can be followed only by e or f. A point in this space is then just a bi-infinite path on the graph shown in Figure 1.2.1 This is an example of a *shift of finite type*. These shifts are the focus of the next chapter. $\square$

Example 1.2.9. X is the set of points in the full shift $\{a, b, c\}^{\mathbb{Z}}$ so that a block of the form $ab^m c^k a$ may occur in the point only if $m = k$. We will refer to this example as the *context-free shift*. $\square$

You can make up infinitely many shift spaces by using different forbidden collections $\mathcal{F}$. Indeed, there are uncountably many shift spaces possible (see Exercise 1.2.12). As subsets of full shifts, these spaces share a common feature called *shift invariance*. This amounts to the observation that the constraints on points are given in terms of forbidden blocks alone, and do not involve the coordinate at which a block might be forbidden. It follows that if x is in $X_{\mathcal{F}}$, then so are its shifts $\sigma(x)$ and $\sigma^{-1}(x)$. This can be neatly expressed as $\sigma(X_{\mathcal{F}}) = X_{\mathcal{F}}$. The *shift map* σ_X on X is the restriction to X of the shift map σ on the full shift.

This shift invariance property allows us to find subsets of a full shift that are not shift spaces. One simple example is the subset X of $\{0, 1\}^{\mathbb{Z}}$

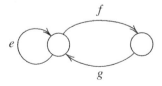

FIGURE 1.2.1. A graph defining a shift space.

consisting of the single point

$$x = \ldots 0101.0101 \ldots = (01)^\infty.$$

Since $\sigma(x) = (10)^\infty \notin X$, we see that X is not shift invariant, so it is not a shift space.

However, shift invariance alone is not enough to have a shift space. What is missing is a sort of "closure" (see Corollary 1.3.5 and Theorem 6.1.21). This is illustrated by the following example.

Example 1.2.10. Let $X \subseteq \{0,1\}^{\mathbb{Z}}$ be the set of points each of which contains exactly one symbol 1 and the rest 0's. Clearly X is shift invariant. If X were a shift space, then no block of 0's could be forbidden. But then the point $0^\infty = \ldots 000.000 \ldots$ would necessarily belong to X, whereas it does not. The set X lacks the "closure" necessary for a shift space.   □

Since a shift space X is contained in a full shift, Definition 1.1.3 serves to define what it means for $x \in X$ to be fixed or periodic under σ_X. However, unlike full shifts and many of the examples we have introduced, there are shift spaces that contain no periodic points at all (Exercise 1.2.13).

EXERCISES

1.2.1. Find a collection $\mathcal{F}$ of blocks over $\{0,1\}$ so that $X_{\mathcal{F}} = \varnothing$.

1.2.2. For Examples 1.2.5 through 1.2.9 find a set of forbidden blocks describing the shift space.

1.2.3. Let X be the subset of $\{0,1\}^{\mathbb{Z}}$ described in Example 1.2.10. Show that $X \cup \{0^\infty\}$ is a shift space.

1.2.4. Find two collections $\mathcal{F}_1$ and $\mathcal{F}_2$ over $\mathcal{A} = \{0,1\}$ with $X_{\mathcal{F}_1} = X_{\mathcal{F}_2} \neq \varnothing$, where $\mathcal{F}_1$ is finite and $\mathcal{F}_2$ is infinite.

1.2.5. Show that $X_{\mathcal{F}_1} \cap X_{\mathcal{F}_2} = X_{\mathcal{F}_1 \cup \mathcal{F}_2}$. Use this to prove that the intersection of two shift spaces over the same alphabet is also a shift space. Extend this to arbitrary intersections.

1.2.6. Show that if $\mathcal{F}_1 \subseteq \mathcal{F}_2$, then $X_{\mathcal{F}_1} \supseteq X_{\mathcal{F}_2}$. What is the relationship between $X_{\mathcal{F}_1} \cup X_{\mathcal{F}_2}$ and $X_{\mathcal{F}_1 \cap \mathcal{F}_2}$?

1.2.7. Let X be the full $\mathcal{A}$-shift.
  (a) Show that if X_1 and X_2 are shift spaces such that $X_1 \cup X_2 = X$, then $X_1 = X$ or $X_2 = X$ (or both).
  (b) Extend your argument to show that if X is the union of any collection $\{X_\alpha\}$ of shift spaces, then there is an α such that $X = X_\alpha$.
  (c) Explain why these statements no longer hold if we merely assume that X is a shift space.

1.2.8. If S is an infinite subset of $\{0, 1, 2, \ldots\}$, show that the collection of all binary sequences of the form

$$x = \ldots 1\, 0^{n-1}\, 1\, 0^{n_0}\, 1\, 0^{n_1}\, 1 \ldots,$$

where each $n_j \in S$, is not a shift space.

1.2.9. Let X_i be a shift over $\mathcal{A}_i$ for $i = 1, 2$. The *product shift* $X = X_1 \times X_2$ consists of all pairs $(x^{(1)}, x^{(2)})$ with $x^{(i)} \in X_i$. If we identify a pair (x, y) of sequences with the sequence $(\ldots (x_{-1}, y_{-1}), (x_0, y_0), (x_1, y_1), \ldots)$ of pairs, we can regard $X_1 \times X_2$ as a subset of $(\mathcal{A}_1 \times \mathcal{A}_2)^{\mathbb{Z}}$. With this convention, show that $X_1 \times X_2$ is a shift space over the alphabet $\mathcal{A}_1 \times \mathcal{A}_2$.

1.2.10. Let X be a shift space, and $N \geqslant 1$. Show that there is a collection $\mathcal{F}$ of blocks, all of which have length at least N, so that $X = X_{\mathcal{F}}$.

1.2.11. For which sets S does the S-gap shift have infinitely many periodic points?

1.2.12. Show there are uncountably many shift spaces contained in the full 2-shift. [*Hint*: Consider S-gap shifts.]

*1.2.13. Construct a nonempty shift space that does not contain any periodic points.

*1.2.14. For a given alphabet $\mathcal{A}$, let

$$X = \{x \in \mathcal{A}^{\mathbb{Z}} : x_{i+n^2} \neq x_i \text{ for all } i \in \mathbb{Z} \text{ and } n \geqslant 1\}.$$

(a) If $|\mathcal{A}| = 2$, prove that $X = \varnothing$.
(b) If $|\mathcal{A}| = 3$, show that $X = \varnothing$. [*Hint*: $3^2 + 4^2 = 5^2$.]

§1.3. Languages

It is sometimes easier to describe a shift space by specifying which blocks are allowed, rather than which are forbidden. This leads naturally to the notion of the language of a shift.

Definition 1.3.1. Let X be a subset of a full shift, and let $\mathcal{B}_n(X)$ denote the set of all n-blocks that occur in points in X. The *language of X* is the collection

$$\mathcal{B}(X) = \bigcup_{n=0}^{\infty} \mathcal{B}_n(X).$$

Example 1.3.2. The full 2-shift has language

$$\{\varepsilon, 0, 1, 00, 01, 10, 11, 000, 001, 010, 011, 100, \ldots\}. \qquad \square$$

Example 1.3.3. The golden mean shift (Example 1.2.3) has language

$$\{\varepsilon, 0, 1, 00, 01, 10, 000, 001, 010, 100, 101, 0000, \ldots\}. \qquad \square$$

The term "language" comes from the theory of automata and formal languages. See [HopU] for a lucid introduction to these topics. Think of the language $\mathcal{B}(X)$ as the collection of "allowed" blocks in X. For a block $u \in \mathcal{B}(X)$, we sometimes use alternative terminology such as saying that u *occurs in X* or *appears in X* or *is in X* or *is allowed in X*.

Not every collection of blocks is the language of a shift space. The following proposition characterizes those which are, and shows that they provide an alternative description of a shift space. In what follows we will denote the complement of a collection $\mathcal{C}$ of blocks over $\mathcal{A}$ relative to the collection of all blocks over $\mathcal{A}$ by $\mathcal{C}^c$.

Proposition 1.3.4.

(1) *Let X be a shift space, and $\mathcal{L} = \mathcal{B}(X)$ be its language. If $w \in \mathcal{L}$, then*

    (a) *every subblock of w belongs to $\mathcal{L}$, and*

    (b) *there are nonempty blocks u and v in $\mathcal{L}$ so that $uwv \in \mathcal{L}$.*

(2) *The languages of shift spaces are characterized by (1). That is, if $\mathcal{L}$ is a collection of blocks over $\mathcal{A}$, then $\mathcal{L} = \mathcal{B}(X)$ for some shift space X if and only if $\mathcal{L}$ satisfies condition (1).*

(3) *The language of a shift space determines the shift space. In fact, for any shift space, $X = X_{\mathcal{B}(X)^c}$. Thus two shift spaces are equal if and only if they have the same language.*

PROOF: (1) If $w \in \mathcal{L} = \mathcal{B}(X)$, then w occurs in some point x in X. But then every subblock of w also occurs in x, so is in $\mathcal{L}$. Furthermore, clearly there are nonempty blocks u and v such that uwv occurs in x, so that $u, v \in \mathcal{L}$ and $uwv \in \mathcal{L}$.

(2) Let $\mathcal{L}$ be a collection of blocks satisfying (1), and X denote the shift space $X_{\mathcal{L}^c}$. We will show that $\mathcal{L} = \mathcal{B}(X)$. For if $w \in \mathcal{B}(X)$, then w occurs in some point of $X_{\mathcal{L}^c}$, so that $w \notin \mathcal{L}^c$, or $w \in \mathcal{L}$. Thus $\mathcal{B}(X) \subseteq \mathcal{L}$. Conversely, suppose that $w = x_0 x_1 \ldots x_m \in \mathcal{L}$. Then by repeatedly applying (1b), we can find symbols x_j with $j > m$ and x_i with $i < 0$ so that by (1a) every subblock of $x = (x_i)_{i \in \mathbb{Z}}$ lies in $\mathcal{L}$. This means that $x \in X_{\mathcal{L}^c}$. Since w occurs in x, we have that $w \in \mathcal{B}(X_{\mathcal{L}^c}) = \mathcal{B}(X)$, proving that $\mathcal{L} \subseteq \mathcal{B}(X)$.

(3) If $x \in X$, no block occurring in x is in $\mathcal{B}(X)^c$ since $\mathcal{B}(X)$ contains all blocks occurring in all points of X. Hence $x \in X_{\mathcal{B}(X)^c}$, showing that $X \subseteq X_{\mathcal{B}(X)^c}$. Conversely, since X is a shift there is a collection $\mathcal{F}$ for which $X = X_{\mathcal{F}}$. If $x \in X_{\mathcal{B}(X)^c}$, then every block in x must be in $\mathcal{B}(X) = \mathcal{B}(X_{\mathcal{F}})$, and so cannot be in $\mathcal{F}$. Hence $x \in X_{\mathcal{F}}$, proving that $X = X_{\mathcal{F}} \supseteq X_{\mathcal{B}(X)^c}$. $\square$

This result shows that although a shift X can be described by different collections of forbidden blocks, there is a largest collection $\mathcal{B}(X)^c$, the complement of the language of X. This is the largest possible forbidden collection that describes X. For a minimal forbidden collection, see Exercise 1.3.8. The proposition also gives a one-to-one correspondence between shifts X and languages $\mathcal{L}$ that satisfy (1). This correspondence can be summarized by the equations

(1–3–1) $$\mathcal{L} = \mathcal{B}(X_{\mathcal{L}^c}), \qquad X = X_{\mathcal{B}(X)^c}.$$

A useful consequence of part (3) above is that to verify that a point x is in a given shift space X, you only need to show that each subblock $x_{[i,j]}$ is in $\mathcal{B}(X)$. In fact, this gives a characterization of shift spaces in terms of "allowed" blocks.

Corollary 1.3.5. *Let X be a subset of the full $\mathcal{A}$-shift. Then X is a shift space if and only if whenever $x \in \mathcal{A}^{\mathbb{Z}}$ and each $x_{[i,j]} \in \mathcal{B}(X)$ then $x \in X$.*

PROOF: The "whenever" condition is equivalent to the condition that $X = X_{\mathcal{B}(X)^c}$. Thus the corollary follows from Proposition 1.3.4(3). $\quad\square$

If X is a shift space, the first part of Proposition 1.3.4 shows that every block $w \in \mathcal{B}(X)$ can be extended on both sides to another block $uwv \in \mathcal{B}(X)$. However, given two blocks u and v in $\mathcal{B}(X)$, it may not be possible to find a block w so that $uwv \in \mathcal{B}(X)$. For example, let $X = \{0^\infty, 1^\infty\} \subseteq \{0,1\}^{\mathbb{Z}}$, and $u = 0$, $v = 1$. Shift spaces for which two blocks can always be "joined" by a third play a special and important role.

Definition 1.3.6. A shift space X is *irreducible* if for every ordered pair of blocks $u, v \in \mathcal{B}(X)$ there is a $w \in \mathcal{B}(X)$ so that $uwv \in \mathcal{B}(X)$.

Note that if u, v is an ordered pair of blocks in $\mathcal{B}(X)$, then so is v, u. Thus to verify that X is irreducible, we must be able to find blocks w_1 and w_2 so that both uw_1v and vw_2u are in $\mathcal{B}(X)$.

The reader should verify that Examples 1.2.2 through 1.2.9 are irreducible. Indeed, most shift spaces we encounter will be irreducible. Those which are not can usually be decomposed into irreducible "pieces," and the theory we develop for irreducible shifts can then be applied to each piece.

There are close connections between symbolic dynamics and the theory of formal languages. For example, special shift spaces called sofic shifts that we will explore in Chapter 3 correspond to regular languages, i.e., those languages accepted by a finite-state automaton. In addition, ideas and techniques from formal languages can sometimes be used in symbolic dynamics. For instance, the idea behind the Pumping Lemma for regular languages is used in Example 3.1.7 to prove that the context-free shift is not sofic. See the notes section of Chapter 3 for more on these ideas.

EXERCISES

1.3.1. Determine the language of the full shift, the even shift (Example 1.2.4), the $(1,3)$ run-length limited shift (Example 1.2.5), and the charge constrained shift (Example 1.2.7).

1.3.2. If $\mathcal{L}_1$ and $\mathcal{L}_2$ are languages satisfying condition 1 of Proposition 1.3.4, show that $\mathcal{L}_1 \cup \mathcal{L}_2$ also satisfies the condition. Use this to prove that the union of two shift spaces is also a shift space. If $\mathcal{L}_1, \mathcal{L}_2, \mathcal{L}_3, \ldots$ are languages over the same alphabet, show that $\bigcup_{n=1}^{\infty} \mathcal{L}_n$ is also a language. Why can't you use this to prove that the union of an infinite number of shift spaces over the same alphabet is also a shift space?

1.3.3. Is the intersection of the languages of two shift spaces also the language of a shift space?

1.3.4. If X and Y are shift spaces, describe the languages of their intersection $X \cap Y$ and of their product $X \times Y$ (defined in Exercise 1.2.9).

1.3.5. Is the intersection of two irreducible shift spaces always irreducible? The product (defined in Exercise 1.2.9)?

1.3.6. Let $\mathcal{A} = \{0,1\}$ and $\mathcal{F} = \{01\}$. Is $X_{\mathcal{F}}$ irreducible?

1.3.7. Let X be an irreducible shift space. Show that for every ordered pair of blocks $u, v \in \mathcal{B}(X)$, there is a *nonempty* block $w \in \mathcal{B}(X)$ such that $uwv \in \mathcal{B}(X)$.

*1.3.8. Let X be a shift space. Call a word w a "first offender" for X if $w \notin \mathcal{B}(X)$, but every proper subword of w is in $\mathcal{B}(X)$. Let $\mathcal{O}$ be the collection of all first offenders for X.

    (a) Prove that $X = X_{\mathcal{O}}$.

    (b) If $X = X_{\mathcal{F}}$, show that for every $w \in \mathcal{O}$ there is a $v \in \mathcal{F}$ such that w is a subword of v, but v contains no other first offenders.

    (c) Use (b) to show that $\mathcal{O}$ is a *minimal* forbidden set, in the sense that if $\mathcal{F} \subseteq \mathcal{O}$ and $X_{\mathcal{F}} = X$, then $\mathcal{F} = \mathcal{O}$.

§1.4. Higher Block Shifts and Higher Power Shifts

One of the basic constructions in symbolic dynamics involves widening our attention from a single symbol to a block of consecutive symbols, and considering such blocks as letters from a new, more elaborate alphabet. This process, which we will call "passing to a higher block shift," is a very convenient technical device, and we will be using it often. It provides an alternative description of the same shift space.

Let X be a shift space over the alphabet $\mathcal{A}$, and $\mathcal{A}_X^{[N]} = \mathcal{B}_N(X)$ be the collection of all allowed N-blocks in X. We can consider $\mathcal{A}_X^{[N]}$ as an alphabet in its own right, and form the full shift $(\mathcal{A}_X^{[N]})^{\mathbb{Z}}$. Define the *Nth higher block code* $\beta_N \colon X \to (\mathcal{A}_X^{[N]})^{\mathbb{Z}}$ by

$$(1\text{--}4\text{--}1) \qquad\qquad (\beta_N(x))_{[i]} = x_{[i, i+N-1]}.$$

Thus β_N replaces the ith coordinate of x with the block of coordinates in x of length N starting at position i. This becomes clearer if we imagine the symbols in $\mathcal{A}_X^{[N]}$ as written vertically. Then the image of $x = (x_i)_{i \in \mathbb{Z}}$ under β_4 has the form

(1–4–2)

$$\beta_4(x) = \ldots \begin{bmatrix} x_0 \\ x_{-1} \\ x_{-2} \\ x_{-3} \end{bmatrix} \begin{bmatrix} x_1 \\ x_0 \\ x_{-1} \\ x_{-2} \end{bmatrix} \begin{bmatrix} x_2 \\ x_1 \\ x_0 \\ x_{-1} \end{bmatrix} . \begin{bmatrix} x_3 \\ x_2 \\ x_1 \\ x_0 \end{bmatrix} \begin{bmatrix} x_4 \\ x_3 \\ x_2 \\ x_1 \end{bmatrix} \begin{bmatrix} x_5 \\ x_4 \\ x_3 \\ x_2 \end{bmatrix} \ldots \in (\mathcal{A}_X^{[4]})^{\mathbb{Z}}.$$

Definition 1.4.1. Let X be a shift space. Then the Nth *higher block shift* $X^{[N]}$ or *higher block presentation* of X is the image $X^{[N]} = \beta_N(X)$ in the full shift over $\mathcal{A}_X^{[N]}$.

Notice that in (1–4–2) consecutive symbols from $\mathcal{A}_X^{[N]}$ overlap. If $u = u_1 u_2 \ldots u_N$ and $v = v_1 v_2 \ldots v_N$ are N-blocks, let us say that u and v *overlap progressively* if $u_2 u_3 \ldots u_N = v_1 v_2 \ldots v_{N-1}$. If the 2-block uv over the alphabet $\mathcal{A}_X^{[N]}$ occurs in some image point $\beta_N(x)$, then a glance at (1–4–2) shows that u and v must overlap progressively. Also observe from

(1–4–2) that by knowing the bottom letter in each symbol of $\beta_N(x)$ we can reconstruct the entire image, as well as the original point x. In this sense $X^{[N]}$ is simply another description of the same shift space X.

Example 1.4.2. Let X be the golden mean shift of Example 1.2.3. Then

$$\mathcal{A}_X^{[2]} = \{a = 00, b = 01, c = 10\},$$

and $X^{[2]}$ is described by the constraints $\mathcal{F} = \{ac, ba, bb, cc\}$. Each of these 2-blocks is forbidden since they fail to overlap progressively. For example, the second symbol of $a = 00$ does not match the first of $c = 10$, so ac is forbidden. Naturally, the block 11 is also forbidden, since it is forbidden in the original shift. This is expressed by its absence from $\mathcal{A}_X^{[2]}$. □

The terminology "higher block shift" implies that it is a shift space. We can verify this as follows.

Proposition 1.4.3. *The higher block shifts of a shift space are also shift spaces.*

PROOF: Let X be a shift space over $\mathcal{A}$, and $N \geqslant 1$. Then there is a collection $\mathcal{F}$ of blocks over $\mathcal{A}$ so that $X = X_{\mathcal{F}}$. Create a new collection $\widetilde{\mathcal{F}}$ by replacing each block u in $\mathcal{F}$ such that $|u| < N$ by all N-blocks over $\mathcal{A}$ containing u. Then clearly $X = X_{\widetilde{\mathcal{F}}}$, and every block in $\widetilde{\mathcal{F}}$ has length $\geqslant N$. (See Exercise 1.2.10.)

For each $w = a_1 a_2 \ldots a_m \in \widetilde{\mathcal{F}}$ let

$$w^{[N]} = (a_1 a_2 \ldots a_N)(a_2 a_3 \ldots a_{N+1}) \ldots (a_{m-N+1} a_{m-N+2} \ldots a_m)$$

be the corresponding $(m - N + 1)$-block over $\mathcal{A}^N$. Let $\mathcal{F}_1$ denote the set of all blocks over the alphabet $\mathcal{A}^N$ of the form $w^{[N]}$ for some $w \in \widetilde{\mathcal{F}}$. This represents one set of constraints on $X^{[N]}$, namely those coming from the constraints on the original shift. It follows that $X^{[N]} \subseteq X_{\mathcal{F}_1}$.

Points in $X^{[N]}$ also satisfy the overlap condition illustrated in (1–4–2). Thus we let

$$\mathcal{F}_2 = \{uv : u \in \mathcal{A}^N, v \in \mathcal{A}^N, \text{ and } u \text{ and } v \text{ do not overlap progressively}\}.$$

Then $X^{[N]} \subseteq X_{\mathcal{F}_2}$, so that by Exercise 1.2.5

$$X^{[N]} \subseteq X_{\mathcal{F}_1} \cap X_{\mathcal{F}_2} = X_{\mathcal{F}_1 \cup \mathcal{F}_2}.$$

Conversely, suppose that $y \in X_{\mathcal{F}_1 \cup \mathcal{F}_2}$, and let x be the point of $\mathcal{A}^{\mathbb{Z}}$ reconstructed from the "bottom" symbols as described after Definition 1.4.1. Then $x \in X = X_{\mathcal{F}}$ since y satisfies the constraints from $\mathcal{F}_1$, and $y = \beta_N(x)$ by the overlap constraints from $\mathcal{F}_2$. This proves that $X^{[N]} \supseteq X_{\mathcal{F}_1 \cup \mathcal{F}_2}$, so that $X^{[N]} = X_{\mathcal{F}_1 \cup \mathcal{F}_2}$ is a shift space. □

The Nth higher block shift of X uses overlapping blocks. The same sort of construction can be made with nonoverlapping blocks, and leads to the notion of the Nth higher power shift of X.

Using the same notation as at the beginning of this section, define the *Nth higher power code* $\gamma_N \colon X \to (\mathcal{A}_X^{[N]})^{\mathbb{Z}}$ by

$$(\gamma_N(x))_{[i]} = x_{[iN, iN+N-1]}.$$

Here γ_N chops up the coordinates of x into consecutive N-blocks and assembles the pieces into a point over $\mathcal{A}_X^{[N]}$. The image of $x = (x_i)_{i \in \mathbb{Z}}$ under γ_4 has the form

(1–4–3)

$$\gamma_4(x) = \ldots \begin{bmatrix} x_{-9} \\ x_{-10} \\ x_{-11} \\ x_{-12} \end{bmatrix} \begin{bmatrix} x_{-5} \\ x_{-6} \\ x_{-7} \\ x_{-8} \end{bmatrix} \begin{bmatrix} x_{-1} \\ x_{-2} \\ x_{-3} \\ x_{-4} \end{bmatrix} . \begin{bmatrix} x_3 \\ x_2 \\ x_1 \\ x_0 \end{bmatrix} \begin{bmatrix} x_7 \\ x_6 \\ x_5 \\ x_4 \end{bmatrix} \begin{bmatrix} x_{11} \\ x_{10} \\ x_9 \\ x_8 \end{bmatrix} \ldots \in (\mathcal{A}_X^{[4]})^{\mathbb{Z}}.$$

Compare this with (1–4–2). Note that the bottom symbols here will usually *not* determine the rest of the symbols since there is no overlapping.

Definition 1.4.4. Let X be a shift space. The Nth *higher power shift* X^N of X is the image $X^N = \gamma_N(X)$ of X in the full shift over $\mathcal{A}_X^{[N]}$.

Example 1.4.5. Let X be the golden mean shift of Example 1.2.3, and $N = 2$. Then $\mathcal{A}_X^{[2]} = \{a = 00, b = 01, c = 10\}$. The 2nd higher power shift X^2 is described by $\mathcal{F} = \{bc\}$, since words containing $bc = 0110$ are the only ones containing the forbidden word 11 of the original shift. □

As before, a higher power shift is also a shift space.

Proposition 1.4.6. *The higher power shifts of a shift space are also shift spaces.*

PROOF: The proof is very similar to that of Proposition 1.4.3, and is left to the reader. □

EXERCISES

1.4.1. For Examples 1.2.3 and 1.2.4 describe explicitly the 3rd higher block shift $X^{[3]}$. To do this, you need to specify the alphabet $\mathcal{A}_X^{[3]}$, and then describe a collection $\mathcal{F}$ of blocks over this alphabet so that $X^{[3]} = X_{\mathcal{F}}$. [*Hint:* Use the proof of Proposition 1.4.3 as a guide.]

1.4.2. If X and Y are shift spaces over the same alphabet, show that

$$(X \cap Y)^{[N]} = X^{[N]} \cap Y^{[N]} \quad \text{and} \quad (X \cup Y)^{[N]} = X^{[N]} \cup Y^{[N]}.$$

1.4.3. If X and Y are shift spaces over possibly different alphabets, show that

$$(X \times Y)^{[N]} = X^{[N]} \times Y^{[N]}$$

1.4.4. If X is a shift space with shift map σ_X, and $X^{[N]}$ is its Nth higher block shift with corresponding shift map $\sigma_{X^{[N]}}$, prove that $\beta_N \circ \sigma_X = \sigma_{X^{[N]}} \circ \beta_N$. [*Hint*: Compute the ith coordinate of each image.]

1.4.5. If X^N is the Nth higher power shift of X, and σ_{X^N} is its shift map, prove that $\gamma_N \circ \sigma_X^N = \sigma_{X^N} \circ \gamma_N$. Here σ_X^N is the N-fold composition of σ_X with itself.

1.4.6. Find an example of a shift space for which the bottom symbols in (1–4–3) of $\gamma_4(x)$ do not determine the rest of the symbols. Find another example for which they do determine the rest.

§1.5. Sliding Block Codes

Suppose that $x = \ldots x_{-1}x_0x_1 \ldots$ is a sequence of symbols in a shift space X over $\mathcal{A}$. We can transform x into a new sequence $y = \ldots y_{-1}y_0y_1 \ldots$ over another alphabet $\mathfrak{A}$ as follows. Fix integers m and n with $-m \leqslant n$. To compute the ith coordinate y_i of the transformed sequence, we use a function Φ that depends on the "window" of coordinates of x from $i - m$ to $i + n$. Here $\Phi \colon \mathcal{B}_{m+n+1}(X) \to \mathfrak{A}$ is a fixed *block map*, called an $(m+n+1)$-*block map* from allowed $(m + n + 1)$-blocks in X to symbols in $\mathfrak{A}$, and so

$$(1\text{--}5\text{--}1) \qquad y_i = \Phi(x_{i-m}x_{i-m+1} \cdots x_{i+n}) = \Phi(x_{[i-m,i+n]}).$$

Definition 1.5.1. Let X be a shift space over $\mathcal{A}$, and $\Phi \colon \mathcal{B}_{m+n+1}(X) \to \mathfrak{A}$ be a block map. Then the map $\phi \colon X \to \mathfrak{A}^{\mathbb{Z}}$ defined by $y = \phi(x)$ with y_i given by (1–5–1) is called the *sliding block code* with *memory* m and *anticipation* n *induced by* Φ. We will denote the formation of ϕ from Φ by $\phi = \Phi_\infty^{[-m,n]}$, or more simply by $\phi = \Phi_\infty$ if the memory and anticipation of ϕ are understood. If not specified, the memory is taken to be 0. If Y is a shift space contained in $\mathfrak{A}^{\mathbb{Z}}$ and $\phi(X) \subseteq Y$, we write $\phi \colon X \to Y$.

Figure 1.5.1 illustrates the action of a sliding block code. The window is slid one coordinate to the right to compute the next coordinate of the image.

The simplest sliding block codes are those with no memory or anticipation, i.e., with $m = n = 0$. Here the ith coordinate of the image of x depends only on x_i. Such sliding block codes are called *1-block codes*. By our convention about memory in Definition 1.5.1, when Φ is a 1-block map, then $\phi = \Phi_\infty$ is taken to be a 1-block code if no memory is specified.

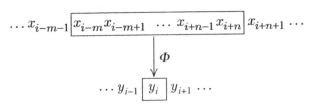

FIGURE 1.5.1. Sliding block code.

Example 1.5.2. Let X be a shift space over an alphabet $\mathcal{A}$, $\mathfrak{A} = \mathcal{A}$, $m = 0$, $n = 1$, and $\Phi(a_0 a_1) = a_1$. Then $\phi = \Phi_\infty^{[0,1]} = \Phi_\infty$ is the shift map σ_X. What happens if we let $\Phi(a_0 a_1) = a_0$ instead?

Now let $m = 1$, $n = 0$, and $\Psi(a_{-1} a_0) = a_{-1}$. Then $\psi = \Psi_\infty^{[-1,0]} = \Psi_\infty$ is the inverse σ_X^{-1} of the shift map, so that $\psi(\phi(x)) = x = \phi(\psi(x))$ for all $x \in X$.

Notice that if $\Theta(a) = a$ for all $a \in \mathcal{A}$, then $\phi = \Theta_\infty^{[1,1]}$ and $\psi = \Theta_\infty^{[-1,-1]}$. So, there may be many ways of representing a mapping between shift spaces as a sliding block code.   □

Example 1.5.3. Let $\mathcal{A} = \{0,1\} = \mathfrak{A}$, $X = \mathcal{A}^{\mathbb{Z}}$, $m = 0$, $n = 1$, and $\Phi(a_0 a_1) = a_0 + a_1 \pmod 2$. Then $\phi = \Phi_\infty$ is the code ϕ discussed near the end of §1.1.   □

Example 1.5.4. Let $\mathcal{A} = \{0,1\}$, $\mathfrak{A} = \{a,b\}$, $m = n = 0$, and $\Phi(0) = a$, $\Phi(1) = b$. Then $\phi = \Phi_\infty$ is a 1-block code from the full 2-shift to the full $\mathfrak{A}$-shift. If $\Psi(0) = \Psi(1) = a$, then $\psi = \Psi_\infty$ collapses the full 2-shift to the single point a^∞.   □

Example 1.5.5. Let X be a shift space over $\mathcal{A}$, $\mathfrak{A} = \mathcal{A}_X^{[N]}$, $m = 0$, $n = N - 1$, $Y = X^{[N]}$, and

$$\Phi(a_0 a_1 \ldots a_{N-1}) = a_0 a_1 \ldots a_{N-1} \in \mathcal{A}_X^{[N]}.$$

Then $\phi = \Phi_\infty \colon X \to Y$ is the Nth higher block code β_N from §1.4.   □

Suppose that $\Phi \colon \mathcal{B}_{m+n+1}(X) \to \mathfrak{A}$ is a block map which induces a sliding block code with memory m and anticipation n. It will sometimes be convenient to imagine Φ as having a larger "window," and ignore the extra coordinates. Thus if $M \geqslant m$ and $N \geqslant n$, define $\widehat{\Phi} \colon \mathcal{B}_{M+N+1}(X) \to \mathfrak{A}$ by

$$\widehat{\Phi}(x_{[-M,N]}) = \Phi(x_{[-m,n]}).$$

Clearly $\widehat{\Phi}_\infty^{[-M,N]} = \Phi_\infty^{[-m,n]}$. The process of passing from Φ to $\widehat{\Phi}$ is called "increasing the window size of Φ," and shows we can assume that a sliding block code is induced by a block map with as large a window as we like.

Let $\Phi \colon \mathcal{B}_{m+n+1}(X) \to \mathfrak{A}$ be a block map. We can extend Φ so that it maps $(m + n + k)$-blocks in X to k-blocks over $\mathfrak{A}$ by sliding its window as follows. If $x_{[-m,n+k-1]}$ is in $\mathcal{B}_{m+n+k}(X)$, put

$$\Phi(x_{[-m,n+k-1]}) = \Phi(x_{[-m,n]})\Phi(x_{[-m+1,n+1]}) \ldots \Phi(x_{[-m+k-1,n+k-1]}).$$

For example, if Φ is the 2-block map of Example 1.5.3, then $\Phi(011010001) = 10111001$.

Example 1.5.6. Let $\mathcal{A} = \mathfrak{A} = \{0,1\}$, X be the golden mean shift of Example 1.2.3, and Y be the even shift of Example 1.2.4. Let Φ be the 2-block map defined by $\Phi(00) = 1$, $\Phi(01) = 0$, and $\Phi(10) = 0$. We do not need to define $\Phi(11)$ since the block 11 does not occur in X. Then we will show that the induced sliding block code $\phi = \Phi_\infty \colon X \to Y$ is onto.

If 10^k1 occurs in $\phi(x)$, it must be the image under Φ of the block $0(01)^r00$, so that $k = 2r$ is even. This shows that $\phi(X) \subseteq Y$. Since each point $y \in Y$ has 1's separated by an even number of 0's, this same observation shows how to construct an $x \in X$ with $\phi(x) = y$, so that ϕ is onto. □

If $\phi \colon X \to Y$ is a sliding block code and $x \in X$, then computing ϕ at the shifted sequence $\sigma_X(x)$ gives the same result as shifting the image $\phi(x)$ using σ_Y. The commuting property is a key feature of sliding block codes.

Proposition 1.5.7. *Let X and Y be shift spaces. If $\phi \colon X \to Y$ is a sliding block code, then $\phi \circ \sigma_X = \sigma_Y \circ \phi$; i.e., the following diagram commutes.*

$$
\begin{array}{ccc}
X & \xrightarrow{\;\sigma_X\;} & X \\[2pt]
\phi \downarrow & & \downarrow \phi \\[2pt]
Y & \xrightarrow{\;\sigma_Y\;} & Y
\end{array}
$$

PROOF: Let ϕ be induced by the block map $\Phi \colon \mathcal{B}_{m+n+1}(X) \to \mathfrak{A}$ and have memory m and anticipation n. For $x \in X$,

$$(\sigma_Y \circ \phi)(x)_{[i]} = \phi(x)_{[i+1]} = \Phi(x_{[i+1-m,i+1+n]}),$$

while

$$
\begin{aligned}
(\phi \circ \sigma_X)(x)_{[i]} &= \phi(\sigma_X(x))_{[i]} \\
&= \Phi(\sigma_X(x)_{[i-m,i+n]}) \\
&= \Phi(x_{[i-m+1,i+n+1]}).
\end{aligned}
$$

Hence the ith coordinates of the images agree for each i, so the images are equal. □

However, shift-commuting is not enough to have a sliding block code (the reader is asked to give a specific example in Exercise 1.5.14). One also needs to know that $\phi(x)_0$ depends only on a central block of x.

Proposition 1.5.8. *Let X and Y be shift spaces. A map $\phi \colon X \to Y$ is a sliding block code if and only if $\phi \circ \sigma_X = \sigma_Y \circ \phi$ and there exists $N \geqslant 0$ such that $\phi(x)_0$ is a function of $x_{[-N,N]}$.*

PROOF: The necessity of the condition is clear from the definition and Proposition 1.5.7. For sufficiency, define the $(2N + 1)$-block map Φ by $\Phi(w) = \phi(x)_0$ where x is any point in X such that $x_{[-N,N]} = w$. It is straightforward to check that $\phi = \Phi_\infty^{[-N,N]}$. $\square$

If a sliding block code $\phi: X \to Y$ is onto, then ϕ is called a *factor code from X onto Y* . A shift space Y is a *factor* of X if there is a factor code from X onto Y. The sliding block codes ϕ in Examples 1.5.2 through 1.5.6 are factor codes. Factor codes are often called "factor maps" in the literature.

If $\phi: X \to Y$ is one-to-one, then ϕ is called an *embedding of X into Y*. The sliding block codes ϕ in Examples 1.5.2 and 1.5.4 are embeddings, as is the higher block code $\beta_N: X \to (\mathcal{A}_X^{[N]})^{\mathbb{Z}}$. The code in Example 1.5.3 is not an embedding since it is two-to-one everywhere.

Sometimes a sliding block code $\phi: X \to Y$ has an *inverse*, i.e., a sliding block code $\psi: Y \to X$ such that $\psi(\phi(x)) = x$ for all $x \in X$ and $\phi(\psi(y)) = y$ for all $y \in Y$. This is the case in Example 1.5.2. If ϕ has an inverse, it is unique (see Exercise 1.5.4), so we can write $\psi = \phi^{-1}$, and we call ϕ *invertible*.

Definition 1.5.9. A sliding block code $\phi: X \to Y$ is a *conjugacy from X to Y*, if it is invertible. Two shift spaces X and Y are *conjugate* (written $X \cong Y$) if there is a conjugacy from X to Y.

If there is a conjugacy from X to Y, we can think of Y as being a "recoded" version of X, sharing all of its properties. Then X and Y are merely different views of the same underlying object. We will explore this idea in greater detail in Chapter 6. Conjugacies are often called "topological conjugacies" in the literature.

Example 1.5.10. Let X be a shift space over $\mathcal{A}$, and $X^{[N]}$ be its Nth higher block shift. According to Example 1.5.5, $\beta_N: X \to X^{[N]}$ is a sliding block code. Define the 1-block map $\Psi: \mathcal{A}_X^{[N]} \to \mathcal{A}$ by $\Psi(a_0 a_1 \dots a_{N-1}) = a_0$, and put $\psi = \Psi_\infty: X^{[N]} \to X$. It is easy to check that $\psi = \beta_N^{-1}$, so that β_N is a conjugacy, and thus $X \cong X^{[N]}$. In this sense, $X^{[N]}$ is a recoded version of X. $\square$

The behavior of periodic points under sliding block codes is described in the following result.

Proposition 1.5.11. *Let $\phi: X \to Y$ be a sliding block code. If $x \in X$ has period n under σ_X, then $\phi(x)$ has period n under σ_Y, and the least period of $\phi(x)$ divides the least period of x. Embeddings, and hence conjugacies, preserve the least period of a point.*

PROOF: If x has period n, then $\sigma_X^n(x) = x$. Hence

$$\sigma_Y^n(\phi(x)) = \phi(\sigma_X^n(x)) = \phi(x),$$

so that $\phi(x)$ has period n. If x has least period n, then $\phi(x)$ has period n, and hence its least period divides n (see Exercise 1.1.5). If ϕ is one-to-one, then $\sigma_X^n(x) = x$ if and only if $\sigma_Y^n(\phi(x)) = \phi(x)$, so that x and $\phi(x)$ must have the same least period. $\qquad\square$

Observe that this proposition shows that for each n, the number of points of period n is the same for all shifts conjugate to a given shift; i.e., it is an *invariant* of the shift. This gives a way to prove that some pairs of shift spaces cannot be conjugate, for example when one has a fixed point and the other doesn't. We shall be meeting several other kinds of invariants in the chapters ahead.

Let $\phi: X \to Y$ be a sliding block code. We show next that we can recode X to a conjugate shift $\widetilde{X}$ so that the corresponding sliding block code $\widetilde{\phi}: \widetilde{X} \to Y$ is a 1-block code. This process, called "recoding ϕ to a 1-block code," is often a starting point in proofs, since 1-block codes are much easier to think about. However, the penalty for making the map simpler is making the alphabet more complicated.

Proposition 1.5.12. *Let $\phi: X \to Y$ be a sliding block code. Then there exist a higher block shift $\widetilde{X}$ of X, a conjugacy $\psi: X \to \widetilde{X}$, and a 1-block code $\widetilde{\phi}: \widetilde{X} \to Y$ so that $\widetilde{\phi} \circ \psi = \phi$; i.e., the following diagram commutes.*

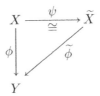

PROOF: Suppose that ϕ is induced by a block map Φ and has memory m and anticipation n. Let $\mathfrak{A} = \mathcal{B}_{m+n+1}(X)$, and define $\psi: X \to \mathfrak{A}^{\mathbb{Z}}$ by $\psi(x)_{[i]} = x_{[i-m,i+n]}$. Then $\psi = \sigma^{-m} \circ \beta_{m+n+1}$. Thus $\widetilde{X} = \psi(X) = X^{[m+n+1]}$ is a shift space, and since σ and β_{m+n+1} are conjugacies, so is ψ. Put $\widetilde{\phi} = \phi \circ \psi^{-1}$. Note that $\widetilde{\phi}$ is a 1-block code. $\qquad\square$

We remark that if a sliding block code ϕ happens to be a conjugacy, then the recoding of ϕ to a 1-block code, given in Proposition 1.5.12, usually does not also recode ϕ^{-1} to a 1-block code (why?). Thus the cost of recoding to "simplify" a sliding block code in one direction is often to make it more "complicated" in the other.

We next show that for any sliding block code $\phi: X \to Y$, $\phi(X)$ is a shift space.

Theorem 1.5.13. *The image of a shift space under a sliding block code is a shift space.*

PROOF: Let X and Y be shift spaces, and $\phi: X \to Y$ be a sliding block code. By Proposition 1.5.12, we can assume that ϕ is a 1-block code. Let Φ be a 1-block map inducing ϕ. Put $\mathcal{L} = \{\Phi(w) : w \in \mathcal{B}(X)\}$. We will show that $\phi(X) = X_{\mathcal{L}^c}$, proving that the image of X is a shift space.

If $x \in X$, then every block in $\phi(x)$ is in $\mathcal{L}$, so that $\phi(x) \in X_{\mathcal{L}^c}$. This proves that $\phi(X) \subseteq X_{\mathcal{L}^c}$.

Suppose now that $y \in X_{\mathcal{L}^c}$. Then for each $n \geqslant 0$ the central $(2n + 1)$-block of y is the image under Φ of the central $(2n + 1)$-block of some point $x^{(n)}$ in X; i.e.,

$$(1\text{--}5\text{--}2) \qquad \Phi(x^{(n)}_{[-n,n]}) = \phi(x^{(n)})_{[-n,n]} = y_{[-n,n]}.$$

We will use the $x^{(n)}$ to find a point $x \in X$ with $\phi(x) = y$.

First consider the 0th coordinates $x^{(n)}_{[0]}$ for $n \geqslant 1$. Since there are only finitely many symbols, there is an infinite set S_0 of integers for which $x^{(n)}_{[0]}$ is the same for all $n \in S_0$. Next, the central 3-blocks $x^{(n)}_{[-1,1]}$ for $n \in S_0$ all belong to the finite set of possible 3-blocks, so there is an infinite subset $S_1 \subseteq S_0$ so that $x^{(n)}_{[-1,1]}$ is the same for all $n \in S_1$. Continuing this way, we find for each $k \geqslant 1$ an infinite set $S_k \subseteq S_{k-1}$ so that all blocks $x^{(n)}_{[-k,k]}$ are equal for $n \in S_k$.

Define x to be the sequence with $x_{[-k,k]} = x^{(n)}_{[-k,k]}$ for all $n \in S_k$ (these blocks are all the same by our construction). Observe that since $S_k \subseteq S_{k-1}$, the central $(2k-1)$-block of $x_{[-k,k]}$ is $x_{[-k+1,k-1]}$, so that x is well-defined. Also observe that every block in x occurs in some $x_{[-k,k]} = x^{(n)}_{[-k,k]} \in \mathcal{B}(X)$, so that $x \in X$ since X is a shift space. Finally, for each $k \geqslant 0$ and $n \in S_k$ with $n \geqslant k$ we have, using (1–5–2), that

$$\Phi(x_{[-k,k]}) = \Phi(x^{(n)}_{[-k,k]}) = \phi(x^{(n)})_{[-k,k]} = y_{[-k,k]},$$

so that $\phi(x) = y$. This proves that $X_{\mathcal{L}^c} \subseteq \phi(X)$, completing the proof. $\square$

This proof repays close study. It uses a version of the *Cantor diagonal argument*, one of the most important and subtle ideas in mathematics. This argument is used, for example, to show that the set of real numbers cannot be arranged in a sequence (i.e., the set of real numbers is uncountable). We will encounter it again in Chapter 6, when we discuss the notion of compactness.

Suppose that $\phi: X \to Y$ is an embedding. By the previous result, $\phi(X)$ is a shift space, and ϕ establishes a one-to-one correspondence between points in X and points in $\phi(X)$. Let $\psi: \phi(X) \to X$ be the reverse correspondence, so that $\psi(y) = x$ whenever $\phi(x) = y$. Then ψ is a mapping, but is it a sliding block code? Another application of the Cantor diagonal argument shows that it is.

Theorem 1.5.14. *A sliding block code that is one-to-one and onto has a sliding block inverse, and is hence a conjugacy.*

PROOF: Let $\phi: X \to Y$ be a sliding block code that is one-to-one and onto. By recoding ϕ if necessary, we can assume that ϕ is a 1-block code. Let Φ be a 1-block map inducing ϕ. Let $\psi: Y \to X$ be the map on points inverse to ϕ, which we will show is a sliding block code.

First observe that if $y = \phi(x)$, then since $\phi(\sigma_X(x)) = \sigma_Y(\phi(x))$, we have that

$$\sigma_X(\psi(y)) = \sigma_X(x) = \psi(\phi(\sigma_X(x)))$$
$$= \psi(\sigma_Y(\phi(x))) = \psi(\sigma_Y(y)),$$

so that $\psi \circ \sigma_Y = \sigma_X \circ \psi$. Hence by Proposition 1.5.8, to show that ψ is a sliding block code, it is enough to find an $n \geqslant 0$ such that the central $(2n+1)$-block of every y determines the 0th coordinate $\psi(y)_{[0]}$ of its image.

If this were not the case, then for every $n \geqslant 0$ there would be two points $y^{(n)}$ and $\tilde{y}^{(n)}$ in Y so that $y^{(n)}_{[-n,n]} = \tilde{y}^{(n)}_{[-n,n]}$ but $\psi(y^{(n)})_{[0]} \neq \psi(\tilde{y}^{(n)})_{[0]}$. Put $x^{(n)} = \psi(y^{(n)})$ and $\tilde{x}^{(n)} = \psi(\tilde{y}^{(n)})$.

Since there are only finitely many symbols in the alphabet of X, there would be distinct symbols $a \neq b$ and an infinite set S_0 of integers so that $x^{(n)}_{[0]} = \psi(y^{(n)})_{[0]} = a$ and $\tilde{x}^{(n)}_{[0]} = \psi(\tilde{y}^{(n)})_{[0]} = b$ for all $n \in S_0$.

Since the number of pairs of possible 3-blocks is finite, there would be an infinite subset $S_1 \subseteq S_0$ so that the $x^{(n)}_{[-1,1]}$ are all equal for $n \in S_1$ and the $\tilde{x}^{(n)}_{[-1,1]}$ are all equal for $n \in S_1$. Continuing this way, for each $k \geqslant 1$ we would find an infinite subset $S_k \subseteq S_{k-1}$ so that the $x^{(n)}_{[-k,k]}$ are all equal for $n \in S_k$, and the $\tilde{x}^{(n)}_{[-k,k]}$ are all equal for $n \in S_k$. As in the proof of Theorem 1.5.13, this would allow us to construct points x and $\tilde{x}$ in X defined by $x_{[-k,k]} = x^{(n)}_{[-k,k]}$ and $\tilde{x}_{[-k,k]} = \tilde{x}^{(n)}_{[-k,k]}$ for $n \in S_k$. Note that $x_{[0]} = a \neq b = \tilde{x}_{[0]}$, so that $x \neq \tilde{x}$. Now if $n \in S_k$ and $n \geqslant k$, then

$$\Phi(x_{[-k,k]}) = \Phi(x^{(n)}_{[-k,k]}) = \phi(x^{(n)})_{[-k,k]} = y^{(n)}_{[-k,k]}$$
$$= \tilde{y}^{(n)}_{[-k,k]} = \phi(\tilde{x}^{(n)})_{[-k,k]} = \Phi(\tilde{x}^{(n)}_{[-k,k]}) = \Phi(\tilde{x}_{[-k,k]}).$$

But this would imply that $\phi(x) = \phi(\tilde{x})$. This contradiction shows that ψ must be a sliding block code. $\square$

If we are given two shift spaces X and Y, it is natural to ask whether Y is conjugate to X, whether Y is a factor of X, or whether Y embeds into X. These questions are very difficult to settle for general shift spaces. Indeed, many of the ideas and results we will encounter originate in attempts to answer these fundamental questions for special classes of shift spaces.

EXERCISES

1.5.1. Suppose that $\phi: X \to Y$ and $\psi: Y \to Z$ are sliding block codes. Show that $\psi \circ \phi: X \to Z$ is also a sliding block code. If ϕ and ψ are factor codes, show that $\psi \circ \phi$ is also a factor code, and similarly for embeddings and conjugacies.

1.5.2. Show that an invertible sliding block code must be one-to-one and onto, so it is simultaneously a factor code and an embedding.

1.5.3. Prove that conjugacy $\cong$ between shift spaces is an equivalence relation; that is, show that (a) $X \cong X$, (b) if $X \cong Y$ then $Y \cong X$, and (c) if $X \cong Y$ and $Y \cong Z$, then $X \cong Z$.

1.5.4. Prove that an invertible sliding block code can have only one inverse.

1.5.5. Does the sliding block code in Example 1.5.3 have an inverse? What about the sliding block codes in Example 1.5.4? Justify your answers.

1.5.6. Let X be a shift space.
   (a) Show that $X^{[1]} = X$.
   (b) Show that $(X^{[N]})^{[2]} \cong X^{[N+1]}$.

1.5.7. Let $X = \{0, 1\}^{\mathbb{Z}}$, and $\Phi: \{0, 1\} \to \{0, 1\}$ be the 1-block map given by $\Phi(0) = 1$ and $\Phi(1) = 0$. Show that $\phi = \Phi_{\infty}: X \to X$ is a conjugacy of the full 2-shift to itself.

1.5.8. Let X be the full 2-shift. Define the block map Φ by

$$\Phi(abcd) = b + a(c + 1)d \pmod 2,$$

and put $\phi = \Phi_{\infty}^{[-1,2]}$.
   (a) Describe the action of ϕ on $x \in X$ in terms of the blocks 1001 and 1101 appearing in x.
   (b) Show that $\phi^2(x) = x$ for all $x \in X$, and hence show that ϕ is a conjugacy of X to itself.
   (c) Use this method to find other conjugacies of the full 2-shift to itself.

1.5.9. Recode Example 1.5.3 to a 1-block code.

1.5.10. Suppose that $X_1 \supseteq X_2 \supseteq X_3 \supseteq \ldots$ are shift spaces whose intersection is X. For each $N \geqslant 1$, use the Cantor diagonal argument to prove that there is a $K \geqslant 1$ such that $\mathcal{B}_N(X_k) = \mathcal{B}_N(X)$ for all $k \geqslant K$.

1.5.11. (a) Is the full 2-shift conjugate to the full 3-shift?
   (b) Find a factor code from the full 3-shift onto the full 2-shift. Can you find infinitely many such factor codes?
   (c) Is there a factor code from the full 2-shift onto the full 3-shift?
   (d) Is the golden mean shift conjugate to a full shift? To the even shift?

1.5.12. Let $\phi: X \to Y$ be a sliding block code, and Z be a shift space contained in Y. Show that $\phi^{-1}(Z) = \{x \in X : \phi(x) \in Z\}$ is a shift space.

1.5.13. (a) Let Z be the full k-shift, and $\phi: X \to Z$ be a sliding block code. If X is a subset of a shift space Y, show that ϕ can be extended to a sliding block code $\psi: Y \to Z$ such that $\psi(x) = \phi(x)$ for all $x \in X$.
   (b) Find an example of shift spaces X, Y, and Z with $X \subset Y$, and a sliding block code $\phi: X \to Z$, such that there is no sliding block code from Y to Z extending ϕ.

1.5.14. Find a point mapping from the full 2-shift to itself that commutes with the shift, but is *not* a sliding block code.

1.5.15. Show that for a forbidden list $\mathcal{F}$, $X_{\mathcal{F}} = \varnothing$ if and only if there exists N such that whenever u and v are blocks with $|u| = N$, then some subblock of uvu belongs to $\mathcal{F}$.

*1.5.16. (a) Show that there is no 1-block or 2-block factor code from the even shift onto the golden mean shift.

  (b) Find a 3-block factor code from the even shift onto the golden mean shift (compare with Example 1.5.6).

*1.5.17. Show that the S-gap shift and the S'-gap shift are conjugate iff either $S = S'$ or for some n, $S = \{0, n\}$ and $S' = \{n, n+1, \dots\}$. Hence, there are uncountably many shifts no pair of which are conjugate.

§1.6. Convolutional Encoders

In symbolic dynamics the term "code" means a mapping from one shift space to another, or more loosely some sort of apparatus or procedure for constructing such a mapping. In the previous section we introduced sliding block codes. Later in this book we consider finite-state codes (Chapter 5), finite-to-one codes (Chapter 8), and almost invertible codes (Chapter 9).

However, in the subject of coding theory the term "code" means something different, namely a set $\mathcal{C}$ of sequences (often finite sequences, but sometimes right-infinite or bi-infinite sequences). The goal is to find "good" error-correcting codes. These are codes $\mathcal{C}$ for which any two distinct sequences in $\mathcal{C}$ differ in a relatively "large" number of coordinates. Thus if the sequences in $\mathcal{C}$ are regarded as messages and transmitted over a "noisy" channel that makes a relatively small number of errors, then these errors can be detected and corrected to recover the original message.

The two broad classes of error-correcting codes that have been studied over the past forty years are block codes and convolutional codes. A *block code* is defined as a finite set of sequences all of the same length over some finite alphabet. *Convolutional codes* are much closer in spirit to symbolic dynamics, and are used in various applications in communications and storage. Such a code is defined as the image of a mapping, called a convolutional encoder, defined below.

Recall that a *finite field* $\mathbb{F}$ is a finite set in which you can add, subtract, multiply and divide so that the basic associative, distributive and commutative laws of arithmetic hold. A good example to keep in mind (and the one that we are mostly concerned with) is the field $\mathbb{F}_2$ with just two elements. Thus $\mathbb{F}_2 = \{0, 1\}$ with the usual additive and multiplicative structure: $0 + 0 = 1 + 1 = 0$, $0 + 1 = 1 + 0 = 1$, $0 \cdot 0 = 0 \cdot 1 = 1 \cdot 0 = 0$, and $1 \cdot 1 = 1$.

A *Laurent polynomial* over a field $\mathbb{F}$ is a polynomial $f(t)$ in the variable t and its inverse t^{-1} whose coefficients are in $\mathbb{F}$. A typical Laurent polynomial looks like

$$f(t) = \sum_{j=-m}^{n} a_j t^j,$$

where the $a_j \in \mathbb{F}$. A *bi-infinite power series* over $\mathbb{F}$ is a series of the form

$$(1\text{--}6\text{--}1) \qquad\qquad f(t) = \sum_{j=-\infty}^{\infty} a_j t^j.$$

Although these series resemble the infinite series studied in calculus, we are using them as a formal algebraic device and are not concerned with convergence. If the coefficients of two series lie in the same field $\mathbb{F}$, then we can add them coefficientwise. Note that the set of all bi-infinite power series with coefficients in $\mathbb{F}$ can be identified with the full shift over $\mathbb{F}$, where the series in (1–6–1) corresponds to the point $\ldots a_{-1}.a_0 a_1 \ldots \in \mathbb{F}^{\mathbb{Z}}$. We can also multiply a bi-infinite power series by a Laurent polynomial using the normal rules of algebra. A *Laurent polynomial matrix* is a (finite-dimensional) rectangular matrix whose entries are Laurent polynomials. For the k-dimensional vector space $\mathbb{F}^k$ over a field $\mathbb{F}$, we identify the full $\mathbb{F}^k$-shift with the set of all k-tuple row vectors of bi-infinite power series with coefficients in $\mathbb{F}$.

Definition 1.6.1. Let $G(t) = [g_{ij}(t)]$ be a $k \times n$ Laurent polynomial matrix. Use $G(t)$ to transform an *input vector* $I(t) = [I_1(t), \ldots, I_k(t)]$, whose components are bi-infinite power series, into an *output vector* $O(t) = [O_1(t), \ldots, O_n(t)]$ via the equation

$$(1\text{--}6\text{--}2) \qquad\qquad O(t) = E(I(t)) = I(t)G(t).$$

A *(k, n)-convolutional encoder* is a mapping E from the full $\mathbb{F}^k$-shift to the full $\mathbb{F}^n$-shift of the form (1–6–2). A *convolutional code* is the image of a convolutional encoder.

The term "convolutional" is used because multiplying a power series by a polynomial is usually called a convolution. In coding theory, what we have defined as a convolutional encoder is often called a "Laurent polynomial convolutional encoder" to distinguish it from a more general class of encoders.

We illustrate these concepts with the following example.

Example 1.6.2. Let $\mathbb{F} = \mathbb{F}_2$ and

$$G(t) = \begin{bmatrix} 1 & 0 & 1+t \\ 0 & t & t \end{bmatrix}.$$

The image of the input vector $I(t) = [I_1(t), I_2(t)]$ under the corresponding convolutional encoder E is

$$E(I(t)) = I(t) \cdot G(t) = [I_1(t), tI_2(t), (1+t)I_1(t) + tI_2(t)].$$

To see how E represents a mapping from $X = (\mathbb{F}_2^2)^{\mathbb{Z}}$ to $Y = (\mathbb{F}_2^3)^{\mathbb{Z}}$, write $I_1(t) = \sum_j a_j t^j$ and $I_2(t) = \sum_j b_j t^j$. We are then identifying $I(t)$ with the point

$$\ldots (a_{-1}, b_{-1}).(a_0, b_0)(a_1, b_1) \ldots \in X,$$

and similarly for $O(t)$ and Y. With these identifications, the jth component of $O(t) = E(I(t))$ is $(a_j, b_{j-1}, a_j + a_{j-1} + b_{j-1})$, whose entries are just the coefficients of t^j in $O_1(t)$, $O_2(t)$, and $O_3(t)$, respectively. Observe that E is actually a sliding block code. Specifically, $E = \Phi_{\infty}^{[-1,0]}$, where Φ is the 2-block map over $\mathbb{F}_2^2$ defined by

$$\Phi\big((a_{-1}, b_{-1})(a_0, b_0)\big) = (a_0,\ b_{-1},\ a_0 + a_{-1} + b_{-1}).$$

The corresponding convolutional code is

$$E(X) = \{\ldots (a_j,\ b_{j-1},\ a_j + a_{j-1} + b_{j-1}) \ldots \in (\mathbb{F}_2^3)^{\mathbb{Z}} : a_j, b_j \in \mathbb{F}_2\}.$$

To describe $E(X)$ more explicitly, let $\mathcal{F}$ be the finite collection of 2-blocks over $\mathbb{F}_2^3$ defined by

$$(1\text{--}6\text{--}3) \qquad \mathcal{F} = \{(c, d, e)(c', d', e') : e' \neq c' + c + d'\}.$$

We leave it to the reader to check that $E(X)$ is the shift space defined using the set $\mathcal{F}$ of forbidden blocks, so that $E(X) = X_{\mathcal{F}}$. $\qquad\qquad\square$

It is not an accident that the convolutional encoder in the previous example is a sliding block code, or that the corresponding convolutional code is a shift space. To prove that this holds generally, consider a convolutional encoder E defined by a Laurent polynomial matrix $G(t) = [g_{ij}(t)]$ over a finite field $\mathbb{F}$. Let M denote the largest power of t that occurs in any of the $g_{ij}(t)$, and N denote the smallest such power. Let g_{ij}^p be the coefficient of t^p in $g_{ij}(t)$. Similarly, if $I(t) = [I_1(t), \ldots, I_k(t)]$ is an input vector of bi-infinite power series over $\mathbb{F}$, let I_i^p denote the coefficient of t^p in $I_i(t)$. Then $I(t)$ is identified with the point

$$\ldots (I_1^{-1}, \ldots, I_k^{-1}).(I_1^0, \ldots, I_k^0)(I_1^1, \ldots I_k^1) \ldots \in (\mathbb{F}^k)^{\mathbb{Z}}.$$

It is then straightforward to check that

$$(1\text{--}6\text{--}4) \qquad\qquad E = \Phi_{\infty}^{[-M,N]},$$

where

$$\Phi\big((I_1^{-M}, \ldots, I_k^{-M}) \ldots (I_1^N, \ldots, I_k^N)\big) =$$
$$\left(\sum_{j=-M}^{N} \sum_{i=1}^{k} I_i^j g_{i,1}^{-j},\ \ldots,\ \sum_{j=-M}^{N} \sum_{i=1}^{k} I_i^j g_{i,n}^{-j} \right).$$

Hence the convolutional code $E\big((\mathbb{F}^k)^{\mathbb{Z}}\big)$, being the image of a full shift under a sliding block code, is a shift space by Theorem 1.5.13.

Note that both $(\mathbb{F}^k)^{\mathbb{Z}}$ and $(\mathbb{F}^n)^{\mathbb{Z}}$ are (infinite-dimensional) vector spaces over $\mathbb{F}$. Also observe from what we just did that a convolutional encoder is a linear transformation. Since the image of a vector space under a linear transformation is again a vector space, it follows that the corresponding convolutional code $E\big((\mathbb{F}^k)^{\mathbb{Z}}\big)$ is a linear subspace of $(\mathbb{F}^n)^{\mathbb{Z}}$, i.e., is a *linear shift space*. Furthermore, it is easy to check that $E\big((\mathbb{F}^k)^{\mathbb{Z}}\big)$ is also irreducible (Exercise 1.6.4). This proves the following result

Theorem 1.6.3.

(1) *Every convolutional encoder is a linear sliding block code.*
(2) *Every convolutional code is a linear irreducible shift space.*

In fact, the converses hold: the convolutional encoders are precisely the linear sliding block codes, and the convolutional codes are precisely the linear irreducible shift spaces (see Exercises 1.6.3 and 1.6.6).

Convolutional encoders are usually viewed as operating on the set of all k-tuples of Laurent series, i.e., objects of the form $\sum_{j=j_0}^{\infty} a_j t^j$, where j_0 may be negative. We have focused on bi-infinite power series instead in order to view these encoders better within the framework of symbolic dynamics. Each formalism has its advantages.

EXERCISES

1.6.1. Verify that in Example 1.6.2, $E(X) = X_{\mathcal{F}}$ where $\mathcal{F}$ is defined by (1–6–3).

1.6.2. Verify (1–6–4).

1.6.3. Let $\mathbb{F}$ be a finite field and E be a mapping from the full shift over $\mathbb{F}^k$ into the full shift over $\mathbb{F}^n$. Show that the following are equivalent:

    (a) E is a convolutional encoder.

    (b) E is a linear sliding block code, i.e., a sliding block code which is linear as a mapping between the vector spaces $(\mathbb{F}^k)^{\mathbb{Z}}, (\mathbb{F}^n)^{\mathbb{Z}}$.

    (c) E is a map from the full F^k-shift to the full F^n-shift of the form $E = \Phi_{\infty}^{[-M,N]}$ where Φ is a linear map $\Phi \colon (\mathbb{F}^k)^{M+N+1} \to \mathbb{F}^n$; here M, N are integers and we identify $(\mathbb{F}^k)^{M+N+1}$ with $\mathbb{F}^{k(M+N+1)}$.

1.6.4. Show that every convolutional code has a fixed point and is irreducible.

*1.6.5. Let $G(t)$ be a Laurent polynomial matrix. Give an algorithm to construct a finite list $\mathcal{F}$ from $G(t)$ such that $X_{\mathcal{F}}$ is the convolutional code defined by $G(t)$.

*1.6.6. Show that a subset of the full $\mathbb{F}^n$-shift is a convolutional code if and only if it is a linear irreducible shift space, i.e., an irreducible shift space that is a linear subspace of $(\mathbb{F}^n)^{\mathbb{Z}}$.

Notes

Symbolic dynamics goes back to Hadamard [Had] (1898) and Morse [Mor1, Mor2] (1921) in modeling of geodesics on surfaces of negative curvature. A notion of shift space described by spelling out an explicit list of restrictions on the allowed sequences was given by Morse and Hedlund [MorH1, pp. 822–824] (1938); see also [MorH2] and Hedlund [Hed4]. A space of sequences was often described by declaring the allowed blocks to be those that appear in a particular bi-infinite sequence; for instance, see Gottschalk and Hedlund [GotH, Chap. 12]. However, in the generality of our textbook, shift spaces were not formally defined until Smale [Sma] (1967); he called them *subshifts*, viewing them as closed, shift invariant subsets of full shifts (in his definition, he also assumed that periodic points are dense). We have chosen the term "shift space" to emphasize the basic nature of these spaces.

See §6.5 and §13.6 for discussions of how shift spaces can be used to model smooth dynamical systems. Shift spaces can also be used to model constraints that naturally occur in data recording; see §2.5. The *Scientific American* article [Mon] gives a lucid explanation of how compact audio disks make use of certain shifts in recording Beethoven symphonies.

Underlying the theory of shift spaces are some fundamental topological notions such as compactness and continuity. We will explain these thoroughly in Chapter 6. But for readers already familiar with these ideas, the following gives a brief account of how they connect with symbolic dynamics.

There is a metric on the full shift such that two points are "close" if and only if they agree in a "large" central block (such a metric is given in Example 6.1.10). With respect to this metric, the full shift is compact and the shift map is continuous. A subset of a full shift is a shift space precisely when it is compact and shift invariant. Sliding block codes are exactly those maps from one shift space to another that are continuous and commute with the shift map. Theorem 1.5.13 then follows from the result that the continuous image of a compact set is compact, while Theorem 1.5.14 follows from the general result that a continuous one-to-one map on a compact metric space has a continuous inverse. The Cantor diagonal argument replaces compactness in our proofs of these results. Other facts are easier to understand from the topological viewpoint. Exercise 1.2.5, for example, translates to the statement that the intersection of compact subsets is compact. We remark that the metric on the full shift mentioned above is compatible with the product topology on the full shift using the discrete topology on its alphabet.

There is a version of shift spaces in which the alphabet is allowed to be countably infinite (see §13.9). However, the requirement of a finite alphabet that we have imposed allows us to take advantage of two very important tools: the Cantor diagonalization argument and, as we shall see in later chapters, finite-dimensional linear algebra. There is also a version of shift spaces where the sequences are one-sided, i.e., indexed over the nonnegative integers rather than the integers (see §5.1 and §13.8). One reason we choose to focus on bi-infinite sequences is to allow for memory as well as anticipation in sliding block codes.

Convolutional encoders and convolutional codes are central objects in coding theory. We refer the reader to [McE], [LinC], [Pir] for further reading on this subject.

Exercise 1.2.14 is due to M. Keane; Exercise 1.3.8 to A. Khayrallah and D. Neuhoff [KhaN1]; Exercise 1.5.16 to E. Coven and M. Paul [CovP2]; Exercise 1.5.15 to V. DeAngelis; and Exercise 1.5.17 to D. A. Dastjerdi and S. Jangjoo [DasJ] and M. Hollander and J. Von Limbach.

CHAPTER 2

SHIFTS OF FINITE TYPE

Shift spaces described by a *finite* set of forbidden blocks are called shifts of finite type. Despite being the "simplest" shifts, they play an essential role in mathematical subjects like dynamical systems. Their study has also provided solutions to important practical problems, such as finding efficient coding schemes to store data on computer disks.

One reason why finite type shifts are so useful is that they have a simple representation using a finite, directed graph. Questions about the shift can often be phrased as questions about the graph's adjacency matrix. Basic results from linear algebra help us take this matrix apart and find the answers.

We begin this chapter by introducing shifts of finite type and giving some typical examples. Next we explain the connections with directed graphs and matrices and introduce the fundamental operation of state splitting. We conclude with a brief account of magnetic storage, indicating why the (1,3) run-length limited shift in Example 1.2.5 is the central feature of one of the earliest methods used for data storage on hard disks of personal computers.

§2.1. Finite Type Constraints

We first define shifts of finite type, and then explain how they have a "Markov" property of finite memory.

Definition 2.1.1. A *shift of finite type* is a shift space that can be described by a finite set of forbidden blocks, i.e., a shift space X having the form $X_{\mathcal{F}}$ for some finite set $\mathcal{F}$ of blocks.

We have already encountered several shifts of finite type.

Example 2.1.2. The full shift $X = \mathcal{A}^{\mathbb{Z}}$ is a shift of finite type; we can simply take $\mathcal{F} = \varnothing$ since nothing is forbidden, so that $X = X_{\mathcal{F}}$. □

Example 2.1.3. The golden mean shift X of Example 1.2.3 is a shift of finite type, since we can use $\mathcal{F} = \{11\}$, and obtain $X = X_{\mathcal{F}}$. □

Example 2.1.4. The shift of Example 1.2.8 has finite type, since we can take $\mathcal{F} = \{eg, fe, ff, gg\}$. □

Note that a shift of finite type X might also be described by an infinite set of forbidden blocks. In fact, Proposition 1.3.4 shows that this can happen whenever X is not the full shift since the complement of the language of X is infinite. Definition 2.1.1 only requires that there be *some* finite $\mathcal{F}$ which works.

Suppose that $X \subseteq \mathcal{A}^{\mathbb{Z}}$ is a shift of finite type, and that $X = X_{\mathcal{F}}$ for a finite set $\mathcal{F}$. Let N be the length of the longest block in $\mathcal{F}$. If we form the collection $\mathcal{F}_N$ of all blocks of length N which contain some subblock in $\mathcal{F}$, then clearly $X_{\mathcal{F}_N} = X_{\mathcal{F}}$, and the blocks in $\mathcal{F}_N$ all have the same length N. For example, if $\mathcal{A} = \{0, 1\}$ and $\mathcal{F} = \{11, 000\}$, then $\mathcal{F}_3 = \{110, 111, 011, 000\}$. It will sometimes be convenient to assume this procedure has already been carried out, and that all forbidden blocks have the same length.

If all blocks in $\mathcal{F}$ have length N, then $x \in \mathcal{A}^{\mathbb{Z}}$ is in $X_{\mathcal{F}}$ exactly when $x_{[i,i+N-1]} \notin \mathcal{F}$ for every $i \in \mathbb{Z}$, or, equivalently, when $x_{[i,i+N-1]} \in \mathcal{B}_N(X_{\mathcal{F}})$ for every $i \in \mathbb{Z}$ (why?). Thus to detect whether or not x is in $X_{\mathcal{F}}$, we need only scan the coordinates of x with a "window" of width N, and check that each block seen through this window is in the allowed collection $\mathcal{B}_N(X)$. This observation is often useful when deciding whether or not a given shift space has finite type.

Example 2.1.5. The even shift in Example 1.2.4 is *not* a shift of finite type. For if it were, there would be an $N \geqslant 1$ and a collection $\mathcal{F}$ of N-blocks so that $X = X_{\mathcal{F}}$. Consider the point $x = 0^{\infty}10^{2N+1}10^{\infty}$. Every N-block in x is in $\mathcal{B}_N(X)$, so we would have that $x \in X_{\mathcal{F}} = X$, contradicting the definition of the even shift. □

There is also a notion of "memory" for a shift of finite type.

Definition 2.1.6. A shift of finite type is *M-step* (or has *memory M*) if it can be described by a collection of forbidden blocks all of which have length $M + 1$.

To motivate this definition, suppose that all forbidden blocks have length $M + 1$. Let $u = a_1 a_2 \dots a_n$ be a block with length n much larger than M. Suppose that a machine reads the symbols of u one at a time from left to right. For this machine to detect whether or not u contains a forbidden block, it only has to remember the previous M symbols read; i.e., it only needs "M steps of memory."

Note that an M-step shift of finite type is also K-step for all $K \geqslant M$. A 0-step shift of finite type is just a full shift, since omitting 1-blocks merely throws out letters from the alphabet. You can think of a 1-step shift of finite type as being given by a collection of symbols together with a description of which symbols can follow which others, similar to a Markov chain but without probabilities (see the end of §2.3). The next section contains many examples of such 1-step shifts.

Proposition 2.1.7. *If X is a shift of finite type, then there is an $M \geqslant 0$ such that X is M-step.*

PROOF: Since X has finite type, $X = X_{\mathcal{F}}$ for a finite $\mathcal{F}$. If $\mathcal{F} = \varnothing$, let $M = 0$. If $\mathcal{F}$ is not empty, let M be one less than the length of the longest block in $\mathcal{F}$. Our discussion above shows that X is described by $\mathcal{F}_{M+1}$, and so is M-step. □

The language of a shift of finite type is characterized by the property that if two words overlap enough, then they can be glued together along their overlap to form another word in the language.

Theorem 2.1.8. *A shift space X is an M-step shift of finite type if and only if whenever $uv, vw \in \mathcal{B}(X)$ and $|v| \geqslant M$, then $uvw \in \mathcal{B}(X)$.*

PROOF: First suppose that X is an M-step shift of finite type, so that $X = X_{\mathcal{F}}$ for a finite list $\mathcal{F}$ consisting of $(M+1)$-blocks. Suppose that $uv, vw \in \mathcal{B}(X)$, where $|v| = n \geqslant M$. Then there are points $x, y \in X$ with $x_{[-k,n]} = uv$ and $y_{[1,l]} = vw$, so that $x_{[1,n]} = y_{[1,n]} = v$. We claim that the point $z = x_{(-\infty,0]} v y_{[n+1,\infty)}$ is in X. For if a word in $\mathcal{F}$ were to occur in z, then it must occur in either $x_{(-\infty,0]} v = x_{(-\infty,n]}$ or in $v y_{[n+1,\infty)} = y_{[1,\infty)}$ since $|v| = n \geqslant M$, contradicting $x \in X$ or $y \in X$. Hence

$$ uvw = x_{[-k,0]} \, v \, y_{[n+1,l]} = z_{[-k,l]} \in \mathcal{B}(X). $$

Conversely, suppose that X is a shift space over $\mathcal{A}$, and that M has the property that if $uv, vw \in \mathcal{B}(X)$ with $|v| \geqslant M$, then $uvw \in \mathcal{B}(X)$. Let $\mathcal{F}$ be the set of all $(M+1)$-blocks over $\mathcal{A}$ that are not in $\mathcal{B}_{M+1}(X)$. We will show that $X = X_{\mathcal{F}}$, verifying that X is an M-step shift of finite type.

If $x \in X$, then no block in $\mathcal{F}$ can occur in x, and so $x \in X_{\mathcal{F}}$. This shows that $X \subseteq X_{\mathcal{F}}$. Now let $x \in X_{\mathcal{F}}$. Then $x_{[0,M]}$ and $x_{[1,M+1]}$ are in $\mathcal{B}(X)$ by our definition of $\mathcal{F}$. Since they overlap in M symbols, $x_{[0,M+1]}$ is also in $\mathcal{B}(X)$ (use $u = x_{[0]}$, $v = x_{[1,M]}$, and $w = x_{[M+1]}$). Now $x_{[2,M+2]}$ is in $\mathcal{B}(X)$, and overlaps in M symbols with $x_{[0,M+1]}$. Thus $x_{[0,M+2]} \in \mathcal{B}(X)$. Repeated application of this argument in each direction shows that $x_{[-k,l]} \in \mathcal{B}(X)$ for all $k, l \geqslant 0$. By Corollary 1.3.5 this implies that $x \in X$, proving that $X_{\mathcal{F}} \subseteq X$. □

Example 2.1.9. This theorem gives another way to see that the even shift X is not a shift of finite type. For if it were, it would be M-step for some $M \geqslant 1$. Since 10^{2M+1} and $0^{2M+1}1$ are allowed blocks in $\mathcal{B}(X)$, we would have $10^{2M+1}1$ in $\mathcal{B}(X)$, which violates the definition of the even shift. □

In Example 1.5.6 we discussed a sliding block code from the golden mean shift onto the even shift. Thus a factor of a shift of finite type does not always have finite type. There are also cases when the inverse image of a shift of finite type does not have finite type. For instance, the 1-point shift $Y = \{0^\infty\}$, which is of finite type, is a factor of any shift X. For a less trivial example, recall from Exercise 1.5.16, that the golden mean shift is a factor of the even shift. These remarks help explain why the proof of the following result, which shows that the property of having finite type is invariant under conjugacy, is more subtle than you might expect.

Theorem 2.1.10. *A shift space that is conjugate to a shift of finite type is itself a shift of finite type.*

PROOF: Suppose that X is a shift space that is conjugate to a shift of finite type Y. Let Φ be a block map that induces a conjugacy from X to Y, and Ψ be a block map that induces its inverse. The idea is to apply Theorem 2.1.8, observing that if two blocks in X overlap sufficiently, then their images under Φ will overlap enough to glue them together to form a block in Y. Applying Ψ to this block yields the original blocks in X glued along their overlap, but shortened at each end. To accommodate this shortening, we need to first extend the original blocks.

According to Theorem 2.1.8, our goal is to find an integer $M \geqslant 1$ such that if $v \in \mathcal{B}(X)$ with $|v| \geqslant M$, and if $uv, vw \in \mathcal{B}(X)$, then $uvw \in \mathcal{B}(X)$.

Let $\phi \colon X \to Y$ be a conjugacy induced by a block map Φ, and $\psi = \phi^{-1} \colon Y \to X$ be its inverse, induced by Ψ. By increasing the window size if necessary, we can assume that ϕ and ψ have the same memory and anticipation, say l. Observe that since $\psi(\phi(x)) = x$ for all $x \in X$, the block map $\Psi \circ \Phi \colon \mathcal{B}_{4l+1}(X) \to \mathcal{B}_1(X)$ merely selects the central symbol in a block.

Since Y has finite type, by Theorem 2.1.8 there is an $N \geqslant 1$ such that two blocks in Y that overlap in at least N places can be glued together along their overlap to form a block in Y.

Set $M = N + 4l$. To verify that this choice satisfies the conditions of Theorem 2.1.8, and thereby prove that X has finite type, let $uv, vw \in \mathcal{B}(X)$ with $|v| \geqslant M$. Figure 2.1.1 sketches the following arguments. By Proposition 1.3.4, there are words $s, t \in \mathcal{B}_{2l}(X)$ such that $suv, vwt \in \mathcal{B}(X)$. Since every $(4l + 1)$-block in $suvwt$ is in $\mathcal{B}(X)$ (although we do not know yet that $suvwt$ is itself in $\mathcal{B}(X)$), the description above of the action of $\Psi \circ \Phi$ shows that $\Psi(\Phi(suvwt)) = uvw$.

Now $\Phi(suv) = u'\Phi(v)$ and $\Phi(vwt) = \Phi(v)w'$, where $u', w' \in \mathcal{B}(Y)$ and $|\Phi(v)| = |v| - 2l \geqslant N$. Hence $u'\Phi(v)$ and $\Phi(v)w'$ can be glued together to form $u'\Phi(v)w' \in \mathcal{B}(Y)$. Then

$$uvw = \Psi(\Phi(suvwt)) = \Psi(u'\Phi(v)w') \in \mathcal{B}(X),$$

proving that X has finite type. $\square$

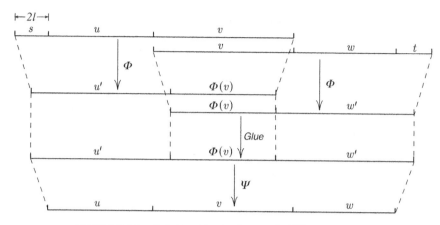

FIGURE 2.1.1.   Relationships in the proof of Theorem 2.1.10.

Let $\mathcal{A}$ be a finite group. Then the full shift $\mathcal{A}^{\mathbb{Z}}$ is a group via coordinate-wise addition, so that

$$(\dots, x_i, \dots) + (\dots, y_i, \dots) = (\dots, x_i + y_i, \dots).$$

Let X be a subshift of $\mathcal{A}^{\mathbb{Z}}$ which is also a subgroup of $\mathcal{A}^{\mathbb{Z}}$; i.e., X is closed under addition and inverses. Such a subshift is called a *group shift*. Recall from §1.6 that since convolutional codes are images of linear sliding block codes, they are linear shift spaces. In particular, any convolutional code is a subshift and subgroup of $\mathcal{A}^{\mathbb{Z}}$, where $\mathbb{F}$ is a finite field and $\mathcal{A} = \mathbb{F}^n$ with the additive group structure. Hence convolutional codes are examples of group shifts.

In Exercise 2.1.11, we outline a proof of the fact that group shifts, in particular convolutional codes, are shifts of finite type. A method of constructing group shifts is outlined in Exercise 2.2.16.

EXERCISES

2.1.1. Decide which of the following examples from Chapter 1 are shifts of finite type. If so, specify a finite collection $\mathcal{F}$ describing the shift.

(a) The $(1, 3)$ run-length limited shift (Example 1.2.5).
(b) The prime gap shift (Example 1.2.6).
(c) The charge constrained shift (Example 1.2.7).
(d) The context-free shift (Example 1.2.9).

2.1.2. Find an infinite collection $\mathcal{F}$ of forbidden blocks over $\mathcal{A} = \{0, 1\}$ so that $X_{\mathcal{F}}$ is the golden mean shift.

2.1.3. Show that if X and Y are shifts of finite type over $\mathcal{A}$, then so is $X \cap Y$. Must $X \cup Y$ also be a shift of finite type?

2.1.4. If X and Y are shifts of finite type, show that their product shift $X \times Y$, as defined in Exercise 1.2.9, also has finite type.

2.1.5. Show that for an M-step shift of finite type X and $n \geqslant M + 1$, a block $u \in \mathcal{B}_n(X)$ if and only if for every subblock v of u with $|v| = M + 1$, we have that $v \in \mathcal{B}_{M+1}(X)$.

2.1.6. Let $\mathcal{A} = \{0, 1\}$ and $\mathcal{F} = \{11, 00000\}$. Form, as in the text, the collection $\mathcal{F}_5$ of 5-blocks such that $X_{\mathcal{F}} = X_{\mathcal{F}_5}$. How many blocks does $\mathcal{F}_5$ contain? Generalize your answer to $\mathcal{F} = \{11, 0^n\}$ for $n \geqslant 1$.

2.1.7. (a) Solve Exercise 1.2.7 when the full $\mathcal{A}$-shift is replaced by an irreducible shift.

    (b) Find a (reducible) shift of finite type that is the union of two shifts of finite type that are proper subsets of it.

    (c) Find a reducible shift of finite type that is *not* the union of two shifts of finite type that are proper subsets of it.

2.1.8. If $\mathcal{F}$ is a collection of blocks for which $X_{\mathcal{F}} = \varnothing$, must there always be a *finite* subcollection $\mathcal{F}_0 \subseteq \mathcal{F}$ such that $X_{\mathcal{F}_0} = \varnothing$?

*2.1.9. Characterize the subsets S of integers such that the S-gap shift is a shift of finite type.

*2.1.10. Show that for any shift of finite type X, the list of first offenders (see Exercise 1.3.8) is the unique forbidden list $\mathcal{F}$ such that $X = X_{\mathcal{F}}$ and $\mathcal{F}$ minimizes $\sum_{w \in \mathcal{F}} |w|$.

*2.1.11. Let X be a group shift over a finite group $\mathcal{A}$. Show that X is a shift of finite type as follows.

    Let e denote the identity element of $\mathcal{A}$, and for a block $u \in \mathcal{B}(X)$, let $F_k(u)$ denote the set of all blocks $v \in \mathcal{B}_k(X)$ such that $uv \in \mathcal{B}(X)$.

    (a) Show that $\mathcal{B}_k(X)$ is a subgroup of $\mathcal{A}^k$.

    (b) Show that for any n, $F_k(e^n)$ is a normal subgroup of $\mathcal{B}_k(X)$.

    (c) Show that for any n, k, the quotient group $\mathcal{B}_k(X)/F_k(e^n) = \{F_k(u) : u \in \mathcal{B}_n(X)\}$.

    (d) Show that for some positive integer N and all $n \geqslant N$, $F_1(e^n) = F_1(e^N)$.

    (e) Show that for this same N, any $n \geqslant N$, and any k, $F_k(e^n) = F_k(e^N)$.

    (f) Show that for this same N, any block $u \in \mathcal{B}(X)$, and any k, $F_k(u) = F_k(v)$ where v is the suffix of u of length N.

    (g) Show that X is a shift of finite type.

§2.2. Graphs and Their Shifts

A fundamental method to construct shifts of finite type starts with a finite, directed graph and produces the collection of all bi-infinite walks (i.e., sequences of edges) on the graph. In a sense that we will make precise in the next section, *every* shift of finite type can be recoded to look like such an edge shift. In this section we introduce edge shifts, and use the adjacency matrix of the graph to answer questions about its shift. Indeed, the reason that an edge shift can be understood far better than a general shift is that we can apply the powerful machinery of linear algebra to its adjacency matrix.

We will work with finite, directed graphs. Only the simplest properties of such graphs will be developed here.

Definition 2.2.1. A *graph* G consists of a finite set $\mathcal{V} = \mathcal{V}_G$ of *vertices* (or *states*) together with a finite set $\mathcal{E} = \mathcal{E}(G)$ of *edges*. Each edge $e \in \mathcal{E}(G)$

starts at a vertex denoted by $i(e) \in \mathcal{V}_G$ and *terminates* at a vertex $t(e) \in \mathcal{V}_G$ (which can be the same as $i(e)$). Equivalently, the edge e has *initial state* $i(e)$ and *terminal state* $t(e)$. There may be more than one edge between a given initial state and terminal state; a set of such edges is called a set of *multiple edges*. An edge e with $i(e) = t(e)$ is called a *self-loop*.

We shall usually shorten $\mathcal{V}_G$ to $\mathcal{V}$ and $\mathcal{E}(G)$ to $\mathcal{E}$ when G is understood. For a state I, $\mathcal{E}_I = \mathcal{E}_I(G)$ denotes the set of outgoing edges, and $\mathcal{E}^I = \mathcal{E}^I(G)$ denotes the set of incoming edges; the number, $|\mathcal{E}_I|$ of outgoing edges from I is called the *out-degree of I*, and likewise, $|\mathcal{E}^I|$ is called the *in-degree of I*.

Be warned that terminology in graph theory is notoriously nonstandard. Edges are sometimes called arcs, and vertices sometimes called nodes. Some authors allow self-loops or multiple edges between a given pair of vertices; others do not. The books [Wil] and [BonM] provide good background reading for general graph theory.

Figure 2.2.1 depicts a typical graph G. For this graph, $\mathcal{V} = \{I, J\}$, and $\mathcal{E}$ contains 8 edges. The edges of a graph are regarded as distinct from each other. You can think of them as carrying different names or colors. The edge e in Figure 2.2.1 has initial state $i(e) = I$ and terminal state $t(e) = J$, while edges f and g are self-loops since they both have initial and terminal state J.

There are natural ways to map graphs to other graphs so that initial states and terminal states are preserved.

Definition 2.2.2. Let G and H be graphs. A *graph homomorphism from G to H* consists of a pair of maps $\partial\Phi\colon \mathcal{V}_G \to \mathcal{V}(H)$ and $\Phi\colon \mathcal{E}(G) \to \mathcal{E}(H)$ such that $i(\Phi(e)) = \partial\Phi(i(e))$ and $t(\Phi(e)) = \partial\Phi(t(e))$ for all edges $e \in \mathcal{E}(G)$. In this case we write $(\partial\Phi, \Phi) : G \to H$.

A graph homomorphism $(\partial\Phi, \Phi)$ is a *graph embedding* if both $\partial\Phi$ and Φ are one-to-one. It is a *graph isomorphism* if both $\partial\Phi$ and Φ are one-to-one and onto, in which case we write $(\partial\Phi, \Phi) : G \cong H$. Two graphs G and H are *graph isomorphic* (written $G \cong H$) if there is a graph isomorphism between them.

This definition says that when two graphs are isomorphic you can obtain one from the other by renaming vertices and edges. We tend to regard isomorphic graphs as being "the same" for all practical purposes.

Next we discuss the adjacency matrix of a graph G. It is convenient to assume that we have listed the vertices in some definite order, say $\mathcal{V} =$

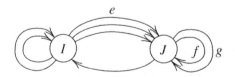

FIGURE 2.2.1. A typical graph.

$\{I_1, I_2, \ldots, I_r\}$. Often we will label the vertices with distinct integers, such as $\mathcal{V} = \{1, 2, \ldots, r\}$ or $\mathcal{V} = \{0, 1\}$, or with distinct letters, as in $\mathcal{V} = \{I, J, K\}$. In such cases, the order of listing is understood to be the usual one, either numerical or alphabetical.

Definition 2.2.3. Let G be a graph with vertex set $\mathcal{V}$. For vertices $I, J \in \mathcal{V}$, let A_{IJ} denote the number of edges in G with initial state I and terminal state J. Then the *adjacency matrix* of G is $A = [A_{IJ}]$, and its formation from G is denoted by $A = A(G)$ or $A = A_G$.

For example, the graph G in Figure 2.2.1 has adjacency matrix

$$A_G = \begin{bmatrix} 2 & 3 \\ 1 & 2 \end{bmatrix}.$$

Notice that the order of listing the vertices in $\mathcal{V}$ simply tells us how to arrange the numbers A_{IJ} into a square array. The effect on A of using a different listing order for $\mathcal{V}$ is easy to describe. To do this, recall that a *permutation matrix* is a square matrix with entries either 0 or 1 that has exactly one 1 in each row and in each column. If $\mathcal{V}$ is listed in a different order, and this order is used to write down an adjacency matrix A', then there is a permutation matrix P such that $A' = PAP^{-1}$ (Exercise 2.2.6).

The information in A_G is all that is needed to reconstruct the original graph G, at least up to graph isomorphism.

Definition 2.2.4. Let $A = [A_{IJ}]$ be an $r \times r$ matrix with nonnegative integer entries. Then the *graph of* A is the graph $G = G(A) = G_A$ with vertex set $\mathcal{V}_G = \{1, 2, \ldots, r\}$, and with A_{IJ} distinct edges with initial state I and terminal state J.

These operations are clearly inverse to one another, in the sense that

(2–2–1) $\qquad A = A(G(A)) \qquad$ and $\qquad G \cong G(A(G))$.

The graph isomorphism here results from the possibility of a different naming of vertices and edges in $G(A(G))$. This correspondence between graphs and their adjacency matrices means that we can use either G or A to specify a graph, whichever is more convenient. Our notation will reflect both possibilities.

Since an $r \times r$ nonnegative integral matrix A is naturally associated to the graph G_A, we will sometimes speak of the indices of A as states and use the notation $\mathcal{V}(A) = \{1, \ldots, r\}$ for the set of indices.

Next, each graph G with corresponding adjacency matrix A gives rise to a shift of finite type.

Definition 2.2.5. Let G be a graph with edge set $\mathcal{E}$ and adjacency matrix A. The *edge shift* X_G or X_A is the shift space over the alphabet $\mathcal{A} = \mathcal{E}$ specified by

$$(2\text{-}2\text{-}2) \quad X_G = X_A = \{\xi = (\xi_i)_{i \in \mathbb{Z}} \in \mathcal{E}^{\mathbb{Z}} : t(\xi_i) = i(\xi_{i+1}) \text{ for all } i \in \mathbb{Z}\}.$$

The shift map on X_G or X_A is called the *edge shift map* and is denoted by σ_G or σ_A.

According to this definition, a bi-infinite sequence of edges is in X_G exactly when the terminal state of each edge is the initial state of the next one; i.e., the sequence describes a *bi-infinite walk* or *bi-infinite trip* on G. For the graph G in Figure 2.2.1, X_G has eight symbols. With edges f and g as indicated, $i(f) = t(f) = i(g) = t(g) = J$, so that any bi-infinite sequence of f's and g's is a point in X_G.

It it easy to see that when G and H are graph isomorphic, X_G and X_H are conjugate (Exercise 2.2.14). In fact, since, in this case, we tend to regard G and H as being the same, we also tend to regard X_G and X_H as being the same.

Our next result shows that edge shifts are indeed shift spaces, and in fact have finite type.

Proposition 2.2.6. *If G is a graph with adjacency matrix A, then the associated edge shift $X_G = X_A$ is a 1-step shift of finite type.*

PROOF: Let $\mathcal{A} = \mathcal{E}$ be the alphabet of X_G. Consider the finite collection

$$\mathcal{F} = \{ef : e, f \in \mathcal{A}, \ t(e) \neq i(f)\}$$

of 2-blocks over $\mathcal{A}$. According to Definition 2.2.5, a point $\xi \in \mathcal{A}^{\mathbb{Z}}$ lies in X_G exactly when no block of $\mathcal{F}$ occurs in ξ. This means that $X_G = X_{\mathcal{F}}$, so that X_G has finite type. Since all blocks in $\mathcal{F}$ have length 2, $X_{\mathcal{F}}$ is 1-step. $\square$

Example 2.2.7. Let $r \geqslant 1$, and A be the 1×1 matrix $A = [r]$. Then G_A has one vertex and r self-loops at that vertex. If we name the edges as $0, 1, \ldots, r - 1$, then X_G is the full r-shift (see Figure 2.2.2), which we therefore will denote by $X_{[r]}$. $\square$

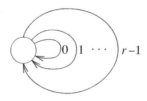

FIGURE 2.2.2.   Graph for the full r-shift.

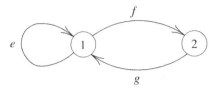

FIGURE 2.2.3. Graph of Example 2.2.8.

Example 2.2.8. Let A be the matrix

$$A = \begin{bmatrix} 1 & 1 \\ 1 & 0 \end{bmatrix}.$$

Its graph $G = G_A$ is shown in Figure 2.2.3. If we name the edges as indicated, then X_G is the same as Example 1.2.8. □

Sometimes certain edges of G can never appear in X_G, and such edges are inessential for the edge shift. For example, if an edge ends at a vertex from which no edges start, it is a "dead end" and cannot occur in any bi-infinite walk on G. Call a vertex $I \in \mathcal{V}$ *stranded* if either no edges start at I or no edges terminate at I. In terms of the adjacency matrix A_G, I is stranded precisely when either the Ith row or the Ith column of A_G contains only 0's.

Definition 2.2.9. A graph is *essential* if no vertex of the graph is stranded.

To remove stranded vertices, we need the notion of a *subgraph* of G, i.e., a graph H with $\mathcal{V}(H) \subseteq \mathcal{V}(G)$, $\mathcal{E}(H) \subseteq \mathcal{E}(G)$, and edges in H starting and terminating at the same vertices as they do in G.

Proposition 2.2.10. *If G is a graph, then there is a unique subgraph H of G such that H is essential and $X_H = X_G$.*

PROOF: Let $\mathcal{E}(H)$ consist of those edges $e \in \mathcal{E}(G)$ contained in some bi-infinite walk on G, and let $\mathcal{V}(H) = \{i(e) : e \in \mathcal{E}(H)\}$ be the set of vertices visited on such walks. Then H is a subgraph of G, and by definition any bi-infinite walk on G is actually a walk on H. Thus H is essential, and $X_H = X_G$.

In order to show that H is unique, first observe that the definition of H shows that it is the largest essential subgraph of G. Thus if H' is an essential subgraph of G and $H' \neq H$, then there is an edge of H not in H'. Since every edge in H occurs in X_H, this proves that $X_{H'} \neq X_H$, establishing the uniqueness of H. □

Since the subgraph H in this proposition contains the only part of G used for symbolic dynamics, we will usually confine our attention to essential graphs.

We shall be using finite paths on a graph as well as the bi-infinite walks used to define the edge shift.

Definition 2.2.11. A *path* $\pi = e_1 e_2 \ldots e_m$ on a graph G is a finite sequence of edges e_i from G such that $t(e_i) = i(e_{i+1})$ for $1 \leqslant i \leqslant m-1$. The *length* of $\pi = e_1 e_2 \ldots e_m$ is $|\pi| = m$, the number of edges it traverses. The path $\pi = e_1 e_2 \ldots e_m$ *starts at* vertex $i(\pi) = i(e_1)$ and *terminates at* vertex $t(\pi) = t(e_m)$, and π is a path *from* $i(\pi)$ *to* $t(\pi)$. A *cycle* is a path that starts and terminates at the same vertex. A *simple cycle* is a cycle that does not intersect itself, i.e., a cycle $\pi = e_1 e_2 \ldots e_m$ such that the states $i(e_1), \ldots, i(e_m)$ are distinct. For each vertex I of G there is also an *empty path* ε_I, having length 0, which both starts and terminates at I.

Notice that if G is essential, then nonempty paths on G correspond to nonempty blocks in its edge shift X_G.

Information about paths on G can be obtained from the adjacency matrix A of G as follows. Let $\mathcal{E}_I^J$ denote the collection of edges in G with initial state I and terminal state J. Then $\mathcal{E}_I^J$ is the collection of paths of length 1 from I to J, and has size A_{IJ}. In particular, A_{II} is the number of self-loops at vertex I. The total number of self-loops is then the sum of the diagonal entries of A, which is the *trace* of A and denoted by $\operatorname{tr} A$. Extending this idea, we can use the matrix powers of A to count longer paths and cycles.

Proposition 2.2.12. *Let G be a graph with adjacency matrix A, and let $m \geqslant 0$.*

(1) *The number of paths of length m from I to J is $(A^m)_{IJ}$, the (I,J)th entry of A^m.*

(2) *The number of cycles of length m in G is $\operatorname{tr}(A^m)$, the trace of A^m, and this equals the number of points in X_G with period m.*

PROOF: (1) This is true for $m = 0$ since the only paths of length 0 are the empty paths at each vertex. If $m \geqslant 1$, then counting the paths of length m from I to J by adding up over the possible sequences of the $m-1$ intervening vertices yields the (I, J)th entry of A^m. Details of this argument are left to the reader.

(2) The first part of the statement follows from (1) and the definition of cycle. For the second part, note that if π is a cycle in G of length m, then π^∞ is a point of period m in X_G, while if $x \in X_G$ has period m, then $x_{[0,m-1]}$ must be a cycle in G of length m. This sets up a one-to-one correspondence between cycles in G of length m and points in X_G of period m. $\qquad \square$

Recall from §1.3 that a shift space is called irreducible if every ordered pair of allowed blocks u and v can be joined with a block w so that uwv is also allowed. The graphs giving rise to irreducible edge shifts are easy to describe.

Definition 2.2.13. A graph G is *irreducible* if for every ordered pair of vertices I and J there is a path in G starting at I and terminating at J.

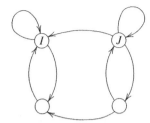

FIGURE 2.2.4. A reducible graph.

Irreducible graphs are sometimes called *strongly connected* in graph theory. The graphs we have looked at thus far are irreducible. Note that for a graph to be irreducible, you need to check that, for any two vertices I and J, there exist a path from I to J *and* a path from J to I. Figure 2.2.4 shows a *reducible* graph, i.e., one that is not irreducible, since there is no path from vertex I to vertex J.

Proposition 2.2.14. *An essential graph is irreducible if and only if its edge shift is irreducible.*

PROOF: Let G be an irreducible graph, and $\pi, \tau \in \mathcal{B}(X_G)$. Suppose that π terminates at vertex I and τ starts at vertex J. By irreducibility of G, there is a path $\omega \in \mathcal{B}(X_G)$ from I to J. Then $\pi\omega\tau$ is a path on G, so that $\pi\omega\tau \in \mathcal{B}(X_G)$.

Conversely, suppose that G is essential, and that X_G is an irreducible shift. Let I and J be vertices of G. Since G is essential, there are edges e and f such that e terminates at I and f starts at J. By irreducibility of X_G, there is a block ω so that $ewf \in \mathcal{B}(X_G)$. Then ω is a path in G from I to J, so that G is an irreducible graph. □

Finally, we discuss how symbolic dynamics treats "time-reversal." For a shift space X, the *transposed shift* X^{T} is the shift space obtained from X by reading the sequences in X backwards. For a graph G, the *transposed graph* G^{T} of G is the graph with the same vertices as G, but with each edge in G reversed in direction. Recall that the *transpose* of a matrix A is the matrix A^{T} obtained from A by switching the (I, J) and (J, I) entries for each I and J. Then $A(G^{\mathsf{T}}) = A(G)^{\mathsf{T}}$ and $(X_G)^{\mathsf{T}} = X_{G^{\mathsf{T}}}$ (provided that we identify the states, edges of G with the states, edges of G^{T} in the obvious way).

EXERCISES

2.2.1. Show that the graph in Figure 2.2.3 is isomorphic to a subgraph of the graph in Figure 2.2.1.

2.2.2. Let G be the graph in Figure 2.2.3.
  (a) For $1 \leqslant n \leqslant 6$ compute the total number of paths in G of length n. Compare your answers to those from Exercise 1.1.4. Can you explain any similarities?

(b) For $1 \leqslant n \leqslant 6$ compute the total number of periodic points in X_G with period n. Can you guess the answer for general n?

2.2.3. Suppose that H is a subgraph of G. What is the relationship between the adjacency matrices A_H and A_G?

2.2.4. For the graph G below, find the unique essential subgraph H guaranteed by Proposition 2.2.10.

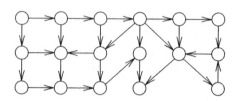

2.2.5. Give a different, and more concrete, proof of Proposition 2.2.10 as follows. Remove from G all stranded vertices and any adjacent edges, obtaining a subgraph. Repeat this process until there are no more stranded vertices. Show that what remains is the required graph H.

2.2.6. Show that if the vertices of G are listed in a different order, and A' is the resulting adjacency matrix, then there is a permutation matrix P so that $A' = PAP^{-1}$.

2.2.7. Prove that, with exactly one exception, an irreducible graph is essential, and find the exception.

2.2.8. Is an edge shift always the union of irreducible edge shifts?

2.2.9. Suppose that G and H are essential graphs, and that $X_G = X_H$. Prove that $G = H$.

2.2.10. Let G be a graph with adjacency matrix A. Show that the following statements are equivalent.
   (a) There is an integer $n \geqslant 1$ such that all paths in G have length less than n.
   (b) A is nilpotent; i.e., $A^n = 0$ for some $n \geqslant 1$.
   (c) $X_G = \varnothing$.

2.2.11. Formulate an algorithm which decides for a given finite list $\mathcal{F}$ of forbidden blocks and a given block w if w occurs in the shift of finite type $X_{\mathcal{F}}$.

2.2.12. How could you use the adjacency matrix of a graph to determine whether or not the graph is irreducible?

2.2.13. For essential graphs G and H, show that a map $\phi: X_G \to X_H$ is a 1-block code if and only if $\phi = \Phi_\infty$ where Φ is the edge mapping of a graph homomorphism from G to H.

2.2.14. Prove that essential graphs G and H are graph isomorphic if and only if there is a 1-block conjugacy $\phi: X_G \to X_H$ whose inverse $\phi^{-1}: X_H \to X_G$ is also 1-block.

*2.2.15. (a) For a sliding block code ϕ whose domain is an edge shift, let W_ϕ denote the set of all pairs of nonnegative integers (m, n) such that ϕ has memory m and anticipation n. Show that there is a unique element $(m, n) \in W_\phi$ such that $m + n$ is minimal.
   (b) Show that part (a) is false if the domain is assumed to be merely a shift of finite type.

*2.2.16. Let $\mathcal{A}$ be a finite group which contains two normal subgroups H and K such that $\mathcal{A}/H$ is isomorphic (as a group) to $\mathcal{A}/K$ via an isomorphism $\phi: \mathcal{A}/H \to \mathcal{A}/K$. Let $p_K: \mathcal{A} \to \mathcal{A}/K$ be the map which sends each element of $\mathcal{A}$ to the coset of K that it belongs to; likewise for $p_H: \mathcal{A} \to \mathcal{A}/H$. Let $G = (\mathcal{V}, \mathcal{E})$ be the graph with vertices $\mathcal{V} = \mathcal{A}/K$, edges $\mathcal{E} = \mathcal{A}$, initial state $i(e) = p_K(e)$ and terminal state $t(e) = \phi \circ p_H(e)$.

    (a) Show that the edge shift X_G is a group shift i.e., a subgroup of the group $\mathcal{A}^{\mathbb{Z}}$.

    (b) Show that any group shift which is also an edge shift arises in this way.

§2.3. Graph Representations of Shifts of Finite Type

The edge shifts introduced in the previous section may seem to be quite special shifts of finite type. For example, they have memory 1 since when walking on a graph, where you can go next only depends on where you are now, not on how you got there. However, we shall show that any shift of finite type can be recoded, using a higher block presentation, to be an edge shift. We will also introduce an analogue for graphs of the higher block presentations, and give an alternative description of shifts of finite type using 0–1 matrices. This section involves some new machinery, but these constructions are basic to coding arguments throughout this book.

The first example shows that not every shift of finite type (in fact, not every 1-step shift of finite type) is an edge shift.

Example 2.3.1. Let $\mathcal{A} = \{0, 1\}$, $\mathcal{F} = \{11\}$, so that $X_\mathcal{F}$ is the golden mean shift of Example 1.2.3. We claim that there is no graph G so that $X_\mathcal{F} = X_G$. For if there were such a G, we could assume by Proposition 2.2.10 that it is essential. Then G would have exactly two edges, named 0 and 1. The only possibilities are that G has just one vertex, in which case X_G is the full 2-shift, or that G has two vertices, so that X_G is just the two points $(01)^\infty$ and $(10)^\infty$. In neither case is X_G the same as $X_\mathcal{F}$. $\square$

Despite this example, the following result shows that any shift of finite type can be recoded to an edge shift. Recall that X is called M-step if $X = X_\mathcal{F}$ for a collection $\mathcal{F}$ of $(M+1)$-blocks.

Theorem 2.3.2. *If X is an M-step shift of finite type, then there is a graph G such that $X^{[M+1]} = X_G$.*

PROOF: First note that if $M = 0$, then X is a full shift, and we can take G to have a single vertex and one edge for each symbol appearing in X. Thus we may assume that $M \geqslant 1$. Define the vertex set of G to be $\mathcal{V} = \mathcal{B}_M(X)$, the allowed M-blocks in X. We define the edge set $\mathcal{E}$ as follows. Suppose that $I = a_1 a_2 \ldots a_M$ and $J = b_1 b_2 \ldots b_M$ are two vertices in G. If $a_2 a_3 \ldots a_M = b_1 b_2 \ldots b_{M-1}$, and if $a_1 \ldots a_M b_M$ $(= a_1 b_1 \ldots b_M)$ is in $\mathcal{B}(X)$, then draw exactly one edge in G from I to J, named $a_1 a_2 \ldots a_M b_M = a_1 b_1 b_2 \ldots b_M$. Otherwise, there is no edge from I

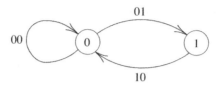

FIGURE 2.3.1.   Graph of the recoded golden mean shift.

to J. From this construction, it is clear that a bi-infinite walk on G is precisely a sequence of $(M+1)$-blocks in $\mathcal{B}_{M+1}(X)$ which overlap progressively. Hence $X_G = X^{[M+1]}$. □

Example 2.3.3. The golden mean shift X in Example 2.3.1 is 1-step, but is not itself an edge shift. Carrying out the process described in the above proof shows that $X^{[2]} = X_G$, where G is the graph in Figure 2.3.1. □

Coding arguments frequently involve considering several symbols grouped together. We have already seen one method to do this, namely the Nth higher block presentation $X^{[N]}$ of X. The following is an analogous construction for graphs.

Definition 2.3.4. Let G be a graph. For $N \geqslant 2$ we define the *Nth higher edge graph* $G^{[N]}$ of G to have vertex set equal to the collection of all paths of length $N - 1$ in G, and to have edge set containing exactly one edge from $e_1e_2 \ldots e_{N-1}$ to $f_1f_2 \ldots f_{N-1}$ whenever $e_2e_3 \ldots e_{N-1} = f_1f_2 \ldots f_{N-2}$ (or $t(e_1) = i(f_1)$ if $N = 2$), and none otherwise. The edge is named $e_1e_2 \ldots e_{N-1}f_{N-1} = e_1f_1f_2 \ldots f_{N-1}$. For $N = 1$ we put $G^{[1]} = G$.

Example 2.3.5. Figure 2.3.2 shows the 2nd and 3rd higher edge graphs of the graph in Figure 2.2.3, including the names of vertices and edges. □

The relation between the edge shift on the Nth higher edge graph of G and the Nth higher block shift of X_G is spelled out as follows.

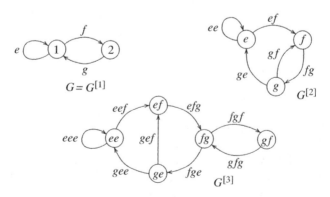

FIGURE 2.3.2.   Higher edge graphs of Figure 2.2.3.

Proposition 2.3.6. *Let G be a graph. Then $(X_G)^{[N]} = X_{G^{[N]}}$.*

PROOF: The symbols for $(X_G)^{[N]}$ are the N-blocks from X_G, which are the paths of length N in G. But these are also the symbols for $X_{G^{[N]}}$. A bi-infinite sequence of these symbols is in either shift precisely when the symbols (i.e., N-blocks) overlap progressively. □

Notice that for $N \geqslant 2$, the adjacency matrix for $G^{[N]}$ contains only 0's and 1's. A matrix whose entries are either 0 or 1 is called a *0–1 matrix*. If G has an adjacency matrix A that is a 0–1 matrix, then between any two vertices there is at most one edge. Hence a walk on G can be equally well described by the sequence of vertices visited. This leads to an alternative shift construction, but one which is only valid for 0–1 matrices.

Definition 2.3.7. Let B be an $r \times r$ matrix of 0's and 1's, or, equivalently, the adjacency matrix of a graph G such that between any two vertices there is at most one edge. The *vertex shift* $\widehat{X}_B = \widehat{X}_G$ is the shift space with alphabet $\mathcal{A} = \{1, 2, \ldots, r\}$, defined by

$$\widehat{X}_B = \widehat{X}_G = \{x = (x_i)_{i \in \mathbb{Z}} \in \mathcal{A}^{\mathbb{Z}} : B_{x_i x_{i+1}} = 1 \text{ for all } i \in \mathbb{Z}\}.$$

The *vertex shift map* is the shift map on $\widehat{X}_B = \widehat{X}_G$ and is denoted $\widehat{\sigma}_B$ or $\widehat{\sigma}_G$.

The reader might think of the hat over the X as an inverted "v" as a reminder that this is a vertex shift.

Example 2.3.8. Let

$$B = \begin{bmatrix} 1 & 1 \\ 1 & 0 \end{bmatrix}$$

Then the vertex shift $\widehat{X}_B$ is the golden mean shift described in Example 2.3.1 above. □

Recall from Proposition 2.2.6 that edge shifts are 1-step shifts of finite type. Vertex shifts are also 1-step shifts of finite type: $\widehat{X}_B$ is the shift defined by the list $\mathcal{F} = \{ij : B_{ij} = 0\}$ of forbidden 2-blocks. The following result establishes the relationship among edge shifts, vertex shifts, and 1-step shifts of finite type. In particular, vertex shifts are essentially the same as 1-step shifts of finite type, and every edge shift is essentially a vertex shift (but not conversely, e.g., Example 2.3.1). Part (3) of this result refines Theorem 2.3.2.

Proposition 2.3.9.

(1) *Up to a renaming of the symbols, the 1-step shifts of finite type are the same as the vertex shifts.*

(2) *Up to a renaming of the symbols, every edge shift is a vertex shift (on a different graph).*

(3) *If X is an M-step shift of finite type, then $X^{[M]}$ is a 1-step shift of finite type, equivalently a vertex shift. In fact, there is a graph G such that $X^{[M]} = \widehat{X}_G$ and $X^{[M+1]} = X_G$.*

PROOF: We have already seen above that every vertex shift is a 1-step shift of finite type. Conversely, every 1-step shift of finite type is, up to a renaming of the symbols, a vertex shift. For if $X = X_{\mathcal{F}}$ where $\mathcal{F}$ consists of 2-blocks, then X may be regarded as the vertex shift $\widehat{X}_B$, where B is the 0–1 matrix indexed by the alphabet of X and $B_{ij} = 0$ if and only if $ij \in \mathcal{F}$. This proves part (1).

Part (2) follows from Part (1) and Proposition 2.2.6.

The first statement of Part (3) follows immediately from the fact that any $(M+1)$-block in X may be regarded as a 2-block in $X^{[M]}$. For the second statement, use the graph constructed in Theorem 2.3.2; we leave the details to the reader. □

Part (2) of the preceding result asserts that every edge shift may be regarded as a vertex shift. To see directly how this works, start with an edge shift X_A having adjacency matrix A, and let $\mathcal{E}$ be the set of edges of the graph. There are $\sum_{I,J} A_{IJ}$ elements in $\mathcal{E}$. Form the 0–1 matrix B indexed by $\mathcal{E}$ defined by

$$B_{ef} = \begin{cases} 1 & \text{if } t(e) = i(f), \\ 0 & \text{if } t(e) \neq i(f). \end{cases}$$

Then $\widehat{X}_B$ is a version of X_A, in which the edges of G_A correspond to the vertices of G_B; essentially $\widehat{X}_B = X_A$ and $G_B = G_A^{[2]}$.

Since 0–1 matrices look easier to deal with than general nonnegative integer matrices, why not just use vertex shifts, instead of edge shifts? There are two reasons. The first is simply economy of expression. To express the edge shift for the adjacency matrix

$$A = \begin{bmatrix} 3 & 2 \\ 2 & 1 \end{bmatrix},$$

as a vertex shift, would require the much larger matrix

$$B = \begin{bmatrix} 1 & 1 & 1 & 1 & 1 & 0 & 0 & 0 \\ 1 & 1 & 1 & 1 & 1 & 0 & 0 & 0 \\ 1 & 1 & 1 & 1 & 1 & 0 & 0 & 0 \\ 0 & 0 & 0 & 0 & 0 & 1 & 1 & 1 \\ 0 & 0 & 0 & 0 & 0 & 1 & 1 & 1 \\ 1 & 1 & 1 & 1 & 1 & 0 & 0 & 0 \\ 1 & 1 & 1 & 1 & 1 & 0 & 0 & 0 \\ 0 & 0 & 0 & 0 & 0 & 1 & 1 & 1 \end{bmatrix}.$$

The second reason is that certain natural operations on matrices, such as taking powers, do not preserve the property of being a 0–1 matrix. This occurs, for example, when using the higher power graph, defined as follows.

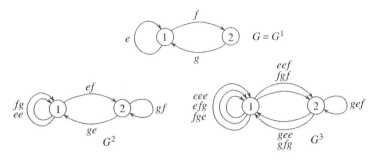

FIGURE 2.3.3. Higher power graphs of Figure 2.2.3.

Definition 2.3.10. Let G be a graph. If $N \geqslant 1$, we define the *Nth higher power graph* G^N of G to have vertex set $\mathcal{V}(G^N) = \mathcal{V}(G)$, and with precisely one edge from I to J for each path in G of length N from I to J.

Thus $G^1 = G$, while G^N usually has many more edges than G for large N.

Example 2.3.11. Let G be the graph shown in Figure 2.2.3. Then the 2nd and 3rd higher power graphs of G are displayed in Figure 2.3.3. $\square$

Proposition 2.3.12. *Let G be a graph with adjacency matrix A_G. Then the adjacency matrix of the Nth power graph G^N is the Nth power of A_G, or in symbols,*

$$A_{G^N} = (A_G)^N.$$

Furthermore, $\mathsf{X}_{G^N} = (\mathsf{X}_G)^N$.

PROOF: By definition, G^N and G have the same vertex set. From Proposition 2.2.12, for vertices I and J in G there are $(A_G^N)_{IJ}$ paths of length N from I to J, and this is the number of edges in G^N from I to J. Hence $A_{G^N} = (A_G)^N$.

Both X_{G^N} and $(\mathsf{X}_G)^N$ have alphabet consisting of paths of length N in G, and a bi-infinite sequence of such paths is in X_{G^N} (equivalently, in $(\mathsf{X}_G)^N$) exactly when it describes a bi-infinite walk on G. Thus $\mathsf{X}_{G^N} = (\mathsf{X}_G)^N$. $\square$

We have just made a case for favoring edge shifts over vertex shifts. In this text, we will use edge shifts much more than vertex shifts. However, there are some places where vertex shifts are more convenient; in particular, there are many useful vertex shifts that are not edge shifts, such as the golden mean shift.

There is a strong connection between the shifts that we have considered in this section and Markov chains, familiar from probability theory [KemS].

Definition 2.3.13. Let S be a finite set. A *Markov chain* μ on S is an assignment of *initial state probabilities* $\mu(I) \geqslant 0$ for $I \in S$ and *conditional probabilities* $\mu(J|I) \geqslant 0$ for $I, J \in S$ such that

$$\sum_{I \in S} \mu(I) = 1$$

and

$$\sum_{J \in S} \mu(J|I) = 1 \quad \text{for all } I \in S.$$

The idea is that S represents the set of outcomes, also called states, of an experiment. The probability that a sequence of outcomes $I_0 I_1 \ldots I_n$ occurs is defined to be

$$\mu(I_0 I_1 \ldots I_n) = \mu(I_0)\mu(I_1|I_0)\mu(I_2|I_1) \cdots \mu(I_n|I_{n-1}).$$

Given that a certain sequence of outcomes $I_0 I_1 \ldots I_{n-1}$ has already occurred with positive probability, the probability that the next outcome will be a particular I_n is

$$\frac{\mu(I_0 I_1 \ldots I_n)}{\mu(I_0 I_1 \ldots I_{n-1})} = \mu(I_n|I_{n-1})$$

and so depends only on the most recent outcome I_{n-1}.

What does this have to do with graphs and vertex shifts? Given a Markov chain, there is an associated graph $G = (\mathcal{V}, \mathcal{E})$ defined as follows. The vertices of G are the states of S with positive probability,

$$\mathcal{V} = \{I \in S : \mu(I) > 0\},$$

and the edges of G are the transitions, from one state to another that have positive conditional probability,

$$\mathcal{E} = \{(I, J) : \mu(J|I) > 0\}.$$

We assume that if $\mu(J|I) > 0$ then both $\mu(I) > 0$ and $\mu(J) > 0$, and if $\mu(J) > 0$, then $\mu(J|I) > 0$ for at least one I. With these assumptions, it follows that G is an essential graph (why?).

In G there is at most one edge between every pair of vertices. Hence sequences of outcomes that have positive probability are precisely the state sequences of paths in the graph G. In other words, they are the blocks that appear in the language of the vertex shift $\widehat{X}_G$. Thus a vertex shift is an "extreme" version of a Markov chain in the sense that it specifies only a set of allowed sequences of outcomes, the sequences that are actually possible, but not the probabilities of these sequences. This is why vertex shifts were originally called *intrinsic Markov chains* or *topological Markov chains*.

Now, consider an arbitrary graph $G = (\mathcal{V}, \mathcal{E})$ which is allowed to have multiple edges between any pair of vertices. If we specify initial state probabilities on the vertices $\mathcal{V}$ and conditional probabilities on the edges $\mathcal{E}$, then we can define probabilities of paths in G viewed as edge sequences.

Definition 2.3.14. A *Markov chain* on a graph $G = (\mathcal{V}, \mathcal{E})$ is an assignment of probabilities $\mu(I) \geqslant 0$ for $I \in \mathcal{V}$ and conditional probabilities $\mu(e|i(e)) \geqslant 0$ for $e \in \mathcal{E}$ such that

$$\sum_{I \in \mathcal{V}} \mu(I) = 1$$

and

$$\sum_{e \in \mathcal{E}_I} \mu(e|I) = 1 \quad \text{for all } I \in \mathcal{V}.$$

Analogous to an ordinary Markov chain, we assume that if $\mu(e|i(e)) > 0$ then both $\mu(i(e)) > 0$ and $\mu(t(e)) > 0$, and if $\mu(J) > 0$, then $\mu(e|i(e)) > 0$ for some edge e with $t(e) = J$.

For a Markov chain on a graph G, we define the probability of a path $e_1 \ldots e_n$ by

$$(2\text{--}3\text{--}1) \qquad \mu(e_1 \ldots e_n) = \mu(i(e_1))\mu(e_1|i(e_1))\mu(e_2|i(e_2)) \cdots \mu(e_n|i(e_n)).$$

Thus the *edges* of G represent the outcomes of the experiment, and the paths in G represent sequences of outcomes. The reader should be careful not to confuse the probability of an edge $\mu(e) = \mu(i(e))\mu(e|i(e))$ with the conditional probability of the edge $\mu(e|i(e))$.

Note that we do not demand that the initial state and conditional probabilities be positive. But if they are all positive, then the sequences that are actually possible are exactly the blocks that appear in the language of the edge shift X_G. Otherwise, the possible sequences are the blocks that appear in the language of the edge shift X_H determined by the subgraph H consisting of states with positive probability and edges with positive conditional probability.

Example 2.3.15. Let G be the graph in Figure 2.2.1. Consider the Markov chain on G defined as follows. The probabilities on states are defined to be $\mu(I) = 1/3, \mu(J) = 2/3$; the conditional probabilities are defined to be $1/8$ on each of the self-loops at state I, $1/2$ on the edge e, $1/8$ on the remaining edges from I to J, $1/2$ on the self-loop f, $1/3$ on the self-loop g, and $1/6$ on the edge from J to I. Then the probability of the path ef is $\mu(ef) = (1/3)(1/2)(1/2) = 1/12$, and the probability of the path eg is $\mu(eg) = (1/3)(1/2)(1/3) = 1/18$. □

For a Markov chain on a graph, it is convenient to assemble the initial state probabilities $\mu(I)$ into a row vector called the *initial state distribution*. This is a vector $\mathbf{p}$ indexed by the states of G defined by

$$p_I = \mu(I).$$

It is also convenient to collect the conditional probabilities $\mu(e|I)$ into a square matrix, called the *conditional probability matrix*. This is the matrix P, also indexed by the states of G, defined by

$$P_{IJ} = \sum_{e \in \mathcal{E}_I^J} \mu(e|I).$$

The vector $\mathbf{p}$ is a *probability vector*, i.e., a vector of nonnegative entries which sum to one. The matrix P is a *stochastic matrix*, i.e., a matrix with nonnegative entries whose rows sum to one.

For Example 2.3.15, we have $\mathbf{p} = [1/3, 2/3]$ and

$$P = \begin{bmatrix} 1/4 & 3/4 \\ 1/6 & 5/6 \end{bmatrix}.$$

We can use $\mathbf{p}$ and P to express certain natural quantities having to do with the Markov chain on G . For instance, $\mathbf{p}P^m$ is the vector whose Ith component is the sum of the probabilities of all the paths of length m in G that end at state I (Exercise 2.3.9).

Observe that if G has no multiple edges between each pair of vertices, then a Markov chain μ on G is essentially an ordinary Markov chain, and in this case the vector $\mathbf{p}$ and matrix P completely describe the Markov chain μ. Otherwise, just as an edge shift can be expressed as a vertex shift, a Markov chain on a graph can be expressed as an ordinary Markov chain (Exercise 2.3.10). But the number of states in the ordinary Markov chain may be much greater.

When we pass from a graph G to a higher edge graph $G^{[N]}$, we can transfer a Markov chain μ on G to a Markov chain $\mu^{[N]}$ on $G^{[N]}$. A state of $G^{[N]}$ is a path $e_1 \ldots e_{N-1}$ of length $N-1$ in G, and we assign the initial state probability of $e_1 \ldots e_{N-1}$ to be the probability $\mu(e_1 \ldots e_{N-1})$ of this path. An edge of $G^{[N]}$ is a path $e_1 \ldots e_N$ of length N in G, and we assign the conditional probability $\mu(e_1 \ldots e_N)/\mu(e_1 \ldots e_{N-1})$ to this edge. A path of length at least N in G can be viewed as a path in either G or $G^{[N]}$, and its μ-probability and $\mu^{[N]}$-probability coincide.

Conversely, any Markov chain on $G^{[N]}$ can be viewed as a "higher order Markov chain" on G where the dependence on initial state is replaced by dependence on an initial subpath of length $N-1$.

EXERCISES

2.3.1. Let G be the graph of the full 2-shift $X_{[2]}$ (see Figure 2.2.2). Draw the graphs $G^{[3]}$ and G^3.

2.3.2. Let $\mathcal{A} = \{0, 1\}$, $\mathcal{F} = \{000, 111\}$, and $X = X_{\mathcal{F}}$.

  (a) Construct a graph G for which $X_G = X_{\mathcal{F}}^{[3]}$.

  (b) Use the adjacency matrix of G to compute the number of points in X with period 5.

2.3.3. Complete the proof of Proposition 2.3.9.

2.3.4. Let X be a 1-step shift of finite type. For each symbol a of X, let $F_X(a)$ denote the set of symbols b such that $ab \in \mathcal{B}_2(X)$. Show that, up to relabeling symbols, X is an edge shift if and only if $F_X(a)$ and $F_X(c)$ are either disjoint or equal whenever a and c are symbols of X.

2.3.5. (a) Let X be a shift of finite type. Show that $X^{[N]}$ and X^N are shifts of finite type.

     (b) Let X be a shift space. If $X^{[N]}$ is a shift of finite type, must X be a shift of finite type? Similarly for X^N.

*2.3.6. (a) Let X be an irreducible shift of finite type and $\phi \colon X \to Y$ a sliding block code. Prove that ϕ is an embedding if and only if it is one-to-one on the periodic points of X.

     (b) Show by example that this can fail if X is only assumed to be a shift space.

2.3.7. Let $w = e_0 e_1 \ldots e_{p-1}$ be a cycle of length p. The *least period* of w is the smallest integer q so that $w = (e_0 e_1 \ldots e_{q-1})^m$ where $m = p/q$ must be an integer. The cycle w is *primitive* if its least period equals its length p. Recall that w is a simple cycle if no proper subblock of w is a cycle – equivalently, if the vertices $i(e_j)$ are all distinct for $0 \leqslant j \leqslant p-1$.

     (a) A simple cycle is clearly primitive. Give an example of a primitive cycle that is not simple.

     (b) If w is a cycle in G, then it corresponds to the cycle $w^{[N]}$ in $G^{[N]}$ given by

$$w^{[N]} = \beta_N(w^\infty)_{[0,p-1]}$$

where we have used Proposition 2.3.6. Show that if w is primitive, then $w^{[N]}$ is simple provided that $N \geqslant p$. Thus a primitive cycle can be "uncoiled" to make it simple by passing to a higher edge graph.

2.3.8. Let v and w be cycles in a graph. Suppose that $v^\infty \neq \sigma^i(w^\infty)$ for any i. Let N be at least twice the length of both v and w. Show that the Nth higher cycles $v^{[N]}$ and $w^{[N]}$ defined in the previous exercise have no edges in common.

2.3.9. Let $\mathbf{p}$ be the initial state distribution and P the conditional probability matrix for a Markov chain on a graph G. Verify that $\mathbf{p}P^m$ is the vector whose Ith component is the sum of the probabilities of all the paths in G of length m that end at state I.

2.3.10. Show how to recode a Markov chain on a graph to an ordinary Markov chain.

§2.4. State Splitting

State splitting is a procedure for constructing new graphs from a given graph. Starting from a partition of the edges, each state is split into a number of derived states (recall that we are using the terms "state" and "vertex" interchangeably). Although the resulting graph may appear to be quite different from the original graph, they have conjugate edge shifts. We will see in Chapter 7 that every conjugacy between edge shifts can be broken down into a finite sequence of state splittings and inverse operations called state amalgamations. This, together with its applications to finite state codes described in Chapter 5, explains the crucial importance of state splitting in symbolic dynamics.

We begin with a description of splitting a single state. Let G be a graph with state set $\mathcal{V}$ and edge set $\mathcal{E}$. Fix a state $I \in \mathcal{V}$, and assume for now that there are no self-loops at I. Recall that $\mathcal{E}_I$ denotes the set of edges in $\mathcal{E}$ starting at I. Partition $\mathcal{E}_I$ into two disjoint subsets $\mathcal{E}_I^1$ and $\mathcal{E}_I^2$ (here the superscripts indicate partition sets, not exponents). We construct a new graph H based on this partition as follows.

The states of H are those of G, except that state I is replaced by two new states called I^1 and I^2. In other words,

$$W = \mathcal{V}(H) = (\mathcal{V} \setminus \{I\}) \cup \{I^1, I^2\}.$$

For each $e \in \mathcal{E}_I^i$, where i is 1 or 2, put an edge in H from I^i to $t(e)$ having the same name e (note that $t(e) \neq I$ since we are assuming that there are no self-loops in G at I, so that $t(e) \in W$). For each $f \in \mathcal{E}$ starting at J and ending at I, put two edges named f^1 and f^2 in H, where f^1 goes from J to I^1 and f^2 goes from J to I^2. The initial and terminal states of all other edges in G remain in W, and copy them verbatim into H, completing its construction. Roughly speaking, we 'split' outgoing edges and 'copy' incoming edges.

Figure 2.4.1 shows how state splitting works. For clarity, graphs in this section are drawn so that edges enter a state on its left and leave on its right.

Example 2.4.1. Let G be the graph in Figure 2.4.2(a). Partition $\mathcal{E}_I = \{a, b, c\}$ into $\mathcal{E}_I^1 = \{a\}$ and $\mathcal{E}_I^2 = \{b, c\}$. The result H of this state splitting is shown in (b).   □

Suppose that H is formed from G by splitting state I as described above. We next construct a conjugacy between their edge shifts X_G and X_H. Recall that $\mathcal{E}^I$ denotes the set of edges in G ending at I. Define the 1-block map $\Psi \colon \mathcal{B}_1(X_H) \to \mathcal{B}_1(X_G)$ by $\Psi(f^i) = f$ if $f \in \mathcal{E}^I$ and $\Psi(e) = e$ if $e \notin \mathcal{E}^I$. In other words, Ψ simply erases any superscripts. It is easy to check that erasing superscripts takes paths in H to paths in G, so that Ψ induces a 1-block code $\psi = \Psi_\infty \colon X_H \to X_G$.

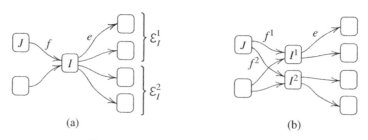

(a)   (b)

FIGURE 2.4.1.   Splitting a single state.

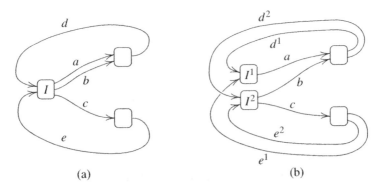

FIGURE 2.4.2. Elementary state splitting.

Going the other way, define a 2-block map $\Phi\colon \mathcal{B}_2(X_G) \to \mathcal{B}_1(X_H)$ by

$$
(2\text{-}4\text{-}1) \qquad \Phi(fe) = \begin{cases} f & \text{if } f \notin \mathcal{E}^I, \\ f^1 & \text{if } f \in \mathcal{E}^I \text{ and } e \in \mathcal{E}^1_I, \\ f^2 & \text{if } f \in \mathcal{E}^I \text{ and } e \in \mathcal{E}^2_I. \end{cases}
$$

Thus Φ "looks ahead" one symbol and may add a superscript, depending on what it sees. The reader should check that Φ takes paths in G to paths in H, so induces a sliding block code $\phi = \Phi_\infty \colon X_G \to X_H$ with memory 0 and anticipation 1.

Since adding and then erasing superscripts has no effect, we see that $\psi(\phi(x)) = x$ for all $x \in X_G$. Conversely, any superscripts are uniquely determined by (2-4-1) since $\mathcal{E}^1_I$ and $\mathcal{E}^2_I$ partition $\mathcal{E}_I$, so that $\phi(\psi(y)) = y$ for all $y \in X_H$. Hence ϕ is a conjugacy from X_G to X_H.

Example 2.4.2. The actions of ϕ and ψ on typical points $x \in X_G$ and $y \in X_H$ from Figure 2.4.2 are shown below.

$$
x = \ldots d\ a\ d\ b\ d\ c\ e\ .\ a\ d\ c\ e\ b\ d \ldots
$$

$$
\phi \Big\downarrow \ \Big\uparrow \psi
$$

$$
y = \ldots d^1\ a\ d^2\ b\ d^2\ c\ e^1\ .\ a\ d^2\ c\ e^2\ b \ldots
$$

Note that the symbol in y below the last d shown in x is not yet determined since it depends on the next symbol in x. $\qquad \square$

The general state splitting procedure extends our previous discussion in two ways: the outgoing edges from a given state can be partitioned into arbitrarily many subsets instead of just two, and the partitioning can occur simultaneously at all of the states instead of just one. This procedure also accommodates self-loops.

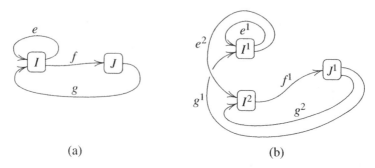

FIGURE 2.4.3.   An example of state splitting.

Definition 2.4.3. Let G be a graph with state set $\mathcal{V}$ and edge set $\mathcal{E}$. For each state $I \in \mathcal{V}$, partition $\mathcal{E}_I$ into nonempty disjoint sets $\mathcal{E}_I^1, \mathcal{E}_I^2, \ldots, \mathcal{E}_I^{m(I)}$, where $m(I) \geqslant 1$. Let $\mathcal{P}$ denote the resulting partition of $\mathcal{E}$, and let $\mathcal{P}_I$ denote the partition $\mathcal{P}$ restricted to $\mathcal{E}_I$. The *state split graph* $G^{[\mathcal{P}]}$ *formed from G using $\mathcal{P}$* has states $I^1, I^2, \ldots, I^{m(I)}$, where I ranges over the states in $\mathcal{V}$, and edges e^j, where e is any edge in $\mathcal{E}$ and $1 \leqslant j \leqslant m(t(e))$. If $e \in \mathcal{E}$ goes from I to J, then $e \in \mathcal{E}_I^i$ for some i, and we define the initial state and terminal state of e^j in $G^{[\mathcal{P}]}$ by $i(e^j) = I^i$ and $t(e^j) = J^j$, that is, e^j goes from I^i to J^j. An *elementary state splitting of G at state I* occurs when $m(I) = 2$ and $m(J) = 1$ for every $J \neq I$.

Our discussion above described an elementary state splitting of a graph at a state with no self-loops, and we suppressed some redundant superscript 1's.

Since the partition elements that define a splitting are all nonempty, any graph formed by splitting an essential graph is still essential, and any graph formed by splitting an irreducible graph is still irreducible (Exercise 2.4.5).

Some examples should help to clarify this procedure.

Example 2.4.4. Let G be the graph in Figure 2.4.3(a), so that $\mathcal{E}_I = \{e, f\}$ and $\mathcal{E}_J = \{g\}$. Partition these sets by $\mathcal{E}_I^1 = \{e\}$, $\mathcal{E}_I^2 = \{f\}$, and $\mathcal{E}_J^1 = \{g\}$, so that $m(I) = 2$ and $m(J) = 1$. Hence $\mathcal{P} = \{\mathcal{E}_I^1, \mathcal{E}_I^2, \mathcal{E}_J^1\}$. The resulting state split graph is drawn in (b). Note in particular that the self-loop e at I splits into a self-loop e^1 at I^1 and another edge e^2.   □

Example 2.4.5. Figure 2.4.4(a) shows a graph for the full 2-shift. There is essentially only one nontrivial partition of $\mathcal{E}_I$, namely $\mathcal{E}_I^1 = \{e\}$ and $\mathcal{E}_I^2 = \{f\}$. The resulting state split graph is shown in (b). Note that the edge shift for (b) is essentially the 2-block presentation of the full 2-shift.   □

Example 2.4.6. For the graph G shown in Figure 2.4.5(a), use the partition $\mathcal{P}$ with $\mathcal{E}_I^1 = \{a\}$, $\mathcal{E}_I^2 = \{b, c\}$, $\mathcal{E}_J^1 = \{d\}$, $\mathcal{E}_K^1 = \{e\}$, and $\mathcal{E}_K^2 = \{f\}$. The state split graph $G^{[\mathcal{P}]}$ is shown in (b).   □

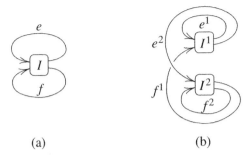

(a)   (b)

FIGURE 2.4.4.   A state splitting of the full 2-shift.

Since the construction of $G^{[\mathcal{P}]}$ uses a partition of the outgoing edges from states, we will sometimes say that $G^{[\mathcal{P}]}$ is the *out-split* graph formed from G. There is a corresponding notion of in-splitting using a partition of the incoming edges.

Definition 2.4.7. Let G be a graph with state set $\mathcal{V}$ and edge set $\mathcal{E}$. For each state $J \in \mathcal{V}$, partition $\mathcal{E}^J$ into nonempty disjoint sets $\mathcal{E}_1^J, \mathcal{E}_2^J, \ldots, \mathcal{E}_{m(J)}^J$, where $m(J) \geqslant 1$. Let $\mathcal{P}$ denote the resulting partition of $\mathcal{E}$. The *in-split graph* $G_{[\mathcal{P}]}$ *formed from* G *using* $\mathcal{P}$ has states $J_1, J_2, \ldots, J_{m(J)}$, where J ranges over the states in $\mathcal{V}$, and edges e_i, where e is any edge in $\mathcal{E}$ and $1 \leqslant i \leqslant m(i(e))$. If $e \in \mathcal{E}$ goes from I to J, then $e \in \mathcal{E}_j^J$ for some j, and we define the initial state and terminal state of e_i in $G_{[\mathcal{P}]}$ by $i(e_i) = I_i$ and $t(e_i) = J_j$.

Example 2.4.8. In-splitting the graph G in Figure 2.4.6(a) using the partition $\mathcal{P}$ with $\mathcal{E}_1^I = \{e\}$, $\mathcal{E}_2^I = \{g\}$, and $\mathcal{E}_1^J = \{f\}$ yields the graph $G_{[\mathcal{P}]}$ shown in (b). □

It will be convenient to regard a graph that is isomorphic to the out-split graph $G^{[\mathcal{P}]}$ as an out-splitting of G (and likewise for in-split graphs). It is also convenient to name the operation inverse to splitting.

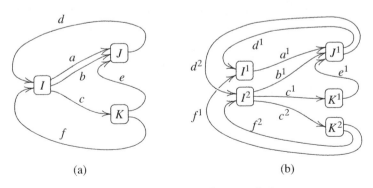

(a)   (b)

FIGURE 2.4.5.   A general state splitting.

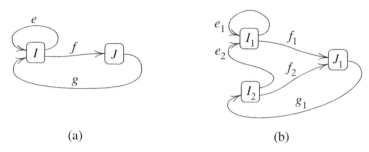

(a)   (b)

FIGURE 2.4.6.   An example of in-splitting.

Definition 2.4.9. A graph H is a *splitting* of a graph G, and G is an *amalgamation* of H, if H is graph isomorphic to the out-split graph $G^{[\mathcal{P}]}$ or the in-split graph $G_{[\mathcal{P}]}$ for some partition $\mathcal{P}$.

If more precision is needed, we will use the terms *out-splitting*, *in-splitting*, *out-amalgamation*, and *in-amalgamation*. Note that graph isomorphism may be regarded as any one of these types of operations.

Just as for elementary state splitting, both in-splitting and out-splitting states produce a graph whose shift is conjugate to that of the initial graph.

Theorem 2.4.10. *If a graph H is a splitting of a graph G, then the edge shifts X_G and X_H are conjugate.*

PROOF: We prove this for out-splittings, the other case being similar.

Using the notation from Definition 2.4.3, we may assume that $H = G^{[\mathcal{P}]}$. Define a 1-block map $\Psi \colon \mathcal{B}_1(X_H) \to \mathcal{B}_1(X_G)$ by $\Psi(e^j) = e$. Observe that if $e^j f^k \in \mathcal{B}_2(X_H)$, then $ef \in \mathcal{B}_2(X_G)$. Hence the image under Ψ of any path in H is a path in G. Thus if $\psi = \Psi_\infty$, then $\psi(X_H) \subseteq X_G$, so that $\psi \colon X_H \to X_G$ is a 1-block code.

Next we define a 2-block map $\Phi \colon \mathcal{B}_2(X_G) \to \mathcal{B}_1(X_H)$. If $fe \in \mathcal{B}_2(X_G)$, then e occurs in a unique $\mathcal{E}_J^j$, and we define $\Phi(fe) = f^j$ (compare with equation (2–4–1)). As before, the image under Φ of a path in G is a path in H. Let $\phi = \Phi_\infty^{[0,1]}$ have memory 0 and anticipation 1. Then $\phi(X_G) \subseteq X_H$. If $x = \ldots e_{-1}e_0e_1 \ldots \in X_G$, then $\phi(x)$ has the form

$$\phi(x) = \ldots e_{-1}^{j_{-1}}.e_0^{j_0}e_1^{j_1}\ldots,$$

so that $\psi(\phi(x)) = x$. Conversely, if

$$y = \ldots e_{-1}^{j_{-1}}.e_0^{j_0}e_1^{j_1}\ldots \in X_H,$$

then $\psi(y) = \ldots e_{-1}e_0e_1\ldots$. Since $e_i^{j_i} e_{i+1}^{j_{i+1}}$ is a 2-block in X_H, it follows that e_{i+1} belongs to $\mathcal{E}_{t(e_i)}^{j_i}$. So, $\Phi(e_ie_{i+1}) = e_i^{j_i}$ for all i, and we conclude that $\phi(\psi(y)) = y$. Hence X_G and X_H are conjugate. $\square$

This proof shows that if H is an out-splitting of G, there is a 1-block code from X_H to X_G called the *out-amalgamation code*. This code has a 2-block inverse, with memory 0 and anticipation 1, called the *out-splitting code*. If instead H is an in-splitting of G, then we have the *in-amalgamation code*, and it has a 2-block inverse, with memory 1 and anticipation 0, called the *in-splitting code*.

It follows from the previous theorem that if a graph H can be obtained from G by a sequence of splittings and amalgamations, then X_H is conjugate to X_G. The conjugacy is a composition of splitting codes and amalgamation codes. One of the main results in Chapter 7 is that *every* conjugacy between shifts of finite type can be decomposed into a composition of these basic types. We emphasize that this includes conjugacies determined by graph isomorphisms. Such conjugacies may appear to be rather trivial, but they turn out to be very important in the theory of automorphisms of shifts of finite type (see §13.2).

How does state splitting affect the adjacency matrix of a graph? Consider Example 2.4.6, depicted in Figure 2.4.5. Let $V = \{I, J, K\}$ and $W = \{I^1, I^2, J^1, K^1, K^2\}$ be the state sets for G and for the out-split graph $H = G^{[\mathcal{P}]}$. We can represent how states in W are derived from states in V by means of the matrix

$$
D = \begin{array}{c} \\ I \\ J \\ K \end{array}
\begin{array}{c}
\begin{array}{ccccc} I^1 & I^2 & J^1 & K^1 & K^2 \end{array} \\
\left[\begin{array}{ccccc}
1 & 1 & 0 & 0 & 0 \\
0 & 0 & 1 & 0 & 0 \\
0 & 0 & 0 & 1 & 1
\end{array} \right]
\end{array}.
$$

Here D is indexed by pairs in $V \times W$, and we have written D with row and column indices added for clarity. We will denote matrix entries with their indices in parentheses rather than as subscripts, so that $D(J, K^2)$ refers to the (J, K^2)-entry in D, etc.

The number of edges in each partition set ending at a given state in V is specified by the matrix

$$
E = \begin{array}{c} \\ I^1 \\ I^2 \\ J^1 \\ K^1 \\ K^2 \end{array}
\begin{array}{c}
\begin{array}{ccc} I & J & K \end{array} \\
\left[\begin{array}{ccc}
0 & 1 & 0 \\
0 & 1 & 1 \\
1 & 0 & 0 \\
0 & 1 & 0 \\
1 & 0 & 0
\end{array} \right]
\end{array},
$$

where, for example, $E(I^1, J)$ is the number of edges in $\mathcal{E}_I^1$ ending at J, in this case 1.

Let $A = A_G$ and $B = A_H$ be the adjacency matrices of G and of H. Direct computation shows that the (I, J)-entry in the matrix product DE

is the sum over partition sets $\mathcal{E}_I^i$ of the number of edges in $\mathcal{E}_I^i$ ending at J. Since each edge in $\mathcal{E}_I$ ending at J occurs in exactly one of the $\mathcal{E}_I^i$, it follows that $(DE)(I,J)$ is the total number of edges in G from I to J, which is by definition $A(I,J)$. Thus

$$DE = A = \begin{array}{c} \\ I \\ J \\ K \end{array} \begin{array}{c} I \quad J \quad K \\ \begin{bmatrix} 0 & 2 & 1 \\ 1 & 0 & 0 \\ 1 & 1 & 0 \end{bmatrix} \end{array}.$$

Consider the product of D and E in reverse order,

$$ED = \begin{array}{c} \\ I^1 \\ I^2 \\ J^1 \\ K^1 \\ K^2 \end{array} \begin{array}{c} I^1 \quad I^2 \quad J^1 \quad K^1 \quad K^2 \\ \begin{bmatrix} 0 & 0 & 1 & 0 & 0 \\ 0 & 0 & 1 & 1 & 1 \\ 1 & 1 & 0 & 0 & 0 \\ 0 & 0 & 1 & 0 & 0 \\ 1 & 1 & 0 & 0 & 0 \end{bmatrix} \end{array}.$$

Since $D(I,K^1) = D(J,K^1) = 0$ and $D(K,K^1) = 1$, the entry $(ED)(I^2,K^1)$ equals $E(I^2,K)$, the number of edges in $\mathcal{E}_I^2$ ending at K. Since incoming edges are duplicated during out-splitting, this number equals the number of edges from I^2 to K^1, which is $B(I^2,K^1)$. Applying this reasoning to each entry shows that $ED = B$.

We now show that this kind of relation between A and B holds in general.

Definition 2.4.11. Let G be a graph, and $H = G^{[\mathcal{P}]}$ be the out-split graph formed from G using $\mathcal{P}$. Let $\mathcal{V} = \mathcal{V}(G)$ and $\mathcal{W} = \mathcal{V}(H)$. The *division matrix D for $\mathcal{P}$* is the $\mathcal{V} \times \mathcal{W}$ matrix defined by

$$D(I, J^k) = \begin{cases} 1 & \text{if } I = J, \\ 0 & \text{otherwise.} \end{cases}$$

The *edge matrix E for $\mathcal{P}$* is the $\mathcal{W} \times \mathcal{V}$ matrix defined by

$$E(I^k, J) = \left| \mathcal{E}_I^k \cap \mathcal{E}^J \right|.$$

The division and edge matrices can be used to compute the adjacency matrices for the graphs involved in an out-splitting.

Theorem 2.4.12. *Let G be a graph, and $H = G^{[\mathcal{P}]}$ be the out-split graph formed from G using the partition $\mathcal{P}$. If D and E are the division and edge matrices for $\mathcal{P}$, then*

$$(2\text{--}4\text{--}2) \qquad\qquad DE = A_G \quad \text{and} \quad ED = A_H.$$

Proof: Using the notation from Definition 2.4.3, we have that

$$(DE)(I, J) = \sum_{i=1}^{m(I)} D(I, I^i) E(I^i, J) = \sum_{i=1}^{m(I)} E(I^i, J)$$

$$= \sum_{i=1}^{m(I)} |\mathcal{E}_I^i \cap \mathcal{E}^J| = \left| \left(\bigcup_{i=1}^{m(I)} \mathcal{E}_I^i \right) \cap \mathcal{E}^J \right|$$

$$= |\mathcal{E}_I \cap \mathcal{E}^J| = A_G(I, J).$$

Hence $DE = A_G$. To prove the other equality, note that

$$(ED)(I^i, J^j) = E(I^i, J) D(J, J^i) = E(I^i, J)$$

$$= |\mathcal{E}_I^i \cap \mathcal{E}^J| = \left| \mathcal{E}_{I^i} \cap \mathcal{E}^{J^j} \right| = A_H(I^i, J^j). \qquad \square$$

Exercise 2.4.6 gives an interpretation of the passage from A_G to A_H, with the edge matrix E as an intermediate step.

Suppose that H is an out-splitting of a graph G; i.e., H is graph isomorphic to (but perhaps not literally equal to) the out-split graph $H' = G^{[\mathcal{P}]}$. Then $A_H = P A_{H'} P^{-1}$ for some permutation matrix P. Thus, from the equations $A_G = DE$ and $A_{H'} = ED$ corresponding to the splitting, we see that $A_G = D'E'$ and $A_H = E'D'$, where $D' = DP^{-1}$ and $E' = PE$, yielding a set of matrix equations similar to (2–4–2). Observe that D' satisfies the following property.

Definition 2.4.13. An (abstract) *division matrix* is a 0–1 rectangular matrix such that each row of D has at least one 1 and each column of D has exactly one 1.

We have the following embellishment of Theorem 2.4.12 which includes a converse.

Theorem 2.4.14. *Let G and H be essential graphs. Then H is an out-splitting of G if and only if there exist a division matrix D and a rectangular matrix E with nonnegative integer entries such that*

$$A_G = DE \quad and \quad A_H = ED.$$

We leave the proof of this result as an exercise for the reader. An analogue of this result for in-splittings is given in Exercise 2.4.9.

EXERCISES

2.4.1. Compute the division and edge matrices for the out-splittings in Examples 2.4.1, 2.4.4, 2.4.5, and 2.4.6, and verify Theorem 2.4.12 for each.

2.4.2. Let G be a graph with edge set $\mathcal{E}$.

  (a) Let $\mathcal{P} = \mathcal{E}$, so that each $\mathcal{E}_I$ is partitioned into singleton sets of individual edges. Show that $G^{[\mathcal{P}]}$ is isomorphic to the 2nd higher edge graph $G^{[2]}$ of G (such a splitting is called the *complete out-splitting* of G).

  (b) For each $N \geqslant 1$, show that you can obtain a graph that is isomorphic to $G^{[N]}$ by using a sequence of out-splittings starting with G.

2.4.3. Suppose we were to allow a partition element $\mathcal{E}_I^i$ in an out-splitting to be the empty set. What would this mean about the derived state I^i?

2.4.4. For each of the two graphs (a) and (b) shown below, draw the out-split graph using the corresponding partitions

  (a) $\mathcal{E}_I^1 = \{a, c\}$, $\mathcal{E}_I^2 = \{b\}$, $\mathcal{E}_I^3 = \{d\}$, $\mathcal{E}_J^1 = \{e, f\}$, $\mathcal{E}_J^2 = \{g\}$;

  (b) $\mathcal{E}_I^1 = \{a, b\}$, $\mathcal{E}_I^2 = \{c\}$, $\mathcal{E}_I^3 = \{d\}$.

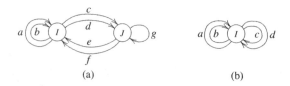

(a)   (b)

2.4.5. Let H be a splitting of an irreducible graph (via a partition whose elements are all nonempty). Show that H is irreducible.

2.4.6. Let $H = G^{[\mathcal{P}]}$ be the out-split graph formed from G using $\mathcal{P}$. Using the division and edge matrices for this splitting, describe how A_H is obtained from A_G by decomposing each row of G into a sum of rows, and duplicating the corresponding columns. Find a similar matrix description of in-splitting.

2.4.7. For each pair of adjacency matrices, decide whether it is possible to obtain the graph of the second from the graph of the first by a finite sequence of splittings and amalgamations.

  (a) $\begin{bmatrix} 1 & 1 \\ 1 & 0 \end{bmatrix}$ and $\begin{bmatrix} 0 & 1 & 1 \\ 1 & 1 & 0 \\ 1 & 0 & 0 \end{bmatrix}$,   (b) $\begin{bmatrix} 1 & 1 \\ 1 & 0 \end{bmatrix}$ and $\begin{bmatrix} 1 & 1 & 0 \\ 0 & 0 & 1 \\ 1 & 1 & 0 \end{bmatrix}$,

  (c) $\begin{bmatrix} 2 \end{bmatrix}$ and $\begin{bmatrix} 1 & 1 & 0 \\ 0 & 0 & 1 \\ 1 & 1 & 1 \end{bmatrix}$,   (d) $\begin{bmatrix} 2 \end{bmatrix}$ and $\begin{bmatrix} 1 & 0 & 1 \\ 1 & 0 & 0 \\ 1 & 1 & 1 \end{bmatrix}$.

  [*Hint*: Consider periodic points.]

2.4.8. Prove Theorem 2.4.14.

2.4.9. Recall the notion of transpose from the end of §2.2.

  (a) Explain why an in-splitting of G corresponds to an out-splitting of G^{T}.

  (b) Use (a) to obtain an analogue of Theorems 2.4.12 and 2.4.14 for in-splittings.

2.4.10. Show that if H is obtained from G by a sequence of in-splittings, then the set of out-degrees of states in H is the same as the set of out-degrees of states in G.

2.4.11. Let G be an irreducible graph.

(a) Show that there is a graph H obtained from G by a sequence of out-splittings such that H has a state with exactly one outgoing edge.

*(b) Show that if G does not consist entirely of a single simple cycle, then for any positive integer m, there is a graph H obtained from G by a sequence of out-splittings such that H has a state with exactly m outgoing edges.

§2.5. Data Storage and Shifts of Finite Type

Computers store data of all kinds as sequences of 0's and 1's, or *bits* (from *binary digits*). You may be familiar with the ASCII code, which converts letters, punctuation, and control symbols into sequences of 8 bits. You might think that these bits are stored *verbatim* on a disk drive. But practical considerations call for more subtle schemes, and these lead directly to shifts of finite type.

We first need to understand something of how magnetic storage devices such as "floppy" and "hard" disk drives actually work. Disk drives contain one or more rapidly spinning platters coated with a magnetic medium. Each platter is divided into concentric circular tracks. An electrical read–write head floats over the platter on a cushion of air about one hundredth the thickness of a human hair. The head moves radially to access different tracks, but we will confine our attention to a single track.

An electrical current passing through the head will magnetize a piece of the track, in effect producing a tiny bar magnet along the track. Reversing the current creates an adjacent bar magnet whose north and south poles are reversed. Data is written by a sequence of current reversals, creating what amounts to a sequence of bar magnets of various lengths and alternating polarity. This is illustrated in the top half of Figure 2.5.1, where N and S show the north and south poles for each magnet.

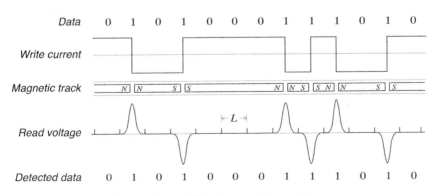

FIGURE 2.5.1. Digital magnetic recording.

Data is read by reversing this process. Polarity reversals along the track generate voltage pulses in the head as they pass by. A scheme called *peak detection* determines whether a pulse occurs within a time interval called the *detection window* having length L. A pulse is interpreted as a 1, while the absence of a pulse represents a 0. Peak detection uses only the reversals in polarity, not the actual polarities; interchanging N and S would yield the same data. The bottom half of Figure 2.5.1 illustrates how data is read.

The most direct method to store a sequence of bits is to divide the track into cells of equal length L, and write a polarity change into each cell storing a 1, while no change is made in cells storing a 0. There are two serious problems with this approach.

The first problem is called *intersymbol interference*. If polarity changes are too close together, the magnetic lines of force tend to cancel out, and the pulses produced in the head are weaker, are harder to detect, and have their peaks shifted (see Figure 2.5.2). Based on the physical characteristics of the magnetic medium and the sensitivity of the electronics, each disk system has a minimum separation distance Δ between polarity changes for reliable operation. In the *verbatim* method above, the detection window length L must be at least Δ, since 1's can occur in adjacent cells. By using schemes in which 1's are separated by at least d 0's, we will be able to use a detection window of length $L = \Delta/(d + 1)$ and pack more data on each track.

The second problem is *clock drift*. The block $10^n 1 = 1000\ldots0001$ is read as two pulses separated by a time interval. The length of this interval is $(n + 1)L$, and so determines the number n of 0's. Any drift in the clocking circuit used to measure this interval could result in an erroneous value for n. Think of measuring the time between lightning and thunder by counting "A thousand one, a thousand two, etc." This is quite accurate for five or ten seconds, but becomes much more error prone for intervals of a minute or two.

Clock drift is corrected by a feedback loop every time a pulse is read. If we can arrange to have pulses occur every so often, then clock drift can never accumulate to dangerous amounts. A typical constraint is that between two successive 1's there are at most k 0's. More accurate clocks

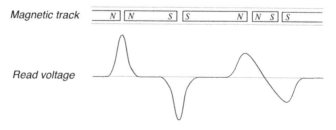

FIGURE 2.5.2.   Intersymbol interference.

allow larger values of k.

We will now describe two schemes to transform an arbitrary sequence of bits into one obeying constraints that control intersymbol interference and clock drift.

The first method is called *frequency modulation* (FM). This scheme controls clock drift with the transparently simple idea of inserting a clock bit $\underline{1}$ between each pair of data bits (0 or 1). If the original data is

$$10001101,$$

then the FM coded data is

$$1\underline{1}1\underline{0}1\underline{0}1\underline{0}1\underline{1}1\underline{1}1\underline{0}1\underline{1}.$$

Although clock bits are underlined here, they are stored identically to data bits. The original data is recovered by ignoring the clock bits. FM coded data obeys the (0,1) run-length limited constraint, and is a block from the shift of finite type $X(0,1)$ in Example 1.2.5. Since adjacent cells can both contain a polarity change, the smallest possible cell length is $L = \Delta$. With the FM scheme, n bits of data are coded to $2n$ bits, so can be stored in a piece of the track of length $2n\Delta$. This scheme was used on some of the earliest disk drives.

We can improve the FM scheme by using two ideas. The first is that if coded sequences have at least d 0's between successive 1's, then we can shrink the detection window and cell size to $L = \Delta/(d+1)$ and still maintain enough separation between pulses to avoid intersymbol interference. The second is that 1's in the data can sometimes double as clocking update bits. One method that uses these ideas is called *modified frequency modulation* (MFM). This scheme is currently used on many disk drives.

The MFM scheme inserts a $\underline{0}$ between each pair of data bits unless both data bits are 0, in which case it inserts a $\underline{1}$. Using the same data sequence as before,

$$10001101,$$

the MFM coded sequence is

$$\underline{0}1\underline{0}0\underline{0}1\underline{0}0\underline{1}0\underline{1}0\underline{0}0\underline{0}1.$$

Here the initial data bit is 1, so the bit inserted to its left must be a $\underline{0}$. If the initial data bit were 0, we would need to know the previous data bit before we could decide which bit to insert to its left. The reader should verify that MFM coded sequences obey the (1,3) run-length limited constraint (Exercise 2.5.1).

Notice that some of the data 1's are used here for clock synchronization. Since $d = 1$, pulses cannot occur in adjacent cells, so we can use a cell

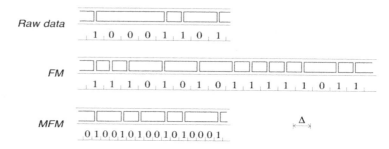

FIGURE 2.5.3.   Comparison of FM and MFM encodings of data 10001101.

length of only $L = \Delta/2$ and maintain enough separation between pulses. With the MFM scheme, n bits of data code to $2n$ bits, and can be stored in length $(2n)(\Delta/2) = n\Delta$, or exactly half the length needed by FM for the same data. Figure 2.5.3 compares the bar magnet layout for these methods with data 10001101; each magnet shown must have length at least Δ. It also shows how the raw data would be stored, but this would be subject to uncontrolled clock drift.

The MFM method stores twice as much data as FM, but this does not come for free. The detection window for MFM is only half that of FM, so that more sensitive electronics is needed. By decreasing the size of the detection window further, even more data can be stored. Carried to an impossible extreme, if pulses could be timed with infinite accuracy, corresponding to $L = 0$, then two pulses could store an infinite amount of data; just use the sequence of 0's and 1's you want to store as the binary expansion of the time between the pulses.

This is just the beginning of increasingly sophisticated schemes for data storage. One popular method is based on a $(2, 7)$ run-length limited constraint. Both encoding and decoding data are more complicated than the simple FM and MFM codes, and can be derived using the theory of symbolic dynamics. We will discuss these applications more thoroughly in Chapter 5. Other storage technologies, such as the familiar compact optical disk or magneto-optical disks, give rise to new sets of constraints. How are codes adapted to these constraints discovered? Is there a systematic way to produce them? What are the theoretical limits to storage density using such codes? In later chapters we will see how symbolic dynamics provides a mathematical foundation for answering these questions.

EXERCISES

2.5.1. Show that every sequence produced by the MFM encoding obeys the $(1, 3)$ run-length limited constraint.

2.5.2. Modify the MFM scheme so that the block 00 is inserted between each pair of data bits unless both are 0, in which case insert 10.

(a) Show that all encoded sequences obey the (2,5) run-length limited constraint.

(b) What is the detection window size for this code? Can you store more data than with the MFM code?

(c) Compare this scheme with the MFM scheme in terms of detection window size and clock drift. Which is preferable?

Notes

While shifts of finite type have been around implicitly for a long time, it wasn't until Parry [Par1] that their systematic study began. He called them *intrinsic Markov chains* since they are the supports of finite-state Markov probability measures. Smale introduced the term *subshift of finite type* [Sma] and showed that they are basic to understanding the dynamics of smooth mappings. Vertex shifts, as models of shifts of finite type, were considered by both Parry and Smale. For reasons of computational convenience, R. Williams [WilR2] allowed multiple edges and thereby introduced edge shifts, although not by that name.

The material on Markov chains in §2.3 is standard; see, for instance, Kemeny and Snell [KemS].

State splitting was introduced to symbolic dynamics by R. Williams [WilR2] in his efforts to solve the conjugacy problem for shifts of finite type. A closely related notion of state splitting appeared earlier in information theory; see Even [Eve1] and Kohavi [Koh1], [Koh2]. Patel [Pat] used state splitting as a tool in his construction of a particular code, called the ZM modulation code, that was actually used in an IBM tape product. The MFM encoding method, discussed in §2.5, was invented by Miller [Mil]. We refer the reader to Chapter 5 for further reading on coding for data recording.

Finally, we mention that there is growing interest in group shifts, introduced at the end of §2.1. In Exercise 2.1.11 the reader was asked to show that every group shift is a shift of finite type. The proof outlined there is essentially due to Kitchens [Kit3]. In fact, it can be shown that every irreducible group shift X is conjugate to a full shift; if X is also a convolutional code, then it is conjugate to a full shift via a linear conjugacy. We refer the reader to Kitchens [Kit3], Forney [For1], [For2], Forney and Trott [ForT], Loeliger and Mittelholzer [LoeM], and Miles and Thomas [MilT] for proofs of various versions of these and related results. "Higher dimensional" versions of group shifts have received a great deal of attention recently (see §13.10).

Exercise 2.1.10 comes from [KhaN1]; Exercise 2.1.11 from Kitchens [Kit3]; Exercise 2.2.15 comes from [AshM] and [KhaN2]; Exercise 2.2.16 is essentially from Forney and Trott [ForT]; and Exercise 2.3.4 is due to N. T. Sindhushayana.

CHAPTER 3

SOFIC SHIFTS

Suppose we label the edges in a graph with symbols from an alphabet $\mathcal{A}$, where two or more edges could have the same label. Every bi-infinite walk on the graph yields a point in the full shift $\mathcal{A}^{\mathbb{Z}}$ by reading the labels of its edges, and the set of all such points is called a *sofic shift*.

Sofic shifts are important for a number of reasons. We shall learn in this chapter that they are precisely the shift spaces that are factors of shifts of finite type. Thus the class of sofic shifts is the smallest collection of shift spaces that contains all shifts of finite type and also contains all factors of each space in the collection. Sofic shifts are analogous to regular languages in automata theory, and several algorithms in §3.4 mimic those for finite-state automata. Shifts of finite type and sofic shifts are also natural models for information storage and transmission; we will study their use in finite-state codes in Chapter 5.

§3.1. Presentations of Sofic Shifts

Sofic shifts are defined using graphs whose edges are assigned labels, where several edges may carry the same label.

Definition 3.1.1. A *labeled graph* $\mathcal{G}$ is a pair $(G, \mathcal{L})$, where G is a graph with edge set $\mathcal{E}$, and the *labeling* $\mathcal{L}: \mathcal{E} \to \mathcal{A}$ assigns to each edge e of G a label $\mathcal{L}(e)$ from the finite alphabet $\mathcal{A}$. The *underlying graph* of $\mathcal{G}$ is G. A labeled graph is *irreducible* if its underlying graph is irreducible. A labeled graph is *essential* if its underlying graph is essential.

Sometimes we refer to a labeled graph as simply a graph, and to states or edges of the underlying graph G of a labeled graph $\mathcal{G}$ as states or edges of $\mathcal{G}$.

The labeling $\mathcal{L}$ can be any assignment of letters from an alphabet $\mathcal{A}$ to edges of G. At one extreme, we could let $\mathcal{A} = \mathcal{E}$ and $\mathcal{L}(e) = e$ for all $e \in \mathcal{E}$, giving a one-to-one labeling of edges by their names. At the other, $\mathcal{A}$ could

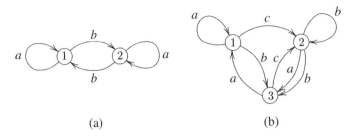

FIGURE 3.1.1.   Typical labeled graphs.

consist of a single letter a and all edges could be labeled with a, so that
$\mathcal{L}(e) = a$ for all $e \in \mathcal{E}$. Usually, $\mathcal{L}$ will be many-to-one, so that several
edges will carry the same label. Although $\mathcal{L}$ need not be onto, any letters
not in the range of $\mathcal{L}$ are not used. We will normally use characters like e,
f, or g for the names of edges, and use a, b, c, or small integers for their
labels. Figure 3.1.1 depicts two typical labeled graphs.

Just as a graph G is conveniently described by its adjacency matrix A_G,
a labeled graph $\mathcal{G}$ has an analogous *symbolic adjacency matrix* $A_{\mathcal{G}}$. The
(I, J)th entry of $A_{\mathcal{G}}$ contains the formal "sum" of the labels of all edges
from I to J, or a "zero" character $\emptyset$ if there are no such edges. For example,
if $\mathcal{G}$ is the labeled graph (a) in Figure 3.1.1 and $\mathcal{H}$ is graph (b), then

$$A_{\mathcal{G}} = \begin{bmatrix} a & b \\ b & a \end{bmatrix} \quad \text{and} \quad A_{\mathcal{H}} = \begin{bmatrix} a & c & b \\ \emptyset & b & a+b \\ a & c & \emptyset \end{bmatrix}.$$

There is an analogue of graph homomorphism for labeled graphs, which
adds the requirement that labels be preserved.

Definition 3.1.2. Let $\mathcal{G} = (G, \mathcal{L}_G)$ and $\mathcal{H} = (H, \mathcal{L}_H)$ be labeled graphs.
A *labeled-graph homomorphism* from $\mathcal{G}$ to $\mathcal{H}$ is a graph homomorphism
$(\partial\Phi, \Phi): G \to H$ such that $\mathcal{L}_H(\Phi(e)) = \mathcal{L}_G(e)$ for all edges $e \in \mathcal{E}(G)$. In
this case, we write $(\partial\Phi, \Phi): \mathcal{G} \to \mathcal{H}$. If both $\partial\Phi$ and Φ are one-to-one and
onto, then $(\partial\Phi, \Phi)$ is a *labeled-graph isomorphism*, written $(\partial\Phi, \Phi): \mathcal{G} \cong \mathcal{H}$.
Two labeled graphs are *labeled-graph isomorphic* (or simply *isomorphic*) if
there is a labeled-graph isomorphism between them.

You should think of isomorphic labeled graphs as being "the same" for
our purposes.

If $\mathcal{G} = (G, \mathcal{L})$ is a labeled graph, then $\mathcal{L}$ can be used to label paths and
bi-infinite walks on the underlying graph G. Define the *label of a path*
$\pi = e_1 e_2 \ldots e_n$ on G to be

$$\mathcal{L}(\pi) = \mathcal{L}(e_1)\mathcal{L}(e_2) \ldots \mathcal{L}(e_n),$$

which is an n-block over $\mathcal{A}$, and which we will sometimes refer to as a *label block*. For each empty path ε_I in G we define $\mathcal{L}(\varepsilon_I) = \varepsilon$, the empty word over $\mathcal{A}$. If $\xi = \ldots e_{-1}e_0e_1 \ldots$ is a bi-infinite walk on G, so that ξ is a point in the edge shift X_G, define the *label of the walk* ξ to be

$$\mathcal{L}_\infty(\xi) = \ldots \mathcal{L}(e_{-1})\mathcal{L}(e_0)\mathcal{L}(e_1)\ldots \in \mathcal{A}^{\mathbb{Z}}.$$

The set of labels of all bi-infinite walks on G is denoted by

$$\begin{aligned}
\mathsf{X}_{\mathcal{G}} &= \{x \in \mathcal{A}^{\mathbb{Z}} : x = \mathcal{L}_\infty(\xi) \text{ for some } \xi \in \mathsf{X}_G\} \\
&= \{\mathcal{L}_\infty(\xi) : \xi \in \mathsf{X}_G\} = \mathcal{L}_\infty(\mathsf{X}_G).
\end{aligned}$$

Thus $\mathsf{X}_{\mathcal{G}}$ is a subset of the full $\mathcal{A}$-shift. For example, if $\mathcal{G}$ is the labeled graph in Figure 3.1.1(a), then $\mathsf{X}_{\mathcal{G}}$ is the full $\{a,b\}$-shift; the labeled graph $\mathcal{H}$ in (b) gives an $\mathsf{X}_{\mathcal{H}}$ which is a proper subset of the full $\{a,b,c\}$-shift (why?). Also note that if $\mathcal{G}$ and $\mathcal{H}$ are isomorphic, then $\mathsf{X}_{\mathcal{G}} = \mathsf{X}_{\mathcal{H}}$.

This chapter concerns shift spaces that arise from labeled graphs.

Definition 3.1.3. A subset X of a full shift is a *sofic shift* if $X = \mathsf{X}_{\mathcal{G}}$ for some labeled graph $\mathcal{G}$. A *presentation* of a sofic shift X is a labeled graph $\mathcal{G}$ for which $\mathsf{X}_{\mathcal{G}} = X$. The shift map on $\mathsf{X}_{\mathcal{G}}$ is denoted $\sigma_{\mathcal{G}}$.

By Proposition 2.2.10, every sofic shift has a presentation by an essential graph.

The term "sofic" was coined by Weiss [Wei], and is derived from the Hebrew word for "finite."

It is important to realize that a given sofic shift will have many different presentations. For example, Figure 3.1.2 shows four presentations of the full 2-shift, none of which are isomorphic (although the underlying graphs of (b) and (c) are graph isomorphic).

If X is a sofic shift presented by $\mathcal{G} = (G, \mathcal{L})$ and w is a block in $\mathcal{B}(X)$, we will say that a path π in G is a *presentation* of w if $\mathcal{L}(\pi) = w$. A

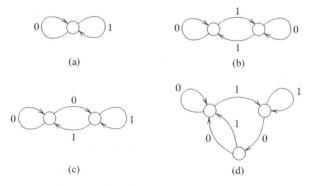

FIGURE 3.1.2.   Presentations of the 2-shift.

given block w can have several different presenting paths. For example, in Figure 3.1.2(d), the block 010001 has three presentations, one starting at each of the three vertices. If $x \in X_{\mathcal{G}}$, we will say that a bi-infinite walk ξ in X_G is a *presentation* of x if $\mathcal{L}(\xi) = x$. As with paths, a given point in $X_{\mathcal{G}}$ may have many different presentations.

Our definition of a sofic shift does not require that it be a shift space. However, it always is.

Theorem 3.1.4. *Sofic shifts are shift spaces.*

PROOF: Let X be a sofic shift over $\mathcal{A}$, and $\mathcal{G} = (G, \mathcal{L})$ be a presentation of X. Then $\mathcal{L}: \mathcal{E} \to \mathcal{A}$ gives a 1-block code $\mathcal{L}_\infty: X_G \to X_{\mathcal{G}}$, so its image $X = \mathcal{L}_\infty(X_G)$ is a shift space by Theorem 1.5.13. □

Notice that edge shifts are sofic, for we can simply use the edge set $\mathcal{E}$ as alphabet and the identity labeling $\mathcal{L}: \mathcal{E} \to \mathcal{E}$ given by $\mathcal{L}(e) = e$. The following result extends this observation to show that shifts of finite type are sofic.

Theorem 3.1.5. *Every shift of finite type is sofic.*

PROOF: Suppose that X is a shift of finite type. By Proposition 2.1.7, X is M-step for some $M \geqslant 0$. The proof of Theorem 2.3.2 constructs a graph G for which $X^{[M+1]} = X_G$. Recall that the vertices of G are the allowed M-blocks in X, and there is an edge e from $a_1 a_2 \dots a_M$ to $b_1 b_2 \dots b_M$ if and only if $a_2 a_3 \dots a_M = b_1 b_2 \dots b_{M-1}$ and $a_1 \dots a_M b_M$ ($= a_1 b_1 \dots b_M$) is in $\mathcal{B}(X)$. In this case, e is named $a_1 a_2 \dots a_M b_M$, and we label e with $\mathcal{L}(e) = a_1$. This yields a labeled graph $\mathcal{G} = (G, \mathcal{L})$ that we will show is a presentation of X.

Let $\beta_{M+1}: X \to X^{[M+1]} = X_G$ be the higher block code from §1.4 given by

$$\beta_{M+1}(x)_{[i]} = x_{[i,i+M]}.$$

Since $\mathcal{L}(x_{[i,i+M]}) = x_i$, we see that $\mathcal{L}_\infty(\beta_{M+1}(x)) = x$ for all $x \in X$, proving that $X \subseteq X_{\mathcal{G}}$. Conversely, every point ξ in $X_G = X^{[M+1]}$ has the form $\xi = \beta_{M+1}(x)$ for some $x \in X$, so that $\mathcal{L}_\infty(\xi) = \mathcal{L}_\infty(\beta_{M+1}(x)) = x \in X$. Hence $X_{\mathcal{G}} = \mathcal{L}_\infty(X_G) \subseteq X$, and so $X = X_{\mathcal{G}}$. □

In this proof, the 1-block code $\mathcal{L}_\infty$ is the inverse of the conjugacy β_{M+1}. A similar situation was described in Example 1.5.10, and is a precise way to say that x can be reconstructed from $\beta_{M+1}(x)$ by reading off the "bottom" letters in (1–4–2).

Not all sofic shifts have finite type. Figure 3.1.3 shows two labeled graphs. The sofic shift presented by graph (a) is the even shift of Example 1.2.4, which we showed in Example 2.1.9 does not have finite type. Similar reasoning applies to graph (b), since 30^M and $0^M 1$ are allowed blocks for every $M \geqslant 1$, but $30^M 1$ is not, showing that this sofic shift also does not have finite type. A sofic shift that does not have finite type is called *strictly sofic*.

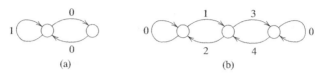

(a)   (b)

FIGURE 3.1.3.   Presentations of two strictly sofic shifts.

Roughly speaking, a shift of finite type uses a finite *length* of memory, while a sofic shift uses a finite *amount* of memory: the even shift only needs to keep track of the parity of the number of 0's since the last 1, although this number can be very large.

The following result gives another perspective on how shifts of finite type differ from strictly sofic shifts.

Proposition 3.1.6. *A sofic shift is a shift of finite type if and only if it has a presentation* $(G, \mathcal{L})$ *such that* $\mathcal{L}_\infty$ *is a conjugacy.*

PROOF: If X is a sofic shift presented by $(G, \mathcal{L})$ and $\mathcal{L}_\infty$ is a conjugacy, then X is conjugate to the shift of finite type, X_G, and is therefore, by Theorem 2.1.10, a shift of finite type itself.

Conversely, if X is a shift of finite type, then it is an M-step shift of finite type for some M, and the labeled graph constructed in Theorem 3.1.5 does the trick. $\square$

It turns out that there are many more shift spaces than sofic shifts. Indeed, while there are only countably many different sofic shifts (Exercise 3.1.5), recall from Exercise 1.2.12 that there are uncountably many different shift spaces.

Here is a specific shift space that we can prove is not sofic.

Example 3.1.7. Let X be the context-free shift in Example 1.2.9, where $\mathcal{A} = \{a, b, c\}$ and $ab^m c^n a$ is allowed exactly when $m = n$. We will show that X is not sofic. For suppose it were, and let $\mathcal{G} = (G, \mathcal{L})$ be a presentation. Let r be the number of states in G. Since $w = ab^{r+1} c^{r+1} a$ is allowed, there is a path π in G presenting w. Let τ be the subpath of π presenting b^{r+1}. Since G has only r states, at least two states in τ must be the same. Hence we can write $\tau = \tau_1 \tau_2 \tau_3$, where τ_2 is a cycle (if the first or last state of τ is one of the repeated states, then τ_1 or τ_3 is an empty path). Then $\tau' = \tau_1 \tau_2 \tau_2 \tau_3$ would also be a path. Replacing τ by τ' in π would give a path π' in G with $\mathcal{L}(\pi') = ab^{r+1+s} c^{r+1} a$, where s is the length of τ_2. But this block is not allowed, so that X cannot be sofic. $\square$

This example uses the so-called Pumping Lemma from automata theory [HopU] which says roughly that every long enough path in a graph has a subpath that can be "pumped" or repeated any number of times. The condition defining the shift space in this example is motivated by the theory of context-free languages. See the Notes section of this chapter for more on the connections between symbolic dynamics and automata theory.

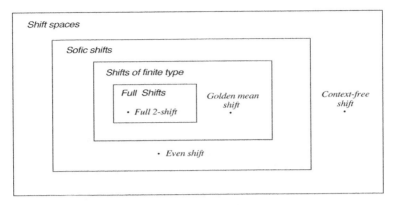

FIGURE 3.1.4. Relations among shift spaces.

Figure 3.1.4 shows how the various types of shift spaces are related.

Finally, we mention that many operations that we have defined on graphs naturally extend to labeled graphs. For instance, consider the notion of state splitting, introduced in §2.4. Let $\mathcal{G} = (G, \mathcal{L})$ be a labeled graph and suppose that H is formed from G by out-splitting. Recall that the edges of H derived from an edge e of G are written e^j. Now, H inherits a labeling, $\mathcal{L}'$, from $\mathcal{L}$: namely, define $\mathcal{L}'(e^j) = \mathcal{L}(e)$. We say that $\mathcal{H} = (H, \mathcal{L}')$ is an *out-split graph formed from* $\mathcal{G}$. We leave it as an exercise for the reader to verify that $\mathcal{G}$ and $\mathcal{H}$ present the same sofic shift (Exercise 3.1.9). Other operations, such as the higher power of a labeled graph, are described in Exercise 3.1.7.

EXERCISES

3.1.1. Show that the sofic shift presented by the labeled graph in Figure 3.1.2(d) is the full 2-shift.

3.1.2. For the labeled graph shown in Figure 3.1.2(d), find a word w so that all paths presenting w end at the same vertex. Can you describe all such words? Do the same for the graph in Figure 3.1.1(b).

3.1.3. Show that the charge-constrained shift from Example 1.2.7 with $c = 3$ is presented by the labeled graph corresponding to the following symbolic adjacency matrix:

$$\begin{bmatrix} \emptyset & 1 & \emptyset & \emptyset \\ -1 & \emptyset & 1 & \emptyset \\ \emptyset & -1 & \emptyset & 1 \\ \emptyset & \emptyset & -1 & \emptyset \end{bmatrix}.$$

Show that this shift is strictly sofic.

3.1.4. Construct an infinite collection of sofic shifts no two of which are conjugate.

3.1.5. Show that there are only countably many sofic shifts contained in the full 2-shift (compare with Exercise 1.2.12).

3.1.6. Denote the collection of all nonempty words over $\mathcal{A}$ by $\mathcal{A}^+$. Let G be a graph, and $\mathcal{W}: \mathcal{E}(G) \to \mathcal{A}^+$ assign a word to every edge in G. Each bi-infinite walk

$\ldots e_{-1}e_0e_1 \ldots$ on G gives rise to a point $\ldots \mathcal{W}(e_{-1}).\mathcal{W}(e_0)\mathcal{W}(e_1) \ldots \in \mathcal{A}^{\mathbb{Z}}$. Show that the collection of all such points and their shifts forms a sofic shift. What happens if we allow $\mathcal{W}(e)$ to be the empty word for some edges e?

3.1.7. (a) Let $\mathcal{G} = (G, \mathcal{L})$ be a labeled graph. Define the *Nth higher power labeled graph* to be $\mathcal{G}^N = (G^N, \mathcal{L}^N)$ where $\mathcal{L}^N(e_1 \ldots e_N) = \mathcal{L}(e_1) \ldots L(e_N)$. Show that $X_{\mathcal{G}^N} = X_{\mathcal{G}}^N$.

      (b) Define a similar notion of higher edge labeled graph and formulate and prove a similar result.

3.1.8. (a) If X is sofic, show that the Nth higher block shift $X^{[N]}$ and the Nth higher power shift X^N are also sofic.

      (b) Let X be a shift space. If $X^{[N]}$ is sofic, must X be sofic? Similarly for X^N.

3.1.9. Let $\mathcal{G}$ and $\mathcal{H}$ be labeled graphs such that $\mathcal{H}$ is an out-split graph formed from $\mathcal{G}$. Show that $\mathcal{G}$ and $\mathcal{H}$ present the same sofic shift.

*3.1.10. Find an explicit necessary and sufficient condition on $S \subseteq \{0, 1, 2, \ldots\}$ in order that the S-gap shift be sofic. Use your condition to prove that the prime gap shift (Example 1.2.6) is not sofic.

§3.2. Characterizations of Sofic Shifts

Suppose we are given a shift space X, specified, say, by a collection of forbidden blocks. How can we tell whether or not X is a sofic shift? In this section we provide two ways to answer this question.

The first is that X is sofic if and only if it is a factor of a shift of finite type. As a consequence, the collection of sofic shifts is the smallest collection of shift spaces containing the finite type shifts and closed under taking factors. This property of sofic shifts was in fact Weiss's original motivation for introducing them.

The second uses the language $\mathcal{B}(X)$ of X. If w is a word in $\mathcal{B}(X)$, then the class of words in $\mathcal{B}(X)$ that can follow w is a subset of $\mathcal{B}(X)$ which depends on w. There could be infinitely many sets as w varies over $\mathcal{B}(X)$, but we will show that X is sofic precisely when there are only finitely many such sets.

We begin with the factor code criterion.

Theorem 3.2.1. *A shift space is sofic if and only if it is a factor of a shift of finite type.*

PROOF: First suppose that X is sofic, and let $\mathcal{G} = (G, \mathcal{L})$ be a presentation of X. The 1-block map $\mathcal{L}$ induces a sliding block code $\mathcal{L}_\infty \colon X_G \to X_{\mathcal{G}}$ that is onto by definition of $X_{\mathcal{G}}$. Hence $X = X_{\mathcal{G}}$ is a factor of the edge shift X_G, which has finite type by Proposition 2.2.6.

Conversely, suppose that X is a shift space for which there is a shift of finite type Y and a factor code $\phi \colon Y \to X$. Let ϕ have memory m and anticipation n. Then ϕ is induced by a block map Φ on $\mathcal{B}_{m+n+1}(Y)$. By increasing m if necessary, we can assume that Y is $(m+n)$-step.

Define $\psi: Y \to Y^{[m+n+1]}$ by

$$\psi(y)_{[i]} = y_{[i-m,i+n]}$$

(ψ is almost the same as the higher block map β_{m+n+1}, except that we use coordinates from y starting at index $i - m$ rather than at i). Since Y is $(m + n)$-step, by Theorem 2.3.2 there is a graph G with edges named by the blocks from $\mathcal{B}_{m+n+1}(Y)$ so that $Y^{[m+n+1]} = X_G$. Define a labeling $\mathcal{L}$ on G by $\mathcal{L}(e) = \Phi(e)$. We will show that $\mathcal{G} = (G, \mathcal{L})$ is a presentation of X, proving that X is sofic.

Observe that the following diagram commutes.

(3–2–1)

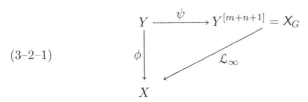

For if $y \in Y$, then $\phi(y)_{[i]} = \Phi(y_{[i-m,i+n]})$, while

$$\mathcal{L}_\infty(\psi(y))_{[i]} = \mathcal{L}(\psi(y)_{[i]}) = \Phi(y_{[i-m,i+n]}).$$

Since ψ is a conjugacy, the images of ϕ and of $\mathcal{L}_\infty$ are equal. Hence $X = \phi(Y) = \mathcal{L}_\infty(X_G)$, so that $\mathcal{G}$ is a presentation of X. $\square$

In this proof we used the conjugacy ψ in (3–2–1) to replace the sliding block code $\phi: Y \to X$ from a general shift of finite type Y with a 1-block code $\mathcal{L}_\infty: Y^{[m+n+1]} \to X$ from a 1-step shift of finite type $Y^{[m+n+1]}$. This replacement is often useful since it is easier to think about 1-block codes and 1-step shifts. We will indicate this replacement with a phrase like "recode $\phi: Y \to X$ so that Y is 1-step and ϕ is 1-block."

Corollary 3.2.2. *A factor of a sofic shift is sofic.*

PROOF: Suppose that $\phi: Y \to X$ is a factor code, and that Y is sofic. By the theorem, there is a factor code $\psi: Z \to Y$ with Z a shift of finite type. Since $\phi \circ \psi: Z \to X$ is a factor code (Exercise 1.5.1), we conclude by the theorem that X is sofic. $\square$

Corollary 3.2.3. *A shift space that is conjugate to a sofic shift is itself sofic.*

PROOF: This follows from Corollary 3.2.2, since a conjugacy is a factor code. $\square$

We now turn to a language-theoretic criterion for sofic shifts.

Definition 3.2.4. Let X be a shift space and w be a word in $\mathcal{B}(X)$. The *follower set* $F_X(w)$ of w in X is the set of all words that can follow w in X; i.e.,

$$F_X(w) = \{v \in \mathcal{B}(X) : wv \in \mathcal{B}(X)\}.$$

The collection of all follower sets in X is denoted by

$$\mathcal{C}_X = \{F_X(w) : w \in \mathcal{B}(X)\}.$$

Each follower set is an infinite set of words, since every word in $\mathcal{B}(X)$ can be extended indefinitely to the right. However, many words can have the same follower set. Examples will show that this collapsing can result in a $\mathcal{C}_X$ with only finitely many different sets.

Example 3.2.5. Let X be the full 2-shift, and $w \in \mathcal{B}(X)$. Since every word can follow w, the follower set of w is the entire language of X, so that

$$F_X(w) = \mathcal{B}(X) = \{0, 1, 00, 01, 10, 11, \dots\}.$$

Thus $\mathcal{C}_X$ contains only one set C, namely $C = \mathcal{B}(X)$. □

Example 3.2.6. Let G be the graph shown in Figure 3.2.1(a), and X be the edge shift X_G, having alphabet $\{e, f, g, h\}$. If π is a path on G (i.e., a word in $\mathcal{B}(X_G)$), then the follower set of π is the set of all paths on G starting at the terminal vertex $t(\pi)$ of π. Let C_0 be the set of paths on G starting at vertex 0, and C_1 be the set of those starting at 1. Then $\mathcal{C}_X = \{C_0, C_1\}$. For example,

$$F_X(e) = F_X(fg) = F_X(hhgee) = C_0,$$

and

$$F_X(f) = F_X(gefh) = F_X(h^{100}) = C_1.$$

The same considerations show that if G is an essential graph having r vertices, then $\mathcal{C}_{X_G}$ contains exactly r sets, one for each vertex. □

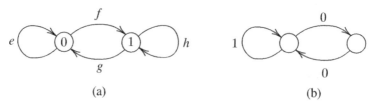

(a)   (b)

FIGURE 3.2.1. Graphs for follower sets.

Example 3.2.7. Let $\mathcal{G}$ be the labeled graph shown in Figure 3.2.1(b), so that $X = X_{\mathcal{G}}$ is the even shift. There are three distinct follower sets, given by

$$C_0 = F_X(0) = \{0, 1, 00, 01, 10, 11, 000, 001, 010, 011, 100, 110, 111, \ldots\},$$
$$C_1 = F_X(1) = \{0, 1, 00, 10, 11, 000, 001, 100, 110, 111, \ldots\},$$
$$C_2 = F_X(10) = \{0, 00, 01, 000, 010, 011, \ldots\}.$$

For $w \in \mathcal{B}(X)$, it is easy to check that

$$F_X(w) = \begin{cases} C_0 & \text{if } w \text{ contains no 1's,} \\ C_1 & \text{if } w \text{ ends in } 10^{2k} \text{ for some } k \geqslant 0, \\ C_2 & \text{if } w \text{ ends in } 10^{2k+1} \text{ for some } k \geqslant 0, \end{cases}$$

so that these three sets are the only ones, and $\mathcal{C}_X = \{C_0, C_1, C_2\}$. □

Example 3.2.8. Let X be the context-free shift from Example 1.2.9. Recall that $\mathcal{A} = \{a, b, c\}$, and that $ab^m c^k a$ is allowed exactly when $m = k$. Hence $c^k a$ is in $F_X(ab^m)$ if and only if $k = m$. This means that for $m = 1, 2, 3, \ldots$ the follower sets $F_X(ab^m)$ are all different from each other, so that $\mathcal{C}_X$ contains infinitely many sets. □

Suppose that X is a shift space over $\mathcal{A}$ such that $\mathcal{C}_X$ is finite. We will construct a labeled graph $\mathcal{G} = (G, \mathcal{L})$ called the *follower set graph* of X as follows. The vertices of G are the elements C in $\mathcal{C}_X$. Let $C = F_X(w) \in \mathcal{V}$ and $a \in \mathcal{A}$. If $wa \in \mathcal{B}(X)$, then $F_X(wa)$ is a set, say $C' \in \mathcal{V}$, and draw an edge labeled a from C to C'. If $wa \notin \mathcal{B}(X)$ then do nothing. Carrying this out for every $C \in \mathcal{V}$ and $a \in \mathcal{A}$ yields the labeled graph $\mathcal{G}$.

For example, let X be the even shift, so that $\mathcal{C}_X = \{C_0, C_1, C_2\}$ as in Example 3.2.7. Since $C_0 = F_X(0)$, the follower set graph contains an edge labeled 0 from C_0 to $F_X(00) = C_0$ and an edge labeled 1 from C_0 to $F_X(01) = C_1$. The entire follower set graph is shown in Figure 3.2.2.

When defining the follower set graph, we glossed over an essential point. Suppose that $C = F_X(w) = F_X(w')$ for another word w', and we had used w' instead of w. Would this change the resulting graph? To see that

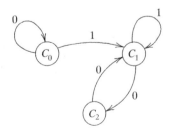

FIGURE 3.2.2. Follower set graph for the even shift.

it would not, first notice that $wa \in \mathcal{B}(X)$ if and only if $w'a \in \mathcal{B}(X)$, since each occurs exactly when $a \in C$. Furthermore, $F_X(wa) = F_X(w'a)$ since each is obtained by taking all words in C starting with the letter a and deleting this letter.

The follower set graph can be used to recover the original shift.

Proposition 3.2.9. *If X is a shift space with a finite number of follower sets, and $\mathcal{G}$ is its follower set graph, then $\mathcal{G}$ is a presentation of X. In particular, X must be sofic.*

PROOF: To show that $X = X_{\mathcal{G}}$, it suffices by Proposition 1.3.4(3) to show they have the same language, i.e., that $\mathcal{B}(X) = \mathcal{B}(X_{\mathcal{G}})$.

First suppose that $u = a_1 a_2 \ldots a_n \in \mathcal{B}(X_{\mathcal{G}})$. Then there is a path π in the underlying graph G with $\mathcal{L}(\pi) = u$. Let π start at vertex $F_X(w)$. The definition of $\mathcal{G}$ shows that $wa_1 \in \mathcal{B}(X)$, next that $wa_1 a_2 \in \mathcal{B}(X)$, and so on, so that $wa_1 a_2 \ldots a_n = wu \in \mathcal{B}(X)$. Hence $u \in \mathcal{B}(X)$, proving that $\mathcal{B}(X_{\mathcal{G}}) \subseteq \mathcal{B}(X)$.

To prove the reverse inclusion, let $u \in \mathcal{B}(X)$. By iteratively applying Proposition 1.3.4(1)(b), there is a word w with $wu \in \mathcal{B}(X)$ and $|w| > |\mathcal{V}(G)|$. By definition of $\mathcal{G}$, there is a path in G labeled by wu. Such a path can be written $\alpha\beta\gamma\pi$, where β is a cycle and π is labeled by u. Since every state of G has at least one outgoing edge, it follows that π can be extended to a bi-infinite path in G, and so $u \in \mathcal{B}(X_{\mathcal{G}})$.   □

Finiteness of the number of follower sets in fact characterizes sofic shifts.

Theorem 3.2.10. *A shift space is sofic if and only if it has a finite number of follower sets.*

PROOF: The "if" part is contained in the previous proposition. To prove the opposite implication, we need to show that if X is sofic, then $\mathcal{C}_X$ is finite.

Let $\mathcal{G} = (G, \mathcal{L})$ be a presentation of X. We may assume that G is essential. For a word $w \in \mathcal{B}(X)$, we will describe how to obtain $F_X(w)$ in terms of $\mathcal{G}$. Consider all paths in G presenting w, and let T be the set of terminal vertices of these paths. Then $F_X(w)$ is the set of labels of paths in G starting at some vertex in T. Hence two words with the same T have the same follower set. Since there are only finitely many subsets T of the vertex set of G, there can be only finitely many follower sets.   □

The proof of the preceding result suggests a procedure for constructing the follower set graph given a presentation $\mathcal{G} = (G, \mathcal{L})$ of a sofic shift. Each follower set corresponds to a subset of vertices. Of course, one needs to identify which subsets actually correspond to follower sets and to identify when two subsets correspond to the same follower set; see Exercise 3.4.9.

In Example 3.2.8 we saw that the context-free shift has an infinite number of follower sets, and so the previous theorem shows that it is not sofic. A more "bare-hands" proof of this is in Example 3.1.7.

Finally, recall from Proposition 3.1.6 that every shift of finite type has a presentation $(G, \mathcal{L})$ such that $\mathcal{L}_\infty$ is a conjugacy. Exercise 3.2.7 shows that the follower set graph is such a presentation. This also holds for a related presentation, called the *future cover*, which is developed in Exercise 3.2.8.

EXERCISES

3.2.1. Let G be an essential graph with labeling $\mathcal{L}(e) = e$ for each edge e in G. Prove that $(G, \mathcal{L})$ is the follower set graph of X_G.

3.2.2. Let $\mathcal{G}$ and $\mathcal{H}$ be the labeled graphs with symbolic adjacency matrices

$$\begin{bmatrix} a & b \\ b & \emptyset \end{bmatrix} \quad \text{and} \quad \begin{bmatrix} a & a \\ b & \emptyset \end{bmatrix}.$$

Show that $X_\mathcal{G}$ is not conjugate to $X_\mathcal{H}$.

3.2.3. For the labeled graphs $\mathcal{G}$ corresponding to each of the following symbolic adjacency matrices, compute the follower set graph of $X_\mathcal{G}$.

$$\text{(a)} \begin{bmatrix} \emptyset & a & a \\ \emptyset & \emptyset & b \\ a & b & \emptyset \end{bmatrix}, \qquad \text{(b)} \begin{bmatrix} a & a+b & a+c \\ a & \emptyset & \emptyset \\ a & \emptyset & \emptyset \end{bmatrix}$$

3.2.4. Let $\mathcal{G}$ be a labeled graph with r vertices. Prove that $X_\mathcal{G}$ has at most $2^r - 1$ follower sets.

3.2.5. (a) Prove that the product of two sofic shifts is also sofic.
(b) Is the union of two sofic shifts also sofic?
*(c) Is the intersection of two sofic shifts also sofic?

3.2.6. Define the predecessor set of a word, and show that a shift space is sofic if and only if the collection of predecessor sets is finite.

3.2.7. Let X be a sofic shift with follower set graph $(G, \mathcal{L})$. Show that X is a shift of finite type if and only if $\mathcal{L}_\infty$ is a conjugacy.

3.2.8. Let X be a shift space. The *future* $F_X(\lambda)$ of a left-infinite sequence λ in X is the collection of all right-infinite sequences ρ such that $\lambda.\rho \in X$. The *future cover* of X is the labeled graph where the vertices are the futures of left-infinite sequences, and there is an edge labeled a from $F_X(\lambda)$ to $F_X(\lambda a)$ whenever λ and λa are left-infinite sequences of X.

(a) Show that the future cover of a sofic shift X is a presentation of X.
(b) Prove that a shift space X is sofic exactly when there are only finitely many futures of left-infinite sequences of X.
(c) Let X be a sofic shift with future cover $\mathcal{G} = (G, \mathcal{L})$. Show that X is a shift of finite type if and only if $\mathcal{L}_\infty$ is a conjugacy.
(d) Let X be a shift of finite type with future cover $\mathcal{G} = (G, \mathcal{L})$ and follower set graph $\mathcal{G}' = (G', \mathcal{L}')$. Show that $X_G = X_{G'}$.
(e) Show that part (d) is false in general for sofic X.

*3.2.9. Find an example of a pair of irreducible sofic shifts X, Y and a sliding block code $\phi\colon X \to Y$ such that ϕ is not one-to-one, but the restriction of ϕ to the periodic points is one-to-one (compare with Exercise 2.3.6).

§3.3. Minimal Right-Resolving Presentations

A labeled graph is called right-resolving if, for each vertex I, the edges starting at I carry different labels. A right-resolving graph is "deterministic," in that given a label block w and a vertex I, there is at most one path starting at I with label w. This section concerns right-resolving presentations of sofic shifts.

Right-resolving presentations are important for a number of reasons. The ambiguity of where different presentations of a label block occur is concentrated in the vertices. For a given irreducible sofic shift, the main result of this section shows that among all of its right-resolving presentations, there is a unique one with fewest vertices. This is much like the canonical form for a matrix, and is useful in both theory and practice. Finally, right-resolving presentations arise in a variety of engineering applications involving encoders and decoders. We will see more on this in Chapter 5.

Recall that for a graph G with edge set $\mathcal{E}$, the set of edges starting at a vertex I is denoted by $\mathcal{E}_I$.

Definition 3.3.1. A labeled graph $\mathcal{G} = (G, \mathcal{L})$ is *right-resolving* if, for each vertex I of G, the edges starting at I carry different labels. In other words, $\mathcal{G}$ is right-resolving if, for each I, the restriction of $\mathcal{L}$ to $\mathcal{E}_I$ is one-to-one. A *right-resolving presentation* of a sofic shift is a right-resolving labeled graph that presents the shift.

For example, the labeled graph shown in Figure 3.1.2(b) is a right-resolving presentation of the full 2-shift, while (c) is not.

We have defined "right-resolving" in terms of the outgoing edges from a vertex. There is a dual property *left-resolving*, in which the incoming edges to each vertex carry different labels.

Does every sofic shift have a right-resolving presentation? The follower set graph from the previous section gives such a presentation. For if X is sofic, then by Theorem 3.2.10 it has a finite number of follower sets. Then Proposition 3.2.9 shows that the resulting follower set graph presents X, and it is right-resolving by its very definition. But this still leaves open the problem of actually *finding* the follower sets. There is an alternative method, called the *subset construction*, to find an explicit right-resolving presentation, given an arbitrary presentation. We use the subset construction in the next proof.

Theorem 3.3.2. *Every sofic shift has a right-resolving presentation.*

PROOF: Let X be sofic over the alphabet $\mathcal{A}$. Then X has a presentation $\mathcal{G} = (G, \mathcal{L})$, so that $X = X_{\mathcal{G}}$. Recall from Proposition 2.2.10 that we may assume that G is essential.

Construct a new labeled graph $\mathcal{H} = (H, \mathcal{L}')$ as follows. The vertices $\mathcal{I}$ of $\mathcal{H}$ are the nonempty subsets of the vertex set $\mathcal{V}(G)$ of G. If $\mathcal{I} \in \mathcal{V}(H)$

and $a \in \mathcal{A}$, let $\mathcal{J}$ denote the set of terminal vertices of edges in G starting at some vertex in $\mathcal{I}$ and labeled a. In other words, $\mathcal{J}$ is the set of vertices reachable from $\mathcal{I}$ using edges labeled a. If $\mathcal{J}$ is nonempty, then $\mathcal{J} \in \mathcal{V}(H)$, and draw an edge in H from $\mathcal{I}$ to $\mathcal{J}$ labeled a; if $\mathcal{J}$ is empty, do nothing. Carrying this out for each $\mathcal{I} \in \mathcal{V}(H)$ and each $a \in \mathcal{A}$ produces the labeled graph $\mathcal{H}$. Note that for each vertex $\mathcal{I}$ in $\mathcal{H}$ there is at most one edge with a given label starting at $\mathcal{I}$, so that $\mathcal{H}$ is right-resolving.

We now verify that $X = X_{\mathcal{G}} = X_{\mathcal{H}}$, so that $\mathcal{H}$ is a right-resolving presentation of X. It suffices to show that $\mathcal{B}(X_{\mathcal{G}}) = \mathcal{B}(X_{\mathcal{H}})$.

First suppose that $w = a_1 a_2 \ldots a_n \in \mathcal{B}(X_{\mathcal{G}})$. Let $\pi = e_1 e_2 \ldots e_n$ be a path in G presenting w. Then π starts at some vertex I in G. Let $\mathcal{I}_0 = \{I\}$, and for $1 \leqslant k \leqslant n$ let $\mathcal{I}_k$ be the set of terminal vertices of paths of length k starting at I and labeled $a_1 a_2 \ldots a_k$. Then $t(e_1 e_2 \ldots e_k) \in \mathcal{I}_k$, so each $\mathcal{I}_k$ is nonempty, and there is an edge in $\mathcal{H}$ labeled a_k from $\mathcal{I}_{k-1}$ to $\mathcal{I}_k$. This shows there is a path in $\mathcal{H}$ labeled w.

For the same reason for any choice of u, v s.t. $uwv \in B(X_{\mathcal{G}})$ there is a path in H labeled uwv. Since $w \in B(X_{\mathcal{G}})$, there are arbitrarily long such u, v, and so $w \in \mathcal{B}(X_{\mathcal{H}})$ (why?). Hence $\mathcal{B}(X_{\mathcal{G}}) \subseteq \mathcal{B}(X_{\mathcal{H}})$.

To prove the reverse inclusion, suppose that $u \in \mathcal{B}(X_{\mathcal{H}})$. Let τ be a path in $\mathcal{H}$ presenting u, starting at $\mathcal{I}$ and ending at $\mathcal{J}$. The construction of $\mathcal{H}$ shows that $\mathcal{J}$ is the set of all vertices in G reachable from a vertex in $\mathcal{I}$ with a path labeled u. Since $\mathcal{J} \neq \varnothing$, this shows there is a path in G labeled u. Since G is essential, $u \in \mathcal{B}(X_{\mathcal{G}})$. Thus $\mathcal{B}(X_{\mathcal{H}}) \subseteq \mathcal{B}(X_{\mathcal{G}})$. $\qquad\square$

Example 3.3.3. Let X be the sofic shift presented by the labeled graph $\mathcal{G}$ shown in Figure 3.3.1(a). Carrying out the subset construction produces the right-resolving labeled graph $\mathcal{H}$ shown in (b). Note that although $\mathcal{G}$ is irreducible, $\mathcal{H}$ is not. However, there is an irreducible subgraph $\mathcal{H}'$ of $\mathcal{H}$, obtained by deleting the vertex $\{b\}$ and the edge from it, which also presents X. $\qquad\square$

If $\mathcal{G}$ is a labeled graph with r states, the subset construction yields a right-resolving labeled graph $\mathcal{H}$ with $2^r - 1$ states. The price of requiring a right-resolving presentation is a large increase in the number of states. Fortunately, often a subgraph of $\mathcal{H}$ presents the same sofic shift, as happened

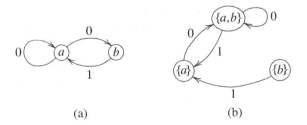

(a)   (b)

FIGURE 3.3.1. The subset construction.

in Example 3.3.3.

We now turn our attention to finding "small" right-resolving presentations of a given sofic shift.

Definition 3.3.4. A *minimal right-resolving presentation* of a sofic shift X is a right-resolving presentation of X having the fewest vertices among all right-resolving presentations of X.

Note that a minimal right-resolving presentation $\mathcal{G}$ is essential; for, if not, one could obtain a smaller right-resolving presentation by passing to the unique essential subgraph H of G such that $X_H = X_G$.

At this point, for all we know there could be several different minimal right-resolving presentations of a given sofic shift. The next series of results leads up to Theorem 3.3.18, which shows that any two minimal right-resolving presentations of an irreducible sofic shift must be isomorphic as labeled graphs. Example 3.3.21 describes a reducible sofic shift with two nonisomorphic minimal right-resolving presentations.

Example 3.3.5. If X is the full 2-shift, then the labeled graph in Figure 3.1.2(a) is a minimal right-resolving presentation since it has only one vertex. The right-resolving graphs shown in (b) and (d) of the same figure are therefore not minimal right-resolving presentations of X. If $\mathcal{G}$ is any minimal right-resolving presentation of X, then $\mathcal{G}$ must have only one vertex, so every edge is a self-loop. Also, $\mathcal{G}$ must have a self-loop labeled 0 and another self-loop labeled 1. Since $\mathcal{G}$ is right-resolving, there cannot be any other self-loops, so that apart from the name of its sole vertex, $\mathcal{G}$ is the labeled graph in (a). Thus any two minimal right-resolving presentations of the full 2-shift are isomorphic. This is a very simple illustration of Theorem 3.3.18.   □

Example 3.3.6. The graph shown in Figure 3.1.3(a) is a minimal right-resolving presentation of the even shift, for any presentation with fewer vertices would have just one vertex and would therefore present a full shift. The reader can verify directly by an ad hoc case-by-case argument that every 2-state right-resolving presentation of the even shift is isomorphic to this graph.   □

An important feature we will establish for minimal right-resolving presentations is that, for each pair of distinct states, there is a label block w such that there is a path labeled w starting from one of the states but not the other.

Definition 3.3.7. Let $\mathcal{G} = (G, \mathcal{L})$ be a labeled graph, and $I \in \mathcal{V}$ be a state of $\mathcal{G}$. The *follower set* $F_{\mathcal{G}}(I)$ of I in $\mathcal{G}$ is the collection of labels of paths starting at I.

Note that if $\mathcal{G}$ is essential, then

$$F_{\mathcal{G}}(I) = \{\mathcal{L}(\pi) : \pi \in \mathcal{B}(X_G) \text{ and } i(\pi) = I\}.$$

We will say that $\mathcal{G}$ is *follower-separated* if distinct states have distinct follower sets; i.e., if $I \neq J$, then $F_{\mathcal{G}}(I) \neq F_{\mathcal{G}}(J)$.

For example, in Figure 3.1.2 the graphs (a) and (c) are follower-separated, while those in (b) and (d) are not (why?).

If a labeled graph is not follower-separated, we show next how to merge states together to obtain a smaller graph, presenting the same sofic shift, that *is* follower-separated.

Let $\mathcal{G}$ be an essential labeled graph. Call two vertices I and J *equivalent* if they have the same follower set, i.e., if $F_{\mathcal{G}}(I) = F_{\mathcal{G}}(J)$. This relation partitions the vertex set of $\mathcal{G}$ into disjoint equivalence classes, which we will denote by $\mathcal{I}_1, \mathcal{I}_2, \ldots, \mathcal{I}_r$. Define a labeled graph $\mathcal{H}$ as follows. The vertex set of $\mathcal{H}$ is $\{\mathcal{I}_1, \mathcal{I}_2, \ldots, \mathcal{I}_r\}$. There is an edge in $\mathcal{H}$ labeled a from $\mathcal{I}$ to $\mathcal{J}$ exactly when there are vertices $I \in \mathcal{I}$ and $J \in \mathcal{J}$ and an edge in the original graph $\mathcal{G}$ labeled a from I to J. We call $\mathcal{H}$ the *merged graph from* $\mathcal{G}$. In the next section we will develop an algorithm to find the classes of equivalent vertices.

Lemma 3.3.8. *Let $\mathcal{G}$ be an essential labeled graph, and $\mathcal{H}$ be the merged graph from $\mathcal{G}$. Then $\mathcal{H}$ is follower-separated and $X_{\mathcal{H}} = X_{\mathcal{G}}$. Furthermore, if $\mathcal{G}$ is irreducible, then so is $\mathcal{H}$, and if $\mathcal{G}$ is right-resolving, then so is $\mathcal{H}$.*

PROOF: Let vertex I in $\mathcal{G}$ have equivalence class $\mathcal{I}$, a vertex in $\mathcal{H}$. We first show that $F_{\mathcal{G}}(I) = F_{\mathcal{H}}(\mathcal{I})$. The definition of $\mathcal{H}$ shows that a path in $\mathcal{G}$ starting at I gives rise to a path in $\mathcal{H}$ starting at $\mathcal{I}$ with the same label, passing through the equivalence classes of vertices on the original path. Hence $F_{\mathcal{G}}(I) \subseteq F_{\mathcal{H}}(\mathcal{I})$. If $F_{\mathcal{G}}(I) \neq F_{\mathcal{H}}(\mathcal{I})$ for some I, let $w = a_1 a_2 \ldots a_m$ be a shortest word in $F_{\mathcal{H}}(\mathcal{I}) \setminus F_{\mathcal{G}}(I)$ among all I in $\mathcal{G}$. There is an edge labeled a_1 from $\mathcal{I}$ to some $\mathcal{J}$, hence an $I' \in \mathcal{I}$, $J \in \mathcal{J}$ and an edge in $\mathcal{G}$ labeled a_1 from I' to J. By minimality of m, $a_2 \ldots a_m \in F_{\mathcal{H}}(\mathcal{J}) \cap \mathcal{A}^{m-1} = F_{\mathcal{G}}(J) \cap \mathcal{A}^{m-1}$. But this gives a path in $\mathcal{G}$ starting at I' labeled w, contradicting $w \notin F_{\mathcal{G}}(I') = F_{\mathcal{G}}(I)$. Hence $F_{\mathcal{H}}(\mathcal{I}) = F_{\mathcal{G}}(I)$.

It is now clear that $\mathcal{H}$ is follower-separated, since the follower sets in $\mathcal{G}$ for inequivalent vertices are different. Since $\mathcal{B}(X_{\mathcal{G}})$ is the union of the follower sets of the vertices in $\mathcal{G}$, and similarly for $\mathcal{H}$, we immediately obtain that $\mathcal{B}(X_{\mathcal{G}}) = \mathcal{B}(X_{\mathcal{H}})$, so that $X_{\mathcal{G}} = X_{\mathcal{H}}$.

Next, assume that $\mathcal{G}$ is irreducible. Let $\mathcal{I}$, $\mathcal{J}$ be distinct vertices of the merged graph $\mathcal{H}$, and let $I \in \mathcal{I}$ and $J \in \mathcal{J}$. Since $\mathcal{G}$ is irreducible, there is a path in $\mathcal{G}$ from I to J. Passing to equivalence classes gives a path in $\mathcal{H}$ from $\mathcal{I}$ to $\mathcal{J}$, and this shows that $\mathcal{H}$ must also be irreducible.

Finally, assume that $\mathcal{G}$ is right-resolving. Consider the situation of an edge in $\mathcal{H}$ labeled a from $\mathcal{I}$ to $\mathcal{J}$. Then there is an edge in $\mathcal{G}$ labeled a from some $I \in \mathcal{I}$ to some $J \in \mathcal{J}$. Since $\mathcal{G}$ is right-resolving, this edge is the *only* edge labeled a starting at I. Thus $F_{\mathcal{G}}(J)$ can be obtained from $F_{\mathcal{G}}(I)$ by removing the first symbol from all words in $F_{\mathcal{G}}(I)$ starting with a. Hence $F_{\mathcal{G}}(I)$ together with a determine $F_{\mathcal{G}}(J)$. Since $F_{\mathcal{H}}(\mathcal{I}) = F_{\mathcal{G}}(I)$ and

$F_{\mathcal{H}}(\mathcal{J}) = F_{\mathcal{G}}(J)$, we see that $F_{\mathcal{H}}(\mathcal{I})$ together with a determine $F_{\mathcal{H}}(\mathcal{J})$. This means that there is at most one edge in $\mathcal{H}$ starting at $\mathcal{I}$ and labeled a, so that $\mathcal{H}$ is right-resolving. $\square$

The merging of states allows us to deduce the main property of minimal right-resolving presentations.

Proposition 3.3.9. *A minimal right-resolving presentation of a sofic shift is follower-separated.*

PROOF: Let $\mathcal{G}$ be a minimal right-resolving presentation of a sofic shift X. If $\mathcal{G}$ is not follower-separated, then the merged graph $\mathcal{H}$ from $\mathcal{G}$ would have fewer states. By Lemma 3.3.8, $\mathcal{H}$ is a right-resolving presentation of X, contradicting minimality of $\mathcal{G}$. $\square$

We next discuss the relationship between irreducibility of a sofic shift and irreducibility of a presentation. First observe that if $\mathcal{G} = (G, \mathcal{L})$ is an irreducible labeled graph, then the sofic shift $X_{\mathcal{G}}$ must be irreducible. For if $u, v \in \mathcal{B}(X_{\mathcal{G}})$, then there are paths π and τ in G with $\mathcal{L}(\pi) = u$ and $\mathcal{L}(\tau) = v$. Since G is irreducible, there is a path ω in G from $t(\pi)$ to $i(\tau)$, so that $\pi\omega\tau$ is a path in G. Let $w = \mathcal{L}(\omega)$. Then

$$\mathcal{L}(\pi\omega\tau) = \mathcal{L}(\pi)\mathcal{L}(\omega)\mathcal{L}(\tau) = uwv \in \mathcal{B}(X_{\mathcal{G}}),$$

establishing irreducibility of $X_{\mathcal{G}}$. This means that a sofic shift presented by an irreducible labeled graph is irreducible.

The converse, however, is not true. For example, let X be irreducible and presented by $\mathcal{G}$. Let $\mathcal{H}$ be the union of two disjoint copies of $\mathcal{G}$. Then although X is irreducible and presented by $\mathcal{H}$, the graph $\mathcal{H}$ is reducible. Nevertheless, we can prove the converse for minimal right-resolving presentations.

Lemma 3.3.10. *Suppose that X is an irreducible sofic shift, and that $\mathcal{G}$ is a minimal right-resolving presentation of X. Then $\mathcal{G}$ is an irreducible graph.*

PROOF: Let $\mathcal{G} = (G, \mathcal{L})$ be a minimal right-resolving presentation of X. Without loss of generality we may assume that G is essential. We first show that for every state $I \in \mathcal{V}(G)$, there is a word $u_I \in \mathcal{B}(X_{\mathcal{G}})$ such that every path in G presenting u_I contains I. Suppose that this is not the case for some I. Then removing I and all edges containing it would create a right-resolving graph $\mathcal{H}$ with fewer states than $\mathcal{G}$ and such that $\mathcal{B}(X_{\mathcal{H}}) = \mathcal{B}(X_{\mathcal{G}})$, so that $X_{\mathcal{H}} = X_{\mathcal{G}}$, contradicting minimality of $\mathcal{G}$.

Now let I and J be distinct vertices in G, and let u_I and u_J be the words described above. Since X is irreducible, there is a word w such that $u_I w u_J \in \mathcal{B}(X_{\mathcal{G}})$. Let π be a path presenting $u_I w u_J$, so $\pi = \tau_I \omega \tau_J$ with $\mathcal{L}(\tau_I) = u_I$, $\mathcal{L}(\omega) = w$, and $\mathcal{L}(\tau_J) = u_J$. But τ_I contains I and τ_J contains J, so there is a subpath of π from I to J. This shows that $\mathcal{G}$ is irreducible. $\square$

Proposition 3.3.11. *A sofic shift is irreducible if and only if it has an irreducible presentation.*

PROOF: If X has an irreducible presentation, then X is irreducible by the discussion preceding Lemma 3.3.10, while that lemma shows that if X is irreducible, then a minimal right-resolving presentation of X is irreducible. □

If X is sofic and presented by $\mathcal{G} = (G, \mathcal{L})$, then a given word $w \in \mathcal{B}(X)$ can be presented by many paths in G. This ambiguity is reduced if $\mathcal{G}$ is right-resolving, since for each vertex in G there is at most one path starting there that presents w. Sometimes there are words such that every presenting path ends at the same vertex, and such paths "focus" to that vertex.

Definition 3.3.12. Let $\mathcal{G} = (G, \mathcal{L})$ be a labeled graph. A word $w \in \mathcal{B}(X_\mathcal{G})$ is a *synchronizing word* for $\mathcal{G}$ if all paths in G presenting w terminate at the same vertex. If this vertex is I, we say that w *focuses to I*.

Example 3.3.13. If e is the only edge labeled a, then a is a synchronizing word, as is any word ending in a. Since edge shifts come from labeled graphs in which each edge is labeled distinctly, every word in an edge shift is synchronizing. □

Example 3.3.14. Consider the four presentations of the full 2-shift shown in Figure 3.1.2. In graph (a), every word is synchronizing since there is only one vertex. In (b) and (c), no word is synchronizing (why?). In (d) the words 00 and 11 are synchronizing, and so any word containing either 00 or 11 is also synchronizing since the graph is right-resolving. This leaves words in which 0's and 1's alternate, such as 0101010, none of which is synchronizing (why?). □

Lemma 3.3.15. *Suppose that w is a synchronizing word for an essential right-resolving labeled graph $\mathcal{G}$. Then any word of the form wu in $\mathcal{B}(X_\mathcal{G})$ is also synchronizing for $\mathcal{G}$. If w focuses to I, then $F_{X_\mathcal{G}}(w) = F_\mathcal{G}(I)$.*

PROOF: Let w focus to I. Any path presenting wu has the form $\pi\tau$, where π must end at I. But there is only one path starting at I labeled u since $\mathcal{G}$ is right-resolving. Thus any path presenting wu must end at $t(\tau)$, showing that wu is synchronizing. Also, any word following w is presented by a path starting at I and conversely, establishing the second statement. □

The next result shows that for some labeled graphs we can find synchronizing words, in fact infinitely many.

Proposition 3.3.16. *Suppose that $\mathcal{G}$ is an essential right-resolving labeled graph that is follower-separated. Then every word $u \in \mathcal{B}(X_\mathcal{G})$ can be extended on the right to a synchronizing word uw.*

PROOF: For any word $v \in \mathcal{B}(X_{\mathcal{G}})$, let $T(v)$ denote the set of terminal vertices of all paths presenting v. If $T(u)$ is just a single vertex then u is synchronizing and w can be any word such that $uw \in \mathcal{B}(X_{\mathcal{G}})$.

Suppose now that $T(u)$ has more than one vertex, and let I and J be distinct vertices in $T(u)$. Since $F_{\mathcal{G}}(I) \neq F_{\mathcal{G}}(J)$, there is a word in one follower set but not in the other, say $v_1 \in F_{\mathcal{G}}(I)$ but $v_1 \notin F_{\mathcal{G}}(J)$. Since $\mathcal{G}$ is right-resolving, there is at most one path labeled v_1 starting at each element of $T(u)$. Since $v_1 \notin F_{\mathcal{G}}(J)$, it follows that $T(uv_1)$ has fewer elements than $T(u)$. We can continue this way to produce ever smaller sets $T(uv_1)$, $T(uv_1v_2), \ldots$, stopping when $T(uv_1v_2 \ldots v_n)$ has only one vertex. Then set $w = v_1v_2 \ldots v_n$. $\qquad \square$

We now come to the main results of this section. These are due to Fischer [Fis1], [Fis2].

Proposition 3.3.17. *If $\mathcal{G}$ and $\mathcal{G}'$ are irreducible, right-resolving presentations of a sofic shift that are also follower-separated, then $\mathcal{G}$ and $\mathcal{G}'$ are isomorphic as labeled graphs.*

PROOF: Let $\mathcal{G} = (G, \mathcal{L})$ and $\mathcal{G}' = (G', \mathcal{L}')$. We will first use the right-resolving and follower-separated properties to find a synchronizing word for $\mathcal{G}$ and $\mathcal{G}'$. This word picks out a vertex from each graph. Irreducibility is then used to extend this correspondence to one-to-one correspondences $\partial\Phi \colon \mathcal{V}(G) \to \mathcal{V}(G')$ and $\Phi \colon \mathcal{E}(G) \to \mathcal{E}(G')$ that respect the labels. A key ingredient is the second statement in Lemma 3.3.15 which asserts that the follower set of a synchronizing word equals the follower set of its focusing vertex.

By Proposition 3.3.16, we can find a synchronizing word u_1 for $\mathcal{G}$ and an extension $w = u_1u_2$ that is synchronizing for $\mathcal{G}'$. By Lemma 3.3.15, w remains synchronizing for $\mathcal{G}$.

We will now use w to define $\partial\Phi \colon \mathcal{V}(G) \to \mathcal{V}(G')$. Since w is synchronizing for both graphs, it focuses to some vertex I_0 in $\mathcal{G}$ and to another vertex I_0' in $\mathcal{G}'$. Define $\partial\Phi(I_0) = I_0'$. If J is any vertex in $\mathcal{G}$, let π be a path in G from I_0 to J. If $u = \mathcal{L}(\pi)$, then wu is synchronizing for both graphs, and focuses to J in $\mathcal{G}$ and to some vertex J' in $\mathcal{G}'$. We define $\partial\Phi(J) = J'$. This does not depend on the particular path π used, since

$$F_{\mathcal{G}}(J) = F_{X_{\mathcal{G}}}(wu) = F_{X_{\mathcal{G}'}}(wu) = F_{\mathcal{G}'}(J')$$

and $\mathcal{G}'$ is follower-separated. This shows that $\partial\Phi$ is one-to-one, and reversing the roles of $\mathcal{G}$ and $\mathcal{G}'$ constructs its inverse. Thus $\partial\Phi$ is a one-to-one correspondence.

We next define $\Phi \colon \mathcal{E}(G) \to \mathcal{E}(G')$. Suppose that an edge e labeled a starts at I and ends at J. As above, let wu be a synchronizing word focusing to I in $\mathcal{G}$. Then wua focuses to J in $\mathcal{G}$, while in $\mathcal{G}'$ let wu and wua focus to I' and J', respectively. Then $\partial\Phi(I) = I'$ and $\partial\Phi(J) = J'$.

Since $a \in \mathcal{F}_{\mathcal{G}}(I) = \mathcal{F}_{\mathcal{G}'}(I')$, there is an edge e' starting at I' labeled a. Put $\Phi(e) = e'$. Since the graphs are right-resolving, Φ is one-to-one. Again reversing the graphs, we can construct the inverse of Φ, so that Φ is a one-to-one correspondence which preserves the labelings. Hence $\mathcal{G}$ and $\mathcal{G}'$ are isomorphic. □

Theorem 3.3.18. *Any two minimal right-resolving presentations of an irreducible sofic shift are isomorphic as labeled graphs.*

PROOF: If $\mathcal{G}$ and $\mathcal{G}'$ are both minimal right-resolving presentations of an irreducible sofic shift, then by Lemma 3.3.10 they are both irreducible graphs, and by Proposition 3.3.9 they are both follower-separated. Apply Proposition 3.3.17. □

By this theorem, we can speak of "the" minimal right-resolving presentation of an irreducible sofic shift X, which we will denote by $\mathcal{G}_X = (G_X, \mathcal{L}_X)$. But how can we describe $\mathcal{G}_X$ directly in terms of X itself, and how can we find $\mathcal{G}_X$ in practice? The answer to the first question involves the follower set graph and "intrinsically synchronizing words," as discussed in Exercise 3.3.4. The answer to the second question uses Corollary 3.3.20 below, starting with an irreducible right-resolving presentation, and is the subject of Theorem 3.4.14 in the next section.

Corollary 3.3.19. *Let X be an irreducible sofic shift with right-resolving presentation $\mathcal{G}$. Then $\mathcal{G}$ is the minimal right-resolving presentation of X if and only if it is irreducible and follower-separated.*

PROOF: If $\mathcal{G}$ is the minimal right-resolving presentation of X, then it is irreducible by Lemma 3.3.10 and follower-separated by Proposition 3.3.9. Conversely, suppose that $\mathcal{G}$ is a right-resolving, irreducible, and follower-separated presentation of X. Let $\mathcal{G}'$ be the minimal right resolving presentation of X. By the first line of this proof, $\mathcal{G}'$ is right-resolving, irreducible and follower-separated. By Proposition 3.3.17, $\mathcal{G}$ and $\mathcal{G}'$ are isomorphic. Thus, $\mathcal{G}$ is the minimal right-resolving presentation of X. □

Corollary 3.3.20. *Let X be an irreducible sofic shift and let $\mathcal{G}$ be an irreducible, right-resolving presentation of X. Then the merged graph of $\mathcal{G}$ is the minimal right-resolving presentation of X.*

PROOF: Apply Lemma 3.3.8 and Corollary 3.3.19. □

The following example, due to N. Jonoska, shows that Theorem 3.3.18 can fail for reducible sofic shifts.

Example 3.3.21. Let $\mathcal{G}$ be graph (a) in Figure 3.3.2, and $\mathcal{H}$ be graph (b). Their symbolic adjacency matrices are

$$A_{\mathcal{G}} = \begin{bmatrix} \emptyset & a & b & \emptyset \\ a & \emptyset & \emptyset & b \\ \emptyset & \emptyset & c & b \\ \emptyset & \emptyset & a & \emptyset \end{bmatrix} \quad \text{and} \quad A_{\mathcal{H}} = \begin{bmatrix} \emptyset & \emptyset & a+c & b \\ b & a & \emptyset & \emptyset \\ \emptyset & \emptyset & c & b \\ \emptyset & \emptyset & a & \emptyset \end{bmatrix}.$$

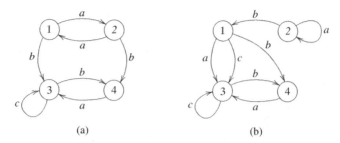

(a)   (b)

FIGURE 3.3.2.   Presentations of a reducible sofic shift.

Since each symbol occurs at most once in each row, both graphs are right-resolving. We will show that these graphs present the same sofic shift X, and that X has no right-resolving presentation with fewer than four states. Note that $\mathcal{G}$ and $\mathcal{H}$ are not isomorphic since $\mathcal{H}$ has a self-loop labeled a and $\mathcal{G}$ does not. Hence $\mathcal{G}$ and $\mathcal{H}$ are nonisomorphic minimal right-resolving presentations of the same (reducible) sofic shift, showing why we need the irreducibility hypothesis in Theorem 3.3.18. Later on, Example 3.4.19 will show that X is actually a 3-step shift of finite type.

We first show that $\mathcal{B}(X_{\mathcal{G}}) = \mathcal{B}(X_{\mathcal{H}})$, so that $X_{\mathcal{G}} = X_{\mathcal{H}} = X$. Any path in $\mathcal{G}$ can be extended to one which cycles through states 1 and 2, then drops down to state 3 or to state 4, and continues. Hence $\mathcal{B}(X_{\mathcal{G}})$ consists of all subwords of the two types of words

$$v_1 = a^k b c^{m_1} b a c^{m_2} b a c^{m_3} b a \dots,$$
$$v_2 = a^k b a c^{n_1} b a c^{n_2} b a c^{n_3} b a \dots,$$

where the exponents k, m_i, n_j can be any integers $\geqslant 0$ (an exponent 0 means that the term is omitted). Similarly, there are three types of paths on $\mathcal{H}$, depending on which edge exits state 1. Thus $\mathcal{B}(X_{\mathcal{H}})$ consists of all subwords of the three types of words

$$w_1 = a^k b a c^{p_1} b a c^{p_2} b a c^{p_3} b a \dots,$$
$$w_2 = a^k b c c^{q_1} b a c^{q_2} b a c^{q_3} b a \dots,$$
$$w_3 = a^k b b a c^{r_1} b a c^{r_2} b a c^{r_3} b a \dots,$$

where again the exponents are any integers $\geqslant 0$. Words of type w_1 correspond to those of type v_2. Type w_2 corresponds to type v_1 with $m_1 = 1 + q_1 \geqslant 1$, while type w_3 corresponds to type v_1 with $m_1 = 0$. This proves that $\mathcal{B}(X_{\mathcal{G}}) = \mathcal{B}(X_{\mathcal{H}})$.

Next we show that no smaller right-resolving graphs can present X. One approach is brute force: write down all right-resolving graphs with three states, and show none of them presents X. To estimate the work involved, note that for each state I ($= 1$, 2, or 3) and each symbol d ($= a$, b, or c), there are 4 choices for an edge labeled d starting at I: it can end at 1 or 2 or 3, or there is no such edge. Hence there are $4^{3\times3} = 262{,}144$ graphs to examine.

A better approach is to use the properties of the language of X. First observe that each of the follower sets $F_X(aa), F_X(c), F_X(cb)$ contains a word that is not in any other. For example $aab \in F_X(aa) \setminus \{F_X(c) \cup F_X(cb)\}$. Thus, if X had a right-resolving presentation $\mathcal{K}$ with only three states, then we can associate the words aa, c, cb to the states 1,2 3 in the sense that

(1) $F_{\mathcal{K}}(1) \subseteq F_X(aa)$,
(2) $F_{\mathcal{K}}(2) \subseteq F_X(c)$, and
(3) $F_{\mathcal{K}}(3) \subseteq F_X(cb)$.

Also, it is straightforward to see that $F_X(aab) = \{F_X(c) \cup F_X(cb)\}$. Thus there must be paths presenting aab, one ending at state 2 and one at state 3. But since $aab \notin \{F_X(c) \cup F_X(cb)\}$, all such paths must start at state 1. This contradicts the fact that $\mathcal{K}$ is right-resolving. $\square$

EXERCISES

3.3.1. Construct the merged graph of the labeled graph whose symbolic adjacency matrix is
$$\begin{bmatrix} \emptyset & b & a \\ b & \emptyset & a \\ a & \emptyset & \emptyset \end{bmatrix}.$$

3.3.2. (a) Find the minimal right-resolving presentation of the even shift from Example 1.2.4 and the charge-constrained shift from Example 1.2.7 with $c = 3$. [*Hint*: Use Corollary 3.3.19.]

   (b) Show that in both cases, the minimal right-resolving presentation is also left-resolving.

   (c) Find an irreducible sofic shift whose minimal right-resolving presentation is not left-resolving.

3.3.3. Let X be an irreducible sofic shift, $\mathcal{G}$ be its minimal right-resolving presentation, and $w \in \mathcal{B}(X)$. Show that w is synchronizing for $\mathcal{G}$ if and only if whenever $uw, wv \in \mathcal{B}(X)$, then $uwv \in \mathcal{B}(X)$.

3.3.4. Let X be an irreducible sofic shift. Call a word w *intrinsically synchronizing* for X if whenever $uw, wv \in \mathcal{B}(X)$, then $uwv \in \mathcal{B}(X)$. Show that the minimal right-resolving presentation of X is the labeled subgraph of the follower set graph formed by using only the follower sets of intrinsically synchronizing words.

3.3.5. Show that a shift space is a shift of finite type if and only if all sufficiently long words are intrinsically synchronizing.

3.3.6. If A is the symbolic adjacency matrix for a labeled graph $\mathcal{G}$, you can think of the entries of A as being sums of "variables" from the alphabet. These variables can be "multiplied" using concatenation, but this multiplication is not commutative since ab is different from ba. Also, a "no symbol" $\emptyset$ obeys $\emptyset a = a\emptyset = \emptyset$. This allows us to multiply A by itself using the usual rules for matrix multiplication, being careful to preserve the correct order in products of variables. For example,

$$
\begin{bmatrix} a+b & a \\ b & \emptyset \end{bmatrix}^2 = \begin{bmatrix} (a+b)(a+b)+ab & (a+b)a \\ b(a+b) & ba \end{bmatrix}
$$

$$
= \begin{bmatrix} a^2+ab+ba+b^2 & a^2+ba \\ ba+b^2 & ba \end{bmatrix}.
$$

How could you use A^k to find all synchronizing words for $\mathcal{G}$ of length k?

3.3.7. Recall that a graph is left-resolving if the edges coming in to each state carry different labels. Prove that every sofic shift has a left-resolving presentation.

3.3.8. The two symbolic adjacency matrices

$$
\begin{bmatrix} b & c & a+d \\ b & c & a+d \\ c+d & \emptyset & a+b \end{bmatrix} \quad \text{and} \quad \begin{bmatrix} a & b & d & c \\ a & b & c & d \\ d & a & b & c \\ d & a & b & c \end{bmatrix}
$$

correspond to two right-resolving labeled graphs. Do they present the same sofic shift?

3.3.9. Find a presentation of the sofic shift in Example 3.3.21 which has three vertices. Why doesn't this contradict the minimality of the two presentations given there?

*3.3.10. Let $\mathcal{G} = (G, \mathcal{L})$ be a labeled graph. Let I and J be states of G with the same follower sets. Define the (I, J)-*merger* $\mathcal{G}'$ to be the labeled graph obtained from $\mathcal{G}$ by deleting all outgoing edges from J, and redirecting into I all remaining edges incoming to J, and finally deleting J.

(a) Show that $\mathcal{G}$ and $\mathcal{G}'$ present the same sofic shift.

(b) Show by example that the (I, J)-merger need not be the same as the (J, I)-merger.

*3.3.11. Let $\mathcal{G} = (G, \mathcal{L})$ be a labeled graph. Let I and J be states with $F_{\mathcal{G}}(I) \subseteq F_{\mathcal{G}}(J)$. Form the (I, J)-merger $\mathcal{G}'$ as in the preceding exercise. Show that $\mathcal{G}'$ presents a subshift of $X_{\mathcal{G}}$.

§3.4. Constructions and Algorithms

Given presentations of two sofic shifts, how can we obtain presentations for their union, intersection, and product? How can we tell whether two labeled graphs present the same sofic shift? How can we find the minimal right-resolving presentation of a given sofic shift? How can we figure out whether a sofic shift has finite type?

This section contains constructions and algorithms to answer these questions. Some of these methods are simple enough that we can "compute" with presentations of moderate size. Others, although of theoretical interest, are unavoidably complicated and create enormous calculations when applied to even small presentations.

We start with several constructions to build new graphs from old.

Let $\mathcal{G}_1 = (G_1, \mathcal{L}_1)$ and $\mathcal{G}_2 = (G_2, \mathcal{L}_2)$ be labeled graphs with alphabets $\mathcal{A}_1$ and $\mathcal{A}_2$. Put $\mathcal{A} = \mathcal{A}_1 \cup \mathcal{A}_2$, and consider each $\mathcal{G}_k$ as having the enlarged alphabet $\mathcal{A}$. By drawing $\mathcal{G}_1$ and $\mathcal{G}_2$ in separate areas of the same piece of paper, we form a new labeled graph denoted by $\mathcal{G}_1 \cup \mathcal{G}_2$, and called the *disjoint union* of $\mathcal{G}_1$ and $\mathcal{G}_2$. There are several technical devices to make this notion precise, but our informal description should suffice.

Proposition 3.4.1. *If $\mathcal{G}_1$ and $\mathcal{G}_2$ are labeled graphs having disjoint union $\mathcal{G}_1 \cup \mathcal{G}_2$, then*

$$X_{\mathcal{G}_1} \cup X_{\mathcal{G}_2} = X_{\mathcal{G}_1 \cup \mathcal{G}_2}.$$

Hence the union of two sofic shifts is also sofic.

PROOF: Let $\mathcal{G} = \mathcal{G}_1 \cup \mathcal{G}_2$, so that $\mathcal{G}_1$ and $\mathcal{G}_2$ are subgraphs of $\mathcal{G}$. Thus any path in each $\mathcal{G}_k$ is a path in $\mathcal{G}$, so that $X_{\mathcal{G}_k} \subseteq X_{\mathcal{G}}$ for $k = 1, 2$, and so $X_{\mathcal{G}_1} \cup X_{\mathcal{G}_2} \subseteq X_{\mathcal{G}_1 \cup \mathcal{G}_2}$. Since there are no edges connecting $\mathcal{G}_1$ and $\mathcal{G}_2$, any path in $\mathcal{G}$ is entirely contained in $\mathcal{G}_1$ or in $\mathcal{G}_2$. Hence $X_{\mathcal{G}} \subseteq X_{\mathcal{G}_1} \cup X_{\mathcal{G}_2}$.

Given two sofic shifts, the disjoint union of their presentations provides a presentation of their union, which is therefore also sofic. $\square$

We next turn to products. Recall that if P and Q are sets, then their *Cartesian product* $P \times Q$ is the set of all ordered pairs (p, q) with $p \in P$ and $q \in Q$. The elements p and q are called the *components* of the pair (p, q). We will use set products to define a product for graphs and also one for labeled graphs.

Definition 3.4.2. Let $G_1 = (\mathcal{V}_1, \mathcal{E}_1)$ and $G_2 = (\mathcal{V}_2, \mathcal{E}_2)$ be graphs. Their *graph product* $G_1 \times G_2$ has vertex set $\mathcal{V} = \mathcal{V}_1 \times \mathcal{V}_2$ and edge set $\mathcal{E} = \mathcal{E}_1 \times \mathcal{E}_2$. An edge $e = (e_1, e_2)$ starts at $i(e) = (i(e_1), i(e_2))$ and terminates at $t(e) = (t(e_1), t(e_2))$.

Example 3.4.3. Figure 3.4.1 shows two graphs and their graph product. Notice that $|\mathcal{V}| = |\mathcal{V}_1| \times |\mathcal{V}_2|$ and $|\mathcal{E}| = |\mathcal{E}_1| \times |\mathcal{E}_2|$, so that the graph product is typically much larger than the original graphs. $\square$

Suppose that X_1 and X_2 are shift spaces with alphabets $\mathcal{A}_1$ and $\mathcal{A}_2$. An element (x, y) in $X_1 \times X_2$ is a pair of sequences

$$(\ldots x_{-1} x_0 x_1 \ldots, \ \ldots y_{-1} y_0 y_1 \ldots),$$

but we can also think of it as a sequence

$$(\ldots, (x_{-1}, y_{-1}), (x_0, y_0), (x_1, y_1), \ldots)$$

of pairs from $\mathcal{A}_1 \times \mathcal{A}_2$. We will therefore view the product $X_1 \times X_2$ as a shift space with alphabet $\mathcal{A}_1 \times \mathcal{A}_2$.

Proposition 3.4.4. *If G_1 and G_2 are graphs, then $X_{G_1} \times X_{G_2} = X_{G_1 \times G_2}$.*

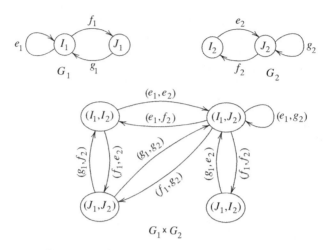

FIGURE 3.4.1.   The graph product of two graphs.

PROOF: A bi-infinite sequence of pairs of edges is in $X_{G_1 \times G_2}$ if and only if the sequence of first components is in X_{G_1} and the sequence of second components is in X_{G_2}. □

There is a similar product for labeled graphs, obtained by taking the label of each component.

Definition 3.4.5. Let $\mathcal{G}_1 = (G_1, \mathcal{L}_1)$ and $\mathcal{G}_2 = (G_2, \mathcal{L}_2)$ be labeled graphs over alphabets $\mathcal{A}_1$ and $\mathcal{A}_2$. Their *graph product* $\mathcal{G} = \mathcal{G}_1 \times \mathcal{G}_2$ has underlying graph $G = G_1 \times G_2$ and labeling $\mathcal{L} = \mathcal{L}_1 \times \mathcal{L}_2 : \mathcal{E}_1 \times \mathcal{E}_2 \to \mathcal{A}_1 \times \mathcal{A}_2$ defined for $e = (e_1, e_2)$ by $\mathcal{L}(e) = (\mathcal{L}_1(e_1), \mathcal{L}_2(e_2))$.

Example 3.4.6. Figure 3.4.2 shows a labeling of the graphs G_1 and G_2 from Figure 3.4.1, and the graph product of these labeled graphs. □

Proposition 3.4.7. *If $\mathcal{G}_1$ and $\mathcal{G}_2$ are labeled graphs, then*

$$X_{\mathcal{G}_1} \times X_{\mathcal{G}_2} = X_{\mathcal{G}_1 \times \mathcal{G}_2}.$$

Hence the product of two sofic shifts is also sofic.

PROOF: Let $\mathcal{G}_1 = (G_1, \mathcal{L}_1)$ and $\mathcal{G}_2 = (G_2, \mathcal{L}_2)$. By Proposition 3.4.4, $X_{G_1} \times X_{G_2} = X_{G_1 \times G_2}$. Because we have identified a pair of sequences with a sequence of pairs, we have that

$$X_{\mathcal{G}_1 \times \mathcal{G}_2} = (\mathcal{L}_1 \times \mathcal{L}_2)_\infty(X_{G_1 \times G_2}) = (\mathcal{L}_1 \times \mathcal{L}_2)_\infty(X_{G_1} \times X_{G_2})$$
$$= (\mathcal{L}_1)_\infty(X_{G_1}) \times (\mathcal{L}_2)_\infty(X_{G_2}) = X_{\mathcal{G}_1} \times X_{\mathcal{G}_2}.$$

This provides a presentation for the product of sofic shifts, which is therefore also sofic. □

The next construction detects whether paths in two labeled graphs have the same label. We will use it to present the intersection of two sofic shifts.

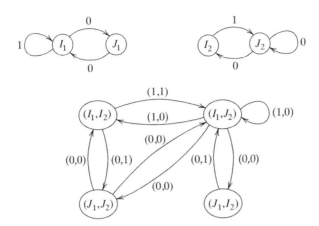

FIGURE 3.4.2. The graph product of two labeled graphs.

Definition 3.4.8. Let $\mathcal{G}_1 = (G_1, \mathcal{L}_1)$ and $\mathcal{G}_2 = (G_2, \mathcal{L}_2)$ be labeled graphs with the same alphabet $\mathcal{A}$, and let their underlying graphs be $G_1 = (\mathcal{V}_1, \mathcal{E}_1)$ and $G_2 = (\mathcal{V}_2, \mathcal{E}_2)$. The *label product* $\mathcal{G}_1 * \mathcal{G}_2$ of $\mathcal{G}_1$ and $\mathcal{G}_2$ has underlying graph G with vertex set $\mathcal{V} = \mathcal{V}_1 \times \mathcal{V}_2$, edge set

$$\mathcal{E} = \{(e_1, e_2) \in \mathcal{E}_1 \times \mathcal{E}_2 : \mathcal{L}_1(e_1) = \mathcal{L}_2(e_2)\},$$

and labeling $\mathcal{L}: G \to \mathcal{A}$ defined for $e = (e_1, e_2) \in \mathcal{E}$ by $\mathcal{L}(e) = \mathcal{L}_1(e_1) = \mathcal{L}_2(e_2)$.

The label product discards all pairs of edges in the product graph whose components have different labels. Notice that, unlike the graph product, the label product has the same alphabet as the individual graphs. If two labeled graphs have alphabets $\mathcal{A}_1$ and $\mathcal{A}_2$, we can think of each of them as having alphabet $\mathcal{A} = \mathcal{A}_1 \cup \mathcal{A}_2$, and then take their label product.

Example 3.4.9. For the two labeled graphs in the top of Figure 3.4.2, their label product is shown in Figure 3.4.3. This example shows that even though $\mathcal{G}_1$ and $\mathcal{G}_2$ do not have stranded vertices, their label product may have such vertices. It also shows that the label product of irreducible graphs is not always irreducible. □

Proposition 3.4.10. *If $\mathcal{G}_1$ and $\mathcal{G}_2$ are labeled graphs having label product $\mathcal{G}_1 * \mathcal{G}_2$, then*

$$X_{\mathcal{G}_1} \cap X_{\mathcal{G}_2} = X_{\mathcal{G}_1 * \mathcal{G}_2}.$$

Hence the intersection of two sofic shifts is also sofic. If $\mathcal{G}_1$ and $\mathcal{G}_2$ are right-resolving, then so is $\mathcal{G}_1 * \mathcal{G}_2$.

PROOF: Let $\mathcal{G}_1 = (G_1, \mathcal{L}_1)$ and $\mathcal{G}_2 = (G_2, \mathcal{L}_2)$, and let $\mathcal{L} = \mathcal{L}_1 \times \mathcal{L}_2$. If $\xi = (\xi_1, \xi_2)$ is a bi-infinite walk on the underlying graph of $\mathcal{G}_1 * \mathcal{G}_2$,

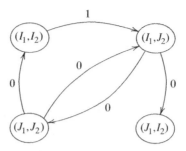

FIGURE 3.4.3.   The label product of the graphs from Figure 3.4.2.

then $\mathcal{L}_\infty(\xi) = (\mathcal{L}_1)_\infty(\xi_1) \in X_{\mathcal{G}_1}$ and $\mathcal{L}_\infty(\xi) = (\mathcal{L}_2)_\infty(\xi_2) \in X_{\mathcal{G}_2}$. Hence $X_{\mathcal{G}_1 * \mathcal{G}_2} \subseteq X_{\mathcal{G}_1} \cap X_{\mathcal{G}_2}$.

On the other hand, if $x \in X_{\mathcal{G}_1} \cap X_{\mathcal{G}_2}$, then there are bi-infinite walks ξ_1 on G_1 and ξ_2 on G_2 such that $(\mathcal{L}_1)_\infty(\xi_1) = x = (\mathcal{L}_2)_\infty(\xi_2)$. Hence, for every i, the ith edge in ξ_1 has the same label as the ith edge in ξ_2, so that (ξ_1, ξ_2) is a path in the underlying graph of $\mathcal{G}_1 * \mathcal{G}_2$. Thus $x = \mathcal{L}_\infty(\xi_1, \xi_2) \in X_{\mathcal{G}_1 * \mathcal{G}_2}$, proving that $X_{\mathcal{G}_1} \cap X_{\mathcal{G}_2} \subseteq X_{\mathcal{G}_1 * \mathcal{G}_2}$.

Suppose now that $\mathcal{G}_1$ and $\mathcal{G}_2$ are right-resolving, and that (e_1, e_2) and (e'_1, e'_2) are distinct edges in $\mathcal{G}_1 * \mathcal{G}_2$ starting at the same vertex and having the same label. If $e_1 \neq e'_1$, then we would have that $\mathcal{L}_1(e_1) = \mathcal{L}(e_1, e_2) = \mathcal{L}(e'_1, e'_2) = \mathcal{L}_1(e'_1)$, contradicting the assumption that $\mathcal{G}_1$ is right-resolving. A similar argument applies if $e_2 \neq e'_2$. Thus $\mathcal{G}_1 * \mathcal{G}_2$ is right-resolving.   □

Example 3.4.11. Let $\mathcal{G}_1$ and $\mathcal{G}_2$ denote the labeled graphs at the top of Figure 3.4.2. Then $X_{\mathcal{G}_1}$ is the even shift, and $X_{\mathcal{G}_2}$ is the golden mean shift (for which consecutive 1's are forbidden). Their intersection is therefore the shift space where 1's are separated by an even, and positive, number of 0's. A glance at Figure 3.4.3 shows that this exactly describes the sofic shift presented by $\mathcal{G}_1 * \mathcal{G}_2$.   □

Having introduced some basic constructions, our next goal is to determine whether two labeled graphs present the same sofic shift. To make sure our algorithm terminates, we will use the following facts about paths in graphs.

Lemma 3.4.12. *Let G be a graph with vertex set $\mathcal{V}$ having r elements. Suppose that $I \in \mathcal{V}$, and that $\mathcal{S}$ is a subset of $\mathcal{V}$ with size s that does not contain I. If there is a path in G from I to a vertex in $\mathcal{S}$, then the shortest such path has length $\leqslant r - s$.*

If A is the adjacency matrix of G, and $B = A + A^2 + \ldots A^{r-s}$, then there is a path in G from I to a vertex in $\mathcal{S}$ if and only if $B_{IJ} > 0$ for some $J \in \mathcal{S}$.

PROOF: Let $\pi = e_1 e_2 \ldots e_n$ be the shortest path in G from I to a vertex in $\mathcal{S}$. Then $i(e_j) \notin \mathcal{S}$ for $1 \leqslant j \leqslant n$, since otherwise there would be a shorter path from I to $\mathcal{S}$. The vertices $i(e_j)$ must all be different for $1 \leqslant j \leqslant n$, since otherwise π would contain a cycle whose removal would produce a shorter path. Hence $\mathcal{V} \setminus \mathcal{S}$ contains n distinct elements, so that

$$|\pi| = n \leqslant |\mathcal{V} \setminus \mathcal{S}| = r - s.$$

The second claim now follows, since by Proposition 2.2.12, the number of paths of length n from I to J is $(A^n)_{IJ}$. $\qquad\square$

We have laid the groundwork for our first algorithm.

Theorem 3.4.13. *There is an algorithm which, given two labeled graphs $\mathcal{G}_1$ and $\mathcal{G}_2$, decides whether $X_{\mathcal{G}_1} = X_{\mathcal{G}_2}$.*

PROOF: For each $\mathcal{G}_k$ we first construct an auxiliary graph $\widehat{\mathcal{G}}_k$ which is used to specify all words that do *not* occur in $X_{\mathcal{G}_k}$. We then form the label product $\widehat{\mathcal{G}} = \widehat{\mathcal{G}}_1 * \widehat{\mathcal{G}}_2$. In $\widehat{\mathcal{G}}$ there are a vertex $\mathcal{I}$ and a subset of vertices $\mathcal{S}$ such that $X_{\mathcal{G}_1} \neq X_{\mathcal{G}_2}$ if and only if there is a path in $\widehat{\mathcal{G}}$ from $\mathcal{I}$ to $\mathcal{S}$. Lemma 3.4.12 shows how to determine whether such a path exists.

As mentioned before, by taking the union of the alphabets involved, we can assume that both labeled graphs use the same alphabet $\mathcal{A}$. Let $\mathcal{G}_k$ have vertex set $\mathcal{V}_k$. We first enlarge $\mathcal{G}_k$ to a new graph $\mathcal{G}'_k$ as follows. Add a new vertex K_k to $\mathcal{V}_k$, which you should think of as a "sink" or "kill state." For each $I \in \mathcal{V}_k$ and $a \in \mathcal{A}$, if there is no edge in $\mathcal{G}_k$ starting at I labeled a, then add an edge labeled a from I to K_k; otherwise, do nothing. Finally, for each $a \in \mathcal{A}$ add a self-loop at K_k labeled a. Enough edges have been added so that for every vertex I in $\mathcal{G}'_k$ and every word w over $\mathcal{A}$, there is at least one path labeled w starting at I. In particular, $X_{\mathcal{G}'_k}$ is the full $\mathcal{A}$-shift. The kill state K_k is a black hole for paths: once they enter K_k they can never leave.

Next, apply the subset construction, described in the proof of Theorem 3.3.2, to $\mathcal{G}'_k$ and obtain a right-resolving graph $\widehat{\mathcal{G}}_k$. Recall from this construction that the vertices of $\widehat{\mathcal{G}}_k$ are the nonempty subsets of $\mathcal{V}'_k = \mathcal{V}_k \cup \{K_k\}$, and that there is an edge labeled a in $\widehat{\mathcal{G}}_k$ from $\mathcal{I}$ to $\mathcal{J}$ exactly when $\mathcal{J}$ is the set of vertices in $\mathcal{G}'_k$ reached by edges labeled a starting at vertices in $\mathcal{I}$.

Let $\mathcal{I}_k$ denote the vertex in $\widehat{\mathcal{G}}_k$ that is the set of vertices from the original graph $\mathcal{G}_k$, so that $\mathcal{I}_k = \mathcal{V}_k$. Denote the vertex $\{K_k\}$ in $\widehat{\mathcal{G}}_k$ by $\mathcal{K}_k$. For every word w over $\mathcal{A}$, there is a unique path in $\widehat{\mathcal{G}}_k$ starting at $\mathcal{I}_k$ and labeled w. The crucial property of $\widehat{\mathcal{G}}_k$ is that this path ends at $\mathcal{K}_k$ if and only if $w \notin \mathcal{B}(X_{\mathcal{G}_k})$. This means that the complement of the language of $X_{\mathcal{G}_k}$ is the set of labels of paths in $\widehat{\mathcal{G}}_k$ from $\mathcal{I}_k$ to $\mathcal{K}_k$. This property follows from the observation that $w \in \mathcal{B}(X_{\mathcal{G}_k})$ exactly when there is at least one path labeled w in $\mathcal{G}_k$, and this occurs if and only if the unique path in $\widehat{\mathcal{G}}_k$ starting at $\mathcal{I}_k$ labeled w ends at a set of vertices containing at least one vertex from the original graph $\mathcal{G}_k$.

Put $\widehat{\mathcal{G}} = \widehat{\mathcal{G}}_1 * \widehat{\mathcal{G}}_2$, let $\mathcal{I} = (\mathcal{I}_1, \mathcal{I}_2)$, and set

$$\mathcal{S}_1 = \{(\mathcal{J}, \mathcal{K}_2) : \mathcal{J} \neq \mathcal{K}_1\}.$$

Since the label product keeps track of pairs of paths with the same labeling, our previous observation shows that a word w is in $X_{\mathcal{G}_1} \setminus X_{\mathcal{G}_2}$ if and only if the unique path in $\widehat{\mathcal{G}}$, starting at $\mathcal{I}$ and labeled w, terminates at a vertex in $\mathcal{S}_1$. Thus $X_{\mathcal{G}_1} \subseteq X_{\mathcal{G}_2}$ if and only if there is no path in $\widehat{\mathcal{G}}$ from $\mathcal{I}$ to $\mathcal{S}_1$. A similar argument shows that $X_{\mathcal{G}_2} \subseteq X_{\mathcal{G}_1}$ if and only if there is no path in $\widehat{\mathcal{G}}$ from $\mathcal{I}$ to $\mathcal{S}_2 = \{(\mathcal{K}_1, \mathcal{J}) : \mathcal{J} \neq \mathcal{K}_2\}$.

Putting these together, we see that $X_{\mathcal{G}_1} = X_{\mathcal{G}_2}$ exactly when there is no path in $\widehat{\mathcal{G}}$ from $\mathcal{I}$ to $\mathcal{S}_1 \cup \mathcal{S}_2$. Lemma 3.4.12 shows how to use the adjacency matrix of $\widehat{\mathcal{G}}$ to answer the latter question. $\qquad\square$

Although this theorem shows that, in principle, we can decide whether two presentations give the same sofic shift, the algorithm given is hopelessly complicated on even moderate examples. For if $\mathcal{G}_k$ has r_k vertices, then $\mathcal{G}'_k$ has $r_k + 1$ vertices, and the subset construction gives graphs $\widehat{\mathcal{G}}_k$ with $2^{r_k+1} - 1$ vertices, so that $\widehat{\mathcal{G}}$ has $(2^{r_1+1} - 1)(2^{r_2+1} - 1)$ vertices. If each $\mathcal{G}_k$ had only nine vertices, the resulting $\widehat{\mathcal{G}}$ would have over one million vertices! However, for irreducible right-resolving presentations, we will soon obtain a much more efficient method.

The next procedure finds minimal right-resolving presentations.

Theorem 3.4.14. *There is an algorithm which, given an irreducible right-resolving presentation of a sofic shift X, constructs the minimal right-resolving presentation of X.*

PROOF: Let $\mathcal{G}$ be an irreducible right-resolving presentation of X. Recall two vertices I and J are called equivalent if they have the same follower sets, i.e., if $F_{\mathcal{G}}(I) = F_{\mathcal{G}}(J)$. This partitions the vertex set of $\mathcal{G}$ into disjoint equivalence classes, which we will denote by $\mathcal{I}_1, \mathcal{I}_2, \ldots, \mathcal{I}_r$. These are the vertices of the merged graph $\mathcal{H}$ of $\mathcal{G}$. By Corollary 3.3.20, $\mathcal{H}$ is the minimal right-resolving presentation of X. Note that once the vertices of $\mathcal{H}$ are found, the definition of $\mathcal{H}$ shows how to obtain its edges and their labels from those of $\mathcal{G}$.

Thus we must find a method to decide whether or not two vertices in $\mathcal{G}$ are equivalent. We do this by introducing an auxiliary graph $\mathcal{G}'$, similar to the first part of the proof of Theorem 3.4.13. Let $\mathcal{V}$ denote the vertex set of $\mathcal{G}$. Add to $\mathcal{V}$ a new "sink" or "kill state" K. For each $I \in \mathcal{V}$ and $a \in \mathcal{A}$, if there is no edge labeled a starting at I, then add an edge labeled a from I to K; otherwise do nothing. Finally, for each $a \in \mathcal{A}$ add a self-loop at K labeled a. As before, K is a black hole for paths. The augmented graph $\mathcal{G}'$ is clearly right-resolving.

Consider the label product $\mathcal{G}' * \mathcal{G}'$. Let

$$\mathcal{S} = (\mathcal{V} \times \{K\}) \cup (\{K\} \times \mathcal{V})$$

be the set of states in $\mathcal{G}' * \mathcal{G}'$ with exactly one component not the kill state. If $I \neq J \in \mathcal{V}$, we claim that $F_{\mathcal{G}}(I) \neq F_{\mathcal{G}}(J)$ if and only if there is a path in $\mathcal{G}' * \mathcal{G}'$ from (I, J) to $\mathcal{S}$. For if $w \in F_{\mathcal{G}}(I) \setminus F_{\mathcal{G}}(J)$, then the unique path in $\mathcal{G}'$ labeled w and starting at I must end at some vertex in $\mathcal{V}$, while that starting at J must end at K. This would give a path in $\mathcal{G}' * \mathcal{G}'$ from (I, J) to $\mathcal{V} \times \{K\}$. Similarly, a word in $F_{\mathcal{G}}(J) \setminus F_{\mathcal{G}}(I)$ would yield a path from (I, J) to $\{K\} \times \mathcal{V}$, establishing half of the claim. For the other half, observe that the label of any path from (I, J) to $\mathcal{S}$ would be a word in one follower set but not the other.

Thus I is equivalent to J exactly when there is no path in $\mathcal{G}' * \mathcal{G}'$ from (I, J) to $\mathcal{S}$. Lemma 3.4.12 gives a method to determine whether or not such a path exists. This method can be used to find the equivalence classes $\mathcal{I}_1, \mathcal{I}_2, \ldots, \mathcal{I}_r$ of vertices in $\mathcal{G}$, and hence construct $\mathcal{H}$. □

In the proof of Theorem 3.4.14, we showed that $F_{\mathcal{G}}(I) \neq F_{\mathcal{G}}(J)$ if and only if there is a path in $\mathcal{G}' * \mathcal{G}'$ from (I, J) to $\mathcal{S}$. Let r denote the number of states in $\mathcal{G}$. Since there are at most $(r+1)^2$ states in $\mathcal{G}' * \mathcal{G}'$, it follows that if $F_{\mathcal{G}}(I) \neq F_{\mathcal{G}}(J)$, then there is some word of length $(r+1)^2$ which belongs to one of $F_{\mathcal{G}}(I), F_{\mathcal{G}}(J)$, but not both. In fact, it turns out that if $F_{\mathcal{G}}(I) \neq F_{\mathcal{G}}(J)$, then they can actually be distinguished by some word of length r (see Exercise 3.4.10).

The practical problem in finding the minimal right-resolving presentation is to determine which states are equivalent to which others. A convenient and efficient way to tabulate state equivalences is described in the following example.

Example 3.4.15. Let $\mathcal{G}$ be the labeled graph shown in Figure 3.4.4. A glance at the figure shows that $\mathcal{G}$ is irreducible and right-resolving. Our goal is to identify which states are equivalent to which others, then merge equivalent states to obtain the minimal right-resolving presentation for $X_{\mathcal{G}}$.

The table shown in Figure 3.4.5(a) has a box for each pair $\{I, J\}$ of distinct vertices. The order of vertices doesn't matter, so we need only list

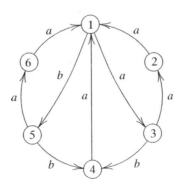

FIGURE 3.4.4.   A presentation to be minimized.

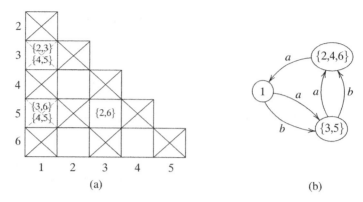

(a)   (b)

FIGURE 3.4.5.   State minimization.

pairs $\{I, J\}$ with $I < J$. Our procedure will eventually mark an "X" in the box for $\{I, J\}$ exactly when I is not equivalent to J. The remaining boxes then tell which vertices are equivalent.

The procedure runs through the pairs one at a time. Here we will go down the columns, and move from the left column to the right one. When we consider pair $\{I, J\}$, each box for a previous pair will either be blank, be marked with an "X," or contain a list of pairs.

Consider a pair of vertices $\{I, J\}$. If there is a symbol c (which in this example is either a or b) such that there is an edge labeled c starting at one vertex but no such edge starting at the other, then I and J are not equivalent, so mark an "X" in box $\{I, J\}$ and also in any box that contains $\{I, J\}$ in its list, and recursively repeat this for any newly marked boxes. Otherwise, make a list for box $\{I, J\}$ as follows. For each symbol c, if there is an edge labeled c from I to I', and also one from J to J', and if $I' \neq J'$, put $\{I', J'\}$ on the list. If no pairs are put on the list, leave box $\{I, J\}$ empty. Move on to the next box.

Let's see how this works on our example. The first pair is $\{1, 2\}$. There is an edge labeled b starting at vertex 1, but none at vertex 2, so we mark an "X" in box $\{1, 2\}$. For $\{1, 3\}$, each vertex has an edge with each symbol, so we form a list. The letter a gives the pair $\{3, 2\} = \{2, 3\}$, and b gives $\{5, 4\} = \{4, 5\}$, so we put the list containing these two pairs in box $\{1, 3\}$.

Next, using b gives an "X" in box $\{1, 4\}$; also, $\{1, 4\}$ does not appear on any previous lists, so nothing else happens. Box $\{1, 5\}$ gets the list indicated, and using b marks box $\{1, 6\}$ with an "X". The next pair is $\{2, 3\}$, which receives an "X" using letter b. Now $\{2, 3\}$ appears in the list for box $\{1, 3\}$, so this box is also marked (we use dotted lines to denote this kind of marking). The newly marked box $\{1, 3\}$ is not on any lists, so recursive marking stops here. Box $\{2, 4\}$ has no elements on its list (why?), so remains blank. The remaining boxes are treated similarly, and the completed table is shown in Figure 3.4.5(a).

The table shows that states 2, 4, and 6 are equivalent, and that states 3 and 5 are also equivalent (even though box $\{3,5\}$ is not blank). Merging equivalent states results in the minimal right-resolving presentation, depicted in Figure 3.4.5(b). □

Suppose we want to determine whether two irreducible right-resolving graphs $\mathcal{G}_1$ and $\mathcal{G}_2$ present the same sofic shift. The method in Example 3.4.15 gives a practical procedure for merging states in each $\mathcal{G}_k$ to obtain the corresponding minimal right-resolving presentation $\mathcal{H}_k$. It follows from Theorem 3.3.18 that $\mathcal{G}_1$ and $\mathcal{G}_2$ present the same sofic shift if and only if $\mathcal{H}_1$ and $\mathcal{H}_2$ are isomorphic. Isomorphism of right-resolving labeled graphs is not difficult to decide, at least for examples of moderate size.

Exercise 3.4.8 gives a cubic upper bound (in terms of the number of states) on the size of the smallest synchronizing word for any right-resolving labeled graph that has a synchronizing word. This bound can probably be improved. In particular, in automata theory, there is the following well-known and long-standing conjecture, due to Cerny [Cer]; see also Pin [Pin].

Conjecture 3.4.16. *Let $\mathcal{G}$ be a right-resolving labeled graph. If there is a synchronizing word for $\mathcal{G}$, then there is a synchronizing word for $\mathcal{G}$ of length at most $(|\mathcal{V}(G)| - 1)^2$.*

Our final procedure determines when an irreducible sofic shift has finite type.

Theorem 3.4.17. *Let $\mathcal{G}$ be a right-resolving labeled graph, and suppose that all words in $\mathcal{B}_N(X_{\mathcal{G}})$ are synchronizing for $\mathcal{G}$. Then $X_{\mathcal{G}}$ is an N-step shift of finite type.*

Conversely, suppose that $\mathcal{G} = (G, \mathcal{L})$ is the minimal right-resolving presentation of an irreducible sofic shift X, and let r be the number of states in $\mathcal{G}$. If X is an N-step shift of finite type, then all words in $\mathcal{B}_N(X)$ are synchronizing for $\mathcal{G}$, and $\mathcal{L}_\infty$ is a conjugacy. If X has finite type, then it must be $(r^2 - r)$-step.

PROOF: Suppose that $\mathcal{G}$ is right-resolving, and that all words in $\mathcal{B}_N(X_{\mathcal{G}})$ are synchronizing for $\mathcal{G}$. By Theorem 2.1.8, it suffices to show that if $uv, vw \in \mathcal{B}(X_{\mathcal{G}})$ with $|v| \geqslant N$, then $uvw \in \mathcal{B}(X_{\mathcal{G}})$. A glance at Figure 3.4.6(a) may help make the following argument clearer. Let $\pi\tau_1$ present uv and $\tau_2\omega$ present vw. Since $|v| \geqslant N$, v is synchronizing for $\mathcal{G}$. Thus τ_1 and τ_2 end at the same vertex. Hence $\pi\tau_1\omega$ is a path in $\mathcal{G}$, so that $uvw = \mathcal{L}(\pi\tau_1\omega) \in \mathcal{B}(X_{\mathcal{G}})$.

For the second part, let $\mathcal{G}$ be the minimal right-resolving presentation of an irreducible sofic shift X, and assume that X is an N-step shift of finite type. Suppose there were a word w of length N that is not synchronizing for $\mathcal{G}$. Then there would be paths π and τ in $\mathcal{G}$ presenting w which end in distinct vertices, say $t(\pi) = I \neq J = t(\tau)$ (see Figure 3.4.6(b)). Since $\mathcal{G}$ is minimal, by Proposition 3.3.9 it is follower-separated. By interchanging

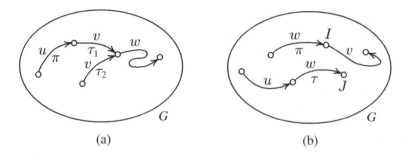

FIGURE 3.4.6.   The proof of Theorem 3.4.17.

π and τ if necessary, we can find a word v in $F_{\mathcal{G}}(I)$ that is not in $F_{\mathcal{G}}(J)$. By Proposition 3.3.16, there is a synchronizing word for $\mathcal{G}$ which, using irreducibility of $\mathcal{G}$ and Lemma 3.3.15, can be extended to a synchronizing word u that focuses to $i(\tau)$. Then $uw, wv \in \mathcal{B}(X_{\mathcal{G}})$, but we claim that $uwv \notin \mathcal{B}(X_{\mathcal{G}})$. This would violate Theorem 2.1.8, and would show that X is not N-step.

To prove our claim, suppose that $\pi = \pi_1 \pi_2 \pi_3$ presents uwv. Since u is synchronizing, $t(\pi_1) = i(\pi_2) = i(\tau)$. Now $\mathcal{L}(\pi_2) = w = \mathcal{L}(\tau)$, and there is a unique path labeled w starting at $i(\pi_2) = i(\tau)$ since $\mathcal{G}$ is right-resolving. Hence $\pi_2 = \tau$. Thus $v \in F_{\mathcal{G}}(t(\tau)) = F_{\mathcal{G}}(J)$, contradicting $v \notin F_{\mathcal{G}}(J)$, and this verifies the claim.

Since all words of length N are synchronizing, $\mathcal{L}_\infty$ is a conjugacy (why?).

Finally, let r be the number of vertices in the minimal right-resolving presentation $\mathcal{G}$. Since X has finite type, it is N-step for some $N \geqslant 1$, so that all words in $\mathcal{B}_N(X)$ are synchronizing for $\mathcal{G}$. Let $\widehat{\mathcal{G}}$ be the labeled graph formed from the label product $\mathcal{G} * \mathcal{G}$ by removing the r vertices of the form (I, I) together with all edges containing them. Then $\widehat{\mathcal{G}}$ has $r^2 - r$ vertices. Since $\mathcal{G} * \mathcal{G}$ keeps track of pairs of paths with equal labels, and since we have removed the pairs (I, I), the labels of paths in $\widehat{\mathcal{G}}$ are exactly the nonsynchronizing words for $\mathcal{G}$. If $\widehat{\mathcal{G}}$ contained a cycle, then $\mathcal{G}$ would have arbitrarily long nonsynchronizing words, which cannot happen since all words in $\mathcal{B}_N(X_{\mathcal{G}})$ are synchronizing. Now $\widehat{\mathcal{G}}$ has only $r^2 - r$ vertices, so any path of length $r^2 - r$ would repeat a vertex and produce a cycle. Thus all paths on $\widehat{\mathcal{G}}$, and hence all nonsynchronizing words for $\mathcal{G}$, have length $\leqslant r^2 - r - 1$. We conclude that all words in $\mathcal{B}_{r^2-r}(X_{\mathcal{G}})$ are synchronizing for $\mathcal{G}$, so the first part of the proof shows that $X_{\mathcal{G}}$ is $(r^2 - r)$-step.   □

One can show that $X_{\mathcal{G}}$ in the preceding result is $((r^2 - r)/2)$-step (see Exercise 3.4.6).

Example 3.4.18. Each of the labeled graphs in Figure 3.1.3 is irreducible, right-resolving, and follower-separated. By Corollary 3.3.19, each is the minimal right-resolving presentation of a sofic shift. For every $N \geqslant 1$, the word 0^N is not synchronizing for either graph. We can therefore conclude using Theorem 3.4.17 that neither sofic shift has finite type, which we had earlier proved with a more "bare hands" argument. □

Example 3.4.19. In Example 3.3.21 we showed that the two labeled graphs shown in Figure 3.3.2 present the same sofic shift X, and that each is a minimal right-resolving presentation. Let $\mathcal{H}$ be graph (b) in the figure. Then every word in $\mathcal{B}_3(X)$ is synchronizing for $\mathcal{H}$. For c, aa, ba, and bb are synchronizing by observation, and all words in $\mathcal{B}_3(X)$ contain at least one of these words (why?). Then Theorem 3.4.17 implies that X is a 3-step shift of finite type.

Let $\mathcal{G}$ be graph (a). Note that for $N \geqslant 1$ the word a^N is not synchronizing for $\mathcal{G}$. Hence although X is a 3-step shift of finite type and $\mathcal{G}$ is a minimal right-resolving presentation of X, there are arbitrarily long nonsynchronizing words for $\mathcal{G}$. This shows why we need to assume that X is irreducible in the second part of Theorem 3.4.17. □

EXERCISES

3.4.1. Let G_1 and G_2 be essential graphs.
   (a) Show that $G_1 \cup G_2$ and $G_1 \times G_2$ are also essential.
   (b) If G_1 and G_2 are irreducible, must $G_1 \times G_2$ also be irreducible?

3.4.2. Find a presentation of the intersection of the two sofic shifts with adjacency matrices

$$\begin{bmatrix} a & b \\ b & \emptyset \end{bmatrix} \quad \text{and} \quad \begin{bmatrix} b & a & \emptyset & \emptyset \\ \emptyset & \emptyset & a & \emptyset \\ b & \emptyset & \emptyset & a \\ b & \emptyset & \emptyset & \emptyset \end{bmatrix}.$$

3.4.3. Find the minimal right-resolving presentation for the sofic shifts presented by

$$\text{(a)} \quad \begin{bmatrix} a & \emptyset & c & b \\ a & c & \emptyset & b \\ a & c & \emptyset & b \\ c & a & \emptyset & b \end{bmatrix} \qquad \text{(b)} \quad \begin{bmatrix} \emptyset & \emptyset & a & b & \emptyset & c \\ a & \emptyset & \emptyset & b & \emptyset & c \\ \emptyset & a & \emptyset & \emptyset & b & c \\ \emptyset & \emptyset & \emptyset & \emptyset & \emptyset & d \\ \emptyset & \emptyset & \emptyset & \emptyset & \emptyset & d \\ \emptyset & c & a & \emptyset & d & \emptyset \end{bmatrix}$$

3.4.4. What is the unique minimal right-resolving presentation of an edge shift?

3.4.5. Let A be the symbolic adjacency matrix of the minimal right-resolving presentation $\mathcal{G}$ of an irreducible sofic shift. Use Exercise 3.3.6 together with Theorem 3.4.17 to give a criterion, using the powers of A, for $X_\mathcal{G}$ to have finite type.

3.4.6. Let $\mathcal{G}$ be the minimal right-resolving presentation of an irreducible sofic shift X, and let r be the number of states in $\mathcal{G}$. Show that if X has finite type, then it actually must be $((r^2 - r)/2)$-step.

3.4.7. Let $\mathcal{G}$ be a right-resolving irreducible labeled graph. Let $\mathcal{H}$ be the (labeled) subgraph obtained from $\mathcal{G} * \mathcal{G}$ by deleting the diagonal states, $\{(I, I)\}$. Show that $\mathcal{G}$ has a synchronizing word if and only if $X_\mathcal{G} \neq X_\mathcal{H}$.

3.4.8. Let $\mathcal{G}$ be an irreducible right-resolving labeled graph which has a synchronizing word. Show that there is a synchronizing word for $\mathcal{G}$ of length at most

$$|\mathcal{V}(G)| \cdot (|\mathcal{V}(G)| - 1)^2 / 2.$$

3.4.9. Give an algorithm for finding the follower set graph given a presentation of a sofic shift [*Hint*: Use the subset construction and the method described in this section for finding the minimal right-resolving presentation.]

*3.4.10. Let $\mathcal{G} = (G, \mathcal{L})$ be a right-resolving labeled graph with r states. Say that $I \sim_n J$ if $F_\mathcal{G}(I)$ and $F_\mathcal{G}(J)$ consist of the exact same words of length n. Let $\mathcal{P}_n$ denote the partition of $\mathcal{V}(G)$ defined by this equivalence relation.

   (a) Show that for each n, $\mathcal{P}_{n+1}$ refines $\mathcal{P}_n$.
   (b) Show that if n is such that $\mathcal{P}_n = \mathcal{P}_{n+1}$, then $F_\mathcal{G}(I) = F_\mathcal{G}(J)$ if and only if I and J belong to the same partition element of $\mathcal{P}_n$.
   (c) Show that if $F_\mathcal{G}(I) \neq F_\mathcal{G}(J)$, then there is some word of length r which belongs to one of $F_\mathcal{G}(I), F_\mathcal{G}(J)$, but not both, and so the words of length r distinguish $F_\mathcal{G}(I)$ from $F_\mathcal{G}(J)$.

*3.4.11. Suppose that X and X' are irreducible sofic shifts with irreducible right-resolving presentations $\mathcal{G}$ and $\mathcal{G}'$. If $X \subseteq X'$, show that for every vertex I in $\mathcal{G}$ there must be a vertex I' in $\mathcal{G}'$ such that $F_\mathcal{G}(I) \subseteq F_{\mathcal{G}'}(I')$.

Notes

Weiss [Wei] asked: what is the smallest collection of shift spaces that includes the shifts of finite type and is closed under factor codes? His answer, sofic shifts, was given in terms of semigroups. It soon became clear that sofic shifts are precisely those shifts presented by labeled graphs. The characterizations of sofic shifts given in §3.2 are due to Weiss.

The uniqueness of the minimal right-resolving presentation of an irreducible sofic shift is due to Fischer [Fis1] [Fis2], and so this presentation is often called the *right Fischer cover* or simply *Fischer cover*. Example 3.3.21, which shows that for reducible sofic shifts there need not be a unique minimal right-resolving presentation, is due to Jonoska [Jon]. The future cover, introduced in Exercise 3.2.8, is often called the *right Krieger cover* or simply *Krieger cover*.

Many of the concepts and constructions in this chapter are very similar to those in automata theory. For instance, a *finite-state automaton* is a labeled graph with a distinguished "initial state" and a distinguished subset of "terminal states." A *language* is a set of words over a finite alphabet. The *language* of a finite-state automaton is the set of all labels of paths that begin at the initial state and end at a terminal state, and a language is called a *regular language* if it is a language of a finite-state automaton. Clearly the language of a sofic shift is regular. It follows from Proposition 1.3.4 that the languages of sofic shifts are precisely the regular languages that are closed under subword and prolongable in the sense that every word in the language can be extended to the left and the right to obtain a longer word in the language (in automata theory, these are called the *factorial, prolongable, regular languages*).

A finite-state automaton whose labeling is right-resolving is called a *deterministic finite-state automaton* or *DFA* for short. In analogy with Theorem 3.3.2, every regular language is the language of a DFA. In fact, the subset construction, used in our proof of that result, was borrowed from automata theory. Also, in analogy with Theorem 3.3.18, every regular language is the language of a unique minimal DFA.

However, Theorem 3.3.18 is a more subtle result because, roughly, a presentation of a sofic shift differs from a DFA by not having a distinguished initial state. Nevertheless, the proof of Theorem 3.3.18 as well as the method given in Example 3.4.15 to compute the minimal right-resolving presentation are both based on a technique in automata theory called the state minimization algorithm. For further reading on automata theory, see Hopcroft and Ullmann [HopU]. For further reading on the connections between automata theory and symbolic dynamics, see Béal [Bea4].

As far as we can tell, W. Krieger was the first to observe the connection between sofic shifts and regular languages.

Exercise 3.3.4 is due to Krieger [Kri8].

CHAPTER 4

ENTROPY

Entropy measures the complexity of mappings. For shifts, it also measures their "information capacity," or ability to transmit messages. The entropy of a shift is an important number, for it is invariant under conjugacy, can be computed for a wide class of shifts, and behaves well under standard operations like factor codes and products. In this chapter we first introduce entropy and develop its basic properties. In order to compute entropy for irreducible shifts of finite type and sofic shifts in §4.3, we describe the Perron–Frobenius theory of nonnegative matrices in §4.2. In §4.4 we show how general shifts of finite type can be decomposed into irreducible pieces and compute entropy for general shifts of finite type and sofic shifts. In §4.5 we describe the structure of the irreducible pieces in terms of cyclically moving states.

§4.1. Definition and Basic Properties

Before we get under way, we review some terminology and notation from linear algebra.

Recall that the *characteristic polynomial* of a matrix A is defined to be $\chi_A(t) = \det(t Id - A)$, where Id is the identity matrix. The *eigenvalues* of A are the roots of $\chi_A(t)$. An *eigenvector* of A corresponding to eigenvalue λ is a vector $\mathbf{v}$, not identically 0, such that $A\mathbf{v} = \lambda\mathbf{v}$.

We say that a (possibly rectangular) matrix A is (strictly) *positive* if each of its entries is positive. In this case we write $A > 0$. Likewise, we say that a matrix A is *nonnegative* if each of its entries is nonnegative and write $A \geqslant 0$. We write $A > B$ (respectively $A \geqslant B$) when A and B have the same size and each entry of A is greater than (respectively greater than or equal to) B. Hence $A > B$ means that $A - B > 0$, and similarly for $A \geqslant B$. Since vectors are special kinds of rectangular matrices, all of these notations apply to vectors as well.

Let X be a shift space. The number $|\mathcal{B}_n(X)|$ of n-blocks appearing in points of X gives us some idea of the complexity of X: the larger the number of n-blocks, the more complicated the space. Instead of using the individual numbers $|\mathcal{B}_n(X)|$ for $n = 1, 2, \ldots$, we can summarize their behavior by computing their growth rate. As we will see a little later, $|\mathcal{B}_n(X)|$ grows approximately like 2^{cn}, and the value of the constant c is the growth rate. This value should be roughly $(1/n)\log_2|\mathcal{B}_n(X)|$ for large n. We will always use the base 2 for logarithms, so that log means $\log_2$. These considerations suggest the following definition.

Definition 4.1.1. Let X be a shift space. The *entropy* of X is defined by

$$(4\text{-}1\text{-}1) \qquad h(X) = \lim_{n\to\infty} \frac{1}{n}\log|\mathcal{B}_n(X)|.$$

In Chapter 6 we will define the entropy of a general continuous map, and show how for a shift map σ_X that definition translates to the growth rate $h(X)$ of n-blocks in X.

The use of the symbol h for entropy has a curious history. Entropy was originally denoted by an upper case Greek letter eta, or H. When variants were needed for different aspects of the mathematical theory, the lower case of the corresponding Roman letter was used.

Note that if X has alphabet $\mathcal{A}$, then $|\mathcal{B}_n(X)| \leqslant |\mathcal{A}|^n$. Hence

$$\frac{1}{n}\log|\mathcal{B}_n(X)| \leqslant \log|\mathcal{A}|$$

for all n, so that $h(X) \leqslant \log|\mathcal{A}|$. Also, if $X \neq \varnothing$, then $|\mathcal{B}_n(X)| \geqslant 1$ for all n, and so $0 \leqslant h(X) < \infty$. Since $\mathcal{B}_n(\varnothing) = 0$ for all n, the definition shows that $h(\varnothing) = -\infty$.

Example 4.1.2. Let $X = X_{[r]}$ be the full r-shift. Then $|\mathcal{B}_n(X)| = r^n$, so that $h(X) = \log r$. □

Example 4.1.3. Let G be an essential graph for which there are exactly r edges starting at each vertex. Since there is a choice of r outgoing edges at each vertex, the number of paths of length n starting at a given vertex is exactly r^n. Hence if G has k vertices, then $|\mathcal{B}_n(X_G)| = k \cdot r^n$. Plugging this into the definition shows that

$$h(X_G) = \lim_{n\to\infty} \frac{1}{n}\log(k \cdot r^n) = \lim_{n\to\infty}\left\{\frac{1}{n}\log k + \log r\right\} = \log r.$$

Example 4.1.2 is the case $k = 1$. □

Example 4.1.4. Let X be the golden mean shift, so that $X = X_{\mathcal{F}}$ with $\mathcal{A} = \{0, 1\}$ and $\mathcal{F} = \{11\}$. Recall that X is the vertex shift $\widehat{X}_G$ where G is the graph in Figure 4.1.1. Now, for $n \geqslant 2$, there is a one-to-one

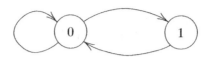

FIGURE 4.1.1.   Graph for golden mean shift.

correspondence between $\mathcal{B}_n(X) = \mathcal{B}_n(\widehat{X}_G)$ and $\mathcal{B}_{n-1}(X_G)$, and we can use Proposition 2.2.12 to count the latter.

The adjacency matrix of G is

$$A = \begin{bmatrix} 1 & 1 \\ 1 & 0 \end{bmatrix}.$$

By Proposition 2.2.12, $|\mathcal{B}_m(X_G)| = \sum_{I,J=1}^{2} (A^m)_{IJ}$. We can compute the entries in A^m by first diagonalizing A. The eigenvalues of A are the roots of its characteristic polynomial

$$\chi_A(t) = \det \begin{bmatrix} t-1 & -1 \\ -1 & t \end{bmatrix} = t^2 - t - 1,$$

which are

$$\lambda = \frac{1+\sqrt{5}}{2} \quad \text{and} \quad \mu = \frac{1-\sqrt{5}}{2}.$$

The corresponding eigenvectors are easily computed to be

$$\begin{bmatrix} \lambda \\ 1 \end{bmatrix} \quad \text{and} \quad \begin{bmatrix} \mu \\ 1 \end{bmatrix}.$$

Thus if we put

$$P = \begin{bmatrix} \lambda & \mu \\ 1 & 1 \end{bmatrix},$$

so that

$$P^{-1} = \frac{1}{\sqrt{5}} \begin{bmatrix} 1 & -\mu \\ -1 & \lambda \end{bmatrix},$$

then

$$P^{-1}AP = \begin{bmatrix} \lambda & 0 \\ 0 & \mu \end{bmatrix}.$$

Hence

$$P^{-1}A^m P = (P^{-1}AP)^m = \begin{bmatrix} \lambda^m & 0 \\ 0 & \mu^m \end{bmatrix},$$

so that

$$A^m = P \begin{bmatrix} \lambda^m & 0 \\ 0 & \mu^m \end{bmatrix} P^{-1}.$$

Using that $\lambda\mu = \det A = -1$, multiplying out the previous equation gives

$$(4\text{-}1\text{-}2) \quad A^m = \frac{1}{\sqrt{5}} \begin{bmatrix} \lambda^{m+1} - \mu^{m+1} & \lambda^m - \mu^m \\ \lambda^m - \mu^m & \lambda^{m-1} - \mu^{m-1} \end{bmatrix} = \begin{bmatrix} f_{m+1} & f_m \\ f_m & f_{m-1} \end{bmatrix},$$

where $f_m = (\lambda^m - \mu^m)/\sqrt{5}$.

Since $\lambda^2 = \lambda + 1$, it follows that $\lambda^{m+2} = \lambda^{m+1} + \lambda^m$, and similarly for μ. Hence $f_{m+2} = f_{m+1} + f_m$, so that the f_m can be successively computed from the starting values $f_1 = 1$ and $f_2 = 1$, producing the familiar Fibonacci sequence $1, 1, 2, 3, 5, 8, 13, 21, 34, \ldots$. This shows, incidentally, that the mth Fibonacci number is given by the formula

$$f_m = \frac{1}{\sqrt{5}} \left[\left(\frac{1 + \sqrt{5}}{2} \right)^m - \left(\frac{1 - \sqrt{5}}{2} \right)^m \right].$$

We conclude that

$$|\mathcal{B}_m(X_G)| = f_{m+1} + f_m + f_m + f_{m-1} = f_{m+2} + f_{m+1} = f_{m+3},$$

so that

$$|\mathcal{B}_n(X)| = |\mathcal{B}_{n-1}(X_G)| = f_{n+2} = \frac{1}{\sqrt{5}}(\lambda^{n+2} - \mu^{n+2}).$$

Since $\lambda \approx 1.61803$ and $\mu \approx -0.61803$, the growth rate of $|\mathcal{B}_n(X)|$ is determined by the size of λ. More explicitly, since $\mu^n/\lambda^n \to 0$,

$$h(X) = \lim_{n \to \infty} \frac{1}{n} \log |\mathcal{B}_n(X)| = \lim_{n \to \infty} \frac{1}{n} \log \left[\frac{1}{\sqrt{5}}(\lambda^{n+2} - \mu^{n+2}) \right]$$

$$= \lim_{n \to \infty} \frac{1}{n} \left[\log \frac{1}{\sqrt{5}} + (n+2) \log \lambda + \log \left(1 - \frac{\mu^{n+2}}{\lambda^{n+2}} \right) \right]$$

$$= \lim_{n \to \infty} \left(1 + \frac{2}{n} \right) \log \lambda = \log \lambda.$$

The number $\lambda = (1 + \sqrt{5})/2$ is called the *golden mean* for geometric reasons (the Greeks felt that rectangle with sides λ and 1 has the most pleasing proportions), and this is why X is called the golden mean shift. $\quad\square$

Remark 4.1.5. In the preceding example, we computed the entropy of the golden mean shift using the edge shift X_G. We did this because, in general, we prefer to work with edge shifts rather than vertex shifts. However, we could have just as easily worked with the vertex shift $\widehat{X}_G$ instead. Note that for any graph G without multiple edges, the vertex shift $\widehat{X}_G$ and the edge shift X_G have the same entropy since $|\mathcal{B}_n(\widehat{X}_G)| = |\mathcal{B}_{n-1}(X_G)|$ for all $n \geqslant 2$. $\quad\square$

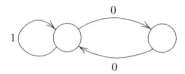

FIGURE 4.1.2. Presentation of the even shift.

Example 4.1.6. Let X be the even shift, so that $X = X_\mathcal{G}$, where $\mathcal{G}$ is the labeled graph shown in Figure 4.1.2. The underlying graph G is isomorphic to the graph in Figure 4.1.1. Now any block in $\mathcal{B}_n(X)$ containing a 1 has a unique presentation in $\mathcal{G}$, while 0^n has exactly two presentations, one starting at each of the two vertices. Hence $|\mathcal{B}_n(X_\mathcal{G})| = |\mathcal{B}_n(X_G)| - 1 = f_{n+3} - 1$ by the previous example. A similar calculation then shows that

$$h(X) = \lim_{n \to \infty} \frac{1}{n} \log |\mathcal{B}_n(X_\mathcal{G})| = \log\left(\frac{1 + \sqrt{5}}{2}\right).$$

Thus the even shift and the golden mean shift have the same entropy. Proposition 4.1.13 will generalize this idea. □

We defined $h(X)$ to be the limit of the numbers $(1/n) \log |\mathcal{B}_n(X)|$. But do these numbers always approach a limit, or could they continually fluctuate and never settle down to a limiting value? Our next goal is to show that the limit always exists. Recall that if $\{b_n\}$ is a sequence of real numbers, its *infimum* $\beta = \inf_{n \geqslant 1} b_n$ is the largest number β less than or equal to all of the b_n.

Lemma 4.1.7. *Let $a_1, a_2, \ldots$ be a sequence of nonnegative numbers such that*

$$a_{m+n} \leqslant a_m + a_n \quad \text{for all} \quad m, n \geqslant 1.$$

Then $\lim_{n \to \infty} a_n/n$ exists and equals $\inf_{n \geqslant 1} a_n/n$.

PROOF: Let $\alpha = \inf_{n \geqslant 1} a_n/n$. By definition, $a_n/n \geqslant \alpha$ for every $n \geqslant 1$. Fix $\epsilon > 0$. We must show that $a_n/n < \alpha + \epsilon$ for all large enough n. Since α is the largest number less than or equal to all of the a_n/n, there is a k for which $a_k/k < \alpha + \frac{1}{2}\epsilon$. Then for $0 \leqslant j < k$ and $m \geqslant 1$ we have that

$$\frac{a_{mk+j}}{mk + j} \leqslant \frac{a_{mk}}{mk + j} + \frac{a_j}{mk + j} \leqslant \frac{a_{mk}}{mk} + \frac{a_j}{mk}$$

$$\leqslant \frac{ma_k}{mk} + \frac{ja_1}{mk} \leqslant \frac{a_k}{k} + \frac{a_1}{m} < \alpha + \frac{\epsilon}{2} + \frac{a_1}{m}.$$

Hence if $n = mk + j$ is large enough so that $a_1/m < \epsilon/2$, then $a_n/n < \alpha + \epsilon$, completing the proof. □

Proposition 4.1.8. *If X is a shift space, then*

$$\lim_{n \to \infty} \frac{1}{n} \log |\mathcal{B}_n(X)|$$

exists, and equals

$$\inf_{n \geqslant 1} \frac{1}{n} \log |\mathcal{B}_n(X)|.$$

PROOF: For integers $m, n \geqslant 1$, a block in $\mathcal{B}_{m+n}(X)$ is uniquely determined by its initial m-block and the subsequent n-block, so that

$$|\mathcal{B}_{m+n}(X)| \leqslant |\mathcal{B}_m(X)| \cdot |\mathcal{B}_n(X)|,$$

from which we see that

$$\log |\mathcal{B}_{m+n}(X)| \leqslant \log |\mathcal{B}_m(X)| + \log |\mathcal{B}_n(X)|.$$

Apply the previous lemma to the sequence $a_n = \log |\mathcal{B}_n(X)|$. $\square$

We next turn to the behavior of entropy under factor codes and embeddings.

Proposition 4.1.9. *If Y is a factor of X, then $h(Y) \leqslant h(X)$.*

PROOF: Let $\phi \colon X \to Y$ be a factor code from X onto Y. Then $\phi = \Phi_\infty^{[-m,k]}$ for a block map Φ and integers m and k. Every block in $\mathcal{B}_n(Y)$ is the image under Φ of a block in $\mathcal{B}_{n+m+k}(X)$ (why?). Hence $|\mathcal{B}_n(Y)| \leqslant |\mathcal{B}_{n+m+k}(X)|$ and

$$h(Y) = \lim_{n \to \infty} \frac{1}{n} |\mathcal{B}_n(Y)| \leqslant \lim_{n \to \infty} \frac{1}{n} \log |\mathcal{B}_{n+m+k}(X)|$$

$$= \lim_{n \to \infty} \left(\frac{n+m+k}{n} \right) \frac{1}{n+m+k} \log |\mathcal{B}_{n+m+k}(X)| = h(X).$$

$\square$

From this it follows that entropy is invariant under conjugacy.

Corollary 4.1.10. *If X is conjugate to Y, then $h(X) = h(Y)$.*

PROOF: Since $X \cong Y$, each is a factor of the other. By the previous proposition, $h(X) \leqslant h(Y)$ and $h(Y) \leqslant h(X)$. $\square$

Example 4.1.11. The full 2-shift is not conjugate to the full 3-shift, since they have different entropies. Similarly, the golden mean shift is not conjugate to a full shift, since the golden mean is not a whole integer. $\square$

Proposition 4.1.12. *If Y embeds into X, then $h(Y) \leqslant h(X)$.*

PROOF: Let $\phi: Y \to X$ be an embedding. Then $\phi(Y)$ is a shift space by Theorem 1.5.13, and ϕ is a conjugacy from Y to $\phi(Y)$ by Theorem 1.5.14. Hence $h(Y) = h(\phi(Y))$ since entropy is invariant under conjugacy, and clearly $h(\phi(Y)) \leqslant h(X)$ since $\mathcal{B}_n(\phi(Y)) \subseteq \mathcal{B}_n(X)$ for all n. □

Our next result generalizes the method we used in Example 4.1.6 to compute the entropy of the even shift.

Proposition 4.1.13. *Let* $\mathcal{G} = (G, \mathcal{L})$ *be a right-resolving labeled graph. Then* $h(X_\mathcal{G}) = h(X_G)$.

PROOF: $\mathcal{L}_\infty$ is a 1-block factor code from X_G onto $X_\mathcal{G}$, so that, by Proposition 4.1.9, $h(X_\mathcal{G}) \leqslant h(X_G)$. Let G have k states. Then any block in $\mathcal{B}_n(X_\mathcal{G})$ has at most k presentations since $\mathcal{G}$ is right-resolving. Hence $|\mathcal{B}_n(X_\mathcal{G})| \geqslant (1/k)|\mathcal{B}_n(X_G)|$. Taking logarithms, dividing by n, and letting $n \to \infty$ show that $h(X_\mathcal{G}) \geqslant h(X_G)$. □

We next want to examine the growth rate of the number of periodic points in a shift.

Definition 4.1.14. For a shift space X, let $p_n(X)$ denote the number of points in X having period n.

To discuss the growth rate of $p_n(X)$, we recall the notion of the *limit superior*, $\limsup_{n \to \infty} b_n$, of a sequence $\{b_n\}$ of real numbers, which is defined as the largest limit point of $\{b_n\}$, or equivalently, the smallest number β such that, for every $\epsilon > 0$, only finitely many of the b_n are greater than $\beta + \epsilon$. If $\{b_n\}$ converges, then $\limsup_{n \to \infty} b_n = \lim_{n \to \infty} b_n$.

Proposition 4.1.15. *Let X be a shift space. Then*

$$\limsup_{n \to \infty} \frac{1}{n} \log p_n(X) \leqslant h(X).$$

PROOF: Each point $x \in X$ of period n is uniquely determined by $x_{[0,n-1]} \in \mathcal{B}_n(X)$, so that $p_n(X) \leqslant |\mathcal{B}_n(X)|$. Take logarithms, divide by n, and take $\limsup$ as $n \to \infty$. □

Example 4.1.16. The numbers $(1/n) \log p_n(X)$ do not always converge to a limit, even for irreducible shifts of finite type. For example, let

$$A = \begin{bmatrix} 0 & 2 \\ 3 & 0 \end{bmatrix}$$

and $X = X_A$. Then $p_n(X) = \operatorname{tr}(A^n)$, and hence

$$p_n(X) = \begin{cases} 2 \cdot 6^{n/2} & \text{for } n \text{ even,} \\ 0 & \text{for } n \text{ odd.} \end{cases}$$ □

EXERCISES

4.1.1. Show that $X_{[m]}$ factors onto $X_{[n]}$ if and only if $m \geqslant n$.

4.1.2. Find an infinite shift of finite type with entropy 0.

4.1.3. Prove that $(1/n) \log p_n(X)$ converges to $h(X)$ when X is (a) a full shift, (b) the golden mean shift, (c) the even shift.

4.1.4. Let $\mathcal{A} = \{0, 1\}$ and $\mathcal{F} = \{111\}$. Explain how you would compute $h(X_{\mathcal{F}})$.

4.1.5. Let X and Y be shift spaces, and $N \geqslant 1$. Show that
   (a) $h(X^{[N]}) = h(X)$,
   (b) $h(X^N) = Nh(X)$,
   (c) $h(X \times Y) = h(X) + h(Y)$,
   (d) $h(X \cup Y) = \max\{h(X), h(Y)\}$

4.1.6. Let f_m denote the mth Fibonacci number. Prove that $f_{m+1}f_{m-1} - f_m^2 = (-1)^m$ for all $m > 1$.

4.1.7. Show directly (without using eigenvalues) that for the golden mean shift X, $|\mathcal{B}_{m+2}(X)| = |\mathcal{B}_{m+1}(X)| + |\mathcal{B}_m(X)|$.

4.1.8. Let G be a graph with adjacency matrix A, let r denote the smallest row sum of A, and let s denote the largest. Prove that $\log r \leqslant h(X_G) \leqslant \log s$.

4.1.9. Find two irreducible shifts of finite type which have the same entropy but are not conjugate.

*4.1.10. For a shift space X, $h(X)$ is the infimum of the sequence $(1/n) \log |\mathcal{B}_n(X)|$. Must this sequence always be monotonic?

*4.1.11. (a) Construct an irreducible shift space X which has no periodic points of any period, but for which $h(X) = 0$.
   (b) Construct an irreducible shift space Y which has no periodic points of any period, but for which $h(Y) > 0$. This shows that the inequality in Proposition 4.1.15 can be strict.

§4.2. Perron–Frobenius Theory

Let G be a graph with an $r \times r$ adjacency matrix A. As in Example 4.1.4, the number of n-blocks in X_G is given by

$$|\mathcal{B}_n(X_G)| = \sum_{I,J=1}^{r} (A^n)_{IJ}.$$

Thus to compute the entropy of X_G, we need to find the growth rate of the entries of A^n. The Perron–Frobenius theory is precisely the tool that will enable us to do this when G is irreducible, and to show that the growth rate is controlled by the largest eigenvalue of A. We will extend this to general graphs in §4.4.

Let A be a (square) nonnegative matrix, and assume that A is not the zero matrix. Suppose for the moment that A has a positive eigenvector $\mathbf{v}$. Then $(A\mathbf{v})_I > 0$ for some I, and so if λ is the eigenvalue for $\mathbf{v}$, then $\lambda \mathbf{v}_I = (A\mathbf{v})_I > 0$; since $\mathbf{v}_I > 0$ it follows that $\lambda > 0$.

Next, applying A to $A\mathbf{v} = \lambda\mathbf{v}$ shows that $A^2\mathbf{v} = \lambda A\mathbf{v} = \lambda^2\mathbf{v}$, and in general that $A^n\mathbf{v} = \lambda^n\mathbf{v}$ for $n \geqslant 1$. Hence, for every I,

$$\sum_{J=1}^{r}(A^n)_{IJ}v_J = \lambda^n v_I.$$

Let

$$c = \min\{v_1, v_2, \ldots, v_r\} \quad \text{and} \quad d = \max\{v_1, v_2, \ldots, v_r\}.$$

Then

$$c\sum_{J=1}^{r}(A^n)_{IJ} \leqslant \sum_{J=1}^{r}(A^n)_{IJ}v_J = \lambda^n v_I \leqslant d\lambda^n.$$

Dividing by c and summing over I shows that

$$\sum_{I,J=1}^{r}(A^n)_{IJ} \leqslant \sum_{I=1}^{r}\frac{d}{c}\lambda^n = \left(\frac{rd}{c}\right)\lambda^n = d_0\,\lambda^n,$$

where $d_0 = rd/c > 0$.

To estimate $\sum_{I,J}(A^n)_{IJ}$ from below, note that, for each I,

$$c\lambda^n \leqslant \lambda^n v_I = \sum_{J=1}^{r}(A^n)_{IJ}v_J \leqslant d\sum_{J=1}^{r}(A^n)_{IJ} \leqslant d\sum_{I,J=1}^{r}(A^n)_{IJ}.$$

Letting $c_0 = c/d > 0$, we conclude that

$$c_0\,\lambda^n \leqslant \sum_{I,J=1}^{r}(A^n)_{IJ}.$$

We can summarize our discussion thus far as follows.

Proposition 4.2.1. *Let $A \neq 0$ be a nonnegative matrix having a positive eigenvector $\mathbf{v}$. Then the corresponding eigenvalue λ is positive, and there are positive constants c_0 and d_0 such that*

(4–2–1) $$c_0\,\lambda^n \leqslant \sum_{I,J=1}^{r}(A^n)_{IJ} \leqslant d_0\,\lambda^n.$$

Hence if G is a graph whose adjacency matrix is A, then $h(X_G) = \log\lambda$.

PROOF: We have already shown all but the last statement. For this, observe that $|\mathcal{B}_n(X_G)|$ equals the central term in (4–2–1). Hence taking logarithms, dividing by n, and letting $n \to \infty$ completes the proof.   □

Our assumption that A has a positive eigenvector implies two additional consequences. First, suppose that $\mathbf{u}$ is another positive eigenvector for A, and that θ is its eigenvalue. Our previous argument applies equally well to $\mathbf{u}$, showing that $h(X_G) = \log\theta$, which forces $\theta = \lambda$. Hence λ is the *only* eigenvalue of A corresponding to a positive eigenvector.

Secondly, suppose that μ is another eigenvalue for A, and let $\mathbf{w}$ be an eigenvector with eigenvalue μ. Note that both μ and $\mathbf{w}$ could involve complex numbers. Define the *sum norm*, abbreviated *norm*, of $\mathbf{w}$ to be $\|\mathbf{w}\| = \sum_{I=1}^{r} |w_I|$. Since $A^n \mathbf{w} = \mu^n \mathbf{w}$ and $A \geqslant 0$,

$$
\begin{aligned}
|\mu|^n \|\mathbf{w}\| = \|\mu^n \mathbf{w}\| &= \|A^n \mathbf{w}\| \\
&= \left| \sum_{J=1}^{r} (A^n)_{1J} w_J \right| + \cdots + \left| \sum_{J=1}^{r} (A^n)_{rJ} w_J \right| \\
&\leqslant \sum_{J=1}^{r} (A^n)_{1J} |w_J| + \cdots + \sum_{J=1}^{r} (A^n)_{rJ} |w_J| \\
&\leqslant \left(\max_{1 \leqslant J \leqslant r} |w_J| \right) \sum_{I,J=1}^{r} (A^n)_{IJ} \\
&\leqslant \|\mathbf{w}\|\, d_0\, \lambda^n.
\end{aligned}
$$

Since $\|\mathbf{w}\| \neq 0$, it follows that $|\mu| \leqslant d_0^{1/n} \lambda$, and letting $n \to \infty$ shows that $|\mu| \leqslant \lambda$. Hence λ is the largest (in absolute value) eigenvalue of A.

The preceding arguments depend very much on the fact that A is nonnegative. However, they do not at all depend on the fact that A has integer entries. Indeed, Perron–Frobenius theory applies to nonnegative (real) matrices, although we will mainly use it when the entries are nonnegative integers.

Definition 4.2.2. A nonnegative matrix A is *irreducible* if for each ordered pair of indices I, J, there exists some $n \geqslant 0$ such that $A_{IJ}^n > 0$. We adopt the convention that for any matrix $A^0 = Id$, and so the 1×1 matrix $[0]$ is irreducible. A nonnegative matrix A is *essential* if none of its rows or columns is zero.

Observe that a graph G is irreducible if and only if its adjacency matrix A_G is irreducible (why?). Note that irreducibility does *not* imply that $A^n > 0$ for some n (see Example 4.1.16).

Observe that a graph G is essential if and only if its adjacency matrix A_G is essential (why?). With the single exception of the 1×1 matrix $[0]$, an irreducible matrix cannot contain any zero rows or columns (i.e., it is essential).

The Perron–Frobenius theory will show that an irreducible matrix always has a positive eigenvector, so that our arguments above always apply.

For the rest of this section, let A be a nonnegative matrix. A real eigenvalue λ for A is *geometrically simple* if its corresponding eigenspace is one dimensional, and is *algebraically simple* if it is a simple root of the characteristic polynomial of A. Recall from linear algebra that algebraic simplicity implies geometric simplicity, but not conversely.

Theorem 4.2.3 (Perron–Frobenius Theorem). *Let $A \neq 0$ be an irreducible matrix. Then A has a positive eigenvector $\mathbf{v}_A$ with corresponding eigenvalue $\lambda_A > 0$ that is both geometrically and algebraically simple. If μ is another eigenvalue for A, then $|\mu| \leqslant \lambda_A$. Any positive eigenvector for A is a positive multiple of $\mathbf{v}_A$.*

For an irreducible matrix A, we call λ_A the *Perron eigenvalue* of A, and $\mathbf{v}_A$ a *Perron eigenvector* of A. The theorem shows that $\mathbf{v}_A$ is unique up to multiplication by a positive scalar.

There are a number of proofs for this fundamental result; see [Gan] or [Sen] for example. Rather than duplicate one of these, we will illustrate the main ideas of proof with the 2×2 case, where we can use elementary versions of more sophisticated arguments. A complete proof of the Perron–Frobenius theorem (in any dimension) is outlined in the exercises.

PROOF: (For the 2-dimensional case.) Let

$$A = \begin{bmatrix} a & b \\ c & d \end{bmatrix}$$

be an irreducible matrix over the nonnegative reals. Irreducibility implies that $b > 0$ and $c > 0$ (why?).

Let Q denote the first quadrant in the plane; i.e., Q consists of those points with nonnegative coordinates. Since $A \geqslant 0$, it maps Q into Q. If $A > 0$, then intuitively it would "compress" the rays in Q starting at the origin, and this would result in some ray being fixed by A (see Corollary 4.5.14). Then any vector pointing along this ray would be mapped by A to a multiple of itself, providing the eigenvector we are after. Here instead, we find a fixed ray for irreducible matrices by using a fixed point theorem from calculus.

For $0 \leqslant t \leqslant 1$ let

$$\mathbf{v}_t = \begin{bmatrix} t \\ 1 - t \end{bmatrix} \in Q.$$

Since $b > 0$ and $c > 0$, it follows that $A\mathbf{v}_t \neq 0$ for $0 \leqslant t \leqslant 1$, and that $\mathbf{v}_0$ and $\mathbf{v}_1$ are not eigenvectors for A. For any vector $\mathbf{w}$ in the plane, recall that $\|\mathbf{w}\| = |w_1| + |w_2|$. Then the $\mathbf{v}_t$ for $0 \leqslant t \leqslant 1$ are exactly the vectors in Q of norm 1.

Now for each $t \in [0, 1]$, clearly $A\mathbf{v}_t / \|A\mathbf{v}_t\|$ is in Q and has norm 1, so there is a unique $f(t) \in [0, 1]$ satisfying

$$\frac{A\mathbf{v}_t}{\|A\mathbf{v}_t\|} = \mathbf{v}_{f(t)}.$$

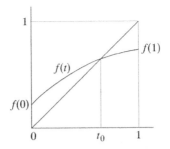

FIGURE 4.2.1. A fixed point $f(t_0) = t_0$ produces an eigenvector.

The function $f: [0, 1] \to [0, 1]$ is continuous. Since $\mathbf{v}_0$ is not an eigenvector for A, it follows that $A\mathbf{v}_0/\|A\mathbf{v}_0\|$ is not $\mathbf{v}_0$, so that $f(0) > 0$. Similarly, $f(1) < 1$. Hence the function $g(t) = f(t) - t$ is continuous on $[0, 1]$, and $g(0) > 0$ while $g(1) < 0$. The Intermediate Value Theorem from calculus shows that there must be a $t_0 \in (0, 1)$ with $g(t_0) = 0$, i.e., $f(t_0) = t_0$ (see Figure 4.2.1).

From $f(t_0) = t_0$ it follows that

$$\frac{A\mathbf{v}_{t_0}}{\|A\mathbf{v}_{t_0}\|} = \mathbf{v}_{t_0}.$$

Since $0 < t_0 < 1$, we conclude that $\mathbf{v}_{t_0}$ is a positive eigenvector for A, and its eigenvalue is $\lambda = \|A\mathbf{v}_{t_0}\|$.

To prove that λ is geometrically simple, put $\mathbf{v} = \mathbf{v}_{t_0}$, and suppose that $\mathbf{w}$ is an eigenvector for λ that is linearly independent of $\mathbf{v}$. The line $\mathbf{v} + s\mathbf{w}$, $-\infty < s < \infty$, must intersect at least one of the positive axes, say at $\mathbf{u}$ (see Figure 4.2.2). Now $\mathbf{u} \neq 0$ since $\mathbf{v}$ and $\mathbf{w}$ are linearly independent. Since $A\mathbf{u} = A\mathbf{v} + s\,A\mathbf{w} = \lambda\mathbf{v} + s(\lambda\mathbf{w}) = \lambda\mathbf{u}$, this would imply that either $\begin{bmatrix} 1 \\ 0 \end{bmatrix}$ or $\begin{bmatrix} 0 \\ 1 \end{bmatrix}$ is an eigenvector for A with eigenvalue λ, contradicting $b > 0$ and $c > 0$.

If $\mathbf{w}$ is another positive eigenvector for A, our discussion on page 109 shows that $\mathbf{w}$ has eigenvalue λ. Geometric simplicity of λ then shows that $\mathbf{w}$ must be a positive multiple of $\mathbf{v}$. The same discussion also shows that $|\mu| \leqslant \lambda$ for the other eigenvalue μ of A.

To show that λ is algebraically simple, consider the characteristic polynomial of A,

$$\chi_A(t) = t^2 - (a + d)t + ad - bc.$$

We can show that λ is a simple root of $\chi_A(t)$ by proving that the derivative $\chi'_A(\lambda) \neq 0$. Since

$$\lambda = \frac{a + d + \sqrt{(a + d)^2 - 4(ad - bc)}}{2} = \frac{a + d + \sqrt{(a - d)^2 + 4bc}}{2},$$

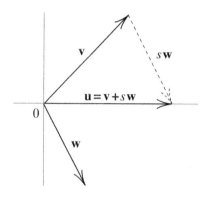

FIGURE 4.2.2.   Proof of geometric simplicity.

and $(a - d)^2 + 4bc > 0$ since $b > 0$ and $c > 0$, we see that $\lambda > (a + d)/2$. Now $\chi'_A(t) = 2t - (a + d)$, so that $\chi'_A(\lambda) = 2\lambda - (a + d) > 0$.   □

We now possess the tools to compute entropy.

EXERCISES

4.2.1. Compute the Perron eigenvalue and eigenvector for each of the following matrices.

$$\text{(a)} \begin{bmatrix} 2 & 1 \\ 1 & 1 \end{bmatrix}, \quad \text{(b)} \begin{bmatrix} 0 & 1 \\ 1 & 0 \end{bmatrix}, \quad \text{(c)} \begin{bmatrix} 0 & 3 \\ 2 & 0 \end{bmatrix}.$$

4.2.2. Show that $\begin{bmatrix} 2 & 0 \\ 0 & 3 \end{bmatrix}$ has no (strictly) positive eigenvectors.

4.2.3. Let A be an irreducible matrix, r denote the smallest row sum of A, and s denote the largest. Show that $r \leqslant \lambda_A \leqslant s$.

4.2.4. Show that if A is irreducible, and $\mathbf{w} \neq \mathbf{0}$ is nonnegative with $A\mathbf{w} = \mu\mathbf{w}$ for any eigenvalue μ of A, then $\mathbf{w}$ is a strictly positive multiple of $\mathbf{v}_A$, and so $\mathbf{w} > 0$.

4.2.5. Compute the function $f(t)$ used in the proof of the Perron–Frobenius Theorem in terms of a, b, c, and d. Use this to show directly that $f(t)$ is continuous, and that $f(0) > 0$ and $f(1) < 1$.

4.2.6. If A is an irreducible matrix with nonnegative real entries, show that

$$\lambda_A = \inf\{\lambda \geqslant 0 : A\mathbf{v} \leqslant \lambda\mathbf{v} \text{ for some } \mathbf{v} > 0\}$$
$$= \sup\{\lambda \geqslant 0 : A\mathbf{v} \geqslant \lambda\mathbf{v} \text{ for some } \mathbf{v} > 0\}.$$

4.2.7. If A is an irreducible matrix over the nonnegative integers, show that we can find a Perron eigenvector $\mathbf{v}_A$ for A having *integer* entries if and only if λ_A is an integer.

In the next five exercises the reader is asked to prove the Perron–Frobenius Theorem (Theorem 4.2.3) for an irreducible $m \times m$ matrix A, by generalizing the arguments in the text.

4.2.8. Use irreducibility to show that any nonnegative eigenvector for A is necessarily positive.

4.2.9. Show that A has a positive eigenvector $\mathbf{v}_A$ with corresponding eigenvalue $\lambda_A >$ 0, using Exercise 4.2.8 and the following result:

Brouwer Fixed Point Theorem: Let

$$\Delta = \{\mathbf{x} \in \mathbb{R}^m : \mathbf{x} \geqslant 0, \; \textstyle\sum_i x_i = 1\},$$

the unit simplex in $\mathbb{R}^m$. Then any continuous map $f: \Delta \to \Delta$ has a fixed point (see [Gre] for a proof of this result; also, see Exercise 6.1.15 for a proof of a special version of the Brouwer Fixed Point Theorem which suffices for this exercise).

4.2.10. Show that λ_A is geometrically simple, by showing that if $\mathbf{v}$ is a positive eigenvector for λ_A and $\mathbf{w}$ is linearly independent of $\mathbf{v}$, then for some value s the vector $\mathbf{u} = \mathbf{v} + s\mathbf{w}$ lies on the boundary of $\{(x_1, \dots, x_m) \geqslant \mathbf{0}\}$ (i.e., u is nonnegative, some component of $\mathbf{u}$ is 0) but $\mathbf{u} \neq \mathbf{0}$.

4.2.11. Using the argument in the text, show that if μ is another eigenvalue for A, then $|\mu| \leqslant \lambda_A$ and any positive eigenvector for A is a positive multiple of $\mathbf{v}_A$.

*4.2.12. In the following steps, show that λ_A is algebraically simple:

(a) Show that an eigenvalue λ of A is algebraically simple if and only if $\frac{d}{dt}\chi_A(t)|_{t=\lambda} \neq 0$.

(b) Define the *adjugate* $\mathrm{adj}(B)$ of a matrix B to be the matrix whose IJth entry is $(-1)^{I+J}$ times the determinant of the matrix obtained by deleting the Jth row and Ith column of B. Show that $B \cdot \mathrm{adj}(B) = \mathrm{adj}(B) \cdot B = det(B) \cdot Id$. [*Hint:* Recall how to compute a determinant by cofactors.]

(c) Show that $\mathrm{adj}(\lambda_A Id - A)$ is not the zero matrix. [*Hint:* Use geometric simplicity of λ_A to prove that the rank of $\lambda_A Id - A$ is $m - 1$.]

(d) Show that the columns (respectively rows) of $\mathrm{adj}(\lambda_A Id - A)$ are right (respectively left) eigenvectors of A corresponding to λ_A.

(e) Show that

$$\frac{d}{dt}\chi_A(t) = \sum_I \mathrm{adj}(tId - A)_{II}.$$

[*Hint:* Differentiate the equation $\mathrm{adj}(tId - A) \cdot (tId - A) = \chi_A(t)Id$.]

(f) Show that λ_A is algebraically simple.

§4.3. Computing Entropy

If X is an irreducible shift of finite type, then it is conjugate to an edge shift with an irreducible adjacency matrix A. We will use our work in the previous two sections to show that $h(X) = \log \lambda_A$. In this section we will also compute the entropy of an irreducible sofic shift and show that the growth rate of its periodic points equals its entropy. We deal with reducible shifts in the next section.

First consider a shift of finite type X. Recall from Theorem 2.3.2 that we can always recode X to an edge shift.

Theorem 4.3.1.

(1) *If G is an irreducible graph, then $h(\mathsf{X}_G) = \log \lambda_{A(G)}$.*

(2) *If X is an irreducible M-step shift of finite type and G is the essential graph for which $X^{[M+1]} = \mathsf{X}_G$, then $h(X) = \log \lambda_{A(G)}$.*

PROOF: Since $A = A(G)$ is irreducible, the Perron–Frobenius Theorem shows that A has a positive eigenvector with a positive eigenvalue λ_A. Hence Proposition 4.2.1 applies, giving (1).

For the second statement, observe that since $X \cong X^{[M+1]}$ and entropy is a conjugacy invariant, it follows that $h(X) = h(X^{[M+1]}) = h(X_G)$. Since X is irreducible, so is $A(G)$, and we can apply the first statement. □

Example 4.3.2. Let $\mathcal{A} = \{0, 1\}$, $\mathcal{F} = \{111\}$, and $X = X_{\mathcal{F}}$. Then X is the space of points in the full 2-shift in which there are no three consecutive 1's. This is a 2-step shift of finite type. The graph G with $X^{[3]} = X_G$ has states 00, 01, 10, and 11, and irreducible adjacency matrix

$$A = \begin{bmatrix} 1 & 1 & 0 & 0 \\ 0 & 0 & 1 & 1 \\ 1 & 1 & 0 & 0 \\ 0 & 0 & 1 & 0 \end{bmatrix}$$

(the transition from 11 to 11 is ruled out since the block 111 is forbidden). Thus $h(X) = \log \lambda_A$. Here $\chi_A(t) = t^4 - t^3 - t^2 - t$, whose roots are $t = 0$, $t \approx -0.41964 \pm 0.60629i$, and $t \approx 1.83929$. Hence $h(X) \approx \log 1.83929 \approx 0.87915$. □

Recall from Remark 4.1.5 that when G has no multiple edges then we have $h(\widehat{X}_G) = h(X_G)$, and so part (1) of the preceding theorem also gives a formula for the entropy of vertex shifts.

We next turn to sofic shifts.

Theorem 4.3.3. Let X be an irreducible sofic shift, and $\mathcal{G} = (G, \mathcal{L})$ be an irreducible right-resolving presentation of X. Then $h(X) = \log \lambda_{A(G)}$.

PROOF: By Proposition 4.1.13, $h(X) = h(X_{\mathcal{G}}) = h(X_G)$, and since G is irreducible, the previous theorem shows that $h(X_G) = \log \lambda_{A(G)}$. □

Of course, Theorem 4.3.3 applies to irreducible shifts of finite type. So, Theorem 4.3.1(2) and Theorem 4.3.3 give two ways to compute the entropy of an irreducible shift of finite type. Usually, the latter is simpler.

Example 4.3.4. Let X be the $(1, 3)$ run-length limited shift. Figure 4.3.1 shows an irreducible right-resolving presentation of X. The adjacency ma-

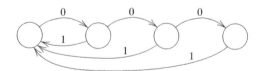

FIGURE 4.3.1. Presentation of $(1, 3)$ run-length limited shift.

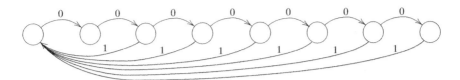

FIGURE 4.3.2. Presentation of $(2,7)$ run-length limited shift.

trix of the underlying graph is

$$A = \begin{bmatrix} 0 & 1 & 0 & 0 \\ 1 & 0 & 1 & 0 \\ 1 & 0 & 0 & 1 \\ 1 & 0 & 0 & 0 \end{bmatrix}.$$

Then $\chi_A(t) = t^4 - t^2 - t - 1$, whose largest root is $\lambda_A \approx 1.46557$. Thus $h(X) = \log \lambda_A \approx 0.55146$. □

Example 4.3.5. The $(2,7)$ run-length limited shift X has the irreducible right-resolving presentation shown in Figure 4.3.2. If A denotes the adjacency matrix of the underlying graph, then $\chi_A(t) = t^8 - t^5 - t^4 - t^3 - t^2 - t - 1$, whose largest root is $\lambda_A \approx 1.43134$. Hence $h(X) = \log \lambda_A \approx 0.51737$. The fact that this number is slightly bigger than $1/2$ plays a significant role in the next chapter. □

We saw in §4.1 that, for a general shift space X, the growth rate of the number $p_n(X)$ of its points of period n is always bounded above by $h(X)$. For many X this rate equals $h(X)$.

Theorem 4.3.6. *If X is an irreducible sofic shift, then*

$$\limsup_{n \to \infty} \frac{1}{n} \log p_n(X) = h(X).$$

PROOF: The inequality

$$\limsup_{n \to \infty} \frac{1}{n} \log p_n(X) \leqslant h(X)$$

holds for all X by Proposition 4.1.15.

We prove the reverse inequality first for irreducible shifts of finite type X. As before, there is an irreducible graph G so that $X \cong X_G$, and so $p_n(X) = p_n(X_G)$ and $h(X) = h(X_G)$. Since G is irreducible, there is an integer N such that for every pair I, J of vertices in G, there is a path from I to J of length $\leqslant N$. Thus if $\pi \in \mathcal{B}_n(X_G)$, there is a path τ of length $\leqslant N$ from $t(\pi)$ to $i(\pi)$. Then $\pi\tau$ is a cycle, and $(\pi\tau)^\infty$ has period $n + |\tau|$, which is between

n and $n + N$. In this way each block in $\mathcal{B}_n(X_G)$ generates a point in X_G whose period is between n and $n + N$, different blocks generating different periodic points. Hence

$$|\mathcal{B}_n(X_G)| \leqslant p_n(X_G) + p_{n+1}(X_G) + \cdots + p_{n+N}(X_G).$$

For each n, there is a largest number among $p_{n+k}(X_G)$ for $0 \leqslant k \leqslant N$, say at $k = k(n)$. Then

$$|\mathcal{B}_n(X_G)| \leqslant (N + 1)p_{n+k(n)}(X_G).$$

Hence

$$h(X) = h(X_G) = \lim_{n \to \infty} \frac{1}{n} \log |\mathcal{B}_n(X_G)|$$

$$\leqslant \limsup_{n \to \infty} \left[\frac{1}{n} \log(N + 1) + \frac{1}{n} \log p_{n+k(n)}(X_G) \right]$$

$$\leqslant \limsup_{m \to \infty} \frac{1}{m} \log p_m(X_G),$$

proving the result for irreducible shifts of finite type.

Next, suppose that X is an irreducible sofic shift. Let $\mathcal{G} = (G, \mathcal{L})$ be an irreducible right-resolving presentation of X. From Proposition 4.1.13 we have that $h(X) = h(X_\mathcal{G}) = h(X_G)$. Let G have r vertices. Since $\mathcal{G}$ is right-resolving, each point in $X_\mathcal{G}$ has at most r presentations in X_G. Since $\mathcal{L}_\infty$ maps points of period n in X_G to points of period n in $X_\mathcal{G}$, it follows that

$$p_n(X_\mathcal{G}) \geqslant \frac{1}{r} p_n(X_G).$$

Since we have verified the result already for shifts of finite type, we see that

$$\limsup_{n \to \infty} \frac{1}{n} \log p_n(X_\mathcal{G}) \geqslant \limsup_{n \to \infty} \left[\frac{1}{n} \log \frac{1}{r} + \frac{1}{n} \log p_n(X_G) \right]$$

$$= \limsup_{n \to \infty} \frac{1}{n} \log p_n(X_G) = h(X_G) = h(X_\mathcal{G}). \qquad \square$$

The same result holds for the growth rate of the number of periodic points of least period n.

Definition 4.3.7. For a shift space X, $q_n(X)$ denotes the number of points in X having least period n.

Corollary 4.3.8. *If X is an irreducible sofic shift with $h(X) > 0$, then*

$$\limsup_{n \to \infty} \frac{1}{n} \log q_n(X) = h(X).$$

PROOF: By Proposition 4.1.15,

$$(4\text{-}3\text{-}1) \qquad \limsup_{n\to\infty} \frac{1}{n}\log q_n(X) \leqslant \limsup_{n\to\infty} \frac{1}{n}\log p_n(X) \leqslant h(X).$$

By Theorem 4.3.6, there is a subsequence $n_i \to \infty$ such that

$$\lim_{i\to\infty} \frac{1}{n_i}\log p_{n_i}(X) = h(X).$$

Let $\epsilon > 0$. Then, for sufficiently large i,

$$(4\text{-}3\text{-}2) \qquad p_{n_i}(X) > 2^{n_i h(X)-n_i\epsilon}.$$

Also, by Theorem 4.3.6, for sufficiently large m and all $j \leqslant m$, we have

$$(4\text{-}3\text{-}3) \qquad p_j(X) < 2^{mh(X)+m\epsilon}.$$

But

$$q_n(X) \geqslant p_n(X) - \sum_{j=1}^{n/2} p_j(X).$$

So, by (4-3-2) and (4-3-3), for sufficiently large i, we have

$$\begin{aligned}
q_{n_i}(X) &\geqslant p_{n_i}(X) - \sum_{j=1}^{n_i/2} p_j(X) \\
&> 2^{n_i h(X)-n_i\epsilon} - (n_i/2)2^{(n_i/2)h(X)+(n_i/2)\epsilon} \\
&= \left[2^{n_i h(X)-n_i\epsilon}\right]\left[1 - (n_i/2)2^{-n_i(h(X)/2-3\epsilon/2)}\right].
\end{aligned}$$

For sufficiently small ϵ, the second factor in brackets tends to 1 as $i \to \infty$, and so

$$\limsup_{i\to\infty} \frac{1}{n_i}\log q_{n_i}(X) \geqslant h(X).$$

This together with (4-3-1) yields the result. □

EXERCISES

4.3.1. For an irreducible graph G, show that $h(X_G) = 0$ if and only if G consists of a single cycle.

4.3.2. (a) If $A = \begin{bmatrix} a & b \\ c & d \end{bmatrix}$ has nonnegative integer entries, show that

$$h(X_A) \geqslant \max\{\log a, \log d, \log \sqrt{bc}\}.$$

(b) If $A = [A_{I,J}]$ is an $m \times m$ matrix, with nonnegative integer entries, show that

$$h(X_A) \geqslant \log((A_{1,2}A_{2,3}A_{3,4}\ldots A_{m-1,m}A_{m,1})^{1/m}).$$

(c) If $A = [A_{I,J}]$ is an $m \times m$ irreducible matrix, with nonnegative integer entries, show that

$$h(X_A) \geqslant \frac{1}{m}\max_{I,J}\log(A_{IJ}).$$

4.3.3. Which of the following numbers is the entropy of an irreducible shift of finite type?

(a) $\log \sqrt{2}$, (b) $\log \frac{3}{2}$, (c) $\log(3 - \sqrt{2})$, (d) $\log \pi$.

[*Hint*: First show that if A is an irreducible integral matrix, then λ_A satisfies a polynomial equation with integer coefficients and whose leading coefficient is 1.]

4.3.4. Let $\mathcal{A} = \{a, b, c\}$ and

$$\mathcal{F} = \{aac, aba, acb, baa, bbb, bca, cac, cba, cca\}.$$

Compute $h(X_{\mathcal{F}})$. [*Hint*: Use Example 4.1.3.]

4.3.5. Find a right-resolving presentation of the shift X in Example 4.3.2 having three states. Use this to compute $h(X)$. Check that your answer agrees with the computed entropy.

4.3.6. (a) Using Examples 4.3.4 and 4.3.5 as a guide, find an irreducible right-resolving presentation of the general (d, k) run-length limited shift, and compute the characteristic polynomial of the adjacency matrix of its underlying graph.

(b) Show that if the entropy of the (d, k) run-length limited shift is $\log \lambda$, then λ satisfies a polynomial with integer coefficients having at most four terms.

*4.3.7. For each subset $S \subseteq \{0, 1, 2, \dots \}$, consider the S-gap shift $X(S)$ introduced in Example 1.2.6.

(a) Show that $h(X(S)) = \log \lambda$, where λ is the unique positive solution to

$$1 = \sum_{n \in S} x^{-n-1}.$$

(b) Verify this when $S = \{0, 2, 4, 6, \dots \}$, where $X(S)$ is the even shift.

(c) Compute the entropy of the prime gap shift to two decimal places.

(d) Show that for every number t between 0 and 1 there is a set S for which $h(X(S)) = t$. This gives another proof that there is an uncountable collection of shift spaces of which no two are conjugate (compare with Exercise 1.5.17).

§4.4. Irreducible Components

Let G be an arbitrary graph. We will see in this section that G contains certain irreducible subgraphs G_i, called its irreducible components, and that the study of X_G reduces for the most part to the study of the irreducible shifts X_{G_i}. For example, we will show that $h(X_G) = \max_i h(X_{G_i})$, and we know that $h(X_{G_i}) = \log \lambda_i$ where λ_i is the Perron eigenvalue of $A(G_i)$.

Matrices in this section have, as usual, nonnegative entries. Let A be such an $m \times m$ matrix. Our goal is to reorganize A by renumbering its states so that it assumes a block triangular form. The square matrices A_i along the diagonal will be the irreducible components of A, and the growth of A^n is controlled by the growth of the A_i^n.

If A has nonnegative integer entries, we have already associated a graph $G(A) = G_A$ to A as the graph with states $\{1, \dots, r\}$ and A_{IJ} distinct edges with initial state I and terminal state J. For an arbitrary nonnegative matrix, the matrix $\lceil A \rceil$, obtained by replacing each entry A_{IJ} of A by its ceiling $\lceil A_{IJ} \rceil$ (i.e., the smallest integer $\geq A_{IJ}$), is a nonnegative integer matrix, and we set $G(A) = G_A = G(\lceil A \rceil)$.

For states I and J, let $I \rightsquigarrow J$ mean that $(A^n)_{IJ} > 0$ for some $n \geq 0$. Thus $I \rightsquigarrow J$ if there is a path in the graph G associated to A from I to J. Note that $I \rightsquigarrow I$ for every I, since $(A^0)_{II} = 1$, or equivalently because the empty path ε_I goes from I to I. Say that I *communicates with* J if $I \rightsquigarrow J$ and $J \rightsquigarrow I$. Communication is an equivalence relation, hence the states of A are partitioned into *communicating classes*. A communicating class is thus a maximal set of states such that each communicates with all others in the class.

Example 4.4.1. Let

$$
A = \begin{array}{c c} & \begin{array}{c c c c c c c} 1 & 2 & 3 & 4 & 5 & 6 & 7 \end{array} \\ \begin{array}{c} 1 \\ 2 \\ 3 \\ 4 \\ 5 \\ 6 \\ 7 \end{array} & \left[\begin{array}{c c c c c c c} 0 & 0 & 2 & 0 & 0 & 1 & 1 \\ 0 & 1 & 0 & 1 & 1 & 0 & 0 \\ 0 & 1 & 0 & 0 & 0 & 1 & 0 \\ 0 & 0 & 0 & 1 & 0 & 0 & 0 \\ 0 & 1 & 0 & 1 & 0 & 0 & 0 \\ 1 & 1 & 0 & 0 & 0 & 0 & 1 \\ 0 & 0 & 0 & 0 & 0 & 0 & 0 \end{array} \right] \end{array},
$$

whose graph is shown in Figure 4.4.1. The communicating classes are $C_1 = \{4\}$, $C_2 = \{2,5\}$, $C_3 = \{7\}$, and $C_4 = \{1,3,6\}$, and these are grouped by the dotted lines in the figure. □

Form a graph $\mathcal{H}$ whose vertex set is the set of communicating classes of G, and for which there is an edge from class C to class D if and only if $C \neq D$ and there is an edge in G from some state in C to some state in D. For Example 4.4.1 the graph $\mathcal{H}$ is shown in Figure 4.4.2.

Note that $\mathcal{H}$ cannot have any cycles, for a cycle would create a larger communicating class. If every state in $\mathcal{H}$ were to have an outgoing edge, following these in succession would eventually lead to a cycle. Hence $\mathcal{H}$

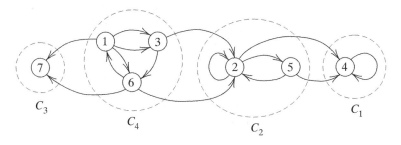

FIGURE 4.4.1. Communicating classes.

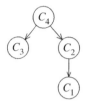

FIGURE 4.4.2.   Graph of communicating classes.

must have at least one state without any outgoing edges, i.e., a *sink*. In Figure 4.4.2, states C_1 and C_3 are sinks. A similar argument using incoming edges shows that $\mathcal{H}$ must also have *sources*, i.e., states with no incoming edges. State C_4 is a source in Figure 4.4.2.

By removing sinks one at a time, we can arrange the communicating classes in an order $C_1, C_2, \ldots, C_k$ so that there can be a path in $\mathcal{H}$ from C_j to C_i only if $j > i$. By using this order on the states, A assumes the block triangular form

$$(4\text{-}4\text{-}1) \qquad A = \begin{array}{c} \\ C_1 \\ C_2 \\ C_3 \\ \vdots \\ C_k \end{array}
\begin{array}{cccccc}
C_1 & C_2 & C_3 & \ldots & C_k \\
\left[\begin{array}{ccccc}
A_1 & 0 & 0 & \ldots & 0 \\
* & A_2 & 0 & \ldots & 0 \\
* & * & A_3 & \ldots & 0 \\
\vdots & \vdots & \vdots & \ddots & \vdots \\
* & * & * & \ldots & A_k
\end{array}\right]
\end{array},$$

where the $*$'s represent possibly nonzero submatrices. Let G_i denote the subgraph of G with vertex set C_i and whose edges are those of G whose initial and terminal states are in C_i. Then each G_i is irreducible (why?) and has adjacency matrix A_i. The G_i are called the *irreducible components* of G, and the A_i are the *irreducible components* of A. We will sometimes refer to the X_{G_i} as irreducible components of X_G. A shift of finite type X naturally inherits irreducible components by recoding to an edge shift (see Exercise 4.4.4).

In Example 4.4.1, using the order C_1, C_2, C_3, C_4 on the components leads to the reorganized matrix

$$A = \begin{array}{c} \\ 4 \\ 2 \\ 5 \\ 7 \\ 1 \\ 3 \\ 6 \end{array}
\begin{array}{c}
\begin{array}{ccccccc} 4 & 2 & 5 & 7 & 1 & 3 & 6 \end{array} \\
\left[\begin{array}{ccccccc}
\boxed{1} & 0 & 0 & 0 & 0 & 0 & 0 \\
1 & \boxed{1 & 1} & 0 & 0 & 0 & 0 \\
1 & 1 & 0 & 0 & 0 & 0 & 0 \\
0 & 0 & 0 & \boxed{0} & 0 & 0 & 0 \\
0 & 0 & 0 & 1 & \boxed{0 & 2 & 1} \\
0 & 1 & 0 & 0 & 0 & 0 & 1 \\
0 & 1 & 0 & 1 & 1 & 0 & 0
\end{array}\right]
\end{array}.$$

In this example we could have used other allowable ways of ordering the communicating classes, for example C_3, C_1, C_2, C_4, and still obtained a block triangular form. The irreducible components would then appear in a different order, but would otherwise be the same.

The characteristic polynomial $\chi_A(t)$ of a matrix A is unchanged by a simultaneous permutation of rows and columns, so we can compute $\chi_A(t)$ from the triangular form (4–4–1). When expanding the determinant, any product involving a term from the $*$ portion below the diagonal must also include a term from the 0 portion above the diagonal. Thus only the blocks on the diagonal matter when computing $\chi_A(t)$, so that

$$\chi_A(t) = \chi_{A_1}(t)\chi_{A_2}(t)\cdots\chi_{A_k}(t).$$

Thus the eigenvalues of A are exactly the eigenvalues of the A_i. In particular, the largest λ_{A_i} (in absolute value) is also an eigenvalue of A. This suggests the following definition.

Definition 4.4.2. Let A be a nonnegative matrix with irreducible components $A_1, \ldots, A_k$. The *Perron eigenvalue* λ_A of A is $\lambda_A = \max\limits_{1 \leqslant i \leqslant k} \lambda_{A_i}$.

Lemma 4.4.3. *For an arbitrary nonnegative matrix A, its Perron eigenvalue λ_A is the largest eigenvalue of A.*

PROOF: The set of eigenvalues of A is the union of the eigenvalues of the components A_i. Thus if μ is an eigenvalue of A, then $|\mu|$ is bounded above by some λ_{A_i}, hence by λ_A. $\qquad\square$

For a general matrix C with complex entries, the *spectral radius* of C is the absolute value of the largest eigenvalue of C. This lemma shows that the spectral radius of a nonnegative matrix A is its Perron eigenvalue λ_A. For this reason we will denote the spectral radius of a general matrix C by λ_C.

In our next result, we show that the Perron eigenvalue controls the growth rate of A^n; this extends Theorem 4.3.1(1) to general graphs.

Theorem 4.4.4. *Let G be a graph having adjacency matrix A. Then $h(X_G) = \log \lambda_A$.*

PROOF: Let A have irreducible components $A_1, A_2, \ldots, A_k$. Then $\lambda_A = \lambda_{A_q}$ for some q. Since $|\mathcal{B}_n(X_G)| \geqslant |\mathcal{B}_n(X_{G_q})|$,

$$h(X_G) \geqslant h(X_{G_q}) = \log \lambda_{A_q} = \log \lambda_A.$$

Note that if $\lambda_A = 0$, then each A_i must be the 1×1 matrix $[0]$. Then G has no bi-infinite walks, and $X_G = \varnothing$, verifying the result in this case. From now on we assume that $\lambda_A > 0$.

We will prove that $h(X_G) \leqslant \log \lambda_A$ by estimating the number of paths of length n in G. Such a path decomposes into subpaths in communicating classes separated by transitional edges between classes. Thus any $\pi \in \mathcal{B}_n(X_G)$ has the form

$$(4\text{–}4\text{–}2) \qquad \pi = \pi_1 e_1 \pi_2 e_2 \ldots \pi_{j-1} e_{j-1} \pi_j,$$

where π_i is a path in some $G_{q(i)}$ and e_i is an edge from a state in $G_{q(i)}$ to one in $G_{q(i+1)}$. Thus $q(1) > q(2) > \cdots > q(j)$, and so $j \leqslant k$. Let M be the total number of transitional edges (i.e., edges whose initial and terminal states lie in different communicating classes). Then there are at most M choices for each e_i in (4–4–2), and at most n places where each could occur. Hence the number of arrangements for the transitional edges is at most $(Mn)^k$. Having fixed such an arrangement of the e_i, the number of ways to choose π_i is at most $|\mathcal{B}_{n(i)}(X_{G_{q(i)}})|$, where $n(i) = |\pi_i|$. By Proposition 4.2.1 and the Perron–Frobenius Theorem, there is a $d > 0$ so that, for every G_q,

$$|\mathcal{B}_m(X_{G_q})| \leqslant d\,\lambda_{A_q}^m \leqslant d\,\lambda_A^m.$$

Hence the number of ways of filling in the π_i's is at most

$$|\mathcal{B}_{n(1)}(X_{G_{q(1)}})| \times \cdots \times |\mathcal{B}_{n(j)}(X_{G_{q(j)}})| \leqslant d^j\,\lambda_A^{n(1)+\cdots+n(j)} \leqslant d^k\,\lambda_A^n.$$

Putting this together, we see that

$$|\mathcal{B}_n(X_G)| \leqslant (Mn)^k\,d^k\,\lambda_A^n.$$

Hence

$$\begin{aligned}
h(X_G) &= \lim_{n \to \infty} \frac{1}{n} \log |\mathcal{B}_n(X_G)| \\
&\leqslant \lim_{n \to \infty} \left[\frac{1}{n} \log(M^k d^k) + k\frac{\log n}{n} + \log \lambda_A \right] \\
&= \log \lambda_A. \qquad\qquad\qquad\qquad\qquad\qquad\quad \square
\end{aligned}$$

Example 4.4.5. Let $A = \begin{bmatrix} 2 & 0 \\ 1 & 2 \end{bmatrix}$. Then A is already in triangular form and

$$A^n = \begin{bmatrix} 2^n & 0 \\ n\,2^{n-1} & 2^n \end{bmatrix},$$

so that $|\mathcal{B}_n(X_A)| = (\frac{1}{2}n + 2)2^n$. Observe that here there is no constant $d > 0$ for which $|\mathcal{B}_n(X_A)| \leqslant d\,2^n$, but that the polynomial factor $\frac{1}{2}n + 2$ has no effect on the growth rate $h(X_A) = \log \lambda_A = \log 2 = 1$. $\qquad \square$

The previous theorem shows how to also compute the entropy of a (possibly reducible) sofic shift X. For if $\mathcal{G} = (G, \mathcal{L})$ is a right-resolving presentation of $X = X_{\mathcal{G}}$, then, according to Proposition 4.1.13, $h(X_{\mathcal{G}}) = h(X_G) = \log \lambda_{A(G)}$. This also shows that every sofic shift contains an irreducible sofic shift of maximal entropy.

How could we find the entropy of an arbitrary shift space X? One approach is to use "outer approximations" of X by shifts of finite type as follows. Let $\mathcal{F} = \{w_1, w_2, \dots\}$ be an infinite collection of forbidden words for which $X = X_{\mathcal{F}}$, and put $\mathcal{F}_k = \{w_1, w_2, \dots, w_k\}$. Then $X_{\mathcal{F}_1} \supseteq X_{\mathcal{F}_2} \supseteq X_{\mathcal{F}_3} \supseteq \dots$ are shifts of finite type, and

$$\bigcap_{k=1}^{\infty} X_{\mathcal{F}_k} = X_{\bigcup_{k=1}^{\infty} \mathcal{F}_k} = X_{\mathcal{F}} = X.$$

We can compute each $h(X_{\mathcal{F}_k})$ by applying the previous theorem to the edge shift X_{G_k} where G_k is the underlying graph of a right-resolving presentation of $X_{\mathcal{F}_k}$. The next result shows that the $h(X_{\mathcal{F}_k})$ approach $h(X_{\mathcal{F}})$.

Proposition 4.4.6. *Let $X_1 \supseteq X_2 \supseteq X_3 \supseteq \dots$ be shift spaces whose intersection is X. Then $h(X_k) \to h(X)$ as $k \to \infty$.*

PROOF: Fix $\epsilon > 0$. We will show that $h(X) \leqslant h(X_k) < h(X) + \epsilon$ for all large enough k.

Since $X \subseteq X_k$, it follows that $h(X) \leqslant h(X_k)$ for all k. By definition, $(1/n) \log |\mathcal{B}_n(X)| \to h(X)$, so there is an $N \geqslant 1$ such that

$$\frac{1}{N} \log |\mathcal{B}_N(X)| < h(X) + \epsilon.$$

Since $\bigcap_{k=1}^{\infty} X_k = X$, there is a $K \geqslant 1$ such that $\mathcal{B}_N(X_k) = \mathcal{B}_N(X)$ for all $k \geqslant K$ (if not, an application of the Cantor diagonal argument, as in Theorem 1.5.13, would produce a point in $\bigcap_{k=1}^{\infty} X_k$ but not in X). By Proposition 4.1.8, for all $k \geqslant K$ we have that

$$h(X_k) = \inf_{n \geqslant 1} \frac{1}{n} \log |\mathcal{B}_n(X_k)| \leqslant \frac{1}{N} \log |\mathcal{B}_N(X_k)|$$

$$= \frac{1}{N} \log |\mathcal{B}_N(X)| < h(X) + \epsilon. \qquad \square$$

Unfortunately, computing $h(X_{\mathcal{F}})$ as the limit of the $h(X_{\mathcal{F}_k})$'s has two serious shortcomings. The first is practical: when $X_{\mathcal{F}_k}$ is presented by a labeled graph, the size of the adjacency matrix can grow rapidly, making it increasingly difficult to compute the Perron eigenvalue accurately. The second is theoretical: we have no idea of how large k must be to make sure that $h(X_{\mathcal{F}_k})$ approximates $h(X_{\mathcal{F}})$ to within a specified tolerance.

One way to deal with the second problem is to find *inside* sofic approximations $Y \subseteq X_{\mathcal{F}}$. For if $h(X_{\mathcal{F}_k}) - h(Y)$ is small, we will have nailed down the value of $h(X_{\mathcal{F}})$ since $h(Y) \leqslant h(X_{\mathcal{F}}) \leqslant h(X_{\mathcal{F}_k})$. But finding inside approximations whose entropy can be computed is not always easy. In Exercise 4.1.11(b) the reader was asked to find a shift space X with $h(X) > 0$ but having no periodic points. Such a shift does not contain any nonempty sofic shifts, much less a sofic shift whose entropy is close to $h(X)$.

Next, we consider how changes in an adjacency matrix affect entropy. If A and B are nonnegative integer matrices and $B \leqslant A$, then the graph $G(B)$ is a subgraph of $G(A)$, so that $\lambda_B \leqslant \lambda_A$. We strengthen this in the next result, which shows that if A is irreducible, and if B has at least one entry strictly less than the corresponding entry in A, then $\lambda_B < \lambda_A$.

Theorem 4.4.7. *Let A be irreducible, $0 \leqslant B \leqslant A$, and $B_{KL} < A_{KL}$ for a pair K, L of indices. Then $\lambda_B < \lambda_A$.*

PROOF: Since A is irreducible there is an N so that for every state I there is an outgoing path of length N which uses an edge from K to L. Thus, for every state I there is a state J such that $(B^N)_{IJ} < (A^N)_{IJ}$.

Let $\mathbf{v} = \mathbf{v}_A > 0$ be the Perron eigenvector for A, and λ_A be the Perron eigenvalue. Then $B^N \mathbf{v} < A^N \mathbf{v} = \lambda_A^N \mathbf{v}$, and thus for some $\epsilon > 0$ we have that

$$B^N \mathbf{v} \leqslant (1 - \epsilon)\lambda_A^N \mathbf{v}.$$

It follows that, for $k \geqslant 1$,

$$(B^N)^k \mathbf{v} \leqslant \left[(1 - \epsilon)\lambda_A^N\right]^k \mathbf{v}.$$

An argument analogous to that of Proposition 4.2.1 shows that there is a constant $d > 0$ such that

$$(4\text{-}4\text{-}3) \qquad \sum_{I,J}(B^{Nk})_{IJ} \leqslant d\left[(1 - \epsilon)\lambda_A^N\right]^k = d\left[(1 - \epsilon)^{1/N}\lambda_A\right]^{Nk}.$$

Now let C be an irreducible component of B with $\lambda_C = \lambda_B$. Then C has a positive Perron eigenvector, and so by Proposition 4.2.1 there is a constant $c > 0$ such that

$$(4\text{-}4\text{-}4) \qquad c\lambda_B^{Nk} \leqslant \sum_{I,J}(C^{Nk})_{IJ} \leqslant \sum_{I,J}(B^{Nk})_{IJ}.$$

Comparing (4-4-3) and (4-4-4), and taking (Nk)th roots and letting $k \to \infty$, shows that

$$\lambda_B \leqslant (1 - \epsilon)^{1/N}\lambda_A < \lambda_A. \qquad \square$$

Example 4.4.8. Let $A = \begin{bmatrix} 2 & 0 \\ 0 & 2 \end{bmatrix}$ and $B = \begin{bmatrix} 1 & 0 \\ 0 & 2 \end{bmatrix}$. Then $B \leqslant A$ and $B_{11} < A_{11}$, yet $\lambda_B = \lambda_A = 2$. This shows that some irreducibility assumption is necessary for the previous theorem. $\qquad\square$

Corollary 4.4.9. *If X is an irreducible sofic shift and Y is a proper subshift of X, then $h(Y) < h(X)$.*

PROOF: First suppose that X has finite type. We may assume that X is an edge shift. Since $\mathcal{B}(Y) \neq \mathcal{B}(X)$, there is a block $w = a_1 \ldots a_N$ appearing in X but not in Y. Let $\beta_N \colon X \to X^{[N]}$ be the conjugacy from X to its Nth higher block shift (from §1.4). Recall from Proposition 2.3.6 that $X^{[N]}$ is also an edge shift, and denote its adjacency matrix by A. Form B from A by replacing the 1 for the transition from $a_1 \ldots a_{N-1}$ to $a_2 \ldots a_N$ by 0. Since A is irreducible, the previous theorem shows that $\lambda_B < \lambda_A$. Hence

$$h(Y) = h(\beta_N(Y)) \leqslant h(X_B) = \log \lambda_B < \log \lambda_A = h(X_A) = h(X).$$

Now suppose that X is sofic, and let $\mathcal{G} = (G, \mathcal{L})$ be an irreducible right-resolving presentation of X. Then $h(X) = h(X_{\mathcal{G}}) = h(X_G)$ by Proposition 4.1.13. Since $\mathcal{L}_\infty \colon X_G \to X_{\mathcal{G}}$ is a factor code, $\widetilde{Y} = \mathcal{L}_\infty^{-1}(Y)$ is a proper subshift of X_G, from which $h(\widetilde{Y}) < h(X_G)$ by the previous paragraph. But entropy cannot increase under factor codes, so that

$$h(Y) \leqslant h(\widetilde{Y}) < h(X_G) = h(X_{\mathcal{G}}) = h(X). \qquad\square$$

EXERCISES

4.4.1. Explain why the irreducible components G_i of a graph are actually irreducible.

4.4.2. For each of the following matrices A, find the communicating classes, draw the communicating class graph, and order the states to put A into block triangular form.

(a) $\begin{bmatrix} 0 & 0 & 0 & 0 & 1 & 1 \\ 0 & 0 & 2 & 0 & 0 & 1 \\ 0 & 0 & 0 & 0 & 0 & 3 \\ 0 & 0 & 0 & 1 & 0 & 0 \\ 1 & 0 & 0 & 2 & 1 & 0 \\ 0 & 0 & 1 & 0 & 0 & 0 \end{bmatrix}$, (b) $\begin{bmatrix} 0 & 0 & 0 & 1 & 0 & 0 \\ 0 & 0 & 1 & 1 & 0 & 0 \\ 0 & 1 & 1 & 0 & 0 & 1 \\ 1 & 0 & 0 & 0 & 0 & 0 \\ 0 & 0 & 0 & 0 & 0 & 0 \\ 1 & 0 & 1 & 0 & 1 & 0 \end{bmatrix}$.

4.4.3. Let X be a (possibly reducible) sofic shift. Prove that

$$\limsup_{n \to \infty} \frac{1}{n} \log p_n(X) = h(X),$$

and so extend Theorem 4.3.6 to the reducible case.

4.4.4. For a shift of finite type X define the irreducible components of X to be the pre-images of the irreducible components of an edge shift X_G via a conjugacy

ϕ from X to X_G. Show that the irreducible components of X are well-defined independent of the choice of edge shift X_G and conjugacy ϕ.

4.4.5. Show that if A is a matrix whose characteristic polynomial cannot be factored into two nonconstant polynomials with integer coefficients, then A is irreducible.

4.4.6. Let A be an irreducible matrix. Use Theorem 4.4.7 to show that each diagonal entry of $\mathrm{adj}(\lambda_A Id - A)$ is strictly positive. Compare this with Exercise 4.2.12.

4.4.7. Let A be a nonnegative matrix with spectral radius λ_A. Prove that A has a nonnegative eigenvector whose eigenvalue is λ_A.

4.4.8. Let A be a nonnegative matrix. Use the communicating class graph of A to determine precisely when A has a (strictly) positive eigenvector.

§4.5. Cyclic Structure

In this section, we consider the cyclic structure of irreducible matrices. We will show that each irreducible matrix has a characteristic period p, and its states can be grouped into classes that move cyclically with period p. Matrices with period one are called aperiodic. We will show how, in some ways, the study of irreducible matrices can be reduced to that of aperiodic irreducible matrices.

Example 4.5.1. Let

$$(4\text{-}5\text{-}1) \qquad A = \begin{bmatrix} 0 & 0 & 1 & 1 & 0 \\ 0 & 0 & 1 & 1 & 0 \\ 0 & 0 & 0 & 0 & 1 \\ 0 & 0 & 0 & 0 & 1 \\ 1 & 1 & 0 & 0 & 0 \end{bmatrix}.$$

Observe that states 1 and 2 can only go to state 3 or 4, each of which can only go to state 5, which can only go back to states 1 and 2. Hence we can group states into the sets $D_0 = \{1,2\}$, $D_1 = \{3,4\}$, and $D_2 = \{5\}$ so that only transitions of the form $D_0 \to D_1 \to D_2 \to D_0$ occur; i.e., the sets D_i are cyclically permuted with period 3. Furthermore,

$$A^3 = \begin{bmatrix} \boxed{\begin{matrix} 2 & 2 \\ 2 & 2 \end{matrix}} & 0 & 0 & 0 \\ & \boxed{\begin{matrix} 2 & 2 \\ 2 & 2 \end{matrix}} & 0 \\ 0 & 0 & & \\ 0 & 0 & 0 & 0 & \boxed{4} \end{bmatrix}$$

has a block diagonal form where each component has no further cyclic decomposition. This is an example of the cyclic structure we are after. $\square$

Definition 4.5.2. Let A be a nonnegative matrix. The *period of a state I*, denoted by $\mathrm{per}(I)$, is the greatest common divisor of those integers $n \geqslant 1$ for which $(A^n)_{II} > 0$. If no such integers exist, we define $\mathrm{per}(I) = \infty$.

The *period* per(A) *of the matrix* A is the greatest common divisor of the numbers per(I) that are finite, or is ∞ if per(I) = ∞ for all I. A matrix is *aperiodic* if it has period 1. The *period* per(G) *of a graph* G is the period of its adjacency matrix.

Observe that per(G) is the greatest common divisor of its cycle lengths. Note that each state in (4–5–1) has period 3. The fact that they all have the same period is not accidental.

Lemma 4.5.3. *If* A *is irreducible, then all states have the same period, and so* per(A) *is the period of any of its states.*

PROOF: Let I be a state, and put p = per(I). If $p = \infty$, then $A = [0]$ since A is irreducible, and we are done. So we may assume that $p < \infty$. Let J be another state. Then there are $r, s \geqslant 1$ for which $(A^r)_{IJ} > 0$ and $(A^s)_{JI} > 0$. Let n be such that $(A^n)_{JJ} > 0$. Then

$$(A^{r+s})_{II} \geqslant (A^r)_{IJ}(A^s)_{JI} > 0,$$

and

$$(A^{r+n+s})_{II} \geqslant (A^r)_{IJ}(A^n)_{JJ}(A^s)_{JI} > 0.$$

Then p divides both $r + s$ and $r + n + s$, hence their difference n. This shows that p divides all n for which $(A^n)_{JJ} > 0$, so that p = per(I) divides their greatest common divisor per(J). Reversing the roles of I and J shows that per(J) divides per(I), proving that per(I) = per(J). □

There is a related notion of period for shift spaces based on periodic points. Recall that $p_n(X)$ denotes the number of points in X of period n.

Definition 4.5.4. Let X be a shift space. The *period* per(X) of X is the greatest common divisor of integers $n \geqslant 1$ for which $p_n(X) > 0$, or is ∞ if no such integers exist.

If $X \cong Y$, then $p_n(X) = p_n(Y)$ for all n, and so per(X) = per(Y), showing that period is a conjugacy invariant. Since a shift of finite type is conjugate to an edge shift, we can find its period by computing the period of an edge shift.

Proposition 4.5.5. *If* G *is a graph, then* per(X_G) = per(G).

PROOF: The periods of the periodic points of X_G are the lengths of the cycles of G. □

Let $A \neq [0]$ be an irreducible matrix, so that p = per(A) $< \infty$. Our next goal is to show that the states of A break up into p cyclically moving classes, as in Example 4.5.1. Say that states I and J are *period equivalent*, denoted by $I \sim J$, if there is a path in $G(A)$ from I to J whose length is divisible by p. To show that $\sim$ is an equivalence relation, first note that ε_I is a path of length 0 from I to I, and since p divides 0 we see that $I \sim I$.

If $I \sim J$, there is a path π from I to J for which p divides $|\pi|$. Since A is irreducible, there is a path τ from J to I. Then $\pi\tau$ is a cycle at I, so that $p = \mathrm{per}(A) = \mathrm{per}(I)$ divides $|\pi\tau| = |\pi| + |\tau|$. This shows that p divides $|\tau|$, so that $J \sim I$. Finally, if $I \sim J$ and $J \sim K$, then concatenating paths shows that $I \sim K$. Thus $\sim$ is an equivalence relation, and therefore it partitions the states of A into *period classes*. In Example 4.5.1 the period classes are $D_0 = \{1,2\}$, $D_1 = \{3,4\}$, and $D_2 = \{5\}$.

Proposition 4.5.6. *Let $A \neq [0]$ be irreducible with period p. Then there are exactly p period classes, which can be ordered as $D_0, D_1, \ldots, D_{p-1}$ so that every edge that starts in D_i terminates in D_{i+1} (or in D_0 if $i = p-1$).*

PROOF: Start with any period class D_0. Inductively define D_{i+1} to be the set of terminal states of edges starting in D_i. A routine application of the definitions shows that each D_i is a period class, that $D_0, D_1, \ldots, D_{p-1}$ partition the states of A, and that every edge starting in D_i ends in D_{i+1}, where $D_p = D_0$. The details of these arguments are left as Exercise 4.5.4. □

By using the period classes of an irreducible matrix A with period p to index states, the matrix can be put into the form

$$
A =
\begin{array}{c}
 \\ D_0 \\ D_1 \\ \vdots \\ D_{p-2} \\ D_{p-1}
\end{array}
\begin{array}{c}
\begin{array}{ccccc} D_0 & D_1 & D_2 & \ldots & D_{p-1} \end{array} \\
\left[
\begin{array}{ccccc}
0 & B_0 & 0 & \ldots & 0 \\
0 & 0 & B_1 & \ldots & 0 \\
\vdots & \vdots & \vdots & \ddots & \vdots \\
0 & 0 & 0 & \ldots & B_{p-2} \\
B_{p-1} & 0 & 0 & \ldots & 0
\end{array}
\right].
\end{array}
$$

Then A^p has the block diagonal form

$$
(4\text{-}5\text{-}2) \qquad A^p =
\begin{array}{c}
 \\ D_0 \\ D_1 \\ D_2 \\ \vdots \\ D_{p-1}
\end{array}
\begin{array}{c}
\begin{array}{ccccc} D_0 & D_1 & D_2 & \ldots & D_{p-1} \end{array} \\
\left[
\begin{array}{ccccc}
A_0 & 0 & 0 & \ldots & 0 \\
0 & A_1 & 0 & \ldots & 0 \\
0 & 0 & A_2 & \ldots & 0 \\
\vdots & \vdots & \vdots & \ddots & \vdots \\
0 & 0 & 0 & \ldots & A_{p-1}
\end{array}
\right],
\end{array}
$$

where $A_i = B_i \ldots B_{p-1} B_0 \ldots B_{i-1}$.

If $I \sim J$, then there is a path in $G(A)$ from I to J whose length is divisible by p, so that each of the diagonal blocks A_i in (4–5–2) is irreducible. Furthermore, each A_i is aperiodic, since otherwise the period of A would be greater than p (why?). Irreducible matrices that are aperiodic are the basic "building blocks" of nonnegative matrices.

Definition 4.5.7. A matrix is *primitive* if it is irreducible and aperiodic. A graph is *primitive* if its adjacency matrix is primitive.

The diagonal blocks in (4–5–2) are hence primitive, as is any strictly positive matrix.

Theorem 4.5.8. *Let A be a nonnegative matrix. The following are equivalent.*

(1) A *is primitive.*
(2) $A^N > 0$ *for some $N \geqslant 1$.*
(3) $A^N > 0$ *for all sufficiently large N.*

PROOF: $(2) \Rightarrow (1)$: Suppose that $A^N > 0$ for some $N \geqslant 1$. Then there is a path of length N in $G(A)$ from any state to any other state, so that A is irreducible. In particular, A has no zero rows, hence $A^{N+1} = A\,A^N > 0$. Thus per(A) divides both N and $N+1$, so that A has period 1 and is hence primitive.

$(1) \Rightarrow (3)$: Suppose that A is primitive. We first show that for each state I there is an $N_I \geqslant 1$ so that $(A^n)_{II} > 0$ for all $n \geqslant N_I$.

Let $R_I = \{n \geqslant 1 : (A^n)_{II} > 0\}$. By definition, the greatest common divisor of the numbers in R_I is per(I) = per(A) = 1. Hence there are numbers $m_i, n_j \in R_I$ and positive integers a_i, b_j such that

$$1 = \sum_{i=1}^{k} a_i m_i - \sum_{j=1}^{l} b_j n_j$$

(see Exercise 4.5.5). Let K denote the first sum and L the second, so that $1 = K - L$. Put $N_I = L^2$. If $n \geqslant N_I$, then $n = cL + d$, where $0 \leqslant d < L$ and $c \geqslant L$. Hence

$$n = cL + d = cL + d(K - L) = (c - d)L + dK,$$

where $c - d \geqslant L - d > 0$ and $d \geqslant 0$. Thus

$$n = \sum_{j=1}^{l} [(c - d)b_j]n_j + \sum_{i=1}^{k} [da_i]m_i$$

is the sum of numbers in R_I, hence in R_I. This shows that R_I contains all $n \geqslant N_I$.

To complete the proof, use irreducibility of A to choose M so that for every pair I, J of states there is a path from I to J of length $\leqslant M$. Define $N = M + \max_I N_I$. Then for any I, J, there is a path of length $\ell \leqslant M$ from I to J, and since $N - \ell \geqslant N_J$ there is a path of length $N - \ell$ from J to itself, hence one of length N from I to J. This proves that $A^N > 0$. But since A is primitive, no row of A can be zero and so $A^n > 0$ for all $n \geqslant N$.

$(3) \Rightarrow (2)$: Obvious. $\qquad \square$

The irreducible graphs correspond to the irreducible shifts of finite type in the following senses: (1) for an essential graph G, the edge shift X_G is irreducible if and only if G is irreducible (Proposition 2.2.14), and (2) any irreducible shift of finite type can be recoded to an irreducible edge shift (Theorem 2.3.2 and Proposition 2.2.14). This raises the question, which class of shifts of finite type correspond to the primitive graphs? The answer is as follows.

Definition 4.5.9. A shift space X is *mixing* if, for every ordered pair $u, v \in \mathcal{B}(X)$, there is an N such that for each $n \geqslant N$ there is word $w \in \mathcal{B}_n(X)$ such that $uwv \in \mathcal{B}(X)$.

Proposition 4.5.10.

(1) *The mixing property is invariant under conjugacy (in fact, a factor of a mixing shift is also mixing).*

(2) *If G is an essential graph, then the edge shift X_G is mixing if and only if G is primitive.*

(3) *A shift of finite type is mixing if and only if it is conjugate to an edge shift X_G where G is primitive.*

(4) *A shift of finite type is mixing if and only if it is irreducible and $\mathrm{per}(X) = 1$, i.e., the greatest common divisor of the periods of its periodic points is 1.*

We leave the proof of this result as an exercise (Exercise 4.5.7) for the reader.

Recall that if A is irreducible, the Perron–Frobenius Theorem says that $|\mu| \leqslant \lambda_A$ for all eigenvalues μ of A. When A is primitive, we can say more.

Theorem 4.5.11. *If A is a primitive matrix and $\mu \neq \lambda_A$ is an eigenvalue of A, then $|\mu| < \lambda_A$.*

PROOF: Put $\lambda = \lambda_A$, and let μ be an eigenvalue of A for which $|\mu| = \lambda$. We will show that $\mu = \lambda$.

Let $\mathbf{w}$ be an eigenvector for A corresponding to μ. Note that $\mathbf{w}$ could have complex numbers as entries. Let $|\mathbf{w}|$ denote the vector obtained from $\mathbf{w}$ by taking the absolute value of each entry. Since $A\mathbf{w} = \mu\mathbf{w}$ and $A \geqslant 0$, applying the triangle inequality in each component shows that

$$(4\text{--}5\text{--}3) \qquad \lambda|\mathbf{w}| = |\mu\mathbf{w}| = |A\mathbf{w}| \leqslant A|\mathbf{w}|.$$

We next show that this inequality is in fact an equality, i.e., that $A|\mathbf{w}| = \lambda|\mathbf{w}|$. For if not, then $A|\mathbf{w}| - \lambda|\mathbf{w}|$ is nonnegative and has at least one positive entry. Since A is primitive, $A^N > 0$ for some $N \geqslant 1$. Hence $A^N(A|\mathbf{w}| - \lambda|\mathbf{w}|) > 0$, so that if we put $\mathbf{u} = A^N|\mathbf{w}|$, then $\mathbf{u} > 0$ and $A\mathbf{u} > \lambda\mathbf{u}$. We may therefore find $\epsilon > 0$ so that $A\mathbf{u} \geqslant (\lambda + \epsilon)\mathbf{u}$, and hence

$A^n \mathbf{u} \geq (\lambda + \epsilon)^n \mathbf{u}$ for all $n \geq 1$. Since $\mathbf{u} > 0$, it then follows as in the proof of Proposition 4.2.1 that there is a $c > 0$ such that

$$c (\lambda + \epsilon)^n \leq \sum_{I,J=1}^{r} (A^n)_{IJ}.$$

But this implies $h(X_A) \geq \log(\lambda + \epsilon)$, a contradiction.

Hence there is equality in (4–5–3), so that $|\mathbf{w}| = t \mathbf{v}_A$ for some $t > 0$, and $|A^N \mathbf{w}| = A^N |\mathbf{w}|$. Thus for every I,

(4–5–4)
$$\left| \sum_{J=1}^{r} (A^N)_{IJ} w_J \right| = \sum_{J=1}^{r} (A^N)_{IJ} |w_J|.$$

Recall that the triangle inequality holds with equality in only very special circumstances: namely, if $z_1, \ldots, z_r$ are complex numbers, then

$$|z_1 + \cdots + z_r| = |z_1| + \cdots + |z_r|$$

if and only if there is a real number θ such that $z_k = e^{i\theta} |z_k|$ for all k (see Exercise 4.5.9). Since $A^N > 0$, we conclude from (4–5–4) that $\mathbf{w} = e^{i\theta} |\mathbf{w}| = t e^{i\theta} \mathbf{v}_A$, showing that the eigenvalue μ for $\mathbf{w}$ is λ_A. □

If A is irreducible but has period > 1, and if $\lambda_A > 1$, then $(A^n)_{IJ}$ oscillates between 0 and exponentially large numbers as $n \to \infty$ (see Example 4.1.16). We will use the previous theorem to show that for primitive A, the entries in A^n grow steadily, and at a predictable rate.

Before we state this result, we introduce some new terminology and a new convention. We will need to use the Perron eigenvector, $\mathbf{w}_A$, for the transpose matrix, A^T, as well as the Perron eigenvector $\mathbf{v}_A$, for A. To avoid using the transpose notation, we consider $\mathbf{w}_A$ to be a row vector and $\mathbf{v}_A$ to be a column vector. We call $\mathbf{w}_A$ the *left Perron eigenvector* to distinguish it from the ordinary Perron eigenvector, $\mathbf{v}_A$, which we call the *right Perron eigenvector*. We will view the matrix A as acting on the right, and so for a subspace U of row vectors, we write UA to denote the image of U via the action of A.

Theorem 4.5.12. *Let A be a primitive matrix with Perron eigenvalue λ. Let $\mathbf{v}, \mathbf{w}$ be right, left Perron eigenvectors for A, i.e., vectors $\mathbf{v}, \mathbf{w} > 0$ such that $A\mathbf{v} = \lambda \mathbf{v}$, $\mathbf{w}A = \lambda \mathbf{w}$, and normalized so that $\mathbf{w} \cdot \mathbf{v} = 1$. Then for each I and J,*

$$(A^n)_{IJ} = [(v_I w_J) + \rho_{IJ}(n)] \lambda^n,$$

where $\rho_{IJ}(n) \to 0$ as $n \to \infty$.

PROOF: Let W be the 1-dimensional subspace of row vectors generated by $\mathbf{w}$, and let $U = \{\mathbf{u} : \mathbf{u} \cdot \mathbf{v} = 0\}$ be the subspace of row vectors orthogonal to the column vector $\mathbf{v}$. Write $WA = \{\mathbf{x}A : \mathbf{x} \in W\}$ and $UA = \{\mathbf{x}A : \mathbf{x} \in U\}$. Then $WA \subseteq W$ since $\mathbf{w}$ is an eigenvector, and $UA \subseteq U$ since if $\mathbf{u} \in U$, then

$$(\mathbf{u}A) \cdot \mathbf{v} = \mathbf{u} \cdot (\lambda \mathbf{v}) = \lambda(\mathbf{u} \cdot \mathbf{v}) = 0.$$

Note that $\lambda^{-1}A$ has Perron eigenvalue 1 with left Perron eigenvector $\mathbf{w}$, while the previous theorem shows that the eigenvalues of $\lambda^{-1}A$ on U all have absolute value < 1. Now, the powers of any such matrix converge to the zero matrix (see Exercise 4.5.10). Hence, as $n \to \infty$, the restriction of $(\lambda^{-1}A)^n$ to U (viewed as a matrix with respect to any basis) converges to the zero matrix. Let m denote the size of A. Now, there is a basis $\{\mathbf{x}^{(1)}, \ldots, \mathbf{x}^{(m)}\}$ for $\mathbb{R}^m$, viewed as row vectors, such that $\mathbf{x}^{(1)} = \mathbf{w}$ and $\{\mathbf{x}^{(2)}, \ldots, \mathbf{x}^{(m)}\}$ is a basis for U. Thus, for any $\mathbf{z} \in \mathbb{R}^m$, we can express $\mathbf{z}$ in terms of this basis as

$$\mathbf{z} = \sum_{i=1}^{m} a_i \mathbf{x}^{(i)},$$

and $\mathbf{z}(\lambda^{-1}A)^n$ converges to $a_1 \mathbf{x}^{(1)} = a_1 \mathbf{w}$. By definition of U and the assumption that $\mathbf{w} \cdot \mathbf{v} = 1$, it follows that $a_1 = \mathbf{z} \cdot \mathbf{v}$. So, $\mathbf{z}(\lambda^{-1}A)^n$ converges to $(\mathbf{z} \cdot \mathbf{v})\mathbf{w}$ (for readers familiar with the notion of "projection", we have just shown that $(\lambda^{-1}A)^n$ converges to the projection to W along U).

Let $\mathbf{e}_I$ denote the Ith standard basis vector (viewed as either a row vector or a column vector). Then

$$(\lambda^{-n}A^n)_{IJ} = \mathbf{e}_I\left[(\lambda^{-1}A)^n\right] \cdot \mathbf{e}_J \to (\mathbf{e}_I \cdot \mathbf{v})\mathbf{w} \cdot \mathbf{e}_J = v_I w_J.$$

Hence $\rho_{IJ}(n) = (\lambda^{-n}A^n)_{IJ} - v_I w_J \to 0$ as $n \to \infty$. □

Exercise 4.5.14 gives a version of Theorem 4.5.12 for irreducible matrices.

The following result is a strengthening of Theorem 4.3.6 and Corollary 4.3.8 in the mixing case.

Corollary 4.5.13.

(1) *For a primitive integral matrix $A \neq [1]$, the following limits all exist and equal $\log \lambda_A$:*

$$(a) \quad \lim_{n\to\infty} \frac{1}{n} \log p_n(X_A),$$

$$(b) \quad \lim_{n\to\infty} \frac{1}{n} \log q_n(X_A),$$

$$(c) \quad \lim_{n\to\infty} \frac{1}{n} \log A_{IJ}^n \text{ for every } I, J.$$

(2) *For a mixing sofic shift X with $h(X) > 0$*

$$\lim_{n\to\infty} \frac{\log p_n(X)}{n} = \lim_{n\to\infty} \frac{\log q_n(X)}{n} = h(X).$$

PROOF: Parts (1a) and (1c) of this result are immediate consequences of Theorem 4.5.12. Part (1b) follows from Part (1a) and the argument in Corollary 4.3.8. For part (2), if X is a mixing shift of finite type, use Proposition 4.5.10(3) to recode X to an edge shift X_G with G primitive. For the mixing sofic case, pass to the minimal right-resolving presentation (see Exercise 4.5.16). □

Recall from calculus that the angle $\theta_{\mathbf{x},\mathbf{y}}$ between two vectors $\mathbf{x}, \mathbf{y} \in \mathbb{R}^m \setminus \{\mathbf{0}\}$ is determined by the equation

$$(4\text{-}5\text{-}5) \qquad \cos(\theta_{\mathbf{x},\mathbf{y}}) = \frac{\mathbf{x} \cdot \mathbf{y}}{\|\mathbf{x}\|_2 \|\mathbf{y}\|_2},$$

where $\|\cdot\|_2$ denotes the *Euclidean norm*: $\|\mathbf{x}\|_2 = \sqrt{x_1^2 + \cdots + x_m^2} = \sqrt{\mathbf{x} \cdot \mathbf{x}}$. The following result is a geometric interpretation of Theorem 4.5.12.

Corollary 4.5.14. *Let A be a primitive $m \times m$ matrix. Let $(\mathbb{R}^m)^+ = \{\mathbf{z} \in \mathbb{R}^m : \mathbf{z} \geqslant 0\}$, the positive orthant in $\mathbb{R}^m$, viewed as row vectors. Let $\mathbf{w}$ be a left Perron eigenvector for A. Then for any vector $\mathbf{z} \in (\mathbb{R}^m)^+ \setminus \{\mathbf{0}\}$, $\theta_{\mathbf{z}A^n,\mathbf{w}}$ converges to zero. Moreover, the convergence is uniform in the sense that given any $\epsilon > 0$, for sufficiently large n, $\theta_{\mathbf{z}A^n,\mathbf{w}} < \epsilon$ for all $\mathbf{z} \in (\mathbb{R}^m)^+ \setminus \{\mathbf{0}\}$.*

In other words, any primitive matrix eventually "squashes" the positive orthant into the single ray, generated by $\mathbf{w}$. A geometrical picture of this idea is shown in Figure 4.5.1.

PROOF: By (4-5-5), it suffices to show that $(\mathbf{z}A^n) \cdot \mathbf{w}/(\|\mathbf{z}A^n\|_2\|\mathbf{w}\|_2)$ converges uniformly to 1. Letting $B = A/\lambda$, this is equivalent to the statement that

$$\frac{(\mathbf{z}B^n) \cdot \mathbf{w}}{\|\mathbf{z}B^n\|_2\|\mathbf{w}\|_2} \quad \text{converges uniformly to 1.}$$

By Theorem 4.5.12, $\mathbf{z}B^n$ converges to $(\mathbf{z} \cdot \mathbf{v})\mathbf{w}$, where $\mathbf{v}$ is a right Perron eigenvector such that $\mathbf{w} \cdot \mathbf{v} = 1$. It then follows that

$$\frac{(\mathbf{z}B^n) \cdot \mathbf{w}}{\|\mathbf{z}B^n\|_2\|\mathbf{w}\|_2} \;\to\; \frac{(\mathbf{z} \cdot \mathbf{v})\mathbf{w} \cdot \mathbf{w}}{\|(\mathbf{z} \cdot \mathbf{v})\mathbf{w}\|_2\|\mathbf{w}\|_2} = 1.$$

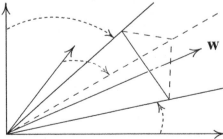

FIGURE 4.5.1.   The positive orthant being "squashed" by a primitive matrix.

The uniformity of convergence is Exercise 4.5.16(b), left to the reader. □

Let us put the material in the last two sections into perspective. In order to compute $h(X_A)$ we need to know the growth rate of the entries of A^n. From this standpoint, primitive matrices are "best," since by the last theorem in this section the entries in A^n all grow regularly and at the same rate. Next "best" are irreducible matrices B, since all states cycle with the same period p, and B^p is block diagonal with primitive blocks. This reduces the study of irreducible edge shifts to those with a primitive adjacency matrix (see Exercise 4.5.6). Finally, a general adjacency matrix C can be reorganized into block triangular form, with irreducible components $C_1, C_2, \ldots, C_k$ on the diagonal. Then bi-infinite walks in X_C can never go from a C_i to a C_j with $j > i$. Hence each walk breaks up into at most k subwalks, each on an irreducible component, with transitional edges in between. This reduces the study of X_C to that of the components X_{C_i}, which we have already understood. For example, this approach underlies our proof that $h(X_C) = \log \lambda_C = \max_i h(X_{C_i})$.

EXERCISES

4.5.1. Compute the periods of the irreducible components of the matrices in Exercise 4.4.2.

4.5.2. For each of the following irreducible matrices A, find its period p, the period classes $D_0, D_1, \ldots, D_{p-1}$, and verify that A^p is block diagonal with primitive blocks.

$$
\text{(a)} \begin{bmatrix} 0 & 0 & 1 & 0 & 0 \\ 2 & 0 & 0 & 0 & 3 \\ 0 & 1 & 0 & 2 & 0 \\ 0 & 0 & 0 & 0 & 1 \\ 0 & 0 & 2 & 0 & 0 \end{bmatrix}, \quad \text{(b)} \begin{bmatrix} 0 & 0 & 0 & 0 & 0 & 1 \\ 0 & 0 & 0 & 0 & 1 & 0 \\ 1 & 0 & 0 & 1 & 0 & 0 \\ 0 & 1 & 0 & 0 & 0 & 0 \\ 0 & 1 & 1 & 0 & 0 & 0 \\ 1 & 0 & 0 & 0 & 1 & 0 \end{bmatrix}.
$$

4.5.3. Let A be an irreducible matrix with period p. Show that A^n is irreducible if and only if n and p are relatively prime.

4.5.4. Complete the proof of Proposition 4.5.6.

4.5.5. Show that if S is a set of positive integers with greatest common divisor d, then for some $k \geqslant 1$ there are $s_1, \ldots, s_k \in S$ and integers $a_1, \ldots, a_k$ such that $d = a_1 s_1 + \cdots + a_k s_k$. [*Hint:* Consider the smallest positive number in the subgroup of the integers generated by S.]

4.5.6. Let A be irreducible with period p, and let $A_1, A_2, \ldots, A_p$ be the primitive blocks in A^p. Prove that the edge shifts X_{A_i} are all conjugate to each other.

4.5.7. Prove Proposition 4.5.10. [*Hint:* For part (2), use Theorem 4.5.8.]

4.5.8. Let X and Y be irreducible shifts. Prove that if either X embeds into Y or X factors onto Y, then per(Y) divides per(X).

4.5.9. In the following steps, show that if $z_1, \ldots, z_r$ are complex numbers, then

$$|z_1 + \cdots + z_r| = |z_1| + \cdots + |z_r|$$

if and only if there is a real number θ such that $z_k = e^{i\theta}|z_k|$ for all k.

   (a) Prove this for $r = 2$ by showing that $|z_1 + z_2|^2 = (|z_1| + |z_2|)^2$ if and only if $z_1 \overline{z_2}$ is nonnegative.

   (b) Use induction to prove this for general r.

4.5.10. (a) Show that if A is a matrix all of whose eigenvalues have absolute value < 1, then A^n converges to the zero matrix. [*Hint*: Use the Jordan form (see §7.4 for a review of the Jordan form) together with an argument similar to that of Theorem 4.4.4.]

   (b) Show that if A is a primitive matrix and $\lambda > \lambda_A$, then $\frac{A^n}{\lambda^n}$ converges to the zero matrix.

4.5.11. A matrix A is *eventually positive* if $A^N > 0$ for all sufficiently large N (the matrix A is allowed to have negative entries). Prove the following extension of the Perron–Frobenius theorem: Let A be an eventually positive matrix. Then A has a positive eigenvector $\mathbf{v}_A$ with corresponding eigenvalue $\lambda_A > 0$ that is both geometrically and algebraically simple. If μ is another eigenvalue for A, then $|\mu| < \lambda_A$. Any positive eigenvector for A is a positive multiple of $\mathbf{v}_A$. Moreover, the conclusion of Theorem 4.5.12 holds.

4.5.12. Recall from §2.3 the conditional probability matrix P and initial state distribution $\mathbf{p}$ for a Markov chain on a graph.

   (a) Show that if P is irreducible, there is a unique vector $\mathbf{q}$ such that $\mathbf{q}P = \mathbf{q}$, $\mathbf{q} \geqslant 0$, and $\sum_i q_i = 1$ (such a vector $\mathbf{q}$ is called a *stationary distribution* for the Markov chain that defines P).

   (b) Show that if P is primitive, then for any probability vector $\mathbf{p}$ we have that $\lim_{m\to\infty} \mathbf{p}P^m = \mathbf{q}$, where $\mathbf{q}$ is the stationary distribution (so, the initial state distribution must evolve into the unique stationary distribution).

4.5.13. Show that if A is an irreducible matrix with period p and λ is an eigenvalue of A, then so is $\lambda \exp(2\pi i/p)$.

4.5.14. Let A be an irreducible matrix with Perron eigenvalue λ and period p. Let $\mathbf{v}, \mathbf{w}$ be right, left Perron eigenvectors for A, i.e., vectors $\mathbf{v}, \mathbf{w} > 0$ such that $A\mathbf{v} = \lambda\mathbf{v}$, $\mathbf{w}A = \lambda\mathbf{w}$, and normalized so that $\mathbf{w} \cdot \mathbf{v} = p$. For each state I, let $c(I)$ denote the period class which contains I, i.e., $I \in D_{c(I)}$. Show that for each I and J,

$$(A^{np+c(J)-c(I)})_{IJ} = [(v_I w_J) + \rho_{IJ}(n)]\lambda^{np+c(J)-c(I)}$$

where $\rho_{IJ}(n) \to 0$ as $n \to \infty$.

4.5.15. Complete the proof of Corollary 4.5.14.

4.5.16. Let X be an irreducible sofic shift with minimal right-resolving presentation $(G_X, \mathcal{L}_X)$.

   (a) Show that X is mixing if and only if G_X is primitive. [*Hint*: Use a synchronizing word for $(G_X, \mathcal{L}_X)$.]

   (b) Complete the proof of Corollary 4.5.13(2) by reducing from the mixing sofic case to the mixing shift of finite type case.

   (c) Show that $\mathrm{per}(G_X) = \mathrm{per}(G_{X^\mathsf{T}})$.

   (d) Give an example to show that $\mathrm{per}(X)$ need not coincide with $\mathrm{per}(G_X)$.

*4.5.17. For each $r \geqslant 1$, find the primitive graph with r vertices having the least possible Perron eigenvalue. Justify your answer.

Notes

In a seminal paper on information theory, Shannon [Sha] introduced the concept that we have called entropy by the name *capacity*. He reserved the term entropy for a probabilistic concept which is now often referred to as *Shannon entropy*. Using Shannon entropy, Kolmogorov [Kol] and Sinai [Sin1] invented *measure-theoretic entropy*, an important invariant of transformations studied in ergodic theory (see §13.4). Motivated by measure-theoretic entropy, Adler, Konheim and McAndrew [AdlKoM] introduced *topological entropy* for continuous maps (see §6.3), although topological entropy for shifts of finite type was effectively given earlier by Parry [Par1]. The notion of entropy introduced in this chapter is a special case of topological entropy, and that is why the term entropy, rather than the term capacity, is used in symbolic dynamics.

Most of the matrix-theoretic and graph-theoretic results in §4.2, §4.4, and §4.5 are standard parts of the Perron–Frobenius theory; see Seneta [Sen] or Gantmacher [Gan]. This theory has a long and extensive history going back to the work of Perron [Per] and Frobenius [Fro].

The proof of Theorem 4.4.4 is due to D. Neuhoff and A. Khayrallah [KhaN1]. J. Smiley supplied Exercise 4.5.17.

CHAPTER 5

FINITE-STATE CODES

In §2.5 we saw why it is useful to transform arbitrary binary sequences into sequences that obey certain constraints. In particular, we used Modified Frequency Modulation to code binary sequences into sequences from the $(1,3)$ run-length limited shift. This gave a more efficient way to store binary data so that it is not subject to clock drift or intersymbol interference.

Different situations in data storage and transmission require different sets of constraints. Thus our general problem is to find ways to transform or encode sequences from the full n-shift into sequences from a preassigned sofic shift X. In this chapter we will describe one encoding method called a finite-state code. The main result, the Finite-State Coding Theorem of §5.2, says that we can solve our coding problem with a finite-state code precisely when $h(X) \geqslant \log n$. Roughly speaking, this condition simply requires that X should have enough "information capacity" to encode the full n-shift.

We will begin by introducing in §5.1 two special kinds of labelings needed for finite-state codes. Next, §5.2 is devoted to the statement and consequences of the Finite-State Coding Theorem. Crucial to the proof is the notion of an approximate eigenvector, which we discuss in §5.3. The proof itself occupies §5.4, where an approximate eigenvector is used to guide a sequence of state splittings that converts a presentation of X into one with out-degree at least n at every state. From this it is short work to construct the required finite-state code. Finally, §5.5 deals with the issue of finding encoders which have sliding block decoders. These have the important practical advantage that errors can only propagate a finite distance.

This chapter uses most of the concepts discussed in the previous four chapters, and it provides concrete and computable solutions to a variety of coding problems.

§5.1. Road Colorings and Right-Closing Labelings

Recall from Chapter 3 that a labeling of a graph is right-resolving if, for each state, the edges starting at that state are labeled differently. The notion of finite-state code uses two variants of the right-resolving property, a stronger one called a road-coloring and a weaker one called right-closing.

A road-coloring is a right-resolving labeling that further requires each symbol be used on the outgoing edges at each state.

Definition 5.1.1. Let G be a graph with edge set $\mathcal{E}$. A *road-coloring* $\mathcal{C}$ of G is a labeling $\mathcal{C} \colon \mathcal{E} \to \mathcal{A}$ that establishes, for each state I of G, a one-to-one correspondence between $\mathcal{A}$ and the edges of G starting at I. A graph is *road-colorable* if it has a road-coloring.

Example 5.1.2. The labeling shown in Figure 5.1.1 is a road-coloring of the underlying graph G using the alphabet $\mathcal{A} = \{0, 1\}$. There are other road-colorings of G as well. In fact, the 2 outgoing edges from each of the 3 states can be colored in 2 ways, so that there are $2 \times 2 \times 2 = 8$ road-colorings of G using $\mathcal{A}$. □

A graph is road-colorable if and only if it has the same number of outgoing edges at each state, i.e., has *constant out-degree*. The equivalent condition on its adjacency matrix is having all row sums equal. If a graph has r states and constant out-degree n, then there are $(n!)^r$ possible road-colorings with a given alphabet of n symbols (why?).

Suppose that $\mathcal{C}$ is a road-coloring of G using alphabet $\mathcal{A}$. Let $w = a_1 a_2 \ldots a_m$ be a word over $\mathcal{A}$, and I be an arbitrary state in G. If we are standing at $I_0 = I$, the symbol a_1 picks out the unique edge e_1 with label a_1. Thinking of a_1 as an instruction to traverse e_1, we walk to a state $I_1 = t(e_1)$. The next symbol a_2 picks out the unique edge e_2 from I_1 with label a_2. We traverse e_2, ending at a state I_2, and so on. In this way, starting at a state I in G, each word w over $\mathcal{A}$ is a sequence of instructions to form the unique path $\pi_{\mathcal{C}}(I, w)$ in G starting at I having label w. In particular, since each word over $\mathcal{A}$ is the label of a path, we see that $(G, \mathcal{C})$ is a presentation of the full $\mathcal{A}$-shift.

At this point, we cannot help but mention a famous result about road-colorings. Recall from §3.3 that a word w over $\mathcal{A}$ is synchronizing for a

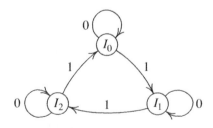

FIGURE 5.1.1. A road-coloring of a graph.

labeled graph if all paths labeled w end up at the same state. For a colored graph $(G, \mathcal{C})$, this means that as I varies over the states of G, the paths $\pi_{\mathcal{C}}(I, w)$ all end at the same state.

The road-coloring in Figure 5.1.1 has no synchronizing words. To see this, notice that a symbol 0 returns you to the same state, while a symbol 1 means that you move to the next. Thus if $w = a_1 a_2 \ldots a_m$ is a word over $\mathcal{A} = \{0, 1\}$, then $\pi_{\mathcal{C}}(I_j, w)$ ends at I_k, where $k \equiv j + a_1 + a_2 + \cdots + a_m \pmod{3}$. Hence every word has 3 presentations, all ending at different states. However, if we switch the colors at state I_0, then there *is* a synchronizing word for the new road-coloring (find one). Can we always find a road-coloring with a synchronizing word? This problem, known as the *Road Problem*, was open for forty years. Trahtman [Tra] showed that the answer is Yes:

Theorem 5.1.3. (**The Road Theorem**) *Every primitive road-colorable graph has a road-coloring for which there is a synchronizing word.*

Imagine that the graph represents a network of one-way roads between various cities. The network is assumed to be primitive and road-colorable with constant out-degree n. The Road Theorem tells the highway department that it is possible to paint the roads of such a network with "colors" $\{0, 1, \ldots, n-1\}$ in a right-resolving way, such that there is a finite sequence of "instructions" (a synchronizing word) that will lead drivers starting at each city to the same place at the same time. The Road Problem originated in an attempt by Adler and Weiss [AdlW] [AdlGW] to factor a shift of finite type with entropy $= \log(n)$ onto the full n-shift in an "almost" one-to-one way. See the discussion around Theorem 9.2.3.

While, strictly speaking, the primitivity assumption on G is not a necessary assumption, it turns out that if G is irreducible and there is a road-coloring for which there is a synchronizing word, then G must be primitive (Exercise 5.1.6). A special case of the Road Theorem, when G is assumed to have a self-loop, is relatively easy to prove (see Exercise 5.1.7).

We now turn to a weaker variant of right-resolving.

Definition 5.1.4. A labeled graph is *right-closing with delay D* if whenever two paths of length $D + 1$ start at the same state and have the same label, then they must have the same initial edge. A labeled graph is *right-closing* if it is right-closing with some delay $D \geqslant 0$.

According to this definition, a labeled graph is right-resolving if and only if it is right-closing with delay 0. Roughly speaking, a labeled graph is right-closing if, starting at a fixed state, we can tell which edge occurs in a path by looking ahead at the symbols of upcoming edges; i.e., it is "eventually right-resolving." We will sometimes call a labeling *right-closing* if the resulting labeled graph is right-closing.

As with right-resolving, we have a notion dual to right-closing: a labeling is *left-closing* if whenever two paths of length $D + 1$ end at the same state and have the same label, then they must have the same terminal edge.

Example 5.1.5. Consider the labeled graphs (a)–(d) shown in Figure 5.1.2. Graph (a) has delay $D = 1$. To see this, consider state I. Since there are two edges labeled a starting at I, the graph is not right-resolving. However, the self-loop at I can only be followed by an edge labeled a, while the edge from I to J can only be followed by an edge labeled b or c. Thus the label of a path of length 2 starting at I determines its initial edge. A similar analysis works at state J, and state K is easier since the edges starting there carry distinct labels. The reader should now verify that graph (b) has delay $D = 1$. In graph (c) there are two paths with different initial edges starting at I and labeled ab, so this graph does not have delay 1. However, the next symbol determines which path is taken, so it is right-closing with delay $D = 2$. In graph (d), for all m, there are two paths which start at state J and present the block ab^m; since these paths have different initial edges, this graph is not right-closing.   □

If a graph $\mathcal{G} = (G, \mathcal{L})$ is right-closing with delay D, then paths of length $D + 1$ starting at the same state and having the same label must "close up" at the initial edge. This is illustrated in Figure 5.1.3.

We can apply this property to longer paths in G, and obtain more closing. Suppose that $k \geqslant D$, and that $e_0 e_1 \ldots e_k$ and $f_0 f_1 \ldots f_k$ are paths in G with the same initial state and same label. Applying the definition to the

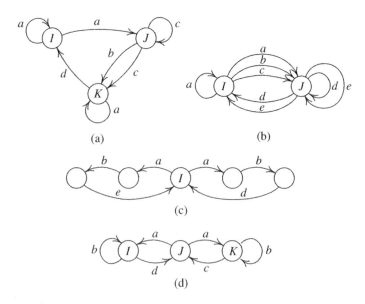

FIGURE 5.1.2.   Examples of right-closing graphs and non-right-closing graphs.

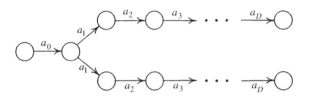

FIGURE 5.1.3.   Closing the initial edge of paths with the same label.

initial $(D + 1)$-blocks, we obtain that $e_0 = f_0$. Hence $e_1 e_2 \ldots e_{D+1}$ and $f_1 f_2 \ldots f_{D+1}$ both start at $t(e_0) = t(f_0)$ and have the same label, so that $e_1 = f_1$. Continuing, we successively conclude that $e_0 = f_0$, $e_1 = f_1$, ..., $e_{k-D} = f_{k-D}$. In other words, given two paths of length $\geqslant D$ having the same initial state and label, the right-closing property acts as a kind of "zipper" to close up the paths through all but the last D edges (see Figure 5.1.4).

Using this idea on paths that extend infinitely into the future, we see that two such paths with the same initial state and label must be equal. The ambiguity due to the closing delay is pushed out to infinity. In order to make this precise, we introduce some notation.

Let X be a shift space over $\mathcal{A}$. Define the *one-sided shift*

$$X^+ = \{x_{[0,\infty)} : x \in X\},$$

the set of right-infinite sequences that appear in X. If G is an essential graph, then X_G^+ is the set of right-infinite trips on G. For a state I of G, let $X_{G,I}^+$ denote the set of those right-infinite trips in X_G^+ starting at I. So a typical element in $X_{G,I}^+$ looks like $e_0 e_1 e_2 \ldots$, where $i(e_0) = I$. If $\mathcal{L}$ is a labeling on G, let $\mathcal{L}^+$ denote its extension to X_G^+ by

$$\mathcal{L}^+(e_0 e_1 e_2 \ldots) = \mathcal{L}(e_0)\mathcal{L}(e_1)\mathcal{L}(e_2) \ldots.$$

If we put $\mathcal{G} = (G, \mathcal{L})$, then $\mathcal{L}^+ \colon X_G^+ \to X_{\mathcal{G}}^+$ is onto. The following result characterizes right-closing labelings in terms of one-sided sequences.

Proposition 5.1.6. *An essential labeled graph $(G, \mathcal{L})$ is right-closing if and only if for each state I of G the map $\mathcal{L}^+$ is one-to-one on $X_{G,I}^+$.*

PROOF: First suppose that $(G, \mathcal{L})$ is right-closing, say with delay D. Let $x = e_0 e_1 e_2 \ldots$ and $y = f_0 f_1 f_2 \ldots$ be in $X_{G,I}^+$ and suppose that $\mathcal{L}^+(x) =$

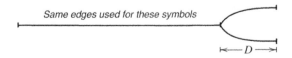

FIGURE 5.1.4.   Closing all but the last D edges of paths with the same label.

$\mathcal{L}^+(y)$. Then $i(e_0) = i(f_0) = I$ and $\mathcal{L}(e_0 e_1 \ldots e_D) = \mathcal{L}(f_0 f_1 \ldots f_D)$, so the definition of right-closing shows that $e_0 = f_0$. Hence $e_1 e_2 \ldots e_{D+1}$ and $f_1 f_2 \ldots f_{D+1}$ have the same initial state and label, so $e_1 = f_1$. We can continue this process indefinitely, showing that $e_k = f_k$ for all $k \geqslant 0$, so that $x = y$. This proves that $\mathcal{L}^+$ is one-to-one on $X_{G,I}^+$.

Conversely, suppose that $(G, \mathcal{L})$ is not right-closing. Then for each $n \geqslant 1$ there must be a state I_n with the property that there are paths π_n and τ_n of length n starting at I_n and having the same label, but with different initial edges. Fix $n \geqslant |\mathcal{V}(G)|^2$. Since there are only $|\mathcal{V}(G)|^2$ distinct pairs of vertices, we can write

$$\pi_n = \alpha\beta\gamma, \ \tau_n = \alpha'\beta'\gamma'$$

where $|\alpha| = |\alpha'|, |\beta| = |\beta'|, |\gamma| = |\gamma'|$, and both β and β' are cycles. Then $\alpha(\beta)^\infty$ and $\alpha'(\beta')^\infty$ are right-infinite trips starting at I_n having the same label sequence. But since they have different initial edges, $\mathcal{L}^+$ is not one-to-one on X_{G,I_n}^+. $\qquad\square$

The following is a situation in which we can show that a labeling is right-closing.

Proposition 5.1.7. *Suppose that* $\mathcal{G} = (G, \mathcal{L})$ *is essential, and that the 1-block code* $\mathcal{L}_\infty \colon X_G \to X_\mathcal{G}$ *is a conjugacy. If* $\mathcal{L}_\infty^{-1}$ *has anticipation* n, *then* $\mathcal{L}$ *is right-closing with delay* n.

PROOF: Suppose that $e_0 e_1 \ldots e_n$ and $f_0 f_1 \ldots f_n$ are paths in G starting at the same state I and having the same label. We must show that $e_0 = f_0$.

Let $\mathcal{L}_\infty^{-1}$ have memory m, so that $\mathcal{L}_\infty^{-1} = \Phi_\infty^{[-m,n]}$ for some block map Φ. Since G is essential, there is a path π in G of length m ending at I. Then $\mathcal{L}(\pi e_0 e_1 \ldots e_n) = \mathcal{L}(\pi f_0 f_1 \ldots f_n)$, and applying Φ shows that $e_0 = f_0$. $\qquad\square$

At the end of §3.1 we remarked that if we out-split the underlying graph of a labeled graph, the labeling carries over to the split graph, resulting in a different presentation of the same sofic shift. If the original labeling is right-closing, then so is the split labeling, but the delay is increased by one.

Proposition 5.1.8. *Let* $\mathcal{G}$ *be right-closing with delay* D. *Any labeled graph obtained from* $\mathcal{G}$ *by out-splitting is right-closing with delay* $D + 1$.

PROOF: Let $\mathcal{G} = (G, \mathcal{L})$, and suppose that H is obtained from G by out-splitting states. Recall that H has states I^i and edges e^i, and that the amalgamation map $e^i \to e$ gives a 1-block conjugacy from X_H to X_G. Recall that $\mathcal{L}$ induces a labeling $\mathcal{L}'$ on H by $\mathcal{L}'(e^i) = \mathcal{L}(e)$. We will show that $(H, \mathcal{L}')$ is right-closing with delay $D + 1$.

Suppose that $e_0^{i_0} e_1^{i_1} \ldots e_{D+1}^{i_{D+1}}$ and $f_0^{j_0} f_1^{j_1} \ldots f_{D+1}^{j_{D+1}}$ are paths in H starting at I^i and having the same $\mathcal{L}'$-label. Applying the amalgamation map, we see that $e_0 e_1 \ldots e_{D+1}$ and $f_0 f_1 \ldots f_{D+1}$ are paths in G starting at I with

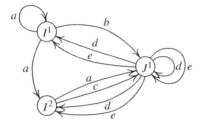

FIGURE 5.1.5. An out-splitting increases the closing delay.

the same $\mathcal{L}$-label. Hence $e_0 e_1 = f_0 f_1$. Recall that i_0 is determined by which partition element contains e_1, and similarly j_0 is determined by f_1. We conclude that $e_0^{i_0} = f_0^{j_0}$, proving that $\mathcal{L}'$ is right-closing with delay $D + 1$. □

Example 5.1.9. Consider the labeled graph in Figure 5.1.2(b), which is right-closing with delay 1. Perform an elementary out-splitting using the sets $\mathcal{E}_I^1$, $\mathcal{E}_I^2$, and $\mathcal{E}_J^1$, where $\mathcal{E}_I^1$ consists of the self-loop at I labeled a and the only edge in $\mathcal{E}_I$ labeled b, $\mathcal{E}_I^2$ is the rest of $\mathcal{E}_I$, and $\mathcal{E}_J^1 = \mathcal{E}_J$. The resulting labeled graph $\mathcal{H}$ is shown in Figure 5.1.5. Since there are two paths in $\mathcal{H}$ starting at I^1 labeled aa using different initial edges, $\mathcal{H}$ is not right-closing with delay 1. However, the proposition does show that $\mathcal{H}$ is right-closing with delay 2 (you should check this directly). □

Proposition 5.1.8 will be used in the proof of the Finite-State Coding Theorem to show that a labeled graph formed by k successive out-splittings of a right-resolving graph is right-closing with delay k.

Recall from Proposition 4.1.13 that right-resolving labelings preserve entropy. The same is true for right-closing labelings:

Proposition 5.1.10. *Suppose that $\mathcal{G}$ is right-closing and has underlying graph G. Then $h(X_{\mathcal{G}}) = h(X_G)$.*

PROOF: Since $X_{\mathcal{G}}$ is a factor of X_G, we have that $h(X_{\mathcal{G}}) \leqslant h(X_G)$. Suppose that $\mathcal{G}$ is right-closing with delay D. Then each state $I \in \mathcal{V}$ and each block $w \in \mathcal{B}_{n+D}(X_{\mathcal{G}})$ determine at most one path of length n in G, and all paths in $\mathcal{B}_n(X_G)$ arise this way. Hence

$$|\mathcal{B}_n(X_G)| \leqslant |\mathcal{V}| \times |\mathcal{B}_{n+D}(X_{\mathcal{G}})|,$$

so that

$$\begin{aligned}
h(X_G) &= \lim_{n \to \infty} \frac{1}{n} \log |\mathcal{B}_n(X_G)| \\
&\leqslant \lim_{n \to \infty} \frac{1}{n} \log |\mathcal{V}| + \lim_{n \to \infty} \frac{1}{n} \log |\mathcal{B}_{n+D}(X_{\mathcal{G}})| \\
&= h(X_{\mathcal{G}}).
\end{aligned}$$

□

Our final result shows that, just as a sliding block code can be recoded to a 1-block code and a shift of finite type can be recoded to an edge shift, a right-closing labeling can be recoded to a right-resolving labeling.

Proposition 5.1.11. *Let Φ be a right-closing labeling on a graph G with delay D. Then there is a graph H and labelings Ψ, Θ of H such that Ψ is right-resolving, $\Theta_\infty \colon X_H \to X_G$ is a conjugacy, and the following diagram commutes.*

$$
\begin{array}{ccc}
X_G & \xleftarrow{\ \Theta_\infty \circ \sigma^D\ } & X_H \\
\Big\downarrow{\scriptstyle \Phi_\infty} & \swarrow{\scriptstyle \Psi_\infty} & \\
\Phi_\infty(X_G) & &
\end{array}
$$

PROOF: If $D = 0$ then Φ is already right-resolving and we can choose $H = G$, $\Psi = \Phi$, and Θ to be the identity. Hence we may assume that $D \geqslant 1$.

Define the graph H as follows. The states of H are pairs of the form $(I, w_1 w_2 \ldots w_D)$, where I is a state of G and $w_1 w_2 \ldots w_D$ is a block that is presented by some path in G starting at I. Suppose that a is a symbol such that $w_1 w_2 \ldots w_D a$ is presented by a path starting at I. The initial edge of all such paths is the same, say e. We then endow H with an edge named $(I, w_1 w_2 \ldots w_D a)$ from state $(I, w_1 w_2 \ldots w_D)$ to state $(t(e), w_2 w_3 \ldots w_D a)$. This defines H. Let Θ assign the label e to the edge $(I, w_1 w_2 \ldots w_D a)$.

By our construction, Θ is a graph homomorphism from H to G (the associated vertex map is defined by $\partial \Theta(I, w_1 w_2 \ldots w_D) = I$). The construction also guarantees that if

$$
\Theta((I^{(1)}, w^{(1)}) \ldots (I^{(D+1)}, w^{(D+1)})) = e_1 \ldots e_{D+1},
$$

then $I^{(1)}$ must be the initial state of e_1 and $w^{(1)}$ must be the Φ-label of $e_1 \ldots e_{D+1}$. Thus Θ_∞^{-1} is a sliding block code with memory 0 and anticipation D, and so Θ_∞ is a conjugacy.

Define the labeling Ψ on H be $\Psi((I, w_1 \ldots w_D a)) = a$. This labeling is right-resolving since the state $(I, w_1 \ldots w_D)$ and label a determine the edge $(I, w_1 \ldots w_D a)$.

Finally, we verify that the diagram commutes. Now

$$
\Phi \circ \Theta(I, w_1 \ldots w_D a) = w_1 \quad \text{and} \quad \Psi(I, w_1 \ldots w_D a) = a.
$$

Thus it suffices to show that if

$$
(I^{(1)}, w^{(1)}) \ldots (I^{(D+1)}, w^{(D+1)})
$$

is a path in H, then the terminal symbol of $w^{(1)}$ is the initial symbol of $w^{(D+1)}$. But this follows directly from the definition of edges in H. $\quad\square$

EXERCISES

5.1.1. Consider the graph shown in Figure 5.1.1.
  (a) Find a road-coloring that has a synchronizing word.
  (b) How many of the 8 possible road-colorings have a synchronizing word?
  (c) Is there a road-coloring such that all sufficiently long words are synchronizing?

5.1.2. Suppose that $h(X_G) > 0$ and $n \geqslant 1$. Show that for all sufficiently large m there is a subgraph of G^m that has constant out-degree n.

5.1.3. Let G be a primitive graph. Prove that G is road-colorable if and only if G^m is road-colorable for some (and hence every) $m \geqslant 1$.

5.1.4. Show that if $\mathcal{G} = (G, \mathcal{L})$ is an irreducible right-resolving presentation of the full n-shift, then $\mathcal{L}$ must be a road-coloring. [*Hint*: Show that the merged-graph of $\mathcal{G}$ (described in §3.3) has only one state.]

5.1.5. If $\mathcal{L}$ is a road-coloring of G with alphabet $\{0, 1, \ldots, n-1\}$, and $\mathcal{L}_\infty \colon X_G \to X_{[n]}$ is the resulting 1-block code, show that each point in $X_{[n]}$ has at most $|\mathcal{V}(G)|$ pre-images. Must every point in $X_{[n]}$ have exactly $|\mathcal{V}(G)|$ pre-images?

5.1.6. Show that if G is an irreducible graph that is not primitive, then G cannot have a road-coloring for which there is a synchronizing word. Thus the primitivity assumption in the Road Problem is needed.

5.1.7. Solve the Road Problem for primitive graphs that have a self-loop. [*Hint*: Construct a "tree" that meets each vertex of the graph and is rooted at a self-loop.]

5.1.8. Show that we can compute the delay of a right-closing graph $\mathcal{G}$ as follows. Let $\mathcal{K}$ denote the label product of $\mathcal{G}$ with itself. Let H be the (unlabeled) subgraph of $\mathcal{K}$ determined by all states of $\mathcal{K}$ which are terminal states of a path with initial state of the form (I, I) and initial edge of the form (e, f), $e \neq f$.
  (a) Show that $\mathcal{G}$ is right-closing if and only if H has no cycles, and that this occurs if and only if $A_H^n = 0$ for some n.
  (b) Show that if $\mathcal{G}$ is right-closing with delay D, then $D - 1$ is the length of the longest path in H.

5.1.9. Proposition 3.3.17 says that two right-resolving graphs which present the same sofic shift are isomorphic if they are irreducible and follower-separated. Is the same true for right-closing graphs?

5.1.10. Let H be the out-split graph formed by out-splitting states of G. Recall from §2.4 that the edges of H are copies e^i of the edges e of G. Let $\mathcal{L} \colon \mathcal{E}(H) \to \mathcal{E}(G)$ be the map $\mathcal{L}(e^i) = e$. Then we can regard $\mathcal{H} = (H, \mathcal{L})$ as a labeled graph which presents X_G. Show that $\mathcal{H}$ is right-closing and left-resolving, but not necessarily right-resolving.

§5.2. Finite-State Codes

We will describe in this section one method for transforming unconstrained sequences into constrained sequences, and give a theoretical result about when this is possible. The proof of this occupies the next two sections, which also provide a practical algorithm for constructing finite-state codes.

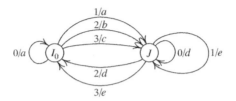

FIGURE 5.2.1.   A finite-state code.

Definition 5.2.1. A *finite-state code* is a triple $(G, \mathcal{I}, \mathcal{O})$, where G is a graph called the *encoder graph*, $\mathcal{I}$ is a road-coloring of G called the *input labeling*, and $\mathcal{O}$ is a right-closing labeling of G called the *output labeling*. If G has constant out-degree n and if X is a shift space containing $\mathcal{O}_\infty(X_G)$, then $(G, \mathcal{I}, \mathcal{O})$ is called a *finite-state (X, n)-code*.

We will draw finite-state codes by marking an edge e with a/b, where $a = \mathcal{I}(e)$ is the input label of e and $b = \mathcal{O}(e)$ is its output label.

Example 5.2.2. The right-closing graph shown in Figure 5.1.2(b) has constant out-degree 4, so provides the output labeling for a finite-state code. Figure 5.2.1 shows one choice of input labeling with alphabet $\{0, 1, 2, 3\}$.   □

Suppose that $(G, \mathcal{I}, \mathcal{O})$ is a finite-state (X, n)-code. We can use this to transform sequences from the full n-shift into sequences from X as follows. Fix a state I_0 in G. Let $x = x_0 x_1 x_2 \ldots \in X_{[n]}^+$ be an infinite sequence from the one-sided full n-shift. Since $\mathcal{I}$ is a road-coloring, there is exactly one infinite path $e_0 e_1 e_2 \ldots \in X_G^+$ starting at I_0 with $\mathcal{I}$-label $x_0 x_1 x_2 \ldots$. Apply $\mathcal{O}$ to this path to create the output $y = \mathcal{O}(e_0)\mathcal{O}(e_1)\mathcal{O}(e_2) \ldots \in X^+$.

Example 5.2.3. If $(G, \mathcal{I}, \mathcal{O})$ is the finite-state code shown in Figure 5.2.1, then, with initial state I_0, the sequence $x = 0231002 \ldots$ is encoded to $y = abeaddd \ldots$.   □

We can imagine the encoding being done by a "machine" whose internal states correspond to the vertices of the encoder graph. The machine starts in state I_0. The input symbol x_0 causes the machine to find the unique edge $e_0 \in \mathcal{E}_{I_0}$ with input label x_0, emit its output label $y_0 = \mathcal{O}(e_0)$, and then change to state $I_1 = t(e_0)$. It then reads the next input symbol x_1, which determines e_1, emits $y_1 = \mathcal{O}(e_1)$, and moves to state $I_2 = t(e_1)$. This process, sketched in Figure 5.2.2, continues as long as there are input symbols.

Naturally, we want to be able to recover the input $x_0 x_1 x_2 \ldots$ from the output $y_0 y_1 y_2 \ldots$. Now $x_0 x_1 x_2 \ldots$ determines a unique path $e_0 e_1 e_2 \ldots$ in X_{G, I_0}^+, and Proposition 5.1.6 tells us that $\mathcal{O}_\infty$ is one-to-one on X_{G, I_0}^+ since $\mathcal{O}$ is right-closing. This shows that we can reconstruct the input from the output.

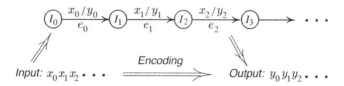

FIGURE 5.2.2. Encoding with a finite-state code.

To be more concrete, suppose that $(G, \mathcal{O})$ is right-closing with delay D. Then I_0 and $y_0 y_1 \ldots y_D$ determine e_0, hence I_1 and $x_0 = \mathcal{I}(e_0)$. Knowing I_1 and $y_1 y_2 \ldots y_{D+1}$ determines e_1, hence I_2 and $x_1 = \mathcal{I}(e_1)$, and so on. Thus from the output $y_0 y_1 \ldots y_k$ we can determine all but the last D input symbols $x_0 x_1 \ldots x_{k-D}$.

Example 5.2.4. Using the finite-state code in Figure 5.2.1, we can decode the output *abeaddd* back to 023100. The final output symbol d cannot yet be decoded, since it could correspond to an edge with input label 0 or one with input label 2. The next output symbol would tell which is correct. □

Since in practice we handle only finite sequences, the loss of the last D input symbols is alarming. One simple solution is to tack on D extra "padding" symbols to the input, which provide the extra output symbols needed. This adds to the encoding, but D is normally microscopic compared to the input length, so the extra work is relatively slight.

When do finite-state codes exist? The next result, the main theorem in this chapter, provides a complete answer.

Theorem 5.2.5 (Finite-State Coding Theorem). *Let X be a sofic shift and $n \geqslant 1$. Then there is a finite-state (X, n)-code if and only if $h(X) \geqslant \log n$.*

The necessity of the entropy condition $h(X) \geqslant \log n$ is an easy consequence of what we already know. For suppose that $(G, \mathcal{I}, \mathcal{O})$ is a finite-state (X, n)-code. By Proposition 5.1.10, right-resolving and right-closing labelings preserve entropy, so that

$$h(\mathcal{I}_\infty(X_G)) = h(X_G) = h(\mathcal{O}_\infty(X_G)).$$

Since $(G, \mathcal{I})$ presents the full n-shift, and $\mathcal{O}_\infty(X_G) \subseteq X$, we obtain that

$$\log n = h(X_{[n]}) = h(\mathcal{I}_\infty(X_G)) = h(\mathcal{O}_\infty(X_G)) \leqslant h(X).$$

We will prove the sufficiency of the entropy condition in §5.4, but this will require some new ideas. In order to see how these ideas will be used, we will briefly sketch the sufficiency proof here. Start with a right-resolving presentation $\mathcal{G}$ of X. Using a vector, called an approximate eigenvector, to guide us, we will perform a sequence of out-splittings on $\mathcal{G}$ to obtain a presentation $\mathcal{H}$ of X such that every state of $\mathcal{H}$ has out-degree at least n. By Proposition 5.1.8, $\mathcal{H}$ is right-closing. We can then prune away edges from $\mathcal{H}$ to create a right-closing graph with constant out-degree n that presents a sofic shift contained in X. Then any road-coloring of this graph gives the finite-state code we are after. In §5.5 we discuss the advantages of some road-colorings over others.

Next, we discuss how to apply the Finite-State Coding Theorem even when the entropy condition is not directly satisfied. Consider a typical situation, in which we seek to encode sequences from the full 2-shift into a sofic shift X over $\mathcal{A} = \{0,1\}$, such as a run-length limited shift. Then $h(X) < \log 2$, so the entropy condition must fail. The way around this difficulty is to encode blocks of symbols instead of individual symbols, and the higher power shifts from Chapter 1 are just the tool that is needed.

Let us divide up the input into blocks of length p, and the output into blocks of length q. This effectively replaces the input shift $X_{[2]}$ by its higher power shift $X_{[2]}^p \cong X_{[2^p]}$, and the output shift X by X^q. The entropy condition for these replacement shifts becomes

$$q\,h(X) = h(X^q) \geqslant \log 2^p = p \log 2 = p,$$

or

(5–2–1) $$h(X) \geqslant \frac{p}{q}.$$

Thus by finding fractions p/q close to, but less than, $h(X)$, we can construct finite-state codes that transform p-blocks from $X_{[2]}$ to q-blocks from X. Such a code is said to have *rate* $p\!:\!q$. When viewed in this way, a finite-state $(X^q, 2^p)$-code is often called a *rate $p\!:\!q$ finite-state code from binary data into X*.

The code rate (really, the ratio p/q) expresses the number of input bits represented by each output bit. When X is a proper subset of the full 2-shift, this ratio will always be less than 1. But the higher the rate, the more efficient the encoding. Thus (5–2–1) tells us the maximum efficiency possible for a finite-state code from binary data into X.

Some examples, whose construction we will describe in §5.4, should help explain these concepts.

Example 5.2.6. Let $X = X(1,3)$ be the $(1,3)$ run-length limited shift. In Example 4.3.4 we computed that $h(X) \approx 0.55$, so we can pick $p = 1$ and $q = 2$. The input alphabet is the set of 1-blocks from $X_{[2]}$, namely $\{0,1\}$, while the output alphabet is the set of 2-blocks from X, namely $\{00, 01, 10\}$. Figure 5.2.3 shows a finite-state $(X^2, 2)$-code, that turns out to be none other than the MFM code from §2.5 (why?). $\square$

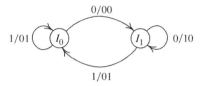

FIGURE 5.2.3. A finite-state code giving the MFM code.

Example 5.2.7. Let $X = X(0, 1)$. Now X is conjugate to the golden mean shift by interchanging 0's and 1's, and we computed the entropy of the latter in Example 4.1.4 to be $h(X) = \log[(1 + \sqrt{5})/2] \approx 0.69424$, so choose $p = 2$ and $q = 3$. The input alphabet is then the set of 2-blocks from $X_{[2]}$, namely $\{00, 01, 10, 11\}$, while the output alphabet is the set of 3-blocks from X, namely $\{010, 011, 101, 110, 111\}$. We show a finite-state $(X^3, 2^2)$-code in Figure 5.2.4.

To illustrate the use of this finite-state code, consider the input sequence 000111010011. We first divide it into 2-blocks, so that it looks like 00 01 11 01 00 11. We then use these 2-blocks as input symbols, starting at state I_0. This produces output triples

$$011\,011\,111\,011\,101\,111,$$

which is a sequence from X^3. □

We conclude this section with some general remarks about coding. First, it would seem better to require that the output labeling be right-resolving, which would reduce the coding delay to 0. Unfortunately, it is not always possible to find such a finite-state code even when the entropy condition is satisfied (see Exercise 5.2.5).

Another approach to transforming arbitrary sequences to constrained sequences is with an embedding, whose inverse would be the decoder. The entropy condition is certainly necessary here, but there are two serious difficulties, one theoretical and the other practical. If the constrained sofic shift X had fewer that n fixed points, then no embedding of $X_{[n]}$ into X is possible (see Exercise 5.2.6). Even if fixed or periodic points did not present a problem, the known constructions for embeddings involve block maps with

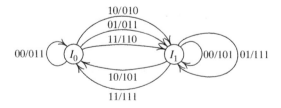

FIGURE 5.2.4. A finite-state code based on the $(0, 1)$ run-length limited shift.

enormous window sizes, and the resulting codes are far too complicated to use.

EXERCISES

5.2.1. Verify that the finite-state code with initial state I_0 shown in Figure 5.2.3 is the MFM code.

5.2.2. Use the finite-state code from Figure 5.2.4 to encode the sequences 0111001011 and 100100011010.

5.2.3. (a) Is the "only if" part of the Finite-State Coding Theorem true for general shifts?

(b) Show that if X is a shift with $h(X) > \log n$ and there are sofic subshifts of X with entropy arbitrarily close to $h(X)$, then there is a finite-state (X, n)-code.

(c) Is the "if" part of the Finite-State Coding Theorem true for general shifts?

5.2.4. Let X be a shift space, and let G be a graph with out-degree n and labelings $\mathfrak{I}$ and $\mathcal{O}$ such that $\mathfrak{I}$ is a road-coloring of G, and X contains $\mathcal{O}_\infty(X_G)$ (but $\mathcal{O}$ is not assumed to be right-closing). Show that even though $(G, \mathfrak{I}, \mathcal{O})$ is not necessarily a finite-state code, it still makes sense to use $(G, \mathfrak{I}, \mathcal{O})$ to encode $X^+_{[n]}$ into X^+ and that this encoding is one-to-one if and only if $\mathcal{O}$ is right-closing.

5.2.5. Let

$$A = \begin{bmatrix} 1 & 3 \\ 1 & 0 \end{bmatrix}.$$

Show that $h(X_A) > \log 2$, but there is no finite-state $(X_A, 2)$-code such that both the input and output labelings are right-resolving.

5.2.6. Let A be the matrix in Exercise 5.2.5. Show that it is impossible to embed $X_{[2]}$ into X_A.

5.2.7. (a) Explain why, in a finite-state code at rate $p\colon q$, it is desirable to have p, q relatively 'small' and to have p/q close to $h(X)$.

(b) Show that if $h(X)$ is irrational and p_n, q_n are pairs of positive integers such that $\lim_{n\to\infty} p_n/q_n = h(X)$, then $\lim_{n\to\infty} q_n = \infty$ (so, in this case, it is impossible to have our cake and eat it too).

(c) Show that for any shift X and any $\epsilon > 0$, there are positive integers p, q such that $|p/q - h(X)| \leqslant \epsilon$ and $q \leqslant \lceil 1/\epsilon \rceil$.

(d) Show that there is a constant C such that for any shift X, there are p_n, q_n satisfying $|p_n/q_n - h(X)| \leqslant C/(q_n)^2$, and $\lim_{n\to\infty} q_n = \infty$. [*Hint:* Read about continued fractions in a number theory book such as [NivZ].]

(e) Explain the significance of parts (c) and (d) for finite-state coding.

§5.3. Approximate Eigenvectors

Let X be a sofic shift satisfying the necessary condition $h(X) \geqslant \log n$ from the Finite-State Coding Theorem. Our goal is to find a finite-state (X, n)-code. Using the subset construction in Theorem 3.3.2, we first find a right-resolving presentation $(H, \mathcal{O})$ for X. Suppose we are lucky, and the out-degree of each state in H is at least n. By selecting edges from H we can form a subgraph G with constant out-degree n, and restrict $\mathcal{O}$ to G, so that $\mathcal{O}_\infty(X_G) \subseteq X$. If $\mathcal{I}$ is any road-coloring of G, then $(G, \mathcal{I}, \mathcal{O})$ is the finite-state (X, n)-code we are after.

Let A denote the adjacency matrix of H. Our assumption on H about out-degrees corresponds to the row sums of A all being at least n. We can express this using the column vector $\mathbf{1} = [1\,1\,\ldots\,1]^\mathsf{T}$ of all 1's as the vector inequality

$$(5\text{--}3\text{--}1) \hspace{3cm} A\mathbf{1} \geqslant n\mathbf{1}.$$

Usually we will not be so lucky with our first presentation of X, and some of its row sums will be $< n$. To handle this situation, we will show in this section that an inequality similar to (5–3–1) still holds, but where $\mathbf{1}$ is replaced by a nonnegative integer vector $\mathbf{v}$ called an approximate eigenvector. This $\mathbf{v}$ is used in the next section to guide a sequence of state splittings designed to modify the original presentation of X to one for which (5–3–1) holds. The discussion at the beginning of this section then shows how to use the modified presentation to construct a finite-state (X, n)-code.

We start by defining an approximate eigenvector.

Definition 5.3.1. Let A be a matrix with nonnegative integer entries, and n be a positive integer. An (A, n)-*approximate eigenvector* is a vector $\mathbf{v} \neq \mathbf{0}$ with nonnegative integer components satisfying $A\mathbf{v} \geqslant n\mathbf{v}$.

Example 5.3.2. Let $A = \begin{bmatrix} 1 & 3 \\ 6 & 1 \end{bmatrix}$ and $n = 5$, which is the adjacency matrix of the graph shown in Figure 5.3.1. Then $\mathbf{v} = \begin{bmatrix} 2 \\ 3 \end{bmatrix}$ is an $(A, 5)$-approximate eigenvector, since

$$A\mathbf{v} = \begin{bmatrix} 11 \\ 15 \end{bmatrix} \geqslant \begin{bmatrix} 10 \\ 15 \end{bmatrix} = 5\mathbf{v}.$$

The reader should verify that $\mathbf{v}$ is the "smallest" $(A, 5)$-approximate eigenvector, in the sense that every other such vector has at least one component > 3. ☐

One useful way to think about an approximate eigenvector $\mathbf{v}$ is that it assigns a nonnegative integer "weight" v_I to each state I. The vector inequality $A\mathbf{v} \geqslant n\mathbf{v}$ then corresponds to the condition that, for each state I, if we add up the weights at the ends of all edges starting at I, the sum should

FIGURE 5.3.1. Graph for approximate eigenvector example.

be at least n times the weight at I. In Example 5.3.2 (see Figure 5.3.1) there are four edges starting at the first state, one ending there (so having weight $v_1 = 2$), and three ending at the other state (each with weight $v_2 = 3$). Their sum is $2 + 3 + 3 + 3 = 11$, and this is at least $5v_1 = 10$, verifying the approximate eigenvector inequality at the first state.

There is another useful way to understand the meaning of approximate eigenvectors. Suppose that A has size r. Then the vector inequality $A\mathbf{v} \geqslant n\mathbf{v}$ is equivalent to r scalar inequalities

$$\sum_{J=1}^{r} A_{IJ}v_J \geqslant n\,v_I \qquad (I = 1, 2, \ldots, r).$$

Each of these inequalities determines a half-space in r-dimensional space, and their intersection will typically be a polyhedral cone (but could have lower dimension). Then any nonnegative integer vector in this cone is an (A, n)-approximate eigenvector.

Example 5.3.3. Consider $A = \begin{bmatrix} 1 & 3 \\ 6 & 1 \end{bmatrix}$ and $n = 5$ from Example 5.3.2. Let us find an $(A, 5)$-approximate eigenvector $\mathbf{v}$ using the geometrical idea just discussed. The inequality $A\mathbf{v} \geqslant 5\mathbf{v}$ is equivalent to the system

$$\begin{cases} v_1 + 3v_2 & \geqslant 5v_1 \\ 6v_1 + \ v_2 & \geqslant 5v_2 \end{cases},$$

which becomes

$$\begin{cases} -4v_1 + 3v_2 & \geqslant 0 \\ 6v_1 - 4v_2 & \geqslant 0 \end{cases}.$$

Figure 5.3.2 shows the two half-planes determined by these inequalities. Their intersection is a wedge in the first quadrant, cross-hatched in the figure. The integer vectors in this wedge are indicated by dots, the first few being located at

$$\begin{bmatrix} 2 \\ 3 \end{bmatrix}, \begin{bmatrix} 3 \\ 4 \end{bmatrix}, \begin{bmatrix} 4 \\ 6 \end{bmatrix}, \begin{bmatrix} 5 \\ 7 \end{bmatrix}, \begin{bmatrix} 6 \\ 8 \end{bmatrix}, \begin{bmatrix} 6 \\ 9 \end{bmatrix}, \begin{bmatrix} 7 \\ 10 \end{bmatrix}, \ldots,$$

and these are precisely the $(A, 5)$-approximate eigenvectors. This confirms that $\begin{bmatrix} 2 \\ 3 \end{bmatrix}$ is the smallest such vector, as mentioned in Example 5.3.2. □

Before coming to the main result of this section, it is first convenient to show how we can reduce to the situation of an irreducible graph and a positive approximate eigenvector.

Lemma 5.3.4. *Let G be a graph, and suppose that $\mathbf{v}$ is an (A_G, n)-approximate eigenvector. There is an irreducible subgraph H of G such that if we form $\mathbf{w}$ by restricting $\mathbf{v}$ to the states in H, then $\mathbf{w} > 0$ and $\mathbf{w}$ is an (A_H, n)-approximate eigenvector.*

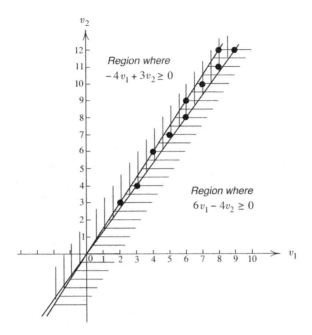

v_2

Region where
$-4v_1 + 3v_2 \geq 0$

Region where
$6v_1 - 4v_2 \geq 0$

v_1

FIGURE 5.3.2.   Geometrical interpretation of approximate eigenvectors.

PROOF: First form the subgraph K of G by removing all states I for which $v_I = 0$ together with all edges incident to such states. Recall from our discussion in §4.4 (see page 120) that K has at least one irreducible component H that is a sink; i.e., H contains the terminal states of all edges that start in H. If $\mathbf{w}$ is the restriction of $\mathbf{v}$ to H, then $\mathbf{w} > 0$ by construction, and use of the "weight" interpretation shows that $\mathbf{w}$ is an (A_H, n)-approximate eigenvector.   □

The next theorem tells us when approximate eigenvectors exist.

Theorem 5.3.5. *Let A be a nonnegative integer matrix and $n \geqslant 1$. Then there is an (A, n)-approximate eigenvector if and only if $\lambda_A \geqslant n$. If we also assume that A is irreducible, then there is a positive (A, n)-approximate eigenvector.*

PROOF: First suppose that there is an (A, n)-approximate eigenvector $\mathbf{v}$. By the previous lemma we can assume that A is irreducible and that $\mathbf{v} > 0$. Then $A\mathbf{v} \geqslant n\mathbf{v}$, so that $A^2\mathbf{v} \geqslant n(A\mathbf{v}) \geqslant n^2\mathbf{v}$, and so $A^k\mathbf{v} \geqslant n^k\mathbf{v}$ for all $k \geqslant 1$. Since $\mathbf{v} > 0$, exactly the same argument we used to prove Proposition 4.2.1 shows that there is a $c > 0$ such that for all $k \geqslant 1$,

$$c\,n^k \leqslant \sum_{I,J=1}^{r} (A^k)_{IJ}.$$

Proposition 4.2.1 also shows that that there is a constant $d > 0$ such that

$$\sum_{I,J=1}^{r} (A^k)_{IJ} \leqslant d\,\lambda_A^k.$$

Hence $(c/d)^{1/k}n \leqslant \lambda_A$ for all $k \geqslant 1$. Letting $k \to \infty$ implies that $\lambda_A \geqslant n$, establishing necessity.

Conversely, suppose that $\lambda_A \geqslant n$. By passing to an irreducible component with maximal spectral radius, we can assume that A is irreducible (why?). Then the Perron eigenvector $\mathbf{v}_A > 0$, and $A\mathbf{v}_A = \lambda_A \mathbf{v}_A \geqslant n\mathbf{v}_A$. Unfortunately, $\mathbf{v}_A$ hardly ever has integer entries, so we must do more work. There are two cases to consider.

First suppose that $\lambda_A = n$, so we are finding an exact eigenvector. We can solve the vector equation $A\mathbf{v} = n\mathbf{v}$ using Gaussian elimination. Since A and n involve only integers, the solution will be a vector $\mathbf{v}$ with rational entries. Now A is irreducible, so the Perron–Frobenius Theorem implies that $\mathbf{v}$ is a multiple of $\mathbf{v}_A$. Multiplying $\mathbf{v}$ by -1 if necessary, we can assume that $\mathbf{v} > 0$. We can then multiply $\mathbf{v}$ by a positive integer to clear fractions, and the result is a positive (A, n)-approximate eigenvector.

Next, suppose that $\lambda_A > n$, so that $A\mathbf{v}_A = \lambda_A \mathbf{v}_A > n\mathbf{v}_A$. Choose a positive rational vector $\mathbf{v}$ so close to $\mathbf{v}_A$ that the inequality $A\mathbf{v} > n\mathbf{v}$ still holds. Multiply $\mathbf{v}$ by a positive integer to clear fractions, obtaining a positive (A, n)-approximate eigenvector. $\qquad\square$

Although this theorem tells us when approximate eigenvectors exist, it does not provide a good way to find them. Fortunately, there is an efficient algorithm to find small approximate eigenvectors, which are crucial to the practicality of finite-state coding.

We need some notation to state this algorithm. If $\mathbf{u}$ and $\mathbf{v}$ are vectors, let $\mathbf{w} = \min\{\mathbf{u}, \mathbf{v}\}$ denote the componentwise minimum, so that $w_I = \min\{u_I, v_I\}$ for each I. For a real number s, let $\lfloor s \rfloor$ denote the integer floor of s, i.e., the largest integer $\leqslant s$.

Theorem 5.3.6 (Approximate Eigenvector Algorithm). *Let A be a nonnegative integer matrix, $n \geqslant 1$, and $\mathbf{z} \neq \mathbf{0}$ be a nonnegative integer vector. Compute*

$$\mathbf{z}' = \min\left\{\mathbf{z}, \left\lfloor \frac{1}{n} A\mathbf{z} \right\rfloor\right\}.$$

If $\mathbf{z}' = \mathbf{z}$, output $\mathbf{z}'$ and stop. If $\mathbf{z}' \neq \mathbf{z}$, replace $\mathbf{z}$ by $\mathbf{z}'$ and repeat. This process eventually stops, and its output is either an (A, n)-approximate eigenvector or $\mathbf{0}$.

PROOF: The process eventually stops since the vectors computed are nonnegative, integral, and monotonically nonincreasing in each component. If the output is $\mathbf{z}$, then $\mathbf{z} \geqslant \mathbf{0}$ and

$$\mathbf{z} = \min \left\{ \mathbf{z}, \left\lfloor \frac{1}{n} A\mathbf{z} \right\rfloor \right\} \leqslant \frac{1}{n} A\mathbf{z},$$

so that $A\mathbf{z} \geqslant n\mathbf{z}$. Thus either $\mathbf{z} = \mathbf{0}$, or $\mathbf{z}$ is an (A, n)-approximate eigenvector. □

A typical use of the Approximate Eigenvector Algorithm starts with initial vector $\mathbf{z} = [1\,1\,\ldots\,1]^{\mathsf{T}}$. If the output is not $\mathbf{0}$, we are done. If it is $\mathbf{0}$, try starting with $\mathbf{z} = [2\,2\,\ldots\,2]^{\mathsf{T}}$. Again, if the output is not $\mathbf{0}$, done; otherwise try $\mathbf{z} = [3\,3\,\ldots\,3]^{\mathsf{T}}$, and so on. This eventually catches the approximate eigenvector with smallest maximal entry (see Exercise 5.3.5).

Example 5.3.7. Let $A = \begin{bmatrix} 1 & 3 \\ 6 & 1 \end{bmatrix}$ and $n = 5$ as in previous examples. Start with $\mathbf{z} = \begin{bmatrix} 1 \\ 1 \end{bmatrix}$. The vectors computed by the algorithm are

$$\begin{bmatrix} 1 \\ 1 \end{bmatrix} \to \begin{bmatrix} 0 \\ 1 \end{bmatrix} \to \begin{bmatrix} 0 \\ 0 \end{bmatrix} \to \begin{bmatrix} 0 \\ 0 \end{bmatrix},$$

so this choice of $\mathbf{z}$ does not work. Next, use $\mathbf{z} = \begin{bmatrix} 2 \\ 2 \end{bmatrix}$, resulting in

$$\begin{bmatrix} 2 \\ 2 \end{bmatrix} \to \begin{bmatrix} 1 \\ 2 \end{bmatrix} \to \begin{bmatrix} 1 \\ 1 \end{bmatrix} \to \begin{bmatrix} 0 \\ 1 \end{bmatrix} \to \begin{bmatrix} 0 \\ 0 \end{bmatrix} \to \begin{bmatrix} 0 \\ 0 \end{bmatrix},$$

another failure. But the third time is the charm:

$$\begin{bmatrix} 3 \\ 3 \end{bmatrix} \to \begin{bmatrix} 2 \\ 3 \end{bmatrix} \to \begin{bmatrix} 2 \\ 3 \end{bmatrix} = \mathbf{v},$$

and we have found the smallest $(A, 5)$-approximate eigenvector. □

Some interesting features and properties of the Approximate Eigenvector Algorithm are discussed in the Exercises.

EXERCISES

5.3.1. Show that the set of (A, n)-approximate eigenvectors is closed under sums, scalar multiplication by positive integers, and componentwise maximum.

5.3.2. Give an example to show that the (componentwise) minimum of two (A, n)-approximate eigenvectors need not be an (A, n)-approximate eigenvector.

5.3.3. Give an explicit upper bound, in terms of the initial vector $\mathbf{z}$, on the number of iterations needed for the Approximate Eigenvector Algorithm to halt.

5.3.4. Use the Approximate Eigenvector Algorithm to find (A, n)-approximate eigenvectors for the following A's and n's. Describe as carefully as you can the set of all the approximate eigenvectors in each case.

(a)
$$A = \begin{bmatrix} 1 & 0 & 1 \\ 1 & 1 & 0 \\ 0 & 1 & 1 \end{bmatrix}, \qquad n = 2.$$

(b)
$$A = \begin{bmatrix} 1 & 2 \\ 5 & 1 \end{bmatrix}, \qquad n = 4.$$

(c)
$$A = \begin{bmatrix} 3 & 2 \\ 2 & 0 \end{bmatrix}, \qquad n = 4.$$

(d)
$$A = \begin{bmatrix} 0 & 0 & 1 & 0 & 0 & 0 & 0 & 0 \\ 1 & 0 & 0 & 1 & 0 & 0 & 0 & 0 \\ 1 & 1 & 0 & 0 & 1 & 0 & 0 & 0 \\ 1 & 1 & 0 & 0 & 0 & 1 & 0 & 0 \\ 1 & 1 & 0 & 0 & 0 & 0 & 1 & 0 \\ 1 & 1 & 0 & 0 & 0 & 0 & 0 & 1 \\ 1 & 1 & 0 & 0 & 0 & 0 & 0 & 0 \\ 0 & 1 & 0 & 0 & 0 & 0 & 0 & 0 \end{bmatrix}, \qquad n = 2.$$

5.3.5. Let A be a nonnegative integer matrix and n a positive integer. For a vector $\mathbf{z}$, let $V_{\mathbf{z}}$ denote the set of all (A, n)-approximate eigenvectors $\mathbf{v}$ that are componentwise less than or equal to $\mathbf{z}$.

(a) Show that if $V_{\mathbf{z}} \neq \varnothing$ then there is a largest (componentwise) element of $V_{\mathbf{z}}$, i.e., one that dominates all the others.

(b) If $\mathbf{z}$ is the initial vector for the Approximate Eigenvector Algorithm, show that the outcome is $\mathbf{0}$ if $V_{\mathbf{z}} = \varnothing$, while it is the largest (componentwise) element in $V_{\mathbf{z}}$ otherwise.

(c) Show that if you apply the Approximate Eigenvector Algorithm to initial vectors $[k\, k\, \ldots\, k]$ with $k = 1, 2, 3, \ldots$, the smallest k for which the algorithm does not output $\mathbf{0}$ will be the maximal component of the smallest (in terms of maximum entry) (A, n)-approximate eigenvector (see Ashley [Ash3] for an explicit estimate on how big k needs to be to guarantee that the output is not $\mathbf{0}$).

(d) Show that if $\lambda_A \geqslant n$, then for all sufficiently large initial vectors $\mathbf{z}$, the output of the Approximate Eigenvector Algorithm is nonzero.

(e) Show that if A is irreducible and $\lambda_A \geqslant n$, then for all sufficiently large initial vectors $\mathbf{z}$, the output of the Approximate Eigenvector Algorithm is a positive (A, n)-approximate eigenvector.

(f) Let
$$A = \begin{bmatrix} 1 & 0 & 1 & 0 \\ 1 & 1 & 0 & 1 \\ 1 & 1 & 0 & 0 \\ 0 & 1 & 0 & 0 \end{bmatrix}.$$

Show that although A is irreducible, its smallest $(A, 2)$-approximate eigenvector has a 0 entry.

5.3.6. (a) Give an example of a reducible nonnegative integer matrix A with $\lambda_A \geqslant n$ such that there is no positive (A, n)-approximate eigenvector.

(b) Formulate a necessary and sufficient condition on such A, n for the existence of a positive (A, n)-approximate eigenvector.

5.3.7. Use approximate eigenvectors to construct all nonnegative adjacency matrices A with $\lambda_A \geqslant n$ from lists of nonnegative integers.

*5.3.8. Recall the adjugate of a matrix, defined in Exercise 4.2.12. Let G be an irreducible graph with $\lambda_{A_G} \geqslant n$. Assume also that for any graph H obtained from G by deleting at least one edge of G, $\lambda_{A_H} < n$. Show that any column of $\mathrm{adj}(nId - A_G)$ is an (A_G, n)-approximate eigenvector.

§5.4. Code Construction

The aim of this section is to complete the proof of the Finite-State Coding Theorem, and to use the ideas of the proof to construct practical codes. Recollect the statement of the theorem: if X is sofic and $n \geqslant 1$, then there is a finite-state (X, n)-code if and only if $h(X) \geqslant \log n$. We showed in §5.2 that the existence of such a code implies that $h(X) \geqslant \log n$. Conversely, if $h(X) \geqslant \log n$, we will construct a finite-state (X, n)-code, and we show how this construction works on some examples.

Suppose we are confronted with a sofic shift X and an integer $n \geqslant 1$. We begin by using the subset construction in Theorem 3.3.2 to find a right-resolving presentation $\mathcal{K}$ of X, although in practice sofic shifts are usually described by right-resolving presentations to begin with. We next use Theorem 4.4.4 to pick out an irreducible component $\mathcal{G} = (G, \mathcal{L})$ of $\mathcal{K}$ with $h(X_\mathcal{G}) = h(X_\mathcal{K}) = h(X) \geqslant \log n$. The heart of the construction is a sequence of elementary state splittings starting with G and ending with a graph H having minimum out-degree at least n. As we indicated in §5.2, from there it is easy to construct the required finite-state code using a subgraph of H as the encoder graph. The following gives a precise statement of the state splitting argument.

Theorem 5.4.1. *Let G be an irreducible graph with adjacency matrix A. If $\lambda_A \geqslant n$, then there is a sequence*

$$G = G_0, \, G_1, \, G_2, \, \ldots, \, G_m = H$$

such that G_j is an elementary state splitting of G_{j-1} for $1 \leqslant j \leqslant m$, and H has minimum out-degree at least n.

If $\mathbf{v}$ is a positive (A, n)-approximate eigenvector and k denotes the sum of its components, then there is such a sequence of m splittings with

$$m \leqslant k - |\mathcal{V}(G)| \quad and \quad |\mathcal{V}(H)| \leqslant k.$$

The proof of this theorem will be given later in this section. But we first use it to complete the proof of the Finite-State Coding Theorem.

PROOF OF THEOREM 5.2.5: Let X be sofic and $n \geqslant 1$ with $h(X) \geqslant \log n$. The discussion at the beginning of this section shows that we may assume that $X = X_{\mathcal{G}}$, where $\mathcal{G} = (G, \mathcal{L})$ is irreducible and right-resolving. Then $h(X_G) = h(X_{\mathcal{G}}) = h(X) \geqslant \log n$. Hence if $A = A_G$, then $\lambda_A \geqslant n$.

According to Theorem 5.4.1, there is a sequence

$$G = G_0, \, G_1, \, G_2, \, \ldots, \, G_m = H$$

of elementary state splittings such that H has minimum out-degree at least n. The labeling $\mathcal{L}$ of G successively carries through these state splittings to a labeling $\mathcal{L}'$ of H such that $\mathcal{L}'_\infty(X_H) = \mathcal{L}_\infty(X_G) = X$. Applying Proposition 5.1.8 m times shows that $(H, \mathcal{L}')$ is right-closing with delay m.

Since H has minimum out-degree at least n, we can choose a subgraph $\widehat{H}$ of H with constant out-degree n. Let $\mathcal{O}$ be the restriction of $\mathcal{L}'$ to $\widehat{H}$, and $\mathcal{I}$ be any road-coloring of $\widehat{H}$. Then $(\widehat{H}, \mathcal{O})$ remains right-closing with delay m, and $\mathcal{O}_\infty(X_{\widehat{H}}) \subseteq \mathcal{L}'_\infty(X_H) = X$. Hence $(\widehat{H}, \mathcal{I}, \mathcal{O})$ is a finite-state (X, n)-code. $\square$

We will prove Theorem 5.4.1 by repeated use of the following, which you might think of as "state splitting by weight splitting."

Proposition 5.4.2. *Let G be an irreducible graph, and $\mathbf{v} > 0$ be an (A_G, n)-approximate eigenvector. Suppose that the minimum out-degree of G is $< n$. Then there exist a state I, an elementary state splitting G' of G at I with split states I^1 and I^2, and an $(A_{G'}, n)$-approximate eigenvector $\mathbf{v}' > 0$ such that $v'_J = v_J$ for $J \neq I$ and $v'_{I^1} + v'_{I^2} = v_I$.*

The proof is based on the following simple lemma.

Lemma 5.4.3. *Let $k_1, k_2, \ldots, k_n$ be positive integers. Then there is a subset $S \subseteq \{1, 2, \ldots, n\}$ such that $\sum_{q \in S} k_q$ is divisible by n.*

PROOF: The partial sums $k_1, \, k_1 + k_2, \, k_1 + k_2 + k_3, \, \ldots, \, k_1 + k_2 + \cdots + k_n$ either are all distinct (mod n), or two are congruent (mod n). In the former case, at least one partial sum must be $\equiv 0$ (mod n). In the latter, there are $1 \leqslant m < p \leqslant n$ such that

$$k_1 + k_2 + \cdots + k_m \equiv k_1 + k_2 + \cdots + k_p \pmod{n}.$$

Hence $k_{m+1} + k_{m+2} + \cdots + k_p \equiv 0 \pmod{n}$. $\square$

We can now turn to the proof of Proposition 5.4.2.

PROOF OF PROPOSITION 5.4.2: Put $p = \max_I v_I$.

We first choose a state I for which $v_I = p$, and such that there is an edge e with initial state I and for which $v_{t(e)} < p$. In order to see that such a state must exist, start with a state J for which $v_J = p$. Observe that there must be a K for which $v_K < p$, for otherwise $\mathbf{1}$ would be an (A_G, n)-approximate eigenvector, and so G would have minimum out-degree at least n. Since G is irreducible, there is a path from J to K, and the last state I along this path for which $v_I = p$ is our choice for I.

We must now partition the set $\mathcal{E}_I$ of outgoing edges from I. The Ith component of the vector inequality $A_G \mathbf{v} \geqslant n\mathbf{v}$ is

$$\sum_{e \in \mathcal{E}_I} v_{t(e)} \geqslant n\, v_I = n\, p.$$

Now each $v_{t(e)}$ is at most p, and at least one is strictly less than p. Hence

$$\sum_{e \in \mathcal{E}_I} v_{t(e)} < p\, |\mathcal{E}_I|,$$

so that if we put $m = |\mathcal{E}_I|$, then $m > n$. Let $\mathcal{E}_I = \{e_1, \ldots, e_n, \ldots, e_m\}$, where e_1 is the special edge e for which $v_{t(e)} < p$. Applying Lemma 5.4.3 to the n positive integers $v_{t(e_1)}, v_{t(e_2)}, \ldots, v_{t(e_n)}$, shows there is a subset $\mathcal{E}_I^1$ of $\{e_1, e_2, \ldots, e_n\}$ for which

$$(5\text{-}4\text{-}1) \qquad \sum_{e \in \mathcal{E}_I^1} v_{t(e)} \equiv 0 \pmod{n}.$$

We set $\mathcal{E}_I^2 = \mathcal{E}_I \setminus \mathcal{E}_I^1$, which is nonempty since $n < m$. Let G' be the resulting state splitting of G using the partition $\mathcal{E}_I^1 \cup \mathcal{E}_I^2$ of $\mathcal{E}_I$.

By (5-4-1), the number

$$q = \frac{1}{n} \sum_{e \in \mathcal{E}_I^1} v_{t(e)}$$

is a positive integer. Define $\mathbf{v}'$ by

$$v'_J = \begin{cases} q & \text{if } J = I^1, \\ v_I - q & \text{if } J = I^2, \\ v_J & \text{otherwise.} \end{cases}$$

We complete the proof by showing that $\mathbf{v}' > 0$, and that $\mathbf{v}'$ is an $(A_{G'}, n)$-approximate eigenvector. Since q and the entries in $\mathbf{v}$ are positive integers, the only component of $\mathbf{v}'$ in doubt is at index $J = I^2$. But

$$q = \frac{1}{n} \sum_{e \in \mathcal{E}_I^1} v_{t(e)} \leqslant \frac{1}{n} \sum_{j=1}^{n} v_{t(e_j)} < p$$

since each $v_{t(e_j)} \leqslant p$ and $v_{t(e_1)} < p$. Hence $v_I - q = p - q > 0$, showing that $\mathbf{v}' > 0$.

To show that $\mathbf{v}'$ is an approximate eigenvector, we must prove that for each state J of G', $\sum_{e\in\mathcal{E}_J} v'_{t(e)} \geqslant n v_J$. For this, first observe that each incoming edge e at I is split into e^1 and e^2, but that the sum $v'_{I^1} + v'_{I^2}$ of the weights of their terminal states equals the weight v_I of the terminal state of e. From this it follows that if J is not I^1 or I^2, then the sums in the approximate eigenvector inequality are unchanged and therefore the inequality remains true. For $J = I^1$ we have that

$$\sum_{e\in\mathcal{E}_{I^1}} v'_{t(e)} = \sum_{e\in\mathcal{E}_I^1} v_{t(e)} = n\,q = n\,v'_{I^1},$$

while for $J = I^2$,

$$\sum_{e\in\mathcal{E}_{I^2}} v'_{t(e)} = \sum_{e\in\mathcal{E}_I\setminus\mathcal{E}_I^1} v_{t(e)} = \left(\sum_{e\in\mathcal{E}_I} v_{t(e)}\right) - n\,q$$

$$\geqslant n\,v_I - n\,q = n(v_I - q) = n\,v'_{I^2}. \qquad \square$$

We can now finish the proofs in this section.

PROOF OF THEOREM 5.4.1: Let $\mathbf{v}$ be a positive (A_G, n)-approximate eigenvector, and k be the sum of its components. If G has minimum out-degree at least n, we are done. If not, then by Proposition 5.4.2 we can state split G to form an irreducible graph G' having one more state, and find an $(A_{G'}, n)$-approximate eigenvector $\mathbf{v}' > 0$ whose components also sum to k. If G' has minimum out-degree n, we are done. Otherwise, keep applying Proposition 5.4.2. To see that this process must eventually stop, note that the approximate eigenvectors $\mathbf{v}, \mathbf{v}', \mathbf{v}'', \ldots$ become taller and taller, while the positive integers in each vector always sum to k. The positive integer weights can only spread so far, and we must stop after at most $k - |\mathcal{V}(G)|$ times, resulting in a graph H with at most k states. $\qquad\square$

The ideas from these proofs can be used in concrete situations to construct codes. We will give some examples of encoding arbitrary binary data into run-length limited shifts.

Example 5.4.4. Let $X = X(1,3)$, with presentation $\mathcal{G}$ shown in Figure 5.4.1(a).

In Example 4.3.4 we found that $h(X) = h(X_{\mathcal{G}}) = 0.55146$. Therefore it is possible to construct a finite-state $(X^2, 2)$-code, i.e., a rate $1\!:\!2$ code (see §5.2). Recall from Exercise 3.1.7 the notion of a higher power labeled graph. The second power graph $\mathcal{G}^2$ presenting X^2 is shown in (b). The adjacency matrix of its underlying graph is

$$A = \begin{bmatrix} 1 & 0 & 1 & 0 \\ 1 & 1 & 0 & 1 \\ 1 & 1 & 0 & 0 \\ 0 & 1 & 0 & 0 \end{bmatrix}.$$

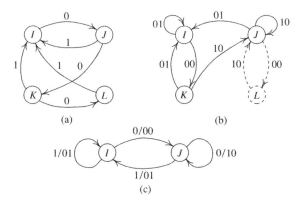

FIGURE 5.4.1. Code construction for the MFM code.

Applying the Approximate Eigenvector Algorithm gives

$$\mathbf{v} = \begin{bmatrix} 1 \\ 1 \\ 1 \\ 0 \end{bmatrix}$$

as an $(A, 2)$-approximate eigenvector. Since $v_4 = 0$, we can eliminate state L and all incident edges, as indicated by the dotted lines in (b). We are in luck, since the remaining graph has out-degree 2 at each state. At this point, we could simply color this graph using input symbols 0 and 1, and be done. However, notice that now states J and K have the same follower set, so we can merge K and J to obtain a smaller graph. The final result, together with an input road-coloring, is shown in (c). This is precisely the MFM code, as described in Example 5.2.6. □

Example 5.4.5. Let $X = X(0, 1)$, with minimal right-resolving presentation $\mathcal{G}$ shown in Figure 5.4.2(a). Note that X is obtained from the golden mean shift by interchanging 0's and 1's, and we computed the entropy of the latter in Example 4.1.4 to be $\log[(1 + \sqrt{5})/2] \approx 0.69424$. Since $2/3$ is less than $h(X)$, there is a rate 2:3 code, i.e., a finite-state $(X^3, 2^2)$-code.

We first compute $\mathcal{G}^3$, which is shown in (b), and has underlying graph with adjacency matrix

$$A = \begin{bmatrix} 3 & 2 \\ 2 & 1 \end{bmatrix}.$$

Now $\mathcal{G}^3$ does not have minimum out-degree 4, so we need to invoke state splitting. To determine which state to split, we first use the Approximate Eigenvector Algorithm to find that $\mathbf{v} = \begin{bmatrix} 2 \\ 1 \end{bmatrix}$ is an $(A, 4)$-approximate eigenvector. State I is the only state for which v_I is maximal, so this is the state to split. There are five edges leaving I, three self-loops each having weight 2,

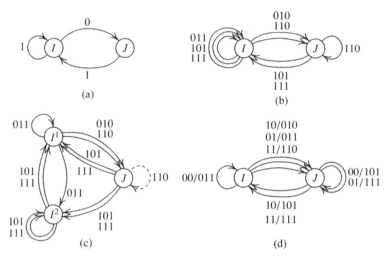

FIGURE 5.4.2.   Code construction for the $(0,1)$ run-length limited shift.

and two edges each having weight 1. We must find a subset $\mathcal{E}_I^1$ of these whose weights sum to a multiple of 4. Among the many ways to do this, we choose for $\mathcal{E}_I^1$ the edges labeled 011, 010, and 110, whose weights add to $2 + 1 + 1 = 4$. The weight of the split state I^1 is therefore $q = 4/4 = 1$, and I^2 gets the remaining weight, in this case $2 - 1 = 1$. The result of this splitting is shown in (c). Since the new approximate eigenvector has only 1's as components, the split graph must have minimum out-degree at least 4, as is indeed the case.

We could trim one edge from state J, road-color the remaining graph, and be done. Notice, however, that if we prune away the self-loop at J, then I^2 and J have the same follower set. We can then merge I^2 into J, which is shown in (d) together with an input road-coloring. This is the finite-state code from Example 5.2.7.   □

These examples illustrate the *State Splitting Algorithm*, which is a recipe to construct a rate $p\!:\!q$ finite-state code from binary data to a given sofic shift. Recall from §5.2 that this is possible whenever $h(X) \geqslant p/q$. The following steps describe the construction.

   (1) Find a right-resolving presentation $\mathcal{K}$ of X with underlying graph K (the minimal right-resolving presentation found in Theorem 3.4.14 may be the easiest starting point).
   (2) Compute $h(X)$ as $\log \lambda_{A(K)}$.
   (3) Choose integers p and q so that $h(X) \geqslant p/q$.
   (4) Construct $\mathcal{K}^q$, observing that $h(X_{\mathcal{K}^q}) \geqslant \log 2^p$.
   (5) Use the Approximate Eigenvector Algorithm to find an $(A(K^q), 2^p)$-approximate eigenvector $\mathbf{v}$. Delete all states I for which $v_I = 0$, and restrict to a sink component $\mathcal{G}$ of this graph (see Lemma 5.3.4).

(6) Keep applying the state splitting method of Proposition 5.4.2 until you obtain another presentation $\mathcal{H}$ of X with minimum out-degree at least 2^p. Theorem 5.4.1 guarantees this will succeed. Observe that $\mathcal{H}$ is right-closing.

(7) Prune edges from $\mathcal{H}$ if necessary to obtain $\widehat{\mathcal{H}} = (\widehat{H}, \mathcal{O})$ with constant out-degree 2^p. Use as input labeling $\mathcal{I}$ a road-coloring of G using binary p-blocks. The resulting $(\widehat{H}, \mathcal{I}, \mathcal{O})$ is the desired finite-state code.

Although we will not discuss complexity issues in any detail, in Remark 5.4.6 we mention a few practical considerations in code construction. One of the primary aims is, naturally, to construct codes that are as "simple" as possible, where simplicity is measured by the number of states in the encoder graph as well as the decoding delay.

Remark 5.4.6. (a) Let X be a sofic shift with $h(X) \geqslant \log(n)$, and let A be the adjacency matrix of a right-resolving presentation of X. Theorem 5.4.1 yields an upper bound on the size of the smallest finite-state (X, n)-code: there is such a code with at most $\min_v \sum_i v_i$ states, where the minimum is taken over all (A, n)- approximate eigenvectors $\mathbf{v}$. Theorem 5.4.1 also gives an upper bound on the smallest closing delay of any finite-state (X, n)-code: $\min_v \sum_i (v_i - 1)$, where again the minimum is taken over all (A, n)-approximate eigenvectors $\mathbf{v}$. For other bounds on these quantities, see [MarR], [AshMR1], [AshMR2], and [AshM].

(b) In the proof of Theorem 5.4.1, we constructed finite-state codes by using a positive approximate eigenvector $\mathbf{v}$. However, it can be simpler to use an approximate eigenvector with some zero components if available, and delete states as in step 5. We did this in Example 5.4.4.

(c) There is no need to restrict to elementary state splittings when applying Proposition 5.4.2. It may be more efficient to split many states at once, and into many pieces. One splits by finding partitions $\mathcal{P}_I = \{\mathcal{E}_I^1, \ldots, \mathcal{E}_I^{m(I)}\}$ of $\mathcal{E}_I$ and decompositions $v_I = \sum_{i=1}^{m(I)} v'_{I,i}$ into positive integers such that for $i = 1, \ldots, m(I)$,

$$\sum_{e \in \mathcal{E}_I^i} v_{t(e)} \geqslant n v'_{I,i}.$$

This will split state I into states $I^1, \ldots, I^{m(I)}$ with new approximate eigenvector components $v'_{I,1}, \ldots, v'_{I,m(I)}$. An out-splitting in which several states may be split simultaneously is often referred to as a *round* of splitting. See Exercises 5.4.1, 5.4.2 and 5.5.2 for examples of this.

(d) At any stage it may be possible to merge several states into one, as we did with Examples 5.4.4 and 5.4.5. This reduces the complexity of the final code. One criterion for state merging is as follows. Let $\mathcal{G} = (G, \mathcal{L})$ be a presentation of a sofic shift, and let $\mathbf{v}$ be an (A_G, n)-approximate eigenvector such that $v_I = v_J$ and $F_{\mathcal{G}}(I) \subseteq F_{\mathcal{G}}(J)$. Then states I and J

can be merged to form a new labeled graph which presents a subshift of X: one deletes J and all outgoing edges from J and then redirects into I all remaining incoming edges to J (see Exercise 3.3.11)). This yields a new graph G' with states $\mathcal{V}(G') = \mathcal{V}(G) \setminus \{J\}$ and the restriction of $\mathbf{v}$ to $\mathcal{V}_G \setminus \{J\}$ is an $(A_{G'}, n)$-approximate eigenvector (Exercise 5.4.6). This can reduce, by $v_I = v_J$, the number of encoder states. If $v_I < v_J$ and $F_{\mathcal{G}}(I) \subset F_{\mathcal{G}}(J)$, then it is natural to try to split state J into two states J_1, J_2 such that states I and J_1 can be merged as above; then for the split graph $\mathcal{G}'$, there will be an approximate eigenvector $\mathbf{v}'$ such that $v'_{J_1} = v_I$, $v'_{J_2} = v_J - v_I$ and $F_{\mathcal{G}'}(I) \subseteq F_{\mathcal{G}'}(J_1)$. Then states I and J_1 could be merged and this would reduce, by v_I, the number of encoder states. While this works often, it does not always work (for details, see [MarS] and [MarSW]).   □

EXERCISES

5.4.1. Find finite-state (X_A, n)-codes for the matrices A in Exercise 5.3.4 ((a), (b) and (c)).

5.4.2. Let X be the sofic shift presented by the labeled graph, whose symbolic adjacency matrix is:

$$\begin{bmatrix} a & b+c+d \\ e+f+g+h+i+j & a \end{bmatrix}.$$

By two rounds of out-splitting, find a finite-state $(X, 5)$-code. [*Hint:* See Remark 5.4.6(c).]

5.4.3. If G is irreducible and $\lambda_{A(G)} \geqslant n$, show that X_G is conjugate to an edge shift X_H where H has minimum out-degree at least n and minimum in-degree at least n.

5.4.4. If G is irreducible and $\lambda_{A(G)} \leqslant n$, show that X_G is conjugate to an edge shift X_H where H has maximum out-degree at most n.

*5.4.5. Show that if G is irreducible and $n \leqslant \lambda_{A(G)} \leqslant n+1$, then X_G is conjugate to an edge shift X_H such that for each state I of H, I has out-degree either n or $n+1$. [*Hint:* Show the following: suppose that n, M and $k_1, \ldots, k_m$ are positive integers, each $k_i \leqslant M$, some $k_i < M$, and $nM \leqslant \sum_{i=1}^{m} k_i < (n+1)M$; then, modulo $\alpha = (\sum_{i=1}^{m} k_i)/M$, two of the partial sums $k_1, k_1 + k_2, \ldots, k_1 + \ldots + k_m$, are within distance 1, i.e., for some $1 < r \leqslant s$, $k_r + \ldots + k_s = \ell\alpha + \theta$ where ℓ is an integer and $|\theta| < 1$.]

5.4.6. Explain why, in Remark 5.4.6 (d), the restriction of $\mathbf{v}$ to $\mathcal{V}_G \setminus \{J\}$ is an $(A_{G'}, n)$-approximate eigenvector.

*5.4.7. The following outlines an alternative proof of the Finite State Coding Theorem in certain cases; this sometimes yields simpler codes.

Let $G = (\mathcal{V}, \mathcal{E})$ be an irreducible graph, and let $\mathcal{G} = (G, \mathcal{L})$ be a right-resolving presentation of an irreducible sofic shift X with $h(X) \geqslant \log(n)$. Let $\mathbf{x}$ be an (A_G, n)-approximate eigenvector. Let H be the graph with states $\mathcal{V}(H) = \{(I, i) : I \in \mathcal{V}(G), 1 \leqslant i \leqslant x_I\}$ and edges $\mathcal{E}(H) = \{(e, j) : e \in \mathcal{E}, 1 \leqslant j \leqslant x_{t(e)}\}$; the edge (e, j) terminates at state $(t(e), j)$, and the initial states of the edges are determined as follows: partition the edges $\{(e, j) \in \mathcal{E}(H) :$

$e \in \mathcal{E}_I\}$ into x_I sets $Q_{(I,i)}, i = 1, \ldots, x_I$ each of size at least n, and declare (I, i) to be the initial state of the edges in $Q_{(I,i)}$.

(a) Explain why it is possible to find such partitions, and show that H has minimum out-degree at least n.

(b) Show that the labeling $\Phi \colon (e, j) \mapsto e$ is left-resolving and induces a factor code $\phi = \Phi_\infty \colon X_H \to X_G$.

(c) Explain how the proof of Theorem 5.4.1 gives an example of this setup where ϕ is a conjugacy.

   Parts (d), (e), and (f) assume that the partitions are constructed in the following way. For each state I, list the edges of $\mathcal{E}_I$ in some order $e_1, \ldots, e_k$; then list the elements $\{(e, j) : e \in \mathcal{E}_I\}$ in the lexicographic order:

$$(e_1, 1) \ldots (e_1, x_{t(e_1)}), (e_2, 1) \ldots (e_2, x_{t(e_2)}), \ldots (e_k, 1) \ldots (e_k, x_{t(e_k)}),$$

   then define $Q_{(I,1)}$ to be the first set of n elements, $Q_{(I,2)}$ to be the next set of n elements, . . ., and $Q_{(I,x_I)}$ to be the remaining set of elements.

(d) Show that if the partitions are constructed in this way, then the factor code ϕ is at most 2-1.

(e) Show how, if the partitions are constructed in this way, this gives a finite-state $(X, n-1)$-code.

(f) Find an example of this where the partitions are constructed as in (d) but Φ is not right-closing and another example where Φ is right-closing.

*5.4.8. The following outlines an alternative proof of the Finite State Coding Theorem in the case that $h(X) > \log(n)$. Let $\mathcal{G} = (G, \mathcal{L})$ be a right-resolving presentation of an irreducible sofic subshift of X with maximal entropy. Let $I \in \mathcal{V}(G)$ and let $\Gamma(I)$ denote the set of all cycles in G which meet I only at their initial states.

(a) Show that there is a finite set $\Gamma' \subseteq \Gamma(I)$ such that $\sum_{\gamma \in \Gamma'} n^{-\ell(\gamma)} > 1$.

(b) Find a list P of n-ary blocks such that (i) every n-ary block either is a prefix of a block in P or has a prefix which is a block in P and (ii) the set of lengths (with multiplicity) of the blocks in P is the same as the set of lengths (with multiplicity) of the paths in Γ'.

(c) Show how to construct a finite-state (X, n)-code by making a length-preserving correspondence between P and Γ'.

§5.5. Sliding Block Decoders

Suppose that we use a finite-state code $(G, \mathcal{I}, \mathcal{O})$ to transform input $x = x_0 x_1 x_2 \ldots$ into output $y = y_0 y_1 y_2 \ldots$ as described in §5.2. In order to recover x, the decoder needs to keep careful track of which state of G it is using at each step. One erroneous symbol in y could cause the decoder to lose track of the current state, forcing it to remain out of "synch" forever. The following is a simple instance of this behavior.

Example 5.5.1. Consider the finite-state code shown in Figure 5.5.1(a). With I_0 as initial state, the input $00000 \ldots$ is encoded to $aaaaa \ldots$. But mistakes happen, and suppose the first output symbol is recorded as b instead of a. The decoder would then reconstruct the input as $11111 \ldots$, not $00000 \ldots$, with every bit wrong. $\square$

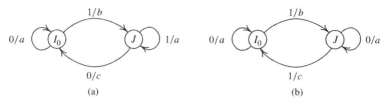

FIGURE 5.5.1. Bad and good input labelings.

This kind of infinite error propagation can be prevented by using a finite-state code with a sliding block decoder.

Definition 5.5.2. Let $(G, \mathfrak{I}, \mathcal{O})$ be a finite-state (X, n)-code. A *sliding block decoder* for $(G, \mathfrak{I}, \mathcal{O})$ is a sliding block code $\phi \colon X \to X_{[n]}$ such that $\phi \circ \mathcal{O}_\infty = \mathfrak{I}_\infty$; i.e., the following diagram commutes.

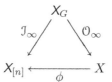

In this definition, we only need ϕ to be defined on the image of $\mathcal{O}_\infty$. But any such mapping can be extended to all of X (see Exercise 1.5.13(a)).

Example 5.5.3. Consider again the finite-state code in Figure 5.5.1(a). Now $\mathcal{O}_\infty$ collapses both fixed points in X_G to a^∞, while $\mathfrak{I}_\infty$ keeps them distinct, so there cannot be a sliding block decoder ϕ with $\mathfrak{I}_\infty = \phi \circ \mathcal{O}_\infty$ for this code.

However, suppose we switch the input labeling on $\mathcal{E}_J$, resulting in (b). This new finite-state code *does* have a sliding block decoder, namely the 1-block code induced by the 1-block map Φ defined by $\Phi(a) = 0$, $\Phi(b) = \Phi(c) = 1$.

Up until now the choice of input road-coloring has been arbitrary. This example shows that some input road-colorings are preferable to others. □

If a finite-state code $(G, \mathfrak{I}, \mathcal{O})$ has a sliding block decoder ϕ, then there is a decoding method that does not need to keep track of the state in G. For suppose that $\phi = \Phi_\infty^{[-m,n]}$, and that $y = y_0 y_1 y_2 \ldots$ is an output sequence. For each $k \geqslant m$, there is a path $e_{k-m} e_{k-m+1} \cdots e_{k+n}$ in G with

$$\mathcal{O}(e_{k-m} e_{k-m+1} \cdots e_{k+n}) = y_{k-m} y_{k-m+1} \cdots y_{k+n}.$$

Since $\phi \circ \mathcal{O}_\infty = \mathfrak{I}_\infty$, it follows that

$$x_k = \mathfrak{I}(e_k) = \Phi(y_{k-m} y_{k-m+1} \cdots y_{k+n}).$$

Thus we can recover all but the first m input symbols by using Φ on blocks of output symbols. The key advantage of using Φ is that an erroneous output symbol can affect decoding only when it occurs in the window for Φ, and so has only finite error propagation.

Many of our previous examples of finite-state codes turn out to have sliding block decoders.

Example 5.5.4. The encoder giving the MFM code, shown in Figure 5.2.3, has a 1-block decoder induced by the 1-block map $\Phi(00) = \Phi(10) = 0$, $\Phi(01) = 1$. □

Example 5.5.5. The rate 2:3 code using the $(0, 1)$ run-length limited shift shown in Figure 5.2.4 has a sliding block decoder of the form $\phi = \Phi_\infty^{[0,1]}$. Here Φ is the mapping from pairs u, v of 3-blocks from $X(0,1)$ to 2-blocks of bits tabulated below, where an entry "$*$" means that any 3-block can appear there.

u	v	$\Phi(uv)$
010	$*$	10
011	010	00
011	011	00
011	110	00
011	101	01
011	111	01
101	101	00
101	111	00
101	011	10
101	010	10
101	110	10
110	$*$	11
111	101	01
111	111	01
111	011	11
111	010	11
111	110	11

□

Since finite-state codes with sliding block decoders are so desirable, when can we be sure of finding them? When the constraint shift has finite type, there is no added difficulty.

Theorem 5.5.6 (Sliding Block Decoding Theorem). *Let X be a shift of finite type with $h(X) \geqslant \log n$. Then there is a finite-state (X, n)-code having a sliding block decoder.*

PROOF: By passing to an irreducible subshift of maximal entropy, we may assume that X is irreducible. Let $\mathcal{G} = (G, \mathcal{L})$ be any presentation of X such that $\mathcal{L}_\infty \colon X_G \to X_\mathcal{G}$ is a conjugacy; for instance, $\mathcal{G}$ could be the minimal right-resolving presentation of X (see Theorem 3.4.17). Since state splittings induce conjugacies, using $\mathcal{G}$ as the starting point in the proof of the Finite-State Coding Theorem, we arrive at a labeled graph $(H, \mathcal{L}')$ for which $\mathcal{L}'_\infty$ is a conjugacy from X_H to X. By pruning edges from H, we get a graph $\widehat{H}$ with constant out-degree n and a finite-state code $(\widehat{H}, \mathcal{I}, \mathcal{O})$, where $\mathcal{O}$ is the restriction of $\mathcal{L}'$ to $\widehat{H}$. By Theorems 1.5.13 and 1.5.14, $\mathcal{O}_\infty^{-1}$ is a sliding block code. Then $\phi = \mathcal{I}_\infty \circ \mathcal{O}_\infty^{-1}$ is also a sliding block code, and clearly $\phi \circ \mathcal{O}_\infty = \mathcal{I}_\infty \circ \mathcal{O}_\infty^{-1} \circ \mathcal{O}_\infty = \mathcal{I}_\infty$, so ϕ is the required sliding block decoder.   □

Remark 5.5.7. (a) Recall that the (d, k) run-length limited shifts have finite type, so Theorem 5.5.6 includes encoding into run-length constraints. In Exercise 5.5.5 we indicate how to extend Theorem 5.5.6 to sofic X when there is strict inequality $h(X) > \log n$. There are sofic X with $h(X) = \log n$ for which there are no finite-state (X, n)-codes with sliding block decoders [KarM].

(b) It may seem that, when constructing finite-state codes, it is always best to use a "small" approximate eigenvector: the smaller the vector, the less splitting one needs to do. Using a "large" approximate eigenvector typically yields finite-state codes with a "large" number of encoder states. However, sometimes using a "large" vector yields "small" decoding window, $m + n + 1$. Examples of this are contained in [Kam], [AshM]. There are examples where state splitting using the smallest (in any sense) approximate eigenvector yields a decoder with window size 2, but using a larger approximate eigenvector yields a decoder with window size only 1. In these examples, there is no 1-block decoder (i.e., $m = n = 0$), but there is a decoder which has memory $m = -1$ and anticipation $n = 1$, so that one decodes by looking into the future and ignoring the present [Imm3].   □

A consequence of Theorem 5.5.6 is the following result which says that among all shifts of finite type with entropy at least $\log n$, the full n-shift is the "simplest" in the sense that all the others factor onto it.

Theorem 5.5.8. *Let X be a shift of finite type with $h(X) \geqslant \log n$. Then X factors onto the full n-shift.*

PROOF: By Theorem 5.5.6, there is a finite-state (X, n)-code $(G, \mathcal{I}, \mathcal{O})$ with sliding block decoder ϕ. Then

$$X_{[n]} = \mathcal{I}_\infty(X_G) = \phi(\mathcal{O}_\infty(X_G)) \subseteq \phi(X) \subseteq X_{[n]},$$

so that ϕ is onto, hence a factor code.   □

EXERCISES

5.5.1. For the sofic shift X in Exercise 5.4.2, find a finite-state $(X, 5)$-code with a sliding block decoder that has memory 0 and anticipation 2.

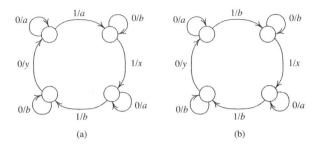

FIGURE 5.5.2.   Graphs for Exercise 5.5.4.

5.5.2. (a) Construct a rate 2:3 finite-state code from binary data into the run-length limited shift $X(1,7)$ that has a sliding block decoder. Use state merging, as in Remark 5.4.6 (d), to try to minimize the number of states in your answer.

  (b) Do the same for a rate 1:2 code from binary data into $X(2,7)$.

5.5.3. Let $(G, \mathcal{I}, \mathcal{O})$ be a finite-state code with G essential. Show that the following are equivalent.

  (a) $(G, \mathcal{I}, \mathcal{O})$ has a sliding block decoder.
  (b) There exist integers $1 \leqslant r \leqslant s$, such that the sequence of output labels of a path of length s uniquely determines its rth input label.
  (c) If $x, y \in X_G$ and $\mathcal{O}_\infty(x) = \mathcal{O}_\infty(y)$, then $\mathcal{I}_\infty(x) = \mathcal{I}_\infty(y)$.

5.5.4. Decide whether the finite-state codes described in Figure 5.5.2 have sliding block decoders.

5.5.5. (a) Show that any sofic shift X contains shifts of finite type with entropy arbitrarily close to $h(X)$. [*Hint*: Find an irreducible sofic subshift $Y \subseteq X$ of maximal entropy, and then use a synchronizing word in the minimal right-resolving presentation of Y to construct the desired shifts of finite type.]

  (b) Show that if a shift X contains sofic subshifts of entropy arbitrarily close to $h(X)$ (in particular, if X is itself sofic) and $h(X) > p/q$, then there is a rate $p : q$ finite-state code from binary data into X with a sliding block decoder.

5.5.6. Show that if X is an irreducible shift of finite type and $h(X) = \log n$, then there is a factor code $\phi : X \to X_{[n]}$ such that each point has only finitely many pre-images under ϕ (this sharpens Theorem 5.5.8).

*5.5.7. Let G be a primitive graph. We say that a finite-state (X, n)-code $(G, \mathcal{I}, \mathcal{O})$ is *noncatastrophic* if whenever x^+, y^+ are right-infinite paths in X_G^+ whose $\mathcal{O}$–labels differ in only finitely many coordinates, then their $\mathcal{I}$–labels differ in only finitely many coordinates.

  (a) Explain the use of the term "noncatastrophic."
  (b) Show that a finite-state code is noncatastrophic if and only if whenever x, y are *periodic* bi-infinite sequences in X_G and $\mathcal{O}_\infty(x) = \mathcal{O}_\infty(y)$, then $\mathcal{I}_\infty(x) = \mathcal{I}_\infty(y)$.
  (c) Show that if a finite-state code has a sliding block decoder, then it is noncatastrophic.
  (d) Of the finite-state codes in Exercise 5.5.4, which are noncatastrophic?
  (e) Show how to view a convolutional encoder (see §1.6) as a finite-state code.

Show that such a finite-state code is noncatastrophic if and only if it has a sliding block decoder.

Notes

Motivated by a classification problem in ergodic theory, Adler, Goodwyn and Weiss [AdlW],[AdlGW] introduced road-colorings. They realized that a factor code induced by a road-coloring is one-to-one "almost everywhere" if the road-coloring has a synchronizing word, and this gave birth to the Road Problem. Before Trahtman's solution to the Road Problem [Tra], special cases were solved by Friedman [Fri2], O'Brien [Obr], Perrin and Schutzenberger [PerS], and Culik, Karhumäki, and Kari [CulKK]. Trahtman's proof built on some of these results.

Right-closing graphs, a natural generalization of right-resolving graphs, were introduced by Kitchens [Kit1] (actually, he introduced a more general notion which will be considered in §8.1). This same concept is known as *lossless of finite order* in automata theory (see Even [Eve1], Kohavi [Koh2], Kurmit [Kur]) and is sometimes called *finite local anticipation* in coding theory. Proposition 5.1.11, which shows that right-closing labelings can be recoded to right-resolving labelings, is due to Kitchens [Kit1].

A finite-state code, sometimes called a *finite-state encoder*, is a standard device used in coding theory to encode sequences for various purposes. Finite-state codes are closely related to *transducers* in automata theory; see [Bea4].

There is a large body of literature on the kinds of coding problems that we have considered in this chapter. Most notable is Franaszek's pioneering work on code construction [Fra1], [Fra2], [Fra3], [Fra4], [Fra5], [Fra6], [Fra7], [Fra8], [Fra9]. Franaszek introduced approximate eigenvectors and what we call the Approximate Eigenvector Algorithm. As mentioned earlier, Patel [Pat] employed state splitting as a tool in his construction of a particular encoder used for storing data on magnetic tape. For more recent developments, we refer the reader to [Imm3], [Imm4], [Hol2], [FraT], [AshMR1], [AshM].

Adler, Coppersmith, and Hassner [AdlCH] introduced the State Splitting Algorithm to construct finite-state codes with sliding block decoders for shifts of finite type. This was partly based on earlier work of Marcus [Mar1], which in turn was inspired by [AdlGW] and the notion of state splitting [WilR2]. For further reading on various aspects of the code construction problem, see the tutorial by Marcus, Siegel, and Wolf [MarSW], the monograph by Khayrallah and Neuhoff [KhaN1], and the book chapter by Marcus, Roth, and Siegel [MarRS].

The state splitting algorithm can also be viewed as a natural outgrowth of work on finite equivalence [Par3] (see Chapter 8) and almost conjugacy [AdlM] (see Chapter 9). The connection between this circle of problems and coding applications was made by Hassner [Has].

The extension of the State Splitting Algorithm to obtain finite-state codes for sofic shifts was straightforward, and this is what we called the Finite-State Coding Theorem in this chapter. As mentioned in the text, it is not always possible to get sliding block decoders for irreducible sofic shifts X when $h(X) = \log(n)$. But it is always possible to get noncatastrophic codes (see Exercise 5.5.7), and under reasonable assumptions one can get sliding block decoders as well [Mar2], [KarM].

Theorem 5.5.8 provides factor codes from certain shifts of finite type onto full shifts. The general problem of deciding when one irreducible shift of finite type factors onto another is treated in §10.3 and Chapter 12.

Exercise 5.3.8 is due to Ashley [Ash3]; Exercise 5.4.5 to Friedman [Fri1]; Exercise 5.4.7 to Ashley, Marcus, and Roth [AshMR1]; and Exercise 5.4.8 to Béal [Bea1].

Erdos credits Lemma 5.4.3 to A. Vazsonyi and M. Sved [AigZ](Chapter 28, part 3).

CHAPTER 6

SHIFTS AS DYNAMICAL SYSTEMS

Symbolic dynamics is part of a larger theory of dynamical systems. This chapter gives a brief introduction to this theory and shows where symbolic dynamics fits in. Certain basic dynamical notions, such as continuity, compactness, and topological conjugacy, have already appeared in a more concrete form. So our development thus far serves to motivate more abstract dynamical ideas. On the other hand, applying basic concepts such as continuity to the symbolic context shows why the objects we have been studying, such as sliding block codes, are natural and inevitable.

We begin by introducing in §6.1 a class of spaces for which there is a way to measure the distance between any pair of points. These are called metric spaces. Many spaces that you are familiar with, such as subsets of 3-dimensional space, are metric spaces, using the usual Euclidean formula to measure distance. But our main focus will be on shift spaces as metric spaces. We then discuss concepts from analysis such as convergence, continuity, and compactness in the setting of metric spaces. In §6.2 we define a dynamical system to be a compact metric space together with a continuous map from the space to itself. The shift map on a shift space is the main example for us. When confronted with two dynamical systems, it is natural to wonder whether they are "the same," i.e., different views of the same underlying process. This leads to the notion of conjugacy, and related ideas of embedding and factor map, which extend the notions we have already encountered for shift spaces.

We have already seen several ways to show that two shift spaces are *not* conjugate. We find an object, such as entropy or the number of fixed points, such that conjugate shifts are assigned the same value of the object. Then if two shifts are assigned different values, we know for sure that they cannot be conjugate (although we cannot necessarily say anything if they are assigned the same value). For example, the full 2-shift cannot be conjugate to the golden mean shift, since they have different entropy, or since they have different numbers of fixed points. Such an assignment is called an invariant of conjugacy. In §6.3 we discuss several invariants for conjugacy, and show how these translate to the symbolic setting. One revealing way to summarize the periodic point information about a transformation is by a function called its zeta function. In §6.4 we define this invariant and show how to compute it for shifts of finite type and sofic shifts. Finally, in §6.5 we discuss how shifts of finite type are used to model smooth dynamical systems.

§6.1. Metric Spaces

Measuring distances is one of the fundamental notions in geometry and analysis. Indeed, the word "geometry" itself means "earth measuring." Many spaces have a natural way to measure distances between pairs of points. With a subset of $\mathbb{R}^n$ such as a hollow sphere, for instance, we can define the distance between two points to be the length of the line segment between them.

The properties we need of such a distance function are captured in the following definition.

Definition 6.1.1. A *metric space* (M, ρ) consists of a set M together with a *metric* (or *distance function*) $\rho \colon M \times M \to [0, \infty)$ such that, for all points $x, y, z \in M$,

(1) $\rho(x, y) = 0$ if and only if $x = y$,
(2) $\rho(x, y) = \rho(y, x)$,
(3) $\rho(x, z) \leqslant \rho(x, y) + \rho(y, z)$.

Condition 3 is called the *triangle inequality* since, when used with the metric on $\mathbb{R}^n$ in Example 6.1.3 below, it says that the length of one side of a triangle is always less than or equal to the sum of the lengths of the other two sides.

For each of the following examples, you should check that conditions (1), (2), and (3) in Definition 6.1.1 are satisfied.

Example 6.1.2. Let $M = \mathbb{R}$ and $\rho(x, y) = |x - y|$. $\qquad\qquad\square$

Example 6.1.3. Let $M = \mathbb{R}^n$. For $\mathbf{x} = (x_1, \ldots, x_n)$ and $\mathbf{y} = (y_1, \ldots, y_n)$ put

$$\rho(\mathbf{x}, \mathbf{y}) = \sqrt{\sum_{j=1}^{n} (x_j - y_j)^2}.$$

We refer to ρ as the *Euclidean metric* on $\mathbb{R}^n$, since it is the one used in Euclidean geometry. $\qquad\qquad\square$

Example 6.1.4. Let $M = \mathbb{R}^n$. For $\mathbf{x} = (x_1, \ldots, x_n)$ and $\mathbf{y} = (y_1, \ldots, y_n)$ put

$$\rho(\mathbf{x}, \mathbf{y}) = \max_{1 \leqslant j \leqslant n} \{|x_j - y_j|\}. \qquad \square$$

The preceding example can be generalized in the following way. A *norm* on a Euclidean space $\mathbb{R}^n$ is an assignment of a number $\|\mathbf{x}\| \geqslant 0$ to each $\mathbf{x} \in \mathbb{R}^n$ such that

(1) $\|\mathbf{x}\| = 0$ if and only if $\mathbf{x} = 0$,
(2) $\|t\mathbf{x}\| = |t| \cdot \|\mathbf{x}\|$ for any scalar $t \in \mathbb{R}$,
(3) $\|\mathbf{x} + \mathbf{y}\| \leqslant \|\mathbf{x}\| + \|\mathbf{y}\|$.

In Chapter 4, we encountered two norms: the sum norm defined by $\|\mathbf{x}\| = |x_1| + \cdots + |x_n|$ and the Euclidean norm defined by $\|\mathbf{x}\| = \sqrt{x_1^2 + \cdots + x_n^2}$.

Example 6.1.5. Let $M = \mathbb{R}^n$ and $\| \cdot \|$ be any norm on M. For $\mathbf{x} = (x_1, \ldots, x_n)$ and $\mathbf{y} = (y_1, \ldots, y_n)$ put

$$\rho(\mathbf{x}, \mathbf{y}) = \|\mathbf{x} - \mathbf{y}\|.$$

Then (M, ρ) is a metric space. Example 6.1.3 corresponds to the case where the norm is the Euclidean norm, and Example 6.1.4 to the max norm defined by $\|\mathbf{x}\| = \max_{1 \leqslant j \leqslant n} \{|x_j|\}$. $\qquad \square$

Example 6.1.6. Let (M, ρ) be an arbitrary metric space, and E be a subset of M. We can define a metric ρ_E on E by restricting ρ to $E \times E$. The resulting metric space (E, ρ_E) is called a *subspace* of (M, ρ). For example, the *unit cube* $[0, 1]^n$ is a subspace of $\mathbb{R}^n$. $\qquad \square$

Example 6.1.7. Let (M_1, ρ_1) and (M_2, ρ_2) be metric spaces. Put $M = M_1 \times M_2$, and define $\rho_1 \times \rho_2$ on $M_1 \times M_2$ by

$$(\rho_1 \times \rho_2)\big((x_1, x_2), (y_1, y_2)\big) = \max\{\rho_1(x_1, y_1), \rho_2(x_2, y_2)\}.$$

We call $\rho_1 \times \rho_2$ the *product metric* on $M_1 \times M_2$ and $(M_1 \times M_2, \rho_1 \times \rho_2)$ the *product space*.

Example 6.1.8. For this example, start with the unit interval $[0, 1]$ and "glue together" the points 0 and 1. You can picture this as bending the interval into a circle and then joining the ends together, as in Figure 6.1.1.

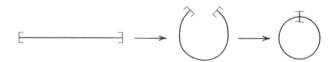

FIGURE 6.1.1.   Identifying endpoints of an interval produces a circle.

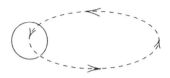 

FIGURE 6.1.2. A torus is formed by sweeping out one circle along another.

To make this precise, we set M to be the half-open interval $[0,1)$. We don't need the endpoint 1 since it is identified with 0. Define

$$\rho(x,y) = \min\{|x-y|, |1-x+y|, |1+x-y|\}.$$

In terms of our circle picture, $\rho(x,y)$ is the length of the shortest arc between x and y. This metric space $([0,1), \rho)$ is commonly denoted by $\mathbb{T}$. $\square$

Example 6.1.9. Let $\mathbb{T} = [0,1)$ be the circle with metric ρ defined in the previous example. The product space $\mathbb{T}^2 = \mathbb{T} \times \mathbb{T}$ with metric $\rho \times \rho$ looks like the surface of a doughnut, for, as depicted in Figure 6.1.2, it is the surface swept out by one circle being rotated along another circle. This surface is called a *torus*, which explains the use of the notation $\mathbb{T}$.

Points in the torus $\mathbb{T}^2 = [0,1) \times [0,1)$ may appear far apart when they are actually close in the torus metric. For example, the point $(.5, .999)$ may look far from $(.5, 0)$, but their distance in $\mathbb{T}^2$ is only .001 since we are identifying $(.5, 1)$ with $(.5, 0)$. This gives another way to view $\mathbb{T}^2$, namely as the unit square $[0,1] \times [0,1]$ with opposite sides glued together. Here $(s, 0)$ is glued to $(s, 1)$ for $0 \leqslant s \leqslant 1$ and $(0, t)$ is glued to $(1, t)$ for $0 \leqslant t \leqslant 1$. Geometrically, first identifying the left and right sides of the unit square gives a vertical cylinder, and then identifying the top and bottom edges of the cylinder gives the surface of a doughnut, again depicted in Figure 6.1.2. $\square$

Example 6.1.10. For symbolic dynamics the most important examples of metric spaces are shift spaces. The metric should capture the idea that points are close when large central blocks of their coordinates agree. Specifically, let $M = \mathcal{A}^{\mathbb{Z}}$, and put

$$\rho(x,y) = \begin{cases} 2^{-k} & \text{if } x \neq y \text{ and } k \text{ is maximal so that } x_{[-k,k]} = y_{[-k,k]}, \\ 0 & \text{if } x = y. \end{cases}$$

In other words, to measure the distance between x and y, we find the largest k for which the central $(2k+1)$-blocks of x and y agree, and use 2^{-k} as the distance (with the conventions that if $x = y$ then $k = \infty$ and $2^{-\infty} = 0$, while if $x_0 \neq y_0$ then $k = -1$). This ρ clearly satisfies conditions (1) and (2) in Definition 6.1.1, so to verify that ρ is a metric we need only show that the triangle inequality (3) holds. If $\rho(x,y) = 2^{-k}$

then $x_{[-k,k]} = y_{[-k,k]}$; likewise, if $\rho(y, z) = 2^{-\ell}$ then $y_{[-\ell,\ell]} = z_{[-\ell,\ell]}$. Put $m = \min\{k, \ell\}$. Then $x_{[-m,m]} = z_{[-m,m]}$, so

$$\rho(x, z) \leqslant 2^{-m} \leqslant 2^{-k} + 2^{-\ell} = \rho(x, y) + \rho(y, z).$$

Hence ρ satisfies the triangle inequality, and so is a metric on M. There are other metrics on the shift space that work just as well (see Exercise 6.1.1).

Using the notion of subspaces described in Example 6.1.6, we obtain a metric on all shift spaces X by restricting ρ to $X \times X$. The interpretation of "closeness" remains the same.   □

It will be convenient to have the following way to measure the size of a subset.

Definition 6.1.11. Let (M, ρ) be a metric space and $E \subseteq M$. The *diameter* of E is the number $\text{diam}(E) = \sup\{\rho(x, y) : x, y \in E\}$.

For example, the diameter of a circle in $\mathbb{R}^2$ using the Euclidean metric is just the usual meaning of diameter.

For sake of brevity, from now on we will usually suppress mention of the metric, and simply refer to "a metric space M" with the understanding that ρ (or ρ_M) is its metric. In the case $M = \mathbb{R}^n$ we will use the Euclidean metric unless we specify otherwise.

One of the basic ideas in calculus is that of convergence. We can extend this idea to general metric spaces as follows.

Definition 6.1.12. Let M be a metric space. A sequence $\{x_n\}_{n=1}^{\infty}$ in M *converges to* x in M if $\rho(x_n, x) \to 0$ as $n \to \infty$. In this case we write $x_n \to x$ as $n \to \infty$, or $\lim_{n \to \infty} x_n = x$.

For the metric space $(\mathbb{R}, \rho)$ in Example 6.1.2, this is the same as the usual idea of convergence for a sequence of real numbers. Note that although the metrics on $\mathbb{R}^n$ defined in Examples 6.1.3 and 6.1.4 are different, they give the same convergent sequences. For the circle $\mathbb{T}$, you need to be careful when sequences get close to 0 or 1, since these points are identified. For instance, the sequence $1/2, 2/3, 1/4, 4/5, 1/6, 6/7, \ldots$ converges to 0 in $\mathbb{T}$.

A sequence of points in a shift space converges exactly when, for each $k \geqslant 0$, the central $(2k + 1)$-blocks stabilize starting at some element of the sequence. Specifically, let X be a shift space, and $x^{(n)}, x \in X$ (we use the notation $x^{(n)}$ for the nth term in a sequence from a shift space since x_n would normally refer to the nth coordinate of the point x). The definition shows that $x^{(n)} \to x$ exactly when, for each $k \geqslant 0$, there is an n_k such that

$$x_{[-k,k]}^{(n)} = x_{[-k,k]}$$

for all $n \geqslant n_k$. For example, if

$$x^{(n)} = (1\,0^n)^{\infty} = \ldots 1\,0^n\,1\,0^n . 1\,0^n\,1\,0^n \ldots,$$

then $x^{(n)} \to \ldots 0000.10000 \ldots$.

It is easy to show that limits are unique.

Lemma 6.1.13. *Let M be a metric space, and suppose that $x_n \to x$ and $x_n \to y$ as $n \to \infty$. Then $x = y$.*

PROOF: By the triangle inequality,

$$0 \leqslant \rho(x, y) \leqslant \rho(x, x_n) + \rho(x_n, y) \to 0 + 0 = 0.$$

Thus $\rho(x, y) = 0$, so that $x = y$. $\square$

Two metrics ρ_1 and ρ_2 on a metric space are called *equivalent* if convergence means the same for both metrics: $\rho_1(x_n, x) \to 0$ implies that $\rho_2(x_n, x) \to 0$, and conversely. It can be shown that the metrics defined by any two norms on $\mathbb{R}^n$ are equivalent (see Exercise 6.1.2).

We next turn to the notion of continuity. Let M and N be metric spaces and $\phi \colon M \to N$ a function. We are interested in those functions that preserve convergence.

Definition 6.1.14. A function $\phi \colon M \to N$ from one metric space to another is *continuous* if, whenever $x_n \to x$ in M, then $\phi(x_n) \to \phi(x)$ in N. If ϕ is continuous, one-to-one, onto, and has a continuous inverse, then we call ϕ a *homeomorphism*.

When $M = N = \mathbb{R}$ with the Euclidean metric, this becomes the usual definition of (sequential) continuity of real-valued functions. This is equivalent to the familiar "ϵ, δ" definition from calculus (see Exercise 6.1.8).

Example 6.1.15. Let $M = X$ and $N = Y$ be shift spaces and $\phi = \Phi_\infty^{[-m,n]}$ a sliding block code. If two points in X are close, they agree on a large central block, hence their images under ϕ also agree on a large (though slightly smaller) central block. Thus ϕ is continuous. In particular, the shift map $\sigma_X \colon X \to X$ is continuous.

Thus sliding block codes are continuous and commute with the respective shift maps. We will see in §6.2 that, conversely, a continuous map from one shift space to another that commutes with the shift maps must be a sliding block code. $\square$

In Chapter 1 we used the Cantor diagonal argument in several crucial places. This argument is a concrete form of the last of the "three C's" in this section: convergence, continuity, and now compactness.

To state the definition, recall that a *subsequence* of $\{x_n\}_{n=1}^\infty$ is a sequence of the form $\{x_{n_k}\}_{k=1}^\infty$, where $n_1 < n_2 < n_3 < \ldots$.

Definition 6.1.16. A metric space M is *compact* if every sequence in M has a convergent subsequence.

Thus to verify that M is compact, we must show that for every sequence $\{x_n\}$ in M there is a subsequence $\{x_{n_k}\}$ and an x in M for which $x_{n_k} \to x$ as $k \to \infty$. Some variant of the Cantor diagonal argument is typically used for this.

Example 6.1.17. The reader should be able to check that the following metric spaces are compact.

(1) A finite set M, with metric ρ defined by

$$\rho(x, y) = \begin{cases} 1 & \text{if } x \neq y, \\ 0 & \text{if } x = y. \end{cases}$$

(2) $M = [0, 1]$ with the Euclidean metric. To see that M is compact, argue as follows. A sequence in M must have an infinite subsequence in at least one of the two half intervals; this subsequence must have a further infinite subsequence in a subinterval of size $1/4$, and so on.

(3) $M = \mathbb{T}$ with the metric in Example 6.1.8.

(4) $M = M_1 \times M_2$, where M_1 and M_2 are compact. Similarly, the product of any finite number of compact metric spaces is compact.

(5) $M = X$, a shift space. To see that M is compact, argue as follows. Given a sequence $x^{(n)}$ in X, construct a convergent subsequence using Cantor diagonalization as in the proof of Theorem 1.5.13. Namely, for $k \geqslant 1$ inductively find a decreasing sequence of infinite subsets S_k of positive integers so that all blocks $x^{(n)}_{[-k,k]}$ are equal for $n \in S_k$. Define x to be the point with $x_{[-k,k]} = x^{(n)}_{[-k,k]}$ for all $n \in S_k$, and inductively define n_k as the smallest element of S_k which exceeds n_{k-1}. Then $x \in X$, and $x^{(n_k)}$ converges to x as $k \to \infty$

On the other hand, the metric space (M, ρ), with $M = \mathbb{R}$ described in Example 6.1.2, is not compact because the sequence $1, 2, 3, \ldots$ has no convergent subsequence. Similarly, the subspace $((0,1), \rho)$ is not compact because the sequence $1/2, 1/3, 1/4 \ldots$ has no convergent subsequence. $\quad\square$

In order to specify which subspaces of a compact metric space are compact, we need the notion of closed set, and the complementary notion of open set.

Definition 6.1.18. Let (M, ρ) be a metric space. For $x \in M$ and $r > 0$, the *open ball of radius* r *about* x is the set $B_r(x) = \{y \in M : \rho(x, y) < r\}$. A subset $U \subseteq M$ is *open* if, for every $x \in U$, there is an $r > 0$ such that $B_r(x) \subseteq U$. A subset $V \subseteq M$ is *closed* if its complement $M \setminus V$ is open.

Roughly speaking, open sets are "fuzzy," while closed sets have "hard edges." For instance, in $\mathbb{R}^n$ the ball $B_1(\mathbf{0})$ is open, while $\{\mathbf{y} \in \mathbb{R}^n : \rho(\mathbf{0}, \mathbf{y}) \leqslant 1\}$ is closed. The reader should verify that the union of any collection of open sets is still open and that the intersection of any *finite* collection of open sets is open. The corresponding statements for closed sets, proved by taking complements, are that the intersection of any collection of closed sets is closed, and the union of any *finite* collection of closed sets is closed.

There are many sets, such as $[0,1) \subset \mathbb{R}$, which are neither open nor closed. Sometimes subsets can be simultaneously open *and* closed. A good example is that of cylinder sets in a shift space, which we now describe. Let X be a shift space, and fix $u \in \mathcal{B}(X)$ and $k \in \mathbb{Z}$. Define the *cylinder set* $C_k(u)$ as

$$C_k(u) = C_k^X(u) = \{x \in X : x_{[k,k+|u|-1]} = u\};$$

i.e., $C_k(u)$ is the set of points in which the block u occurs starting at position k. To see that cylinder sets are open, first observe that for any $x \in X$ and nonnegative integer n we have that

$$C_{-n}(x_{[-n,n]}) = B_{2^{-(n-1)}}(x).$$

Hence if $x \in C_k(u)$ and $n = \max\{|k|, |k+|u|-1|\}$, then $B_{2^{-(n-1)}}(x) \subseteq C_k(u)$, proving that $C_k(u)$ is open. To see that cylinder sets are also closed, observe that the complement of a cylinder set is a finite union of cylinder sets (why?), which is open since the union of open sets is open.

We say that a subset E of a metric space (M, ρ) is compact when the subspace $(E, \rho|_E)$ is compact. We will prove two assertions about compact sets, one characterizing the compact sets in $\mathbb{R}^n$, and the other characterizing shift spaces as the compact, shift-invariant subsets of a full shift. Both proofs make use of the following assertion.

Proposition 6.1.19. *Let M be a compact metric space. A subset $E \subseteq M$ is compact if and only if E is closed.*

PROOF: First, suppose that $E \subseteq M$ is compact. If E were not closed, then $M \setminus E$ would not be open. Hence there would be $x \in M \setminus E$ such that $B_{1/n}(x) \cap E \neq \varnothing$ for all $n \geqslant 1$. For each n, choose $x_n \in B_{1/n}(x) \cap E$. Then $\{x_n\}$ is a sequence in E, so by compactness it has a convergent subsequence $x_{n_k} \to y \in E$. But clearly $x_{n_k} \to x$ as well, so by uniqueness of limits we see that $y = x \notin E$, a contradiction. Thus E is closed. (Note that this part did not use compactness of M, i.e., compact sets are always closed.)

Conversely, suppose that E is closed, and let $\{x_n\}$ be a sequence in E. Since $\{x_n\}$ is also a sequence in the compact metric space M, it has a subsequence $\{x_{n_k}\}$ that converges to some $y \in M$. We need to show that y is actually in E. Suppose not. Since E is closed, $M \setminus E$ is open. Hence there is an $r > 0$ such that $B_r(y) \subseteq M \setminus E$. By definition of convergence, $\rho(y, x_{n_k}) < r$ for large enough k, so that $x_{n_k} \notin E$. This contradiction shows that E is compact. $\qquad\square$

Note in particular that since cylinder sets are closed, they are compact as well.

Our first use of this proposition is for subsets of $\mathbb{R}^n$. Let us call a set $E \subseteq \mathbb{R}^n$ *bounded* if E is contained in a ball of positive radius centered at $\mathbf{0}$.

Theorem 6.1.20 (Bolzano–Weierstrass Theorem). *A subset of* $\mathbb{R}^n$ *is compact if and only if it is closed and bounded.*

PROOF: We showed that compact sets are always closed when proving Proposition 6.1.19. If a compact set $E \subseteq \mathbb{R}^n$ were not bounded, then for each n there would be an $x_n \in E$ with $\rho(\mathbf{0}, x_n) \geqslant n$. Such a sequence clearly could not have a convergent subsequence (why?), contradicting compactness. Thus compact subsets of $\mathbb{R}^n$ are closed and bounded.

Conversely, suppose that $E \subseteq \mathbb{R}^n$ is closed and bounded. Then there is an $r > 0$ such that $E \subseteq [-r, r]^n$, and the latter set, being a product of (compact) intervals, is compact. Hence E is a closed subset of a compact space, thus compact by Proposition 6.1.19. □

Our next result characterizes shift spaces.

Theorem 6.1.21. *A subset of* $\mathcal{A}^{\mathbb{Z}}$ *is a shift space if and only if it is shift-invariant and compact.*

PROOF: Shift spaces are shift-invariant, and also compact according to Example 6.1.17(5).

Conversely, suppose that $X \subseteq \mathcal{A}^{\mathbb{Z}}$ is shift-invariant and compact. By Proposition 6.1.19, X is closed, so that $\mathcal{A}^{\mathbb{Z}} \setminus X$ is open. Hence, for each $y \in \mathcal{A}^{\mathbb{Z}} \setminus X$ there is a $k = k(y)$ such that if u_y denotes the block $y_{[-k,k]}$, then $C_{-k}^{\mathcal{A}^{\mathbb{Z}}}(u_y) \subseteq \mathcal{A}^{\mathbb{Z}} \setminus X$. Let $\mathcal{F} = \{u_y : y \in \mathcal{A}^{\mathbb{Z}} \setminus X\}$. It is easy to verify using shift invariance of X that $X = X_{\mathcal{F}}$, so that X is a shift space. □

Exercise 6.1.9 suggests an alternative proof, using compactness, of Theorem 1.5.13, which states that the image of a shift space under a sliding block code is again a shift space.

There is another way to characterize compactness, this time in terms of coverings by collections of open sets. This characterization is sometimes even used as the definition of compactness (see [ProM] for a proof).

Theorem 6.1.22 (Heine–Borel Theorem). *A subset* E *of a metric space* (M, ρ) *is compact if and only if whenever* E *is contained in the union of a collection of open subsets of* M, *then* E *is also contained in the union of a finite subcollection of these sets.*

We conclude this brief introduction to metric spaces by discussing the notions of interior and closure of sets.

What stops a subset E of a metric space from being open? The definition shows that there must be a point x in E such that every ball about x contains points in the complement of E. Let us say that x is an *interior point* of E if there is an $r > 0$ for which $B_r(x) \subseteq E$, and denote by $\text{int}(E)$ the set of all interior points of E. Then E is open if and only if all points in E are interior points, or equivalently $E = \text{int}(E)$.

What stops a subset E from being closed? There must be a point x in the complement of E such that there are balls around x of arbitrarily small radius that intersect E. It follows that there is a sequence $\{x_n\}$ of points in E converging to x. Define a *limit point* of E to be a point that is the limit of a sequence of points in E. The *closure* of E is defined to be the union of E and all of its limit points. We denote the closure of E by $\overline{E}$. Our remarks show that E is closed if and only if it contains all of its limit points, or equivalently $E = \overline{E}$.

The notions of closure and interior are complementary in the sense that the complement of the closure of a set E is the interior of the complement of E (Exercise 6.1.13).

We say that a set E is *dense* in a set F if $F \subseteq \overline{E}$. For instance, the set of rationals in $[0, 1]$ is dense in $[0, 1]$, and the set of periodic points in an irreducible shift of finite type X is dense in X (Exercise 6.1.12). The intersection of two dense subsets need not itself be dense. For instance, both the set of rationals in $[0, 1]$ and the set of irrationals in $[0, 1]$ are dense in $[0, 1]$, yet their intersection is empty and so certainly is not dense. However, if the sets are both dense and open, then their intersection is also dense.

Proposition 6.1.23. *Let M be a metric space. Suppose that E_1, E_2, $\ldots$, E_n are each open and dense in M. Then $\bigcap_{k=1}^{n} E_k$ is open and dense in M.*

PROOF: Since the intersection of a finite number of open sets is open, by induction it suffices to prove the proposition for $n = 2$. Let E and F be open dense sets in M. To prove that $E \cap F$ is dense, let $x \in M$ and $\epsilon > 0$ be arbitrary. We will find a point z in $E \cap F$ such that $\rho(x, z) < \epsilon$. Since E is dense, there is a $y \in E$ such that $\rho(x, y) < \epsilon/2$. Since E is open, there is a $\delta > 0$ with $\delta < \epsilon/2$ such that $B_\delta(y) \subseteq E$. Since F is dense, there is a $z \in F$ such that $\rho(y, z) < \delta$. Then $z \in E \cap F$ and $\rho(x, z) \leqslant \rho(x, y) + \rho(y, z) < \epsilon/2 + \delta < \epsilon$. $\qquad\square$

Although the intersection of an infinite collection of open sets need no longer be open, the density part of the previous proposition carries over.

Theorem 6.1.24 (Baire Category Theorem). *Let M be a compact metric space, and suppose that E_1, E_2, E_3, $\ldots$ are each open and dense in M. Then $\bigcap_{k=1}^{\infty} E_k$ is dense in M.*

PROOF: Let $x \in M$ and $\epsilon > 0$ be arbitrary. We need to find some $y \in \bigcap_{k=1}^{\infty} E_k$ such that $\rho(x, y) < \epsilon$.

Set $x_0 = x$ and $\delta_0 = \epsilon/2$. Use Proposition 6.1.23 to find, for each $n \geqslant 1$,

$x_n \in M$ and $\delta_n > 0$ such that

$$\overline{B_{\delta_n}(x_n)} \subset \left(\bigcap_{k=1}^{n} E_k \right) \cap B_{\delta_{n-1}}(x_{n-1}).$$

Since M is compact, the sequence $\{x_n\}$ has a subsequence that converges to some $y \in M$. Now the closed sets

$$\overline{B_{\delta_0}(x_0)} \supseteq \overline{B_{\delta_1}(x_1)} \supseteq \overline{B_{\delta_2}(x_2)} \supseteq \cdots$$

are nested, hence $y \in \overline{B_{\delta_n}(x_n)} \subseteq E_n$ for all $n \geqslant 0$. Thus $y \in \bigcap_{k=1}^{\infty} E_k$. Since $y \in \overline{B_{\delta_0}(x_0)} \subseteq B_\epsilon(x)$, it follows that $\rho(x, y) < \epsilon$.   □

EXERCISES

6.1.1. Show that the following three metrics on the full r-shift $\{0, \dots, r-1\}^{\mathbb{Z}}$ are all equivalent:

(a) The metric given in Example 6.1.10.

(b) $\rho(x, y) = 1/(k+2)$, where k is maximal such that x and y agree in their central $(2k+1)$-block (we put $1/\infty = 0$ by convention, and $k = -1$ when $x_0 \neq y_0$).

(c)

$$\rho(x, y) = \sum_{k=-\infty}^{\infty} \frac{|x_k - y_k|}{2^{|k|}}.$$

6.1.2. (a) Show that the two metrics on $\mathbb{R}^n$ in Examples 6.1.3 and 6.1.4 are equivalent.

*(b) Show that in fact the metrics defined by any two norms on $\mathbb{R}^n$ are equivalent [*Hint:* Use continuity and compactness to show that for any two norms, $\| \cdot \|_1$, $\| \cdot \|_2$, on $\mathbb{R}^n$, there are constants $a, b > 0$ such that $a < \|\mathbf{x}\|_1 / \|\mathbf{x}\|_2 < b$ for all $\mathbf{x} \neq 0$.]

(c) Let S be the hollow unit sphere in $\mathbb{R}^3$. For $x, y \in S$, let $\rho_1(x, y)$ be the length of the line segment joining x and y, and $\rho_2(x, y)$ be the length of the shortest arc of a great circle on S joining x and y. Show that ρ_1 and ρ_2 are equivalent metrics on S.

6.1.3. Complete the verification in Example 6.1.17(5) that shift spaces are compact.

6.1.4. Let M be a compact metric space, and $E_1 \supseteq E_2 \supseteq E_3 \supseteq \dots$ be a decreasing sequence of nonempty compact sets. Prove that $\bigcap_{n=1}^{\infty} E_n \neq \varnothing$. Use this to solve Exercise 2.1.8.

6.1.5. Let M be a metric space.

(a) Suppose that M is compact and $M = \bigcup_{n=1}^{\infty} A_n$ where each A_n is an open set. Show that finitely many of the A_n cover M, i.e. $M = \bigcup_{n=1}^{m} A_n$ for some positive integer m.

*(b) Prove the Heine–Borel Theorem 6.1.22 for M.

6.1.6. Let E and F be compact subsets of a metric space. Show that both $E \cup F$ and $E \cap F$ are compact. Use this to solve Exercise 1.2.5.

6.1.7. (a) Show that a subset of a shift space is open if and only if it is a union of cylinder sets.

(b) Show that a subset of a shift space is both open and closed if and only if it is a finite union of cylinder sets.

(c) Show that a subset of a shift space is both open and compact if and only if it is a finite union of cylinder sets.

6.1.8. Let M and N be metric spaces.

(a) Let $\phi\colon M \to N$. Show that ϕ is continuous if and only if, for every $x \in M$ and every $\epsilon > 0$, there is a $\delta = \delta(\epsilon, x) > 0$ such that if $\rho_M(x, y) < \delta$, then $\rho_N(\phi(x), \phi(y)) < \epsilon$. Thus continuity can be defined by the standard "ϵ, δ" definition from calculus. If M is compact, show that δ can be chosen independent of x, i.e., that ϕ is *uniformly continuous*.

(b) Show that $\phi\colon M \to N$ is continuous if and only if the inverse image under ϕ of every open set in N is open in M.

(c) Show that $\phi\colon M \to N$ is continuous if and only if the inverse image under ϕ of every closed set in N is closed in M.

6.1.9. Let M and N be metric spaces and $\phi\colon M \to N$ be continuous. Suppose that E is a compact subset of M. Show that $\phi(E)$ is compact. Use this to give an alternative proof of Theorem 1.5.13.

6.1.10. Let M and N be metric spaces with M compact, and let $\phi\colon M \to N$ be continuous. Show that if E is a compact subset of N, then $\phi^{-1}(E)$ is compact.

6.1.11. Let E and F be compact, disjoint subsets of a metric space M. Show that there is a $\delta > 0$ such that $\rho(x, y) \geqslant \delta$ whenever $x \in E$ and $y \in F$.

6.1.12. Show that the set of periodic points in an irreducible shift of finite type X is dense in X.

6.1.13. Show that the complement of the closure of a set E is the interior of the complement of E.

6.1.14. Let M and N be metric spaces and $\phi\colon M \to N$ continuous.

(a) Show that $\phi(\overline{A}) \subseteq \overline{\phi(A)}$.

(b) Show that if M is compact, then $\phi(\overline{A}) = \overline{\phi(A)}$.

*6.1.15. (Special case of Brouwer Fixed Point Theorem). Define the sum norm $\|\mathbf{x}\| = |x_1| + |x_2| + \ldots + |x_n|$ on $\mathbb{R}^n$. Let $\Delta = \{\mathbf{x} \in \mathbb{R}^n : \mathbf{x} \geqslant 0, \ \|\mathbf{x}\| = 1\}$. Let A be an irreducible matrix. Define the map $\bar{A}\colon \Delta \to \Delta$ by

$$\bar{A}(\mathbf{x}) = \frac{A\mathbf{x}}{\|A\mathbf{x}\|}.$$

In the following steps, show that $\bar{A}$ has a fixed point in Δ, which is therefore a nonnegative eigenvector for A. In your proof, use compactness and the following notions of convexity, but do not use the Perron–Frobenius theory from §4.2.

A *convex combination* of points $x^{(1)}, \ldots, x^{(m)}$ in a vector space is a point $\sum_i t_i x^{(i)}$, where each $t_i \geqslant 0$ and $\sum_i t_i = 1$. A convex combination is *nontrivial* if $t_i \neq 0$ and $t_j \neq 0$ for some $i \neq j$. The *convex hull* of a set S in a vector space is the set of all convex combinations of finite subsets of S. A *polytope* is the convex hull of a finite set. An *extreme point* of a convex set S is a point

$x \in S$ which cannot be expressed as a nontrivial convex combination of points in S.

(a) Show that any polytope has only finitely many extreme points and that it is the convex hull of its extreme points.

(b) Let $\Delta^m = \bar{A}^m(\Delta)$ and $\Delta^\infty = \bigcap_{m=0}^\infty \Delta^m$. Show that Δ^∞ is a polytope.

(c) Show that

$$\bar{A}(\Delta^\infty) = \Delta^\infty.$$

(d) Let E be the (finite) set of extreme points of Δ^∞. Show that $\bar{A}(E) = E$.

(e) Show by explicit computation that any matrix of the form

$$\begin{bmatrix} 0 & a_1 & 0 & \cdots & 0 \\ 0 & 0 & a_2 & \cdots & 0 \\ \vdots & \vdots & \vdots & \vdots & \vdots \\ 0 & 0 & 0 & \cdots & a_{k-1} \\ a_k & 0 & 0 & \cdots & 0 \end{bmatrix},$$

where each $a_i > 0$, has a positive eigenvector.

(f) Show that $\bar{A}$ has a fixed point.

*6.1.16. Let A be an irreducible matrix, and let Δ and $\bar{A}$ be as in Exercise 6.1.15. In the following steps, show, without using Perron–Frobenius theory, that $\bar{A}$ has a unique fixed point and that the fixed point is strictly positive (and so A has a unique positive eigenvector up to a scalar multiple).

(a) For $\mathbf{x}, \mathbf{y} \in \Delta$, let

$$K(\mathbf{x}, \mathbf{y}) = \sup\{\lambda \in \mathbb{R} : \mathbf{x} - \lambda\mathbf{y} \geq 0\}$$

and

$$\rho(\mathbf{x}, \mathbf{y}) = -\log K(\mathbf{x}, \mathbf{y}) - \log K(\mathbf{y}, \mathbf{x}).$$

Show that ρ is a metric on $\text{int}(\Delta)$, the set of all positive vectors in Δ.

(b) Show that if A is positive, then $\bar{A}$ strictly contracts distances between distinct points, i.e.,

$$\rho(\bar{A}(\mathbf{x}), \bar{A}(\mathbf{y})) < \rho(\mathbf{x}, \mathbf{y})$$

for every pair $\mathbf{x}, \mathbf{y}$ of distinct points in $\text{int}(\Delta)$.

(c) Show that any continuous map on a compact metric space which strictly contracts distances between distinct points has a unique fixed point.

(d) Show that if A is primitive, then $\bar{A}$ has a unique fixed point and that the fixed point is strictly positive.

(e) Do part (d) when A is merely irreducible. [*Hint:* Consider $A + Id$.]

§6.2. Dynamical Systems

Physical systems can usually be described by giving the values of a finite number of measurements. A swinging pendulum, for example, is determined by its angle from the vertical and its angular velocity. A gas is described by giving the position and momentum of each molecule. As time evolves, these values change. Let M be the set of all possible values for a system. Imagine taking a film of the system, with one frame each second. Each frame depends on the previous one, namely how the system evolved during an interval of one second, and we will assume that the laws governing change do not change with time. This dependence of each frame on the previous one gives a function $\phi\colon M \to M$, usually continuous. Thus if $x \in M$ describes the system at time $t = 0\,\mathrm{sec}$, then $\phi(x)$ describes it at time $t = 1\,\mathrm{sec}$, $\phi^2(x)$ at $t = 2\,\mathrm{sec}$, and so on. Thus to study how the system behaves through time, we need to study how the iterates x, $\phi(x)$, $\phi^2(x)$, ... behave. These considerations motivate the following notion.

Definition 6.2.1. A *dynamical system* (M, ϕ) consists of a compact metric space M together with a continuous map $\phi\colon M \to M$. If ϕ is a homeomorphism we call (M, ϕ) an *invertible dynamical system*.

We often abbreviate (M, ϕ) by ϕ to emphasize the dynamics.

Several of the following examples use the notion of the fractional part of a real number. Each $r \in \mathbb{R}$ can be written uniquely as $r = n + s$, where $n \in \mathbb{Z}$ and $s \in [0, 1)$. In this case we write $r \equiv s \pmod 1$ and call s the *fractional part* of r.

Example 6.2.2. Let $M = \mathbb{T}$, fix $\alpha \in \mathbb{T}$, and define $\phi_\alpha\colon \mathbb{T} \to \mathbb{T}$ by $\phi_\alpha(x) \equiv x + \alpha \pmod 1$. Thinking of $\mathbb{T}$ as a circle (see Example 6.1.8), ϕ_α acts by rotating $\mathbb{T}$ by an angle of $2\pi\alpha$ radians, and so ϕ_α is called a *circle rotation*. □

Example 6.2.3. Let $M = \mathbb{T}$, and $\phi(x) = 2x \pmod 1$. The graph of ϕ is shown in Figure 6.2.1.

Note that although this appears to be discontinuous at $x = 0$ and at $x = 1/2$, it is actually continuous there since 0 is identified with 1 in $\mathbb{T}$. Note also that ϕ maps each of the subintervals $[0, 1/2)$ and $[1/2, 1)$ one-to-one and onto the entire circle. For this reason, ϕ is called the *doubling map*. □

Example 6.2.4. Let $M = \mathbb{T}^2$ from Example 6.1.9 and $\phi(x, y) \equiv (x + y, x) \pmod 1$. You should verify that, as in the previous example, ϕ is continuous. This is an example of a *toral automorphism*, defined in §6.5. □

Example 6.2.5. Let $M = X$, a shift space, and $\phi = \sigma_X$. We will call this a *shift dynamical system*, and is the motivation for the title of this chapter.

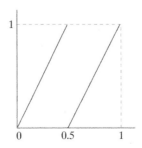

FIGURE 6.2.1. Graph of the doubling map.

In symbolic dynamics, we often abbreviate (X, σ_X) by X instead of σ_X, since the shift map is understood to be the underlying map that provides the dynamics.   □

Example 6.2.6. Let $M = X$, a shift space, and $\phi = \Phi_\infty : X \to X$ be a sliding block code. Continuity of ϕ was established in Example 6.1.15, so that (M, ϕ) is a dynamical system.   □

The study of dynamical systems is vast and rapidly expanding. We can barely touch on some of its problems and ideas. For thorough accounts of the basic theory, you can consult [Dev], [Edg], and the more advanced [Wal].

One set of problems involves the study of a single dynamical system and the behavior of points and sets under iteration. For a dynamical system (M, ϕ), the *orbit* of a point $x \in M$ is the the set of iterates $\{\phi^n(x)\}_{n \in \mathbb{Z}}$ when ϕ is invertible, and $\{\phi^n(x)\}_{n \geqslant 0}$ otherwise. A *periodic point* is a point $x \in M$ such that $\phi^n(x) = x$ for some $n > 0$. An orbit $\{\phi^n(x)\}$ is called a *periodic orbit* if x is a periodic point. Some typical questions about a given (M, ϕ) are:

(1) For each point $x \in M$, are there periodic points arbitrarily close to x?

(2) How are the points of an orbit distributed throughout M?

(3) Given a subset U of M, how do the sets $\phi^n(U)$ spread throughout M as $n \to \infty$? In particular, if U is open and V is another open set, must there be an n for which $\phi^n(U) \cap V \neq \varnothing$?

Another set of problems involves the relationship between one dynamical system and another. To make these questions precise, let us define a *homomorphism* $\theta : (M, \phi) \to (N, \psi)$ from one dynamical system to another to be a continuous function $\theta : M \to N$ satisfying the commuting property that $\psi \circ \theta = \theta \circ \phi$, so that the following diagram commutes.

$$
\begin{array}{ccc}
M & \xrightarrow{\phi} & M \\
\theta \downarrow & & \downarrow \theta \\
N & \xrightarrow{\psi} & N
\end{array}
$$

In this case we say that θ *intertwines* ϕ and ψ. Note that when (M, ϕ) and (N, ψ) are shift dynamical systems, a homomorphism is simply a continuous shift-commuting mapping.

Definition 6.2.7. Let (M, ϕ) and (N, ψ) be dynamical systems, and let $\theta : (M, \phi) \to (N, \psi)$ be a homomorphism. Then θ is called an *embedding* if it is one-to-one, a *factor map* if it is onto, and a *topological conjugacy* if it is both one-to-one and onto and its inverse map is continuous. If θ is a topological conjugacy we write $\theta : (M, \phi) \cong (N, \psi)$. Two dynamical systems are *topologically conjugate* if there is a topological conjugacy between them.

The next result shows that although we have required continuity of the inverse in defining topological conjugacy, this property already follows from the others.

Lemma 6.2.8. *Let M and N be compact metric spaces, and $\theta \colon M \to N$ be continuous, one-to-one, and onto. Then $\theta^{-1} \colon N \to M$ is also continuous.*

PROOF: If θ^{-1} were not continuous, then there would be y_n and y in N with $y_n \to y$, but $\theta^{-1}(y_n) \not\to \theta^{-1}(y)$. By passing to a subsequence, we may assume that there is an $\epsilon > 0$ such that

$$\rho_M\big(\theta^{-1}(y_n), \theta^{-1}(y)\big) \geqslant \epsilon.$$

Since M is compact, there is a subsequence $\theta^{-1}(y_{n_k})$ that converges to some $x \in M$. Taking limits shows that $\rho_M(x, \theta^{-1}(y)) \geqslant \epsilon$, so that $x \neq \theta^{-1}(y)$. But $y_{n_k} \to y$, and since θ is continuous we also have

$$y_{n_k} = \theta(\theta^{-1}(y_{n_k})) \to \theta(x).$$

Uniqueness of limits implies that $\theta(x) = y$, contradicting $x \neq \theta^{-1}(y)$. $\square$

The following result characterizes the homomorphisms from one shift dynamical system to another as being exactly the sliding block codes.

Theorem 6.2.9 (Curtis–Hedlund–Lyndon Theorem). *Suppose that (X, σ_X) and (Y, σ_Y) are shift dynamical systems, and that $\theta \colon X \to Y$ is a (not necessarily continuous) function. Then θ is a sliding block code if and only if it is a homomorphism.*

PROOF: We saw in Example 6.1.15 that sliding block codes are continuous, and they are clearly shift-commuting. So, any sliding block code is a homomorphism.

Conversely, suppose that θ is a homomorphism. Let $\mathcal{A}$ be the alphabet of X and $\mathfrak{A}$ that of Y. For each $b \in \mathfrak{A}$ let $C_0(b)$ denote the cylinder set $\{y \in Y : y_0 = b\}$. The sets $C_0(b)$ for $b \in \mathfrak{A}$ are disjoint and compact, so by Exercise 6.1.10 their inverse images $E_b = \theta^{-1}(C_0(b))$ are also disjoint and compact in X. Hence by Exercise 6.1.11, there is a $\delta > 0$ such that points in different sets E_b are at least δ apart in distance. Choose n with $2^{-n} < \delta$. Then any pair of points $x, x' \in X$ such that $x_{[-n,n]} = x'_{[-n,n]}$ must lie in the same set E_b, so that $\theta(x)_0 = b = \theta(x')_0$. Thus the 0th coordinate of $\theta(x)$ depends only on the central $(2n+1)$-block of x, so that, by Proposition 1.5.8, θ is a sliding block code. $\square$

Thus for shift dynamical systems (X, σ_X) and (Y, σ_Y), a factor map from σ_X to σ_Y is the same as a factor code from X to Y, and a topological conjugacy from σ_X to σ_Y is the same as a conjugacy from X to Y.

Combining the Curtis–Hedlund–Lyndon Theorem with Lemma 6.2.8 provides another proof of Theorem 1.5.14, that a sliding block code that is one-to-one and onto has a sliding block inverse. Note the use of compactness to replace the Cantor diagonal argument.

The notions of embedding, factor map, and topological conjugacy lead to a set of classification problems in dynamical systems. Just as for shifts, when given two dynamical systems, it is natural to ask when one can be embedded in the other, when one can be factored onto the other, and, most important, when they are topologically conjugate. These questions can be extremely difficult to answer. The reader should stop now to ponder them for the edge shifts based on adjacency matrices

$$\begin{bmatrix} 1 & 3 \\ 2 & 1 \end{bmatrix} \quad \text{and} \quad \begin{bmatrix} 1 & 6 \\ 1 & 1 \end{bmatrix}.$$

In the next chapter we investigate whether these particular edge shifts are topologically conjugate, and the general question of topological conjugacy for shifts of finite type.

Finally, we mention that the study of dynamics is not limited to continuous maps on compact metric spaces. The theory of dynamical systems was motivated by examples of "continuous-time" (as opposed to "discrete-time") dynamics, often on noncompact spaces. See the Notes for this chapter and §13.6.

EXERCISES

6.2.1. Find a mapping $\phi \colon X_{[2]} \to X_{[2]}$ that is continuous, but that is *not* a sliding block code.

6.2.2. Let $M = [0, 1)$ with the Euclidean metric, and $N = \mathbb{T}$. Define $\theta \colon M \to N$ by $\theta(x) = x$. Show that θ is continuous, one-to-one, and onto, but that θ^{-1} is *not* continuous. Why doesn't this contradict Lemma 6.2.8?

6.2.3. Consider the map ϕ_r on the unit interval, $[0, 1]$, defined by $\phi_r(x) = r\,x(1-x)$.
  (a) Show that for $0 \leqslant r \leqslant 4$, $\phi_r([0, 1]) \subseteq [0, 1]$.
  (b) Show that for $0 \leqslant r \leqslant 1$ and every $x \in [0, 1]$, the iterates $\phi_r^n(x)$ converge to 0.
  (c) Show that for $r = 2$ and every $x \in (0, 1)$, the iterates $\phi_r^n(x)$ converge to $1/2$.

6.2.4. Let ϕ_α be the circle rotation defined in Example 6.2.2
  (a) Show that if α is rational, then every orbit is periodic.
  (b) Show that if α is irrational, then every orbit is dense. [*Hint*: Use the fact that for any $\epsilon > 0$, there is a rational number p/q such that $|\alpha - p/q| < \epsilon/q$. See [HarW, Theorem 164] for an explanation.]

6.2.5. (a) Show that for the doubling map in Example 6.2.3 the periodic points are the rational points in $\mathbb{T}$ of the form p/q where p is an integer and q is an odd integer.
  (b) Show that for the toral automorphism in Example 6.2.4, the periodic points are exactly the rational points, i.e., the points (x, y) such that both x and y are rational.

§6.3.   Invariants

When confronted with two dynamical systems, how can we tell whether or not they are topologically conjugate? There is one strategy, which we have already seen in several guises, to show that systems are *not* topologically conjugate. Suppose we can attach to each dynamical system an "object," which could be a real number, or finite set, or group, or other mathematical structure, so that topologically conjugate dynamical systems are assigned the same value of the object. Such an assignment is called a *conjugacy invariant*, or simply an *invariant*, since it does not vary when applied to topologically conjugate dynamical systems. If two systems are assigned different values, then we know for sure they cannot be topologically conjugate.

A simple example of an invariant is the number of fixed points. This invariant clearly distinguishes the full 2-shift from the golden mean shift.

It is important to realize that if two systems are assigned the same value of an invariant, then we may be no wiser about whether or not they are topologically conjugate. We need to look for more sensitive invariants to try to tell them apart.

There is a similar approach at work in linear algebra when trying to decide whether two complex matrices A and B are similar, i.e., whether there is an invertible matrix P such that $A = PBP^{-1}$. An invariant here is the Jordan canonical form of the matrix (see §7.4). Moreover, this invariant is the best possible kind, since it tells the complete story: two complex matrices are similar if and only if they have the same Jordan canonical form. Such invariants are called *complete*. The goal of classification problems in many areas of mathematics is to find a computable, complete invariant. This goal has certainly not been realized for dynamical systems in general, but that should not stop us from trying to find invariants that are useful in detecting differences in dynamical behavior or from finding complete invariants for special classes of dynamical systems.

For an invariant to be useful, it should be computable, especially for the sorts of finitely described dynamical systems like shifts of finite type, and we should be able to tell when two values of the invariant are equal. We will give some examples of invariants below. In the next section, we will assemble the periodic point behavior of a dynamical system into a single function called its zeta function and show how to compute this invariant for shifts of finite type and sofic shifts. In the next chapter we will develop another invariant for shifts of finite type called the dimension triple that is very sensitive for detecting when two such shifts are not conjugate.

Example 6.3.1. Let (M, ϕ) be a dynamical system. For $n \geqslant 1$ let $p_n(\phi)$ denote the number of periodic points of period n, i.e.,

$$p_n(\phi) = |\{x \in M : \phi^n(x) = x\}|.$$

Then p_n takes values in $\{0, 1, 2, \ldots, \infty\}$. To show that p_n is an invariant, observe that if $\theta: (M, \phi) \cong (N, \psi)$, then $\phi^n(x) = x$ if and only if

$$\theta(x) = \theta(\phi^n(x)) = \psi^n(\theta(x)).$$

Thus θ establishes a one-to-one correspondence between the fixed points of ϕ^n and those of ψ^n, so that $p_n(\phi) = p_n(\psi)$.

A point $x \in M$ is said to have *least period* n under ϕ if $\phi^n(x) = x$ but $\phi^k(x) \neq x$ for $0 < k < n$. Let $q_n(\phi)$ denote the number of points in M having least period n under ϕ. An argument similar to that in the preceding paragraph shows that q_n is also an invariant.

If X is a shift space, then our previous notations $p_n(X)$ and $q_n(X)$ in Chapter 4 are equivalent here to $p_n(\sigma_X)$ and $q_n(\sigma_X)$.

Proposition 2.2.12 shows how to compute $p_n(\sigma_G)$ for an edge shift σ_G. Since every shift of finite type is conjugate to an edge shift, we can compute p_n for these as well. The next section gives an explicit procedure for computing p_n for sofic shifts. See Exercise 6.3.1 for the relationship between the p_n's and q_n's. $\square$

Example 6.3.2. Let (M, ϕ) be a dynamical system. We call ϕ *topologically transitive* if, for every ordered pair U, V of nonempty open sets in M, there is an $n > 0$ for which $\phi^n(U) \cap V \neq \varnothing$. Thus a topologically transitive system moves each open set enough so that it eventually intersects every other open set. In fact, we can find $n_i \to \infty$ such that $\phi^{n_i}(U) \cap V \neq \varnothing$ (see Exercise 6.3.4).

Since a topological conjugacy $\theta \colon (M, \phi) \cong (N, \psi)$ establishes a one-to-one correspondence between the open sets of M with those of N, it is clear that ϕ is topologically transitive if and only if ψ is. In this sense topological transitivity is an invariant with just two values, "yes" and "no."

We claim that for a shift dynamical system (X, σ_X), topological transitivity of σ_X is the same as irreducibility of X.

To see this, first suppose that X is irreducible. We must show that for every ordered pair U, V of nonempty open sets in X there is an $n > 0$ such that $\sigma_X^n(U) \cap V \neq \varnothing$. It is instructive to first consider a pair $C_0(u), C_0(v)$ of cylinder sets defined by blocks u, v. Since X is irreducible, there is a block w such that uwv is an allowed block in X. Hence if $n = |uw|$, then $C_0(u) \cap \sigma_X^{-n}(C_0(v)) \neq \varnothing$, or equivalently $\sigma_X^n(C_0(u)) \cap C_0(v) \neq \varnothing$, confirming topological transitivity for pairs of cylinder sets. Now suppose that U, V is a pair of nonempty open sets. Then U contains a cylinder set $C_k(u) = \sigma_X^{-k}(C_0(u))$, and V contains $C_\ell(v) = \sigma_X^{-\ell}(C_0(v))$. Irreducibility of X implies that there are blocks $w^{(i)}$ such that $uw^{(i)}v$ are allowed blocks in X and $|w^{(i)}| \to \infty$ (why?). Putting $n_i = |uw^{(i)}| + k - \ell$, we have that $\sigma_X^{n_i}(U) \cap V \neq \varnothing$ and $n_i > 0$ for sufficiently large i.

Conversely, suppose that σ_X is topologically transitive. Then given an ordered pair u, v of allowed blocks, it follows that there is an $n > |u|$ with $\sigma_X^n(C_0(u)) \cap C_0(v) \neq \varnothing$, or equivalently $C_0(u) \cap \sigma_X^{-n}(C_0(v)) \neq \varnothing$. If z is in the latter set, then $z_{[0,n+|v|-1]} = u\, z_{[|u|,n-1]}\, v = uwv$ is allowed in X, proving that X is irreducible. $\square$

Example 6.3.3. Again consider a dynamical system (M, ϕ). Call ϕ *topo-*

logically mixing if, for each ordered pair U, V of nonempty open sets, there is an n_0 such that $\phi^n(U) \cap V \neq \varnothing$ for *all* $n \geqslant n_0$. For the same reasons as in Example 6.3.2, topological mixing is an invariant. This agrees with our notion of mixing for shift spaces given in Definition 4.5.9 (Exercise 6.3.5). □

Our last example shows how to define the entropy of an arbitrary dynamical system.

Example 6.3.4. Let (M, ϕ) be a dynamical system. For $\epsilon > 0$ we will say that a set $E \subseteq M$ is (n, ϵ)-*spanning for* ϕ if, for every $x \in M$, there is a $y \in E$ such that $\rho(\phi^j(x), \phi^j(y)) \leqslant \epsilon$ for $0 \leqslant j < n$. Continuity of ϕ and compactness of M show that there are always finite (n, ϵ)-spanning sets for every $\epsilon > 0$ and $n \geqslant 1$ (Exercise 6.3.9). Denote by $r_n(\phi, \epsilon)$ the size of the (n, ϵ)-spanning set for ϕ with fewest number of elements. Put

$$r(\phi, \epsilon) = \limsup_{n \to \infty} \frac{1}{n} \log r_n(\phi, \epsilon),$$

the growth rate of $r_n(\phi, \epsilon)$ as $n \to \infty$. It is easy to see that $r(\phi, \epsilon)$ is nondecreasing as ϵ decreases to 0, so we may define

$$h(\phi) = \lim_{\epsilon \to 0} r(\phi, \epsilon).$$

The quantity $h(\phi)$ is called the *topological entropy* of ϕ, and is either a nonnegative real number or ∞.

Despite its technical appearance, this definition has the following simple interpretation. Define the n-partial orbit of a point x in a dynamical system (M, ϕ) to be the set $\{x, \phi(x), \phi^2(x), \ldots, \phi^{n-1}(x)\}$. Then $r(\phi, \epsilon)$ is the growth rate of the number of n-partial orbits as seen by someone whose vision is only good enough to tell when two points are ϵ apart.

It turns out that topological entropy is an invariant, but we shall not prove this here (see [Wal]). If X is a shift space with shift map σ_X, then the topological entropy $h(\sigma_X)$ equals the entropy $h(X)$ of the shift space X that we studied in Chapter 4 (see Exercise 6.3.8). □

EXERCISES

6.3.1. (a) Show that

$$p_n(\phi) = \sum_{k|n} q_k(\phi),$$

where the notation $k|n$ means that the sum is over all divisors k of n. For example, $p_6(\phi) = q_1(\phi) + q_2(\phi) + q_3(\phi) + q_6(\phi)$. Hence the q_n's determine the p_n's.

(b) Show that $q_1(\phi) = p_1(\phi)$ and

$$q_n(\phi) = p_n(\phi) - \sum_{\substack{k|n \\ k<n}} q_k(\phi).$$

Hence the p_n's inductively determine the q_n's.

*(c) (Möbius Inversion Formula) Define the Möbius function μ by $\mu(1) = 1$ and

$$\mu(m) = \begin{cases} (-1)^k & \text{if } m \text{ is the product of } k \text{ distinct primes,} \\ 0 & \text{if } m \text{ is divisible by a perfect square.} \end{cases}$$

Prove that

$$q_n(\phi) = \sum_{k \mid n} \mu\left(\frac{n}{k}\right) p_k(\phi).$$

For example, this says that $q_6(\phi) = p_6(\phi) - p_3(\phi) - p_2(\phi) + p_1(\phi)$. This gives a formula for the q_n's in terms of the p_n's all at once, rather than the inductive process of Exercise 6.3.1(b).

6.3.2. Let $X = X_{[2]}$.

(a) For the map $\phi \colon X \to X$ defined by $\phi(x)_j = x_j + x_{j+1} \pmod 2$, compute $p_n(\phi)$ and $q_n(\phi)$ for all $n \geqslant 1$.

(b) For $\psi \colon X \to X$ defined by $\psi(x)_j = x_j + x_{j+1}x_{j+2} \pmod 2$, compute $p_n(\psi)$ and $q_n(\psi)$ for $1 \leqslant n \leqslant 4$.

6.3.3. (a) Show that a shift space X is irreducible if and only if for some $x \in X$ the set $\{\sigma_X^n(x) : n > 0\}$ is dense in X.

(b) Show that a dynamical system (M, ϕ) for which ϕ is onto is topologically transitive if and only if for some $x \in M$, $\{\phi^n(x) : n > 0\}$ is dense in M. [*Hint*: Use the Heine-Borel Theorem 6.1.22 and the Baire Category Theorem 6.1.24.]

6.3.4. Show that a dynamical system (M, ϕ) is topologically transitive if and only if for each ordered pair U, V of nonempty open sets, there exist $n_i \to \infty$ such that $\phi^{n_i}(U) \cap V \neq \varnothing$. [*Hint*: First show that if n is any positive integer, then $\phi^n(U) \cap V \neq \varnothing$ if and only if $U \cap \phi^{-n}(V) \neq \varnothing$.]

6.3.5. Show that for shift spaces, topological mixing agrees with the notion of mixing given in Definition 4.5.9.

*6.3.6. Let $M = \mathbb{T}$ and ϕ_α be the circle rotation as described in Example 6.2.2.

(a) For which α is ϕ_α topologically transitive?

(b) For which α is ϕ_α topologically mixing?

*6.3.7. Let $M = \mathbb{T}$ and let ϕ be the doubling map described in Example 6.2.3.

(a) Is ϕ topologically transitive?

(b) Is ϕ topologically mixing?

6.3.8. Let X be a shift space with the metric in Example 6.1.10. Let $\phi = \sigma_X$. In the notation of Example 6.3.4, show that

$$r_n(\phi, 2^{-k}) = |\mathcal{B}_{n+2k}(X)|.$$

Use this to verify that $h(\sigma_X) = h(X)$.

6.3.9. Fill in the following details to show that the topological entropy of a dynamical system is well-defined:

(a) Show that for any $\epsilon > 0$ and positive integer n, there is a finite (n, ϵ)-spanning set. [*Hint*: Use the Heine–Borel Theorem 6.1.22 and Exercise 6.1.8.]

(b) Show that $r(\phi, \epsilon)$ is nondecreasing as ϵ decreases to 0.

6.3.10. The following gives an alternative definition of topological entropy. Let (M, ϕ) be a dynamical system. For $\epsilon > 0$ and a positive integer n, say that a set $E \subseteq M$ is (n, ϵ)-*separated for* ϕ if, for every pair x, y of distinct points in E, $\rho(\phi^j(x), \phi^j(y)) > \epsilon$ for some $0 \leqslant j \leqslant n - 1$. Let $s_n(\phi, \epsilon)$ be the size of the (n, ϵ)-separated set for ϕ with the most number of points. Put

$$s(\phi, \epsilon) = \limsup_{n \to \infty} \frac{1}{n} \log s_n(\phi, \epsilon),$$

the growth rate of $s_n(\phi, \epsilon)$ as $n \to \infty$.

(a) Show that for all ϵ and n, $r_n(\phi, \epsilon) \leqslant s_n(\phi, \epsilon) \leqslant r_n(\phi, \epsilon/2)$.

(b) Show that $h(\phi) = \lim_{\epsilon \to 0} s(\phi, \epsilon)$.

*6.3.11. Let $M = \mathbb{T}$ and ϕ_α be the circle rotation as described in Example 6.2.2. Compute $h(\phi_\alpha)$.

*6.3.12. Let $M = \mathbb{T}$ and let ϕ be the doubling map described in Example 6.2.3. Compute $h(\phi)$.

6.3.13. A dynamical system (M, ϕ) is called *expansive* if for some number $c > 0$, whenever $x \neq y$, there is an integer n such that $\rho(\phi^n(x), \phi^n(y)) \geqslant c$.

(a) Show that any shift space, as a dynamical system, is expansive.

(b) Show that if ϕ is expansive, then $p_n(\phi) < \infty$ for all n and $h(\phi) \geqslant \limsup_{n \to \infty} (1/n) \log |p_n(\phi)|$, thereby generalizing Proposition 4.1.15.

§6.4. Zeta Functions

Let (M, ϕ) be a dynamical system. For each $n \geqslant 1$ we have the periodic point invariant $p_n(\phi)$. There is a convenient way to combine all of these into a single invariant called the zeta function of ϕ. Although the definition will appear strange at first, it is designed to bring out important features of the dynamical system.

For notational ease, we will sometimes use $\exp(x)$ for e^x. We will assume for this section that the reader is familiar with the rudiments of the theory of power series and Taylor series. This includes the meaning of the radius of convergence of a series and what is meant by saying that the Taylor series

$$(6\text{-}4\text{-}1) \qquad \exp(x) = e^x = 1 + \frac{x}{1!} + \frac{x^2}{2!} + \frac{x^3}{3!} + \cdots = \sum_{n=0}^{\infty} \frac{x^n}{n!}$$

has radius of convergence ∞.

Definition 6.4.1. Let (M, ϕ) be a dynamical system for which $p_n(\phi) < \infty$ for all $n \geqslant 1$. The *zeta function* $\zeta_\phi(t)$ is defined as

$$\zeta_\phi(t) = \exp\left(\sum_{n=1}^{\infty} \frac{p_n(\phi)}{n} t^n\right).$$

We can compute the Taylor series of $\zeta_\phi(t)$ from the numbers $p_n(\phi)$ by simply plugging in the series

$$\sum_{n=1}^{\infty} \frac{p_n(\phi)}{n} t^n$$

for x in equation (6–4–1) and expanding out powers of the series formally. Doing this gives the first few terms of ζ_ϕ to be

$$\zeta_\phi(t) = 1 + p_1(\phi)t + \frac{1}{2}\big[p_2(\phi) + p_1(\phi)^2\big]t^2$$
$$+ \frac{1}{6}\big[2p_3(\phi) + 3p_1(\phi)p_2(\phi) + p_1(\phi)^3\big]t^3 + \dots .$$

Conversely, assuming that $\zeta_\phi(t)$ has radius of convergence > 0, we can compute the $p_n(\phi)$ from $\zeta_\phi(t)$. One way is to observe that

$$\ln \zeta_\phi(t) = \sum_{n=1}^{\infty} \frac{p_n(\phi)}{n} t^n,$$

where ln denotes the natural logarithm base e. By Taylor's formula,

$$\frac{d^n}{dt^n} \ln \zeta_\phi(t)\Big|_{t=0} = n! \frac{p_n(\phi)}{n} = (n-1)!\, p_n(\phi).$$

Another, perhaps more traditional, way to assemble the numbers $p_n(\phi)$ into a function is by using the *periodic point generating function* (or simply the *generating function*) of ϕ, defined as

$$g_\phi(t) = \sum_{n=1}^{\infty} p_n(\phi)\, t^n.$$

Both the zeta function and the generating function of ϕ are determined by the same information, namely the numbers $p_n(\phi)$. However, the zeta function has certain conceptual and formal advantages over the generating function. For instance, the zeta function of a shift of finite type is the reciprocal of a polynomial (Theorem 6.4.6), and Exercise 6.4.4 shows that the zeta function of ϕ can be written as an infinite product over the periodic orbits of ϕ. The latter property hints at a deep analogy with the Euler product formula for the Riemann zeta function on which the study of the distribution of prime numbers is based [Edw].

Example 6.4.2. Let M have just one point and ϕ be the identity map. Then $p_n(\phi) = 1$ for all $n \geqslant 1$. Thus

$$\zeta_\phi(t) = \exp\Big(\sum_{n=1}^{\infty} \frac{t^n}{n}\Big).$$

Now the Taylor series of $-\ln(1-t)$ is

$$(6\text{–}4\text{–}2) \qquad -\ln(1-t) = \frac{t}{1} + \frac{t^2}{2} + \frac{t^3}{3} + \dots = \sum_{n=1}^{\infty} \frac{t^n}{n},$$

as the reader should readily verify. Thus

$$\zeta_\phi(t) = \exp(-\ln(1-t)) = \frac{1}{1-t}.$$

The Taylor series of this function has radius of convergence 1. $\qquad\square$

Example 6.4.3. Let $M = X_{[2]}$ and $\phi = \sigma_{[2]}$. Then $p_n(\phi) = 2^n$, and using the Taylor series (6–4–2) again shows that

$$\zeta_\phi(t) = \exp\Big(\sum_{n=1}^{\infty} \frac{2^n}{n} t^n\Big) = \exp\Big(\sum_{n=1}^{\infty} \frac{(2t)^n}{n}\Big)$$

$$= \exp\big(-\ln(1 - 2t)\big) = \frac{1}{1 - 2t}. \qquad \square$$

Example 6.4.4. Let $A = \begin{bmatrix} 1 & 1 \\ 1 & 0 \end{bmatrix}$, $M = X_A$, and $\phi = \sigma_A$. Then by Proposition 2.2.12 and Example 4.1.4,

$$p_n(\phi) = \operatorname{tr} A^n = \lambda^n + \mu^n,$$

where

$$\lambda = \frac{1 + \sqrt{5}}{2} \quad \text{and} \quad \mu = \frac{1 - \sqrt{5}}{2}$$

are the roots of $\chi_A(t) = t^2 - t - 1 = 0$. Thus

$$\zeta_\phi(t) = \exp\Big(\sum_{n=1}^{\infty} \frac{\lambda^n + \mu^n}{n} t^n\Big) = \exp\Big(\sum_{n=1}^{\infty} \frac{(\lambda t)^n}{n} + \sum_{n=1}^{\infty} \frac{(\mu t)^n}{n}\Big)$$

$$= \exp\big(-\ln(1 - \lambda t) - \ln(1 - \mu t)\big) = \frac{1}{(1 - \lambda t)(1 - \mu t)}$$

$$= \frac{1}{1 - t - t^2}. \qquad \square$$

Example 6.4.5. Let M be the even shift X and $\phi = \sigma_X$. We will use the usual right-resolving presentation $\mathcal{G} = (G, \mathcal{L})$ of X shown in Figure 6.4.1. Our approach is to first show that $p_n(\sigma_{\mathcal{G}}) = p_n(\sigma_G) - (-1)^n$, then use the calculation of $p_n(\sigma_G)$ from the previous example to compute $\zeta_\phi(t)$.

Observe that the 1-block code $\mathcal{L}_\infty \colon X_G \to X_{\mathcal{G}}$ is one-to-one except over the point 0^∞, where it is two-to-one. If n is even, then every point in $X_{\mathcal{G}}$ of period n is the image of exactly one point in X_G of the same period, except 0^∞, which is the image of two points of period n. Hence $p_n(\sigma_{\mathcal{G}}) = p_n(\sigma_G) - 1$ when n is even. If n is odd, then 0^∞ is not the image of any point in X_G

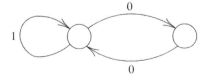

FIGURE 6.4.1. Right-resolving presentation of the even shift.

with period n, so that $p_n(\sigma_{\mathcal{G}}) = p_n(\sigma_G) + 1$ for odd n. This together with the previous example shows that

$$p_n(\phi) = p_n(\sigma_{\mathcal{G}}) = p_n(\sigma_G) - (-1)^n = \lambda^n + \mu^n - (-1)^n.$$

Therefore

$$\zeta_\phi(t) = \exp\left(\sum_{n=1}^\infty \frac{\lambda^n + \mu^n - (-1)^n}{n} t^n\right)$$

$$= \exp\left(\sum_{n=1}^\infty \frac{(\lambda t)^n}{n} + \sum_{n=1}^\infty \frac{(\mu t)^n}{n} - \sum_{n=1}^\infty \frac{(-t)^n}{n}\right)$$

$$= \exp\left(-\ln(1 - \lambda t) - \ln(1 - \mu t) + \ln(1 + t)\right)$$

$$= \frac{1 + t}{(1 - \lambda t)(1 - \mu t)} = \frac{1 + t}{1 - t - t^2}. \qquad \square$$

In all of our examples thus far, the zeta function has been rational, i.e., the ratio of two polynomials. Our next goal is to show that this is always true for sofic shifts. We first treat the easier case of shifts of finite type.

We know by Theorem 2.3.2 that if X is a shift of finite type, then there is an edge shift X_A conjugate to X. Thus the following result is enough to compute the zeta function of any shift of finite type.

Theorem 6.4.6. *Let A be an $r \times r$ nonnegative integer matrix, $\chi_A(t)$ its characteristic polynomial, and σ_A the associated shift map. Then*

$$\zeta_{\sigma_A}(t) = \frac{1}{t^r \, \chi_A(t^{-1})} = \frac{1}{\det(Id - tA)}.$$

Thus the zeta function of a shift of finite type is the reciprocal of a polynomial.

PROOF: Denote the roots of $\chi_A(t)$ by $\lambda_1, \dots, \lambda_r$, listed with multiplicity. By Proposition 2.2.12,

$$p_n(\sigma_A) = \operatorname{tr} A^n = \lambda_1^n + \lambda_2^n + \cdots + \lambda_r^n.$$

Thus

$$\zeta_{\sigma_A}(t) = \exp\left(\sum_{n=1}^\infty \frac{\lambda_1^n + \cdots + \lambda_r^n}{n} t^n\right)$$

$$= \exp\left(\sum_{n=1}^\infty \frac{(\lambda_1 t)^n}{n} + \cdots + \sum_{n=1}^\infty \frac{(\lambda_r t)^n}{n}\right)$$

$$= \frac{1}{1 - \lambda_1 t} \times \cdots \times \frac{1}{1 - \lambda_r t}.$$

Since

$$\chi_A(u) = (u - \lambda_1) \ldots (u - \lambda_r) = u^r(1 - \lambda_1 u^{-1}) \ldots (1 - \lambda_r u^{-1})$$
$$= u^r \det(Id - u^{-1}A),$$

substituting t^{-1} for u shows that

$$\chi_A(t^{-1}) = t^{-r}(1 - \lambda_1 t) \ldots (1 - \lambda_r t) = t^{-r} \det(Id - tA).$$

Hence

$$\zeta_{\sigma_A}(t) = \frac{1}{t^r \chi_A(t^{-1})} = \frac{1}{\det(Id - tA)}. \qquad \Box$$

This result allows us to give four equivalent ways, stated below, to view the information contained in the zeta function of an edge shift X_A. One of these ways involves the nonzero eigenvalues of A, and we use this opportunity to introduce some convenient notation and terminology as follows. By a *list* of complex numbers we mean a collection of complex numbers where the order of listing is irrelevant, but multiplicity counts. Thus, the list $\{2, 1\}$ is different from $\{2, 1, 1, 1\}$, but is the same as the list $\{1, 2\}$. The *spectrum* of a matrix A is the list of eigenvalues of A (with repeated eigenvalues listed according to their multiplicity). Let $\mathrm{sp}^\times(A)$ denote the list of nonzero eigenvalues of A (again, with repeated eigenvalues listed according to their multiplicity). We call $\mathrm{sp}^\times(A)$ the *nonzero spectrum of A*.

Corollary 6.4.7. *Let A be a nonnegative integer matrix. Then each of the following determines the other three.*

(1) $\zeta_{\sigma_A}(t)$,
(2) $\mathrm{sp}^\times(A)$,
(3) $\{p_n(\sigma_A)\}_{n \geqslant 1}$,
(4) $\{q_n(\sigma_A)\}_{n \geqslant 1}$.

PROOF: According to Exercise 6.3.1, the p_n's and q_n's determine one another, and the p_n's clearly determine the zeta function. By Theorem 6.4.6, the zeta function can be expressed as

$$\zeta_{\sigma_A}(t) = \frac{1}{\det(Id - tA)} = \frac{1}{\displaystyle\prod_{\lambda \in \mathrm{sp}^\times(A)} (1 - \lambda t)}.$$

So the zeta function of σ_A determines $\mathrm{sp}^\times(A)$. Finally, $\mathrm{sp}^\times(A)$ determines the p_n's because

$$p_n(\sigma_A) = \mathrm{tr}\, A^n = \sum_{\lambda \in \mathrm{sp}^\times(A)} \lambda^n. \qquad \Box$$

Counting periodic points for sofic shifts is trickier than for shifts of finite type, since the same periodic point may have several presentations and the underlying paths of these presentations may have larger periods. We have already seen this with the even shift, where the fixed point 0^∞ is presented by a point of period 2 in the edge shift of the underlying graph.

We first need to build some machinery for a more sophisticated way to count. Let $\mathcal{G} = (G, \mathcal{L})$ be a right-resolving labeled graph, where $\mathcal{L} \colon \mathcal{E} \to \mathcal{A}$. We let r denote the number of vertices in G and so may assume that $\mathcal{V} = \{1, 2, \ldots, r\}$. Let B denote the symbolic adjacency matrix of $\mathcal{G}$ and A the adjacency matrix of the underlying graph G. Therefore A is obtained from B by setting all the symbols equal to 1.

To define some auxiliary graphs, we need to recall some properties of permutations of a finite set (see [Her] for a more complete explanation). If $F = \{f_1, \ldots, f_j\}$, we can represent a permutation π of F by an expression $(f_{i_1}, \ldots, f_{i_j})$, which means that $\pi(f_1) = f_{i_1}, \ldots, \pi(f_j) = f_{i_j}$. Permutations come in two flavors: even or odd, depending on the parity of the number of interchanges (or transpositions) needed to generate the permutation. We define the *sign* of a permutation π, denoted by $\mathrm{sgn}(\pi)$, to be 1 if π is even and -1 if π is odd. The sign of a permutation turns out to be well-defined, i.e., independent of how it is represented in terms of transpositions. The sign of a cyclic permutation on a set of size n is $(-1)^{n+1}$. Every permutation π can be expressed as a composition of cyclic permutations on disjoint sets, and the sign of π is the product of the signs of these cyclic permutations.

For each j with $1 \leqslant j \leqslant r = |\mathcal{V}|$ we will construct a labeled graph $\mathcal{G}_j$ with alphabet $\{\pm a : a \in \mathcal{A}\}$. The vertex set of $\mathcal{G}_j$ is the set $\mathcal{V}_j$ of all subsets of $\mathcal{V}$ having j elements. Thus $|\mathcal{V}_j| = \binom{r}{j}$. Fix an ordering on the states in each element of $\mathcal{V}_j$.

Since $\mathcal{G}$ is right-resolving, for each $I \in \mathcal{V}$ there is at most one edge labeled a starting at I. If such an edge exists, denote its terminal state by $a(I)$; otherwise $a(I)$ is not defined. Let $\mathcal{I} = \{I_1, \ldots, I_j\}$, $\mathcal{J} = \{J_1, \ldots, J_j\} \in \mathcal{V}_j$, and $a \in \mathcal{A}$. In $\mathcal{G}_j$ there is an edge labeled a from $\mathcal{I}$ to $\mathcal{J}$ provided all the $a(I_i)$ are defined, and $(a(I_1), \ldots, a(I_j))$ is an even permutation of $\mathcal{J}$. If the permutation is odd, we assign the label $-a$. Otherwise, there is no edge with label $\pm a$ from $\mathcal{I}$ to $\mathcal{J}$. For example, $\mathcal{G}_1$ reduces to the original $\mathcal{G}$.

Let B_j denote the symbolic adjacency matrix of $\mathcal{G}_j$. Thus each entry of B_j is a signed combination of symbols in $\mathcal{A}$. Let A_j be obtained from B_j by setting all the symbols in $\mathcal{A}$ equal to 1. We call A_j the *jth signed subset matrix of* $\mathcal{G}$. With these preparations we can now describe how to compute $\zeta_{\sigma_\mathcal{G}}(t)$.

Theorem 6.4.8. *Let $\mathcal{G}$ be a right-resolving labeled graph with r vertices, and let A_j be its jth signed subset matrix. Then*

$$\zeta_{\sigma_\mathcal{G}}(t) = \prod_{j=1}^{r} \left[\det(Id - tA_j) \right]^{(-1)^j}.$$

Hence the zeta function of a sofic shift is a rational function.

PROOF: Our calculation of $\zeta_{\sigma_g}(t)$ is based on showing that

$$(6\text{-}4\text{-}3) \qquad p_n(\sigma_g) = \sum_{j=1}^{r} (-1)^{j+1} \operatorname{tr}(A_j^n),$$

which we now do.

Consider the symbolic counterpart to the sum in (6-4-3),

$$(6\text{-}4\text{-}4) \qquad \sum_{j=1}^{r} (-1)^{j+1} \operatorname{tr}(B_j^n).$$

Let $u = a_1 a_2 \ldots a_n \in \mathcal{A}^n$. Since $\mathcal{G}$ is right-resolving, for each $I \in \mathcal{V}$ there is at most one path labeled u starting at I. If it exists, let $u(I)$ denote its terminal state; otherwise $u(I)$ is not defined. Suppose that $u^\infty \notin X_{\mathcal{G}}$. Then there can be no subset $E \subseteq \mathcal{V}$ on which u acts as a permutation, for otherwise there would be a point labeled u^∞ in $X_{\mathcal{G}}$. Thus u cannot appear in $\operatorname{tr}(B_j^n)$ for $1 \leqslant j \leqslant r$, and so u does not appear in the symbolic expression (6-4-4). We can therefore write

$$(6\text{-}4\text{-}5) \qquad \sum_{j=1}^{r} (-1)^{j+1} \operatorname{tr}(B_j^n) = \sum_{u \in \mathcal{A}^n;\, u^\infty \in X_{\mathcal{G}}} c_u \cdot u,$$

where the coefficients $c_u \in \mathbb{Z}$. We will show that $c_u = 1$ for every $u \in \mathcal{A}^n$ with $u^\infty \in X_{\mathcal{G}}$.

Suppose that $u^\infty \in X_{\mathcal{G}}$. There must be at least one subset of $\mathcal{V}$ on which u acts as a permutation. If two subsets have this property, then so does their union. Hence there is a largest subset $F \subseteq \mathcal{V}$ on which u acts as a permutation. At this point, we pause to prove a combinatorial lemma, in which the notation $\pi|_E$ means the restriction of a permutation π to a set E.

Lemma 6.4.9. *Let π be a permutation of a finite set F and $\mathcal{C} = \{E \subseteq F : E \neq \varnothing, \pi(E) = E\}$. Then*

$$\sum_{E \in \mathcal{C}} (-1)^{|E|+1} \operatorname{sgn}(\pi|_E) = 1.$$

PROOF: Recall that F decomposes under π into disjoint cycles, say $C_1, C_2,$ $\ldots, C_m$. Thus each $\pi|_{C_k}$ is a cyclic permutation, and so

$$\operatorname{sgn}(\pi|_{C_k}) = (-1)^{1+|C_k|}.$$

The nonempty sets $E \subseteq F$ for which $\pi(E) = E$ are exactly the nonempty unions of subcollections of $\{C_1, \ldots, C_m\}$. Thus

$$\sum_{E \in \mathcal{C}} (-1)^{|E|+1} \operatorname{sgn}(\pi|_E) = \sum_{\varnothing \neq K \subseteq \{1,\ldots,m\}} (-1)^{1+|\cup_{k \in K} C_k|} \operatorname{sgn}(\pi|_{\cup_{k \in K} C_k})$$

$$= \sum_{\varnothing \neq K \subseteq \{1,\ldots,m\}} (-1)^{1+\sum_{k \in K}|C_k|} \prod_{k \in K} (-1)^{1+|C_k|}$$

$$= \sum_{\varnothing \neq K \subseteq \{1,\ldots,m\}} (-1)^{|K|+1+2\sum_{k \in K}|C_k|}$$

$$= \sum_{j=1}^{m} (-1)^{j+1} \binom{m}{j} = 1 - (1-1)^m = 1. \qquad \square$$

Returning to the proof of the theorem, suppose that $u \in \mathcal{A}^n$ and $u^\infty \in X_{\mathcal{G}}$, and let F be the largest subset of $\mathcal{V}$ on which u acts as a permutation. Then the coefficient c_u of u in equation (6–4–5) is, by the definition of the B_j and the lemma,

$$c_u = \sum_{\varnothing \neq E \subseteq F; \, u(E)=E} (-1)^{|E|+1} \operatorname{sgn}(u|_E) = 1.$$

Hence

$$\sum_{j=1}^{r} (-1)^{j+1} \operatorname{tr}(B_j^n) = \sum_{u \in \mathcal{A}^n; \, u^\infty \in X_{\mathcal{G}}} u.$$

Setting all symbols in $\mathcal{A}$ to 1 proves equation (6–4–3).

To conclude the proof, first observe that our proof of Theorem 6.4.6 shows that, for any matrix A,

$$\exp\left(\sum_{n=1}^{\infty} \frac{\operatorname{tr}(A^n)}{n} t^n\right) = \left[\det(Id - tA)\right]^{-1}.$$

Thus

$$\zeta_{\sigma_{\mathcal{G}}}(t) = \exp\left(\sum_{n=1}^{\infty} \frac{1}{n} \left\{\sum_{j=1}^{r} (-1)^{j+1} \operatorname{tr}(A_j^n)\right\} t^n\right)$$

$$= \prod_{j=1}^{r} \left[\exp\left(\sum_{n=1}^{\infty} \frac{\operatorname{tr}(A_j^n)}{n} t^n\right)\right]^{(-1)^{j+1}}$$

$$= \prod_{j=1}^{r} \left[\det(Id - tA_j)\right]^{(-1)^j}. \qquad \square$$

Example 6.4.10. Let $\mathcal{G}$ be the labeled graph with symbolic adjacency matrix

$$B = \begin{bmatrix} a & b \\ b & \emptyset \end{bmatrix},$$

so that $X_\mathcal{G}$ is essentially the even shift of Example 6.4.5 (we have replaced the symbol 1 by a and the symbol 0 by b). In the above notation, $\mathcal{G}_2$ has a single vertex, and one edge labeled $-b$ from this vertex to itself, since b induces an odd permutation on $\mathcal{V}$. The corresponding auxiliary matrices are

$$B_1 = \begin{bmatrix} a & b \\ b & \emptyset \end{bmatrix}, \quad A_1 = \begin{bmatrix} 1 & 1 \\ 1 & 0 \end{bmatrix}, \quad B_2 = [-b], \quad A_2 = [-1].$$

According to the theorem,

$$\zeta_{\sigma_\mathcal{G}}(t) = \frac{\det(Id - tA_2)}{\det(Id - tA_1)} = \frac{1+t}{1-t-t^2},$$

which agrees with the calculation in Example 6.4.5.   □

EXERCISES

6.4.1. For a matrix A, let $f_A(t) = \det(Id - tA)$. Show that the generating function for σ_A is

$$g_{\sigma_A}(t) = -\frac{f'_A(t)}{f_A(t)}\, t.$$

6.4.2. Compute the zeta functions of the sofic shifts presented by the labeled graphs with the following symbolic adjacency matrices.

$$\text{(a)} \begin{bmatrix} \emptyset & a & b \\ a & \emptyset & b \\ a & b & \emptyset \end{bmatrix}, \quad \text{(b)} \begin{bmatrix} a & b & \emptyset \\ \emptyset & \emptyset & b \\ b & \emptyset & a \end{bmatrix}$$

6.4.3. Let $\mathcal{G} = (G, \mathcal{L})$ be a labeled graph. A *graph diamond* for $\mathcal{G}$ is a pair of distinct paths in G with the same initial state, terminal state and label.

(a) Show that a right-closing (in particular, right-resolving) labeled graph has no graph diamonds.

(b) Assume that $\mathcal{G}$ has no graph diamonds. Let $\{I_1, \ldots, I_j\}$ be a collection of distinct states in $\mathcal{V}$, and let w be a block. Show that if there is a permutation π of $\{1, \ldots, j\}$ and paths labeled w from I_i to $I_{\pi(i)}$, then π is unique.

(c) Assume that $\mathcal{G}$ has no graph diamonds. For each j with $1 \leqslant j \leqslant r = |\mathcal{V}|$, construct the following labeled graph $\mathcal{G}_j$, which generalizes the construction in this section: the vertex set of $\mathcal{G}_j$ is, as before, the set of all subsets of $\mathcal{V}$ having j elements; for each pair $\mathcal{I} = \{I_1, \ldots, I_j\}, \mathcal{J} = \{J_1, \ldots, J_j\}$ of states in $\mathcal{G}_j$, each label $a \in \mathcal{A}$, and each permutation π of $\{1, \ldots, j\}$ such that there are edges labeled a from I_i to $J_{\pi(i)}$, endow $\mathcal{G}_j$ with exactly one edge, labeled $(\text{sgn } \pi)a$, from $\mathcal{I}$ to $\mathcal{J}$. Let B_j denote the corresponding symbolic adjacency matrix of $\mathcal{G}_j$, and let A_j denote the corresponding signed subset matrix, i.e., the matrix obtained from B_j by setting each symbol $a \in \mathcal{A}$ to 1. Show that the zeta function formula of Theorem 6.4.8 still holds.

6.4.4. Let (M, ϕ) be a dynamical system for which $p_n(\phi) < \infty$ for all n. Verify the following "product formula" for the zeta function:

$$\zeta_\phi(t) = \prod_\gamma (1 - t^{|\gamma|})^{-1},$$

where the product is taken over all periodic orbits γ of ϕ, $|\gamma|$ denotes the length (i.e., the number of elements) of γ, and $|t|$ is less than the radius of convergence of $\sum_{n=1}^\infty p_n(\phi)t^n/n$.

§6.5. Markov Partitions

One of the main sources of interest in symbolic dynamics is its use in representing other dynamical systems. This section describes how such representations arise and why they are useful.

Let us begin with the rough idea of symbolic representations. Suppose we wish to study a dynamical system (M, ϕ). We first consider the case where ϕ is invertible so that all iterates of ϕ, positive and negative, are used. To describe the orbits $\{\phi^n(y) : n \in \mathbb{Z}\}$ of points $y \in M$, we can try to use an "approximate" description constructed in the following way. Divide M into a finite number of pieces E_0, E_1, ..., E_{r-1}, which you should think of as small and localized. We can track the orbit of a point y by keeping a record of which of these pieces $\phi^n(y)$ lands in. This yields a corresponding point $x = \ldots x_{-1}x_0x_1 \ldots \in \{0, 1, \ldots, r - 1\}^{\mathbb{Z}} = X_{[r]}$ defined by

$$\phi^n(y) \in E_{x_n} \quad \text{for } n \in \mathbb{Z}.$$

So to every $y \in M$ we get a symbolic point x in the full r-shift, and the definition shows that the image $\phi(y)$ corresponds to the shifted point $\sigma(x)$.

For many invertible dynamical systems (M, ϕ) this correspondence is one-to-one (or "almost" one-to-one), so that the dynamics of ϕ on M is faithfully recorded by the shift map on a subset of the full shift. However, the set of all points in the full shift arising from points in M need not be a shift space; we will deal with this problem shortly.

Example 6.5.1. (a) Let M be a shift space over $\mathcal{A} = \{0, 1, \ldots, r - 1\}$, let $\phi = \sigma_M$ be the shift map on M, and $E_j = \{y \in M : y_0 = j\}$ for $0 \leqslant j \leqslant r - 1$. Then the symbolic representation of $y \in M$ using this subdivision of M is just the sequence y itself.

(b) Again let M be a shift space with alphabet $\mathcal{A} = \{0, 1, \ldots, r - 1\}$, ϕ be the shift map on M, and $\mathcal{B}_2(M)$ be the set of allowed 2-blocks occurring in M. For $ab \in \mathcal{B}_2(M)$ let $E_{ab} = \{y \in M : y_0 y_1 = ab\}$. It is easy to check that the symbolic representation of (M, ϕ) using this subdivision is just the 2nd higher block presentation $M^{[2]}$ of M. □

There is a one-sided version of symbolic representations that applies to (not necessarily invertible) dynamical systems. If (M, ϕ) is a dynamical system and $\{E_0, E_1, \ldots, E_{r-1}\}$ subdivides M as before, then we can record which pieces the forward iterates $\phi^n(y)$ land in for $n \geqslant 0$ (negative iterates do not necessarily make sense since ϕ need not be invertible). This produces a point x in the one-sided full r-shift $X_{[r]}^+$ (see §5.1 for the definition) defined by $\phi^n(y) \in E_{x_n}$ for $n \geqslant 0$. The action of ϕ on y corresponds to the action of the one-sided shift map on x.

Example 6.5.2. Let $M = \mathbb{T}$ be the circle described in Example 6.1.8 and $\phi \colon M \to M$ be defined by $\phi(y) = 10y \pmod 1$. Subdivide M into ten equal subintervals $E_j = [j/10, (j+1)/10)$ for $j = 0, 1, \ldots, 9$, so the alphabet here is $\mathcal{A} = \{0, 1, \ldots, 9\}$. For $y \in M$ and $n \geqslant 0$ we define a symbol $x_n \in \mathcal{A}$ by $\phi^n(y) \in E_{x_n}$. The resulting point $x = x_0 x_1 x_2 \ldots$ lies in the one-sided shift space $X_{[10]}^+$. In this case y corresponds precisely to the sequence of digits in its decimal expansion. The action of ϕ corresponds to the (one-sided) shift map; i.e., multiplying a real number by 10 shifts the decimal digits to the left by one and deletes the left-most digit since our operations are modulo 1. □

In the last example, there is some ambiguity when a point lands on the common boundary of two subdivisions. This reflects the ambiguity in decimal expansions that, for example, $.200000 \ldots$ represents the same real number as $.199999 \ldots$. One consequence is that the set of all points in the shift space corresponding to points in M is not a shift space because it may not be compact. For instance, the sequence $190000 \ldots$, $199000 \ldots$, $199900 \ldots$, of points in this set has the limiting sequence $199999 \ldots$ that is not in the set. This sort of ambiguity is the reason that symbolic representation of general dynamical systems may appear more complicated than might be first expected.

Let us now turn to the actual method used to represent dynamical systems symbolically.

Definition 6.5.3. A *topological partition* of a metric space M is a finite collection $\mathcal{P} = \{P_0, P_1, \ldots, P_{r-1}\}$ of disjoint open sets whose closures $\overline{P}_j$ together cover M in the sense that $M = \overline{P}_0 \cup \cdots \cup \overline{P}_{r-1}$.

For instance, $\mathcal{P} = \{(0, 1/10), (1/10, 2/10), \ldots, (9/10, 1)\}$ is a topological partition of $M = \mathbb{T}$. Note that a topological partition is not necessarily a partition in the usual sense, since the union of its elements need not be the whole space.

Suppose that (M, ϕ) is a (not necessarily invertible) dynamical system and that $\mathcal{P} = \{P_0, \ldots, P_{r-1}\}$ is a topological partition of M. Let $\mathcal{A} = \{0, 1, \ldots, r-1\}$. Say that a word $w = a_1 a_2 \ldots a_n$ is *allowed* for $\mathcal{P}, \phi$ if $\bigcap_{j=1}^{n} \phi^{-j}(P_{a_j}) \neq \varnothing$, and let $\mathcal{L}_{\mathcal{P}, \phi}$ be the collection of all allowed words for $\mathcal{P}, \phi$. It can be checked easily that $\mathcal{L}_{\mathcal{P}, \phi}$ is the language of a shift space (see Exercise 6.5.1). Hence by Proposition 1.3.4 there is a unique shift space $X_{\mathcal{P}, \phi}$ whose language is $\mathcal{L}_{\mathcal{P}, \phi}$. If (M, ϕ) is invertible, we call

$X_{\mathcal{P},\phi}$ the *symbolic dynamical system corresponding to* $\mathcal{P},\phi$. If (M,ϕ) is not necessarily invertible, then the one-sided shift space $X_{\mathcal{P},\phi}^+$ is called the *one-sided symbolic dynamical system corresponding to* $\mathcal{P},\phi$.

Example 6.5.4. Let (M,ϕ) be the dynamical system from Example 6.5.2, $\mathcal{P} = \{(0,1/10),(1/10,2/10),\ldots,(9/10,1)\}$, and $\mathcal{A} = \{0,1,\ldots,9\}$. It is easy to verify that the language $\mathcal{L}_{\mathcal{P},\phi}$ is the set of all words over $\mathcal{A}$, so that the one-sided symbolic dynamical system $X_{\mathcal{P},\phi}^+$ is the full one-sided 10-shift $X_{[10]}^+$.   □

Example 6.5.5. Let us use the same space M and topological partition $\mathcal{P}$ as in the previous example, but replace the mapping ϕ by the identity mapping Id on M. Then $\mathcal{L}_{\mathcal{P},Id}$ contains only words of the form a^n where $a \in \mathcal{A}$ and $n \geqslant 0$. Hence $X_{\mathcal{P},\phi}$ consists of just the ten points 0^∞, 1^∞, $\ldots$, 9^∞.   □

In the last example, $X_{\mathcal{P},\phi}$ is not a particularly accurate description of the original dynamical system since it is a finite set. We next investigate circumstances under which the symbolic dynamical system better models the dynamical system used to define it.

First consider the case of an invertible dynamical system (M,ϕ). Let $\mathcal{P} = \{P_0, P_1, \ldots, P_{r-1}\}$ be a topological partition of M. For each $x \in X_{\mathcal{P},\phi}$ and $n \geqslant 0$ there is a corresponding nonempty open set

$$D_n(x) = \bigcap_{k=-n}^{n} \phi^{-k}(P_{x_k}) \subseteq M.$$

The closures $\overline{D}_n(x)$ of these sets are compact and decrease with n, so that $\overline{D}_0(x) \supseteq \overline{D}_1(x) \supseteq \overline{D}_2(x) \supseteq \ldots$. It follows that $\bigcap_{n=0}^{\infty} \overline{D}_n(x) \neq \varnothing$ (see Exercise 6.1.4). In order for points in $X_{\mathcal{P},\phi}$ to correspond to points in M, this intersection should contain only one point. This leads to the following definition.

Definition 6.5.6. Let (M,ϕ) be an invertible dynamical system. A topological partition $\mathcal{P} = \{P_0, P_1, \ldots, P_{r-1}\}$ of M gives a *symbolic representation* of (M,ϕ) if for every $x \in X_{\mathcal{P},\phi}$ the intersection $\bigcap_{n=0}^{\infty} \overline{D}_n(x)$ consists of exactly one point. We call $\mathcal{P}$ a *Markov partition* for (M,ϕ) if $\mathcal{P}$ gives a symbolic representation of (M,ϕ) and furthermore $X_{\mathcal{P},\phi}$ is a shift of finite type.

There are natural versions of these notions for not necessarily invertible dynamical systems, replacing $D_n(x)$ with $D_n^+(x) = \bigcap_{k=0}^{n} \phi^{-k}(P_{x_k})$ and $X_{\mathcal{P},\phi}$ with $X_{\mathcal{P},\phi}^+$.

Definition 6.5.7. Let (M,ϕ) be a general dynamical system. A topological partition $\mathcal{P} = \{P_0, P_1, \ldots, P_{r-1}\}$ of M gives a *one-sided symbolic representation* of (M,ϕ) if for every $x \in X_{\mathcal{P},\phi}^+$ the intersection $\bigcap_{n=0}^{\infty} \overline{D_n^+}(x)$

consists of exactly one point. We call $\mathcal{P}$ a *one-sided Markov partition* for (M, ϕ) if $\mathcal{P}$ gives a one-sided symbolic representation of (M, ϕ) and furthermore $X_{\mathcal{P},\phi}$ is a shift of finite type, so that $X_{\mathcal{P},\phi}^+$ is a one-sided shift of finite type.

For instance, the topological partition $\mathcal{P} = \{(0, 1/10), \ldots, (9/10, 1)\}$ in Example 6.5.4 is a one-sided Markov partition for that dynamical system.

There are a number of variants on the definition of Markov partition in the literature, some involving the underlying geometry of the mapping ϕ. Ours is simple to state, but is somewhat weaker than other variants.

Suppose that $\mathcal{P}$ gives a symbolic representation of the invertible dynamical system (M, ϕ). Then there is a well-defined mapping π from $X = X_{\mathcal{P},\phi}$ to M which maps a point $x = \ldots x_{-1} x_0 x_1 \ldots$ to the unique point $\pi(x)$ in the intersection $\bigcap_{n=0}^{\infty} \overline{D}_n(x)$. We call x a *symbolic representation of* $\pi(x)$. Using that

$$D_{n+1}(\sigma x) \subseteq \phi(D_n(x)) \subseteq D_{n-1}(\sigma x),$$

the reader can verify that the definitions are set up so that the following diagram commutes.

(6–5–1)

$$
\begin{array}{ccc}
X_{\mathcal{P},\phi} & \xrightarrow{\ \sigma\ } & X_{\mathcal{P},\phi} \\
\pi \downarrow & & \downarrow \pi \\
M & \xrightarrow[\ \phi\ ]{} & M
\end{array}
$$

Of course there is a one-sided version of this as well.

The following result shows that π is also continuous and onto and therefore a factor map from $(X_{\mathcal{P},\phi}, \sigma)$ to (M, ϕ). We leave it to the reader to state and prove an analogous result for the not necessarily invertible case.

Proposition 6.5.8. *Let $\mathcal{P}$ give a symbolic representation of the invertible dynamical system (M, ϕ), and let $\pi \colon X_{\mathcal{P},\phi} \to M$ be the map defined above. Then π is a factor map from $(X_{\mathcal{P},\phi}, \sigma)$ to (M, ϕ).*

PROOF: We have already observed above that $\pi \circ \sigma = \phi \circ \pi$. It remains only to show that π is continuous and onto.

To prove continuity of π at $x \in X_{\mathcal{P},\phi}$, first observe that since $\mathcal{P}$ gives a symbolic representation of (M, ϕ), the sets $\overline{D}_0(x) \supseteq \overline{D}_1(x) \supseteq \ldots$ intersect to a single point. Hence diam $\overline{D}_n(x) \to 0$ as $n \to \infty$ (Exercise 6.5.2). It follows that if $y \in X_{\mathcal{P},\phi}$ and $y_{[-n,n]} = x_{[-n,n]}$, then $\pi(x), \pi(y) \in \overline{D}_n(x)$ and so $\rho(\pi(x), \pi(y)) \leqslant$ diam $\overline{D}_n(x)$. This shows that if y is close to x then $\pi(y)$ is close to $\pi(x)$, so that π is continuous.

To show that π is onto recall the Baire Category Theorem 6.1.24, which asserts that in a compact metric space the intersection of a countable collection of open dense sets is dense and in particular nonempty. Let

$$U = \bigcup_{i=0}^{r-1} P_i,$$

which is an open set. We claim that U is dense. To see this, simply observe that

$$\overline{U} = \overline{\bigcup_{i=0}^{r-1} P_i} = \overline{P}_0 \cup \cdots \cup \overline{P}_{r-1} = M.$$

Recall from Proposition 6.1.23 that finite intersections of open and dense sets are again open and dense. Since ϕ is a homeomorphism, the ϕ-images and the ϕ^{-1}-images of open and dense sets are also open and dense. It then follows that for each n, the set

$$U_n = \bigcap_{k=-n}^{n} \phi^{-k}(U)$$

is open and dense. Thus by the Baire Category Theorem, the set

$$U_\infty = \bigcap_{n=0}^{\infty} U_n$$

is dense.

Now if $y \in U_\infty$, then $y \in \bigcap_{n=0}^\infty D_n(x)$ for some $x \in X_{\mathcal{P},\phi}$. Since $\mathcal{P}$ gives a symbolic representation of (M, ϕ), we have that

$$\{y\} = \bigcap_{n=0}^{\infty} \overline{D}_n(x) = \bigcap_{n=0}^{\infty} D_n(x),$$

so that $y = \pi(x)$. Thus the image of π contains the dense set U_∞. Since the image of a compact set via a continuous map is again compact (Exercise 6.1.9) and therefore closed (Proposition 6.1.19), it follows that the image of π is all of M, as desired. □

The following result shows how we can use symbolic representations to understand the dynamical systems that they represent: establish the properties for the symbolic representation and then "push them down" to the dynamical system.

Proposition 6.5.9. *Suppose that $\mathcal{P}$ gives a symbolic representation of the invertible dynamical system (M, ϕ). For each of the following properties, if $(X_{\mathcal{P},\phi}, \sigma)$ has the property, then so does (M, ϕ).*

    (1) *Topological transitivity.*

    (2) *Topological mixing.*

    (3) *The set of periodic points is dense.*

PROOF: By Proposition 6.5.8, we know that there is a factor map π from $(X_{\mathcal{P},\phi}, \sigma)$ to (M, ϕ). Suppose that $(X_{\mathcal{P},\phi}, \sigma)$ is topologically transitive. Let U, V be nonempty open subsets of M. Then $\pi^{-1}(U)$ and $\pi^{-1}(V)$ are nonempty open sets, so that $\sigma^n(\pi^{-1}(U)) \cap \pi^{-1}(V) \neq \varnothing$ for some $n > 0$. Applying π shows that $\phi^n(U) \cap V \neq \varnothing$, proving that (M, ϕ) is topologically transitive. A similar argument can be used for topological mixing. We leave the proof for density of the set of periodic points as an exercise for the reader. □

Note that the proof of the preceding result really shows that the properties listed there are preserved by all factor maps.

We remind the reader that for shift spaces, viewed as symbolic dynamical systems, topological transitivity is the same as irreducibility and topological mixing is the same as mixing. These properties are particularly easy to check for shifts of finite type, as is density of the set of periodic points (see Exercise 6.5.5(a) and (b)). So the preceding results give a good method for establishing dynamical properties of dynamical systems that have Markov partitions.

If $\mathcal{P}$ is a Markov partition for (M, ϕ), then we obtain a description of the dynamics of (M, ϕ) in terms of a shift of finite type. As pointed out earlier, there may be some ambiguity in this symbolic representation at the boundaries of topological partition elements, but these boundaries are often "small" in various senses and usually do not impair the usefulness of the symbolic representation. For instance, in Example 6.5.2 the only points with ambiguous symbolic representation are the rational numbers with finite decimal expansions, and these form a countable set.

Which dynamical systems have Markov partitions? How do we find them? The major breakthrough was made by Adler and Weiss [AdlW] in the late 1960's. They discovered that certain mappings of the two-dimensional torus that expand in one direction and contract in another always have Markov partitions, and in fact the elements of these partitions are parallelograms. The idea that such partitions might exist for a wide collection of dynamical systems, and a general method to construct them, led to a burst of activity in the 1970's in developing the theory of Markov partitions as a general tool in dynamics. Some of this work is discussed briefly in the Notes for this chapter.

We will content ourselves with giving one example of a Markov partition for a particular mapping of the two-dimensional torus.

Example 6.5.10. Let $M = \mathbb{T}^2$ be the two-dimensional torus described in Example 6.1.9. Recall that the torus was described there as the closed unit square $[0, 1] \times [0, 1]$ with opposite sides identified, so that the point $(0, t)$ is glued to $(1, t)$ for $0 \leqslant t \leqslant 1$ and $(s, 0)$ is glued to $(s, 1)$ for $0 \leqslant s \leqslant 1$. The mapping $\phi: \mathbb{T}^2 \to \mathbb{T}^2$ defined by $\phi(s, t) = (s + t, s) \pmod 1$ has a continuous inverse given by $\phi^{-1}(s, t) = (t, s - t) \pmod 1$. Hence $(\mathbb{T}^2, \phi)$ is an invertible dynamical system.

Using coordinate-wise addition modulo 1 as a group operation, $\mathbb{T}^2$ becomes a group. Then ϕ preserves the group operation on $\mathbb{T}^2$ and so is a group automorphism. Such an automorphism is called a (two-dimensional) *toral automorphism*.

It will be helpful to view the torus in the following alternative way. We identify the torus $\mathbb{T}^2$ with the quotient group $\mathbb{R}^2/\mathbb{Z}^2$ by identifying a point (s,t) in the torus with the coset $(s,t) + \mathbb{Z}^2$. In this way the torus is formed from the plane $\mathbb{R}^2$ by identifying points (s,t) and (s',t') whenever they differ by an integer lattice point, i.e., $(s,t) - (s',t') \in \mathbb{Z}^2$. It is straightforward to see that via this identification the group operation on the torus that we described above coincides with addition of cosets in $\mathbb{R}^2/\mathbb{Z}^2$.

Let $q \colon \mathbb{R}^2 \to \mathbb{T}^2$ denote the natural quotient mapping given by $q((s,t)) = (s,t) + \mathbb{Z}^2$. We can use q to "push down" objects in the plane to the torus; that is, for a subset E of the plane, $q(E)$ is the corresponding object in the torus. For example, pushing down the closed unit square $S = [0,1] \times [0,1] \subseteq \mathbb{R}^2$ under q amounts to identifying the sides of S to form $\mathbb{T}^2$, as described in Example 6.1.9. Sets other than S can also be used to describe $\mathbb{T}^2$ this way, as we shall shortly see.

Observe that if $A = \begin{bmatrix} 1 & 1 \\ 1 & 0 \end{bmatrix}$ is considered as a linear transformation of $\mathbb{R}^2$ to itself, then the following diagram commutes.

$$
\begin{array}{ccc}
\mathbb{R}^2 & \xrightarrow{\ A\ } & \mathbb{R}^2 \\
\downarrow{\scriptstyle q} & & \downarrow{\scriptstyle q} \\
\mathbb{T}^2 & \xrightarrow[\ \phi\ ]{} & \mathbb{T}^2
\end{array}
$$

In this way, we can use algebraic properties of the linear transformation $\mathbf{v} \mapsto A\mathbf{v}$ to understand the toral automorphism ϕ.

Now A has a one-dimensional eigenspace L_1 corresponding to its Perron eigenvalue $\lambda = (1 + \sqrt{5})/2 \approx 1.618$ and another one-dimensional eigenspace L_2 corresponding to $\mu = (1 - \sqrt{5})/2 = -\lambda^{-1} \approx -0.6180$ (see Figure 6.5.1(a)). Since these eigenspaces are one-dimensional, we call them *eigenlines*. Note that $\lambda > 1$ while $|\mu| < 1$, so that A expands in the direction L_1 but it contracts and flips in the direction L_2. For this reason, we call any line segment parallel to L_1 an *expanding segment* and any line segment parallel to L_2 a *contracting segment*.

We will construct a Markov partition consisting of three open rectangles in $\mathbb{R}^2$; the closures of the q-images of these rectangles cover the torus. These rectangles P_1, P_2, and P_3 are shown in Figure 6.5.1(b). We do not include the boundaries as part of these rectangles, so the P_I are open sets. Note that the boundaries of the P_I consist of contracting and expanding segments.

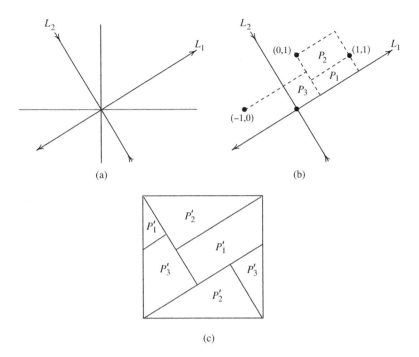

FIGURE 6.5.1. Constructing a Markov partition.

Consider the images $P'_I = q(P_I)$ of these rectangles in $\mathbb{T}^2$, shown in Figure 6.5.1(c). Although each image may look disconnected, each is in reality connected because of the identifications on the square used to form $\mathbb{T}^2$. For example, the top and bottom edges of the square are glued together, and this joins the two parts of P'_2 into a single connected piece. The reader should verify using basic geometry that P'_1, P'_2, and P'_3 fit together as shown, so that these give a topological partition of $\mathbb{T}^2$. This shows that we can equally well describe $\mathbb{T}^2$ as the union Ω of the closed rectangles $\overline{P}_1$, $\overline{P}_2$, and $\overline{P}_3$ with edges identified using q, and we shall use this description in what follows since it is better adapted to the mapping ϕ.

Next, we form a matrix B from $\mathcal{P}$ and ϕ defined by

$$B_{IJ} = \begin{cases} 1 \text{ if } \phi(P_I) \cap P_J \neq \varnothing, \\ 0 \text{ if } \phi(P_I) \cap P_J = \varnothing. \end{cases}$$

We can obtain pictures of the $\phi(P_I)$ by translating the $A(P_I)$ into Ω. As shown in Figure 6.5.2, $\phi(P_1)$ meets P_2 and P_3 but not P_1, $\phi(P_2)$ meets P_1 and P_3 but not P_2, and $\phi(P_3)$ meets only P_2. Hence

$$(6\text{--}5\text{--}2) \qquad\qquad B = \begin{bmatrix} 0 & 1 & 1 \\ 1 & 0 & 1 \\ 0 & 1 & 0 \end{bmatrix}.$$

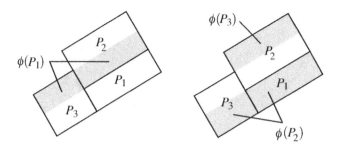

FIGURE 6.5.2.   Images of the Markov rectangles.

We claim that $\mathcal{P}$ is a Markov partition for ϕ. To verify this, we need to check that

(1) $\mathcal{P}$ gives a symbolic representation of $(\mathbb{T}^2, \phi)$, and
(2) the corresponding symbolic dynamical system $X_{\mathcal{P},\phi}$ is a shift of finite type, in our case the vertex shift $\widehat{X}_B$.

The key to understanding both (1) and (2) is the following property.

Intersection Property: Whenever $\phi(P_I)$ meets P_J, then the intersection $\phi(P_I) \cap P_J$ is a single connected strip parallel to L_1, which cuts completely through P_J in the L_1 direction and does not contain an expanding boundary segment of P_J.

In other words, the intersection $\phi(P_I) \cap P_J$ is a single connected strip that looks like Figure 6.5.3, not like those in Figure 6.5.4. This is immediately evident from Figure 6.5.2. But there is a more conceptual way to see this as follows.

As the reader can verify, the union of the boundaries of the rectangles is the image under q of the union of two line segments: the expanding segment E contained in L_1 and the contracting segment C contained in L_2, as shown in Figure 6.5.5. Now since $\lambda > 0$ and A expands L_1, it follows that

$$(6\text{--}5\text{--}3) \qquad\qquad A(E) \supset E.$$

Since $\mu < 0$, the linear map A flips and contracts L_2 by a factor of $|\mu|$. It is an exercise in geometry using Figure 6.5.1(b) to show that the part of C

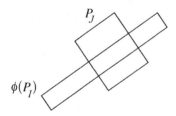

FIGURE 6.5.3.   Good rectangle intersection.

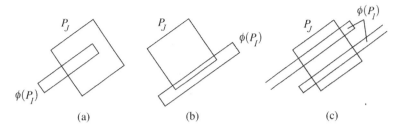

FIGURE 6.5.4. Bad rectangle intersections.

in the second quadrant is shorter than the part in the fourth quadrant by exactly a factor of $|\mu|$, so it follows that

(6-5-4) $$A(C) \subset C.$$

Now if the intersection $\phi(P_I) \cap P_J$ looked like Figure 6.5.4(a), then $A(C)$ would not be contained in C, contradicting (6–5–4). Similarly if the intersection looked like Figure 6.5.4(b), then $A(E)$ would not contain E, contradicting (6–5–3). Finally if the intersection looked like Figure 6.5.4(c), then the image $\phi(P_I)$ would have "wrapped around" the torus; but if the rectangles are sufficiently small (and ours are), this cannot happen.

Let (a_I, b_I) denote the dimensions of the rectangle P_I. That is, a_I denotes the dimension of P_I in the expanding (or L_1) direction and b_I denotes the dimension of P_I in the contracting (or L_2) direction. Since A contracts by a uniform factor of $|\mu| = \lambda^{-1}$ in the L_2 direction, it follows from the Intersection Property that whenever $\phi(P_I)$ meets P_J, the intersection $\phi(P_I) \cap P_J$ is a single strip within P_J of dimensions $(a_J, \lambda^{-1}b_I)$. Hence $P_I \cap \phi^{-1}(P_J)$ is a single strip in P_I with dimensions $(\lambda^{-1}a_J, b_I)$. Repeating this argument shows that $\phi^2(P_I) \cap \phi(P_J) \cap P_K$ is a strip in P_K with dimensions $(a_K, \lambda^{-2}b_I)$ (see Figure 6.5.6), so that $P_I \cap \phi^{-1}(P_J) \cap \phi^{-2}(P_K)$ has dimensions $(\lambda^{-2}a_K, b_I)$. Extending this to multiple intersections shows

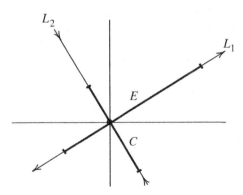

FIGURE 6.5.5. Segments forming the boundary of a Markov partition.

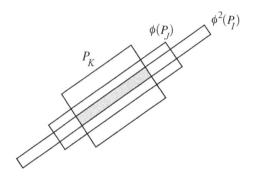

FIGURE 6.5.6.   Iterated intersections.

that whenever $x \in X_{\mathcal{P},\phi}$, then $\bigcap_{k=0}^{n} \phi^{-k}(P_{x_k})$ is a single strip in P_{x_0} with dimensions $(\lambda^{-n} a_{x_n}, b_{x_0})$. Applying this argument to ϕ^{-1} instead of ϕ shows that $\bigcap_{k=-n}^{0} \phi^{-k}(P_{x_k})$ is a single strip in P_{x_0} with dimensions $(a_{x_0}, \lambda^{-n} b_{x_{-n}})$. It follows that $D_n(x) = \bigcap_{k=-n}^{n} \phi^{-k}(P_{x_k})$ is a rectangle of dimensions $(\lambda^{-n} a_{x_n}, \lambda^{-n} b_{x_{-n}})$. Since diam $D_n(x) = $ diam $\overline{D}_n(x)$, we see that $\overline{D}_0(x) \supseteq \overline{D}_1(x) \supseteq \overline{D}_2(x) \supseteq \ldots$ is a decreasing sequence of compact sets whose diameters shrink to zero, so that $\bigcap_{n=0}^{\infty} \overline{D}_n(x)$ consists of a single point (see Exercise 6.5.2). This proves (1).

It remains to prove (2). By the Intersection Property, we see that if $\phi(P_I) \cap P_J \neq \varnothing$ and $\phi(P_J) \cap P_K \neq \varnothing$, then $\phi^2(P_I) \cap \phi(P_J) \cap P_K \neq \varnothing$ (see Figure 6.5.6). Hence if IJ and JK are allowed blocks in $X_{\mathcal{P},\phi}$, then so is IJK. This argument extends to arbitrary intersections, showing that the language $\mathcal{L}_{\mathcal{P},\phi}$ is exactly the set of words over $\{1, 2, 3\}$ whose subblocks of length two are allowed in $\widehat{X}_B$. In other words, $X_{\mathcal{P},\phi} = \widehat{X}_B$, and so we have (2).

This proves that $\mathcal{P}$ is indeed a Markov partition for (M, ϕ).   □

In the preceding example it turns out that the set of points with ambiguous symbolic representation is exactly the union $q(L_1) \cup q(L_2)$ of the images of two lines in $\mathbb{R}^2$ (Exercise 6.5.6). Although this set is dense, it is "small" in a number of senses (just as the set of points whose decimal expansion is ambiguous is also small).

Since the matrix B in (6–5–2) is primitive, the vertex shift $\widehat{X}_B$ is mixing. It follows from Proposition 6.5.9 that the toral automorphism ϕ is topologically mixing, and therefore also topologically transitive. Also, since any mixing shift of finite type has a dense set of periodic points (Exercise 6.5.5(b)), ϕ has this property as well. While it is possible to establish these properties directly for ϕ (for instance, see Exercise 6.2.5(b)), the symbolic representation gives a unified way of establishing several dynamical properties all within one framework.

There are higher-dimensional versions of toral automorphisms. We define

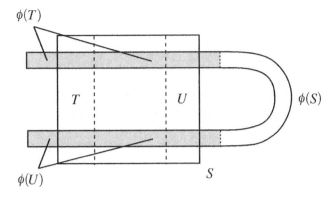

FIGURE 6.5.7. The horseshoe dynamical system.

an (n-dimensional) *toral automorphism* to be a group automorphism ϕ of the n-dimensional torus $\mathbb{T}^n = \mathbb{R}^n/\mathbb{Z}^n$. Any such mapping is given by an $n \times n$ integer matrix A with determinant ± 1 via $\phi(\mathbf{x} + \mathbb{Z}^n) = A\mathbf{x} + \mathbb{Z}^n$. The determinant condition is imposed so that ϕ is invertible (see Exercise 6.5.7). A toral automorphism is called *hyperbolic* if it has no eigenvalues with absolute value one. Thus for a hyperbolic toral automorphism ϕ there are natural directions in which ϕ expands and others where it contracts, and these directions span all n dimensions. There are Markov partitions for such automorphisms ([Sin2], [Bow6]). The example that we have worked out in this section suggests that the elements of such a Markov partition might be n-dimensional "parallelepipeds" in $\mathbb{T}^n$. However, it turns out that for $n \geqslant 3$ the boundaries of the partition elements are typically fractals (see [Bow8],[Caw],[Bed]).

At the end of §9.3, we briefly indicate how the symbolic dynamics framework can be used to give a meaningful classification of hyperbolic toral automorphisms.

The construction of Markov partitions has gone far beyond hyperbolic toral automorphisms. Markov partitions have been constructed for a wide class of dynamical systems, called Axiom A systems, that share the expanding and contracting properties of hyperbolic toral automorphisms. One such example is the *horseshoe* described as follows. Let S be a square in the plane, and let $\phi : S \to \mathbb{R}^2$ be a continuous one-to-one mapping whose image looks like that shown in Figure 6.5.7: the images of the subsets T and U of S are as shown, and both restrictions $\phi|_T$ and $\phi|_U$ are linear mappings which contract in the vertical direction and expand in the horizontal direction. This ϕ is called the *horseshoe mapping*. As it stands, (S, ϕ) is not a dynamical system since $\phi(S)$ is not contained in S. However, restricting our attention to the compact set

$$(6\text{-}5\text{-}5) \qquad\qquad M = \bigcap_{n=-\infty}^{\infty} \phi^{-n}(S)$$

of those points in S whose orbit remains in S gives an invertible dynamical system $(M, \phi|_M)$. One can show that the topological partition $\mathcal{P} = \{T \cap M, U \cap M\}$ is a Markov partition for this dynamical system and that $X_{\mathcal{P}, \phi|_M}$ is the full 2-shift. The factor map $\pi \colon X_{\mathcal{P}, \phi|_M} \to M$ used to give the symbolic representation is one-to-one here, so $(M, \phi|_M)$ is topologically conjugate to the full 2-shift (Exercise 6.5.8).

EXERCISES

6.5.1. Let (M, ϕ) be a dynamical system and $\mathcal{P}$ be a topological partition of M. Show that $\mathcal{L}_{\mathcal{P}, \phi}$ is the language of a shift space.

6.5.2. Let $E_0 \supseteq E_1 \supseteq E_2 \supseteq \dots$ be compact subsets of a metric space. Show that $\bigcap_{n=0}^{\infty} E_n$ consists of exactly one point if and only if $\mathrm{diam}(E_n) \to 0$ as $n \to \infty$.

6.5.3. Verify that the diagram in (6–5–1) commutes.

6.5.4. Use the geometry of Figure 6.5.1(b) to verify (6–5–4).

6.5.5. (a) Show that if G is essential, then the edge shift X_G has dense set of periodic points if and only if every edge of G is contained in an irreducible component of G.

    (b) Show that for a shift of finite type the set of periodic points is dense if and only if it is the disjoint union of irreducible shifts of finite type. (In particular, as in Exercise 6.1.12, an irreducible shift of finite type has dense periodic points.)

    (c) Show that density of the set of periodic points is preserved under factor maps.

    (d) Use the symbolic representation constructed in Example 6.5.10 to show that the toral automorphism ϕ has dense periodic points (see also Exercise 6.2.5(b)).

6.5.6. Let $\pi \colon X_{\mathcal{P}, \phi} \to \mathbb{T}^2$ be the mapping from Example 6.5.10.

    (a) Show that a point of $\mathbb{T}^2$ has more than one pre-image under π if and only if it is contained in $q(L_1) \cup q(L_2)$.

    (b) What is the maximal number of π-pre-images of any point in the torus?

6.5.7. Let A be an $n \times n$ integral matrix. Show that the linear mapping $\mathbf{x} \mapsto A\mathbf{x}$ defines a homeomorphism of the n-dimensional torus $\mathbb{T}^n = \mathbb{R}^n/\mathbb{Z}^n$ if and only if $\det(A) = \pm 1$.

*6.5.8. For the horseshoe mapping ϕ and M as in (6–5–5), show that

    (a) The restriction of ϕ to M is a homeomorphism.

    (b) $\{T \cap M, U \cap M\}$ is a Markov partition for $(M, \phi|_M)$.

    (c) $(M, \phi|_M)$ is topologically conjugate to the full 2-shift.

    (d) M is the largest ϕ-invariant set contained in S.

6.5.9. For a dynamical system (M, ϕ) and $x \in M$, let

$$W^s(x) = \{ y \in M : \lim_{n \to +\infty} \rho(\phi^n(x), \phi^n(y)) = 0 \}.$$

(a) Show that if (M, ϕ) is a mixing shift of finite type, then each $W^s(x)$ is dense.

(b) Show that if (M, ϕ) is a dynamical system, $\mathcal{P}$ is a Markov partition, and $X_{\mathcal{P}, \phi}$ is mixing, then $W^s(x)$ is dense in M for every $x \in M$.

Notes

The material on metric spaces in §6.1 is standard. More detail can be found in most textbooks that include metric spaces, for instance [Sim], [Edg]. The Baire Category Theorem is valid in the more general context of complete metric spaces.

The theory of dynamical systems has its roots in physics, going back to Galileo and Newton and continuing with the work of many prominent nineteenth century mathematicians. The general mathematical theory of dynamical systems was developed by Poincaré [Poi] and Birkhoff [Bir1], [Bir2] in the late nineteenth and early twentieth centuries. There are now several modern textbooks on dynamical systems, for instance Arrowsmith and Place [ArrP], Nitecki [Nit] and Ruelle [Rue2]; see also Devaney [Dev] for a treatment at the undergraduate level.

The dynamical systems that we have considered in this chapter are often called "discrete-time dynamical systems": time is discrete because an orbit $\{\phi^n(x)\}_{n \in \mathbb{Z}}$ is viewed as tracking the evolution of a point x at the discrete-time steps $1, 2, 3, \ldots$. However, "continuous-time" dynamical systems, where orbits are parameterized by the set of real numbers, are also of great importance. Such systems arise naturally from differential equations [HirS], [Har]. The theory of dynamical systems grew out of efforts to understand continuous-time dynamical systems, such as geodesic flows on surfaces of negative curvature (see §13.6 for more on this).

While we have assumed compactness for our dynamical systems, there is also great interest in dynamics on noncompact spaces – for instance, geodesic flows on noncompact spaces of negative curvature and shifts over countably-infinite alphabets (see §13.9).

Topological transitivity is an old notion in dynamical systems. According to Exercise 6.3.3(b) every topologically transitive dynamical system (M, ϕ) is the closure of the orbit of some point $x \in M$, and so (M, ϕ) may be regarded as the "orbit closure" of x. Before the notion of shift space was formally defined by a list of forbidden blocks, a topologically transitive shift space was quite often defined as the orbit closure of a bi-infinite sequence [GotH, Chap. 12]. The fundamental notion of ergodicity, studied in ergodic theory, is a variant of transitivity (see §13.4 and the textbooks Peterson [Pet3] and Walters [Wal]). In the 1950's Gottschalk and Hedlund [GotH] gave an exposition of various notions of transitivity and mixing.

Topological entropy was introduced in [AdlKoM], based on earlier notions in information theory [Sha] and ergodic theory [Kol], [Sin1], [Par1]. The definition that we have given here is due to Bowen [Bow3]. The formula for the zeta function of an edge shift is due to Bowen and Lanford [BowL], and the formula for the zeta function of a sofic shift is due to Manning [Man] and Bowen [Bow7]; for an exposition, see Béal [Bea4]. Lind [Lin8] calculates the effect on entropy and the zeta function when an additional block is forbidden from a shift of finite type. For further reading on zeta functions of dynamical systems, see Parry and Pollicott [ParP] and Ruelle [Rue3].

The symbolic dynamics approach that we have described in §6.5 has been extended considerably. For instance, hyperbolic toral automorphisms fit into the more general framework of Axiom A dynamical systems, and Markov partitions for these systems were constructed by Adler-Weiss [AdlW], Berg [Berg], Sinai [Sin2], Ratner [Rat], Bowen [Bow1], [Bow2], [Bow6], [Bow7], [Bow5], [Bow4], and Series [Ser], among others. More general notions of Markov partition for more general classes of dynamical systems have since been developed by Fried [Frie] and Adler [Adl2], and infinite Markov partitions have been used to understand certain maps of the interval [Hof1], [Hof2], [Hof3], [Hof4], and billiard flows [KruT], [BunS], [BunSC].

Toral automorphisms that have some eigenvalues with absolute value one cannot have Markov partitions [Lin1]. However, if the eigenvalues of absolute value one are not roots of unity, then the automorphism retains many of the dynamical properties of a hyperbolic automorphism [Lin2].

Symbolic dynamics has been used extensively as a tool in the study of continuous maps of the interval. The Markov partition for the map in Example 6.5.4 is a simple example. Markov partitions for piecewise monotonic continuous maps of the interval have been used to compute entropy and numbers of periodic points as well as to investigate other dynamical properties; see Parry [Par2], as well as the expositions by Block and Coppel [BloC], Devaney [Dev], and Collet and Eckmann [ColE].

In §6.2, we observed that a sliding block code from a shift space to itself may be considered as a dynamical system in its own right. These systems, often called cellular automata, have very rich, interesting, and complicated dynamical behavior; see, for instance, Gilman [Gil], Hurd, Kari and Culik [HurKC], and Lind [Lin5], as well as the collections of articles [DemGJ], [FarTW].

Exercise 6.1.16 comes from Furstenberg [Fur1, pp. 91–93]; the Möbius Inversion Formula can be found in [NivZ]); and Exercise 6.4.3 essentially comes from [Man] and [Bow7].

CHAPTER 7

CONJUGACY

When are two shifts of finite type conjugate? Since shifts of finite type are conjugate to edge shifts, it suffices to know when $X_A \cong X_B$ for nonnegative integer matrices A and B. Concretely, let

$$(7\text{--}0\text{--}1) \qquad A = \begin{bmatrix} 1 & 3 \\ 2 & 1 \end{bmatrix} \quad \text{and} \quad B = \begin{bmatrix} 1 & 6 \\ 1 & 1 \end{bmatrix}.$$

Are X_A and X_B conjugate?

Invariants such as entropy and zeta function can tell some shifts apart. A simple calculation shows that A and B in (7--0--1) have the same eigenvalues, and so these invariants do not help. In fact, all invariants introduced so far agree on X_A and X_B. This chapter tackles the conjugacy question.

Recall from Chapter 2 that state splittings and amalgamations give rise to conjugacies called splitting codes and amalgamation codes. In §7.1 we prove a basic decomposition result: every conjugacy between edge shifts can be broken down into a finite succession of splitting and amalgamation codes. Such decompositions are exploited in §7.2 to motivate an equivalence relation between matrices called strong shift equivalence, and to show that $X_A \cong X_B$ if and only if A and B are strong shift equivalent.

So strong shift equivalence appears to answer our original question. But, unfortunately, there is no general algorithm known for deciding whether two matrices are strong shift equivalent, even for 2×2 matrices. Even innocent-looking pairs such as the A and B above pose extraordinary difficulties.

Faced with this unsatisfactory answer, we introduce in §7.3 a weaker equivalence relation called shift equivalence. As the names suggest, strong shift equivalence implies shift equivalence. Hence invariants of shift equivalence are automatically conjugacy invariants. In §7.4 we define two invariants of shift equivalence: the Jordan form away from zero and the Bowen–Franks group. Together they provide much more sensitive ways to detect nonconjugacy.

The matrix equations defining shift equivalence look, at first, bizarre and mysterious. However, they are the natural matrix formulations of an underlying invariant of shift equivalence called the dimension triple. In §7.5 we define this invariant and show that it completely captures shift equivalence. We also show how the dimension triple can be defined "intrinsically" in terms of certain compact subsets of the shift space. This point of view shows why the dimension triple must automatically be a conjugacy invariant.

There are analogues for symbolic matrices of the two matrix equivalence relations. We will briefly mention how these are used in the conjugacy problem for sofic shifts.

The diagram at the end of §7.5 gives a handy summary of the main implications of this chapter.

§7.1. The Decomposition Theorem

The graph operations of state splitting and amalgamation described in §2.4 give rise to basic conjugacies called splitting codes and amalgamation codes. The main result of this section is the Decomposition Theorem, which says that every conjugacy between edge shifts can be broken into a composition of basic conjugacies. Before discussing this theorem, let us briefly review state splitting and associated codes.

Let G be a graph with edge set $\mathcal{E}$ and state set $\mathcal{V}$. For each $I \in \mathcal{V}$ recall that $\mathcal{E}_I$ denotes the set of edges starting at I. Let $\mathcal{P}$ be a partition of $\mathcal{E}$ refining $\{\mathcal{E}_I : I \in \mathcal{V}\}$, so that $\mathcal{P}$ divides each $\mathcal{E}_I$ into subsets $\mathcal{E}_I^1$, $\mathcal{E}_I^2$, $\ldots$, $\mathcal{E}_I^{m(I)}$. According to Definition 2.4.3, the out-split graph $G^{[\mathcal{P}]}$ formed from G using $\mathcal{P}$ has states I^i, where $I \in \mathcal{V}$ and $1 \leqslant i \leqslant m(I)$, and edges e^j, where $e \in \mathcal{E}$ and $1 \leqslant j \leqslant m(t(e))$. If $e \in \mathcal{E}_I^i$ goes from I to J, then e^j goes from I^i to J^j. There is an analogous notion of in-split graph $G_{[\mathcal{P}]}$, where now $\mathcal{P}$ refines the partition of $\mathcal{E}$ by incoming edges to states.

For a graph G recall that we defined H to be an out-splitting of G if H is graph isomorphic to some out-split graph $G^{[\mathcal{P}]}$ of G. The same condition defines when G is an out-amalgamation of H. For example, recall from Exercise 2.4.2, the complete out-splitting of a graph G is defined by taking $\mathcal{P} = \mathcal{E}$. In this case, $G^{[\mathcal{P}]}$ is graph isomorphic to the 2nd-higher edge graph of G via the map $e^f \mapsto ef$. In this sense, $G^{[2]}$ is an out-splitting of G.

We saw in §2.4 that if H is an out-splitting of G, then there is a natural out-splitting code $\psi_{GH} : X_G \to X_H$ with memory 0 and anticipation 1. Its inverse $\alpha_{HG} : X_H \to X_G$ is 1-block and called an out-amalgamation code. Similar remarks apply to in-splitting codes and in-amalgamation codes. A *splitting code* is either an out-splitting code or an in-splitting code. Similarly for *amalgamation codes*.

Let G and H be essential graphs. Then a 1-block code $\phi : X_G \to X_H$ having a 1-block inverse has the form $\phi = \Phi_\infty$, where Φ is the edge mapping of a graph isomorphism from G to H (Exercise 2.2.14). Since this amounts to relabeling the edges of G, we call such codes *relabeling codes*. Note that by definition a relabeling code is simultaneously an in- and out-splitting code and an in- and out-amalgamation code.

Our main goal is to show that every conjugacy between edge shifts is decomposable into a succession of splitting codes and amalgamation codes. Let us first consider a simple example.

Example 7.1.1. Let G be a graph, and $\phi = \sigma_G \colon X_G \to X_G$. We are therefore considering the shift map as a conjugacy from X_G to itself. We will represent ϕ as an out-splitting code, followed by a relabeling code, followed by an in-amalgamation code.

Let H be the complete out-splitting of G, so that the out-splitting code $\psi_{GH} = (\Psi_{GH}^{[0,1]})_\infty$, where $\Psi_{GH}(ef) = e^f$. Let K be the complete in-splitting of G, so that the in-splitting code $\psi_{GK} = (\Psi_{GK}^{[-1,0]})_\infty$, where $\Psi_{GK}(fg) = g_f$. The in-amalgamation code $\alpha_{KG} = \psi_{GK}^{-1}$ is given by the 1-block code $g_f \mapsto g$. Finally, it is easy to check that $\Theta(f^g) = g_f$ is a graph isomorphism from H to K, hence induces a relabeling code $\theta \colon X_H \to X_K$. The following diagram shows what happens to a point in X_G when we apply ψ_{GH}, then θ, and then α_{KG}.

$$
\begin{array}{ccc}
\cdots\; c\; d\; e\,.f\; g\; h\; i\; \cdots & \xrightarrow{\;\psi_{GH}\;} & \cdots\; c^d d^e\, e^f.\, f^g g^h\, h^i\; \cdots \\[4pt]
\Big\downarrow{\scriptstyle \phi\,=\,\sigma_G} & & \Big\downarrow{\scriptstyle \theta} \\[6pt]
\cdots\; d\; e\; f\,.g\; h\; i\; \cdots & \xleftarrow{\;\alpha_{KG}\;} & \cdots\; d_c\, e_d\, f_e\,.\, g_f\, h_g\, i_h\; \cdots
\end{array}
$$

This diagram shows clearly that $\phi = \alpha_{KG} \circ \theta \circ \psi_{GH}$, providing the desired decomposition. $\qquad\square$

We now state the Decomposition Theorem.

Theorem 7.1.2 (Decomposition Theorem). *Every conjugacy from one edge shift to another is the composition of splitting codes and amalgamation codes.*

Before giving the formal proof, we describe its outline. Let $\phi \colon X_G \to X_H$ be a conjugacy. Proposition 1.5.12 shows that we can recode ϕ to be a 1-block code $\tilde{\phi}$ from a higher block shift of X_G to X_H. This recoding amounts to writing ϕ as the composition of $\tilde{\phi}$ with a succession of splitting codes, obtained from complete splittings. If $\tilde{\phi}^{-1}$ were also 1-block, we would be done, since then $\tilde{\phi}$ would be a relabeling code. Thus we need a way, using splitting codes and amalgamation codes, to reduce the memory or anticipation of the inverse of a 1-block code, while keeping the resulting code 1-block. The following lemma serves this purpose.

Lemma 7.1.3. *Let* $\phi\colon X_G \to X_H$ *be a 1-block conjugacy whose inverse has memory* $m \geqslant 0$ *and anticipation* $n \geqslant 1$. *Then there are out-splittings* $\widetilde{G}$ *of* G *and* $\widetilde{H}$ *of* H *such that*

$$
\begin{array}{ccc}
X_G & \xrightarrow{\ \psi_{G\widetilde{G}}\ } & X_{\widetilde{G}} \\
{\scriptstyle \phi}\Big\downarrow & & \Big\downarrow{\scriptstyle \widetilde{\phi}} \\
X_H & \xleftarrow{\ \alpha_{\widetilde{H}H}\ } & X_{\widetilde{H}}
\end{array}
$$

commutes, where $\widetilde{\phi}$ *is a 1-block conjugacy whose inverse has memory* m *and anticipation* $n - 1$.

PROOF: Let $\widetilde{H}$ be the complete out-splitting of H, so the edges of $\widetilde{H}$ have the form h^k, where $h, k \in \mathcal{E}(H)$ and k follows h in H. Let $\phi = \Phi_\infty$, where $\Phi\colon \mathcal{E}(G) \to \mathcal{E}(H)$. Partition the outgoing edges at each state of G according to their Φ-images, so that for each $I \in \mathcal{V}_G$ and $h \in \mathcal{E}(H)$ we define $\mathcal{E}_I^h = \{g \in \mathcal{E}_I : \Phi(g) = h\}$. Let $\widetilde{G}$ be the resulting out-split graph. Finally, define $\widetilde{\Phi}\colon \mathcal{E}(\widetilde{G}) \to \mathcal{E}(\widetilde{H})$ by $\widetilde{\Phi}(g^h) = \Phi(g)^h$. It is simple to check that $\widetilde{\Phi}$ induces a 1-block code $\widetilde{\phi} = \widetilde{\Phi}_\infty\colon X_{\widetilde{G}} \to X_{\widetilde{H}}$. The following diagram shows the actions of these codes.

$$
\begin{array}{ccc}
\cdots\, g_{-3}\, g_{-2}\, g_{-1} \cdot g_0\, g_1\, g_2\, \cdots & \xrightarrow{\ \psi_{G\widetilde{G}}\ } & \cdots\, g_{-3}^{h_{-2}}\, g_{-2}^{h_{-1}}\, g_{-1}^{h_0} \cdot g_0^{h_1}\, g_1^{h_2}\, g_2^{h_3}\, \cdots \\[2mm]
{\scriptstyle \phi = \Phi_\infty}\Big\downarrow & & \Big\downarrow{\scriptstyle \widetilde{\phi} = \widetilde{\Phi}_\infty} \\[2mm]
\cdots\, h_{-3}\, h_{-2}\, h_{-1} \cdot h_0\, h_1\, h_2\, \cdots & \xleftarrow{\ \alpha_{\widetilde{H}H}\ } & \cdots\, h_{-3}^{h_{-2}}\, h_{-2}^{h_{-1}}\, h_{-1}^{h_0} \cdot h_0^{h_1}\, h_1^{h_2}\, h_2^{h_3}\, \cdots
\end{array}
$$

To show that $\widetilde{\phi}^{-1}$ has the required memory and anticipation, we must show that whenever $\widetilde{y} \in X_{\widetilde{H}}$, the coordinates $-m$ through $n-1$ of $\widetilde{y}$ determine the 0th coordinate of $\widetilde{x} = \widetilde{\phi}^{-1}(\widetilde{y})$. For this, write $y = \alpha_{\widetilde{H}H}(\widetilde{y})$, $x = \phi^{-1}(y)$ and observe that

$$
\widetilde{x}_0 = x_0^{y_1}.
$$

Since $\widetilde{y}_{[-m, n-1]}$ determines $y_{[-m,n]}$ and since $y_{[-m,n]}$ determines x_0, it follows that $\widetilde{y}_{[-m,n-1]}$ determines $\widetilde{x}_0$. $\square$

Repeated application of this lemma proves the Decomposition Theorem.

PROOF OF DECOMPOSITION THEOREM: As mentioned above, by Proposition 1.5.12, we can recode and so assume that the conjugacy ϕ is 1-block. Let ϕ^{-1} have memory m and anticipation n. By increasing the window size if necessary (see §1.5), we may also assume that $m \geqslant 0$ and $n \geqslant 0$. If $m = n = 0$, then ϕ is 1-block with a 1-block inverse, hence a relabeling code, and we are done. So suppose that $n \geqslant 1$. By Lemma 7.1.3, there is a sequence of out-splitting codes ψ_j, out-amalgamation codes α_j, and 1-block conjugacies $\tilde{\phi}_j$, where $1 \leqslant j \leqslant n$, for which $\tilde{\phi}_j^{-1}$ has memory m and anticipation $n - j$, as shown below.

$$
\begin{array}{ccccccc}
X_G & \xrightarrow{\psi_1} & X_{G_1} & \xrightarrow{\psi_2} & X_{G_2} \longrightarrow \cdots & \xrightarrow{\psi_n} & X_{G_n} \\
\phi \downarrow & & \tilde{\phi}_1 \downarrow & & \tilde{\phi}_2 \downarrow & & \tilde{\phi}_n \downarrow \\
X_H & \xleftarrow{\alpha_1} & X_{H_1} & \xleftarrow{\alpha_2} & X_{H_2} \longleftarrow \cdots \longleftarrow & \xleftarrow{\alpha_n} & X_{H_n}
\end{array}
$$

The inverse of the resulting $\tilde{\phi}_n$ has memory m and anticipation 0.

Lemma 7.1.3 can be used to decrease memory by applying it to the transposed graph (defined at the end of §2.2). Hence there is a further sequence of in-splitting codes ψ_{n+k}, in-amalgamation codes α_{n+k} and 1-block conjugacies $\tilde{\phi}_{n+k}$ for $1 \leqslant k \leqslant m$ such that $\tilde{\phi}_{n+k}^{-1}$ has memory $m - k$ and anticipation 0, as shown below.

$$
\begin{array}{ccccccc}
X_{G_n} & \xrightarrow{\psi_{n+1}} & X_{G_{n+1}} & \xrightarrow{\psi_{n+2}} & X_{G_{n+2}} \longrightarrow \cdots & \xrightarrow{\psi_{n+m}} & X_{G_{n+m}} \\
\tilde{\phi}_n \downarrow & & \tilde{\phi}_{n+1} \downarrow & & \tilde{\phi}_{n+2} \downarrow & & \tilde{\phi}_{n+m} \downarrow \\
X_{H_n} & \xleftarrow{\alpha_{n+1}} & X_{H_{n+1}} & \xleftarrow{\alpha_{n+2}} & X_{H_{n+2}} \longleftarrow \cdots \longleftarrow & \xleftarrow{\alpha_{n+m}} & X_{H_{n+m}}
\end{array}
$$

In particular, $\tilde{\phi}_{n+m}^{-1}$ has memory 0 and anticipation 0, so it is a relabeling code.

Paste these diagrams together along their common vertical edge to get the following diagram.

$$
\begin{array}{ccccccccc}
X_G & \xrightarrow{\psi_1} & X_{G_1} & \xrightarrow{\psi_2} \cdots \xrightarrow{\psi_n} & X_{G_n} & \xrightarrow{\psi_{n+1}} \cdots \xrightarrow{\psi_{n+m}} & X_{G_{n+m}} \\
\phi \downarrow & & \tilde{\phi}_1 \downarrow & & \tilde{\phi}_n \downarrow & & \tilde{\phi}_{n+m} \downarrow \\
X_H & \xleftarrow{\alpha_1} & X_{H_1} & \xleftarrow{\alpha_2} \cdots \xleftarrow{\alpha_n} & X_{H_n} & \xleftarrow{\alpha_{n+1}} \cdots \xleftarrow{\alpha_{n+m}} & X_{H_{n+m}}
\end{array}
$$

Traversing this diagram along the top, down the right side, and then along the bottom shows that ϕ is a composition of conjugacies of the required type. $\square$

Let us see how this works on an example.

Example 7.1.4. Consider the graph G shown in Figure 7.1.1(a). We label its edges a, b, ..., f. The parenthetical symbols (0) or (1) on edges indicate a 1-block code $\phi = \Phi_\infty \colon X_G \to X_{[2]}$, where $\Phi(a) = 0$, $\Phi(b) = 0$, $\Phi(c) = 1$, etc. It turns out that ϕ is a 1-block conjugacy, and that $\phi^{-1} = \theta = \Theta_\infty$ has memory 1 and anticipation 1. The 3-block map Θ is given below.

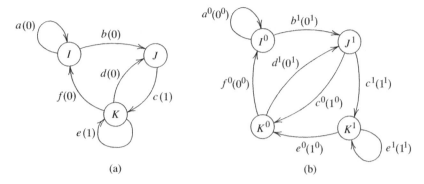

FIGURE 7.1.1.   Reducing the anticipation of the inverse of a 1-block code.

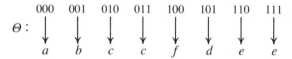

Let us apply Lemma 7.1.3 to obtain a new 1-block conjugacy with memory 1 and anticipation 0. We out-split G according to the Φ-image of edges, resulting in the graph G_1 shown in Figure 7.1.1(b). Here there is only one split state I^0 from I, since all edges leaving I have Φ-image 0; similarly there is only one split state J^1 from J. On the other hand, state K is split into K^0 and K^1. The labeling on G_1 follows the proof of Lemma 7.1.3, so that $\widetilde{\Phi}(a^0) = \Phi(a)^0 = 0^0$, etc. and H_1 is the 2nd higher edge graph of $G([2])$. Let $\widetilde{\phi} = \widetilde{\Phi}_\infty$. Then $\widetilde{\phi}^{-1}$ has memory 1 and anticipation 0 since $\widetilde{\phi}^{-1} = \widetilde{\Theta}_\infty$ where $\widetilde{\Theta}$ is the 2-block map below.

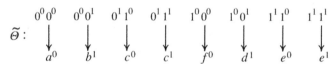

At this stage, the proof of the Decomposition Theorem says to in-split G_1 according to the $\widetilde{\Phi}$-images of incoming edges, producing a new graph G_2. However, note from Figure 7.1.1(b) that in this case for every state in G_1 the $\widetilde{\Phi}$-images of all incoming edges to the state are the same. Thus the in-split graph G_2 is isomorphic to G_1, but its vertices and edges have new labels that reflect the in-splitting operation, as shown in Figure 7.1.2(a). The notational complexity can be reduced by observing that parts of the labels are redundant. State labels have the form $L^s_{r^s}$, which can be shortened to L^s_r; edge labels have the form $k^t_{r^s}$ where $\Phi(k) = s$, so this edge label can be replaced by k^t_r without losing information; parenthetical labels have the form $r^s_{t^r}$, which can be replaced by r^s_t. Figure 7.1.2(b) uses this simplified labeling, which becomes even more useful if further rounds of splitting are

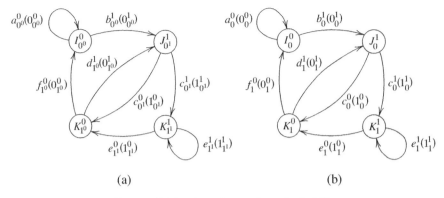

FIGURE 7.1.2. A conjugacy from the 2-shift.

needed. Identifying a parenthetical label r_t^s with the 3-block trs gives a graph isomorphism Φ_2 from G_2 to H_2, the 3rd higher edge graph of $G([2])$. The decomposition of ϕ is shown below.

$$\begin{array}{ccccc}
X_G & \xrightarrow{\psi_1} & X_{G_1} & \xrightarrow{\psi_2} & X_{G_2} \\
{\scriptstyle\phi}\Big\downarrow & & & & \Big\downarrow{\scriptstyle\tilde{\phi}_2} \\
X_{[2]} & \xleftarrow{\alpha_1} & X_{H_1} & \xleftarrow{\alpha_2} & X_{H_2}
\end{array}$$

Here $\psi_1^{-1} \circ \psi_2^{-1} \circ \tilde{\phi}_2^{-1}$ is induced by the block map Θ shown above, so that Θ can be read off directly from Figure 7.1.2(b) (again, identifying the parenthetical label r_t^s with trs). $\qquad\square$

We remark that the procedure given in the proof of the Decomposition Theorem is often not the simplest way to decompose a conjugacy into splitting and amalgamation codes.

Our next result follows directly from the Decomposition Theorem together with the fact that for any edge shift X there is a unique essential graph G such that $X = X_G$ (see Exercise 2.2.9).

Corollary 7.1.5. *Let G and H be essential graphs. The edge shifts X_G and X_H are conjugate if and only if G is obtained from H by a sequence of out-splittings, in-splittings, out-amalgamations, and in-amalgamations.*

The matrix consequences of this corollary will be explored in the next section. We conclude this section with some remarks about the Decomposition Theorem. The theorem makes no claim about the uniqueness of decomposition, because typically there are infinitely many. For example, we could at any point throw in a succession of splitting codes followed by their inverse amalgamation codes. The Decomposition Theorem uses four types of conjugacies: in- and out-splittings and in- and out-amalgamations (recall that relabeling codes are special kinds of splitting and amalgamation codes). Exercise 7.1.7 shows that, by composing with a power of the shift, any conjugacy can be decomposed into just two types: out-splittings followed by in-amalgamations.

Finally, there is a notion of splitting and amalgamation for general shift spaces, in which relabeling codes correspond to renaming letters of the alphabet. Exercise 7.1.10 explains this, and outlines a proof of the Decomposition Theorem valid for conjugacies between arbitrary shift spaces.

EXERCISES

7.1.1. Let

$$A_G = \begin{bmatrix} 3 \end{bmatrix} \quad \text{and} \quad A_H = \begin{bmatrix} 0 & 0 & 1 & 0 & 0 \\ 1 & 1 & 0 & 1 & 1 \\ 1 & 1 & 1 & 1 & 1 \\ 1 & 1 & 0 & 0 & 0 \\ 0 & 0 & 1 & 1 & 1 \end{bmatrix}.$$

Find an explicit conjugacy from X_G to X_H and its inverse. [*Hint*: If A_H has a repeated row or column, then H is a splitting of another graph.]

7.1.2. Let

$$A = \begin{bmatrix} 1 & 1 & 0 \\ 0 & 0 & 1 \\ 1 & 1 & 1 \end{bmatrix} \quad \text{and} \quad B = \begin{bmatrix} 2 \end{bmatrix}.$$

Example 7.1.4 gives a conjugacy from X_A to X_B and shows how to decompose it into splitting and amalgamation codes. Show that $G_A^{[m]}$ is never graph isomorphic to $G_B^{[n]}$ for all $m, n \geqslant 1$. This shows that in the Decomposition Theorem complete splittings and their inverses will not suffice.

7.1.3. Let G be the graph below, where parenthetical labels indicate a conjugacy ϕ from X_G to $X_{[2]}$.

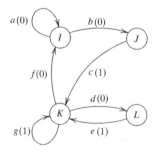

Decompose ϕ into splitting and amalgamation codes.

7.1.4. Let $\mathcal{L}$ be a right-closing labeling on a graph G with delay $D \geqslant 1$ (see §5.1), and $X = X_{(G,\mathcal{L})}$. Show that there exist an out-splitting $\widetilde{G}$ of G and a labeling $\widetilde{\mathcal{L}}$ of $\widetilde{G}$ presenting $X^{[2]}$ such that

(a) the following diagram commutes

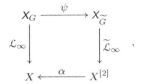

where $\psi\colon X_G \to X_{\widetilde{G}}$ denotes the out-splitting code and $\alpha\colon X^{[2]} \to X$ is the amalgamation code determined by $x_0 x_1 \mapsto x_0$,

(b) $\widetilde{\mathcal{L}}$ is right-closing with delay $D - 1$, and

(c) if $\mathcal{L}$ is left-closing, then so is $\widetilde{\mathcal{L}}$ and with the same delay as $\widetilde{\mathcal{L}}$.

Using properties (a) and (b), give another proof of Proposition 5.1.11.

7.1.5. Let G and H be irreducible graphs. Show that a mapping $\phi\colon X_G \to X_H$ is an in-amalgamation code if and only if ϕ is a 1-block conjugacy and ϕ^{-1} has memory 1 and anticipation 0. [*Hint for* "*if*": Write $\phi = \Phi_\infty$ where Φ is the edge mapping of a graph homomorphism. Show that (G, Φ) is an irreducible right-resolving presentation of X_H and use Corollary 3.3.20 to show that for each state I of G, the restriction of Φ to $\mathcal{E}_I(G)$ is a bijection onto $\mathcal{E}_{\partial\Phi(I)}(H)$.]

7.1.6. Let $\phi\colon X_G \to X_H$ be a conjugacy. Show that for k sufficiently large, $\phi\circ\sigma^{-k}$ has anticipation 0 and its inverse has memory 0.

7.1.7. Let $\phi\colon X_A \to X_B$ be a conjugacy such that ϕ has anticipation 0 and that ϕ^{-1} has memory 0. Let ϕ have memory m and ϕ^{-1} have anticipation n. Define a graph G whose edges are all pairs of the form $(x_{[-m,0]}, y_{[0,n]})$, where $x \in X_A$ and $y = \phi(x)$; the initial state of this edge is the pair $(x_{[-m,-1]}, y_{[0,n-1]})$ and its terminal state is $(x_{[-m+1,0]}, y_{[1,n]})$.

Let $\theta_A\colon X_G \to X_A$ be the 1-block code induced by $\Theta_A(x_{[-m,0]}, y_{[0,n]}) = x_{-m}$, and $\theta_B\colon X_G \to X_B$ be the 1-block code induced by $\Theta_B(x_{[-m,0]}, y_{[0,n]}) = y_n$.

(a) Show that θ_A and θ_B are conjugacies and that $\phi = \sigma^{-(n+m)}\circ\theta_B\circ\theta_A^{-1}$.

(b) Give explicit decompositions of θ_A as a sequence of out-amalgamation codes and θ_B as a sequence of in-amalgamation codes. [*Hint:* for θ_A construct, by out-splitting, a graph with edges $(x_{[-m,0]}, y_{[0,i]})$ from a graph with edges $(x_{[-m,0]}, y_{[0,i-1]})$.]

(c) Use this to give an alternative proof of the Decomposition Theorem, showing in fact that every conjugacy is a composition of a power of the shift, out-splitting codes, and in-amalgamation codes.

7.1.8. Let $X = X_{[2]}$ and $u = 1001$, $v = 1101$.

(a) Define $\phi\colon X \to X$ by replacing u with v and v with u wherever they occur. Show that ϕ is a well-defined conjugacy, and find ϕ^{-1} (see Exercise 1.5.8).

(b) Use the procedure in the previous exercise to outline a decomposition of ϕ into splitting codes and amalgamation codes.

7.1.9. Let $\phi\colon X_G \to X_M$ be an out-amalgamation code and $\psi\colon X_H \to X_M$ be an in-amalgamation code. Let

$$Z = \{(x,y) \in X_G \times X_H : \phi(x) = \psi(y)\}.$$

Show that (up to a renaming of symbols) Z is an edge shift X_K, where K is an in-splitting of G and an out-splitting of H.

7.1.10. (Generalization of the Decomposition Theorem to arbitrary shift spaces) Let X be a shift space, let $\mathcal{P}$ be a partition of its alphabet $\mathcal{A}$, and let $[a]$ denote the partition element that contains the symbol a. Define the shift space $\overline{X}$ over the alphabet $\{a[b] : ab \in \mathcal{B}_2(X)\}$ as the image of the 2-block code induced by the mapping $\Phi(ab) = a[b]$. We say that $\overline{X}$, and more generally any shift obtained from $\overline{X}$ by simply renaming symbols, is an *out-splitting* of X. Similarly, the notions of out-amalgamation, in-splitting, in-amalgamation, splitting codes, and amalgamation codes generalize.

    (a) Show that these notions of splitting and amalgamation agree with our notions of splitting and amalgamation for edge shifts.

    (b) Show that any conjugacy between shift spaces can be decomposed as a composition of splitting codes and amalgamation codes.

§7.2. Strong Shift Equivalence

In this section we translate the Decomposition Theorem for edge shift conjugacies into a matrix condition on the corresponding adjacency matrices. This leads to the notion of strong shift equivalence for matrices.

Let H be a splitting or an amalgamation of a graph G, and $A = A_G$, $B = A_H$. Recall the matrix formulation of state splitting from §2.4; in particular, there are nonnegative integer matrices D and E so that $A = DE$ and $ED = B$. This includes the case where H is graph isomorphic to G. To see this, first note that in this case there is a permutation matrix P such that $A = P^{-1}BP$. Then letting $D = P^{-1}$ and $E = BP$, we obtain the same matrix equations $A = DE$ and $ED = B$.

It therefore follows from the Decomposition Theorem (really Corollary 7.1.5) that if A and B are essential matrices and $X_A \cong X_B$, then there is a sequence (D_0, E_0), (D_1, E_1), ..., (D_ℓ, E_ℓ) of pairs of nonnegative integral matrices such that

$$A = D_0 E_0, \quad E_0 D_0 = A_1$$
$$A_1 = D_1 E_1, \quad E_1 D_1 = A_2$$
$$A_2 = D_2 E_2, \quad E_2 D_2 = A_3$$
$$\vdots$$
$$A_\ell = D_\ell E_\ell, \quad E_\ell D_\ell = B.$$

Definition 7.2.1. Let A and B be nonnegative integral matrices. An *elementary equivalence from A to B* is a pair (R, S) of rectangular nonnegative integral matrices satisfying

$$A = RS \qquad \text{and} \qquad B = SR.$$

In this case we write $(R, S)\colon A \approx B$.

A *strong shift equivalence of lag ℓ from A to B* is a sequence of ℓ elementary equivalences

$$(R_1, S_1)\colon A = A_0 \approx A_1, (R_2, S_2) : A_1 \approx A_2, \ldots, (R_\ell, S_\ell)\colon A_{\ell-1} \approx A_\ell = B.$$

In this case we write $A \approx B$ (lag ℓ). Say that A is strong shift equivalent to B (and write $A \approx B$) if there is a strong shift equivalence of some lag from A to B.

Note that the matrices A and B in this definition need not have the same size. The use of "strong" in "strong shift equivalence" is to distinguish this from another notion called shift equivalence that we will encounter in the next section.

Example 7.2.2. Let

$$A = \begin{bmatrix} 1 & 1 & 0 \\ 0 & 0 & 1 \\ 1 & 1 & 1 \end{bmatrix}, \quad B = \begin{bmatrix} 1 & 1 \\ 1 & 1 \end{bmatrix}, \quad \text{and} \quad C = [\, 2 \,].$$

It is easy to check that if

$$R = \begin{bmatrix} 1 & 0 \\ 0 & 1 \\ 1 & 1 \end{bmatrix} \quad \text{and} \quad S = \begin{bmatrix} 1 & 1 & 0 \\ 0 & 0 & 1 \end{bmatrix},$$

then $(R, S)\colon A \approx B$. Similarly, if

$$T = \begin{bmatrix} 1 \\ 1 \end{bmatrix} \quad \text{and} \quad U = [\, 1 \quad 1 \,],$$

then $(T, U)\colon B \approx C$. Hence A is strong shift equivalent to C with lag 2, or in symbols $A \approx C$ (lag 2).

Observe, however, that there is no elementary equivalence from A to C. For if $(V, W)\colon A \approx C$, then W is 1×3, so that $A = VW$ would have rank $\leqslant 1$, contradicting that A has rank 2. $\square$

This example shows that, in spite of its name, elementary equivalence is *not* an equivalence relation. The next result shows that strong shift equivalence is the equivalence relation that results from making $\approx$ transitive.

Proposition 7.2.3. *Strong shift equivalence is an equivalence relation on nonnegative integral matrices.*

PROOF: First note that $(A, I)\colon A \approx A$, so that $A \approx A$. Also, if $(R, S)\colon A \approx B$, then $(S, R)\colon B \approx A$. Hence any chain of elementary equivalences from A to B can be reversed, so that $\approx$ is symmetric. Finally, a chain of elementary equivalences from A to B can be followed by another chain from B to C, so that $\approx$ is transitive. $\square$

Our comments at the beginning of this section show that if $X_A \cong X_B$ then $A \approx B$. Our main goal in this section is to prove the converse. The fact that conjugacy of X_A and X_B is equivalent to strong shift equivalence of A and B is a fundamental result of R. Williams [WilR2]

In order to prove the converse, we first show how an elementary equivalence $(R, S) \colon A \approx B$ can be used to produce a conjugacy $\gamma_{R,S} \colon X_A \to X_B$.

First construct a graph $G_{R,S}$ as follows. Start with the disjoint union of G_A and G_B. Edges in G_A are called *A-edges*, and those in G_B are called *B-edges*. For each state I in G_A and state J in G_B, add R_{IJ} edges from I to J (called *R-edges*) and S_{JI} edges from J to I (called *S-edges*), completing the construction of $G_{R,S}$. By an *R,S-path* we mean an R-edge followed by an S-edge forming a path in $G_{R,S}$; an *S,R-path* is defined similarly.

Let I and I' be states in G_A. Since $A = RS$, there is a one-to-one correspondence (or *bijection*) between the A-edges from I to I' and the R,S-paths from I to I'. We will denote a fixed choice of bijection by $a \leftrightarrow \mathsf{r}(a)\mathsf{s}(a)$, where $\mathsf{r}(a)$ is an R-edge and $\mathsf{s}(a)$ is an S-edge, or going the other way by $a = \mathsf{a}(rs)$. Similarly, since $SR = B$, for each pair of states J, J' in G_B there is a bijection between the edges from J to J' and the S,R-paths from J to J'. We denote a choice of bijection by $b \leftrightarrow \mathsf{s}(b)\mathsf{r}(b)$, or by $b = \mathsf{b}(sr)$.

Example 7.2.4. Let

$$A = \begin{bmatrix} 3 & 2 \\ 2 & 0 \end{bmatrix}, \quad B = \begin{bmatrix} 1 & 3 \\ 2 & 2 \end{bmatrix}, \quad R = \begin{bmatrix} 1 & 1 \\ 0 & 1 \end{bmatrix}, \quad \text{and} \quad S = \begin{bmatrix} 1 & 2 \\ 2 & 0 \end{bmatrix}.$$

Then $(R, S) \colon A \approx B$. The graph $G_{R,S}$ is depicted in Figure 7.2.1.

One choice of bijection between A-edges and R,S-paths is the following.

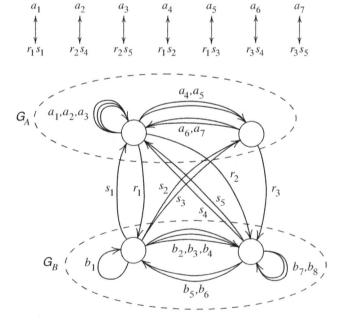

FIGURE 7.2.1. Graph to construct a conjugacy from an elementary equivalence.

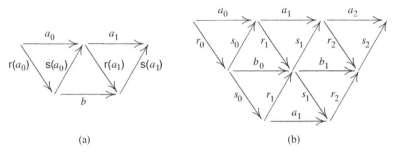

(a)   (b)

FIGURE 7.2.2.   The 2-block code from an elementary equivalence.

In our bijection notation, we have $r(a_1) = r_1$, $s(a_1) = s_1$, $r(a_2) = r_2$, $s(a_2) = s_4$, and so on, or going the other way $a(r_1 s_1) = a_1$, $a(r_2 s_4) = a_2$, etc. There are other possible bijections, for example by permuting the first three bottom entries in the previous display.

Similarly, one choice of bijection between B-edges and S,R-paths is

$$
\begin{array}{cccccccc}
b_1 & b_2 & b_3 & b_4 & b_5 & b_6 & b_7 & b_8 \\
\updownarrow & \updownarrow & \updownarrow & \updownarrow & \updownarrow & \updownarrow & \updownarrow & \updownarrow \\
s_1 r_1 & s_1 r_2 & s_2 r_3 & s_3 r_3 & s_4 r_1 & s_5 r_1 & s_4 r_2 & s_5 r_2
\end{array}
$$

With this choice our bijection functions have values $s(b_1) = s_1$, $r(b_1) = r_1$, $s(b_2) = s_1$, $r(b_2) = r_2$, and so on, while $b(s_1 r_1) = b_1$, $b(s_1 r_2) = b_2$, etc. Again, there are many choices for the bijection. □

We can now define a 2-block map $\Gamma_{R,S} \colon \mathcal{B}_2(X_A) \to \mathcal{B}_1(X_B)$ using the chosen bijections by putting

$$
\Gamma_{R,S}(a_0 a_1) = b(s(a_0) r(a_1)).
$$

What is happening here is that the double A-path $a_0 a_1$ corresponds to a double R,S-path $r(a_0) s(a_0) r(a_1) s(a_1)$. The middle part of this, $s(a_0) r(a_1)$, is an S,R-path that corresponds to a B-edge $b = b(s(a_0) r(a_1))$, the value of $\Gamma_{R,S}$ on the block $a_0 a_1$. Figure 7.2.2(a) gives one way to visualize this situation. With a fixed choice of bijections, we define

$$
\gamma_{R,S} = (\Gamma_{R,S})_{\infty}^{[0,1]}.
$$

Of course, if A and B are 0–1 matrices, then there is only one possible choice of bijections.

Proposition 7.2.5. *If the same bijections are used to define $\gamma_{R,S} \colon X_A \to X_B$ and $\gamma_{S,R} \colon X_B \to X_A$, then*

$$
\gamma_{S,R} \circ \gamma_{R,S} = \sigma_A \qquad \text{and} \qquad \gamma_{R,S} \circ \gamma_{S,R} = \sigma_B.
$$

In particular, the sliding block codes $\gamma_{R,S}$ and $\gamma_{S,R}$ are conjugacies.

PROOF: The definitions immediately show that if $a_0 a_1 a_2 \in \mathcal{B}_3(X_A)$, then $\Gamma_{R,S}(a_1 a_2)$ follows $\Gamma_{R,S}(a_0 a_1)$ in G_B, so that $\gamma_{R,S}(X_A) \subseteq X_B$. Similarly, $\gamma_{S,R}(X_B) \subseteq X_A$. Figure 7.2.2(b) shows why $\gamma_{S,R} \circ \gamma_{R,S} = \sigma_A$. Another way to visualize this is shown in the diagram below.

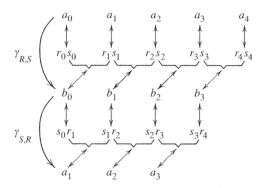

This argument can also be expressed in terms of our bijection notation:

$$\Gamma_{S,R} \circ \Gamma_{R,S}(a_0 a_1 a_2) = \Gamma_{S,R}(\mathsf{b}(\mathsf{s}(a_0)\mathsf{r}(a_1))\mathsf{b}(\mathsf{s}(a_1)\mathsf{r}(a_2)))$$
$$= \mathsf{a}(\mathsf{r}(a_1)\mathsf{s}(a_1)) = a_1.$$

Notice that the second equality is valid because we used the *same* bijections to define $\Gamma_{S,R}$ as $\Gamma_{R,S}$, so that $\mathsf{r}(\mathsf{b}(\mathsf{s}(a_0)\mathsf{r}(a_1))) = \mathsf{r}(a_1)$, etc.

A similar argument shows that $\gamma_{S,R} \circ \gamma_{R,S} = \sigma_B$. Since σ_A and σ_B are conjugacies, it follows immediately that $\gamma_{R,S}$ and $\gamma_{S,R}$ must also be conjugacies. □

Example 7.2.6. For A, B, R, and S as in Example 7.2.4 and the choice of bijections there, the actions of $\gamma_{R,S}$ and $\gamma_{S,R}$ are shown below.

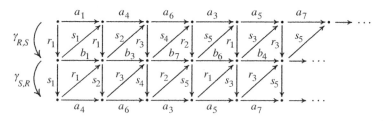

Note that a different choice of bijections would yield different conjugacies $\gamma_{R,S}$ and $\gamma_{S,R}$. □

We have now laid the groundwork for Williams' celebrated criterion for the conjugacy of edge shifts [WilR2].

Theorem 7.2.7 (Classification Theorem). *The edge shifts X_A and X_B are conjugate if and only if the matrices A and B are strong shift equivalent.*

PROOF: Our discussion of the Decomposition Theorem at the beginning of this section shows that if A and B are essential matrices and $X_A \cong X_B$, then $A \approx B$. In fact, this holds even when A or B is inessential (see Exercise 7.2.7). Conversely, if $A \approx B$ via a sequence (R_j, S_j) of elementary equivalences, then the composition of the conjugacies γ_{R_j, S_j} is a conjugacy from X_A to X_B. □

Example 7.2.8. Consider the conjugacy in Example 7.1.4. Here

$$A = A_G = \begin{bmatrix} 1 & 1 & 0 \\ 0 & 0 & 1 \\ 1 & 1 & 1 \end{bmatrix} \qquad \text{and} \qquad B = [\,2\,].$$

Following the notation and the use of the Decomposition Theorem there, we obtain a chain of elementary equivalences as follows, where $A_1 = A_{G_1} = A_{G_2} = A_{H_2}$ and $A_2 = A_{H_1}$.

$$A = \begin{bmatrix} 1 & 0 & 0 & 0 \\ 0 & 1 & 0 & 0 \\ 0 & 0 & 1 & 1 \end{bmatrix} \begin{bmatrix} 1 & 1 & 0 \\ 0 & 0 & 1 \\ 1 & 1 & 0 \\ 0 & 0 & 1 \end{bmatrix} = R_1 S_1, \quad S_1 R_1 = \begin{bmatrix} 1 & 1 & 0 & 0 \\ 0 & 0 & 1 & 1 \\ 1 & 1 & 0 & 0 \\ 0 & 0 & 1 & 1 \end{bmatrix} = A_1$$

$$A_1 = \begin{bmatrix} 1 & 0 \\ 0 & 1 \\ 1 & 0 \\ 0 & 1 \end{bmatrix} \begin{bmatrix} 1 & 1 & 0 & 0 \\ 0 & 0 & 1 & 1 \end{bmatrix} = R_2 S_2, \quad S_2 R_2 = \begin{bmatrix} 1 & 1 \\ 1 & 1 \end{bmatrix} = A_2$$

$$A_2 = \begin{bmatrix} 1 \\ 1 \end{bmatrix} [\,1 \quad 1\,] = R_3 S_3, \quad S_3 R_3 = [\,2\,] = B. \qquad \square$$

The strong shift equivalence in this example has lag three. Example 7.2.2 gives another strong shift equivalence with lag two.

Since every shift of finite type is conjugate to an edge shift by passing to a higher block presentation, Theorem 7.2.7 gives an algebraic criterion for conjugacy of shifts of finite type. But how practical is this criterion?

For given essential matrices A and B, there is a simple way to determine whether or not there is an elementary equivalence from A to B. For if $(R, S): A \approx B$, then the sizes of R and S are fixed and the entries in R and S must be bounded by the maximum of the entries in A and B, so there are only finitely many possibilities for R and S. However, when trying to decide strong shift equivalence two problems immediately arise. There is no bound on the lag known in advance, and consequently no bound on the sizes of the intermediate matrices. How would one proceed with the 2×2 matrices in (7–0–1) of the introduction? We will return to this question later.

It is convenient to name the conjugacies used in the proof of Theorem 7.2.7.

Definition 7.2.9. An *elementary conjugacy* is a conjugacy of the form $\gamma_{R,S}$ or $\gamma_{R,S}^{-1}$.

The following proposition, whose proof is left as an exercise, lists some examples of elementary conjugacies.

Proposition 7.2.10. *Let A be a nonnegative integral matrix, and Id denote the identity matrix of the same size. With an appropriate choice of bijections, the following hold.*

(1) *$\gamma_{Id,A}$ is the identity map from X_A to X_A.*

(2) *$\gamma_{A,Id} = \sigma_A$, so that σ_A and σ_A^{-1} are elementary conjugacies.*

(3) *If P is a permutation matrix then $\gamma_{P,P^{-1}A}$ is a relabeling code corresponding to a graph isomorphism determined by P.*

The following proposition, whose proof is also left as an exercise, shows how elementary conjugacies are related to splitting codes and amalgamation codes.

Proposition 7.2.11.

(1) *Splitting codes and amalgamation codes are elementary conjugacies. More precisely, out-splitting and in-amalgamation codes have the form $\gamma_{R,S}$ while in-splitting and out-amalgamation codes have the form $\gamma_{R,S}^{-1}$.*

(2) *Every $\gamma_{R,S}$ is the composition of an out-splitting code and an in-amalgamation code.*

We conclude this section by describing how conjugacy between sofic shifts can be recast as a condition on the corresponding symbolic adjacency matrices.

Let $\mathcal{C}$ be a finite alphabet. Denote by $\mathcal{S}_{\mathcal{C}}$ the set of all finite formal sums of elements of $\mathcal{C}$. For alphabets $\mathcal{C}$ and $\mathcal{D}$ put $\mathcal{CD} = \{cd : c \in \mathcal{C}, d \in \mathcal{D}\}$. If $u = \sum_j c_j \in \mathcal{S}_{\mathcal{C}}$ and $v = \sum_k d_k \in \mathcal{S}_{\mathcal{D}}$, define $uv = \sum_j \sum_k c_j d_k \in \mathcal{S}_{\mathcal{CD}}$. Thus if R is an $m \times n$ matrix with entries in $\mathcal{S}_{\mathcal{C}}$ and S is $n \times m$ with entries in $\mathcal{S}_{\mathcal{D}}$, their product RS is a symbolic matrix with entries in $\mathcal{S}_{\mathcal{CD}}$. If C and D are square matrices over $\mathcal{S}_{\mathcal{C}}$ and $\mathcal{S}_{\mathcal{D}}$, respectively, write $C \leftrightarrow D$ if there is a bijection from $\mathcal{C}$ to $\mathcal{D}$ that transforms C into D.

Let A and B be symbolic adjacency matrices over alphabets $\mathcal{A}$ and $\mathcal{B}$, respectively. A *symbolic elementary equivalence from A to B* is a pair (R, S) of matrices, where R has entries in an alphabet $\mathcal{C}$ and S has entries in an alphabet $\mathcal{D}$, such that $A \leftrightarrow RS$ and $SR \leftrightarrow B$. Say that A and B are *strong shift equivalent* if there is a sequence of symbolic adjacency matrices

$$A = A_0, A_1, A_2, \ldots, A_{\ell-1}, A_\ell = B$$

such that for $1 \leqslant j \leqslant \ell$ there is a symbolic elementary equivalence from A_{j-1} to A_j.

The following result, whose proof is outlined in Exercise 7.2.9, is an analogue of Theorem 7.2.7.

Theorem 7.2.12. *Let X and Y be irreducible sofic shifts, and let A and B be the symbolic adjacency matrices of the minimal right-resolving presentations of X and Y, respectively. Then X and Y are conjugate if and only if A and B are strong shift equivalent.*

A version of this result for general (i.e., possibly reducible) sofic shifts can be found in Nasu [Nas4]; see also Krieger [Kri8].

EXERCISES

7.2.1. Let $R = \begin{bmatrix} 1 & 1 \\ 0 & 1 \end{bmatrix}$, and let A be a 2×2 nonnegative integral matrix.

(a) Show that $R^{-1}A$ is nonnegative if and only if the first row of A is at least as large, entry-by-entry, as the second row of A. Show that if $R^{-1}A \geqslant 0$, then $(R, R^{-1}A)$ defines an elementary equivalence between A and $R^{-1}AR$.

(b) Show that there is an elementary equivalence from $\begin{bmatrix} 3 & 2 \\ 1 & 1 \end{bmatrix}$ to each of the following matrices:

$$\begin{bmatrix} 2 & 3 \\ 1 & 2 \end{bmatrix}, \quad \begin{bmatrix} 1 & 2 \\ 1 & 3 \end{bmatrix}, \quad \begin{bmatrix} 2 & 1 \\ 3 & 2 \end{bmatrix}, \quad \text{and} \quad \begin{bmatrix} 3 & 1 \\ 2 & 1 \end{bmatrix},$$

but not to $\begin{bmatrix} 3 & 1 \\ 1 & 2 \end{bmatrix}$ or $\begin{bmatrix} 3 & 2 \\ 2 & 1 \end{bmatrix}$.

(c) Assuming $R^{-1}A \geqslant 0$, explain how $G_{R^{-1}AR}$ can be obtained from G_A by an out-splitting followed by an in-amalgamation.

7.2.2. Find an elementary conjugacy from X_A to X_B, where

(a) $A = \begin{bmatrix} 1 & 1 & 1 \\ 1 & 0 & 0 \\ 2 & 1 & 1 \end{bmatrix}$ and $B = \begin{bmatrix} 1 & 1 \\ 3 & 1 \end{bmatrix}$,

(b) $A = \begin{bmatrix} 3 & 2 \\ 1 & 1 \end{bmatrix}$ and $B = \begin{bmatrix} 2 & 3 \\ 1 & 2 \end{bmatrix}$.

7.2.3. Show that there is no elementary equivalence between the matrices

$$A = \begin{bmatrix} 1 & 3 \\ 2 & 1 \end{bmatrix} \quad \text{and} \quad B = \begin{bmatrix} 1 & 6 \\ 1 & 1 \end{bmatrix}$$

in (7–0–1) at the beginning of the chapter. [*Hint:* Observe that if $(R, S): A \approx B$, then $AR = RB$.]

7.2.4. Verify the assertions in Proposition 7.2.10.

7.2.5. Verify the assertions in Proposition 7.2.11.

7.2.6. Let $(R, S): A \approx B$ and put $C = \begin{bmatrix} 0 & R \\ S & 0 \end{bmatrix}$. Show that X_C is the disjoint union of two σ_C^2-invariant sets Y and Z such that, as dynamical systems, (Y, σ_C^2) is conjugate to (X_A, σ_A), (Z, σ_C^2) is conjugate to (X_B, σ_B), and σ_C is a conjugacy from (Y, σ_C^2) to (Z, σ_C^2). This gives another way of seeing how an elementary equivalence from A to B induces a conjugacy from X_A to X_B.

7.2.7. Show that any nonnegative integral matrix is strong shift equivalent to an essential nonnegative integral matrix. [*Hint*: Show that for any $n \times n$ nonnegative integral matrix A, we have $A \approx B$ where B is the $(n+1) \times (n+1)$ matrix obtained from A by adding a row of n arbitrary nonnegative integers and then a column of $n+1$ zeros.]

7.2.8. Let A and B be essential matrices.
  (a) Find an explicit upper bound $M = M(A, B)$ such that if $(R, S): A \approx B$, then the entries in R and S are all less than or equal to M.
  (b) Find explicit bounds $N = N(A)$ and $d = d(A)$ such that if C is essential with $A \approx C$, then C must have size $\leqslant d$ and all entries $\leqslant N$.
  (c) For fixed ℓ, find an algorithm to decide whether or not two matrices are strong shift equivalent with lag ℓ.

*7.2.9. Prove Theorem 7.2.12 as follows. Let A and B be the symbolic adjacency matrices of presentations $(G, \mathcal{L})$ and $(H, \mathcal{M})$ of sofic shifts X and Y. For a symbolic adjacency matrix A, let $|A|$ denote the nonnegative integer matrix obtained by setting all the symbols equal to 1.

  (a) Suppose that (R, S) is a symbolic elementary equivalence from A to B. Show that there is a choice of bijections such that the conjugacy $\psi = \gamma_{|R|, |S|} \colon X_{|A|} \to X_{|B|}$ induces a conjugacy $\phi \colon X \to Y$ in the sense that the following diagram commutes.

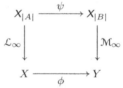

  (b) Show that if A and B are strong shift equivalent (as symbolic adjacency matrices), then X and Y are conjugate.
  In parts (c) and (d), assume that X and Y are irreducible, and that $(G, \mathcal{L})$ and $(H, \mathcal{M})$ are the minimal right-resolving presentations of X and Y.
  (c) Let $\phi \colon X \to Y$ be a splitting code or an amalgamation code, in the sense of Exercise 7.1.10. Show that there is an elementary conjugacy $\psi \colon X_{|A|} \to X_{|B|}$ such that the above diagram commutes.
  (d) Use part (c) and Exercise 7.1.10 to show that if X and Y are conjugate, then A and B are strong shift equivalent.

§7.3. Shift Equivalence

Although strong shift equivalence gives a matrix answer to the conjugacy problem for edge shifts, we seem to have merely traded one hard problem for another. In this section we will show that a weaker relation called shift equivalence follows from strong shift equivalence. The advantage of shift equivalence is that it is easier to decide while still being a very sensitive invariant.

There are several ways to define shift equivalence. We will use a down-to-earth matrix formulation for now. The matrix definition is concrete, but looks at first glance to be somewhat mysterious. However, just as strong shift equivalence is a matrix formulation of the natural notion of conjugacy, we will see in §7.5 that shift equivalence is a matrix version of another intrinsic natural idea, the dimension triple.

Definition 7.3.1. Let A and B be nonnegative integral matrices and $\ell \geqslant 1$. A *shift equivalence of lag ℓ from A to B* is a pair (R, S) of rectangular nonnegative integral matrices satisfying the four shift equivalence equations

$$(7\text{–}3\text{–}1) \qquad \text{(i)} \quad AR = RB, \qquad \text{(ii)} \quad SA = BS,$$

$$(7\text{–}3\text{–}2) \qquad \text{(i)} \quad A^\ell = RS, \qquad \text{(ii)} \quad B^\ell = SR.$$

We denote this situation by $(R, S)\colon A \sim B$ (lag ℓ). We say that A is *shift equivalent* to B, written $A \sim B$, if there is a shift equivalence from A to B of some lag.

A diagram capturing all four matrix equations (with lag $\ell = 2$) is shown below.

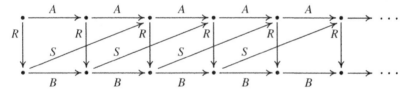

The first square shows that $AR = RB$, the first tilted parallelogram that $SA = BS$, the top left triangle that $A^2 = RS$, and the complementary triangle that $B^2 = SR$. It will be useful in proofs to think of this "ladder" as extending indefinitely to the right.

As a mnemonic, you might think of $\sim$ as one sideways S for <u>s</u>hift equivalence, while $\approx$ is two sideways S's for <u>s</u>trong <u>s</u>hift equivalence.

We first establish some basic properties of shift equivalence.

Proposition 7.3.2. *Let A and B be nonnegative integral matrices.*

(1) *A shift equivalence with lag $\ell = 1$ is the same as an elementary equivalence.*

(2) *If $A \sim B$ (lag ℓ), then $A \sim B$ (lag ℓ') for all $\ell' \geqslant \ell$.*

(3) *If $A \sim B$ (lag ℓ) and $B \sim C$ (lag ℓ'), then $A \sim C$ (lag $\ell + \ell'$).*

(4) *$\sim$ is an equivalence relation.*

PROOF: (1) When $\ell = 1$, equations (7–3–2) imply (7–3–1).

(2) If $(R, S)\colon A \sim B$ (lag ℓ) and $n \geqslant 0$, then

$$(R, SA^n)\colon A \sim B \text{ (lag } \ell + n).$$

Roughly speaking, we can always go further along the shift equivalence ladder in the above diagram.

(3) If (R, S): $A \sim B$ (lag ℓ) and (R', S'): $B \sim C$ (lag ℓ'), then it is routine to check that $(RR', S'S)$: $A \sim C$ (lag $\ell + \ell'$).

(4) Clearly (A, Id): $A \sim A$ (lag 1), and if (R, S): $A \sim B$ (lag ℓ) then (S, R): $B \sim A$ (lag ℓ). Transitivity of $\sim$ follows from (3). □

The next result shows what the names imply.

Theorem 7.3.3. *Strong shift equivalence implies shift equivalence. More precisely, if $A \approx B$ (lag ℓ), then $A \sim B$ (lag ℓ).*

PROOF: If $A \approx B$ (lag ℓ), then there is a chain of ℓ elementary equivalences from A to B. By Proposition 7.3.2(1), each of these elementary equivalences is a shift equivalence with lag 1. Then ℓ applications of Proposition 7.3.2(3) show that $A \sim B$ (lag ℓ). □

This theorem shows that the shift equivalence class of a matrix is a conjugacy invariant of the corresponding edge shift. In the next section we will see that this invariant contains all other invariants we have so far used, including the zeta function.

Example 7.3.4. Let

$$A = \begin{bmatrix} 4 & 1 \\ 1 & 0 \end{bmatrix} \quad \text{and} \quad B = \begin{bmatrix} 3 & 2 \\ 2 & 1 \end{bmatrix}.$$

Then $\chi_A(t) = t^2 - 4t - 1 = \chi_B(t)$, so that X_A and X_B have the same zeta function, hence the same entropy. Both A and B are also primitive. Therefore using only invariants from previous chapters, we are unable to tell whether or not $X_A \cong X_B$. But the following shows that A is not shift equivalent to B, so we can conclude that $X_A \not\cong X_B$.

For suppose that (R, S): $A \sim B$ (lag ℓ). Then R and S are both integral 2×2 matrices. Since $RS = A^\ell$ and $\det A = -1$, it follows that $\det R = \pm 1$. Let

$$R = \begin{bmatrix} a & b \\ c & d \end{bmatrix}.$$

The equation $AR = RB$ is

$$\begin{bmatrix} 4a + c & 4b + d \\ a & b \end{bmatrix} = \begin{bmatrix} 3a + 2b & 2a + b \\ 3c + 2d & 2c + d \end{bmatrix}.$$

Hence $a = 3c + 2d$ and $b = 2c + d$, so that

$$\det R = \det \begin{bmatrix} 3c + 2d & 2c + d \\ c & d \end{bmatrix} = 2cd + 2d^2 - 2c^2.$$

This contradicts $\det R = \pm 1$, proving that $A \not\sim B$. Note that this argument uses only two of the four shift equivalence equations. □

An invariant of shift equivalence called the Bowen–Franks group, introduced in the next section, makes this example tick.

In Example 7.3.4 we made no use of nonnegativity of R and S, and so
we have actually shown something more: there are no integral matrices
(nonnegative or not) satisfying the shift equivalence equations $(7\text{–}3\text{–}1)(i)$
and $(7\text{–}3\text{–}2)(i)$. By removing the nonnegativity assumptions on R and S,
we arrive at a purely algebraic notion.

Definition 7.3.5. Let A and B be integral matrices. Then A and B are
shift equivalent over $\mathbb{Z}$ *with lag* ℓ, written $A \sim_{\mathbb{Z}} B$ (lag ℓ), if there are rectangular integral matrices R and S satisfying the shift equivalence equations
$(7\text{–}3\text{–}1)$ and $(7\text{–}3\text{–}2)$. In this case we write $(R, S)\colon A \sim_{\mathbb{Z}} B$ (lag ℓ).

If two nonnegative integral matrices are shift equivalent (over $\mathbb{Z}^{+}$), then
they are clearly shift equivalent over $\mathbb{Z}$ since nonnegative integral matrices
are of course integral. The converse is true (and quite useful) for primitive
matrices.

Theorem 7.3.6. Let A and B be primitive integral matrices. Then $A \sim B$
if and only if $A \sim_{\mathbb{Z}} B$.

The proof uses the following two results. For a general matrix A let us
use the notation λ_A for the spectral radius of A, i.e., the absolute value
of the largest eigenvalue of A. When A is nonnegative, then λ_A is just
the Perron eigenvalue of A (see §4.4), but in general λ_A need not be an
eigenvalue of A (why?).

Proposition 7.3.7. Let A and B be integral matrices, and suppose that R
and S are rectangular integer matrices such that $(R, S)\colon A \sim_{\mathbb{Z}} B$. If $\mu \neq 0$
and $\mathbf{v} \neq \mathbf{0}$ are such that $A\mathbf{v} = \mu\mathbf{v}$, then $\mathbf{w} = S\mathbf{v} \neq \mathbf{0}$ and $B\mathbf{w} = \mu\mathbf{w}$. Hence
A and B have the same set of nonzero eigenvalues, so that $\lambda_A = \lambda_B$.

PROOF: Suppose that $(R, S)\colon A \sim_{\mathbb{Z}} B$ (lag ℓ). Assume that $\mu \neq 0$ and
$\mathbf{v} \neq \mathbf{0}$ such that $A\mathbf{v} = \mu\mathbf{v}$. Then $RS\mathbf{v} = A^{\ell}\mathbf{v} = \mu^{\ell}\mathbf{v} \neq \mathbf{0}$, so that
$\mathbf{w} = S\mathbf{v} \neq \mathbf{0}$. Since

$$B\mathbf{w} = B(S\mathbf{v}) = (BS)\mathbf{v} = (SA)\mathbf{v} = \mu(S\mathbf{v}) = \mu\mathbf{w},$$

we see that μ is also an eigenvalue of B. This proves that a nonzero eigenvalue of A is also an eigenvalue of B. Reversing the roles of A and B shows
that A and B have the same set of nonzero eigenvalues. Hence $\lambda_A = \lambda_B$
by the definition of spectral radius. □

This proposition shows in particular that if A and B are nonnegative and
shift equivalent over $\mathbb{Z}$, then $h(X_A) = h(X_B)$. Thus entropy is an invariant
of shift equivalence over $\mathbb{Z}$, hence of shift equivalence.

Lemma 7.3.8. Let A be primitive and $\mathbf{v}$ be a right Perron eigenvector
for A. If $\mathbf{u}$ is a row vector, then $\mathbf{u}A^{k}$ is eventually positive if and only if
$\mathbf{u} \cdot \mathbf{v} > 0$.

PROOF: Since A is primitive, its Perron eigenvalue $\lambda = \lambda_A > 0$. So if $\mathbf{u}A^k > 0$ for some k, then $0 < (\mathbf{u}A^k) \cdot \mathbf{v} = \lambda^k(\mathbf{u} \cdot \mathbf{v})$, hence $\mathbf{u} \cdot \mathbf{v} > 0$.

Conversely, suppose that $\mathbf{u} \cdot \mathbf{v} > 0$. By Theorem 4.5.12, there is a left Perron eigenvector $\mathbf{w} = [w_1, \dots, w_r] > 0$ for A such that for every I and J

$$(A^k)_{IJ} = v_I w_J \lambda^k + \rho_{IJ}(k)\lambda^k,$$

where $\rho_{IJ}(k) \to 0$ as $k \to \infty$. Thus

$$\mathbf{u}A^k = (\mathbf{u} \cdot \mathbf{v})\lambda^k \mathbf{w} + \xi(\mathbf{u}, k),$$

where $\xi(\mathbf{u}, k)/\lambda^k \to \mathbf{0}$ as $k \to \infty$. Since $\mathbf{u} \cdot \mathbf{v} > 0$, it follows that $\mathbf{u}A^k$ is positive for all large enough k. $\square$

PROOF OF THEOREM 7.3.6: Shift equivalence clearly implies shift equivalence over $\mathbb{Z}$.

Conversely, suppose that $A \sim_\mathbb{Z} B$, and let $(R, S)\colon A \sim_\mathbb{Z} B$ (lag ℓ). It is easy to verify that if $k \geqslant 0$, then $(RB^k, SA^k)\colon A \sim_\mathbb{Z} B$ (lag $\ell + 2k$). We claim that RB^k and SA^k are either both eventually positive or both eventually negative. In the former case we are done, and in the latter case we can simply replace (R, S) with $(-R, -S)$ to complete the proof.

To verify our claim, first observe from Proposition 7.3.7 that $\lambda_A = \lambda_B$ and if $\mathbf{v}$ is a right Perron eigenvector for A, then either $\mathbf{w} = S\mathbf{v}$ or its negative must be a right Perron eigenvector for B. Combining the inequality

$$(7\text{--}3\text{--}3) \qquad 0 < \lambda_A^\ell v_I = \mathbf{e}_I A^\ell \mathbf{v} = \mathbf{e}_I RS\mathbf{v} = \mathbf{e}_I R\mathbf{w}$$

with Lemma 7.3.8 shows that by applying large enough powers of B to the rows $\mathbf{e}_I R$ of R, they become either all positive or all negative. Thus RB^k is either eventually positive or eventually negative. A similar conclusion applies to SA^k. If the eventual signs of RB^k and SA^k were opposite, we would have

$$0 > (RB^k)(SA^k) = RS\, A^{2k} = A^{2k+\ell},$$

contradicting $A \geqslant 0$. This establishes our claim, completing the proof. $\square$

We remark that Theorem 7.3.6 fails if A and B are merely irreducible (see Boyle [Boy2]).

One reason that $\sim_\mathbb{Z}$ is useful is that the lags are often much smaller than for $\sim$. This is the case for primitive matrices that are similar over $\mathbb{Z}$.

Definition 7.3.9. An integral matrix P is *invertible over $\mathbb{Z}$* if it is nonsingular and its inverse is also integral. Two integral matrices A and B are *similar over $\mathbb{Z}$* if there is an integral matrix P that is invertible over $\mathbb{Z}$ such that $A = P^{-1}BP$.

It is not hard to see that an integral matrix P is invertible over $\mathbb{Z}$ if and only if $\det(P) = \pm 1$ (Exercise 7.3.5).

Proposition 7.3.10. *Let A and B be primitive integral matrices that are similar over $\mathbb{Z}$. Then $A \sim_{\mathbb{Z}} B$ (lag 1), hence $A \sim B$.*

PROOF: There is an integral matrix P invertible over $\mathbb{Z}$ such that $A = P^{-1}BP$. Put $R = P^{-1}$ and $S = BP$. Then $RS = A$ and $SR = B$, so that $(R, S): A \sim_{\mathbb{Z}} B$ (lag 1). Theorem 7.3.6 now shows that $A \sim B$. □

Example 7.3.11. Let

$$A = \begin{bmatrix} 1 & 3 \\ 2 & 1 \end{bmatrix} \quad \text{and} \quad B = \begin{bmatrix} 1 & 6 \\ 1 & 1 \end{bmatrix}$$

be the matrices in (7–0–1). If we put

$$P = \begin{bmatrix} 2 & 3 \\ 1 & 1 \end{bmatrix}, \quad \text{then} \quad P^{-1} = \begin{bmatrix} -1 & 3 \\ 1 & -2 \end{bmatrix}$$

and $A = P^{-1}BP$, so that A and B are similar over $\mathbb{Z}$. By Proposition 7.3.10, we know that A and B must be shift equivalent. In particular, $R = P^{-1}B^3$, $S = P$ defines a shift equivalence of lag 3 from A to B. However, there are *no* shift equivalences of lag 1 or lag 2 from A to B (Exercise 7.3.4). □

So the pair of matrices A and B in the preceding example are at least shift equivalent. But are they *strong* shift equivalent? This question was answered affirmatively by K. Baker (email communication) using a computer search.

Example 7.3.12. The following is a strong shift equivalence of lag 7 between

$$A = \begin{bmatrix} 1 & 3 \\ 2 & 1 \end{bmatrix} \quad \text{and} \quad B = \begin{bmatrix} 1 & 6 \\ 1 & 1 \end{bmatrix}.$$

$$A = \begin{bmatrix} 1 & 3 \\ 2 & 1 \end{bmatrix} = \begin{bmatrix} 0 & 1 & 1 \\ 1 & 0 & 0 \end{bmatrix} \begin{bmatrix} 2 & 1 \\ 1 & 2 \\ 0 & 1 \end{bmatrix},$$

$$\begin{bmatrix} 2 & 1 \\ 1 & 2 \\ 0 & 1 \end{bmatrix} \begin{bmatrix} 0 & 1 & 1 \\ 1 & 0 & 0 \end{bmatrix} = \begin{bmatrix} 1 & 2 & 2 \\ 2 & 1 & 1 \\ 1 & 0 & 0 \end{bmatrix} = A_1$$

$$A_1 = \begin{bmatrix} 1 & 2 & 2 \\ 2 & 1 & 1 \\ 1 & 0 & 0 \end{bmatrix} = \begin{bmatrix} 1 & 0 & 2 & 0 \\ 0 & 1 & 1 & 1 \\ 0 & 1 & 0 & 0 \end{bmatrix} \begin{bmatrix} 1 & 0 & 2 \\ 1 & 0 & 0 \\ 0 & 1 & 0 \\ 1 & 0 & 1 \end{bmatrix},$$

$$\begin{bmatrix} 1 & 0 & 2 \\ 1 & 0 & 0 \\ 0 & 1 & 0 \\ 1 & 0 & 1 \end{bmatrix} \begin{bmatrix} 1 & 0 & 2 & 0 \\ 0 & 1 & 1 & 1 \\ 0 & 1 & 0 & 0 \end{bmatrix} = \begin{bmatrix} 1 & 2 & 2 & 0 \\ 1 & 0 & 2 & 0 \\ 0 & 1 & 1 & 1 \\ 1 & 1 & 2 & 0 \end{bmatrix} = A_2$$

$$A_2 = \begin{bmatrix} 1 & 2 & 2 & 0 \\ 1 & 0 & 2 & 0 \\ 0 & 1 & 1 & 1 \\ 1 & 1 & 2 & 0 \end{bmatrix} = \begin{bmatrix} 2 & 0 & 0 & 1 \\ 0 & 2 & 0 & 1 \\ 1 & 0 & 1 & 0 \\ 1 & 1 & 0 & 1 \end{bmatrix} \begin{bmatrix} 0 & 1 & 1 & 0 \\ 0 & 0 & 1 & 0 \\ 0 & 0 & 0 & 1 \\ 1 & 0 & 0 & 0 \end{bmatrix},$$

$$\begin{bmatrix} 0 & 1 & 1 & 0 \\ 0 & 0 & 1 & 0 \\ 0 & 0 & 0 & 1 \\ 1 & 0 & 0 & 0 \end{bmatrix} \begin{bmatrix} 2 & 0 & 0 & 1 \\ 0 & 2 & 0 & 1 \\ 1 & 0 & 1 & 0 \\ 1 & 1 & 0 & 1 \end{bmatrix} = \begin{bmatrix} 1 & 2 & 1 & 1 \\ 1 & 0 & 1 & 0 \\ 1 & 1 & 0 & 1 \\ 2 & 0 & 0 & 1 \end{bmatrix} = A_3$$

$$A_3 = \begin{bmatrix} 1 & 2 & 1 & 1 \\ 1 & 0 & 1 & 0 \\ 1 & 1 & 0 & 1 \\ 2 & 0 & 0 & 1 \end{bmatrix} = \begin{bmatrix} 0 & 1 & 1 & 0 \\ 0 & 0 & 0 & 1 \\ 0 & 1 & 0 & 0 \\ 1 & 0 & 0 & 0 \end{bmatrix} \begin{bmatrix} 2 & 0 & 0 & 1 \\ 1 & 1 & 0 & 1 \\ 0 & 1 & 1 & 0 \\ 1 & 0 & 1 & 0 \end{bmatrix},$$

$$\begin{bmatrix} 2 & 0 & 0 & 1 \\ 1 & 1 & 0 & 1 \\ 0 & 1 & 1 & 0 \\ 1 & 0 & 1 & 0 \end{bmatrix} \begin{bmatrix} 0 & 1 & 1 & 0 \\ 0 & 0 & 0 & 1 \\ 0 & 1 & 0 & 0 \\ 1 & 0 & 0 & 0 \end{bmatrix} = \begin{bmatrix} 1 & 2 & 2 & 0 \\ 1 & 1 & 1 & 1 \\ 0 & 1 & 0 & 1 \\ 0 & 2 & 1 & 0 \end{bmatrix} = A_4$$

$$A_4 = \begin{bmatrix} 1 & 2 & 2 & 0 \\ 1 & 1 & 1 & 1 \\ 0 & 1 & 0 & 1 \\ 0 & 2 & 1 & 0 \end{bmatrix} = \begin{bmatrix} 0 & 1 & 1 & 1 \\ 1 & 0 & 1 & 1 \\ 1 & 0 & 0 & 0 \\ 0 & 1 & 0 & 0 \end{bmatrix} \begin{bmatrix} 0 & 1 & 0 & 1 \\ 0 & 2 & 1 & 0 \\ 0 & 0 & 1 & 0 \\ 1 & 0 & 0 & 0 \end{bmatrix},$$

$$\begin{bmatrix} 0 & 1 & 0 & 1 \\ 0 & 2 & 1 & 0 \\ 0 & 0 & 1 & 0 \\ 1 & 0 & 0 & 0 \end{bmatrix} \begin{bmatrix} 0 & 1 & 1 & 1 \\ 1 & 0 & 1 & 1 \\ 1 & 0 & 0 & 0 \\ 0 & 1 & 0 & 0 \end{bmatrix} = \begin{bmatrix} 1 & 1 & 1 & 1 \\ 3 & 0 & 2 & 2 \\ 1 & 0 & 0 & 0 \\ 0 & 1 & 1 & 1 \end{bmatrix} = A_5$$

$$A_5 = \begin{bmatrix} 1 & 1 & 1 & 1 \\ 3 & 0 & 2 & 2 \\ 1 & 0 & 0 & 0 \\ 0 & 1 & 1 & 1 \end{bmatrix} = \begin{bmatrix} 1 & 0 & 1 \\ 0 & 1 & 0 \\ 0 & 0 & 1 \\ 1 & 0 & 0 \end{bmatrix} \begin{bmatrix} 0 & 1 & 1 & 1 \\ 3 & 0 & 2 & 2 \\ 1 & 0 & 0 & 0 \end{bmatrix},$$

$$\begin{bmatrix} 0 & 1 & 1 & 1 \\ 3 & 0 & 2 & 2 \\ 1 & 0 & 0 & 0 \end{bmatrix} \begin{bmatrix} 1 & 0 & 1 \\ 0 & 1 & 0 \\ 0 & 0 & 1 \\ 1 & 0 & 0 \end{bmatrix} = \begin{bmatrix} 1 & 1 & 1 \\ 5 & 0 & 5 \\ 1 & 0 & 1 \end{bmatrix} = A_6$$

$$A_6 = \begin{bmatrix} 1 & 1 & 1 \\ 5 & 0 & 5 \\ 1 & 0 & 1 \end{bmatrix} = \begin{bmatrix} 1 & 0 \\ 0 & 5 \\ 0 & 1 \end{bmatrix} \begin{bmatrix} 1 & 1 & 1 \\ 1 & 0 & 1 \end{bmatrix},$$

$$\begin{bmatrix} 1 & 1 & 1 \\ 1 & 0 & 1 \end{bmatrix} \begin{bmatrix} 1 & 0 \\ 0 & 5 \\ 0 & 1 \end{bmatrix} = \begin{bmatrix} 1 & 6 \\ 1 & 1 \end{bmatrix} = A_7 = B.$$

This is the smallest lag known for a strong shift equivalence between A and B.   □

Thus our original pair of matrices has turned out to be strong shift equivalent. But the following example shows that this pair is on the edge of what is known.

Example 7.3.13. Let

$$A_k = \begin{bmatrix} 1 & k \\ k-1 & 1 \end{bmatrix} \quad \text{and} \quad B_k = \begin{bmatrix} 1 & k(k-1) \\ 1 & 1 \end{bmatrix}.$$

Then $A_k \sim B_k$ for all $k \geqslant 1$ (Exercise 7.3.3), but it is not known whether $A_k \approx B_k$ when $k \geqslant 4$. The matrices in the previous example are A_3 and B_3. □

It is natural to wonder whether shift equivalence implies strong shift equivalence. The problem, known as the *Shift Equivalence Problem*, was open for more than twenty years. Kim and Roush [KimR12] solved the Shift Equivalence Problem in the negative by constructing two irreducible shifts of finite type that are shift equivalent but not strong shift equivalent. Earlier they found a reducible example [KimR11]. Their solution depends on a new invariant known as the sign-gyration homomorphism that they developed in collaboration with Wagoner [KimRW1], and whose origins can be traced back to work of Boyle and Krieger [BoyK1].

Like all stimulating problems, the Shift Equivalence Problem has generated a large body of methods, ideas, and techniques, some of which we will encounter in later chapters.

EXERCISES

7.3.1. Show that if $A \sim B$ then $A^{\mathsf{T}} \sim B^{\mathsf{T}}$, and that a similar statement holds for $\approx$.

7.3.2. For each of the following pairs A, B of matrices, decide whether $A \sim B$.

(a) $A = \begin{bmatrix} 2 & 3 \\ 2 & 1 \end{bmatrix}$ and $B = \begin{bmatrix} 3 & 5 \\ 1 & 0 \end{bmatrix}$.

(b) $A = \begin{bmatrix} 1 & 2 \\ 2 & 3 \end{bmatrix}$ and $B = \begin{bmatrix} 3 & 4 \\ 1 & 1 \end{bmatrix}$.

(c) $A = \begin{bmatrix} 3 & 5 \\ 2 & 3 \end{bmatrix}$ and $B = \begin{bmatrix} 1 & 3 \\ 2 & 5 \end{bmatrix}$.

(d) $A = \begin{bmatrix} 2 & 1 \\ 1 & 2 \end{bmatrix}$ and $B = \begin{bmatrix} 1 & 2 \\ 2 & 1 \end{bmatrix}$.

[*Hint*: Consider Exercise 7.2.1, Proposition 7.3.7, and the argument in Example 7.3.4.]

7.3.3. Let k be a positive integer, and let

$$A = \begin{bmatrix} 1 & k \\ k-1 & 1 \end{bmatrix} \quad \text{and} \quad B = \begin{bmatrix} 1 & k(k-1) \\ 1 & 1 \end{bmatrix}.$$

(a) Show that $A \sim B$.
(b) Show that $A \approx A^{\mathsf{T}}$ and $B \approx B^{\mathsf{T}}$.

7.3.4. Let

$$A = \begin{bmatrix} 1 & 3 \\ 2 & 1 \end{bmatrix} \quad \text{and} \quad B = \begin{bmatrix} 1 & 6 \\ 1 & 1 \end{bmatrix}.$$

Show that there is a shift equivalence between A and B of lag 3, but not one of lag 1 or one of lag 2. [*Hint*: If $A \sim B$ (lag 2), then $A^2 \sim B^2$ (lag 1).]

7.3.5. Show that an integral matrix P is invertible over $\mathbb{Z}$ if and only if $\det(P) = \pm 1$.

7.3.6. Let A and B be integral matrices of the same size with $\det A = \det B = \pm 1$. Show that A is similar over $\mathbb{Z}$ to B if and only if $A \sim_{\mathbb{Z}} B$.

7.3.7. We say that two symbolic adjacency matrices A and B are shift equivalent if there are matrices R, S such that the four shift equivalence equations (7–3–1) and (7–3–2) are satisfied with equality replaced by $\leftrightarrow$ (see the end of §7.2). For example, "equation" (7–3–1)(i) means that there is a bijection from the alphabet of AR to the alphabet of RB which transforms AR into RB. Show that if two symbolic adjacency matrices are strong shift equivalent, then they are shift equivalent. The converse is known to be false [KimR8] even when A and B are the symbolic adjacency matrices of the minimal right-resolving presentations of irreducible sofic shifts.

§7.4. Invariants for Shift Equivalence

There are theoretical procedures that will decide whether or not two matrices are shift equivalent [KimR3], but they are usually difficult to apply. So it is useful to have a collection of invariants for shift equivalence that are easy to compute. Since strong shift equivalence implies shift equivalence, every invariant of shift equivalence is automatically a conjugacy invariant for the associated edge shifts. For example, in this section we show that if $A \sim B$, then $\zeta_{\sigma_A} = \zeta_{\sigma_B}$.

This section focuses on two invariants for shift equivalence: the Jordan form away from zero $J^\times$ and the *Bowen–Franks group* BF. Roughly speaking, $J^\times$ captures the invertible part of the rational linear algebra of a matrix while BF reflects integrality.

Before discussing these new invariants, we first show that two familiar properties of essential nonnegative integral matrices are preserved by shift equivalence.

Proposition 7.4.1. *For essential nonnegative integral matrices, both irreducibility and primitivity are invariant under shift equivalence.*

PROOF: Suppose that $(R, S): A \sim B$ (lag ℓ), and let J, K be states for B. Since $SR = B^\ell$ is essential, there is an I for which $S_{JI} \geqslant 1$. Similarly, $RS = A^\ell$ is essential, so there is an L for which $R_{LK} \geqslant 1$. Hence for every $k \geqslant 0$,

$$(B^{k+\ell})_{JK} = (B^k SR)_{JK} = (SA^k R)_{JK} \geqslant S_{JI}(A^k)_{IL}R_{LK} \geqslant (A^k)_{IL}.$$

It now follows easily that if A is irreducible, then so is B, and if A is primitive, then so is B. □

Simple examples show that irreducibility and primitivity are *not* invariants of shift equivalence over $\mathbb{Z}$ (see Exercise 7.4.2(a)) nor are they invariants of shift equivalence when the assumption of essential is dropped (see Exercise 7.4.1).

We next turn to the linear algebra of an $r \times r$ integral matrix A. Since the rational field $\mathbb{Q}$ is the smallest one containing the integers, we will use $\mathbb{Q}$ as the scalar field. Thus linearity, subspaces, linear combinations, and so on, will always be with respect to the field $\mathbb{Q}$.

Let A have size r. Then A defines a linear transformation $A \colon \mathbb{Q}^r \to \mathbb{Q}^r$. It will be convenient for several reasons to have A act on the *right*, and to consider elements of $\mathbb{Q}^r$ as row vectors. Thus if $\mathbf{v} = [v_1, v_2, \ldots, v_r] \in \mathbb{Q}^r$, the image of $\mathbf{v}$ under A is $\mathbf{v}A$. Similarly, if $E \subseteq \mathbb{Q}^r$, its image under A is denoted by EA. Readers familiar with the theory of Markov chains will understand that by writing matrices and vectors this way, matrix equations representing transitions take on a more natural form.

There are two A-invariant subspaces of $\mathbb{Q}^r$ that are particularly important.

Definition 7.4.2. Let A be an $r \times r$ integral matrix. The *eventual range* $\mathcal{R}_A$ of A is the subspace of $\mathbb{Q}^r$ defined by

$$\mathcal{R}_A = \bigcap_{k=1}^{\infty} \mathbb{Q}^r A^k.$$

The *eventual kernel* $\mathcal{K}_A$ of A is the subspace of $\mathbb{Q}^r$ defined by

$$\mathcal{K}_A = \bigcup_{k=1}^{\infty} \ker(A^k).$$

where $\ker(A) = \{\mathbf{v} \in \mathbb{Q}^r : \mathbf{v}A = 0\}$ is the kernel of A viewed as a linear transformation.

Example 7.4.3. (a) Let

$$A = \begin{bmatrix} 1 & 1 \\ 1 & 1 \end{bmatrix}.$$

Then $\mathbb{Q}^2 A = \{[q, q] : q \in \mathbb{Q}\} = \mathbb{Q} \cdot [1, 1] = \mathcal{U}$, say, and on this 1-dimensional subspace A is multiplication by 2. Hence

$$\mathbb{Q}^2 \supsetneq \mathbb{Q}^2 A = \mathbb{Q}^2 A^2 = \mathbb{Q}^2 A^3 = \ldots,$$

so that $\mathcal{R}_A = \mathcal{U}$. Since $A^k = 2^{k-1}A$, it follows that $\ker(A^k) = \ker(A) = \mathbb{Q} \cdot [1, -1] = \mathcal{V}$, so that $\mathcal{K}_A = \mathcal{V}$. Note that $\mathbb{Q}^2 = \mathcal{R}_A \oplus \mathcal{K}_A$.

(b) Let

$$B = \begin{bmatrix} 0 & 0 & 0 \\ 1 & 0 & 0 \\ 0 & 0 & 1 \end{bmatrix},$$

and, as usual, let $\mathbf{e}_j$ denote the jth standard basis vector in $\mathbb{Q}^r$. Then

$$\mathbb{Q}^3 \supsetneq \mathbb{Q}^3 B \supsetneq \mathbb{Q}^3 B^2 = \mathbb{Q}^3 B^3 = \cdots = \mathbb{Q}\mathbf{e}_3,$$

so that $\mathcal{R}_B = \mathbb{Q}\mathbf{e}_3$. Also,

$$0 \subsetneqq \ker(B) \subsetneqq \ker(B^2) = \ker(B^3) = \cdots = \mathbb{Q}\mathbf{e}_1 \oplus \mathbb{Q}\mathbf{e}_2,$$

so that $\mathcal{K}_B = \mathbb{Q}\mathbf{e}_1 \oplus \mathbb{Q}\mathbf{e}_2$. Again note that $\mathbb{Q}^3 = \mathcal{R}_B \oplus \mathcal{K}_B$. □

Remark 7.4.4. (1) If $\mathbb{Q}^r A^k = \mathbb{Q}^r A^{k+1}$, then A is invertible on $\mathbb{Q}^r A^k$, so that $\mathbb{Q}^r A^k = \mathbb{Q}^r A^{k+n}$ for all $n \geqslant 0$. So once equality occurs in the nested sequence

$$\mathbb{Q}^r \supseteq \mathbb{Q}^r A \supseteq \mathbb{Q}^r A^2 \supseteq \mathbb{Q}^r A^3 \supseteq \ldots$$

of subspaces, all further inclusions are equalities. Since proper subspaces must have strictly smaller dimension, there can be at most r strict inclusions before equality occurs. Hence

$$\mathcal{R}_A = \mathbb{Q}^r A^r.$$

A similar argument shows that

$$\mathcal{K}_A = \ker(A^r).$$

(2) The eventual range of A is the largest subspace of $\mathbb{Q}^r$ on which A is invertible. Similarly, the eventual kernel of A is the largest subspace of $\mathbb{Q}^r$ on which A is nilpotent, i.e., on which some power of A is the zero transformation. Hence both $\mathcal{R}_A$ and $\mathcal{K}_A$ are A-invariant subspaces of $\mathbb{Q}^r$.

(3) For every A we claim that $\mathbb{Q}^r = \mathcal{R}_A \oplus \mathcal{K}_A$. For if $\mathbf{v} \in \mathcal{R}_A \cap \mathcal{K}_A$, then $\mathbf{v}A^r = 0$, so that $\mathbf{v} = \mathbf{0}$ since A is invertible on $\mathcal{R}_A$. Hence $\mathcal{R}_A \cap \mathcal{K}_A = 0$. Let $\mathbf{v} \in \mathbb{Q}^r$. Then $\mathbf{v}A^r \in \mathcal{R}_A$. Since A is invertible on $\mathcal{R}_A$, there is a $\mathbf{u} \in \mathcal{R}_A$ such that $\mathbf{v}A^r = \mathbf{u}A^r$. Then $\mathbf{v} - \mathbf{u} \in \mathcal{K}_A$, so that $\mathbf{v} = \mathbf{u} + (\mathbf{v} - \mathbf{u}) \in \mathcal{R}_A + \mathcal{K}_A$, establishing the claim. □

Definition 7.4.5. Let A be an integral matrix. The *invertible part* $A^\times$ of A is the linear transformation obtained by restricting A to its eventual range, i.e., $A^\times \colon \mathcal{R}_A \to \mathcal{R}_A$ is defined by $A^\times(\mathbf{v}) = \mathbf{v}A$.

Next we examine the relation between the invertible parts of matrices that are shift equivalent over $\mathbb{Z}$. For this we need the following standard notions from linear algebra.

We say that a linear transformation is a *linear isomorphism* if it is one-to-one and onto. Note that $A^\times$ is a linear isomorphism. We say that two linear transformations $f \colon V \to V$ and $g \colon W \to W$ are *isomorphic* if there is a linear isomorphism $h \colon V \to W$ such that $h \circ f = g \circ h$.

Theorem 7.4.6. *If $A \sim_{\mathbb{Z}} B$, then $A^\times$ and $B^\times$ are isomorphic as linear transformations.*

PROOF: Suppose that A is $m \times m$, B is $n \times n$, and $(R, S): A \sim_{\mathbb{Z}} B$ (lag ℓ). Put $q = \max\{m, n\}$. Define $\widetilde{R}: \mathcal{R}_A \to \mathbb{Q}^n$ by $\widetilde{R}(\mathbf{v}) = \mathbf{v}R$. Similarly, define $\widetilde{S}: \mathcal{R}_B \to \mathbb{Q}^m$ by $\widetilde{S}(\mathbf{v}) = \mathbf{v}S$. Now,

$$\widetilde{R}(\mathcal{R}_A) = \mathcal{R}_A R = \mathbb{Q}^m A^q R = \mathbb{Q}^m R B^q \subseteq \mathcal{R}_B$$

and

$$\widetilde{S}(\mathcal{R}_B) = \mathcal{R}_B S = \mathbb{Q}^n B^q S = Q^n S A^q \subseteq \mathcal{R}_A.$$

Since $A^\times$ is a linear isomorphism and $RS = A^\ell$, it follows that $\widetilde{R}$ is one-to-one. Similarly $\widetilde{S}$ is one-to-one. Hence

(7-4-1) $\dim \mathcal{R}_A = \dim(\widetilde{R}(\mathcal{R}_A)) \leqslant \dim \mathcal{R}_B = \dim(\widetilde{S}(\mathcal{R}_B)) \leqslant \dim \mathcal{R}_A$

and therefore all dimensions in (7-4-1) are equal. This shows that $\widetilde{R}$ maps $\mathcal{R}_A$ onto $\mathcal{R}_B$, and so is a linear isomorphism between them. The relation $AR = RB$ implies $\widetilde{R} \circ A^\times = B^\times \circ \widetilde{R}$, so that $A^\times$ and $B^\times$ are isomorphic as linear transformations. □

In order to tell whether two linear transformations are isomorphic, we will use their Jordan canonical forms. We briefly review this concept (see [Her] for a detailed account).

For a complex number λ and integer $m \geqslant 1$, define the $m \times m$ *Jordan block for λ* to be

$$J_m(\lambda) = \begin{bmatrix} \lambda & 1 & 0 & \cdots & 0 & 0 \\ 0 & \lambda & 1 & \cdots & 0 & 0 \\ 0 & 0 & \lambda & \cdots & 0 & 0 \\ \vdots & \vdots & \vdots & \ddots & \vdots & \\ 0 & 0 & 0 & \cdots & \lambda & 1 \\ 0 & 0 & 0 & \cdots & 0 & \lambda \end{bmatrix}.$$

If a matrix

$$A = \begin{bmatrix} \boxed{A_1} & & & \\ & \boxed{A_2} & & \\ & & \ddots & \\ & & & \boxed{A_k} \end{bmatrix}$$

has block diagonal form, we will write A more compactly as a direct sum:

$$A = A_1 \oplus A_2 \oplus \cdots \oplus A_k.$$

The basic theorem on Jordan canonical form says that every matrix is similar over the complex numbers $\mathbb{C}$ to a direct sum of Jordan blocks, and that this representation is unique apart from order of the blocks. Thus if A

is a matrix, then there are complex numbers $\lambda_1, \lambda_2, \ldots, \lambda_k$ (not necessarily distinct) and integers $m_1, m_2, \ldots, m_k$ such that A is similar to

$$(7\text{-}4\text{-}2) \qquad J(A) = J_{m_1}(\lambda_1) \oplus J_{m_2}(\lambda_2) \oplus \cdots \oplus J_{m_k}(\lambda_k).$$

The matrix $J(A)$ is called the *Jordan form* of A. Furthermore, two matrices are similar over $\mathbb{C}$ if and only if they have the same Jordan form (up to order of blocks), so that the Jordan form is a complete invariant for similarity over $\mathbb{C}$. The Jordan form of a linear transformation f is is defined to be the Jordan form of the matrix of f with respect to any basis. Since the Jordan form of a matrix is a complete invariant for similarity over $\mathbb{C}$, the Jordan form of a linear transformation is well-defined and is a complete invariant for isomorphism.

Note that if A has Jordan form (7–4–2), then the eigenvalues of A are the λ_j (repeated m_j times) and the characteristic polynomial of A is

$$\chi_A(t) = (t - \lambda_1)^{m_1}(t - \lambda_2)^{m_2} \ldots (t - \lambda_k)^{m_k}.$$

Also note that if the roots of $\chi_A(t)$ are distinct, then all Jordan blocks must be 1×1 and the Jordan form of A is then a diagonal matrix.

Example 7.4.7. Let $A = \begin{bmatrix} 1 & 1 \\ 1 & 1 \end{bmatrix}$, so that $\chi_A(t) = t(t - 2)$. The roots of $\chi_A(t)$ are distinct, therefore

$$J(A) = J_1(2) \oplus J_1(0) = \begin{bmatrix} \boxed{2} & \\ & \boxed{0} \end{bmatrix}. \qquad \square$$

Example 7.4.8. (a) Let

$$A = \begin{bmatrix} 3 & 1 & 1 \\ 2 & 2 & 1 \\ 1 & 2 & 2 \end{bmatrix}.$$

A computation shows that $\chi_A(t) = (t - 5)(t - 1)^2$. Since 5 is a simple eigenvalue, its Jordan block must have size 1. Since 1 is a double root, there are two possibilities:

$$J_2(1) = \begin{bmatrix} 1 & 1 \\ 0 & 1 \end{bmatrix} \quad \text{or} \quad J_1(1) \oplus J_1(1) = \begin{bmatrix} 1 & 0 \\ 0 & 1 \end{bmatrix}.$$

Now rank$(A - Id) = 2$, ruling out the second possibility. Hence

$$J(A) = J_1(5) \oplus J_2(1) = \begin{bmatrix} \boxed{5} & \\ & \boxed{\begin{matrix} 1 & 1 \\ 0 & 1 \end{matrix}} \end{bmatrix}.$$

(b) Let

$$B = \begin{bmatrix} 3 & 1 & 1 \\ 2 & 2 & 1 \\ 2 & 1 & 2 \end{bmatrix}.$$

Then $\chi_B(t) = (t-5)(t-1)^2 = \chi_A(t)$. Again, 5 is a simple eigenvalue. However, $\text{rank}(B - Id) = 1$, ruling out $J_2(1)$, so that

$$J(B) = J_1(5) \oplus J_1(1) \oplus J_1(1) = \begin{bmatrix} \boxed{5} & & \\ & \boxed{1} & \\ & & \boxed{1} \end{bmatrix}.$$

Definition 7.4.9. Let A be an integral matrix. The *Jordan form away from zero* of A, written $J^\times(A)$, is the Jordan form of the invertible part $A^\times$ of A.

Thus in Example 7.4.7, $J^\times(A) = [2]$, while in Example 7.4.8, $J^\times(A) = J(A)$ and $J^\times(B) = J(B)$ since all eigenvalues are nonzero.

Theorem 7.4.10. *If $A \sim_{\mathbb{Z}} B$ then $J^\times(A) = J^\times(B)$.*

PROOF: By Theorem 7.4.6, $A^\times$ and $B^\times$ are isomorphic. The result then follows from the fact that isomorphic linear transformations have the same Jordan form. ☐

Example 7.4.11. Consider the matrices

$$A = \begin{bmatrix} 3 & 1 & 1 \\ 2 & 2 & 1 \\ 1 & 2 & 2 \end{bmatrix} \quad \text{and} \quad B = \begin{bmatrix} 3 & 1 & 1 \\ 2 & 2 & 1 \\ 2 & 1 & 2 \end{bmatrix}$$

from Example 7.4.8. Since $\chi_A(t) = \chi_B(t)$, the zeta functions of X_A and X_B agree. However, we saw that $J^\times(A) \neq J^\times(B)$, so that X_A and X_B are not conjugate. This shows that $J^\times$ can distinguish (some) edge shifts with the same zeta function. ☐

The following corollaries use invariance of $J^\times$ under $\sim_{\mathbb{Z}}$ to show that other quantities we have previously considered are also invariant under $\sim_{\mathbb{Z}}$; note that the second corollary below was proved earlier (see the remarks following the proof of Proposition 7.3.7). Recall from §6.4 the notions of nonzero spectrum and zeta function.

Corollary 7.4.12. *Let A and B be integral matrices. If $A \sim_{\mathbb{Z}} B$ then their nonzero spectra $\text{sp}^\times(A) = \text{sp}^\times(B)$ coincide, and hence when A and B are nonnegative $\zeta_{\sigma_A}(t) = \zeta_{\sigma_B}(t)$.*

PROOF: By Theorem 7.4.10, we see that $J^\times(A) = J^\times(B)$. It follows that A and B have the same nonzero eigenvalues (counted with the same multiplicity). Corollary 6.4.7 then shows that $\zeta_{\sigma_A}(t) = \zeta_{\sigma_B}(t)$. ☐

Corollary 7.4.13. *Let A and B be nonnegative integral matrices. If $A \sim_{\mathbb{Z}} B$ then $h(X_A) = h(X_B)$.*

PROOF: Since $J^{\times}(A) = J^{\times}(B)$, it follows that A and B have the same nonzero eigenvalues, hence the same Perron eigenvalue.   □

Corollary 7.4.14. *Let A and B be nonnegative integral matrices. If $A \sim_{\mathbb{Z}} B$ then $\mathrm{per}(A) = \mathrm{per}(B)$.*

PROOF: The period $\mathrm{per}(A)$ of A from Definition 4.5.2 is determined by the numbers $p_n(X_A)$ of points in X_A of period n for $n \geqslant 1$. These in turn are determined by $\zeta_{\sigma_A}(t)$ (see Corollary 6.4.7). The result then follows from Corollary 7.4.12.   □

Our work so far has the following implications:

$$A \sim B \Longrightarrow A \sim_{\mathbb{Z}} B \Longrightarrow J^{\times}(A) = J^{\times}(B)$$
$$\Longrightarrow \zeta_{\sigma_A}(t) = \zeta_{\sigma_B}(t) \Longleftrightarrow \mathrm{sp}^{\times}(A) = \mathrm{sp}^{\times}(B).$$

By Theorem 7.3.6, the first implication can be reversed when A and B are primitive. Example 7.3.4 shows that the second implication cannot be reversed, and Example 7.4.11 shows that the third cannot be reversed.

We now turn to a different kind of invariant, one whose values are not numbers or functions, but finitely-generated abelian groups. For a good introduction to such groups, the reader could consult [Kap].

Let $\mathbb{Z}_d$ denote the cyclic group of order d, so that $\mathbb{Z}_d = \mathbb{Z}/d\mathbb{Z}$. Then $\mathbb{Z}_1$ is just the trivial group of one element. It is convenient to let $\mathbb{Z}_0$ denote $\mathbb{Z}$. The fundamental theorem on finitely-generated abelian groups states that every such group is isomorphic to a direct sum of cyclic groups, finite or infinite. In other words, if Γ is a finitely generated abelian group, then there are nonnegative integers $d_1, d_2, \ldots, d_k$ such that

$$\Gamma \simeq \mathbb{Z}_{d_1} \oplus \mathbb{Z}_{d_2} \oplus \cdots \oplus \mathbb{Z}_{d_k},$$

where $\simeq$ denotes group isomorphism. The theorem goes on to say that there is a unique choice for the d_j for which $d_j \neq 1$ and d_j divides d_{j+1} for $1 \leqslant j \leqslant k-1$ (with the convention that every integer divides 0); the d_j's for this choice are called the *elementary divisors* of Γ. The theorem concludes by saying that two finitely generated abelian groups are isomorphic if and only if they have the same set of elementary divisors; i.e., the set of elementary divisors is a *complete* invariant of isomorphism for such groups.

Definition 7.4.15. Let A be an $r \times r$ integral matrix. The *Bowen–Franks group* of A is

$$BF(A) = \mathbb{Z}^r / \mathbb{Z}^r (Id - A),$$

where $\mathbb{Z}^r (Id - A)$ is the image of $\mathbb{Z}^r$ under the matrix $Id - A$ acting on the right.

In order to "compute" the Bowen–Franks group, we will use the Smith form (defined below) of an integral matrix. Define the *elementary operations over* $\mathbb{Z}$ on integral matrices to be:

(1) Exchange two rows or two columns.

(2) Multiply a row or column by -1.

(3) Add an integer multiple of one row to another row, or of one column to another column.

The matrices corresponding to these operations are invertible over $\mathbb{Z}$. Hence if C is obtained from B by an elementary operation over $\mathbb{Z}$, then $\mathbb{Z}^r/\mathbb{Z}^r C \simeq \mathbb{Z}^r/\mathbb{Z}^r B$ (see Exercise 7.4.10).

Every integral matrix can be transformed by a sequence of elementary operations over $\mathbb{Z}$ into a diagonal matrix

$$(7\text{–}4\text{–}3) \qquad \begin{bmatrix} d_1 & & & \\ & d_2 & & \\ & & \ddots & \\ & & & d_r \end{bmatrix},$$

where the $d_j \geqslant 0$ and d_j divides d_{j+1}. This is called the *Smith form* of the matrix (see [New]). If we put $Id - A$ into its Smith form $(7\text{–}4\text{–}3)$, then

$$BF(A) \simeq \mathbb{Z}_{d_1} \oplus \mathbb{Z}_{d_2} \oplus \cdots \oplus \mathbb{Z}_{d_r}.$$

By our convention, each summand with $d_j = 0$ is $\mathbb{Z}$, while summands with $d_j > 0$ are finite cyclic groups. Since $\mathbb{Z}_1$ is the trivial group, the elementary divisors of $BF(A)$ are the diagonal entries of the Smith form of $Id - A$ which are not 1.

Example 7.4.16. Let

$$A = \begin{bmatrix} 4 & 1 \\ 1 & 0 \end{bmatrix} \qquad \text{and} \qquad B = \begin{bmatrix} 3 & 2 \\ 2 & 1 \end{bmatrix}.$$

The following sequence of elementary operations over $\mathbb{Z}$ puts $Id - A$ into its Smith form.

$$Id - A = \begin{bmatrix} -3 & -1 \\ -1 & 1 \end{bmatrix} \to \begin{bmatrix} -4 & 0 \\ -1 & 1 \end{bmatrix} \to \begin{bmatrix} -4 & 0 \\ 0 & 1 \end{bmatrix} \to \begin{bmatrix} 4 & 0 \\ 0 & 1 \end{bmatrix} \to \begin{bmatrix} 1 & 0 \\ 0 & 4 \end{bmatrix}.$$

Thus $BF(A) \simeq \mathbb{Z}_1 \oplus \mathbb{Z}_4 \simeq \mathbb{Z}_4$. Similarly,

$$Id - B = \begin{bmatrix} -2 & -2 \\ -2 & 0 \end{bmatrix} \to \begin{bmatrix} -2 & -2 \\ 0 & -2 \end{bmatrix} \to \begin{bmatrix} -2 & 0 \\ 0 & -2 \end{bmatrix} \to \begin{bmatrix} 2 & 0 \\ 0 & 2 \end{bmatrix},$$

so that $BF(B) \simeq \mathbb{Z}_2 \oplus \mathbb{Z}_2$. Therefore $BF(A) \not\simeq BF(B)$. $\qquad\square$

Our motivation for introducing the Bowen–Franks groups is its invariance under shift equivalence. It is even invariant for shift equivalence over $\mathbb{Z}$.

Theorem 7.4.17. *If $A \sim_{\mathbb{Z}} B$ then $BF(A) \simeq BF(B)$.*

PROOF: Let m denote the size of A and n the size of B. Suppose that $(R, S): A \sim_{\mathbb{Z}} B$ (lag ℓ). Let $\mathbf{v} \in \mathbb{Z}^m$. Then

$$\mathbf{v}(Id - A)R = \mathbf{v}R - \mathbf{v}AR = \mathbf{v}R - \mathbf{v}RB = (\mathbf{v}R)(Id - B).$$

This shows that $(\mathbb{Z}^m(Id - A))R \subseteq \mathbb{Z}^n(Id - B)$. Hence R induces a well-defined map $\widehat{R}$ on quotients,

$$\widehat{R}: \mathbb{Z}^m/\mathbb{Z}^m(Id - A) \to \mathbb{Z}^n/\mathbb{Z}^n(Id - B),$$

or $\widehat{R}: BF(A) \to BF(B)$. Similarly, S induces $\widehat{S}: BF(B) \to BF(A)$. Now the map induced by A on the quotient $\mathbb{Z}^m/\mathbb{Z}^m(Id - A)$ is just the identity map, since the cosets of $\mathbf{v}$ and $\mathbf{v}A$ modulo $\mathbb{Z}^m(Id - A)$ are equal. Since $\widehat{S} \circ \widehat{R}: BF(A) \to BF(A)$ is induced by A^ℓ, it follows that $\widehat{S} \circ \widehat{R}$ is the identity on $BF(A)$. Similarly, $\widehat{R} \circ \widehat{S}$ is the identity on $BF(B)$. Thus $\widehat{R}$ is an isomorphism from $BF(A)$ to $BF(B)$.   □

By this theorem, the matrices in Example 7.4.16 are not shift equivalent over $\mathbb{Z}$. We have seen this pair before, in Example 7.3.4, where we gave a "bare-hands" argument that they are not shift equivalent.

There are polynomials in A other than $Id - A$ that also provide invariants of shift equivalence over $\mathbb{Z}$.

Definition 7.4.18. Let $\mathbb{Z}[t]$ denote the set of polynomials in the variable t with integer coefficients. Let A be an $n \times n$ integral matrix, and $p(t) \in \mathbb{Z}[t]$ such that $p(0) = \pm 1$. Define the *generalized Bowen–Franks group* $BF_p(A)$ to be $\mathbb{Z}^n/\mathbb{Z}^n p(A)$.

The Bowen–Franks group from Definition 7.4.15 corresponds to the choice $p(t) = 1 - t$. The generalization of Theorem 7.4.17 holds: If $A \sim_{\mathbb{Z}} B$ then $BF_p(A) \simeq BF_p(B)$ for all $p(t)$ with $p(0) = \pm 1$ (Exercise 7.4.8). For example, using $p(t) = 1 + t$ shows that $\mathbb{Z}^n/\mathbb{Z}^n(Id + A)$ is also an invariant of shift equivalence, and such further invariants can be useful (see Exercise 7.4.12).

Let A be an integral matrix. Then A and A^{T} have the same Jordan form away from zero (Exercise 7.4.5). Furthermore $BF_p(A) \simeq BF_p(A^{\mathsf{T}})$ for all $p(t)$ with $p(0) = \pm 1$ (Exercise 7.4.9). The following example shows, however, that A is not always shift equivalent (even over $\mathbb{Z}$) to A^{T}.

Example 7.4.19. Let $A = \begin{bmatrix} 19 & 5 \\ 4 & 1 \end{bmatrix}$. Using results in number theory about Pell's equation, Kollmer showed (see Parry and Tuncel [ParT2, pp. 80–81]) that A is not shift equivalent over $\mathbb{Z}$ to A^{T}. For another instance of this, see Example 12.3.2.   □

EXERCISES

7.4.1. Let $A = \begin{bmatrix} 1 & 1 \\ 1 & 1 \end{bmatrix} = \begin{bmatrix} 1 & 0 \\ 1 & 0 \end{bmatrix} \begin{bmatrix} 1 & 1 \\ 1 & 1 \end{bmatrix} = RS$ and $B = \begin{bmatrix} 2 & 0 \\ 2 & 0 \end{bmatrix} = SR$. Then A and B are shift equivalent, while A is primitive and B is not primitive. Why doesn't this contradict Proposition 7.4.1?

7.4.2.
(a) Show that $\begin{bmatrix} 5 & 1 \\ 0 & 1 \end{bmatrix}$ and $\begin{bmatrix} 4 & 1 \\ 3 & 2 \end{bmatrix}$ are similar over $\mathbb{Z}$ and hence shift equivalent over $\mathbb{Z}$ (so irreducibility is *not* invariant for shift equivalence over $\mathbb{Z}$).

(b) Show that if A is irreducible, B is primitive and $A \sim_{\mathbb{Z}} B$, then A is primitive.

7.4.3.
(a) Show that $\begin{bmatrix} 2 & 1 & 0 \\ 1 & 1 & 1 \\ 0 & 1 & 2 \end{bmatrix}$ is shift equivalent over $\mathbb{Z}$ to $\begin{bmatrix} 3 & 0 \\ 0 & 2 \end{bmatrix}$.

(b) Show that $\begin{bmatrix} 2 & 1 & 0 \\ 1 & 1 & 1 \\ 0 & 1 & 2 \end{bmatrix}$ is not shift equivalent (over $\mathbb{Z}^+$) to any matrix with nonzero determinant.

7.4.4. Show that if $p(t)$ is a polynomial in one variable with integral coefficients and constant term 1, and if $A \sim_{\mathbb{Z}} B$, then $\det p(A) = \det p(B)$. What if p has constant term -1?

7.4.5.
(a) Let λ be an eigenvalue of a matrix A. For each $j \geqslant 1$ let n_j denote the number of Jordan blocks for λ of size j appearing in the Jordan form of A, and let d_j denote the dimension of $\ker(A - \lambda Id)^j$. Show that $n_j = 2d_j - d_{j+1} - d_{j-1}$. Using this, show that the dimensions of $\ker(A - \lambda Id)^j$ determine the Jordan form of A.

(b) Show that a matrix and its transpose have the same Jordan form.

7.4.6. Decide whether $A \sim B$ where $A = \begin{bmatrix} 5 & 4 \\ 2 & 1 \end{bmatrix}$ and $B = \begin{bmatrix} 4 & 11 \\ 1 & 2 \end{bmatrix}$.

7.4.7. Show that matrices A and B in Example 7.4.8 have different Bowen–Franks groups, hence giving an alternative reason why they are not shift equivalent.

7.4.8. Show that the generalized Bowen–Franks group $BF_p(A)$ is an invariant of shift equivalence over $\mathbb{Z}$. [*Hint*: Show that if $p \in \mathbb{Z}[t]$ with $p(0) = \pm 1$ and L is a positive integer, then there is a polynomial $q \in \mathbb{Z}[t]$ such that p divides $t^L q - 1$.]

7.4.9. Show that if A is an integral matrix and $p(t)$ is a polynomial with integral coefficients and constant term ± 1, then $BF_p(A) = BF_p(A^{\mathsf{T}})$.

7.4.10. Verify that if E and F are integral $n \times n$ matrices and $\det E = \pm 1$, then

$$\mathbb{Z}^n/(\mathbb{Z}^n F) \simeq \mathbb{Z}^n/(\mathbb{Z}^n FE) \simeq \mathbb{Z}^n/(\mathbb{Z}^n EF).$$

7.4.11.
(a) Show that if $\det(Id - A) = \pm 1$, then $BF(A)$ is the trivial group.

(b) Show that if $\det(Id - A) = \det(Id - B) = \pm p$ where p is prime, then $BF(A) \simeq BF(B)$.

7.4.12. Show that $A = \begin{bmatrix} 5 & 3 \\ 3 & 5 \end{bmatrix}$ and $B = \begin{bmatrix} 6 & 8 \\ 1 & 4 \end{bmatrix}$ have the same Jordan form away from zero and the same Bowen–Franks group, but that $\mathbb{Z}^n/\mathbb{Z}^n(Id + A)$ is not isomorphic to $\mathbb{Z}^n/\mathbb{Z}^n(Id + B)$, so that A and B are not shift equivalent (by Exercise 7.4.8).

7.4.13. For an $n \times n$ integral matrix A, let δ_j be the greatest common divisor of all $j \times j$ minors of $Id - A$ (or 0 if all the minors vanish). Let $d_1 = \delta_1$ and $d_j = \delta_j/\delta_{j-1}$ for $2 \leqslant j \leqslant n$ (with the convention that $0/0 = 0$). Show that

$$BF(A) \simeq \mathbb{Z}_{d_1} \oplus \mathbb{Z}_{d_2} \oplus \cdots \oplus \mathbb{Z}_{d_n}.$$

Verify this fact when $A = \begin{bmatrix} 4 & 1 \\ 1 & 0 \end{bmatrix}$ and when $A = \begin{bmatrix} 3 & 2 \\ 2 & 1 \end{bmatrix}$.

7.4.14. Show that $A \sim [n]$ if and only if $\zeta_{\sigma_A}(t) = 1/(1 - nt)$. [*Hint*: Show that the zeta function condition holds if and only if A^k has rank one for sufficiently large k.]

7.4.15. Suppose that $A \sim B$. Show directly that there is an n_0 such that $\operatorname{tr} A^n = \operatorname{tr} B^n$ for all $n \geqslant n_0$. Conclude from this that $\operatorname{tr} A^n = \operatorname{tr} B^n$ for all n. Hence give an alternative proof that the zeta function is an invariant of shift equivalence. [*Hint*: Use the fact that the exponential function has no zeros.]

§7.5. Shift Equivalence and the Dimension Group

In the previous section we described two readily computable invariants of shift equivalence. In this section we will introduce an algebraic invariant called the dimension triple that completely captures shift equivalence. While it is not easy to compute this object in general, we will see that it can be easily computed in some special cases. We will also show how this object arises from the "internal structure" of an edge shift, ultimately showing that the matrix equations for shift equivalence follow from a natural idea.

We will associate to each integral matrix A an additive subgroup Δ_A of the eventual range $\mathcal{R}_A$ of A and an automorphism δ_A of Δ_A that is simply the restriction of A to Δ_A. If A is nonnegative, then we will also associate to A a "positive part" $\Delta_A^+ \subseteq \Delta_A$ that is closed under addition and preserved by δ_A. The triple $(\Delta_A, \Delta_A^+, \delta_A)$ completely captures shift equivalence: A and B are shift equivalent if any only if there is a group isomorphism from Δ_A to Δ_B that maps Δ_A^+ onto Δ_B^+ and intertwines δ_A with δ_B. In the proof of this, δ_A takes care of the necessary linear algebra, Δ_A takes care of integrality, and Δ_A^+ takes care of positivity.

Definition 7.5.1. Let A be an $r \times r$ integral matrix. The *dimension group* of A is

$$\Delta_A = \{\mathbf{v} \in \mathcal{R}_A : \mathbf{v}A^k \in \mathbb{Z}^r \quad \text{for some } k \geqslant 0\}.$$

The *dimension group automorphism* δ_A of A is the restriction of A to Δ_A, so that $\delta_A(\mathbf{v}) = \mathbf{v}A$ for $\mathbf{v} \in \Delta_A$. The *dimension pair* of A is (Δ_A, δ_A).

If A is also nonnegative, then we define the *dimension semigroup* of A to be

$$\Delta_A^+ = \{\mathbf{v} \in \mathcal{R}_A : \mathbf{v}A^k \in (\mathbb{Z}^+)^r \quad \text{for some } k \geqslant 0\}.$$

The *dimension triple* of A is $(\Delta_A, \Delta_A^+, \delta_A)$.

We will use group isomorphisms that preserve all the relevant structure. Thus when we write $\theta: (\Delta_A, \delta_A) \simeq (\Delta_B, \delta_B)$, we mean that $\theta: \Delta_A \to \Delta_B$ is a group isomorphism such that $\theta \circ \delta_A = \delta_B \circ \theta$ (i.e., θ intertwines δ_A with δ_B). An isomorphism $\theta: (\Delta_A, \Delta_A^+, \delta_A) \simeq (\Delta_B, \Delta_B^+, \delta_B)$ of triples adds the condition that $\theta(\Delta_A^+) = \Delta_B^+$.

Since A is integral, once $\mathbf{v}A^k \in \mathbb{Z}^r$ it follows that $\mathbf{v}A^n \in \mathbb{Z}^r$ for all $n \geqslant k$. This observation shows that the sum and difference of two elements in Δ_A is again in Δ_A, so that Δ_A is an additive group. If A is nonnegative as well, once $\mathbf{v}A^k \in (\mathbb{Z}^+)^r$, then $\mathbf{v}A^n \in (\mathbb{Z}^+)^r$ for all $n \geqslant k$. This shows that Δ_A^+ is closed under addition. It also contains $\mathbf{0}$ by its definition. Recall that $A^\times$ is defined to be the restriction of A to $\mathcal{R}_A$, and is invertible. For every $\mathbf{v} \in \Delta_A$, it is easy to check that $\mathbf{v}(A^\times)^{-1} \in \Delta_A$, and that $\delta_A(\mathbf{v}(A^\times)^{-1}) = \mathbf{v}$, so that δ_A maps Δ_A onto itself. Since δ_A is the restriction of the one-to-one map $A^\times$, it follows that δ_A is indeed an automorphism of Δ_A. It also follows directly that $\delta_A(\Delta_A^+) = \Delta_A^+$.

Example 7.5.2. (a) Let $A = [2]$. Then $v \in \Delta_A$ if and only if $vA^k = 2^k v \in \mathbb{Z}$ for some $k \geqslant 0$. This occurs exactly when $v = p/2^k$, i.e., when v is a *dyadic rational*. We denote the set of dyadic rationals by $\mathbb{Z}[1/2]$. Then $\Delta_A = \mathbb{Z}[1/2]$, $\Delta_A^+ = \mathbb{Z}[1/2] \cap \mathbb{R}^+$, and δ_A is multiplication by 2.

(b) Let $B = \begin{bmatrix} 1 & 1 \\ 1 & 1 \end{bmatrix}$. Then $\mathcal{R}_B = \mathbb{Q} \cdot [1,1]$, and on $\mathcal{R}_B$ the action of B is multiplication by 2. Using reasoning similar to (a), we obtain that $\Delta_B = \mathbb{Z}[1/2] \cdot [1,1]$, $\Delta_B^+ = \Delta_B \cap (\mathbb{R}^+)^2$, and δ_B is multiplication by 2.

Note that the map $\theta: \Delta_A \to \Delta_B$ defined by $\theta(v) = [v,v]$ defines an isomorphism $\theta: (\Delta_A, \Delta_A^+, \delta_A) \simeq (\Delta_B, \Delta_B^+, \delta_B)$. This is no accident: here $A \sim B$, and we will see that the dimension triple is invariant under shift equivalence. □

Example 7.5.3. Let $A = \begin{bmatrix} 1 & 1 \\ 1 & 0 \end{bmatrix}$. Since $\det A = -1$, we see that A is invertible over $\mathbb{Z}$. Hence if $\mathbf{v}A^k \in \mathbb{Z}^2$, then $\mathbf{v} \in \mathbb{Z}^2$. Thus $\Delta_A = \mathbb{Z}^2$.

A right Perron eigenvector for A is

$$\mathbf{z} = \begin{bmatrix} \lambda_A \\ 1 \end{bmatrix}.$$

Lemma 7.3.8 shows that $\mathbf{v}A^k$ is eventually positive if and only if $\mathbf{v} \cdot \mathbf{z} > 0$ (note that the notation in that lemma is different). Similarly, $\mathbf{v}A^k$ is eventually negative if and only if $\mathbf{v} \cdot \mathbf{z} < 0$. Since there are no rational vectors orthogonal to $\mathbf{z}$ except $\mathbf{0}$, it follows that

$$\Delta_A^+ = \{\mathbf{v} \in \mathbb{Z}^2 : \mathbf{v} \cdot \mathbf{z} > 0\} \cup \{\mathbf{0}\}.$$

We can picture Δ_A^+ as the set of lattice points in a half-plane (see Figure 7.5.1). Finally, δ_A is just the restriction of A to $\mathbb{Z}^2$. □

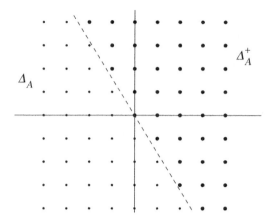

FIGURE 7.5.1.   The dimension group of the golden mean shift.

Remark 7.5.4. The previous example is a special case of the observation that if A is an $r \times r$ integral matrix with $\det A = \pm 1$, then $\Delta_A = \mathbb{Z}^r$ and $\delta_A = A$.   □

Example 7.5.5. Let

$$A = \begin{bmatrix} 2 & 1 & 0 \\ 1 & 1 & 1 \\ 0 & 1 & 2 \end{bmatrix}.$$

The eigenvalues of A are 3, 2, and 0, with corresponding eigenvectors $\mathbf{u} = [1, 1, 1]$, $\mathbf{v} = [1, 0, -1]$, and $\mathbf{w} = [1, -2, 1]$. Hence $\mathcal{R}_A = \mathbb{Q}\mathbf{u} \oplus \mathbb{Q}\mathbf{v}$. Let $s, t \in \mathbb{Q}$. Then

$$(s\mathbf{u} + t\mathbf{v})A^k = 3^k s\mathbf{u} + 2^k t\mathbf{v} = [3^k s + 2^k t, \, 3^k s, \, 3^k s - 2^k t]$$

is in $\mathbb{Z}^3$ if and only if $3^k s \in \mathbb{Z}$ and $2^k t \in \mathbb{Z}$. Hence $s\mathbf{u} + t\mathbf{v} \in \Delta_A$ if and only if $s \in \mathbb{Z}[1/3]$ and $t \in \mathbb{Z}[1/2]$. Thus

$$\Delta_A = \mathbb{Z}[1/3]\mathbf{u} \oplus \mathbb{Z}[1/2]\mathbf{v} = \Big\{ [s + t, s, s - t] : s \in \mathbb{Z}[1/3] \text{ and } t \in \mathbb{Z}[1/2] \Big\}.$$

Since $\mathbf{u}^\mathsf{T}$ is a right Perron eigenvector for A and $\mathbf{v} \cdot \mathbf{u}^\mathsf{T} = 0$, it follows from Lemma 7.3.8 that

$$\Delta_A^+ = \{ s\mathbf{u} + t\mathbf{v} \in \Delta_A : s > 0 \} \cup \{\mathbf{0}\}.$$

Finally, $\delta_A(s\mathbf{u} + t\mathbf{v}) = 3s\mathbf{u} + 2t\mathbf{v}$.   □

In the preceding example, Δ_A turned out to be the direct sum of two groups, each of which lies in an eigenspace. Other dimension groups can be more complicated; for instance, see Exercise 7.5.2(a).

Next, we establish some elementary properties of the dimension triple.

Proposition 7.5.6. *Let A and B be integral matrices.*

(1) Δ_A *spans* $\mathcal{R}_A$ *as a rational vector space.*

(2) *Every group homomorphism from* Δ_A *to* Δ_B *extends uniquely to a linear transformation from* $\mathcal{R}_A$ *to* $\mathcal{R}_B$.

(3) *If A is nonnegative, then* Δ_A^+ *generates* Δ_A *as a group.*

PROOF: (1) Let A have size r, and $\mathbf{v} \in \mathcal{R}_A$. Then there is an integer n such that $n\mathbf{v} \in \mathbb{Z}^r$. Hence $n\mathbf{v} \in \Delta_A$, so that $\mathbf{v}$ is a rational multiple of an element in Δ_A.

(2) Let $\theta: \Delta_A \to \Delta_B$ be a group homomorphism. By (1), we can pick elements of Δ_A that form a rational basis for $\mathcal{R}_A$. Extend θ to $\mathcal{R}_A$ using linearity over $\mathbb{Q}$. It is then routine to show that the extension is well-defined and unique.

(3) Let A have size r, and $\mathbf{v} \in \Delta_A$. Choose k so that $\mathbf{v}A^k \in \mathbb{Z}^r$. Write $\mathbf{v}A^k = \mathbf{u} - \mathbf{w}$, where $\mathbf{u}, \mathbf{w} \in (\mathbb{Z}^+)^r$. Then $\mathbf{v}A^{k+r} = \mathbf{u}A^r - \mathbf{w}A^r$, and $\mathbf{u}A^r, \mathbf{w}A^r \in \Delta_A^+$, so we have expressed $\mathbf{v}A^{k+r}$ as a difference of elements in Δ_A^+. Since Δ_A^+ is δ_A-invariant, applying δ_A^{-k-r} gives the result. □

The next result shows that the dimension pair captures shift equivalence over $\mathbb{Z}$.

Theorem 7.5.7. *Let A and B be integral matrices. Then* $A \sim_{\mathbb{Z}} B$ *if and only if* $(\Delta_A, \delta_A) \simeq (\Delta_B, \delta_B)$.

PROOF: First suppose that $A \sim_{\mathbb{Z}} B$. Let m denote the size of A and n the size of B. Let $(R, S): A \sim_{\mathbb{Z}} B$ (lag ℓ). The proof of Theorem 7.4.6 shows that the maps $\widetilde{R}: \mathcal{R}_A \to \mathcal{R}_B$ and $\widetilde{S}: \mathcal{R}_B \to \mathcal{R}_A$ defined by $\widetilde{R}(\mathbf{v}) = \mathbf{v}R$ and $\widetilde{S}(\mathbf{w}) = \mathbf{w}S$ are linear isomorphisms intertwining $A^\times$ with $B^\times$, so that the following diagram commutes.

$$
\begin{array}{ccc}
\mathcal{R}_A & \xrightarrow{\ A^\times\ } & \mathcal{R}_A \\
{\scriptstyle\widetilde{R}}\Big\downarrow & & \Big\downarrow{\scriptstyle\widetilde{R}} \\
\mathcal{R}_B & \xrightarrow{\ B^\times\ } & \mathcal{R}_B \\
{\scriptstyle\widetilde{S}}\Big\downarrow & & \Big\downarrow{\scriptstyle\widetilde{S}} \\
\mathcal{R}_A & \xrightarrow{\ A^\times\ } & \mathcal{R}_A
\end{array}
$$

Suppose that $\mathbf{v} \in \Delta_A$. Then there is a $k > 0$ such that $\mathbf{v}A^k \in \mathbb{Z}^m$. Since R is integral, we see that

$$(\mathbf{v}R)B^k = \mathbf{v}A^k R \in \mathbb{Z}^n,$$

so that $\widetilde{R}(\mathbf{v}) \in \Delta_B$. Hence $\widetilde{R}(\Delta_A) \subseteq \Delta_B$, and similarly $\widetilde{S}(\Delta_B) \subseteq \Delta_A$. But $\widetilde{S} \circ \widetilde{R} = (A^\times)^\ell$ and $\widetilde{R} \circ \widetilde{S} = (B^\times)^\ell$ are linear isomorphisms. Hence $\widetilde{R}$ is a group isomorphism intertwining δ_A with δ_B.

Conversely, suppose that $\theta\colon (\Delta_A, \delta_A) \simeq (\Delta_B, \delta_B)$. Since $\mathbf{e}_I A^m \in \Delta_A$, it follows that

$$\theta(\mathbf{e}_I A^m) \in \Delta_B.$$

Hence there is a k such that for $1 \leqslant I \leqslant m$ we have

$$\theta(\mathbf{e}_I A^m) B^k \in \mathbb{Z}^n.$$

Let R be the $m \times n$ matrix whose Ith row is $\theta(\mathbf{e}_I A^m) B^k$. Similarly, there is an ℓ such that for each J, $\theta^{-1}(\mathbf{e}_J B^n) A^\ell \in \mathbb{Z}^m$. Let S be the $n \times m$ matrix whose Jth row is $\theta^{-1}(\mathbf{e}_J B^n) A^\ell$. It is now routine to check that $(R, S)\colon A \sim_\mathbb{Z} B$. $\qquad\square$

By including the positive part of the dimension group, we capture shift equivalence (over $\mathbb{Z}^+$).

Theorem 7.5.8. *Let A and B be nonnegative integral matrices. Then $A \sim B$ if and only if $(\Delta_A, \Delta_A^+, \delta_A) \simeq (\Delta_B, \Delta_B^+, \delta_B)$.*

PROOF: Let $(R, S)\colon A \sim B$. Since R and S are nonnegative, it is easy to verify that the isomorphism $\widetilde{R}\colon \Delta_A \to \Delta_B$ defined in the proof of Theorem 7.5.7 maps Δ_A^+ into Δ_B^+, and similarly that $\widetilde{S}(\Delta_B^+) \subseteq \Delta_A^+$. Now $\widetilde{R} \circ \widetilde{S} = \delta_B^\ell$ maps Δ_B^+ onto itself, so that $\widetilde{R}$ must map Δ_A^+ onto Δ_B^+. Hence $\widetilde{R}\colon (\Delta_A, \Delta_A^+, \delta_A) \simeq (\Delta_B, \Delta_B^+, \delta_B)$.

Conversely, suppose that $(\Delta_A, \Delta_A^+, \delta_A) \simeq (\Delta_B, \Delta_B^+, \delta_B)$. If, in the proof of Theorem 7.5.7, we use large enough powers of A and B in the construction of the shift equivalence $(R, S)\colon A \sim_\mathbb{Z} B$, then the R and S can be made nonnegative. $\qquad\square$

A simple but useful consequence of this is the following.

Corollary 7.5.9. *Let A and B be primitive integral matrices. Then $A \sim B$ if and only if $(\Delta_A, \delta_A) \simeq (\Delta_B, \delta_B)$.*

PROOF: The previous theorem implies that if $A \sim B$, then $(\Delta_A, \delta_A) \simeq (\Delta_B, \delta_B)$. Conversely, if $(\Delta_A, \delta_A) \simeq (\Delta_B, \delta_B)$, then we see by Theorem 7.5.7 that $A \sim_\mathbb{Z} B$, from which it follows that $A \sim B$ by Theorem 7.3.6. $\qquad\square$

Observe that Theorem 7.5.8 implies that conjugate edge shifts must have isomorphic dimension triples. We will see that this is a natural consequence of the following alternative, more "internal," way to define the dimension triple.

Let X_A be an edge shift and σ_A its associated shift map. To avoid some trivial technicalities, we assume that A is irreducible and $\lambda_A > 1$. We will construct a semigroup D_A^+ from certain compact subsets of X_A, and D_A will be the group that it generates. The action of the shift map on these subsets will define an automorphism $d_A\colon D_A \to D_A$ that preserves D_A^+.

An *m-ray* is a subset of X_A of the form

$$R(x, m) = \{y \in X_A : y_{(-\infty, m]} = x_{(-\infty, m]}\},$$

where $x \in X_A$ and $m \in \mathbb{Z}$. An *m-beam* is a finite union of m-rays. Since m-rays are either disjoint or equal, we will always assume that the m-rays in an m-beam are disjoint. Note that an m-beam is also an n-beam for all $n \geqslant m$. For example, if $x = 0^\infty \in X_{[2]}$, then

$$R(x, 0) = R(x^{(1)}, 2) \cup R(x^{(2)}, 2) \cup R(x^{(3)}, 2) \cup R(x^{(4)}, 2),$$

where $x^{(1)} = 0^\infty.000^\infty$, $x^{(2)} = 0^\infty.010^\infty$, $x^{(3)} = 0^\infty.100^\infty$, $x^{(4)} = 0^\infty.110^\infty$, which expresses a 0-beam as a 2-beam. A *ray* is an m-ray for some m, and a *beam* is an m-beam for some m.

Let r be the size of A, and let

$$U = \bigcup_{j=1}^{k} R(x^{(j)}, m)$$

be an m-beam. Let $\mathbf{v}_{U,m} \in \mathbb{Z}^r$ be the vector whose Ith component is the number of $x^{(j)}$ with $t(x_m^{(j)}) = I$, i.e., the number of $x^{(j)}$ with $x^{(j)}_{(-\infty, m]}$ ending at state I. Writing U as an $(m + 1)$-beam shows that

$$\mathbf{v}_{U,m+1} = \mathbf{v}_{U,m} A,$$

and more generally that

(7–5–1) $$\mathbf{v}_{U,m+k} = \mathbf{v}_{U,m} A^k$$

for all $k \geqslant 0$.

Also, observe that for all m,

$$\mathbf{v}_{U,m+1} = \mathbf{v}_{\sigma_A(U),m}.$$

Declare two beams U and V to be equivalent if $\mathbf{v}_{U,m} = \mathbf{v}_{V,m}$ for some m. In this case, equation (7–5–1) shows that $\mathbf{v}_{U,n} = \mathbf{v}_{V,n}$ for all $n \geqslant m$. The reader should verify that this is a legitimate equivalence relation. Let $[U]$ denote the equivalence class of a beam U. Note that, typically, each class contains uncountably many equivalent beams, so passing from beams to their equivalence classes represents an enormous collapse.

Let D_A^+ denote the set of equivalence classes of beams. Define an addition operation on D_A^+ as follows. Since X_A is irreducible and $\lambda_A > 1$, G_A contains a cycle with an incoming edge not on the cycle; it follows that there are infinitely many distinct rays with the same terminal state (why?). Thus, whenever U and V are beams, there are beams U' equivalent to U and V' equivalent to V such that $U' \cap V' = \varnothing$, and we put

$$[U] + [V] = [U' \cup V'].$$

The reader should check that this operation is well-defined. Just as the group of integers can be obtained from the nonnegative integers $\{0, 1, 2, \ldots\}$ by using formal differences, we can construct a group D_A from D_A^+. The elements of D_A are formal differences $[U] - [V]$, where we consider two such differences $[U] - [V]$ and $[U'] - [V']$ to be equal if $[U] + [V'] = [U'] + [V]$. Finally, we define a map $d_A \colon D_A \to D_A$ by

$$d_A([U] - [V]) = [\sigma_A(U)] - [\sigma_A(V)].$$

It is routine to check that d_A is a well-defined automorphism of D_A that preserves D_A^+.

Example 7.5.10. Let $A = [2]$. If R is an m-ray in X_A, then $\mathbf{v}_{R,m} = 1$, so all m-rays are equivalent. Also, $d_A([R]) = [\sigma_A(R)] = 2[R]$ since $\sigma_A(R)$ is an $(m-1)$-ray that is the union of two m-rays each equivalent to R. Since beams are unions of rays, we see that $d_A([U]) = 2[U]$ for all beams U.

If U is an m-beam that is the disjoint union of p m-rays, define $\theta([U]) = p/2^m$. It is easy to check that this gives a well-defined group homomorphism $\theta \colon D_A \to \mathbb{Z}[1/2]$ such that $\theta(D_A^+) = \mathbb{Z}[1/2] \cap \mathbb{R}^+$. Furthermore, θ intertwines d_A with the map $\times 2$ that multiplies a number by 2. Because all m-rays are equivalent, we see that θ is an isomorphism from (D_A, D_A^+, d_A) to $(\mathbb{Z}[1/2], \mathbb{Z}[1/2] \cap \mathbb{R}^+, \times 2)$, the latter being the dimension triple of A.   □

Example 7.5.11. Let $A = \begin{bmatrix} 1 & 1 \\ 1 & 1 \end{bmatrix}$. Every m-ray R in X_A can be expressed as the union of two $(m+1)$-rays, one ending at each state in G_A. Thus $\mathbf{v}_{R,m+1} = [1, 1]$, so all m-rays are equivalent. Furthermore, $d_A([R]) = 2[R]$ as in the previous example. If U is a disjoint union of p m-rays, define $\theta([U]) = (p/2^m) \cdot [1, 1]$. The same arguments as before show that θ is an isomorphism between

$$(D_A, D_A^+, d_A) \quad \text{and} \quad (\mathbb{Z}[1/2] \cdot [1,1], (\mathbb{Z}[1/2] \cdot [1,1]) \cap (\mathbb{R}^+)^2, \times 2),$$

the latter being the dimension triple of A.   □

Example 7.5.12. Let $A = \begin{bmatrix} 1 & 1 \\ 1 & 0 \end{bmatrix}$. Let $x, y \in X_A$ such that $x_{(-\infty, 0]}$ ends at state 1 and that $y_{(-\infty, 0]}$ ends at state 2. Define two basic rays $R_1 = R(x, 0)$ and $R_2 = R(y, 0)$. Then every 0-ray is equivalent to either R_1 or R_2. Observe that $\mathbf{v}_{R_1, 0} = [1, 0]$ and $\mathbf{v}_{R_2, 0} = [0, 1]$. The equation (7-5-1) combined with the fact that A^m has distinct rows (see equation (4-1-2)) shows that R_1 and R_2 are not equivalent. The class of each 0-beam is a nonnegative integral combination of $[R_1]$ and $[R_2]$. Since A is invertible over $\mathbb{Z}$, any element of D_A^+ (i.e., equivalence class of m-beams) may be identified with an integral combination $a[R_1] + b[R_2]$ where $(a, b) \in \cup_{n=0}^{\infty} A^{-n}((\mathbb{Z}^+)^2)$ and

$$D_A = (\mathbb{Z} \cdot [R_1]) + (\mathbb{Z} \cdot [R_2]) \cong \mathbb{Z}^2.$$

The map $\theta: D_A \to \mathbb{Z}^2 = \Delta_A$ defined by $\theta(a \cdot [R_1] + b \cdot [R_2]) = [a, b]$ gives an isomorphism that intertwines d_A with A and maps D_A^+ to Δ_A^+. Thus $\theta: (D_A, D_A^+, d_A) \simeq (\Delta_A, \Delta_A^+, \delta_A)$. $\qquad\square$

The following result shows that, as suggested by the preceding examples, our "internal" description of the dimension triple is accurate:

Theorem 7.5.13. *Let A be an irreducible integral matrix with $\lambda_A > 1$. Then*
$$(D_A, D_A^+, d_A) \simeq (\Delta_A, \Delta_A^+, \delta_A).$$

PROOF: Let U be an m-beam in X_A, and $\mathbf{v}_{U,m} \in \mathbb{Z}^r$ its associated vector. Then $\mathbf{v}_{U,m} A^r \in \Delta_A^+$, and so $\delta_A^{-r}(\mathbf{v}_{U,m} A^r) \in \Delta_A^+$. We define $\theta: D_A^+ \to \Delta_A^+$ by $\theta([U]) = \delta_A^{-r-m}(\mathbf{v}_{U,m} A^r)$. Note that $\theta([U]) = \delta_A^{-s-m}(\mathbf{v}_{U,m} A^s)$ for all $s \geqslant r$. Extend θ to D_A by $\theta([U] - [V]) = \theta([U]) - \theta([V])$. It is routine to check that θ is well-defined. Now

$$\theta(d_A([U])) = \theta([\sigma(U)]) = \delta_A^{-r-m}(\mathbf{v}_{\sigma(U),m} A^r) = \delta_A^{-r-m}(\mathbf{v}_{U,m} A^{r+1})$$
$$= \delta_A(\delta_A^{-(r+m+1)}(\mathbf{v}_{U,m} A^{r+1})) = \delta_A(\theta([U])),$$

so that θ intertwines d_A with δ_A. Every vector in Δ_A^+ is clearly the image of a beam, so that $\theta(D_A^+) = \Delta_A^+$. Since Δ_A is generated by Δ_A^+ from Proposition 7.5.6(3), it follows that θ is onto. Finally, the definition of equivalence of beams shows that θ is one-to-one. $\qquad\square$

Let X_A and X_B be conjugate edge shifts. Then $A \sim B$, so that Theorems 7.5.8 and 7.5.13 imply that

$$(7\text{-}5\text{-}2) \qquad\qquad (D_A, D_A^+, d_A) \simeq (D_B, D_B^+, d_B).$$

There is a more direct "internal" way to see why the isomorphism (7–5–2) must hold. Let $\phi: X_A \to X_B$ be a conjugacy. An argument similar to the proof of Theorem 2.1.10 shows that if U is a beam in X_A, then $\phi(U)$ is a beam in X_B and ϕ preserves equivalence between beams (Exercise 7.5.9). Then the map $[U] \mapsto [\phi(U)]$ gives the required isomorphism. This points up the advantage of our "internal" description: invariance of the dimension triple under conjugacy is a natural consequence, not merely a computational accident.

Since matrices that are strong shift equivalent have isomorphic dimension triples, we can define the dimension triple of a shift of finite type X to be $(\Delta_A, \Delta_A^+, \delta_A)$ for any matrix A for which X_A is conjugate to X. But Theorem 7.5.8 is needed to make sure this does not depend on the particular A used. On the other hand, it is possible to define an "internal" dimension triple for an arbitrary shift space (in particular, a shift of finite type) that is a conjugacy invariant and agrees, up to isomorphism, with the dimension triple of an edge shift (Exercise 7.5.10).

Suppose that $A \sim B$, and let $(R, S): A \sim B$ (lag ℓ). Then for $m \geqslant 0$ we have $(R, SA^m): A^{\ell+m} \approx B^{\ell+m}$. Thus if $A \sim B$, then X_A^n is conjugate to X_B^n for all large enough n. Let us give this property a name.

Definition 7.5.14. Shift spaces X and Y are *eventually conjugate* if X^n is conjugate to Y^n for all sufficiently large n.

For shift spaces X and Y, any conjugacy from X to Y naturally defines a conjugacy from X^n to Y^n for all n (why?). Thus conjugate shifts are eventually conjugate. But eventual conjugacy does not always imply conjugacy, even for irreducible shifts of finite type [KimR12].

Our discussion above shows that if $A \sim B$, then X_A and X_B are eventually conjugate. Our final goal in this section is to prove the converse.

Theorem 7.5.15. *Edge shifts X_A and X_B are eventually conjugate if and only if their defining matrices A and B are shift equivalent.*

PROOF: We have already shown that if $A \sim B$, then X_A and X_B are eventually conjugate.

To prove the converse, suppose that X_A and X_B are eventually conjugate, so there is an n_0 such that $X_A^n \cong X_B^n$ for all $n \geqslant n_0$. Since the dimension triple is a conjugacy invariant, we see that

$$(\Delta_{A^n}, \Delta_{A^n}^+, \delta_{A^n}) \simeq (\Delta_{B^n}, \Delta_{B^n}^+, \delta_{B^n})$$

for $n \geqslant n_0$. A moment's reflection shows that $\Delta_{A^n} = \Delta_A$, $\Delta_{A^n}^+ = \Delta_A^+$, and $\delta_{A^n} = \delta_A^n$, so for each $n \geqslant n_0$ there is an isomorphism

$$\theta_n: (\Delta_A, \Delta_A^+, \delta_A^n) \simeq (\Delta_B, \Delta_B^+, \delta_B^n).$$

Let $\psi_n = \theta_n^{-1} \circ \delta_B \circ \theta_n$. Then δ_A and ψ_n are automorphisms of Δ_A such that $\delta_A^n = \psi_n^n$. By Proposition 7.5.6(2) we can extend both δ_A and ψ_n to $\mathcal{R}_A$, and the uniqueness of this extension shows that $\delta_A^n = \psi_n^n$ on $\mathcal{R}_A$. We would like to be able to conclude that $\delta_A = \psi_n$, for then θ_n would be an isomorphism from $(\Delta_A, \Delta_A^+, \delta_A)$ to $(\Delta_B, \Delta_B^+, \delta_B)$, and A would then be shift equivalent to B. To prove that $\delta_A = \psi_n$ for some $n \geqslant n_0$, we use the following result from linear algebra.

Lemma 7.5.16. *Let $n \geqslant 1$ and C, D be nonsingular matrices such that $C^n = D^n$. Suppose that whenever λ is an eigenvalue of C and μ is an eigenvalue of D such that $\lambda^n = \mu^n$, we have that $\lambda = \mu$. Then $C = D$.*

PROOF: Suppose that $\mathbf{v}$ is a left eigenvector for C with eigenvalue λ, so that $\mathbf{v}C = \lambda\mathbf{v}$. Let $\omega = \exp(2\pi i/n)$, and

$$q(t) = \prod_{j=1}^{n-1} (t - \omega^j \lambda) = \frac{t^n - \lambda^n}{t - \lambda}.$$

Then

$$0 = \mathbf{v}(C^n - \lambda^n) = \mathbf{v}(D^n - \lambda^n) = \mathbf{v}(D - \lambda)\left(\prod_{j=1}^{n-1}(D - \omega^j \lambda)\right).$$

Now our assumption on eigenvalues shows that for $1 \leqslant j \leqslant n-1$ each $\omega^j \lambda$ is not an eigenvalue of D, so that $D - \omega^j \lambda$ is invertible. It follows that $\mathbf{v}D = \lambda\mathbf{v}$, so that C and D agree on the subspace spanned by $\mathbf{v}$. The argument up to now suffices to prove the lemma when C is diagonalizable, i.e., has a basis of eigenvectors.

The proof is more intricate when the Jordan form of C contains blocks larger than size one. We illustrate the basic idea for a Jordan block of size two, leaving the general case as Exercise 7.5.5.

If the Jordan form of C contains $J_2(\lambda)$, there are vectors $\mathbf{v}$ and $\mathbf{w}$ such that $\mathbf{v}C = \lambda\mathbf{v}$ and $\mathbf{w}C = \lambda\mathbf{w} + \mathbf{v}$. Our previous argument shows that $\mathbf{v}D = \lambda\mathbf{v}$. We will prove that $\mathbf{w}D = \lambda\mathbf{w} + \mathbf{v}$, hence that C agrees with D on the subspace spanned by $\mathbf{v}$ and $\mathbf{w}$.

Note that since $\mathbf{w}(C - \lambda) = \mathbf{v}$, applying $q(C)$ gives

$$\mathbf{w}(C^n - \lambda^n) = \mathbf{w}(C - \lambda)q(C) = \mathbf{v}q(C).$$

Since $C^n = D^n$, we see that

$$\mathbf{w}(D - \lambda)q(D) = \mathbf{w}(D^n - \lambda^n) = \mathbf{w}(C^n - \lambda^n) = \mathbf{v}q(C).$$

As before, $q(D)$ is invertible, so that

$$\mathbf{w}(D - \lambda) = \mathbf{v}q(C)q(D)^{-1}.$$

But on the subspace spanned by $\mathbf{v}$, both C and D act as multiplication by λ. Hence

$$\mathbf{v}q(C)q(D)^{-1} = q(\lambda)q(\lambda)^{-1}\mathbf{v} = \mathbf{v}.$$

This shows that $\mathbf{w}D = \lambda\mathbf{w} + \mathbf{v}$, completing the proof of the lemma (when the Jordan blocks are of size two). □

We can now complete the proof of Theorem 7.5.15. As linear transformations on $\mathcal{R}_A$, the spectrum of δ_A is the nonzero spectrum of A, while the spectrum of ψ_n is the nonzero spectrum of B since ψ_n is similar to δ_B. Let λ and μ be nonzero eigenvalues of A and B respectively. If $\lambda \neq \mu$, then $\{k : \lambda^k = \mu^k\}$ is either $\{0\}$ or $m_{\lambda,\mu}\mathbb{Z}$, where $m_{\lambda,\mu} \geqslant 2$. Let M be the product of the $m_{\lambda,\mu}$ over all distinct pairs λ, μ for which $\{k : \lambda^k = \mu^k\} \neq \{0\}$. If there is no such pair, set $M = 1$. Put $n = n_0 M + 1 \geqslant n_0$. Then for each pair λ, μ of nonzero eigenvalues of A, B, either $\lambda = \mu$ or $\lambda^n \neq \mu^n$. We can then apply Lemma 7.5.16 to the extensions C of δ_A and D of ψ_n to conclude that $\delta_A = \psi_n$. □

We can summarize the implications of this chapter in the following diagram.

$$\begin{array}{ccc}
 & \text{Thm 2.7} & \\
A \approx B & \Longleftrightarrow & X_A \cong X_B
\end{array}$$

$$\Big\Downarrow \text{Thm 3.3}$$

$$\begin{array}{ccccc}
 & \text{Thm 5.8} & & \text{Thm 5.15} & \\
(\Delta_A, \Delta_A^+, \delta_A) \simeq (\Delta_B, \Delta_B^+, \delta_B) & \Longleftrightarrow & A \sim B & \Longleftrightarrow & X_A, X_B \; \begin{array}{l}\text{eventually}\\ \text{conjugate}\end{array}
\end{array}$$

$$\Big\Downarrow \quad \begin{array}{l}\text{if } A, B \text{ primitive}\\ \text{Thm 3.6}\end{array}$$

$$\begin{array}{ccc}
 & \text{Thm 5.7} & \\
(\Delta_A, \delta_A) \simeq (\Delta_B, \delta_B) & \Longleftrightarrow & A \sim_{\mathbb{Z}} B
\end{array}$$

$$\begin{array}{ccc}
\text{Thm 4.17} & & \text{Thm 4.10}\\
BF(A) \simeq BF(B) & & J^\times(A) = J^\times(B)
\end{array}$$

$$\begin{array}{ccc}
 & \text{Cor 6.4.7} & \\
\mathrm{sp}^\times(A) = \mathrm{sp}^\times(B) & \Longleftrightarrow & \zeta_{\sigma_A}(t) = \zeta_{\sigma_B}(t)
\end{array}$$

$$\mathrm{per}(A) = \mathrm{per}(B) \qquad\qquad h(X_A) = h(X_B)$$

EXERCISES

7.5.1. Find the dimension triples of the following matrices.

(a) $\begin{bmatrix} 4 & 1 \\ 1 & 0 \end{bmatrix}$, (b) $\begin{bmatrix} 8 & 2 \\ 2 & 0 \end{bmatrix}$.

7.5.2. (a) Show that for $A = \begin{bmatrix} 2 & 1 \\ 1 & 2 \end{bmatrix}$,

$$\Delta_A = \{[x, x+y] : x \in \mathbb{Z}[1/3], y \in \mathbb{Z}\} = \{s\mathbf{u} + t\mathbf{v} : t \in \mathbb{Z}/2, \; s - t \in \mathbb{Z}[1/3]\},$$

where $\{\mathbf{u}, \mathbf{v}\}$ is a basis of eigenvectors for A, and explain why Δ_A is *not* a direct sum of subsets of eigenspaces.

(b) Find the dimension groups of

(i) $\begin{bmatrix} 1 & 2 \\ 2 & 1 \end{bmatrix}$, (ii) $\begin{bmatrix} 3 & 2 \\ 4 & 1 \end{bmatrix}$.

7.5.3. Show that the addition $[U] + [V]$ of equivalence classes of beams is well-defined and that d_A is a well-defined automorphism of D_A that preserves D_A^+.

7.5.4. Fill in the details of the proofs of

(a) Theorem 7.5.7.

(b) Theorem 7.5.8.

(c) Theorem 7.5.13.

7.5.5. Prove Lemma 7.5.16 when there may be Jordan blocks of arbitrary size.

7.5.6. The *direct limit* L_A of an $n \times n$ integral matrix A is defined to be the set $\mathbb{Z}^n \times \mathbb{Z}$ modulo the equivalence relation generated by declaring $(\mathbf{v}, j)$ to be equivalent to $(\mathbf{v}A, j+1)$. Let $[(\mathbf{v}, j)]$ denote the equivalence class of $(\mathbf{v}, j)$, and endow L_A with the group structure defined by $[(\mathbf{v}, j)] + [(\mathbf{w}, i)] = [(\mathbf{v}A^i + \mathbf{w}A^j, i+j)]$.

(a) Show that L_A is isomorphic to Δ_A.

(b) The isomorphism in part (a) carries Δ_A^+ onto a subset of the direct limit, and δ_A onto an automorphism of the direct limit. Explicitly describe this subset and automorphism.

7.5.7. Let A be an $n \times n$ integral matrix. $\mathbb{Z}[t]$, the set of polynomials in t with integer coefficients, can be viewed as a group where the group operation is addition of polynomials.

(a) Show that the group $\mathbb{Z}[t]^n / \mathbb{Z}[t]^n (Id - tA)$ is isomorphic to Δ_A.

(b) The isomorphism in part (a) carries Δ_A^+ onto a subset of $\mathbb{Z}[t]^n / \mathbb{Z}[t]^n (Id - tA)$ and δ_A onto an automorphism of $\mathbb{Z}[t]^n / \mathbb{Z}[t]^n (Id - tA)$. Explicitly describe this subset and automorphism.

7.5.8. For a set $\mathbb{E}$ of real numbers, say that two matrices with entries in $\mathbb{E}$ are shift equivalent over $\mathbb{E}$ (and write $A \sim_{\mathbb{E}} B$) if there are matrices R and S with entries in $\mathbb{E}$ satisfying the shift equivalence equations (7–3–1) and (7–3–2). Let A and B be rational matrices.

(a) Show that $A \sim_{\mathbb{R}} B$ if and only if $A \sim_{\mathbb{Q}} B$ if and only if $J^{\times}(A) = J^{\times}(B)$.

(b) Show that if A and B are primitive, then $A \sim_{\mathbb{Q}+} B$ if and only if $A \sim_{\mathbb{Q}} B$ if and only if $J^{\times}(A) = J^{\times}(B)$.

7.5.9. Show that a conjugacy between edge shifts maps beams to beams and preserves equivalence of beams. [*Hint*: Consider the proof of Theorem 2.1.10.]

7.5.10. Formulate a notion of dimension triple for shift spaces in general and show that it is invariant under conjugacy.

7.5.11. (a) Show that if (7–3–1)(i), (7–3–2)(i), and (7–3–2)(ii) hold, then $A \sim B$ (with possibly a different choice of R and S).

(b) Find examples to show that none of the matrix equations in (7–3–1) and (7–3–2) is redundant, i.e., that no three of them imply the remaining one.

Notes

The problem of classifying shifts of finite type up to conjugacy comes from smooth dynamics. In that subject, shifts of finite type are used to model hyperbolic dynamical systems (such as hyperbolic toral automorphisms) via Markov partitions. Williams [WilR1] introduced the notions of shift equivalence and strong shift equivalence for certain one-dimensional systems called 1-dimensional Axiom A attractors. He adapted these notions to the symbolic setting partly in response to the question, raised by Rufus Bowen, of whether the zeta function is a complete invariant for conjugacy of shifts of finite type (see Example 7.3.4). Williams proved the Decomposition Theorem and the Classification Theorem, but realized that shift equivalence is easier to decide than strong shift equivalence; Kim and Roush [KimR3] later showed that shift equivalence is decidable. Using work of Boyle and Krieger [BoyK1], Wagoner [Wag6] as well as [KimRW1] and [KimRW2], Kim and Roush [KimR12] showed that shift equivalence does not imply strong shift equivalence even for irreducible matrices. However, in other settings, it does.

One such setting is one-dimensional Axiom A attractors. Another is integral matrices: we have already defined shift equivalence over $\mathbb{Z}$, and with the obvious definition of strong shift equivalence over $\mathbb{Z}$ it turns out that shift equivalence over $\mathbb{Z}$ does imply (and therefore is equivalent to) strong shift equivalence over $\mathbb{Z}$ (see [WilR1], [WilR3]).

The ideas of elementary equivalence and strong shift equivalence were generalized to sofic shifts by Nasu [Nas4]; for the generalization of shift equivalence to sofic shifts, see Boyle and Krieger [BoyK2]. Our treatment of the Decomposition Theorem for edge shifts partly follows Nasu's treatment for sofic shifts. Theorem 7.3.6 as well as the invariance of the Jordan form away from zero are due to Parry and Williams [ParW] (see also Effros [Eff]). The invariance of the Bowen–Franks group is due to Bowen and Franks [BowF] who discovered it in their study of continuous-time analogues of shifts of finite type (see suspension flows in §13.6). Generalized Bowen–Franks groups were introduced in Kitchens [Kit4].

The dimension group perspective, introduced by Krieger [Kri3], [Kri4], is based on ideas in the theory of C^*-algebras [Eff]. The matrix-oriented definition that we have given here was communicated to us by Handelman. The fact that shift equivalence completely captures eventual conjugacy of edge shifts (Theorem 7.5.15) is due to Williams [WilR2] and Kim and Roush [KimR1].

The methods used for the Decomposition Theorem can also be adapted to conjugacies of one-sided shifts (see [WilR2]).

Exercise 7.1.7 is essentially in [BoyFK]; Exercise 7.2.6 is essentially in Franks [Fran1]; Exercise 7.2.9 is in Nasu [Nas4]; Exercise 7.4.14 is in Williams [WilR2]; Exercise 7.5.6 is in Krieger [Kri3] and [Kri4]; Exercise 7.5.7 is an observation of Wagoner; and part of Exercise 7.5.8 comes from [ParW]. Exercise 7.5.11(a) comes from [BoyK2].

CHAPTER 8

FINITE-TO-ONE CODES AND FINITE EQUIVALENCE

Recall from Chapter 6 that we regard two dynamical systems as being "the same" if they are conjugate and otherwise "different." In Chapter 7 we concentrated on conjugacy for edge shifts, and found that two edge shifts are conjugate if and only if the adjacency matrices of their defining graphs are strong shift equivalent. We also saw that it can be extremely difficult to decide whether two given matrices are strong shift equivalent.

Thus it makes sense to ask if there are ways in which two shifts of finite type can be considered "nearly the same," and which can be more easily decided. This chapter investigates one way called finite equivalence, for which entropy is a complete invariant. Another, stronger way, called almost conjugacy, is treated in the next chapter, where we show that entropy together with period form a complete set of invariants.

In §8.1 we introduce finite-to-one codes, which are codes used to describe "nearly the same." Right-resolving codes are basic examples of finite-to-one codes, and in §8.2 we describe a matrix formulation for a 1-block code from one edge shift to another to be finite-to-one. In §8.3 we introduce the notion of finite equivalence between sofic shifts, and prove that entropy is a complete invariant. A stronger version of finite equivalence is discussed and characterized in §8.4.

§8.1. Finite-to-One Codes

We begin this chapter by introducing finite-to-one sliding block codes. These are sliding block codes $\phi: X \to Y$ for which there is a uniform upper bound on the number of points in $\phi^{-1}(y)$ as y varies over Y. We will focus mainly on the case when the domain X is a shift of finite type. We will state extensions to the case of sofic domains, mostly leaving the proofs as exercises.

Definition 8.1.1. A sliding block code $\phi\colon X \to Y$ is *finite-to-one* if there is an integer M such that $\phi^{-1}(y)$ contains at most M points for every $y \in Y$.

We will see later in this section that if X is sofic and $\phi\colon X \to Y$ is a sliding block code for which $\phi^{-1}(y)$ is finite for every $y \in Y$, then ϕ is finite-to-one; i.e., there is a *uniform* upper bound on the number of points in $\phi^{-1}(y)$ over all $y \in Y$. This no longer holds if X is allowed to be an arbitrary shift space (see Exercise 8.1.15).

Example 8.1.2. Conjugacies and embeddings are finite-to-one, since they are one-to-one. □

Example 8.1.3. Let $\phi\colon X_{[2]} \to X_{[2]}$ be given by the 2-block map $\Phi(ab) = a + b \pmod 2$ (see Example 1.5.3). Then each point in $X_{[2]}$ has exactly two pre-images, so ϕ is finite-to-one. Similarly, each point in $X_{[2]}$ has 2^n pre-images under ϕ^n, so ϕ^n is also finite-to-one. □

Example 8.1.4. If $\phi\colon X \to Y$ and $\psi\colon Y \to Z$ are finite-to-one codes, then so is their composition $\psi \circ \phi\colon X \to Z$. For if $|\phi^{-1}(y)| \leqslant M$ for all $y \in Y$ and $|\psi^{-1}(z)| \leqslant N$ for all $z \in Z$, then clearly $|(\psi \circ \phi)^{-1}(z)| \leqslant MN$ for all $z \in Z$. □

Example 8.1.5. Let $\phi = \Phi_\infty\colon X_{[3]} \to X_{[2]}$ be given by the 1-block map Φ, where $\Phi(0) = 0$, $\Phi(1) = 1$, and $\Phi(2) = 1$. If $y \in X_{[2]}$ contains only finitely many 1's then $\phi^{-1}(y)$ is finite, while if y contains infinitely many 1's then $\phi^{-1}(y)$ is infinite, indeed uncountable. Thus ϕ is not finite-to-one, although $\phi^{-1}(y)$ is finite for some points $y \in Y$. □

Example 8.1.6. Let Φ be a right-resolving labeling on a graph G. Put $X = X_G$, $Y = X_{(G,\Phi)}$, and $\phi = \Phi_\infty\colon X \to Y$. We claim that ϕ is finite-to-one.

Since Φ is right-resolving, each presentation of a block in $\mathcal{B}(Y)$ is determined by its initial vertex. Hence every block in $\mathcal{B}(Y)$ has at most $|\mathcal{V}(G)|$ inverse images under Φ. Now suppose there were a point $y \in Y$ with $|\phi^{-1}(y)| > |\mathcal{V}(G)| = M$, say. Then there would be distinct points $x^{(1)}, \ldots, x^{(M+1)}$ in $\phi^{-1}(y)$. Choose k large enough so that the central $(2k+1)$-blocks of the $x^{(j)}$ are all distinct. This would give a block $y_{[-k,k]}$ with $M + 1$ inverse images under Φ, contradicting what we just proved. Hence $|\phi^{-1}(y)| \leqslant M$ for all $y \in Y$, so that ϕ is indeed finite-to-one. This argument is used in a more general setting in Lemma 8.1.10 below. □

Let us generalize the notion of right-resolving to arbitrary 1-block codes.

Definition 8.1.7. Let X and Y be shift spaces, and $\phi = \Phi_\infty\colon X \to Y$ be a 1-block code. Then ϕ is *right-resolving* if whenever ab and ac are 2-blocks in X with $\Phi(b) = \Phi(c)$, then $b = c$. Similarly, $\phi = \Phi_\infty$ is *left-resolving* if whenever ba and ca are 2-blocks in X with $\Phi(b) = \Phi(c)$, then $b = c$.

The reader should pause to see that when $X = X_G$ is an edge shift and Φ is a labeling of G, this definition is equivalent to Definition 3.3.1.

There is also a notion of right-closing for arbitrary sliding block codes.

Definition 8.1.8. A pair x, x' of points in a shift space X is *left-asymptotic* if there is an integer N for which $x_{(-\infty,N]} = x'_{(-\infty,N]}$. A sliding block code $\phi: X \to Y$ is *right-closing* if whenever x, x' are left-asymptotic and $\phi(x) = \phi(x')$, then $x = x'$. Similarly, we have the notions of *right-asymptotic points* and *left-closing codes*.

In other words, a sliding block code ϕ on an arbitrary shift space is right-closing if distinct points in the pre-image of a point must disagree infinitely often to the left. Using the following concrete characterization of right-closing codes, the reader can verify that when $X = X_G$ and $\phi = \Phi_\infty$ is given by a labeling Φ of G, then ϕ is right-closing in this sense precisely when the labeling Φ is right-closing in the sense of Definition 5.1.4.

The following result characterizes right-closing codes in terms of blocks.

Proposition 8.1.9. *Let* $\phi = \Phi_\infty^{[-m,n]}: X \to Y$ *be a sliding block code. Then* ϕ *is right-closing if and only if there is a* K *such that whenever* $\Phi(x_{[-K,K]}) = \Phi(x'_{[-K,K]})$ *and* $x_{[-K,0]} = x'_{[-K,0]}$, *then* $x_1 = x'_1$.

PROOF: First suppose that Φ and K satisfy the conditions of the proposition. Suppose that x and x' are left-asymptotic, say $x_{(-\infty,N]} = x'_{(-\infty,N]}$, and $\phi(x) = \phi(x')$. Since $x_{[N-K,N]} = x'_{[N-K,N]}$ and $\Phi(x_{[N-K,N+K]}) = \Phi(x'_{[N-K,N+K]})$, we can conclude that $x_{N+1} = x'_{N+1}$. Continuing inductively, we obtain that $x_{N+j} = x'_{N+j}$ for all $j \geqslant 1$, so that $x = x'$. This proves that ϕ is right-closing.

Conversely, suppose that $\phi = \Phi_\infty$ is right-closing. If Φ did not satisfy the conclusion, then for all $k \geqslant 1$ there would be points $x^{(k)}$ and $y^{(k)}$ in X with $\Phi(x^{(k)}_{[-k,k]}) = \Phi(y^{(k)}_{[-k,k]})$ and $x^{(k)}_{[-k,0]} = y^{(k)}_{[-k,0]}$, but $x^{(k)}_1 \neq y^{(k)}_1$. By compactness (equivalently the Cantor diagonal argument as in the proof of Theorem 1.5.13), we could then find points x and y in X with $x_{(-\infty,0]} = y_{(-\infty,0]}$, $\phi(x) = \phi(y)$, but $x_1 \neq y_1$. This contradicts the assumption that ϕ is right-closing.   □

In order to prove that right-closing codes are finite-to-one, we first make the following observation.

Lemma 8.1.10. *Let* Φ *be a 1-block map inducing* $\phi = \Phi_\infty: X \to Y$. *If there is an* M *for which* $|\Phi^{-1}(u)| \leqslant M$ *for all* $u \in \mathcal{B}(Y)$, *then* ϕ *is finite-to-one.*

PROOF: Suppose that for some $y \in Y$ the pre-image $\phi^{-1}(y)$ contains $M+1$ points, say $x^{(1)}, x^{(2)}, \ldots, x^{(M+1)}$. Since these are distinct there is a $k \geqslant 0$ such that the central $(2k+1)$-blocks of the $x^{(j)}$ are all distinct. But these blocks are all in $\Phi^{-1}(y_{[-k,k]})$, contradicting the assumption on Φ.   □

Proposition 8.1.11. *Right-closing codes and left-closing codes on arbitrary shift spaces are finite-to-one.*

PROOF: Proposition 8.1.9 shows that if $\phi = \Phi_\infty : X \to Y$ is a right-closing sliding block code then there is a K such that if $m \geqslant 2K$, $u = u_0 u_1 \ldots u_m$ and $v = v_0 v_1 \ldots v_m$ are blocks in X with $\Phi(u) = \Phi(v)$ and $u_{[0,K]} = v_{[0,K]}$, then $u_0 u_1 \ldots u_{m-K} = v_0 v_1 \ldots v_{m-K}$. It follows that the conditions of Lemma 8.1.10 are satisfied, so that ϕ is finite-to-one. A symmetric argument applies to left-closing codes.   □

The following is a reformulation of a result proved in Chapter 5 and provides further examples of finite-to-one codes.

Proposition 8.1.12. *Let X be an irreducible shift of finite type such that $h(X) = \log n$, where $n \geqslant 1$ is an integer. Then there is a right-closing factor code from X onto $X_{[n]}$.*

PROOF: In the case that $h(X) = \log n$, the factor code $\phi : X \to X_{[n]}$ constructed in the proof of Theorem 5.5.8 is the composition of a conjugacy from X to an edge shift, splitting codes, and a right-resolving code (given by a road-coloring). Each of these is right-closing, hence ϕ is also.   □

Having seen some important examples of finite-to-one codes, our next goal is to determine whether or not a given sliding block code is finite-to-one. The key idea for this is the notion of a *diamond*. We will describe this notion first for edge shifts, then for arbitrary shift spaces.

Definition 8.1.13. Let Φ be a labeling of a graph G. A *graph diamond for Φ* is a pair of distinct paths in G having the same Φ-label, the same initial state, and the same terminal state.

Note that since the paths in a graph diamond have the same label, they must have the same length. Figure 8.1.1(a) shows a graph diamond that earns its name, while a more typical graph diamond is shown in (b).

Distinct paths may have some common edges. For example, if the pair π, τ is a graph diamond for Φ and ω is a path starting at the terminal state $t(\pi) = t(\tau)$, then $\pi\omega, \tau\omega$ is also a graph diamond for Φ. Clearly right-closing labelings and left-closing labelings on essential graphs do not have graph diamonds. We will see in this section that this is consistent with their being finite-to-one. The next example shows that a labeling on a graph having no graph diamonds need not be right-closing or left-closing.

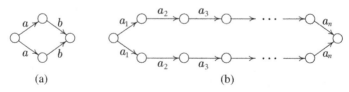

(a)   (b)

FIGURE 8.1.1.   A graph diamond.

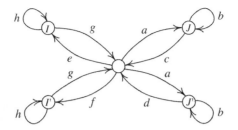

FIGURE 8.1.2. A graph with no graph diamonds.

Example 8.1.14. Consider the labeled graph (G, Φ) depicted in Figure 8.1.2. If (G, Φ) had a graph diamond, then it would have a graph diamond π, τ where π and τ have distinct initial edges and distinct terminal edges (why?). It follows that both π and τ begin and end at the central state K. But then the label of one of these paths would have a prefix of the form $ab^m c$ and the other would have a prefix of the form $ab^m d$; this contradicts the fact that π and τ have the same label. □

Notice that a labeling Φ of a graph G has a graph diamond if and only if there are distinct points x, x' in X_G differing in only finitely many coordinates and such that $\Phi_\infty(x) = \Phi_\infty(x')$. This motivates the following notion of diamond for arbitrary sliding block codes.

Definition 8.1.15. Let $\phi \colon X \to Y$ be a sliding block code. A *point diamond* (or simply *diamond*) *for* ϕ is a pair of distinct points in X differing in only finitely many coordinates with the same image under ϕ.

Figure 8.1.3 pictures a point diamond and shows why point diamonds are sometimes called "bubbles."

A moment's reflection shows that if G is an essential graph, then a 1-block code $\phi = \Phi_\infty$ on X_G has a (point) diamond if and only if Φ has a graph diamond.

We can now formulate several conditions to determine if a sliding block code on an irreducible shift of finite type is finite-to-one; these include the absence of diamonds and the preservation of entropy.

Theorem 8.1.16. *Let X be an irreducible shift of finite type and $\phi \colon X \to Y$ a sliding block code. Then the following are equivalent.*

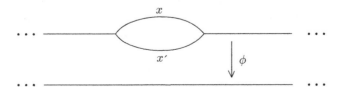

FIGURE 8.1.3. A point diamond.

(1) ϕ is finite-to-one.

(2) For every $y \in Y$, the pre-image $\phi^{-1}(y)$ is finite.

(3) For every $y \in Y$, the pre-image $\phi^{-1}(y)$ is at most countable.

(4) ϕ has no diamonds.

(5) $h(X) = h(\phi(X))$.

PROOF: All five conditions on ϕ are preserved under recoding to higher blocks. Hence using Proposition 1.5.12 and Theorem 2.3.2 we may assume that $\phi = \Phi_\infty$ is a 1-block code induced by a labeling Φ on an irreducible graph G (and $X = X_G$). We may also replace Y with $\phi(X)$ and so assume that $\phi(X) = Y$, i.e., that ϕ is a factor code.

Clearly (1) $\Rightarrow$ (2) $\Rightarrow$ (3). We will show (3) $\Rightarrow$ (4) $\Rightarrow$ (5) $\Rightarrow$ (4) and (4) $\Rightarrow$ (1).

(3) $\Rightarrow$ (4). We prove the contrapositive, i.e., that if ϕ has a diamond, then there is at least one point $y \in Y$ for which $\phi^{-1}(y)$ is uncountable.

Suppose that ϕ has a diamond, so that Φ has a graph diamond π, τ. By appending to π and to τ a path in G from their common terminal state to the common initial state, we may assume that π and τ are distinct cycles based at the same state. Let $u = \Phi(\pi) = \Phi(\tau)$. Then $y = u^\infty$ has pre-images of the form

$$\ldots \omega_{-3}\omega_{-2}\omega_{-1}.\omega_0\omega_1\omega_2 \ldots \,,$$

where each ω_j can be either π or τ. This gives uncountably many points in $\phi^{-1}(y)$.

(4) $\Rightarrow$ (5). Suppose that $\phi = \Phi_\infty$ has no diamonds. Then for each pair $I, J \in \mathcal{V}(G)$ and block $w \in \mathcal{B}(Y)$ there is at most one path π in G from I to J with $\Phi(\pi) = w$. Since ϕ is a factor code, each block in Y is the image of some path in G. Hence for all $n \geqslant 1$ we have

$$|\mathcal{B}_n(Y)| \leqslant |\mathcal{B}_n(X)| \leqslant |\mathcal{B}_n(Y)| \cdot |\mathcal{V}(G)|^2.$$

Applying the definition of entropy, we obtain $h(Y) = h(X)$.

(5) $\Rightarrow$ (4). We again prove the contrapositive, i.e., that if ϕ has a diamond, then $h(X) > h(Y)$. Suppose that ϕ has a diamond, so that Φ has a graph diamond. We will show that diamonds in G create additional entropy in X.

Let π, τ be a diamond with length k. By appending a common path to π and τ, we may assume that k is relatively prime to the period of G. Hence the kth power graph G^k is also irreducible (see Exercise 4.5.3) and π and τ are edges in G^k with the same initial state and same terminal state. Let H be the graph obtained from G^k by omitting the edge π. By Theorem 4.4.7, $h(X_{G^k}) > h(X_H)$. Now Φ induces a factor code $\psi \colon X_{G^k} \to Y^k$, and removal of π from G^k does not affect the image, so $\psi(X_H) = Y^k$. Hence

$$h(X) = h(X_G) = \frac{1}{k}h(X_{G^k}) > \frac{1}{k}h(X_H) \geqslant \frac{1}{k}h(Y^k) = h(Y).$$

(4) ⇒ (1). The proof of (4) ⇒ (5) shows that for every block $u \in \mathcal{B}(Y)$ we have that $|\Phi^{-1}(u)| \leqslant |\mathcal{V}(G)|^2$. Then Lemma 8.1.10 shows that ϕ is finite-to-one.   □

The following example shows that the irreducibility assumption in Theorem 8.1.16 is needed.

Example 8.1.17. Let $\mathcal{G} = (G, \Phi)$ be the labeled graph defined by the symbolic adjacency matrix $\begin{bmatrix} a & a \\ \emptyset & a \end{bmatrix}$. Then $X = X_G$ is countably infinite while $Y = X_{\mathcal{G}} = \{a^\infty\}$ is a single point. Since $h(X) = 0 = h(Y)$, this shows that without assuming irreducibility of X in Theorem 8.1.16, we cannot conclude that (5) implies (4), (2), or (1).   □

When given an explicit sliding block code on an irreducible shift of finite type, you can recode it to a 1-block code on an edge shift. Exercise 8.1.6 gives an algorithm for deciding whether or not such a code has a graph diamond.

We can now give a partial converse to Lemma 8.1.10.

Corollary 8.1.18. *Let X be an irreducible shift of finite type and $\phi = \Phi_\infty: X \to Y$ a finite-to-one code. Then there is an M such that $|\Phi^{-1}(u)| \leqslant M$ for all blocks $u \in \mathcal{B}(Y)$.*

PROOF: Recode ϕ to a 1-block code on an irreducible edge shift X_G. The proof of the implication (4) ⇒ (5) in Theorem 8.1.16 shows that $|\Phi^{-1}(u)| \leqslant |\mathcal{V}(G)|^2$ for all $u \in \mathcal{B}(Y)$.   □

The right-closing codes and left-closing codes provide a rich class of finite-to-one codes. Example 8.1.14 gives a finite-to-one code that is neither right-closing nor left-closing, although it is not hard to see that it is the composition of right-closing codes and left-closing codes (first identify I with I', then J with J'). We can enlarge this class further by forming compositions of right-closing codes and left-closing codes. Do we obtain all finite-to-one codes this way? This is not an unreasonable idea. After all, a finite-to-one code has the "no diamond" property that a path is determined by its label plus initial and terminal states, while for right-resolving codes a path is determined by its label plus initial state, and for left-resolving codes requires its label plus terminal state.

However, not all finite-to-one codes on edge shifts can be decomposed into right-closing codes and left-closing codes (see Exercise 8.1.13). In later chapters we discuss two related results. In an exercise in Chapter 9 (Exercise 9.1.12), we suggest a proof to show that if X is an irreducible shift of finite type and $\phi: X \to Y$ is a finite-to-one code, then there is a factor code $\psi: Z \to X$ with Z an irreducible shift of finite type such that $\phi \circ \psi$ is the composition of a right-resolving factor code followed by a left-resolving factor code. In §12.2 we show that there is an "eventual" sense in which every finite-to-one code can be replaced by a composition of a right-closing code and a left-closing code. Further information about this is contained in [Tro6], [Boy7].

We now turn to extending Theorem 8.1.16 to irreducible sofic domains. The proof is an easy consequence of that theorem and is left as an exercise.

Theorem 8.1.19. *Let X be an irreducible sofic shift, $\phi \colon X \to Y$ a sliding block code, and $\psi = \mathcal{L}_\infty \colon X_G \to X$ the 1-block code induced by the minimal right-resolving presentation $(G, \mathcal{L})$ of X. Then the following are equivalent.*

(1) *ϕ is finite-to-one.*
(2) *For every $y \in Y$, the pre-image $\phi^{-1}(y)$ is finite.*
(3) *For every $y \in Y$, the pre-image $\phi^{-1}(y)$ is at most countable.*
(4) *$\phi \circ \psi$ has no diamonds.*
(5) *$h(X) = h(\phi(X))$.*

Exercise 8.1.9(b) gives an example of an irreducible sofic X and a sliding block code $\phi \colon X \to Y$ that is finite-to-one but has a diamond. This is why part (4) of Theorem 8.1.19 differs from part (4) of Theorem 8.1.16.

The following result spells out a useful relationship among three properties of sliding block codes on irreducible sofic shifts.

Corollary 8.1.20. *Let X and Y be irreducible sofic shifts and $\phi \colon X \to Y$ be a sliding block code. Then any two of the following conditions imply the third.*

(1) *ϕ is finite-to-one.*
(2) *ϕ is onto.*
(3) *$h(X) = h(Y)$.*

PROOF: Suppose that (1) and (2) hold. Then $h(X) = h(\phi(X)) = h(Y)$.

Suppose that (1) and (3) hold. Then $h(\phi(X)) = h(X) = h(Y)$. Since Y is irreducible, it follows from Corollary 4.4.9 that $\phi(X) = Y$.

Finally, suppose that (2) and (3) hold. Then $h(X) = h(Y) = h(\phi(X))$, so that ϕ is finite-to-one by Theorem 8.1.19. □

We conclude this section by describing a connection between finite-to-one codes and a fundamental notion from automata theory and information theory called a uniquely decipherable set.

Uniquely decipherable sets are those collections of words with a "unique factorization" property.

Definition 8.1.21. A (finite or infinite) collection $\mathcal{U}$ of words over an alphabet $\mathcal{A}$ is *uniquely decipherable* if whenever $u_1 u_2 \ldots u_m = v_1 v_2 \ldots v_n$ with $u_i, v_j \in \mathcal{U}$, then $m = n$ and $u_j = v_j$ for $1 \leqslant j \leqslant n$.

Example 8.1.22. (a) $\mathcal{U} = \{0, 10\}$ is uniquely decipherable. For suppose that $w = w_1 w_2 \ldots w_n$ is a concatenation of words from $\mathcal{U}$. If the first symbol in w is 0 then $w_1 = 0$, while if the first symbol is 1 then $w_1 = 10$. Delete this word, and repeat the argument on the remaining part, to show that we can recover the individual words w_j from w.

(b) The collection $\{0, 01, 10\}$ is not uniquely decipherable, since $010 = (01)0 = 0(10)$ has two distinct "factorizations." $\qquad\square$

Let G be a graph and I be a state in G. Denote by $\mathcal{C}_I$ the set of cycles that start and end at I but do not visit I in between. Let $\mathcal{C}_I^*$ denote the set of all cycles in G that start and end at I. Equivalently, $\mathcal{C}_I^*$ is the set of all concatenations of blocks from $\mathcal{C}_I$ (why?). The following result relates finite-to-one codes and uniquely decipherable sets.

Proposition 8.1.23. *Let Φ be a labeling of an irreducible graph G. Then the following are equivalent.*

(1) *$\phi = \Phi_\infty$ is finite-to-one.*
(2) *For every state I of G, the restriction of Φ to $\mathcal{C}_I$ is one-to-one and the image $\Phi(\mathcal{C}_I)$ is uniquely decipherable.*
(3) *For every state I of G, the restriction of Φ to $\mathcal{C}_I^*$ is one-to-one.*

PROOF: (1) $\Rightarrow$ (2). Since ϕ is finite-to-one, the labeling Φ has no graph diamonds, implying (2).

(2) $\Rightarrow$ (3). Suppose that $\pi_i, \tau_j \in \mathcal{C}_I$ with

$$\Phi(\pi_1 \pi_2 \dots \pi_m) = \Phi(\tau_1 \tau_2 \dots \tau_n).$$

Since $\Phi(\mathcal{C}_I)$ is uniquely decipherable, it follows that $m = n$ and that $\Phi(\pi_j) = \Phi(\tau_j)$ for $1 \leqslant j \leqslant n$. Since the restriction of Φ to $\mathcal{C}_I$ is one-to-one, $\pi_j = \tau_j$ for $1 \leqslant j \leqslant n$. Thus Φ is one-to-one on $\mathcal{C}_I^*$.

(3) $\Rightarrow$ (1). Suppose that Φ has a graph diamond π, τ. As before, we could extend these to distinct cycles at the same state with the same Φ-label, contradicting our assumption on Φ. $\qquad\square$

Example 8.1.24. (a) For the labeled graph (G, Φ) depicted in (a) of Figure 8.1.4, we see that $\Phi(\mathcal{C}_I) = \{0, 10\}$ is uniquely decipherable and

$$\Phi(\mathcal{C}_J) = \{01, 001, 0001, \dots\}$$

is uniquely decipherable, consistent with the fact that $\phi = \Phi_\infty$ is finite-to-one.

(b) Let (G, Φ) be the labeled graph in Figure 8.1.4(b). Then

$$\Phi(\mathcal{C}_I) = \{0, 10, 110, 1110, \dots\} \quad \text{and} \quad \Phi(\mathcal{C}_J) = \{1, 01, 001, 0001, \dots\}$$

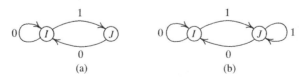

FIGURE 8.1.4. Uniquely decipherable graphs.

are uniquely decipherable, again consistent with the fact that $\phi = \Phi_\infty$ is finite-to-one. □

Exercise 8.1.11 describes those labeled graphs which produce uniquely decipherable sets that are finite.

EXERCISES

8.1.1. Verify that for 1-block codes on edge shifts, i.e., codes induced by labeled graphs, the definitions of right-resolving and right-closing given in this section agree with the definitions given in §3.3 and §5.1.

8.1.2. Show that every conjugacy is right-closing.

8.1.3. Show that every right-closing code $\phi : X \to Y$ can be "recoded" to a right-resolving code in the following sense: there are a shift space X' and a conjugacy $\theta: X' \to X$ such that $\phi \circ \theta$ is right-resolving. [*Hint*: Use Proposition 8.1.9 to modify the proof of Proposition 5.1.11.]

8.1.4. For each of the following symbolic adjacency matrices, decide whether or not the 1-block code determined by the labeling is finite-to-one.

$$\text{(a)} \quad \begin{bmatrix} a & b & b \\ d & c & \emptyset \\ d & \emptyset & c \end{bmatrix}, \qquad \text{(b)} \quad \begin{bmatrix} a & b & b \\ d & c & \emptyset \\ e & \emptyset & c \end{bmatrix}.$$

8.1.5. Formulate a notion of "diamond" for a block map Φ on an arbitrary N-step shift of finite type which generalizes the notion of graph diamond for labelings on graphs and which satisfies the following: Φ has no diamonds if and only if Φ_∞ has no diamonds.

8.1.6. Let $\mathcal{G} = (G, \Phi)$ be an irreducible labeled graph, and let $\phi = \Phi_\infty$. Recalling §3.4, let $\mathcal{K}$ denote the diagonal subgraph of the label product $\mathcal{G} * \mathcal{G}$ (i.e., the labeled subgraph consisting of all edges which begin and end at states of the form (I, I)).

    (a) Show that ϕ is a finite-to-one code if and only if any path of length at most $|\mathcal{V}(\mathcal{G} * \mathcal{G})|$ which begins at a state of $\mathcal{K}$ and ends at a state of $\mathcal{K}$ lies completely within $\mathcal{K}$.

    (b) Formulate an equivalent condition involving the symbolic adjacency matrices of $\mathcal{G} * \mathcal{G}$ and $\mathcal{K}$.

8.1.7. Let X be a possibly reducible shift of finite type, and $\phi: X \to Y$ a sliding block code. Which implications between the five properties listed in Theorem 8.1.16 remain true?

8.1.8. (a) Find an example of a sliding block code from one irreducible shift of finite type to another which has some points with infinitely many pre-images and some points with only finitely many pre-images.

    (b) Show that for an infinite-to-one factor code (i.e., a factor code which is not finite-to-one) on an irreducible shift of finite type, the set of points in the range shift with infinitely many pre-images is dense.

8.1.9. (a) Prove Theorem 8.1.19.

    (b) Show that the sliding block code from the even shift onto the golden mean shift induced by the 3-block map

$$\Phi(abc) = \begin{cases} 1 & \text{if } abc = 001 \text{ or } 110, \\ 0 & \text{otherwise,} \end{cases}$$

is finite-to-one but has a diamond.

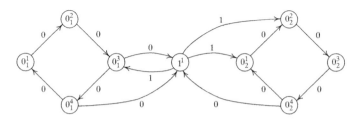

FIGURE 8.1.5. Graph for Exercise 8.1.12.

8.1.10. Show that if ϕ is a finite-to-one 1-block code on an irreducible edge shift, $x \neq x'$ and $\phi(x) = \phi(x')$, then exactly one of the following must hold:

(1) the pair $\{x, x'\}$ is left-asymptotic,
(2) the pair $\{x, x'\}$ is right-asymptotic,
(3) for all sufficiently large n, $x_n \neq x'_n$ and $x_{-n} \neq x'_{-n}$.

8.1.11. Let $(G, \mathcal{L})$ be an irreducible labeled graph and assume that $\mathcal{L}_\infty$ is finite-to-one. Let I be a state of G and let $\mathcal{C}_I$ denote the set of first returns to I as in Proposition 8.1.23. Show that $\mathcal{C}_I$ is finite if and only if $\Phi(\mathcal{C}_I)$ is finite if and only if every cycle in G passes through I.

8.1.12. Let $\mathcal{G} = (G, \mathcal{L})$ be the labeled graph shown in Figure 8.1.5.

(a) Show that $\mathcal{G}$ has no graph diamonds and so $\mathcal{L}_\infty$ is finite-to-one.
(b) Show that $X_{\mathcal{G}} = \mathcal{L}_\infty(X_G)$ is the golden mean shift.

8.1.13. Let $\mathcal{G} = (G, \mathcal{L})$ be the labeled graph in Figure 8.1.5. The following steps outline a proof that $\mathcal{L}_\infty$ is not the composition of right-closing and left-closing codes.

(a) Let $\mathcal{L}'$ denote the mapping on states of G defined by erasing subscripts and superscripts, e.g., $\mathcal{L}'(0_1^2) = 0$ and $\mathcal{L}'(1^1) = 1$. Since G has no multiple edges, the edge shift X_G is conjugate to the vertex shift $\widehat{X}_G$ by the conjugacy $\theta = \Theta_\infty : X_G \to \widehat{X}_G$, where $\Theta(e) = i(e)$. Show that $\mathcal{L}'_\infty \circ \theta = \mathcal{L}_\infty$. (In this way, $\mathcal{L}'_\infty$ is essentially the same as $\mathcal{L}_\infty$.)

(b) Explain why it suffices to show that whenever $\mathcal{L}'_\infty = \psi \circ \phi$ and ψ is right-closing or left-closing, then ψ is a conjugacy. (The remaining items in this exercise outline a proof of this statement where, without loss of generality, we assume that ψ is right-closing.)

(c) Suppose that $\mathcal{L}'_\infty = \psi \circ \phi$ and ψ is right-closing. Let

$$x = (0_1^3 1^1)^\infty . (0_1^3 0_1^4 0_1^1 0_1^2)^\infty,$$

$$y = (0_1^3 1^1)^\infty . (0_2^1 0_2^2 0_2^3 0_2^4)^\infty,$$

$$z = (0_1^3 1^1)^\infty . (0_2^2 0_2^3 0_2^4 0_2^1)^\infty.$$

Show that $\phi(x) = \phi(y) = \phi(z)$.

(d) Let

$$\bar{x} = (0_1^3 0_1^4 0_1^1 0_1^2)^\infty . (0_1^3 0_1^4 0_1^1 0_1^2)^\infty,$$

$$\bar{y} = (0_2^1 0_2^2 0_2^3 0_2^4)^\infty . (0_2^1 0_2^2 0_2^3 0_2^4)^\infty,$$

$$\bar{z} = (0_2^2 0_2^3 0_2^4 0_2^1)^\infty . (0_2^2 0_2^3 0_2^4 0_2^1)^\infty.$$

Show that $\phi(\bar{x}) = \phi(\bar{y}) = \phi(\bar{z})$.

(e) Show that $\phi(\bar{x})$ is a fixed point.

(f) Conclude that ψ is a conjugacy. [*Hint*: See Exercise 2.3.6.]

8.1.14. Let ϕ be a sliding block code on an irreducible shift of finite type X and assume that the memory of ϕ is zero. Then ϕ naturally defines a mapping ϕ^+ on the one-sided shift X^+ (see §5.1). Show the following.

(a) ϕ is onto if and only if ϕ^+ is onto.

(b) ϕ is finite-to-one if and only if ϕ^+ is finite-to-one, i.e., there is a uniform upper bound on the sizes of ϕ^+-pre-images.

(c) If ϕ^+ is one-to-one, then so is ϕ.

(d) The converse of (c) is false.

8.1.15. Let X be the subset of $\{0, 1, 2\}^{\mathbb{Z}}$ consisting of all shifts of points of the form $0^{\infty}.a_1 0^n a_2 0^n \cdots a_n 0^n 0^{\infty}$, $0^{\infty}.a0^{\infty}$, or 0^{∞}, where $n = 1, 2, 3, \ldots$ and $a = 1, 2$, $a_i = 1, 2$. Let Y be the subset of $\{0, 1\}^{\mathbb{Z}}$ consisting of all shifts of points of the form $0^{\infty}.(10^n)^n 0^{\infty}$, $0^{\infty}.10^{\infty}$, or 0^{∞}.

(a) Show that X and Y are shift spaces.

(b) Define the 1-block map Φ by $\Phi(0) = 0$, $\Phi(1) = 1$, and $\Phi(2) = 1$. Show that $\phi = \Phi_{\infty} : X \to Y$ has the property that $\phi^{-1}(y)$ is finite for every $y \in Y$ yet ϕ is *not* a finite-to-one code.

*8.1.16. (a) Show that finite-to-one codes on any shift space preserve entropy. [*Hint:* Use the definition of entropy from Example 6.3.4.]

(b) Show that in fact a sliding block code on any shift space such that each point in the image has only countably many pre-images must preserve entropy.

§8.2.   Right-Resolving Codes

In the last section we saw that left-resolving codes and right-resolving codes together with left-closing codes and right-closing codes give a rich class of finite-to-one codes. Here we will focus mainly on right-resolving codes from one edge shift to another.

Let $\phi = \Phi_{\infty} : X_G \to X_H$ be a 1-block code, where G and H are essential. Then Φ is the edge map of a graph homomorphism from G to H. Recall that there is an associated vertex map $\partial\Phi$ satisfying $\partial\Phi(i(e)) = i(\Phi(e))$ and $\partial\Phi(t(e)) = t(\Phi(e))$ for all $e \in \mathcal{E}(G)$. We will sometimes refer to Φ itself as a graph homomorphism, with $\partial\Phi$ understood as above.

A graph homomorphism $\Phi : G \to H$ maps $\mathcal{E}_I(G)$ into $\mathcal{E}_{\partial\Phi(I)}(H)$ for each state I of G. Thus $\phi = \Phi_{\infty}$ is right-resolving if and only if for every state I of G the restriction Φ_I of Φ to $\mathcal{E}_I(G)$ is one-to-one. One part of Proposition 8.2.2 below shows that if G and H are irreducible and ϕ is a right-resolving code from X_G *onto* X_H, then each Φ_I must be a bijection. We formulate this property into a definition.

Definition 8.2.1. A graph homomorphism $\Phi : G \to H$ is *right-covering* if, for each state I of G, the restriction Φ_I of Φ to $\mathcal{E}_I(G)$ is a bijection from $\mathcal{E}_I(G)$ to $\mathcal{E}_{\partial\Phi(I)}(H)$. A 1-block code ϕ from one edge shift to another is *right-covering* if there is a right-covering graph homomorphism Φ such that $\phi = \Phi_{\infty}$. *Left-covering* is defined similarly using the set $\mathcal{E}^I(G)$ of incoming edges to I.

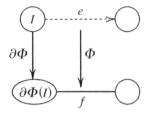

FIGURE 8.2.1. The unique lifting property for right-covering codes.

Note that a right-covering code Φ is automatically right-resolving, but not conversely. The right-covering property can be viewed as a "unique lifting property": for every state I of G and every edge $f \in \mathcal{E}_{\partial\Phi(I)}(H)$, there exists a unique "lifted" edge $e \in \mathcal{E}_I(G)$ such that $\Phi(e) = f$. See Figure 8.2.1. This lifting property inductively extends to paths: for every state I of G and every path w in H starting at $\partial\Phi(I)$, there is a unique path π in G starting at I such that $\Phi(\pi) = w$.

Exercise 8.2.4 shows how to lift a right Perron eigenvector via a right-covering code.

Proposition 8.2.2. *Let G and H be irreducible graphs, and suppose that $\phi = \Phi_\infty : X_G \to X_H$ is a right-resolving code. Then the following are equivalent.*

(1) *ϕ is onto.*
(2) *ϕ is right-covering.*
(3) *$h(X_G) = h(X_H)$.*

In particular, a 1-block code from one irreducible edge shift to another is a right-resolving factor code if and only if it is a right-covering code.

PROOF: $(1) \Rightarrow (2)$ Since $\phi = \Phi_\infty$ is right-resolving, the restriction

$$\Phi_I : \mathcal{E}_I(G) \to \mathcal{E}_{\partial\Phi(I)}(H)$$

is one-to-one. We argue by contradiction. Suppose that there is a state I for which Φ_I is not onto. Put $J = \partial\Phi(I)$. Let f be an edge in $\mathcal{E}_J(H)$ that is not in the image of Φ_I. Since ϕ is onto, there is a state K in G such that $\partial\Phi(K) = t(f)$. Create a new graph G^* by adding to G an edge e that starts at I and ends at K. Extend Φ to a labeling Φ^* on G^* by defining $\Phi^*(e) = f$.

Observe that Φ^* is a graph homomorphism from G^* to H that is right-resolving as a labeling. Hence $\phi^* = \Phi_\infty^*$ is a right-resolving code. Since ϕ is onto, so is ϕ^*. By Proposition 4.1.13 right-resolving codes preserve entropy, so that

$$h(X_{G^*}) = h(X_H) = h(X_G).$$

But the adjacency matrix A_{G^*} is obtained from A_G by increasing one entry by 1. Since G^* is clearly irreducible, Theorem 4.4.7 implies that $h(X_{G^*}) >$

$h(X_G)$. This contradiction proves that each Φ_I is a bijection, so ϕ is right-covering.

(2) $\Rightarrow$ (1) We first claim that $\partial\Phi: \mathcal{V}(G) \to \mathcal{V}(H)$ is onto. To see this, first observe that there is some $J \in \mathcal{V}(H)$ in the image of $\partial\Phi$, say $J = \partial\Phi(I)$. Let J' be another vertex in H. Since H is irreducible, there is a path w in H from J to J'. By the path lifting property of right-covering maps, there is a path π in G starting at I such that $\Phi(\pi) = w$. Hence $J' = \partial\Phi(t(\pi))$, proving that $\partial\Phi$ is onto as claimed. Since $\partial\Phi$ is onto, the path lifting property now shows that every block in $\mathcal{B}(X_H)$ is the image of a block in $\mathcal{B}(X_G)$. Hence $\phi(X_G)$ and X_H are shift spaces with the same language, so they must be equal by Proposition 1.3.4(3).

(1) $\Longleftrightarrow$ (3) Since right-resolving codes are finite-to-one, both directions follow from Corollary 8.1.20. $\qquad\square$

Exercise 8.2.9 suggests a different proof that (1) $\Rightarrow$ (2). Exercise 8.2.8 shows how to determine when $\Phi_\infty(X_G)$ is an edge shift for a labeling Φ on a graph G.

The following observation will be useful later.

Remark 8.2.3. The implication (2) $\Rightarrow$ (1) in Proposition 8.2.2 does not require that G be irreducible. $\qquad\square$

We next show how the existence of right-resolving or right-covering codes between edge shifts induces matrix inequalities or equalities for the corresponding adjacency matrices. For this we first need to introduce two special kinds of matrices.

Definition 8.2.4. A *subamalgamation matrix* is a rectangular 0–1 matrix with exactly one 1 in each row. An *amalgamation matrix* is a rectangular 0–1 matrix with exactly one 1 in each row and at least one 1 in each column.

For example,

$$\begin{bmatrix} 1 & 0 & 0 \\ 0 & 1 & 0 \\ 0 & 1 & 0 \\ 1 & 0 & 0 \end{bmatrix}$$

is a subamalgamation matrix, but not an amalgamation matrix. Note that an amalgamation matrix is just the transpose of a division matrix as defined in Definition 2.4.13.

Let G and H be graphs and $\phi = \Phi_\infty: X_G \to X_H$ be a right-resolving code. Then for each state I of G the restriction Φ_I of Φ to $\mathcal{E}_I(G)$ is one-to-one. We can break up Φ_I as follows. For each state J of H let $\mathcal{F}_{IJ}$ denote the set of edges in G starting at I and ending at a state which maps to J under $\partial\Phi$. In symbols,

$$\mathcal{F}_{IJ} = \mathcal{E}_I(G) \cap \Phi^{-1}(\mathcal{E}^J(H)).$$

Let Φ_{IJ} denote the restriction of Φ to $\mathcal{F}_{IJ}$. Since $\mathcal{E}_I(G)$ is partitioned into the sets $\mathcal{F}_{IJ}$ as J varies over $\mathcal{V}(H)$, we can think of Φ_I as being decomposed into the maps Φ_{IJ}.

Now each Φ_{IJ} is one-to-one on $\mathcal{F}_{IJ}$ and has image contained in $\mathcal{E}^J_{\partial\Phi(I)}(H)$, so that

$$(8\text{--}2\text{--}1) \qquad |\mathcal{F}_{IJ}| \leqslant |\mathcal{E}^J_{\partial\Phi(I)}(H)|.$$

Let S be the matrix with rows indexed by $\mathcal{V}(G)$, columns indexed by $\mathcal{V}(H)$, and entries given by

$$S_{IJ} = \begin{cases} 1 & \text{if } J = \partial\Phi(I), \\ 0 & \text{if } J \neq \partial\Phi(I). \end{cases}$$

This S is just a matrix description of $\partial\Phi$ and clearly S is a subamalgamation matrix. Now

$$(A_G S)_{IJ} = \sum_{K \in \partial\Phi^{-1}(J)} (A_G)_{IK} = |\mathcal{F}_{IJ}|,$$

while

$$(SA_H)_{IJ} = (A_H)_{\partial\Phi(I)J} = |\mathcal{E}^J_{\partial\Phi(I)}(H)|.$$

Applying the inequality (8–2–1) shows that

$$(8\text{--}2\text{--}2) \qquad A_G S \leqslant SA_H.$$

Thus the existence of a right-resolving code from X_G to X_H implies the existence of a subamalgamation matrix S satisfying the matrix inequality (8–2–2).

Figure 8.2.2 may help to clarify (8–2–2). The top part belongs to G and the bottom part to H. Here $\partial\Phi^{-1}(J) = \{K_1, K_2, K_3\}$. The matrix S is used to construct an "S-edge" from $I \in \mathcal{V}(G)$ to $J \in \mathcal{V}(H)$ if $S_{IJ} = 1$, similar to the construction following the proof of Proposition 7.2.3. Since Φ_I is injective, the number of "top-right" paths in the figure must be less than or equal to the number of "left-bottom" paths – this is precisely the inequality $(A_G S)_{IJ} \leqslant (SA_H)_{IJ}$.

Conversely, if S is a subamalgamation matrix satisfying (8–2–2), then we can use S to construct a right-resolving code from X_G to X_H as follows. For each $I \in \mathcal{V}(G)$ define $\partial\Phi(I)$ to be the unique $J \in \mathcal{V}(H)$ for which $S_{IJ} = 1$. The inequality $A_G S \leqslant SA_H$ means that for each pair $I \in \mathcal{V}(G)$ and $J \in \mathcal{V}(H)$ we can find a one-to-one map

$$(8\text{--}2\text{--}3) \qquad \bigcup_{K \in \partial\Phi^{-1}(J)} \mathcal{E}^K_I(G) \to \mathcal{E}^J_{\partial\Phi(I)}(H).$$

Since the domains of these maps are disjoint, they define a graph homomorphism $\Phi: G \to H$ that is a right-resolving labeling. This gives a right-resolving code from X_G to X_H.

This discussion can be summarized as follows.

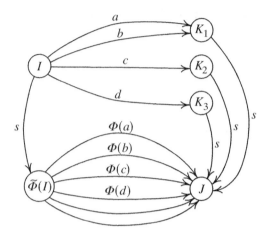

FIGURE 8.2.2.   Illustration of a matrix inequality.

Proposition 8.2.5. *Let G and H be graphs. There is a right-resolving code from X_G to X_H if and only if there is a subamalgamation matrix S such that $A_G S \leqslant S A_H$.*

When the graphs are irreducible and the right-resolving code is onto, the matrix inequality becomes a matrix equality.

Proposition 8.2.6. *Let G and H be irreducible graphs. Then there is a right-resolving factor code from X_G to X_H if and only if there is an amalgamation matrix S such that $A_G S = S A_H$. If this condition holds, then $h(X_G) = h(X_H)$.*

PROOF: Suppose there is a right-resolving factor code $\Phi_\infty : X_G \to X_H$. The proof of (1) $\Rightarrow$ (2) in Proposition 8.2.2 shows that each of the inequalities (8–2–1) must be an equality, so $A_G S = S A_H$. Since Φ_∞ and hence $\partial \Phi$ is onto, the subamalgamation matrix S is actually an amalgamation matrix.

Conversely, if $A_G S = S A_H$ for an amalgamation matrix S, the maps in (8–2–3) are bijections, giving a right-covering code $\phi : X_G \to X_H$ which must therefore be onto by Proposition 8.2.2.

The remaining claim is true since right-resolving factor codes always preserve entropy.   □

Example 8.2.7. Let A be an $r \times r$ matrix all of whose rows sum to 2 and let $B = [\,2\,]$. If S is the $r \times 1$ matrix of all 1's, then S is an amalgamation matrix and $AS = SB$, showing that there is a right-resolving factor code ϕ from X_A onto the full 2-shift X_B. This is a special case of Proposition 8.1.12. In this case, ϕ is given by a road-coloring of G_A.   □

Example 8.2.8. Let (G, Φ) be the labeled graph shown in Figure 8.2.3(a), and H the graph with $\mathcal{E}(H) = \{a, b, c\}$ shown in (b).

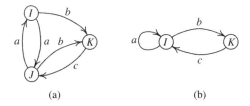

FIGURE 8.2.3. A labeling gives a right-resolving code.

The labeling Φ induces a right-resolving code $\phi = \Phi_\infty \colon X_G \to X_H$ whose corresponding amalgamation matrix is

$$
S = \begin{bmatrix} 1 & 0 \\ 1 & 0 \\ 0 & 1 \end{bmatrix}.
$$

Then

$$
A_G S = \begin{bmatrix} 0 & 1 & 1 \\ 1 & 0 & 1 \\ 0 & 1 & 0 \end{bmatrix} \begin{bmatrix} 1 & 0 \\ 1 & 0 \\ 0 & 1 \end{bmatrix} = \begin{bmatrix} 1 & 0 \\ 1 & 0 \\ 0 & 1 \end{bmatrix} \begin{bmatrix} 1 & 1 \\ 1 & 0 \end{bmatrix} = S A_H,
$$

giving a matrix verification that ϕ is onto. $\qquad\square$

There is a dual version of Proposition 8.2.6 for left-resolving codes: for irreducible graphs G and H, there exists a left-resolving factor code from X_H to X_G if and only if there is a division matrix D such that $A_G D = D A_H$.

Actually, we have already seen a special case of this. Recall from Theorem 2.4.12 that if H is an out-splitting of G, then $A_G = DE$ and $A_H = ED$ where D is a division matrix. Our dual version of Proposition 8.2.6 shows that in this case there must be a left-resolving factor code from X_H onto X_G. The out-amalgamation code (under which e^j maps to e) is just such a factor code, since for each edge e in G the only edge in H that terminates at $t(e)^j$ and maps to e is e^j. Similarly, in-amalgamation codes are right-resolving.

EXERCISES

8.2.1. For each of the following pairs of adjacency matrices A_G, A_H, find an amalgamation matrix S satisfying $A_G S = S A_H$ and hence a right-covering code $\phi \colon X_G \to X_H$.

(a)

$$
A_G = \begin{bmatrix} 1 & 1 & 0 & 0 \\ 0 & 0 & 1 & 1 \\ 1 & 0 & 0 & 1 \\ 0 & 1 & 1 & 0 \end{bmatrix} \quad \text{and} \quad A_H = \begin{bmatrix} 2 \end{bmatrix},
$$

(b)

$$A_G = \begin{bmatrix} 0 & 1 & 0 & 0 & 1 \\ 0 & 0 & 0 & 1 & 0 \\ 1 & 0 & 0 & 0 & 0 \\ 0 & 0 & 1 & 0 & 1 \\ 0 & 0 & 1 & 0 & 0 \end{bmatrix} \quad \text{and} \quad A_H = \begin{bmatrix} 0 & 1 & 0 \\ 1 & 0 & 1 \\ 1 & 0 & 0 \end{bmatrix}.$$

8.2.2. Let H be the graph consisting of one state and n self-loops. Characterize the irreducible graphs G such that X_H is a right-resolving factor of X_G.

8.2.3. Let H be the graph with adjacency matrix $\begin{bmatrix} 1 & 1 \\ 1 & 0 \end{bmatrix}$. Find all irreducible graphs G (up to graph isomorphism) with at most four states such that X_H is a right-resolving factor of X_G.

8.2.4. Let G and H be irreducible graphs, and let $\Phi\colon G \to H$ be a right-covering graph homomorphism. Let $\mathbf{v}$ be a right Perron eigenvector for A_H. Define $\mathbf{x}$ to be the vector, indexed by states of G, with $\mathbf{x}_I = \mathbf{v}_{\partial\Phi(I)}$. Show that $\mathbf{x}$ is a right Perron eigenvector for A_G (so right Perron eigenvectors "lift" under right-covering codes).

8.2.5. Let G and H be irreducible graphs. Assume that the out-degrees of states of H are all distinct and that H has no multiple edges. Show that there is at most one right-covering code $\phi\colon X_G \to X_H$.

8.2.6. Let X be an irreducible sofic shift with minimal right-resolving presentation $(G, \mathcal{L})$.
  (a) Let $(G', \mathcal{L}')$ be an irreducible right-resolving presentation of X. Show that there is graph homomorphism $\Phi\colon G' \to G$ which is a right-covering such that $\mathcal{L}'_\infty = \mathcal{L}_\infty \circ \Phi_\infty$. [*Hint:* Use Corollary 3.3.20.]
  (b) Let Z be an irreducible shift of finite type. Show that any right-closing factor code $\psi\colon Z \to X$ *factors through* $\mathcal{L}_\infty$ in the sense that $\psi = \mathcal{L}_\infty \circ \phi$ for some right-closing factor code $\phi\colon Z \to X_G$.

8.2.7. Let $\mathcal{G}$ be the labeled graph whose symbolic adjacency matrix is

$$\begin{bmatrix} \emptyset & a & b+e \\ a & \emptyset & b \\ c & d & \emptyset \end{bmatrix}.$$

  (a) Show that $\mathcal{G}$ is the minimal right-resolving presentation of $X = X_\mathcal{G}$.
  (b) Show that the minimal left-resolving presentation $\mathcal{G}' = (G', \mathcal{L}')$ of X has a self-loop labeled a.
  (c) Conclude that $\mathcal{L}'_\infty$ does not factor through $\mathcal{L}_\infty$ and explain why this does not contradict the result of Exercise 8.2.6.

8.2.8. For an irreducible labeled graph $(G, \mathcal{L})$, give a procedure to decide if the image $\mathcal{L}_\infty(X_G)$ is an edge shift. [*Hint:* See Exercise 2.3.4.]

8.2.9. Prove the implication $(1) \Rightarrow (2)$ of Proposition 8.2.2 using the uniqueness of the minimal right-resolving presentation of an irreducible sofic shift but not using entropy.

8.2.10. Let G and H be graphs. Show that if we assume only that H is irreducible (but not necessarily G), then the implications $(2) \Rightarrow (1)$ and $(1) \Longleftrightarrow (3)$ in Proposition 8.2.2 are still valid, but that $(1) \Rightarrow (2)$ no longer holds. Show

that if we assume G is irreducible (but not necessarily H), the implications
(1) $\Rightarrow$ (2) and (1) $\Rightarrow$ (3) in Proposition 8.2.2 are still valid, but no others.

8.2.11. (a) Show that if X is an irreducible sofic shift, Y an irreducible shift of finite type, and Y a right-closing factor of X, then X must be a shift of finite type.

(b) Show that the result of part (a) need not hold if Y is merely a finite-to-one factor of X.

§8.3. Finite Equivalence

In this section we consider a coding relation between shift spaces called finite equivalence. This relation is considerably weaker than conjugacy, yet it and its variants have interesting consequences (see §8.4). We will prove that, unlike conjugacy, finite equivalence is completely determined by the simple numerical invariant entropy.

Definition 8.3.1. Shift spaces X and Y are *finitely equivalent* if there is a shift of finite type W together with finite-to-one factor codes $\phi_X \colon W \to X$ and $\phi_Y \colon W \to Y$. We call W a *common extension*, and ϕ_X, ϕ_Y the *legs*. The triple (W, ϕ_X, ϕ_Y) is a *finite equivalence* between X and Y.

A finite equivalence has the following diagram which explains the terminology "legs."

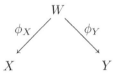

If shift spaces are finitely equivalent, they must be sofic since the definition requires that the common extension have finite type. We require this strong assumption about the common extension since the constructions of this section give finite-type common extensions, and weaker assumptions on the common extension result in the same equivalence relation (Exercise 8.3.6).

For sofic shifts conjugacy implies finite equivalence. To see this, let X and Y be sofic shifts, $\phi \colon X \to Y$ a conjugacy and $(G, \mathcal{L})$ a right-resolving presentation of X. Then $(X_G, \mathcal{L}_\infty, \phi \circ \mathcal{L}_\infty)$ is a finite equivalence between X and Y.

Suppose that X is an irreducible sofic shift with $h(X) = \log n$. Consider a finite-state (X, n)-code $(G, \mathfrak{I}, \mathfrak{O})$ as constructed in Chapter 5; that is, G has constant out-degree n, $\mathfrak{I}$ is a road-coloring of G, and $\mathfrak{O}$ is right-closing. Then $\mathfrak{I}_\infty(X_G) = X_{[n]}$ and $\mathfrak{O}_\infty(X_G) \subseteq X$. Since

$$\log n = h(X_G) = h(\mathfrak{O}_\infty(X_G)) \leqslant h(X) = \log n$$

and X is irreducible, Corollary 4.4.9 shows that $\mathfrak{O}_\infty(X_G) = X$. Hence $(X_G, \mathfrak{I}_\infty, \mathfrak{O}_\infty)$ is a finite equivalence between $X_{[n]}$ and X. We can regard the Finite-State Coding Theorem of Chapter 5 as a strengthening, in a particular case, of the main result of this section, the Finite Equivalence Theorem 8.3.7. Indeed, attempts to better understand the Finite Equivalence Theorem eventually led to the Finite-State Coding Theorem.

It follows immediately from the definition that finite equivalence is reflexive and symmetric, but transitivity is harder to show. The following construction is used to verify transitivity.

Definition 8.3.2. Let X, Y, and Z be shift spaces, and $\psi_X \colon X \to Z$ and $\psi_Y \colon Y \to Z$ be sliding block codes. The *fiber product* of (ψ_X, ψ_Y) is the triple (W, ϕ_X, ϕ_Y), where

$$W = \{(x, y) \in X \times Y : \psi_X(x) = \psi_Y(y)\},$$

$\phi_X(x, y) = x$, and $\phi_Y(x, y) = y$. The maps ϕ_X and ϕ_Y are the *projections* on the fiber product.

The fiber product of (ψ_X, ψ_Y) has the following diagram.

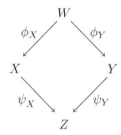

This diagram commutes since

$$\psi_X \circ \phi_X(x, y) = \psi_X(x) = \psi_Y(y) = \psi_Y \circ \phi_Y(x, y).$$

The fiber product generalizes the label product of labeled graphs (introduced in §3.4) in the special case where the sliding block codes ψ_X and ψ_Y are 1-block codes induced by labelings with the same alphabet.

Some specific properties of the maps ψ_X and ψ_Y are inherited by their "opposites" in the fiber product.

Proposition 8.3.3. *Let the fiber product of (ψ_X, ψ_Y) be (W, ϕ_X, ϕ_Y). For each of the following properties, if ψ_X (respectively ψ_Y) satisfies the property, then so does ϕ_Y (respectively ϕ_X).*

(1) *One-to-one.*
(2) *Onto.*
(3) *Right-resolving.*
(4) *Right-closing.*
(5) *Finite-to-one.*

PROOF: We verify (2), the rest being similar. Let $y \in Y$. Since ψ_X is onto, there is an $x \in X$ such that $\psi_X(x) = \psi_Y(y)$. Then $(x, y) \in W$ and $\phi_Y(x, y) = y$, proving that ϕ_Y is onto. $\square$

We can now show that finite equivalence is transitive.

Proposition 8.3.4. *Finite equivalence is an equivalence relation.*

PROOF: Suppose that (W, ϕ_X, ϕ_Y) is a finite equivalence between X and Y and (V, ϕ'_Y, ϕ'_Z) is a finite equivalence between Y and Z. Let $(\widetilde{W}, \phi_W, \phi_V)$ be the fiber product of (ϕ_Y, ϕ'_Y) as shown in the following diagram.

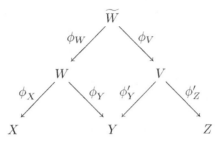

Since W and V are shifts of finite type, so is $W \times V$. The condition $\phi_Y(w) = \phi'_Y(v)$ on $W \times V$ depends on only finitely many coordinates, so that $\widetilde{W}$ is again a shift of finite type. Furthermore, ϕ_W is a finite-to-one factor code since its opposite ϕ'_Y is, and similarly for ϕ_V. Since the composition of finite-to-one factor codes is again a finite-to-one factor code, we see that $(\widetilde{W}, \phi_X \circ \phi_W, \phi'_Z \circ \phi_V)$ is a finite equivalence between X and Z. $\square$

Proposition 8.3.3(5) together with the proof of the preceding result shows that when two shifts of finite type have a common finite-to-one factor, then they are finitely equivalent. The following is an example where two shifts of finite type have a common finite-to-one factor and hence are finitely equivalent, but neither factors onto the other.

Example 8.3.5. Let

$$A = \begin{bmatrix} 0 & 2 \\ 2 & 0 \end{bmatrix}, \quad B = \begin{bmatrix} 0 & 2 & 0 \\ 0 & 0 & 2 \\ 2 & 0 & 0 \end{bmatrix},$$

$X = X_A$, and $Y = X_B$. Then X and Y both factor onto the full 2-shift by road-colorings. So, by the remarks above, they are finitely equivalent. Now X has a point x of period two. Since the image of x under a sliding block code must have either period one or two, and Y has no such points, it follows that there is no sliding block code from X to Y. A similar argument, using a point of period three in Y, shows that there is also no sliding block code from Y to X. For a mixing example, see Exercise 8.3.1(b). $\square$

When two irreducible sofic shifts are finitely equivalent, the following shows that we can arrange that the common extension is irreducible.

Proposition 8.3.6. *Let X and Y be irreducible sofic shifts. If (W, ϕ_X, ϕ_Y) is a finite equivalence between X and Y, then for every irreducible shift of finite type V contained in W with $h(V) = h(W)$ the triple $(V, \phi_X|_V, \phi_Y|_V)$ is also a finite equivalence between X and Y. Hence, if two irreducible sofic shifts are finitely equivalent, then there is a finite equivalence with an irreducible common extension.*

PROOF: Let V be an irreducible shift of finite type of maximal entropy contained in W. Such V's exist by the decomposition of a general adjacency matrix into irreducible components (see §4.4). Then

$$h(\phi_X(V)) = h(V) = h(W) = h(X),$$

and since X is irreducible it follows that $\phi_X(V) = X$. Similarly, we see that $\phi_Y(V) = Y$. Since $\phi_X|_V$ and $\phi_Y|_V$ are clearly finite-to-one, it follows that

$$(V, \phi_X|_V, \phi_Y|_V)$$

is a finite equivalence between X and Y with an irreducible common extension. □

We are now ready for the main result of this section.

Theorem 8.3.7 (Finite Equivalence Theorem). *Two irreducible sofic shifts are finitely equivalent if and only if they have the same entropy.*

Since finite-to-one codes preserve entropy, equality of entropy is clearly necessary for finite equivalence. For sufficiency, observe first that by passing to irreducible right-resolving presentations we may assume that the shifts are irreducible edge shifts, say X_G and X_H. The proof uses two main ideas. The first is that since $\lambda_{A_G} = \lambda_{A_H}$, there is a nonnegative integer matrix $F \neq 0$ such that $A_G F = F A_H$. The second idea is to use this F to construct, in a concrete manner, a finite equivalence between X_G and X_H (observe that a special case of this was done in Proposition 8.2.6). This construction is reminiscent of our construction of conjugacies from elementary equivalences between matrices in §7.2. We will actually prove somewhat more.

Theorem 8.3.8. *Let G and H be irreducible graphs. Then the following are equivalent.*

(1) *X_G and X_H are finitely equivalent.*
(2) *$h(X_G) = h(X_H)$.*
(3) *There is a positive integer matrix F such that $A_G F = F A_H$.*
(4) *There is a nonnegative integer matrix $F \neq 0$ such that $A_G F = F A_H$.*
(5) *There is a finite equivalence (X_M, ϕ_G, ϕ_H) between X_G and X_H for which $\phi_G: X_M \to X_G$ is left-resolving and $\phi_H: X_M \to X_H$ is right-resolving.*

In particular, for irreducible sofic shifts, entropy is a complete invariant of finite equivalence. Moreover, the common extension guaranteed in parts (1) and (5) can be chosen to be irreducible.

PROOF: (1) ⇒ (2). Finite-to-one factors preserve entropy. Thus if W is the common extension of a finite equivalence between X_G and X_H, then $h(X_G) = h(W) = h(X_H)$.

(2) ⇒ (3). Let $A = A_G$, $B = A_H$, and $\lambda = \lambda_A = \lambda_B$ be the common Perron eigenvalue. Let $\mathbf{v} > 0$ be a right Perron eigenvector for A and $\mathbf{w} > 0$ a left Perron eigenvector for B, so that $A\mathbf{v} = \lambda\mathbf{v}$ and $\mathbf{w}B = \lambda\mathbf{w}$. Put $E = \mathbf{v}\mathbf{w}$, which is a positive rectangular matrix.

Note that

$$AE = (A\mathbf{v})\mathbf{w} = (\lambda\mathbf{v})\mathbf{w} = \lambda E$$

and

$$EB = \mathbf{v}(\mathbf{w}B) = \mathbf{v}(\lambda\mathbf{w}) = \lambda E,$$

so that E is a positive real matrix with $AE = EB$. If E were integral, then we could put $F = E$ and be done. Although E is typically *not* integral, we will show that the existence of a positive solution E to $AE = EB$ implies the existence of a positive integral solution F to $AF = FB$.

Let m denote the size of A and n the size of B. The equation $AF = FB$ is equivalent to the set of mn equations

$$(8\text{--}3\text{--}1) \qquad \sum_{K=1}^{m} A_{IK}F_{KJ} - \sum_{L=1}^{n} F_{IL}B_{LJ} = 0,$$

where $1 \leqslant I \leqslant m$, $1 \leqslant J \leqslant n$, and the A_{IK}, B_{LJ} are regarded as fixed integer coefficients while the F_{KJ} are the unknowns. Considering a matrix $F = [F_{IJ}]$ as an element of $\mathbb{R}^{mn}$, we know that the subspace $V \subseteq \mathbb{R}^{mn}$ of solutions to the mn equations (8–3–1) contains the positive element E. Now the coefficients of the equations (8–3–1) are integers, and it is a consequence of the Gaussian elimination method that V has a basis consisting of rational solutions, say $F_1, F_2, \ldots, F_r$ (see Exercise 8.3.10 for a proof of this). Hence E can be expressed as a linear combination $t_1F_1 + \cdots + t_rF_r$ for some $t_j \in \mathbb{R}$. By choosing rational numbers q_j close enough to the corresponding t_j we obtain a rational matrix solution $F_0 = q_1F_1 + \cdots + q_rF_r \in V$ that is positive. Multiplying F_0 by the least common multiple of the denominators of its entries produces a positive integral matrix F with $AF = FB$, as required.

(3) ⇒ (4). This is immediate.

(4) ⇒ (5). We use the matrix F to construct the required common extension. First, we introduce an auxiliary graph.

Let $A = A_G$ and $B = A_H$. We form the auxiliary graph T as follows. The vertex set $\mathcal{V}(T)$ of T is the disjoint union of $\mathcal{V}(G)$ and $\mathcal{V}(H)$. There are three types of edges in T: A-edges, B-edges, and F-edges. The A-edges are simply all edges in G, now regarded as having initial and terminal states in $\mathcal{V}(T)$. The B-edges are defined similarly using H. The F-edges are new edges between G and H; specifically, for each pair $I \in \mathcal{V}(G)$ and $J \in \mathcal{V}(H)$, we form F_{IJ} edges in T from I to J.

Note that for each $I \in \mathcal{V}(G)$ and $J \in \mathcal{V}(H)$, a path in T from I to J of length two is either a *left-bottom path*, by which we mean an F-edge followed by a B-edge, or a *top-right path*, i.e., an A-edge followed by an F-edge. These can be pictured as follows.

$$
\begin{array}{ccc}
I & & \\
\ \ \downarrow f & & \\
& \xrightarrow{\ \ b\ \ } & J
\end{array}
\qquad\qquad
\begin{array}{ccc}
I & \xrightarrow{\ a\ } & \\
& & \downarrow f' \\
& & J
\end{array}
$$

Let $LB(I, J)$ denote the set of all left-bottom paths from I to J, and $TR(I, J)$ the set of all top-right paths from I to J.

For each pair of states $I \in \mathcal{V}(G)$ and $J \in \mathcal{V}(H)$, the equation $(AF)_{IJ} = (FB)_{IJ}$ is equivalent to

$$|LB(I, J)| = |TR(I, J)|.$$

Thus for each I, J we can choose a bijection

$$\Gamma_{IJ} \colon LB(I, J) \to TR(I, J).$$

We assemble all these bijections to form a bijection

$$\Gamma \colon \bigcup_{\substack{I \in \mathcal{V}(G) \\ J \in \mathcal{V}(H)}} LB(I, J) \to \bigcup_{\substack{I \in \mathcal{V}(G) \\ J \in \mathcal{V}(H)}} TR(I, J).$$

Each left-bottom path fb is paired by Γ to a top-right pair af', and each such pair can be put together to form a "box" as follows.

$$
\begin{array}{ccc}
I & & \\
\ \ \downarrow f & & \\
& \xrightarrow{\ \ b\ \ } & J
\end{array}
\quad \overset{\Gamma}{\longleftrightarrow} \quad
\begin{array}{ccc}
I & \xrightarrow{\ a\ } & \\
& & \downarrow f' \\
& & J
\end{array}
\quad \Longrightarrow \quad
\begin{array}{ccc}
I & \xrightarrow{\ a\ } & \\
\ \ \downarrow f & & \downarrow f' \\
& \xrightarrow{\ \ b\ \ } & J
\end{array}
$$

We denote the resulting box by $\Box(f, a, b, f')$. Such a box is determined by its left-bottom path fb and also by its top-right path af'.

Using T we now construct the required common extension using a new graph $M = M(F, \Gamma)$. The vertex set $\mathcal{V}(M)$ is the set of all F-edges in T. The edge set $\mathcal{E}(M)$ is the set of all boxes $\Box(f, a, b, f')$. The initial state of $\Box(f, a, b, f')$ is $f \in \mathcal{V}(M)$ and its terminal state is $f' \in \mathcal{V}(M)$.

Now define $\Phi_H : \mathcal{E}(M) \to \mathcal{E}(H)$ by

$$\Phi_H\big(\square(f,a,b,f')\big) = b.$$

It is clear that Φ_H is the edge map of a graph homomorphism from M to H, since if $\square(f,a,b,f')$ is followed by $\square(f',a',b',f'')$ in M, then $t(b) = t(f') = i(b')$, so that b is followed by b' in H.

Hence $\phi_H = (\Phi_H)_\infty$ defines a 1-block code from X_M to X_H. We claim that ϕ_H is a right-resolving factor code. Suppose that $\square(f,a,b,f')$ is an edge in M with initial state f and Φ_H-image b. This information tells us the left-bottom part fb and hence the entire box $\square(f,a,b,f')$, verifying that Φ_H is right-resolving. In fact, once f is fixed, Γ establishes a bijection between boxes starting at f and edges in H starting at $t(f)$, so that ϕ_H is actually a right-covering code. Since X_H is irreducible, by using Remark 8.2.3 we can conclude that ϕ_H is onto.

Let $\Phi_G : \mathcal{E}(M) \to \mathcal{E}(G)$ be defined by $\Phi_G(\square(f,a,b,f')) = a$. A similar argument shows that $\phi_G = (\Phi_G)_\infty$ is a left-resolving factor code from X_M onto X_G.

(5) $\Rightarrow$ (1). This is immediate.

The existence of an irreducible common extension in parts (1) and (5) follows from Proposition 8.3.6 $\qquad\square$

We can use Theorem 8.3.8 to refine Theorem 8.3.7.

Theorem 8.3.9. *If two irreducible sofic shifts have the same entropy, then there is a finite equivalence between them for which one leg is left-resolving and the other is right-resolving.*

PROOF: Apply Theorem 8.3.8 to an irreducible left-resolving presentation of one of the sofic shifts and an irreducible right-resolving presentation of the other. $\qquad\square$

According to Exercise 7.1.7, if X_G and X_H are conjugate then we can pass from G to H by a sequence of out-splittings followed by a sequence of in-amalgamations. Recall from the end of §8.2 that out-amalgamation codes are left-resolving and in-amalgamation codes are right-resolving. Thus we can conclude the following.

Proposition 8.3.10. *Let G and H be irreducible graphs. Then X_G and X_H are conjugate if and only if there are an irreducible graph M, a left-resolving conjugacy $X_M \to X_G$, and a right-resolving conjugacy $X_M \to X_H$.*

Observe that the equivalence of statements (1) and (5) in Theorem 8.3.8 is an analogue of the preceding result.

Our proof of the Finite Equivalence Theorem treated the graphs symmetrically. There is another way to construct a finite equivalence, called *filling in the tableau*, which provides an effective notation for the finite equivalence but does so in an unsymmetric way. We illustrate this method with an example.

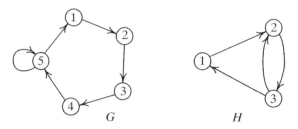

FIGURE 8.3.1.   Graphs for filling in tableau.

Example 8.3.11. Let G and H be the graphs shown in Figure 8.3.1. The corresponding adjacency matrices are

$$A = A_G = \begin{bmatrix} 0 & 1 & 0 & 0 & 0 \\ 0 & 0 & 1 & 0 & 0 \\ 0 & 0 & 0 & 1 & 0 \\ 0 & 0 & 0 & 0 & 1 \\ 1 & 0 & 0 & 0 & 1 \end{bmatrix} \quad \text{and} \quad B = A_H = \begin{bmatrix} 0 & 1 & 0 \\ 0 & 0 & 1 \\ 1 & 1 & 0 \end{bmatrix}.$$

Since both G and H do not have multiple edges, we can denote edges in each with the notation $I \to I'$. For example, $1 \to 2$ in G is the unique edge from state 1 to state 2.

Let

$$F = \begin{bmatrix} 0 & 0 & 1 \\ 1 & 1 & 0 \\ 0 & 1 & 1 \\ 1 & 1 & 1 \\ 1 & 2 & 1 \end{bmatrix}.$$

It is easy to verify that $AF = FB$. We use F to construct a finite equivalence between X_G and X_H. There happens to be a smaller F that also satisfies $AF = FB$, but this choice of F better illustrates the ideas of the proof.

For this we need a convenient notation for F-edges. For $I \in \mathcal{V}(G)$ and $J \in \mathcal{V}(H)$ there are F_{IJ} F-edges in the auxiliary graph T from I to J, and we name them $(I, J, 1)$, $(I, J, 2)$, $\ldots$, (I, J, F_{IJ}). At this point, you may find it helpful to sketch T with the F-edges labeled. The vertices of M are the F-edges of T, which in this example are as follows.

$(1,3,1)$	$(3,2,1)$	$(4,2,1)$	$(5,2,1)$
$(2,1,1)$	$(3,3,1)$	$(4,3,1)$	$(5,2,2)$
$(2,2,1)$	$(4,1,1)$	$(5,1,1)$	$(5,3,1)$

We will describe the edges of M by a table of allowed transitions called the *tableau*. We begin by writing down a partially filled-in version of this table called the *skeleton*. The left column of the skeleton contains each F-edge (I, J, k) followed by an arrow $\to$. We list the F-edges in lexicographic order. To the right of $(I, J, k) \to$ we place triples of the form $(\cdot, J', \cdot)$, one for each edge $J \to J'$ in H. The resulting skeleton is shown below.

$$
\begin{array}{lll}
(1,3,1) & \longrightarrow & (\cdot,1,\cdot) \quad (\cdot,2,\cdot) \\
(2,1,1) & \longrightarrow & (\cdot,2,\cdot) \\
(2,2,1) & \longrightarrow & (\cdot,3,\cdot) \\
(3,2,1) & \longrightarrow & (\cdot,3,\cdot) \\
(3,3,1) & \longrightarrow & (\cdot,1,\cdot) \quad (\cdot,2,\cdot) \\
(4,1,1) & \longrightarrow & (\cdot,2,\cdot) \\
(4,2,1) & \longrightarrow & (\cdot,3,\cdot) \\
(4,3,1) & \longrightarrow & (\cdot,1,\cdot) \quad (\cdot,2,\cdot) \\
(5,1,1) & \longrightarrow & (\cdot,2,\cdot) \\
(5,2,1) & \longrightarrow & (\cdot,3,\cdot) \\
(5,2,2) & \longrightarrow & (\cdot,3,\cdot) \\
(5,3,1) & \longrightarrow & (\cdot,1,\cdot) \quad (\cdot,2,\cdot)
\end{array}
$$

Next, we "fill in the tableau," i.e., assign values to the dots. Fix states $I \in \mathcal{V}(G)$ and $J' \in \mathcal{V}(H)$. We restrict our attention to the set $\mathcal{R}_I$ of rows in the skeleton whose left triple starts with I. For each edge $I \to I'$ in G and $1 \leqslant k' \leqslant F_{I'J'}$, replace exactly one appearance of $(\cdot, J', \cdot)$ in $\mathcal{R}_I$ by (I', J', k'). Doing this for each pair I, J' then defines the filled-in tableau.

For example, consider $I = 4$ and $J' = 2$. The relevant part of the skeleton is

$$
\begin{array}{lll}
(4,1,1) & \longrightarrow & (\cdot,2,\cdot) \\
(4,3,1) & \longrightarrow & (\cdot,2,\cdot)
\end{array}
$$

Since $4 \to 5$ is the only edge in G starting at 4, the initial dots of both triples must be filled in with 5. There are exactly two F-edges of the form $(5,2,\cdot)$, and filling the terminal dots gives this part of the tableau to be

$$
\begin{array}{lll}
(4,1,1) & \longrightarrow & (5,2,1) \\
(4,3,1) & \longrightarrow & (5,2,2)
\end{array}
$$

For another example, consider $I = 5$ and $J' = 3$. This part of the skeleton is

$$
\begin{array}{lll}
(5,2,1) & \longrightarrow & (\cdot,3,\cdot) \\
(5,2,2) & \longrightarrow & (\cdot,3,\cdot)
\end{array}
$$

The edges $5 \to 1$ and $5 \to 5$ in G are used to fill in this part of the tableau, giving

$$
\begin{array}{lll}
(5,2,1) & \longrightarrow & (1,3,1) \\
(5,2,2) & \longrightarrow & (5,3,1)
\end{array}
$$

In our example, the full tableau is as follows.

$$
\begin{array}{lll}
(1,3,1) & \longrightarrow & (2,1,1) \quad (2,2,1) \\
(2,1,1) & \longrightarrow & (3,2,1)
\end{array}
$$

$$
\begin{array}{lll}
(2,2,1) & \longrightarrow & (3,3,1) \\
(3,2,1) & \longrightarrow & (4,3,1) \\
(3,3,1) & \longrightarrow & (4,1,1) \quad (4,2,1) \\
(4,1,1) & \longrightarrow & (5,2,1) \\
(4,2,1) & \longrightarrow & (5,3,1) \\
(4,3,1) & \longrightarrow & (5,1,1) \quad (5,2,2) \\
(5,1,1) & \longrightarrow & (5,2,1) \\
(5,2,1) & \longrightarrow & (1,3,1) \\
(5,2,2) & \longrightarrow & (5,3,1) \\
(5,3,1) & \longrightarrow & (5,1,1) \quad (5,2,2)
\end{array}
$$

This completes our description of filling in the tableau.

We claim that this filling-in process is really the same as making a choice Γ of bijections, as in the proof of the implication (4) $\Rightarrow$ (5) in Theorem 8.3.8, and that the graph determined by the transitions in the tableau is isomorphic to the graph $M = M(F, \Gamma)$ constructed there.

To see this, fix states $I \in \mathcal{V}(G)$ and $J' \in \mathcal{V}(H)$. Each left-bottom path from I to J' in the auxiliary graph T corresponds to a choice of row (I, J, k) (the left edge) and entry $(\cdot, J', \cdot)$ on the right of the arrow in the skeleton (the bottom edge $J \to J'$), giving the picture below.

$$
\begin{array}{c}
I \\
(I,J,k)\Big\downarrow \qquad\qquad \\
J \xrightarrow[\;(\cdot,\,J',\,\cdot)\;]{} J'
\end{array}
$$

This shows that the left-bottom paths in T correspond to entries $(\cdot, J', \cdot)$ in the skeleton yet to be filled in. Each top-right path from I to J' corresponds to a choice of edge $I \to I'$ (the top edge) and a choice of state (I', J', k') (the right edge), as follows.

$$
\begin{array}{c}
I \xrightarrow{\qquad\qquad} I' \\
\qquad\qquad \Big\downarrow (I',J',k') \\
\qquad\qquad J'
\end{array}
$$

Thus the top-right paths provide the ways to fill in the appearances of $(\cdot, J', \cdot)$ in the skeleton. This gives the part $\Gamma_{IJ'}$ of the bijection Γ, and together these define the graph $M = M(F, \Gamma)$. The edge $(I, J, k) \to (I', J', k')$ in this graph corresponds to the following box.

$$
\begin{array}{c}
I \xrightarrow{\qquad\qquad} I' \\
(I,J,k)\Big\downarrow \qquad\qquad \Big\downarrow (I',J',k') \\
J \xrightarrow[\;(\cdot,\,J',\,\cdot)\;]{} J'
\end{array}
$$

For instance, the part Γ_{53} of the bijection can be described as follows.

$$
\begin{array}{cccc}
f & b & \longleftrightarrow & a \qquad f' \\
(5,2,1) & (2 \to 3) & & (5 \to 1) \quad (1,3,1) \\
(5,2,2) & (2 \to 3) & & (5 \to 5) \quad (5,3,1)
\end{array}
$$

To complete our construction of the finite equivalence between X_H and X_G, observe that the 1-block code $\phi_H = (\Phi_H)_\infty$ is defined by

$$\Phi_H\big((I, J, k) \to (I', J', k')\big) = (J \to J')$$

and $\phi_G = (\Phi_G)_\infty$ by

$$\Phi_G\big((I, J, k) \to (I', J', k')\big) = (I \to I').$$

You should now verify, as in the proof of (4) $\Rightarrow$ (5) in Theorem 8.3.8, that ϕ_G is a left-resolving factor code onto X_G and that ϕ_H is a right-resolving factor code onto X_H. You should also check that M is already irreducible in this example, so there is no need to pass to an irreducible component. $\square$

EXERCISES

8.3.1. Exhibit a finite equivalence between X_A and X_B for the following pairs of matrices by using the indicated matrix F.

(a)

$$
A = \begin{bmatrix} 1 & 1 & 0 & 0 & 0 \\ 1 & 0 & 1 & 0 & 0 \\ 0 & 0 & 0 & 1 & 0 \\ 0 & 1 & 0 & 0 & 1 \\ 0 & 0 & 0 & 1 & 0 \end{bmatrix}, \quad
B = \begin{bmatrix} 0 & 1 & 0 & 1 \\ 1 & 0 & 1 & 1 \\ 1 & 0 & 0 & 0 \\ 1 & 0 & 0 & 0 \end{bmatrix},
$$

$$
\text{and} \quad F = \begin{bmatrix} 1 & 0 & 0 & 1 \\ 0 & 1 & 0 & 0 \\ 0 & 0 & 1 & 0 \\ 1 & 0 & 0 & 0 \\ 0 & 0 & 0 & 1 \end{bmatrix}.
$$

(b)

$$
A = \begin{bmatrix} 0 & 1 & 0 & 0 & 1 & 0 & 0 \\ 1 & 0 & 0 & 1 & 0 & 0 & 0 \\ 1 & 0 & 0 & 0 & 0 & 0 & 0 \\ 0 & 0 & 0 & 0 & 0 & 1 & 0 \\ 0 & 0 & 0 & 0 & 0 & 0 & 1 \\ 0 & 0 & 1 & 0 & 0 & 0 & 0 \\ 0 & 0 & 1 & 0 & 0 & 0 & 0 \end{bmatrix}, \quad
B = \begin{bmatrix} 0 & 1 & 0 & 1 & 0 & 0 \\ 0 & 0 & 1 & 0 & 0 & 0 \\ 1 & 0 & 0 & 0 & 0 & 0 \\ 0 & 0 & 0 & 0 & 0 & 1 \\ 1 & 1 & 1 & 0 & 0 & 0 \\ 0 & 0 & 0 & 0 & 1 & 0 \end{bmatrix},
$$

$$
\text{and} \quad F = \begin{bmatrix} 1 & 1 & 1 & 0 & 0 & 0 \\ 1 & 1 & 1 & 0 & 0 & 0 \\ 0 & 0 & 0 & 0 & 1 & 0 \\ 0 & 0 & 0 & 1 & 0 & 0 \\ 0 & 0 & 0 & 1 & 0 & 0 \\ 0 & 0 & 0 & 0 & 0 & 1 \\ 0 & 0 & 0 & 0 & 0 & 1 \end{bmatrix}.
$$

For (b), show that neither X_A nor X_B factors onto the other.

8.3.2. Modify the tableau notation for the case where the graphs G and H have multiple edges and exhibit a finite equivalence between X_A and X_B for the following pairs of matrices.

(a)
$$A = \begin{bmatrix} 0 & 1 & 1 \\ 1 & 0 & 1 \\ 1 & 1 & 0 \end{bmatrix} \quad \text{and} \quad B = \begin{bmatrix} 1 & 2 \\ 1 & 0 \end{bmatrix}.$$

(b)
$$A = \begin{bmatrix} 3 & 2 \\ 2 & 1 \end{bmatrix} \quad \text{and} \quad B = \begin{bmatrix} 4 & 1 \\ 1 & 0 \end{bmatrix}.$$

8.3.3. Let X_A and X_B be irreducible shifts of finite type with $h(X_A) \leqslant h(X_B)$.

(a) Show that there is a positive integral solution F to the matrix inequality $AF \leqslant FB$.

(b) Use part (a) to construct a finite equivalence between X_A and some sofic subshift of X_B.

8.3.4. Show that entropy is not a complete invariant of finite equivalence for reducible shifts of finite type.

8.3.5. (a) Complete the proof of Proposition 8.3.3.

(b) Show that if ψ_Y is onto and ϕ_Y has any of the properties (1), (3), (4) or (5), then so does ψ_X.

8.3.6. (a) Show that, even if arbitrary shift spaces are allowed as common extensions, entropy is a complete invariant of finite equivalence for irreducible sofic shifts. [*Hint*: Exercise 8.1.16(a).]

(b) Show that, even if arbitrary shift spaces are allowed as common extensions, entropy is not a complete invariant of finite equivalence for irreducible shifts. [*Hint*: Exercise 4.1.11(a).]

8.3.7. (a) Show that the matrix E arising in the proof of (2) $\Rightarrow$ (3) of Theorem 8.3.8 has rank 1.

(b) Let
$$A = \begin{bmatrix} 3 & 2 \\ 2 & 1 \end{bmatrix} \quad \text{and} \quad B = \begin{bmatrix} 4 & 1 \\ 1 & 0 \end{bmatrix}.$$

Show that no integral solution to $AF = FB$ can have rank 1.

8.3.8. Show that if A and B are irreducible and $F \neq 0$ is a nonnegative matrix satisfying $AF = FB$, then no row or column of F is zero.

8.3.9. Let G, H and M be irreducible graphs such that there are a left-covering code $\phi_G: X_M \to X_G$ and a right-covering code $\phi_H: X_M \to X_H$. Show directly how to construct from ϕ_G and ϕ_H a nonnegative integral solution $F \neq 0$ to the equation $A_G F = F A_H$.

8.3.10. For any homogeneous linear system of equations with rational coefficients, show that the vector space of real solutions has a basis consisting of rational solutions. [*Hint*: Any such system can be viewed as an equation $Cx = 0$ where C is a rectangular rational matrix. Observe that the rational kernel of C is contained in the real kernel of C and the rational image of C is contained in the real image of C, and use the following fact from linear algebra: the dimensions of the kernel and image of a linear transformation sum to the dimension of the domain (over any field).]

8.3.11. Show that the transitivity of the relation of finite equivalence for irreducible sofic shifts is a direct consequence of Theorem 8.3.7. (Thus it was not necessary to prove Proposition 8.3.4.)

8.3.12. Show that if $X_1, \ldots, X_k$ are irreducible sofic shifts with the same entropy, then there are an irreducible edge shift X_M and finite-to-one factor codes $X_M \to X_1, X_M \to X_2, \ldots, X_M \to X_k$.

8.3.13. Complete the steps of the following alternate proof of the implication (2) $\Rightarrow$ (3) of Theorem 8.3.8.

(a) For any positive matrix M and any $\epsilon > 0$, show that there is a positive integer q such that $qM = N + P$ where N is a positive integer matrix and each entry of P is at most ϵ in absolute value. [*Hint*: Simultaneously approximate all of the entries of M by rational numbers; i.e., for any finite set $\alpha_1, \ldots, \alpha_k$ of positive numbers and $\epsilon > 0$ there are positive integers $p_1, \ldots, p_k$ and q such that $|\alpha_i - p_i/q| < \epsilon/q$ (see [HarW, Thm. 36]).]

(b) Show that if $M = E$, the matrix constructed in the proof of Theorem 8.3.8, and if ϵ is sufficiently small, then $AN = NB$, where N is as in part (a).

8.3.14. (a) Let Φ be a labeling of a graph G. For each state I in G and each block w in the language of $\Phi_\infty(X_G)$ let $S(I, w)$ denote the set of all terminal states of paths in G that begin at I and present w. Define a labeled graph (G_R, Φ_R) as follows. The states of G_R are the sets $S(I, w)$. For each $S = S(I, w) \in \mathcal{V}(G_R)$ and symbol a that is the label of an edge in G starting at some element of S, endow G_R with exactly one edge, labeled a, from S to $S(I, wa)$. Show that $\phi_R = (\Phi_R)_\infty$ is a right-resolving code with the same image as $\phi = \Phi_\infty$.

(b) Suppose that Φ is a right-closing labeling of an irreducible graph and let $\phi = \Phi_\infty$. Let (W, θ_1, θ_2) be the fiber product of ϕ, ϕ_R (where ϕ_R is as in part (a)). Show that there is an irreducible component W' of W such that $\theta_1|_{W'}$ is a right-resolving conjugacy and $\theta_2|_{W'}$ is a left-resolving conjugacy.

§8.4. Right-Resolving Finite Equivalence

According to Theorem 8.3.8 if two irreducible edge shifts X_G and X_H are finitely equivalent, then we can find a finite equivalence such that one leg is left-resolving, the other is right-resolving and the common extension X_M is an irreducible edge shift. If both legs could be chosen to be right-resolving, then we could build a "machine" to encode blocks from X_G into blocks from X_H in a way similar to a finite-state code (see §5.2). Namely, let $(\Phi_G)_\infty, (\Phi_H)_\infty$ denote the legs and choose an arbitrary initial state I_0 in M. Then "lift" a path in G starting at $\partial\Phi_G(I_0)$ to a path in M starting at I_0 and "push down" to a path in H via Φ_H. This specifies an encoder, and decoding is done in exactly the same way.

Let us call a finite equivalence between sofic shifts in which both legs are right-resolving a *right-resolving finite equivalence*. It is natural to wonder if equality of entropy, which is clearly necessary, is a sufficient condition for such a finite equivalence – equivalently if a finite equivalence between irreducible edge shifts can always be replaced by a right-resolving finite equivalence. This is not always possible. In this section we will see why, and we will find a simple necessary and sufficient criterion for the existence of a right-resolving finite equivalence.

Suppose that X_G and X_H are irreducible edge shifts with the same entropy and that (W, ϕ_G, ϕ_H) is a right-resolving finite equivalence between them. We may assume by Proposition 8.3.6 that the common extension W is an irreducible edge shift X_M. Let Φ_G and Φ_H be the edge labelings of M corresponding to ϕ_G and ϕ_H. By Proposition 8.2.2 both Φ_G and Φ_H are right-coverings. Hence the set of all out-degrees of states in M must equal the set of all out-degrees of states in G (although for a fixed n there may be more states in M than in G with out-degree n). A similar statement holds for H. Hence the set of out-degrees for G must equal the set of out-degrees for H. But equality of entropy, or even conjugacy, does not imply such a strong relationship between the defining graphs. For example, it is possible to out-split a typical graph and change the set of out-degrees. The following gives a simple example of this.

Example 8.4.1. Let G and H be the graphs defined by adjacency matrices

$$A_G = \begin{bmatrix} 1 & 1 \\ 1 & 1 \end{bmatrix} \quad \text{and} \quad A_H = \begin{bmatrix} 0 & 0 & 1 \\ 1 & 1 & 0 \\ 1 & 1 & 1 \end{bmatrix}.$$

Then H is formed from G by out-splitting, and so X_H is conjugate to X_G; in particular, X_G and X_H are finitely equivalent. But they cannot be right-resolving finitely equivalent since the set of out-degrees of G is $\{2\}$ while the set of out-degrees of H is $\{1, 2, 3\}$.   □

Using state splittings, it is possible for us to construct an infinite family of edge shifts all conjugate to each other, but no pair of which are right-resolving finitely equivalent (Exercise 8.4.3).

The rest of this section is devoted to describing a simple necessary and sufficient condition for the existence of a right-resolving finite equivalence. It is based on our observation above that the set of out-degrees is an invariant of this relation.

We first recall some terminology and introduce some notation. A *partially ordered set* $(R, \preceq)$ is a set R together with an ordering $\preceq$ on R that satisfies the following conditions:

(1) $x \preceq x$
(2) $x \preceq y$ and $y \preceq x$ imply $x = y$
(3) $x \preceq y$ and $y \preceq z$ imply $x \preceq z$.

An element $x \in R$ is called *smallest* if $x \preceq y$ for every $y \in R$. Not every partially ordered set has a smallest element, but when it does the smallest element is unique (why?).

For irreducible graphs G and H, we write $H \preceq G$ if X_H is a right-resolving factor of X_G (equivalently if there is a right-covering code from X_G onto X_H). Let $\mathcal{R}_G$ be the collection of graph-isomorphism classes of graphs H for which $H \preceq G$. The ordering $\preceq$ naturally determines an ordering, which we still call $\preceq$ on $\mathcal{R}_G$. Since the composition of right-resolving factor codes is also a right-resolving factor code and since $H \preceq G$ and $G \preceq H$ imply $G \cong H$, it follows that $(\mathcal{R}_G, \preceq)$ is a partially ordered set. Note also that $\mathcal{R}_G$ is finite, since if $H \preceq G$ then $|\mathcal{V}(H)| \leqslant |\mathcal{V}(G)|$ and $|\mathcal{E}(H)| \leqslant |\mathcal{E}(G)|$. We will sometimes abuse terminology by referring to elements of $\mathcal{R}_G$ as graphs rather than graph-isomorphism classes, but the meaning should be clear from context.

Theorem 8.4.2. *Let G be an irreducible graph. The partially ordered set $(\mathcal{R}_G, \preceq)$ has a smallest element M_G.*

PROOF: Let $\mathcal{V} = \mathcal{V}(G)$ be the set of vertices of G. We define a nested sequence of equivalence relations $\sim_n$ on $\mathcal{V}$ for $n \geqslant 0$. The partition of $\mathcal{V}$ into $\sim_n$ equivalence classes is denoted by $\mathcal{P}_n$.

To define $\sim_n$, first declare that $I \sim_0 J$ for all $I, J \in \mathcal{V}$. For $n \geqslant 1$, declare that $I \sim_n J$ if and only if for each class (or *atom*) $P \in \mathcal{P}_{n-1}$ the total number of edges from I to states in P equals the total number from J to states in P. For example, $\sim_1$ partitions states by their out-degree.

Next, we inductively show that the partitions $\mathcal{P}_n$ are nested; i.e., that each atom in $\mathcal{P}_n$ is a union of atoms in $\mathcal{P}_{n+1}$. Since $\mathcal{P}_0 = \{\mathcal{V}\}$, clearly $\mathcal{P}_1$ refines $\mathcal{P}_0$. Suppose that $\mathcal{P}_n$ refines $\mathcal{P}_{n-1}$. To show that $\mathcal{P}_{n+1}$ refines $\mathcal{P}_n$, we fix states I and J with $I \sim_{n+1} J$ and we prove that $I \sim_n J$. For $Q \in \mathcal{P}_n$, the total number of edges from I to states in Q equals the total number from J to states in Q. Now an atom P in $\mathcal{P}_{n-1}$ is the disjoint union of atoms Q in $\mathcal{P}_n$. It follows that the total number of edges from I to states in P equals the total number from J to states in P. Hence $I \sim_n J$ as claimed.

Since the $\mathcal{P}_n$ are nested and $\mathcal{V}$ is finite, it follows that the $\mathcal{P}_n$ are equal for all large enough n. Let $\mathcal{P}$ denote this limiting partition. We next use $\mathcal{P}$ to define the graph M_G.

The vertex set of M_G is defined to be $\mathcal{P}$. Since $\mathcal{P} = \mathcal{P}_n = \mathcal{P}_{n+1}$ for all large enough n, it follows that for each pair $P, Q \in \mathcal{P}$ there is a k such that for every $I \in P$ there are exactly k edges in G from I to states in Q. We then declare there to be exactly k edges in M_G from P to Q.

We show that M_G is the smallest element of $\mathcal{R}_G$ by showing that if $H \preceq G$, then $M_G \preceq H$. Let Φ be the edge map of a right-resolving factor code from X_G to X_H and $\partial \Phi$ its associated map on states. Define an equivalence relation $\sim$ on $\mathcal{V} = \mathcal{V}(G)$ by $I \sim I'$ if and only if $\partial \Phi(I) = \partial \Phi(I')$. We inductively show that $I \sim I'$ implies $I \sim_n I'$ for all $n \geqslant 0$. This holds for $n = 0$ by definition. Next, assume that $I \sim I'$ implies that $I \sim_n I'$. Let $I \sim I'$. Since ϕ is a right-covering code, to each edge e with initial state I there corresponds a unique edge e' with initial state I' and $\Phi(e) = \Phi(e')$.

Let $J = t(e)$ and $J' = t(e')$. Now $J \sim J'$ and so $J \sim_n J'$ by hypothesis. Thus to every element P of $\mathcal{P}_n$ and every edge from I to a state of P there corresponds an edge from I' to perhaps some other state of P. This shows that $I \sim_{n+1} I'$, completing the inductive step.

It follows from the previous paragraph that for each state J of H, the inverse image $\partial\Phi^{-1}(J)$ is completely contained in a single element of $\mathcal{P}$. We denote this element by $P(J)$.

We now define a graph homomorphism $\Psi \colon H \to M_G$ as follows. For state $J \in \mathcal{V}(H)$ let $\partial\Psi(J) = P(J)$. To define Ψ on edges, fix a state J in H and an element $Q \in \mathcal{P}$. Let $I \in \partial\Phi^{-1}(J)$. Since Φ is right-covering, the map $e \mapsto \Phi(e)$ gives a bijection between edges in G starting at I with edges in H starting at J. In particular, there is a bijection between the edges in G starting at I and ending at states in Q with edges in H starting at J and ending at states J' with $P(J') = Q$. Let $P = P(J)$, so that $I \in P$. By the definition of M_G there is a bijection between the set of edges of M_G from $P(J)$ to Q and the set of edges in H from J to states J' with $P(J') = Q$. These bijections define the edge map Ψ.

Since the restriction of Ψ to the set of edges starting at a given vertex of H is a bijection, the 1-block code $\Psi_\infty \colon X_H \to X_{M_G}$ is right-covering and thus by Proposition 8.2.2 is a right-resolving factor code. Hence $M_G \preceq H$ as claimed. □

Exercise 8.4.4 outlines another proof of Theorem 8.4.2, but the proof we have given here is more algorithmic.

Corollary 8.4.3. *If G and H are irreducible graphs with $H \preceq G$, then $M_G \cong M_H$.*

PROOF: Since $M_H \preceq H \preceq G$, by Theorem 8.4.2 we see that $M_G \preceq M_H$. Thus $M_G \preceq M_H \preceq H$. Another application of Theorem 8.4.2 then shows that $M_H \preceq M_G$. Since $\preceq$ is a partial ordering, we obtain $M_G \cong M_H$. □

Example 8.4.4. Let G_1, G_2, G_3, and G_4 be the graphs shown in Figure 8.4.1. Then $M_{G_1} = M_{G_2} = G_1$. Since all out-degrees of G_3 are distinct, $M_{G_3} = G_3$. The reader can verify that for G_4 the procedure in Theorem 8.4.2 terminates at $n = 2$, giving $M_{G_4} = G_4$. □

The minimal element M_G in $\mathcal{R}_G$ is closely related to the total amalgamation of a graph, used by R. Williams in the classification of one-sided shifts of finite type [WilR2] (see §13.8).

We now come to the characterization of right-resolving finite equivalence.

Theorem 8.4.5. *Suppose that G and H are irreducible graphs. Then X_G and X_H are right-resolving finite equivalent if and only if $M_G \cong M_H$. If this is the case, then the common extension can be chosen to be irreducible.*

PROOF: First suppose that X_G and X_H are right-resolving finitely equivalent. Let X_K be the common extension. By Proposition 8.3.6 we may

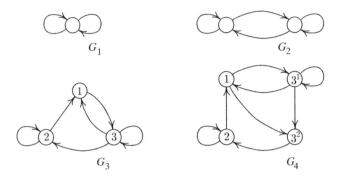

FIGURE 8.4.1.   Graphs for Example 8.4.4.

assume that K is irreducible. Then $G \preceq K$ and $H \preceq K$. By Corollary 8.4.3 we can conclude that $M_G \cong M_K \cong M_H$.

Conversely, suppose that $M \cong M_G \cong M_H$ and let $\Phi_G \colon G \to M$ and $\Phi_H \colon H \to M$ be graph homomorphisms that induce right-resolving factor codes. Then the fiber product of $(\Phi_G)_\infty$ and $(\Phi_H)_\infty$ is a right-resolving finite equivalence of X_G and X_H. The common extension can be chosen to be irreducible by Proposition 8.3.6. □

Example 8.4.6. Consider the graphs G_3 and G_4 in Figure 8.4.1. Now X_{G_3} is conjugate to X_{G_4} (and therefore finitely equivalent) since G_4 is obtained from G_3 by out-splitting state 3; also G_3 and G_4 have the same set of out-degrees. However, X_{G_3} and X_{G_4} are not right-resolving finitely equivalent since $M_{G_3} \not\cong M_{G_4}$. □

The following is a sofic version of Theorem 8.4.5.

Theorem 8.4.7. *Suppose that X and Y are irreducible sofic shifts. Let G_X and G_Y denote the underlying graphs of their minimal right-resolving presentations. Then X and Y are right-resolving finitely equivalent if and only if $M_{G_X} \cong M_{G_Y}$. Moreover, the common extension can be chosen to be irreducible.*

PROOF: Suppose that X and Y are right-resolving finitely equivalent. As before, we may assume that the common extension is an edge shift X_K with K irreducible. By Exercise 8.2.6(a), $G_X \preceq K$ and $G_Y \preceq K$. By Corollary 8.4.3 we conclude that $M_{G_X} \cong M_{G_Y}$. Conversely, if $M_{G_X} \cong M_{G_Y}$, then by Theorem 8.4.5 we see that X_{G_X} and X_{G_Y} are right-resolving finitely equivalent and hence so are X and Y. □

If there is a right-resolving finite equivalence between irreducible edge shifts X_G and X_H, or equivalently if $M_G \cong M_H$, then clearly $\lambda_{A(G)} = \lambda_{A(H)}$. However, the condition $M_G \cong M_H$ is much stronger than $\lambda_{A(G)} = \lambda_{A(H)}$. Thus it is much harder to have a right-resolving finite equivalence than just a finite equivalence.

There is an intermediate relation between right-resolving finite equivalence and finite equivalence: namely, finite equivalence with right-*closing* legs. Such a relation is attractive from a coding point of view: one leg could be recoded to be right-resolving, leading to a "machine" for encoding and decoding very much like a finite-state code (see §5.2). It turns out that neither the condition $\lambda_{A(G)} = \lambda_{A(H)}$ (for finite equivalence) nor the condition $M_G \cong M_H$ (for right-resolving finite equivalence) characterizes right-closing finite equivalence, but we will see in Chapter 12 that there is a complete characterization of right-closing finite equivalence in terms of an ideal class associated with a right Perron eigenvector.

EXERCISES

8.4.1. Find M_G for the graphs G with the following adjacency matrices.

(a) $\begin{bmatrix} 1 & 1 & 0 \\ 0 & 1 & 1 \\ 1 & 0 & 1 \end{bmatrix}$, (b) $\begin{bmatrix} 1 & 0 & 0 & 1 & 0 \\ 0 & 0 & 1 & 1 & 0 \\ 0 & 1 & 0 & 0 & 1 \\ 0 & 0 & 1 & 0 & 0 \\ 1 & 0 & 0 & 0 & 0 \end{bmatrix}$, (c) $\begin{bmatrix} 0 & 0 & 1 & 0 & 0 & 0 \\ 1 & 0 & 0 & 0 & 0 & 0 \\ 0 & 1 & 0 & 1 & 0 & 0 \\ 0 & 0 & 0 & 0 & 1 & 0 \\ 1 & 0 & 0 & 0 & 0 & 1 \\ 0 & 0 & 0 & 1 & 0 & 0 \end{bmatrix}$,

(d) $\begin{bmatrix} 0 & 0 & 0 & 0 & 0 & 1 & 0 & 0 & 0 & 1 \\ 0 & 0 & 1 & 0 & 0 & 0 & 0 & 1 & 0 & 1 \\ 0 & 1 & 0 & 1 & 0 & 0 & 1 & 1 & 0 & 1 \\ 0 & 0 & 1 & 0 & 0 & 1 & 1 & 0 & 0 & 0 \\ 0 & 0 & 0 & 0 & 0 & 1 & 0 & 1 & 0 & 0 \\ 1 & 0 & 0 & 1 & 1 & 0 & 1 & 0 & 0 & 0 \\ 0 & 0 & 1 & 1 & 0 & 1 & 0 & 1 & 0 & 1 \\ 0 & 1 & 1 & 0 & 1 & 0 & 1 & 0 & 1 & 0 \\ 0 & 0 & 0 & 0 & 0 & 0 & 0 & 1 & 0 & 1 \\ 1 & 1 & 1 & 0 & 0 & 0 & 1 & 0 & 1 & 0 \end{bmatrix}$.

8.4.2. Show that for irreducible graphs G and H, the following are equivalent.
   (1) X_G and X_H are right-resolving finitely equivalent.
   (2) X_G and X_H have a common right-resolving (sofic) factor.
   (3) X_G and X_H have a common right-resolving edge shift factor.
   (4) $M_G \cong M_H$.

8.4.3. For a given edge shift X_G of positive entropy, find an infinite collection of edge shifts each conjugate to X_G, but such that no two of them are right-resolving finitely equivalent.

8.4.4. Let G and H be irreducible graphs. For a right-covering code $\phi = \Phi_\infty : X_G \to X_H$, let $\mathcal{P}_\Phi$ denote the partition of $\mathcal{V}$ defined by $\partial\Phi$; i.e., I and J belong to the same element of $\mathcal{P}_\Phi$ if and only if $\partial\Phi(I) = \partial\Phi(J)$.
   (a) Let P and Q be elements of $\mathcal{P}_\Phi$ and let I and J be states in P. Show that the number of edges from I to Q is the same as the number of edges from J to Q.
   (b) Let $\mathcal{U}$ denote a finite collection of partitions of a set S. The *intersection* of $\mathcal{U}$ is defined to be the finest partition coarser than each partition in $\mathcal{U}$. Show that two elements $x, y \in S$ belong to the same element of the

intersection of $\mathcal{U}$ if and only if there is a sequence of subsets $A_1, \ldots, A_k$ such that each A_i is an element of some partition in $\mathcal{U}$, $A_i \cap A_{i+1} \neq \varnothing$ for $1 \leqslant i \leqslant k-1$, $x \in A_1$, and $y \in A_k$.

(c) Explicitly describe M_G as a graph whose states are elements of the intersection of the collection $\mathcal{U} = \{ \mathcal{P}_\Phi : \Phi_\infty \text{ is a right-covering code on } X_G \}$.

8.4.5. Let G be an irreducible graph.

   (a) Show that if H is an in-amalgamation of G, then X_H is a right-resolving factor of X_G.

   (b) Show that if H is an in-amalgamation of G, then X_{M_G} is a right-resolving factor of X_H.

   (c) Give an example to show that M_G need not be an in-amalgamation of G.

   *(d) A *total amalgamation* of G is a graph H such that (1) H is obtained from G by a sequence of in-amalgamations and (2) any in-amalgamation of H is graph isomorphic to H. Show that up to graph isomorphism G has a unique total amalgamation.

Notes

Theorem 8.1.16, which asserts the equivalence of the finite-to-one condition and various other conditions, has its roots in the work of Hedlund [Hed5] and Coven and Paul [CovP1]. The former established that endomorphisms of a full shift (i.e., factor codes from the full shift to itself) are finite-to-one, and the latter generalized this to endomorphisms of irreducible shifts of finite type. For 1-block codes on edge shifts (equivalently, labelings of graphs), the finite-to-one condition is known in the information theory literature as *lossless* – signifying that no "information" (or no "entropy") is lost [Koh2]; in automata theory, it is known as *nonambiguous* [Bea4]. Proposition 8.1.23 is due to J. Ashley and D. Perrin.

The matrix interpretation of right-resolving codes, given in Proposition 8.2.6, is essentially in Williams [WilR2] and was made more explicit in later papers such as Parry and Tuncel [ParT1].

The Finite Equivalence Theorem was proved by Parry [Par3]. Of critical importance was the construction of the positive integral matrix F (used to prove the implication (2) $\Rightarrow$ (3) of Theorem 8.3.8), which was first established by Furstenberg using the proof outlined in Exercise 8.3.13. The special case of the Finite Equivalence Theorem where the entropy is the logarithm of a positive integer was treated earlier by Adler, Goodwyn and Weiss [AdlGW]. The state splitting algorithm described in Chapter 5 was an outgrowth of the Finite Equivalence Theorem, but really derives more from [AdlGW].

The material in §8.4 on right-resolving finite equivalence is mostly taken from Ashley–Marcus–Tuncel [AshMT], but Theorem 8.4.2 is really a much older result in computer science; see Cordon and Crochemore [CorC] and Corneil and Gottlieb [CorG]. While right-resolving finite equivalence is not an invariant of conjugacy, it is an invariant of one-sided conjugacy.

Exercise 8.1.10 is essentially in [CovP1] and [Hed5]; Exercises 8.1.12 and 8.1.13 are in [KitMT] (originally due to Kitchens); Exercise 8.2.11 comes from Kitchens (cited in [BoyMT]); Exercise 8.3.13 is due to Furstenberg (cited in [Par3]); Exercise 8.3.14 is due to Nasu [Nas2] and [AshM]; Exercise 8.4.1(d) is taken from [CorG]; and the total amalgamation from Exercise 8.4.5(d) is due to R. Williams [WilR2].

CHAPTER 9

DEGREES OF CODES AND
ALMOST CONJUGACY

In Chapters 7 and 8 we studied two notions of equivalence for shifts of finite type and sofic shifts: conjugacy and finite equivalence. Conjugacy is the stronger and more fundamental notion, while finite equivalence is more decidable. In this chapter, we introduce an intermediate concept of equivalence called almost conjugacy, which was motivated by constructions of codes in ergodic theory.

For a finite-to-one factor code on an irreducible shift of finite type or sofic shift, there is, by definition, a uniform upper bound on the number of pre-images of points in the image. Thus, the number of pre-images can vary through a finite set of positive integers. In §9.1 we show that certain points which are "representative" of the range shift all have the same number of pre-images. This number is called the degree of the code. In §9.2 we focus on codes of degree one, called almost invertible codes, and show how under certain circumstances finite-to-one codes can be replaced by almost invertible codes without changing the domain and range. In §9.3 we introduce the notion of almost conjugacy, which by definition is a finite equivalence in which both legs are almost invertible. We show that entropy and period form a complete set of almost conjugacy invariants for irreducible shifts of finite type and for irreducible sofic shifts. Finally, in §9.4 we conclude with results which assert that the "representative" points are really "typical" points in a probabilistic sense.

§9.1. The Degree of a Finite-to-One Code

Some points of a shift space X may be more "representative" of X than others. A fixed point $x = \ldots aaa \ldots$ says little about the shift it is sitting in (unless X consists solely of x). On the other hand, points in which every block appears over and over again both to the left and to the right are more representative of the shift. We give the following name to such points.

Definition 9.1.1. A point x in a shift space X is *doubly transitive* if every block in X appears in x infinitely often to the left and to the right.

Are there any doubly transitive points in X, and if so how would you find one? If X has a doubly transitive point, then clearly X must be irreducible. Conversely, if X is irreducible, we can construct a doubly transitive point in the following explicit way. List all the blocks in X in some order, say $u^{(1)}, u^{(2)}, u^{(3)}, \ldots$. Set $w^{(1)}$ to be a block of odd length which contains $u^{(1)}$. We define $w^{(n)}$ inductively. Assuming $w^{(n-1)}$ has been defined, use irreducibility of X repeatedly to find blocks s and t such that

(1) both s and t contain each of $u^{(1)}, \ldots, u^{(n)}$,
(2) s and t have the same length, and
(3) $sw^{(n-1)}t$ is a block in X.

Set $w^{(n)} = sw^{(n-1)}t$. Thus $|w^{(n)}|$ is odd. Let $x^{(n)}$ be any point in X which contains $w^{(n)}$ as a central block (and therefore $w^{(k)}$ as a central block for $k \leqslant n$). Then the sequence $\{x^{(n)}\}$ converges to a point $x \in X$ which contains each block $w^{(n)}$, so x is doubly transitive. Thus, the set of doubly transitive points of a shift space X is nonempty if and only if X is irreducible. A more abstract proof of this result is suggested in Exercise 9.1.7.

The condition that a point be doubly transitive seems to be strong. You might think that even an irreducible shift of finite type may not have many doubly transitive points. But in §9.4 we will see that for an irreducible sofic shift the set of doubly transitive points is the complement of a set which is negligible in a probabilistic sense. Hence the doubly transitive points are not only "representative" of an irreducible sofic shift but also "typical."

The remainder of this section will be devoted to showing that for a finite-to-one factor code on an irreducible shift of finite type the doubly transitive points in the range shift all have the same number of pre-images. Thus "almost" all points have the same number of pre-images, motivating the following definition.

Definition 9.1.2. Let X be a shift space and $\phi \colon X \to Y$ a factor code. If there is a positive integer d such that every doubly transitive point of Y has exactly d pre-images under ϕ, then we call d the *degree* of ϕ and define $d_\phi = d$.

Note that the definition is meaningless if Y is reducible, since such a Y has no doubly transitive points. Even when Y is irreducible, the degree of a factor code need not be defined. In order to show that the degree of a finite-to-one factor code on an irreducible shift of finite type is well-defined, we introduce another quantity d_ϕ^*, defined concretely in terms of (finite) blocks, and prove that all doubly transitive points have exactly d_ϕ^*

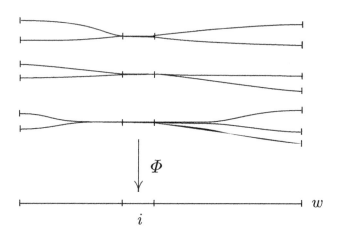

FIGURE 9.1.1.   Meaning of $d_\phi^*(w, i)$.

pre-images (and so $d_\phi = d_\phi^*$). We will only need to define d_ϕ^* for 1-block factor codes on edge shifts, i.e., codes defined by labeled graphs.

Definition 9.1.3. Let (G, Φ) be a labeled graph and $\phi = \Phi_\infty$. Let $w = w_1 \ldots w_m$ be an m-block in $Y = \phi(X_G)$ and $1 \leqslant i \leqslant m$. We define $d_\phi^*(w, i)$ to be the number of distinct edges that you can see at coordinate i in pre-images of the block w; i.e., $d_\phi^*(w, i)$ is the number of edges e in G such that there is a path $u = u_1 u_2 \ldots u_m$ in G with $\Phi(u) = w$ and $u_i = e$.

Figure 9.1.1 depicts the situation when $d_\phi^*(w, i) = 3$.

Minimizing $d_\phi^*(w, i)$ over all w and i, we obtain a quantity depending only on ϕ.

Definition 9.1.4. With the notations of the previous definition, put

$$d_\phi^* = \min\{d_\phi^*(w, i) : w \in \mathcal{B}(Y),\ 1 \leqslant i \leqslant |w|\}.$$

A *magic word* is a block w such that $d_\phi^*(w, i) = d_\phi^*$ for some i. Such an index i is called a *magic coordinate*. We sometimes abbreviate d_ϕ^* to d^*.

The definition will have meaning for us only when ϕ is a finite-to-one code.

Example 9.1.5. In Figure 9.1.2, we have displayed several examples of labeled graphs which define finite-to-one 1-block codes.

In (a) we see that $d_\phi^*(b, 1) = 1$ and thus $d_\phi^* = 1$. In fact every block w that contains a b is a magic word, and every coordinate $i = 1, \ldots, |w|$ is a magic coordinate. On the other hand, $d_\phi^*(a^n, i) = 2$ for $1 \leqslant i \leqslant n$.

In (b) we see that $d_\phi^*(b, 1) = 1$ and thus $d_\phi^* = 1$. Note that $d_\phi^*(w, |w|) = 1$ for all blocks w of length $n \geqslant 2$, and thus all such blocks are magic words.

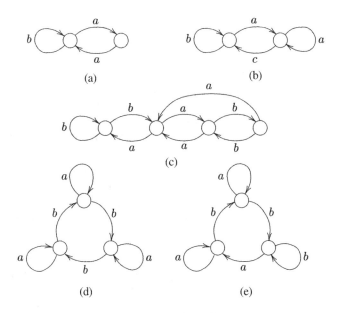

FIGURE 9.1.2.   Edge shifts with magic words.

In (c) again $d_\phi^* = 1$, but the shortest magic word has length three: bab is such a magic word (why?).

In (d) and (e), we revisit an example that appeared in §5.1. In (d) each binary block is presented by a unique path starting from each state. If we imagine tracking the path generated by a block starting at some state, each time that we append an a to the block, we stay wherever we happen to be, and each time that we append a b, we rotate by 120 degrees. So the paths that present a given block occupy distinct edges at each time. Thus, $d_\phi^*(w,i) = 3$ for all w,i, and thus $d_\phi^* = 3$. In (e) we have modified the labeling of the underlying graph in (d) so that $d_\phi^* = 1$.   □

The following shows that $d_\phi = d_\phi^* = 1$ for two prominent classes of codes.

Proposition 9.1.6. *Let (G, Φ) be a labeled graph and $\phi = \Phi_\infty$.*

(1) *If ϕ is a conjugacy, then $d_\phi = d_\phi^* = 1$.*
(2) *If (G, Φ) is the minimal right-resolving presentation of an irreducible sofic shift, then $d_\phi = d_\phi^* = 1$.*

PROOF: (1) Since ϕ is a conjugacy, each point (in particular, each doubly transitive point) has only one pre-image and thus $d_\phi = 1$. Now ϕ^{-1} is a sliding block code which has memory m and anticipation n for some m, n. So for any block $w = w_1 \ldots w_{m+n+1}$ of length $m + n + 1$, all paths which present w must occupy the same edge at coordinate $m + 1$. Thus $d_\phi^*(w, m + 1) = 1$ and so $d_\phi^* = 1$.

(2) By Proposition 3.3.9 and Proposition 3.3.16, the minimal right-resolving presentation has a synchronizing word, i.e., a block w such that all paths that present w end at the same state. Extend w one symbol to the right to obtain an allowed block wa. Since Φ is right-resolving, we see that $d_\phi^*(wa, |wa|) = 1$. Hence wa is a magic word and $d_\phi^* = 1$.

To see that $d_\phi = 1$, observe that for every doubly transitive point y the synchronizing word w appears in y infinitely often to the left. Thus if $\phi(x) = y = \phi(x')$, then $i(x_i) = i(x_i')$ for infinitely many $i < 0$. But since ϕ is right-resolving, we have $x = x'$, and so y has only one pre-image. □

In the same way that it is convenient to recode sliding block codes as 1-block codes, shifts of finite type as 1-step shifts of finite type, and right-closing labelings as right-resolving labelings, it is also convenient to recode so that a magic word becomes a *magic symbol*, i.e., a magic word of length one.

Proposition 9.1.7. *Let X be an edge shift and $\phi\colon X \to Y$ be a finite-to-one 1-block factor code. Then there exist an edge shift $\widetilde{X}$, a higher block shift $\widetilde{Y}$ of Y, a 1-block factor code $\psi\colon \widetilde{X} \to \widetilde{Y}$, and conjugacies $\alpha\colon \widetilde{X} \to X$, $\beta\colon \widetilde{Y} \to Y$ such that*

(1) $\psi = \beta^{-1} \circ \phi \circ \alpha$
(2) $d_\psi^* = d_\phi^*$, *and*
(3) ψ *has a magic symbol.*

PROOF: Let $w = w_1 \ldots w_m$ be a magic word for ϕ and i be a magic coordinate. Write $\phi = \Phi_\infty$. Define two $(m-1)$-blocks $u_1 \ldots u_{m-1}$ and $v_1 \ldots v_{m-1}$ in X to be equivalent if $\Phi(u_1 \ldots u_{m-1}) = \Phi(v_1 \ldots v_{m-1})$ and $i(u_i) = i(v_i)$. Similarly, define two m-blocks $u_1 \ldots u_m$ and $v_1 \ldots v_m$ in X to be equivalent if $\Phi(u_1 \ldots u_m) = \Phi(v_1 \ldots v_m)$ and $u_i = v_i$. The second equivalence relation is illustrated in Figure 9.1.3.

Denote the equivalence class of an $(m-1)$-block $u_1 \ldots u_{m-1}$ in X by $[u_1 \ldots u_{m-1}]$, and that of an m-block $u_1 \ldots u_m$ by $[u_1 \ldots u_m]$. Let $\mathcal{V}$ denote the set of all equivalence classes of $(m-1)$-blocks, $\mathcal{E}$ the set of all equivalence classes of m-blocks, and $\widetilde{G}$ the graph with vertices $\mathcal{V}$ and edges $\mathcal{E}$. In $\widetilde{G}$ the initial state of the edge $[u_1 \ldots u_m]$ is $[u_1 \ldots u_{m-1}]$, and its terminal state is $[u_2 \ldots u_m]$. Put $\widetilde{X} = X_{\widetilde{G}}$.

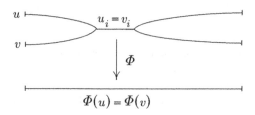

FIGURE 9.1.3. Equivalence relation on m-blocks.

Let $\alpha: \widetilde{X} \to X$ be the 1-block code induced by the map

$$[u_1 \ldots u_m] \mapsto u_i.$$

Then α does map $\widetilde{X}$ into X, since whenever $[u_1 \ldots u_m][v_1 \ldots v_m]$ is a 2-block in $\widetilde{X}$, then $u_i v_i$ is a 2-block in X. Furthermore, α is a conjugacy since the m-block code induced by $u_1 \ldots u_m \mapsto [u_1 \ldots u_m]$ is an inverse of α modulo a power of the shift.

Put $\widetilde{Y} = Y^{[m]}$ and let β be the code induced by $y_1 \ldots y_m \to y_i$. Define

$$\psi = \beta^{-1} \circ \phi \circ \alpha: \widetilde{X} \to \widetilde{Y}.$$

It is straightforward to check that ψ is the 1-block factor code induced by the map

$$[u_1 \ldots u_m] \mapsto \Phi(u_1 \ldots u_m)$$

(regarding $\Phi(u_1 \ldots u_m)$ as a symbol of $\widetilde{Y}$). Then d_ψ^*-values on blocks of $\widetilde{Y}$ agree with the d_ϕ^*-values on corresponding blocks in Y. Hence $d_\psi^* \geqslant d_\phi^*$. Since $d_\psi^*(w, 1) = d_\phi^*(w, i)$, we have $d_\psi^* = d_\phi^*$, and so w is a magic symbol for ψ. □

Our next result gives us a picture of the set of pre-images of blocks that begin and end with a magic symbol. An important part of this picture involves the following notion.

Definition 9.1.8. Let $\pi^{(1)} = e_1^{(1)} e_2^{(1)} \ldots e_m^{(1)}, \ldots, \pi^{(k)} = e_1^{(k)} e_2^{(k)} \ldots e_m^{(k)}$ be paths of length m in a graph. The set $\{\pi^{(1)}, \ldots, \pi^{(k)}\}$ is *mutually separated* if, for each $i = 1, 2, \ldots, m$, the initial states of $e_i^{(1)}, e_i^{(2)}, \ldots, e_i^{(k)}$ are all distinct and also the terminal states of $e_i^{(1)}, e_i^{(2)}, \ldots, e_i^{(k)}$ are all distinct (in particular, the edges $e_i^{(1)}, e_i^{(2)}, \ldots, e_i^{(k)}$ are all distinct).

In other words, a set of paths is mutually separated if they never occupy the same state or edge at the same time.

Proposition 9.1.9. *Let X be an irreducible edge shift and $\phi = \Phi_\infty: X \to Y$ be a finite-to-one 1-block factor code with a magic symbol b. Write $\Phi^{-1}(b) = \{a_1, \ldots, a_{d^*}\}$. Let v be a block in Y which begins and ends with b. Then*

(1) *v has exactly $d^* = d_\phi^*$ pre-images under Φ.*
(2) *There is a permutation τ_v of $\{a_1, \ldots, a_{d^*}\}$ such that, for $1 \leqslant i \leqslant d^*$, exactly one of the pre-images of v begins with a_i and ends with $\tau_v(a_i)$.*
(3) *The set of pre-images of v is mutually separated.*

The meaning of this proposition is depicted in Figure 9.1.4.

PROOF: (1) This holds for $v = b$ by definition. Thus we may assume that $v \neq b$.

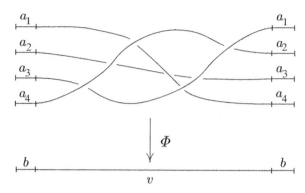

FIGURE 9.1.4. Permutation induced by mutually separated pre-images.

Observe that each a_i is the initial symbol of some pre-image of v since otherwise we would have $d^*(v, 1) < d^*$, contrary to the definition of d^*. Hence it suffices to show that no two pre-images have the same initial symbol. Suppose to the contrary that there were two distinct pre-images of v with the same initial symbol a_{i_0}.

Write $v = b\bar{v}b$. We claim that

$$(9\text{--}1\text{--}1) \qquad |\Phi^{-1}(b(\bar{v}b)^{m+1})| > |\Phi^{-1}(b(\bar{v}b)^m)|$$

for each $m \geq 1$. This would show that there are blocks in Y with an arbitrarily large number of pre-images, contradicting that ϕ is finite-to-one (see Corollary 8.1.18).

To verify (9–1–1), let u be a pre-image of $b(\bar{v}b)^m$. Then u ends in some $a_k \in \Phi^{-1}(b)$. By our remarks above, there is a word $u^{(k)} \in \Phi^{-1}(\bar{v}b)$ such that $a_k u^{(k)}$ is a block in X and $a_k u^{(k)} \in \Phi^{-1}(b\bar{v}b)$; since X is an edge shift, $u u^{(k)}$ is a block in X and $u u^{(k)} \in \Phi^{-1}(b(\bar{v}b)^{m+1})$ (see Figure 9.1.5). Thus every pre-image of $b(\bar{v}b)^m$ can be extended to the right to form a pre-image of $b(\bar{v}b)^{m+1}$. Now some pre-image u of $b(\bar{v}b)^m$ ends with the symbol a_{i_0}, and since there are two pre-images of $v = b\bar{v}b$ starting with a_{i_0} we see that u can be extended to the right to give two pre-images of $b(\bar{v}b)^{m+1}$. This verifies (9–1–1) and therefore proves (1).

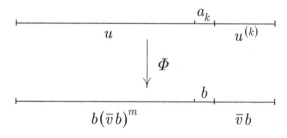

FIGURE 9.1.5. Extending a pre-image of $b(\bar{v}b)^m$.

(2) In the course of proving (1) we have shown that for each i there is a unique pre-image of v that begins with a_i. Define $\tau_v(a_i)$ to be the terminal symbol of this unique pre-image of v. Since each pre-image of b is the terminal symbol of some pre-image of v, then τ_v maps $\{a_1, \ldots, a_{d^*}\}$ onto $\{a_1, \ldots, a_{d^*}\}$ and is therefore a permutation (see Figure 9.1.4).

(3) Suppose that $u = u_1 \ldots u_m$ and $u' = u'_1 \ldots u'_m$ are distinct pre-images of v. If $i(u_i) = i(u'_i)$ for some $i \geqslant 2$, then we can splice together u and u' to give two pre-images of v with the same initial symbol but different terminal symbols, contradicting (2). Suppose $i(u_1) = i(u'_1)$, and let w be any path whose initial edge is a pre-image of b and which terminates at the common initial state of u_1 and u'_1. Then wu and wu' are distinct pre-images of a block that begins and ends with b, and wu and wu' have the same initial symbol, again contradicting (2). The argument for terminal states is similar.   □

Recall from Proposition 8.2.2 that for an irreducible right-resolving labeled graph (G, Φ) which presents an edge shift X_H, the factor code Φ_∞ is right-covering. Consequently, if $\Phi(a) = b$ and bb' is a 2-block in X_H, then there is an edge a' such that aa' is a 2-block in X_G and $\Phi(aa') = bb'$. The following shows that this holds whenever Φ_∞ is merely finite-to-one and b is a magic symbol.

Corollary 9.1.10. *Let X be an irreducible edge shift and let $\phi = \Phi_\infty \colon X \to Y$ be a finite-to-one 1-block factor code with a magic symbol b. Suppose that $\Phi(a) = b$.*

(1) *If bb' is a 2-block in Y then there is a (not necessarily unique) 2-block aa' in X such that $\Phi(aa') = bb'$.*

(2) *If $b'b$ is a 2-block in Y then there is a (not necessarily unique) 2-block $a'a$ in X such that $\Phi(a'a) = b'b$.*

PROOF: Since Y is irreducible, there is a block v in Y which begins with bb' and ends with b. By Proposition 9.1.9(2), some pre-image u of v begins with a. Letting a' denote the second symbol of u, we have that aa' is a 2-block of X and $\Phi(aa') = bb'$. The second statement follows similarly.   □

We now come to the main result of this section.

Theorem 9.1.11. *Let X be an irreducible shift of finite type and $\phi \colon X \to Y$ be a finite-to-one factor code. Then*

(1) *d_ϕ is well-defined, i.e., the doubly transitive points of Y all have the same number of pre-images.*

(2) *d_ϕ is the minimal number of ϕ-pre-images of points in Y.*

(3) *When X is an irreducible edge shift and ϕ is a 1-block factor code, then $d_\phi = d_\phi^*$.*

PROOF: Clearly d_ϕ (if well-defined) is invariant under the standard recoding, via higher blocks, from a sliding block code on a shift of finite type to a 1-block code on an edge shift. Thus, it suffices to show that for a 1-block factor code ϕ on an irreducible edge shift X, $d_\phi = d_\phi^*$ and that d_ϕ is the minimal number of ϕ-pre-images. By Proposition 9.1.7 we may assume that ϕ has a magic symbol.

Let $d^* = d_\phi^*$. Observe that, by definition of d^*, if $y \in Y$, then for each m, there are at least d^* distinct symbols that appear at coordinate 0 in the pre-images of $y_{[-m,m]}$. Thus, by a Cantor diagonal argument (compactness), y has at least d^* pre-images. Hence it suffices to show that every doubly transitive point has at most (and therefore exactly) d^* pre-images.

Let $\{a_1, \ldots, a_{d^*}\}$ be the set of pre-images of a magic symbol w. Let $y \in Y$ be doubly transitive. By Proposition 9.1.9(1), for any $n, m > 0$ such that $y_{-m} = w = y_n$, the block $y_{[-m,n]}$ has exactly d^* pre-images. Since w appears in y infinitely often to the left and to the right, it follows that y has at most d^* pre-images as desired. $\square$

The reader should verify directly that $d_\phi = d_\phi^*$ for the examples shown in Figure 9.1.2.

For a block w in a sofic shift Y, let $I_Y(w)$ denote the set of points $y \in Y$ such that w appears in y infinitely often to the left and to the right. We saw in the proof of Theorem 9.1.11 that for any magic symbol w, each point in $I_Y(w)$ has exactly d_ϕ pre-images; in fact this holds for any magic word w (actually, an examination of the proof shows that it suffices to have some magic word appearing infinitely often to the left and some other magic word appearing infinitely often to the right). Since the set $I_Y(w)$ is generally larger than the set of doubly transitive points, we have really proved more than we stated. However, $I_Y(w)$ depends on w and therefore on the particular code ϕ. The set of doubly transitive points is the intersection of all of the $I_Y(w)$, and so is a canonical set that does not depend on the code ϕ.

Theorem 9.1.11 shows how to compute the degree of a sliding block code ϕ on an irreducible shift of finite type: recode ϕ to a 1-block code ψ on an irreducible edge shift and compute d_ψ^*. In terms of ϕ this amounts to the following result. We leave the proof to the reader.

Proposition 9.1.12. *Suppose that X is an irreducible N-step shift of finite type. Let $\phi = \Phi_\infty : X \to Y$ be a finite-to-one k-block factor code and $M = \max\{N, k\}$. Then d_ϕ is the minimum over all blocks $w = w_1 \ldots w_{|w|}$ in Y and $1 \leqslant i \leqslant |w| - M + 1$ of the number of M-blocks $v \in \mathcal{B}_M(X)$ that you see beginning at coordinate i among the Φ-pre-images of w.*

At this point, we are in a position to show that the degree of a finite-to-one code on an irreducible sofic shift is well-defined (Corollary 9.1.14 below). For this, we will use the following fact.

Lemma 9.1.13. *Let X be an irreducible sofic shift, $\phi = \Phi_\infty : X \to Y$ a finite-to-one factor code, and $x \in X$. Then x is doubly transitive if and* *only if* $\phi(x)$ *is.*

PROOF: Suppose that x is doubly transitive. Let v be a block in Y. Since ϕ is a factor code, there is a block u in X such that $\Phi(u) = v$. Since $x \in I_X(u)$, we see that $\phi(x) \in I_Y(v)$. Hence $\phi(x)$ is doubly transitive.

Conversely, suppose that $\phi(x)$ is doubly transitive, but x is not. Without loss of generality, we may assume that there is a block w in X which does not occur infinitely often to the right in x. Let Z be the shift space obtained by forbidding the block w from X.

Any block v in Y appears infinitely often to the right in $\phi(x)$. Hence at least one of the pre-images $u \in \Phi^{-1}(v)$ appears infinitely often to the right in x. By a Cantor diagonal argument (compactness), there is a limit point $z \in X$ of $\{\sigma^n(x) : n \geqslant 0\}$ which contains the block u. But z cannot contain the block w. Thus $z \in Z$ and so v is a block in $\phi(Z)$. Since v was arbitrary, this shows that $\phi(Z) = Y$. Now ϕ is finite-to-one, so

$$h(X) \geqslant h(Z) \geqslant h(Y) = h(X).$$

Therefore X would contain a proper subshift Z of full entropy, contradicting the irreducibility of X (see Corollary 4.4.9). ☐

Notice that the "only if" direction of the preceding result does not require ϕ to be finite-to-one, only that ϕ be onto.

Corollary 9.1.14. *Let X be an irreducible sofic shift with minimal right-resolving presentation $(G, \mathcal{L})$ and $\phi \colon X \to Y$ a finite-to-one factor code. Then each doubly transitive point in Y has exactly $d_{\phi \circ \mathcal{L}_\infty}$ pre-images (via ϕ). In particular, d_ϕ is well-defined, and $d_\phi = d_{\phi \circ \mathcal{L}_\infty}$.*

PROOF: By Proposition 9.1.6(2), $d_{\mathcal{L}_\infty} = 1$. Thus, every doubly transitive point of X has exactly one pre-image via $\mathcal{L}_\infty$. Together with Lemma 9.1.13, this shows that each doubly transitive point of Y has the same number of $(\phi \circ \mathcal{L}_\infty)$-pre-images as ϕ-pre-images. ☐

For finite-to-one codes on irreducible sofic shifts (in contrast to irreducible shifts of finite type), the degree need not coincide with the minimum number of pre-images, as shown in the following example.

Example 9.1.15. Recall the code ϕ on the full 2-shift $X_{[2]}$ defined by the 2-block map $\Phi(x_0 x_1) = x_0 + x_1 \pmod 2$. This code has degree two since every point has exactly two pre-images. Let W be the (irreducible) sofic shift obtained from $X_{[2]}$ by identifying the two fixed points; that is, W is the sofic shift presented by the labeled graph shown in Figure 9.1.6, whose underlying graph is that of the 2-block presentation of $X_{[2]}$. Since ϕ agrees on the two fixed points, it determines a finite-to-one code $\psi \colon W \to X_{[2]}$ (explicitly, ψ is induced by the 1-block map $a \mapsto 0, b \mapsto 1, c \mapsto 1$). This code has degree two because every point other than 0^∞ has two pre-images. But 0^∞ has only one pre-image. ☐

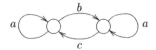

FIGURE 9.1.6.   A sofic shift used to illustrate degree.

In the following, we make use of Lemma 9.1.13 to understand how degree behaves under composition.

Proposition 9.1.16. *Let X be an irreducible sofic shift and $\phi: X \to Y$, $\psi: Y \to Z$ be finite-to-one factor codes. Then $d_{\psi \circ \phi} = d_\psi \cdot d_\phi$.*

PROOF: Since X is irreducible, so are Y and Z. Let $z \in Z$ be doubly transitive. Then z has exactly d_ψ pre-images $z^{(i)}$ for $1 \leqslant i \leqslant d_\psi$ via ψ, and by Lemma 9.1.13, each $z^{(i)}$ is doubly transitive. But then each $z^{(i)}$ has exactly d_ϕ pre-images via ϕ. Thus, z has exactly $d_\psi \cdot d_\phi$ pre-images via $\psi \circ \phi$ and so we have $d_{\psi \circ \phi} = d_\psi \cdot d_\phi$ as desired. □

In Definition 9.1.3, we defined $d_\phi^*(w, i)$ by counting edges. An alternative would be to count states instead. Specifically, for a labeled graph (G, Φ) presenting a sofic shift Y, an m-block $w = w_1 \ldots w_m$ in Y, and $0 \leqslant i \leqslant m$, let $d_\phi'(w, i)$ be the number of states I in G such that there is a path u in G with $\Phi(u) = w$ and $s_i = I$, where $s_0 s_1 \ldots s_m$ is the *state sequence* of u (i.e., $s_i = i(u_{i+1})$ for $0 \leqslant i \leqslant m - 1$ and $s_m = t(u_m)$). Then define

$$d_\phi' = \min\{d_\phi'(w, i) : w \in \mathcal{B}(Y), \ 0 \leqslant i \leqslant m\}.$$

It is not hard to show that if G is irreducible and $\phi = \Phi_\infty$ is finite-to-one, then $d_\phi' = d_\phi^*$ (Exercise 9.1.13). The advantage of counting states rather than edges is that d_ϕ' is often achieved by a block which is shorter (by one symbol) than a block that achieves d_ϕ^*. We chose to count edges rather than states in order to better stay within the framework of edge shifts, (edge) labelings, etc.

EXERCISES

9.1.1. Find the degrees of the 1-block codes defined by the following symbolic adjacency matrices.

$$
\text{(a)} \quad
\begin{bmatrix}
\emptyset & b & a & \emptyset & \emptyset \\
b & \emptyset & a & \emptyset & \emptyset \\
\emptyset & \emptyset & \emptyset & \emptyset & b \\
\emptyset & b & \emptyset & \emptyset & \emptyset \\
\emptyset & \emptyset & \emptyset & a & b
\end{bmatrix},
\quad
\text{(b)} \quad
\begin{bmatrix}
\emptyset & a & b \\
b & \emptyset & a \\
a+b & \emptyset & \emptyset
\end{bmatrix}.
$$

9.1.2. Show that if ϕ is a factor code on an irreducible shift of finite type, then either ϕ is finite-to-one (and hence has a well-defined (finite) degree) or each doubly transitive point has infinitely many pre-images.

9.1.3. Let X_G be an irreducible edge shift. We say that a set of points in X_G is mutually separated if they occupy distinct states at each time (compare with Definition 9.1.8). Let $\phi \colon X_G \to Y$ be a 1-block finite-to-one factor code.

(a) Show that the pre-image of each point in Y contains a set of d_ϕ mutually separated points.

(b) Show that for all doubly transitive points $y \in Y$ the set of pre-images of y is mutually separated.

(c) Show that for all periodic points $y \in Y$ the set of pre-images of y is mutually separated.

9.1.4. Let $\phi = \Phi_\infty \colon X_G \to Y$ be a finite-to-one 1-block factor code on an irreducible edge shift. Define the following sequence of labeled graphs $(H^{(1)}, \Phi^{(1)})$, $(H^{(2)}, \Phi^{(2)})$, ... The set of vertices of $H^{(k)}$ is the set of subsets of size k of $\mathcal{V}(G)$. For each pair $V = \{v_1, \dots, v_k\}$, $V' = \{v'_1, \dots, v'_k\}$ of such subsets and each symbol $b \in \mathcal{A}(Y)$, endow $H^{(k)}$ with an edge from V to V' labeled b if there is a permutation τ of $\{1, \dots, k\}$ such that b is the Φ-label of an edge from v_i to $v'_{\tau(i)}$ for each $1 \leqslant i \leqslant k$. Let $\phi^{(k)} = \Phi^{(k)}_\infty$.

(a) Show that $\phi^{(k)}$ maps onto Y if and only if $k \leqslant d_\phi$.

(b) Show that $d_{\phi^{(k)}} = \binom{d_\phi}{k}$ for $1 \leqslant k \leqslant d_\phi$. In particular, $d_{\phi^{(k)}} = 1$ for $k = d_\phi$.

9.1.5. A sliding block code from one shift space into another is called *bi-resolving* if it is both right-resolving and left-resolving and is called *bi-closing* if it is both right-closing and left-closing.

(a) Show how any bi-closing code $\phi \colon X \to Y$ on an irreducible shift of finite type X can be "recoded" to a bi-resolving code $\psi \colon \widetilde{X} \to \widetilde{Y}$, in the sense that there are conjugacies $\alpha \colon X \to \widetilde{X}$ and $\beta \colon Y \to \widetilde{Y}$ such that $\psi = \beta \circ \phi \circ \alpha^{-1}$.

(b) Let X and Y be irreducible shifts of finite type. Show that a factor code $\phi \colon X \to Y$ is *constant-to-one* (i.e., the points in the image all have the same number of pre-images) if and only if ϕ is bi-closing.

(c) Let X and Y be irreducible shifts of finite type. Show that a factor code $\phi \colon X \to Y$ is a conjugacy if and only if ϕ is bi-closing and $d_\phi = 1$.

(d) Show that (b) and (c) are false if Y is allowed to be strictly sofic.

(e) Show that the image of a constant-to-one code on an irreducible shift of finite type must be an irreducible shift of finite type.

9.1.6. Let X be an irreducible shift of finite type, and $\phi \colon X \to Y$ be a finite-to-one factor code. Assume that ϕ is not constant-to-one. Let B_ϕ denote the set of points in Y which have more than d_ϕ pre-images.

(a) Show that B_ϕ must contain a periodic point.

(b) Show that if Y is a shift of finite type, then B_ϕ must contain a point with a pair of either left-asymptotic or right-asymptotic pre-images.

(c) Show that if Y is a shift of finite type, then B_ϕ must be dense.

(d) Show that the conclusions of parts (b) and (c) are not valid when Y is strictly sofic.

9.1.7. Use the Baire Category Theorem (Theorem 6.1.24) to show that any irreducible shift space contains a doubly transitive point.

9.1.8. Let X be an irreducible shift of finite type and $\phi \colon X \to Y$ be a finite-to-one code.

(a) Show that the minimum number of pre-images is always attained by a periodic point in Y.

(b) Show that if ϕ is not one-to-one, and each point in Y has at most two pre-images via ϕ, then some periodic point in the range has two pre-images.

(c) Find an example to show that the maximum number of pre-images need not be attained by a periodic point in Y.

9.1.9. Let $(G, \mathcal{L})$ be a labeled graph such that $\phi = \mathcal{L}_\infty$ is finite-to-one. Find an explicit upper bound on the length of the smallest magic word for ϕ.

9.1.10. Show that Lemma 9.1.13 is false without the finite-to-one assumption.

*9.1.11. Prove Lemma 9.1.13 without using entropy.

9.1.12. Show that for any finite-to-one factor code ϕ from an irreducible shift of finite type onto a sofic shift, there is a finite-to-one factor code ψ such that $\phi \circ \psi$ is both the composition of a right-resolving code followed by a left-resolving code and the composition of a left-resolving code followed by a right-resolving code. [*Hint*: Use Exercise 8.3.14(a) and the fiber product.]

9.1.13. Show that if G is irreducible, Φ is a labeling of G, and $\phi = \Phi_\infty \colon X_G \to Y$ is finite-to-one, then $d'_\phi = d^*_\phi$.

§9.2. Almost Invertible Codes

In §9.1 we showed that a finite-to-one factor code on an irreducible shift of finite type has a well-defined (finite) degree. In this section, we study such codes having degree one. We view such codes as being "almost invertible."

Definition 9.2.1. A factor code ϕ is *almost invertible* if every doubly transitive point has exactly one pre-image, i.e., $d_\phi = 1$.

We first show that almost invertible codes on irreducible shifts of finite type are automatically finite-to-one (see also Exercise 9.1.2).

Proposition 9.2.2. *An almost invertible code on an irreducible shift of finite type is finite-to-one.*

PROOF: Let X be an irreducible shift of finite type and $\phi \colon X \to Y$ be almost invertible. By recoding, we may assume that $\phi = \Phi_\infty$ is a 1-block code on an irreducible edge shift X_G. If Φ had a graph diamond $\{\pi, \tau\}$, then since any doubly transitive point $x \in X_G$ contains π, $\phi(x)$ must have at least two pre-images (simply replace an appearance of π by τ). But $\phi(x)$ is doubly transitive since x is. Hence Φ has no graph diamond, and so ϕ is finite-to-one. □

Conjugacies (the "truly invertible codes") are of course almost invertible. So almost invertibility sits strictly in between conjugacy and finite-to-one.

We use the term "almost invertible" since, by Theorem 9.1.11 and Lemma 9.1.13, such a code gives a bijection on the set of doubly transitive points, and, as we shall see in §9.4, the set of doubly transitive points exhausts "most" of the space.

Let us consider the special situation of right-resolving factor codes, i.e., factor codes induced by right-resolving labelings. In this case a magic word w is a block such that the number of terminal *edges* that you see among the paths that present w is minimal (among all possible blocks). If a factor code is right-resolving and almost invertible, then a magic word is simply a block w such that all paths that present w have the same terminal edge. In this setting any magic word is a synchronizing word. Conversely, any synchronizing word for a right-resolving labeling can be extended, by adding just one more symbol on the right, to a magic word (we used this fact in the proof of Proposition 9.1.6(2)). Hence a right-resolving factor code induced by a right-resolving labeling Φ is almost invertible if and only if Φ has a synchronizing word. So for right-resolving labelings almost invertibility is really a synchronizing property.

This should remind you of the Road Problem introduced in §5.1. Indeed, since the condition that a primitive graph G have out-degree n is equivalent to the existence of a right-resolving factor code from X_G to $X_{[n]}$ we can restate the Road Theorem (Theorem 5.1.3) as follows (see Exercise 5.1.4).

Theorem 9.2.3. *Given a primitive graph G such that there is a right-resolving factor code $X_G \to X_{[n]}$, there exists an almost invertible, right-resolving factor code $X_G \to X_{[n]}$.*

Recall from Proposition 4.5.10 that that if G is essential then X_G is mixing if and only if G is primitive. So, Theorem 9.2.3 is an instance where a right-resolving factor code between mixing edge shifts can be replaced with an almost invertible, right-resolving factor code. But in general, one cannot always do this; in fact, one cannot always do this even if you are allowed to pass to a higher block version of the domain. In the following example, we exhibit two primitive graphs G, H and a right-resolving factor code $X_G \to X_H$, and we show that for all N there is no almost invertible, right-resolving factor code $X_{G^{[N]}} \to X_H$.

Example 9.2.4. Let G be the 4-state graph with edges $a_0, a_1, b_0, b_1, c_0, c_1$ and H be the 2-state graph with edges a, b, c shown in Figure 9.2.1

The labeled graph (G, Φ) defined by erasing subscripts yields a right-resolving factor code (equivalently, right-covering) $\phi \colon X_G \to X_H$. We claim that ϕ is the only right-covering code from $X_G \to X_H$. This is a consequence

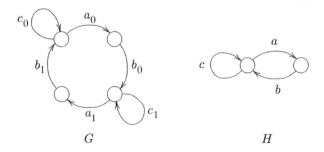

G H

FIGURE 9.2.1. Graphs for Example 9.2.4.

of the following fact: if G, H are irreducible graphs, H has no multiple edges, and the out-degrees of the vertices of H are distinct, then there is at most one right-covering code from $X_G \rightarrow X_H$ (Exercise 8.2.5). This fact also shows that the right-covering code $\phi^{[N]} \colon X_{G^{[N]}} \rightarrow X_H$ defined by

$$\phi^{[N]} = (\Phi^{[N]})_\infty, \quad \text{where} \quad \Phi^{[N]}(e_1 \ldots e_N) = \Phi(e_N),$$

is the only right-covering code from $X_{G^{[N]}} \rightarrow X_H$. Now ϕ has degree 2 since each point in X_H has, by symmetry, exactly two pre-images. But $\phi^{[N]}$ is simply a version of ϕ on a higher block shift, and so it also has degree 2. So there is no almost invertible, right-resolving factor code from a higher block shift $X_{G^{[N]}}$ to X_H.   □

Nevertheless, there is a version of Theorem 9.2.3 which holds in greater generality, namely Theorem 9.2.5 below. This result shows that for irreducible shifts of finite type X, Y, if Y is a *right-closing* factor of X, then Y is also an almost-invertible *right-closing* factor of X provided that the periods of X and Y coincide (recall that the period, per(X), of a shift space X is the greatest common divisor of the periods of its periodic points). In particular, for mixing shifts of finite type, any right-closing factor code ϕ can be replaced by an almost invertible right-closing factor code ψ. In the proof ψ is obtained from ϕ by switching the images of two carefully selected blocks wherever they occur. For Example 9.2.4, a specific replacement is suggested in Exercise 9.2.5.

Theorem 9.2.5 (Replacement Theorem). *Let X and Y be irreducible shifts of finite type. Suppose there is a finite-to-one factor code $\phi \colon X \rightarrow Y$. Then there is an almost invertible factor code $\psi \colon X \rightarrow Y$ if and only if* per$(X) =$ per(Y). *Moreover, if ϕ is right-closing, then ψ can be taken to be right-closing.*

The assumption in the Replacement Theorem that there be a finite-to-one factor code from X to Y is not always easy to check. Of course, equality of entropy is a necessary condition (and is therefore an implicit assumption of the theorem). But there are other necessary conditions involving periodic points and the dimension group; see §12.2. There are also some sufficient conditions known; for instance, in Proposition 8.1.12 we saw that if X has entropy exactly $\log n$ and Y is the full n-shift, then X factors onto Y via a finite-to-one (in fact, right-closing) code. There is a version of this result for more general shifts of finite type, but the conditions also involve periodic points and the dimension group; again, see §12.2.

In §9.3 we will make use of the Replacement Theorem to strengthen the Finite Equivalence Theorem by replacing the legs of a finite equivalence (which are finite-to-one factor codes) with almost invertible factor codes.

PROOF OF REPLACEMENT THEOREM: As usual, we may assume that $X = X_G, Y = X_H$ are irreducible edge shifts. Recall that the period of an edge shift X_G coincides with per(G), the greatest common divisor of cycle lengths in G.

First, we check the necessity of the period condition. For this, we may assume by recoding that $\psi = \Psi_\infty$ is a 1-block code. Then Ψ maps each cycle in G to a cycle in H of the same length (not necessarily of the same least period). So per(H) divides per(G).

Recall from Lemma 4.5.3 that for an irreducible graph and any state in that graph, the period of the graph is the greatest common divisor of the lengths of the cycles that pass through that state. Applying this fact to a higher edge graph of H, we see that for any magic word w for ψ, there is a finite collection of cycles $\gamma^{(1)}, \ldots, \gamma^{(k)}$ each containing w such that per(H) is the greatest common divisor of the lengths of these cycles. Since ψ has no diamonds, each periodic point $(\gamma^{(i)})^\infty$ has exactly one ψ-pre-image $(\delta^{(i)})^\infty$, where $\delta^{(i)}$ has the same length as $\gamma^{(i)}$. So G has a collection of cycles whose lengths have greatest common divisor equal to per(H). Thus per(G) divides per(H), and so per(G) = per(H).

The proof of sufficiency of the period condition is somewhat more involved, but the idea is fairly simple. We will assume that the common period is one, i.e., that G and H are primitive, and leave as an exercise the (routine) reduction to this situation. Assume also that G and H are not single self-loops (otherwise the Replacement Theorem is trivial). We may also assume that $\phi = \Phi_\infty \colon X_G \to X_H$ is a 1-block finite-to-one factor code and has a magic symbol b. Let $d = d_\phi$. The symbol b has exactly d pre-images, say $a_1, \ldots, a_d$. Of course, since X_G and X_H are edge shifts, the symbol b is really an edge in the graph H and the symbols $a_1, \ldots, a_d$ are really edges in the graph G. We may assume that $d \geqslant 2$, since otherwise we can take $\psi = \phi$.

We will need the following lemma, which we leave as an exercise for the reader (Exercise 9.2.2).

Lemma 9.2.6. *Assume the notation and assumptions in the preceding paragraph. By replacing G and H with higher edge graphs if necessary, and with an appropriate choice of magic symbol b, we can find blocks $u, \bar{u}$ in X_G satisfying the following properties:*

(1) *u begins and ends with a_1.*
(2) *$\bar{u}$ begins with a_1 and ends with a_2.*
(3) *$|u| = |\bar{u}|$.*
(4) *$v = \Phi(u), \bar{v} = \Phi(\bar{u})$ overlap themselves or each other only at their initial and terminal symbols; that is, if there are blocks α, β, γ such that $v = \alpha\beta$ and $\bar{v} = \beta\gamma$, then $|\beta| \leqslant 1$, and the same holds if we replace v by $\bar{v}$ and/or $\bar{v}$ by v.*

Note that in Lemma 9.2.6 we have $|u| = |\bar{u}| = |v| = |\bar{v}|$.

Now we define a new code ψ by switching the images of u and $\bar{u}$. More precisely, we define ψ as follows.

(i) If $x_{[i,j]} = u$ then $\psi(x)_{[i,j]} = \bar{v}$.

(ii) If $x_{[i,j]} = \bar{u}$ then $\psi(x)_{[i,j]} = v$.

(iii) If coordinate m does not lie within an interval $[i,j]$ for which $x_{[i,j]} = u$ or $\bar{u}$, then $\psi(x)_m = \phi(x)_m$.

We claim that ψ is a well-defined sliding block code. To see this, first observe that since v and $\bar{v}$ overlap themselves or each other only at their initial and terminal symbols, the same holds for u and $\bar{u}$. This, together with the fact that both v and $\bar{v}$ begin and end with b, implies that ψ is a well-defined mapping from X_G into X_H. Now ψ is shift-commuting by definition. To determine $\psi(x)_0$, we first need to know whether $x_{[i,j]} = u$ or $\bar{u}$ for some $[i,j]$ containing 0. If so, then $\psi(x)_0$ is determined by i and $x_{[i,j]}$; if not, then $\psi(x)_0$ is determined by only x_0. Hence ψ is a sliding block code with both memory and anticipation equal to $|u| - 1 = |v| - 1$.

We will show that ψ is a finite-to-one factor code $X_G \to X_H$, that $d_\psi < d = d_\phi$, and that ψ is right-closing whenever ϕ is. Thus an inductive application of the procedure which passes from ϕ to ψ keeps reducing the degree until we arrive at a right-closing factor code $X_G \to X_H$ of degree one.

In passing from ϕ to ψ, why must the degree decrease? For this, we first show that

$$(9\text{--}2\text{--}1) \qquad \psi(x)_{[i,j]} = v \iff x_{[i,j]} \in (\Phi^{-1}(v) \setminus \{u\}) \cup \{\bar{u}\}.$$

The "$\Leftarrow$" implication of this is clear from the definition. For the converse, suppose that $\psi(x)_{[i,j]} = v$. If an appearance of u or $\bar{u}$ overlaps $x_{[i,j]}$ in more than one symbol, then by Property (4) of Lemma 9.2.6, we have $x_{[i,j]} = \bar{u}$. If neither u nor $\bar{u}$ overlaps $x_{[i,j]}$ in more than one symbol, then $\Phi(x_{[i,j]}) = v$ and $x_{[i,j]} \neq u$. This gives (9–2–1) (see Figure 9.2.2).

Since v begins and ends with a magic symbol, we have by Proposition 9.1.9(2) that the elements of $\Phi^{-1}(v)$ have distinct terminal symbols

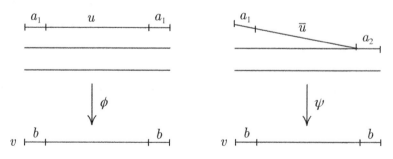

FIGURE 9.2.2.   Switching blocks to define ψ.

$a_1, \ldots, a_d$. Since u ends with the symbol a_1 and $\bar{u}$ does not, a_1 does not appear as a terminal symbol of any element of $(\Phi^{-1}(v) \setminus \{u\}) \cup \{\bar{u}\}$. Thus if $\psi(x)_{[i,j]} = v$, then $x_j \in \{a_2, \ldots, a_d\}$. If ψ were a 1-block code, then we would have $d_\psi^*(v, |v|) < d$, and so according to Theorem 9.1.11 we would have $d_\psi = d_\psi^* < d$. But ψ is not necessarily a 1-block code, and so we need to apply Proposition 9.1.12. In order to do so, we will use the following key lemma.

We assume the notation we have been using in the proof so far. For a block w in a shift space X we use the notation $[w]$ to denote the cylinder set $C_1(w) = \{x \in X : x_{[1,|w|]} = w\}$.

Lemma 9.2.7. *Let w be a block of X_H which begins with v or $\bar{v}$ and ends with v or $\bar{v}$. Then*

$$(9\text{-}2\text{-}2) \qquad \psi^{-1}([w]) = \bigcup_{k=1}^{d} [w^{(k)}]$$

where $w^{(1)}, \ldots, w^{(d)}$ are blocks of X_G all of length $|w|$ such that
 (i) the initial symbol of each $w^{(k)}$ is a_k, and
 (ii) for some $i \neq j$, $w^{(i)}$ and $w^{(j)}$ agree except for at most their first $|v| - 1$ symbols (see Figure 9.2.3).

PROOF: We prove this by induction on $|w|$. The shortest blocks w which can satisfy the hypothesis are $w = v$ and $w = \bar{v}$. We first prove the lemma for $w = v$.

Write

$$(9\text{-}2\text{-}3) \qquad (\Phi^{-1}(v) \setminus \{u\}) \cup \{\bar{u}\} = \{u^{(1)}, \ldots, u^{(d)}\}$$

ordered so that the initial symbol of $u^{(k)}$ is a_k. From (9–2–1) we have

$$(9\text{-}2\text{-}4) \qquad \psi^{-1}([v]) = \bigcup_{k=1}^{d} [u^{(k)}],$$

and so we have (i). Now in this case (ii) simply asserts that there are two distinct elements of $\{u^{(1)}, \ldots, u^{(d)}\}$ which have the same terminal symbol.

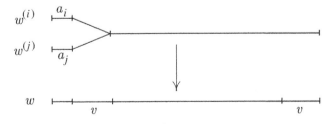

FIGURE 9.2.3. Blocks used in Lemma 9.2.7.

But as explained above, since u ends with a_1 and the terminal symbols of $\Phi^{-1}(v)$ exhaust all of the a_i, $\bar{u}$ has the same terminal symbol, a_2, as some element of $\Phi^{-1}(v) \setminus \{u\}$. By (9–2–3), this gives (ii). This proves the lemma for $w = v$. The proof for $w = \bar{v}$ is similar.

For the inductive step, we write

$$w = \alpha\beta\gamma$$

where α is the prefix of w which ends with the next to last appearance of v or $\bar{v}$, and $\gamma = v$ or $\gamma = \bar{v}$. For definiteness, we assume that α begins and ends with v and that $\gamma = v$. Observe that the only appearances of v or $\bar{v}$ in $v\beta v$ are at the beginning and end.

By the inductive assumption,

$$\psi^{-1}([\alpha]) = \bigcup_{k=1}^{d} [\alpha^{(k)}]$$

where $\alpha^{(1)}, \ldots, \alpha^{(d)}$ are blocks of X_G with initial symbols $a_1, \ldots, a_d$ such that for some i, j, the blocks $\alpha^{(i)}$ and $\alpha^{(j)}$ agree except for at most their first $|v| - 1$ symbols. It suffices to show that each $\alpha^{(k)}$ can be extended to a block $w^{(k)}$ so that (9–2–2) holds and that $w^{(k)}$ is completely determined by the terminal symbol of $\alpha^{(k)}$.

Fix $k \in \{1, \ldots, d\}$. Let $x \in \psi^{-1}([w])$ such that

$$x_{[1,|\alpha|]} = \alpha^{(k)}.$$

Then we may write

$$x_{[1,|w|]} = \alpha^{(k)} x_\beta x_\gamma$$

where

$$|x_\beta| = |\beta|, \quad |x_\gamma| = |\gamma|.$$

Let a_α denote the terminal symbol of $\alpha^{(k)}$ and a_γ the initial symbol of x_γ.

We show that $x_\beta x_\gamma$ is determined by a_α. For this, first note that since the only appearances of v or $\bar{v}$ in $v\beta v$ are at the beginning and end,

$$b\beta b = \Phi(a_\alpha x_\beta a_\gamma).$$

So by Proposition 9.1.9(2), $a_\alpha x_\beta a_\gamma$ must be the unique block in $\Phi^{-1}(b\beta b)$ which begins with the symbol a_α. Now since we have already proved the current lemma for the block v, it follows that x_γ must be the unique element of $(\Phi^{-1}(v) \setminus \{u\}) \cup \{\bar{u}\}$ which begins with a_γ. So $x_\beta x_\gamma$ is completely determined by the terminal symbol a_α of $\alpha^{(k)}$ as desired.

Define $w^{(k)} = \alpha^{(k)} x_\beta x_\gamma$. Since for any $x \in \psi^{-1}([w])$, we have that $x_{[1,|\alpha|]} = \alpha^{(k)}$ for some k; so $x_{[1,|w|]} = w^{(k)}$, and we conclude

$$\psi^{-1}([w]) \subseteq \bigcup_{k=1}^{d} [w^{(k)}].$$

The reverse inclusion follows from the definition of ψ. □

We continue the proof of the Replacement Theorem by showing that ψ is a finite-to-one factor code with $d_\psi < d$ and that if ϕ is right-closing, then so is ψ.

Since ϕ is a finite-to-one factor code, $h(X_G) = h(X_H)$. Thus, to show that ψ is a finite-to-one factor code, it suffices by Corollary 8.1.20 to show that ψ is onto. Since X_H is irreducible, any block w' of X_H is a subblock of a block w that begins and ends with v. So by Lemma 9.2.7, w and hence w' belongs to $\mathcal{B}(\psi(X_G))$. Thus $\psi(X_G) = X_H$ and ψ is a finite-to-one factor code.

Next, we show that $d_\psi < d$. Let w be a block of X_H which begins and ends with v. Let $\{w^{(1)}, \ldots, w^{(d)}\}$ be the blocks in X_G produced by Lemma 9.2.7 for the block w. If $\psi(x)_{[1,|w|]} = w$, then $x_{[1,|w|]} = w^{(k)}$ for some k. But by (ii) of Lemma 9.2.7 the set $\{w^{(1)}_{[|v|,|w|]}, \ldots, w^{(d)}_{[|v|,|w|]}\}$ consists of at most $d - 1$ distinct blocks. These blocks are of length $|w| - |v| + 1$. So for large enough $|w|$ the number of distinct $(2|v| - 1)$-blocks that we see beginning at coordinate $|v|$ among the Ψ-pre-images of w is at most $d - 1$. Since ψ is a $(2|v| - 1)$-block code, it follows from Proposition 9.1.12 that $d_\psi \leqslant d - 1$ as desired.

Finally, we show that if ϕ is right-closing, then so is ψ. Suppose that $\psi(x) = \psi(x')$ and the pair $x, x' \in X$ is left asymptotic, i.e., $x_{(-\infty,m]} = x'_{(-\infty,m]}$ for some m. We must show $x = x'$.

By changing a common left-infinite tail of x and x' and shifting x and x' as necessary, we may assume

$$\psi(x)_{(-|v|,0]} = \psi(x')_{(-|v|,0]} = v$$

and

$$x_{(-\infty,0]} = x'_{(-\infty,0]}.$$

We claim that if for some $0 \leqslant i < j$, $\psi(x)_{[i,j]} = v$ or $\psi(x)_{[i,j]} = \bar{v}$, then $x_{(-\infty,j]} = x'_{(-\infty,j]}$. To see this, apply Lemma 9.2.7 to $w = \psi(x)_{(-|v|,j]}$, a block which begins with v and ends with v or $\bar{v}$. Since the blocks $x_{(-|v|,j]}$ and $x'_{(-|v|,j]}$ have the same initial symbol, (i) of Lemma 9.2.7 tells us that they must coincide. So we get $x_{(-\infty,j]} = x'_{(-\infty,j]}$, as desired.

If blocks from the set $\{v, \bar{v}\}$ appear infinitely often in $\psi(x)_{[0,\infty)}$, the preceding shows that $x = x'$. Otherwise, the preceding shows that for some

$j \geqslant 0$, $x_{(-\infty,j]} = x'_{(-\infty,j]}$ and all appearances of u or $\bar{u}$ in x or x' are entirely confined to $x_{(-\infty,j]}, x'_{(-\infty,j]}$. So, by definition of ψ,

$$\phi(x)_{[j+1,\infty)} = \psi(x)_{[j+1,\infty)} = \psi(x')_{[j+1,\infty)} = \phi(x')_{[j+1,\infty)}.$$

Since $x_{[j+1,\infty)}$ and $x'_{[j+1,\infty)}$ have the same initial state and ϕ is a right-closing 1-block code, it follows that $x_{[j+1,\infty)} = x'_{[j+1,\infty)}$, and so $x = x'$. Thus, ψ is right-closing. This completes the proof of the Replacement Theorem. □

Finally, we have the following version of the Replacement Theorem for sofic shifts.

Theorem 9.2.8. *Let X, Y be irreducible sofic shifts with minimal right-resolving presentations $(G_X, \mathcal{L}_X), (G_Y, \mathcal{L}_Y)$. Suppose that there is a finite-to-one factor code $\phi\colon X \to Y$. Then there is an almost invertible factor code $\psi\colon X \to Y$ if and only if $\mathrm{per}(G_X) = \mathrm{per}(G_Y)$.*

We leave the proof as an exercise.

EXERCISES

9.2.1. Assume the notation in this section. The code $\eta\colon X_H \to X_H$ defined by replacing v by $\bar{v}$ and vice versa whenever they appear is a well-defined sliding block code. Explain why ψ is *not* the composition $\eta \circ \phi$.

9.2.2. Prove Lemma 9.2.6 in the following steps.

(a) Show that, by replacing G and H by higher edge graphs, we may assume that there is an edge b in H which is not a self-loop and is a magic symbol for ϕ and that $t(b)$ has at least two outgoing edges and $i(b)$ has at least two incoming edges.

(b) Let $e_2 \ldots e_L$ be a simple path from $t(b)$ to $i(b)$. Let $f \neq e_2$ be an outgoing edge from $t(b)$. Let $g \neq e_L$ be an incoming edge to $i(b)$. Let $a^{(1)}$ and $a^{(2)}$ be pre-images of b. Show that there are edges $g^{(1)}$ and $g^{(2)}$ in G such that $t(g^{(1)}) = i(a^{(1)})$, $t(g^{(2)}) = i(a^{(2)})$, and $\Phi(g^{(1)}) = \Phi(g^{(2)}) = g$.

(c) Choose N_0 such that for any two states I, J of G there is a path of length N_0 from I to J, and choose a positive integer p such that $pL \geqslant N_0 + 2$. Show that there is a pre-image η of the path $(be_2 \ldots e_L)^p bf$ whose initial symbol is $a^{(1)}$.

(d) Let $\gamma^{(1)}$ be a path in G from $t(\eta)$ to $i(g^{(1)})$, and $\gamma^{(2)}$ be a path in G from $t(\eta)$ to $i(g^{(2)})$, both of length N_0. Let

$$u = \eta\gamma^{(1)}g^{(1)}a^{(1)}, \quad \bar{u} = \eta\gamma^{(2)}g^{(2)}a^{(2)}.$$

Show that $u, \bar{u}$ satisfy Properties (1)–(4) of Lemma 9.2.6.

9.2.3. Complete the proof of the Replacement Theorem by reducing to the mixing case.

9.2.4. Let X_G, X_H be irreducible shifts of finite type. Suppose that $h(X_G) = h(X_H)$ and there is a (finite-to-one) factor code $\phi\colon X_G \to X_H$. Show that $\mathrm{per}(X_H)$ divides $\mathrm{per}(X_G)$ and that there is a finite-to-one factor code $\phi'\colon X_G \to X_H$ with $d_{\phi'} = \mathrm{per}(X_G)/\mathrm{per}(X_H)$.

9.2.5. Verify that, in Example 9.2.4, the code obtained from the factor code $\phi\colon X_G \to X_H$ by switching the images of $a_0b_0a_1b_1c_0a_0$ and $a_0b_0c_1c_1c_1a_1$ is a right-closing, almost invertible factor code $X_G \to X_H$.

9.2.6. Is the "Moreover" of the Replacement Theorem true with both appearances of the term "right-closing" replaced by the term "bi-closing"? [*Hint:* Consider Exercise 9.1.5.]

9.2.7. Outline a proof of the Replacement Theorem that reduces the degree of ϕ to one in a single stroke, instead of inductively.

9.2.8. Prove Theorem 9.2.8.

9.2.9. A point x in a shift space X is called *left-transitive* if every block in X appears in $x_{(-\infty,0]}$. Show that we can add the following to Theorem 9.2.8:

> If ϕ does not identify any pair of distinct left-asymptotic, left-transitive points (i.e., whenever x, x' are distinct left-asymptotic, left-transitive points, $\phi(x) \neq \phi(x')$), then neither does ψ.

§9.3. Almost Conjugacy

In this section, we strengthen the relation of finite equivalence by requiring the legs to be almost invertible.

Definition 9.3.1. Shift spaces X and Y are *almost conjugate* if there is a shift of finite type W and almost invertible factor codes $\phi_X\colon W \to X$, $\phi_Y\colon W \to Y$. We call (W, ϕ_X, ϕ_Y) an *almost conjugacy* between X and Y.

Just as for finite equivalence, the definition makes sense only for sofic shifts, indeed only for irreducible sofic shifts, since a reducible sofic shift cannot be the image of an almost invertible code. As in Proposition 8.3.6, by passing to an irreducible component of maximal entropy we may take the common extension of an almost conjugacy to be irreducible.

We note that the finite equivalence given in Example 8.3.11 is actually an almost conjugacy (see Exercise 9.3.1).

Since conjugacies are almost invertible and since almost invertible codes are finite-to-one, we have the following implications for irreducible sofic shifts:

$$\text{conjugacy} \Rightarrow \text{almost conjugacy} \Rightarrow \text{finite equivalence.}$$

We now state the result which classifies irreducible shifts of finite type up to almost conjugacy.

Theorem 9.3.2. *Let X and Y be irreducible shifts of finite type. Then X and Y are almost conjugate if and only if $h(X) = h(Y)$ and $\operatorname{per}(X) = \operatorname{per}(Y)$. Moreover, if X and Y are almost conjugate, then there is an almost conjugacy of X and Y in which one leg is left-resolving and the other leg is right-resolving (and the common extension is irreducible).*

In particular, for mixing shifts of finite type entropy is a complete invariant of almost conjugacy.

Our proof of this result is not terribly constructive, partly because it relies on the Replacement Theorem. However, in the special case that X and Y have fixed points, there is an alternative construction that is very algorithmic (see [AdlM]).

Since entropy is an invariant of finite equivalence, it is also an invariant of almost conjugacy. Since irreducibility of the common extension comes for free, the "only if" part of the Replacement Theorem (Theorem 9.2.5) shows that the period is an invariant of almost conjugacy. To see that entropy and period form a complete set of invariants for almost conjugacy, we will show how a finite equivalence can be modified to be an almost conjugacy. This will be done by applying the "if" part of the Replacement Theorem and the lemmas below, the first of which depends on the following concept.

Definition 9.3.3. A graph is *nonwandering* if every edge belongs to an irreducible component.

Thus a nonwandering graph is one in which there are no (one-way) connections between irreducible components. For instance, an irreducible graph is clearly nonwandering, while the graph with adjacency matrix $\begin{bmatrix} 1 & 1 \\ 0 & 1 \end{bmatrix}$ is not (why?).

Lemma 9.3.4. *Let G and H be irreducible graphs. Let M be a graph and (X_M, ϕ_G, ϕ_H) be a finite equivalence between X_G and X_H where ϕ_G is left-covering and ϕ_H is right-covering. Then M is nonwandering and the restriction of both legs to each irreducible component of X_M is onto.*

For the second lemma, recall from §8.3 that $M = M(F, \Gamma)$ is a graph constructed as in the proof of (4) $\Rightarrow$ (5) of Theorem 8.3.8 using a nonnegative integral solution $F \neq 0$ to the matrix equation $A_G F = F A_H$ and choice of bijection Γ.

Lemma 9.3.5. *Let G, H be primitive graphs and $A = A_G$, $B = A_H$. Let F be a positive integer matrix satisfying $AF = FB$. If the entries of F are sufficiently large, then the graph $M = M(F, \Gamma)$ can be constructed so that it contains a set of cycles (not necessarily simple) which pass through a common state and have relatively prime lengths.*

Before we prove the lemmas, we first show how to use them to prove the "if" part of Theorem 9.3.2 in the mixing case (the irreducible case follows in the usual way).

Let X, Y be mixing shifts of finite type having the same entropy. By Proposition 4.5.10, we may assume that $X = X_G, Y = X_H$ for primitive graphs G, H. By Theorem 8.3.8 there is a finite equivalence with common extension X_M, left-resolving leg $\phi_G: X_M \to X_G$ and right-resolving leg

$\phi_H \colon X_M \to X_H$. In fact, by the construction used in proving (4) $\Rightarrow$ (5) in Theorem 8.3.8, we may assume that $M = M(F, \Gamma)$ for some positive F and bijection Γ and also that ϕ_G is left-covering and ϕ_H is right-covering.

By Lemma 9.3.5, we may construct M so that it has an irreducible component K which has cycles with relatively prime lengths; just make sure that the entries of the solution F are sufficiently large (any positive integer multiple of a solution F to $A_G F = F A_H$ is still a solution). Thus X_M has a mixing component X_K. It follows by Lemma 9.3.4 that $\phi_G|_{X_K}$ is still a left-resolving factor code onto X_G and $\phi_H|_{X_K}$ is still a right-resolving factor code onto X_H. Now since X_K, X_G, X_H are all mixing, we can apply the Replacement Theorem to $\phi_G|_{X_K}$ and $\phi_H|_{X_K}$ to obtain almost invertible factor codes from X_K onto X_G and from X_K onto X_H. Thus X_G, X_H are almost conjugate. Moreover, the Replacement Theorem guarantees that the replacements can be chosen to be left-closing and right-closing, respectively. We leave it as an exercise for the reader to show that we can recode so that the left-closing leg becomes left-resolving and the right-closing leg becomes right-resolving. This completes the proof of Theorem 9.3.2, except for the two lemmas. $\qquad\square$

PROOF OF LEMMA 9.3.4: We will first show that if K is a sink of M, i.e, every edge in M which originates in K must belong to K, then

(1) K is also a source; i.e., every edge which terminates in K must belong to K,

(2) $\phi_H|_{X_K}$ is a right-resolving factor code onto X_H,

(3) $\phi_G|_{X_K}$ is a left-resolving factor code onto X_G.

To see this, we argue as follows. Since K is a sink and since ϕ_H is right-covering, so is $\phi_H|_{X_K}$. Hence, by Proposition 8.2.2, ϕ_H maps X_K *onto* X_H, and we have (2) above.

Since ϕ_H is finite-to-one, we have $h(X_K) = h(X_H) = h(X_G)$. Since ϕ_G is finite-to-one, $\phi_G(X_K)$ is a subshift of X_G with full entropy. Since G is irreducible, $\phi_G(X_K) = X_G$, and we have (3) above.

We claim that this forces K to be a source. If not, then there is an incoming edge c to some state I of K such that c is not an edge of K. Now Φ_G maps this edge to some edge g in G. But $\phi_G|_{X_K}$ is left-covering onto X_G by Proposition 8.2.2. So there is an edge c' in K that terminates at I and maps to g via ϕ_G. Since c' is in K and c is not, we have $c \neq c'$, and this contradicts the fact that ϕ_G is left-resolving. Thus K is a source and we have (1) above. So, we have established (1), (2) and (3).

Now let K' be an irreducible component of M. Then there is a path from K' to some sink K. On the other hand, we have shown that every sink K is a source. The only possible explanation for this is that $K' = K$. Thus, every irreducible component of M is a sink and a source; it is straightforward to show that this means that M is nonwandering. Finally, since we have shown that the restriction of both legs to each of the sinks is still a factor

code onto the appropriate edge shift X_G, X_H, we have that the same holds
for all irreducible components.   □

PROOF OF LEMMA 9.3.5: Before proceeding, the reader should review the
construction of $M(F, \Gamma)$ given in (4) $\Rightarrow$ (5) of Theorem 8.3.8, in particular
the notions of A-edge, B-edge, F-edge and boxes.

First, we dispose of a trivial case. If $A = B = [1]$, then the solutions F
are the 1×1 matrices $[k]$. In this case, we have only one A-edge, which we
call a, and only one B-edge, which we call b, and there are exactly k F-
edges. We construct $M = M(F, \Gamma)$ according to the bijection $\Gamma(fb) = af$
for all F-edges f. Then M consists of exactly k disjoint self-loops. Any
one of these self-loops constitutes the desired set of cycles.

Hence we may assume $A \neq [1]$ and $B \neq [1]$. Thus, by primitivity, H
must have a state J with at least two distinct outgoing edges b, b', and G
must have a state I with at least two distinct incoming edges a, a'.

Again by primitivity, for all sufficiently large integers m, there are cycles
$\beta = b_1 \ldots b_m$ of length m and $\beta' = b'_1 \ldots b'_{m+1}$ of length $m + 1$ in H such
that

(1) both β and β' begin (and end) at J, and
(2) $b_1 = b$, $b'_1 = b'$.

Likewise, for all sufficiently large m, there are cycles $\alpha = a_1 \ldots a_m$ of
length m and $\alpha' = a'_1 \ldots a'_{m+1}$ of length $m + 1$ in G such that

(1) both α and α' begin (and end) at I, and
(2) $a_m = a$, $a'_{m+1} = a'$.

Fix m sufficiently large that β, β', α, α' are all defined. Let

$$I_i = i(a_i), \ I'_i = i(a'_i), \ J_i = i(b_i), \ J'_i = i(b'_i).$$

We will show that if the entries of F are all at least $2m$ then the bijection
Γ (used in the construction of $M(F, \Gamma)$) can be chosen so that there are
cycles γ, γ' in M with lengths $m, m + 1$ both passing through a common
state. We construct γ, γ' by "joining together" β with α and β' with α' as
follows.

Since the entries of F are all at least m, we may choose distinct F-edges
f_i such that $i(f_i) = I_i$ and $t(f_i) = J_i$ for $i = 1, \ldots, m$. Set $f_{m+1} = f_1$.
This situation is depicted as follows.

$$
\begin{array}{ccccccccc}
I_1 & \xrightarrow{a_1} & I_2 & \xrightarrow{a_2} & I_3 & \xrightarrow{a_3} & \cdots & \xrightarrow{a_{m-1}} & I_m & \xrightarrow{a_m} & I_1 \\
\downarrow{f_1} & & \downarrow{f_2} & & \downarrow{f_3} & & & & \downarrow{f_m} & & \downarrow{f_{m+1} = f_1} \\
J_1 & \xrightarrow{b_1} & J_2 & \xrightarrow{b_2} & J_3 & \xrightarrow{b_3} & \cdots & \xrightarrow{b_{m-1}} & J_m & \xrightarrow{b_m} & J_1
\end{array}
$$

For $i = 1, \ldots, m + 1$, we may also choose F-edges f'_i such that $i(f'_i) = I'_i$, $t(f'_i) = J'_i$, $f'_1 = f_1$, and the F-edges $f_1, \ldots, f_m, f'_2, \ldots, f'_{m+1}$ are all
distinct (since the entries of F are at least $2m$). Set $f'_{m+2} = f'_1 = f_1$.

$$I'_1 \xrightarrow{a'_1} I'_2 \xrightarrow{a'_2} I'_3 \xrightarrow{a'_3} \cdots \xrightarrow{a'_m} I'_{m+1} \xrightarrow{a'_{m+1}} I'_1$$

$$f_1 = f'_1 \Big\downarrow \qquad \Big\downarrow f'_2 \qquad \Big\downarrow f'_3 \qquad \qquad \Big\downarrow f'_{m+1} \qquad \Big\downarrow f'_{m+2} = f_1$$

$$J'_1 \xrightarrow{b'_1} J'_2 \xrightarrow{b'_2} J'_3 \xrightarrow{b'_3} \cdots \xrightarrow{b'_m} J'_{m+1} \xrightarrow{b'_{m+1}} J'_1$$

We declare all of the boxes that you see in the previous two pictures to be boxes in the sense of the construction in Theorem 8.3.8($4 \Rightarrow 5$); to be precise, recalling the notation Γ_{IJ} from that proof, we declare

$$(9\text{--}3\text{--}1) \qquad \Gamma_{I_i J_{i+1}}(f_i b_i) = a_i f_{i+1} \quad \text{for } i = 1, \dots, m$$

and

$$(9\text{--}3\text{--}2) \qquad \Gamma_{I'_i J'_{i+1}}(f'_i b'_i) = a'_i f'_{i+1}, \quad \text{for } i = 1, \dots, m+1$$

(where we put $J_{m+1} = J_1$ and $J'_{m+2} = J'_1$).

Since

$$f_1 b_1, \dots, f_m b_m, f'_1 b'_1, \dots, f'_{m+1} b'_{m+1}$$

are all distinct, and since

$$a_1 f_2, \dots, a_m f_{m+1}, a'_1 f'_2, \dots, a'_{m+1} f'_{m+2}$$

are all distinct, there are no contradictions in these assignments.

Extend the definitions of the bijections in any way that is consistent with (9–3–1) and (9–3–2). This defines $M = M(F, \Gamma)$. Let γ be the path in M defined by the sequence of boxes

$$\Box(f_1, b_1, a_1, f_2)\Box(f_2, b_2, a_2, f_3) \dots \Box(f_m, b_m, a_m, f_{m+1}).$$

Then γ is a cycle in M which begins (and ends) at the state $f_{m+1} = f_1$ and has length m. Let γ' be the path in M defined by the sequence of boxes

$$\Box(f'_1, b'_1, a'_1, f'_2)\Box(f_2,' b'_2, a'_2, f'_3) \dots \Box(f'_{m+1}, b'_{m+1}, a'_{m+1}, f'_{m+2}).$$

Then γ' is a cycle in M which begins (and ends) at the state $f'_{m+2} = f'_1 = f_1$ and has length $m + 1$. $\qquad \Box$

By passing to the minimal right-resolving presentations, we get the following extension of Theorem 9.3.2 to the irreducible sofic case. We leave the proof as an exercise for the reader.

Theorem 9.3.6. *Let X, Y be irreducible sofic shifts with minimal right-resolving presentations $(G_X, \mathcal{L}_X)$, $(G_Y, \mathcal{L}_Y)$. Then X, Y are almost conjugate if and only if $h(X) = h(Y)$ and $\mathrm{per}(G_X) = \mathrm{per}(G_Y)$. Moreover, if X and Y are almost conjugate, then there is an almost conjugacy of X and Y in which one leg is left-resolving and the other leg is right-resolving (and the common extension is irreducible).*

Recall from Theorem 8.4.7 that irreducible sofic shifts X, Y are right-resolving finitely equivalent if and only if $M_{G_X} = M_{G_Y}$ where G_X, G_Y are the underlying graphs of their minimal right-resolving presentations. In view of Theorem 9.3.6, it would be natural to conjecture that a right-resolving finite equivalence can always be replaced by a *right-resolving almost conjugacy* (i.e., an almost conjugacy in which both legs are right-resolving) if $\text{per}(G_X) = \text{per}(G_Y)$. This turns out to be false. In fact, Example 9.2.4 is a counterexample (Exercise 9.3.8). Nevertheless, there is an algorithm to decide when two irreducible sofic shifts are right-resolving almost conjugate [AshMT].

Since Theorem 9.3.6 shows that almost conjugacy is completely captured by the numerical invariants, entropy and period, it follows that almost conjugacy is an equivalence relation. This fact can also be proved directly, using the fiber product construction, as in Proposition 8.3.4; we leave this to the reader.

Finally, we state a consequence of Theorem 9.3.2 which classifies the hyperbolic toral automorphisms introduced in §6.5. For this, we first need to define an analogue of doubly transitive point for invertible dynamical systems.

Definition 9.3.7. Let (M, ϕ) be an invertible dynamical system. A point $x \in M$ is *doubly transitive* if both half-orbits $\{\phi^n(x) : n \geqslant 0\}$ and $\{\phi^n(x) : n \leqslant 0\}$ are dense in M.

When (M, ϕ) is a symbolic dynamical system (i.e., M is a shift space and ϕ is the shift mapping), this definition agrees with our definition of doubly transitive in §9.1 (Exercise 9.3.9). Also, an application of the Baire Category Theorem shows that an invertible dynamical system has a doubly transitive point if and only if it is topologically transitive (Exercise 9.3.10).

Now we extend the notions of almost invertibility and almost conjugacy to invertible dynamical systems.

Definition 9.3.8. Let (M_1, ϕ_1) and (M_2, ϕ_2) be topologically transitive, invertible dynamical systems, and let π be a factor map from (M_1, ϕ_1) to (M_2, ϕ_2). We say that π is *almost invertible* if each doubly transitive point for (M_2, ϕ_2) has exactly one pre-image under π in M_1. An *almost conjugacy* is a triple (W, π_1, π_2) where W, the *common extension*, is a shift of finite type and π_i is an almost invertible factor map from (W, σ) to (M_i, ϕ_i) for $i = 1, 2$.

Recall from §6.5 that hyperbolic toral automorphisms have Markov partitions and thus are factors of shifts of finite type. It turns out that these shifts of finite type can be chosen to be mixing and that these factor maps can be chosen to be almost invertible (for the specific Markov partition that we constructed in Example 6.5.10, see Exercise 9.3.11). It can also be shown that topological entropy is an invariant of almost conjugacy. Thus from Theorem 9.3.2 we obtain the following result.

Theorem 9.3.9. *Two hyperbolic toral automorphisms are almost conjugate if and only if they have the same topological entropy. Moreover, the common extension can be chosen to be mixing.*

See [AdlM] for a proof of this result.

EXERCISES

9.3.1. Show that the finite equivalence given in Example 8.3.11 is an almost conjugacy by showing that for ϕ_G the edge $5 \to 1$ is a magic word and for ϕ_H the path $3 \to 2 \to 3 \to 1 \to 2 \to 3$ is a magic word.

9.3.2. (a) Show that the fiber product construction preserves degree in the following sense: let $\psi_X : X \to Z$ and $\psi_Y : Y \to Z$ be finite-to-one factor codes on irreducible shifts of finite type X and Y, and let (W, ϕ_X, ϕ_Y) denote the fiber product of ψ_X, ψ_Y. Show that $d_{\phi_X} = d_{\psi_Y}$ and $d_{\phi_Y} = d_{\psi_X}$.
   (b) Use this to show directly that almost conjugacy is an equivalence relation for irreducible sofic shifts.
   (c) With the notation in part (a), give an example where for each irreducible component V of W, $d_{\phi_X|_V} < d_{\psi_Y}$ and $d_{\phi_Y|_V} < d_{\psi_X}$.

9.3.3. Show that for a graph G, the following are equivalent.
   (a) G is nonwandering.
   (b) Every irreducible component of G is a sink and a source.
   (c) Every irreducible component of G is a sink.
   (d) Every irreducible component of G is a source.
   (e) Every edge of G belongs to a cycle.

9.3.4. Let X, Y, Z be irreducible shifts of finite type. Show that given a left-closing factor code $Z \to X$ with degree d_1 and a right-closing factor code $Z \to Y$ with degree d_2, then there exist a shift of finite type Z' conjugate to Z, a left-resolving factor code $Z' \to X$ with degree d_1 and a right-resolving factor code $Z' \to Y$ with degree d_2 (this completes the proof of Theorem 9.3.2 in the mixing case). [*Hint:* Use Proposition 5.1.11.]

9.3.5. Reduce the proof of Theorem 9.3.2 to the mixing case.

9.3.6. Construct almost conjugacies for the examples in Exercises 8.3.1 and 8.3.2.

9.3.7. Prove Theorem 9.3.6.

9.3.8. Show that for the graphs G, H in Example 9.2.4 X_G and X_H are almost conjugate but not right-resolving almost conjugate.

9.3.9. Show that for a symbolic dynamical system (M, ϕ) (i.e., M is a shift space and ϕ is the shift mapping), the definition of doubly transitive for invertible dynamical systems agrees with our definition of doubly transitive in §9.1.

9.3.10. Show that an invertible dynamical system has a doubly transitive point if and only if it is topologically transitive.

9.3.11. Show that for the specific toral automorphism and the specific Markov partition constructed in Example 6.5.10, the factor map π is almost invertible. [*Hint:* See Exercise 6.5.6(a).]

§9.4. Typical Points According to Probability

In this section, we justify why the set of doubly transitive points should be viewed as a "large" set or as a set of "typical" points. Probability is our point of view.

In §2.3 we defined Markov chains on graphs. We noted that these are slightly different from ordinary Markov chains. Recall that a Markov chain μ on a graph G is defined by initial state probabilities $\mu(I)$ for $I \in \mathcal{V}(G)$, and conditional probabilities $\mu(e|i(e))$ for $e \in \mathcal{E}(G)$. Formula (2–3–1) shows how, using the initial state and conditional probabilities, a Markov chain assigns probabilities on paths by putting

$$\mu(e_1 \ldots e_n) = \mu(i(e_1))\mu(e_1|i(e_1)) \ldots \mu(e_n|i(e_n)).$$

Recall also that the initial state probabilities are assembled into a probability vector, called the initial state distribution $\mathbf{p}$, where

$$p_I = \mu(I),$$

and the conditional probabilities are assembled into a stochastic matrix, called the conditional probability matrix P, defined by

$$P_{IJ} = \sum_{e \in \mathcal{E}_I^J} \mu(e|I).$$

Since edge shifts are stationary in the sense that they are preserved by the shift map, it is natural to consider Markov chains that are stationary in some sense. Stationarity for Markov chains is commonly expressed as follows.

Definition 9.4.1. A Markov chain on a graph with initial state distribution $\mathbf{p}$ and conditional probability matrix P is *stationary* if

$$(9\text{–}4\text{–}1) \qquad\qquad\qquad \mathbf{p}P = \mathbf{p}.$$

Such a vector $\mathbf{p}$ is called a *stationary distribution*.

Equation (9–4–1) means that, for each state I, the probability of I coincides with the sum of the probabilities of the *incoming* edges to I, or in symbols

$$(9\text{–}4\text{–}2) \qquad \mu(I) = \sum_{e \in \mathcal{E}^I} \mu(i(e))\mu(e|i(e)) = \sum_{e \in \mathcal{E}^I} \mu(e).$$

It follows from the definition of Markov chain that for each state I, the probability of I coincides with the sum of the probabilities of the *outgoing* edges from I, or

$$(9\text{–}4\text{–}3) \qquad \mu(I) = \sum_{e \in \mathcal{E}_I} \mu(I)\mu(e|I) = \sum_{e \in \mathcal{E}_I} \mu(e).$$

Hence stationarity means that for each state I the sum of the probabilities of incoming edges equals the sum of the probabilities of outgoing edges.

It follows inductively from equations (9–4–2) and (9–4–3) that for a stationary Markov chain on a graph G and a path w in G

$$(9\text{–}4\text{–}4) \qquad \mu(w) = \sum_{e \in \mathcal{E}^{i(w)}} \mu(ew)$$

and

$$(9\text{–}4\text{–}5) \qquad \mu(w) = \sum_{e \in \mathcal{E}_{t(w)}} \mu(we).$$

The following result shows that with a modest assumption the conditional probability matrix for a stationary Markov chain on a graph uniquely determines its initial state distribution.

Proposition 9.4.2. *Let* **p** *be the initial state distribution and* P *be the conditional probability matrix of a stationary Markov chain* μ *on an irreducible graph* G *such that* $\mu(e|i(e)) > 0$ *for each edge* e. *Then* **p** *is the unique probability vector such that* $\mathbf{p}P = \mathbf{p}$.

PROOF: Since P is stochastic, the column vector of all 1's is a positive right eigenvector corresponding to eigenvalue 1. Since the assumptions guarantee that P is irreducible, we have, by the Perron–Frobenius Theorem, that $\lambda_P = 1$. Thus, up to scale, there is a unique positive vector $\mathbf{w}$ satisfying $\mathbf{w}P = \mathbf{w}$. Such a vector becomes unique if we require it to be a probability vector. Therefore the initial state distribution is this unique vector. ☐

Given a Markov chain on a graph, we can define "measures" of cylinder sets as follows. Recall the notion of cylinder set from Chapter 6:

$$C_i(w) = \{x \in X : x_{[i,i+|w|)} = w\}.$$

Definition 9.4.3. For a Markov chain μ on a graph G, a path w of length n in G, and $i \in \mathbb{Z}$, we define the *measure* of the cylinder set $C_i(w)$ to be the probability of the path w:

$$\mu(C_i(w)) = \mu(w).$$

For any set K which is a finite union of cylinder sets, express K as a finite *disjoint* union of cylinder sets C_i, $i = 1, \ldots k$ in any way and define

$$\mu(K) = \sum_{i=1}^{k} \mu(C_i).$$

In order for this definition to make sense, we need to verify that (a) any finite union of cylinder sets can be expressed as a disjoint finite union of cylinder sets and (b) the definition of $\mu(K)$ does not depend on the particular expression. Now (a) holds because whenever $K = \bigcup_{i=1}^{k} C_i$ and the C_i are cylinder sets, then for some n, each C_i can be expressed as a disjoint union of "central" cylinder sets i.e., cylinder sets of the form $C_{-n}(w)$ where w is a $(2n+1)$-block. As we will see in our next result, (b) is a consequence of the definition of Markov chain and stationarity.

Proposition 9.4.4. *For a stationary Markov chain on a graph G, the assignment of measures to finite unions of cylinder sets of $X = X_G$ above satisfies the following properties.*

(1) *(Consistency) If a set K is expressed in any way as a finite disjoint union of cylinder sets $K = \bigcup_{i=1}^{k} C_i$, then $\sum_{i=1}^{k} \mu(C_i)$ depends only on K.*

(2) *(Probability) The measure of X is 1.*

(3) *(Stationarity) A finite union of cylinder sets and its shift have the same measure.*

PROOF: The crux of the proof is the Consistency Condition (1). Once we have this, the assignment of measures to cylinder sets is then well-defined. Then (2) follows from the fact that X_G is the disjoint union of the cylinder sets $C_0(e)$ for $e \in \mathcal{E}(G)$:

$$\mu(X) = \sum_{e \in \mathcal{E}(G)} \mu(C_0(e)) = \sum_{e \in \mathcal{E}(G)} \mu(i(e))\mu(e|i(e)) = 1.$$

And (3) follows from the fact that the measure of a cylinder set $C_i(w)$ depends only on the path w that defines it.

To verify (1), first observe that it follows inductively from equations (9–4–4) and (9–4–5) that for any cylinder set $C = C_i(w)$ and any positive integer m,

$$(9\text{–}4\text{–}6) \qquad \mu(C) = \sum_{\{\text{paths } \gamma \,:\, |\gamma|=m, t(\gamma)=i(w)\}} \mu(C_{i-m}(\gamma w))$$

and

$$(9\text{–}4\text{–}7). \qquad \mu(C) = \sum_{\{\text{paths } \gamma \,:\, |\gamma|=m, i(\gamma)=t(w)\}} \mu(C_i(w\gamma))$$

Now, suppose that a cylinder set C is expressed as the disjoint union of an arbitrary finite collection of cylinder sets C_ℓ. There is a positive integer m such that each C_ℓ is a disjoint union of cylinder sets, each of the form $C_{-m}(u)$, where u is a $(2m+1)$-block. From (9–4–6) and (9–4–7), we get

$$\sum_{\ell} \mu(C_\ell) = \sum_{\{u \,:\, |u|=2m+1, \, C_{-m}(u) \subseteq C\}} \mu(C_{-m}(u)) = \mu(C).$$

Hence the consistency condition (1) holds whenever $K = C$ is a cylinder set.

Now if a set K is decomposed as a finite disjoint union of cylinder sets in two different ways, say

$$K = \bigcup_i C_i = \bigcup_j C'_j,$$

then by what we have already proved,

$$\sum_i \mu(C_i) = \sum_{i,j} \mu(C_i \cap C'_j) = \sum_j \mu(C'_j),$$

and so we get the same value for $\mu(K)$ for each decomposition. This verifies the Consistency Condition, and completes the proof. $\square$

Probability theory gives a way to define measures of much more general sets than finite unions of cylinder sets. Here, we will only need to know what it means for a set to have measure zero. These sets are called null sets.

Definition 9.4.5. Let G be a graph and $S \subseteq X_G$. We say that S is a *null set* for a Markov chain μ on G if, for every $\epsilon > 0$, there is a collection $\mathcal{C}$ of cylinder sets such that

(9–4–8) $$S \subseteq \bigcup_{C \in \mathcal{C}} C$$

and

(9–4–9) $$\sum_{C \in \mathcal{C}} \mu(C) < \epsilon.$$

Note that the sum in (9–4–9) is automatically countable since there are only countably many cylinder sets. If a property holds for all points in the complement of a null set, then we regard the property as holding on almost all points, or on most of the space.

Next, we establish some basic properties of null sets.

Proposition 9.4.6.

(1) *The shift of any null set is again a null set.*
(2) *A countable union of null sets is a null set.*

PROOF: (1) If $\mathcal{C}$ works for S and ϵ, then $\{\sigma(C) : C \in \mathcal{C}\}$ works for $\sigma(S)$ and ϵ.

(2) If $\mathcal{C}_i$ works for S_i and $\epsilon/2^i$, then $\cup_{i \in \mathbb{Z}^+} \mathcal{C}_i$ works for $\cup_{i \in \mathbb{Z}^+} S_i$ and ϵ. $\square$

If the complement of a null set could also be a null set, then this notion would be worthless. Fortunately, this cannot happen.

Proposition 9.4.7. *The complement of a null set cannot be a null set.*

PROOF: Suppose that both S and S^c are null. Then by part (2) of Proposition 9.4.6, X itself would be null. We will prove that X is not null by showing that whenever a collection $\mathcal{C}$ of cylinder sets covers X, we have

$$(9\text{-}4\text{-}10) \qquad\qquad 1 \leqslant \sum_{C \in \mathcal{C}} \mu(C).$$

Recall from §6.1 that cylinder sets are open sets. Recall also from the Heine–Borel Theorem (Theorem 6.1.22) that in a compact metric space X every collection of open sets that covers X has a finite subcollection that also covers X. Thus there is a finite subcollection $\mathcal{C}'$ of $\mathcal{C}$ that covers X. It suffices to show that

$$(9\text{-}4\text{-}11) \qquad\qquad 1 \leqslant \sum_{C' \in \mathcal{C}'} \mu(C').$$

Write $\mathcal{C}' = \{C_1', \dots, C_n'\}$, and let

$$C_1'' = C_1', C_2'' = C_2' - C_1'', \dots, C_n'' = C_n' \setminus \left(\bigcup_{i=1}^{n-1} C_i'' \right).$$

Observe that X is the disjoint union of the sets C_i'', $1 \leqslant i \leqslant n$, each C_i'' is a finite union of cylinder sets, and each $C_i'' \subseteq C_i'$. By Proposition 9.4.4, we have

$$(9\text{-}4\text{-}12) \qquad\qquad 1 = \mu(X) = \sum_{i=1}^{n} \mu(C_i'').$$

The Consistency Condition of Proposition 9.4.4 also implies that $\mu(C_i') = \mu(C_i'') + \mu(C_i' \setminus C_i'')$, and so

$$\mu(C_i'') \leqslant \mu(C_i').$$

This together with (9–4–12) gives (9–4–11) as desired.   □

The main result of this section, Theorem 9.4.9 below, shows that for an irreducible edge shift the set of doubly transitive points is the complement of a null set. We separate out a lemma which contains the crucial idea.

Lemma 9.4.8. *Let G be an irreducible graph. Fix a Markov chain on G which assigns positive conditional probability to every edge. For any path w in G, the probability that a path of length m does not contain w approaches zero as m tends to infinity (in fact, exponentially fast as $m \to \infty$).*

PROOF: Recall from the end of §2.3 that we may pass to a higher block presentation $G^{[N]}$ and transfer the Markov chain on G to a Markov chain on $G^{[N]}$. In this way, we may assume that w is an edge of G.

Let P denote the conditional probability matrix of the Markov chain, let $\mathbf{p}$ denote the initial state distribution and let $\mathbf{1}$ denote the column vector of all 1's. The assumptions guarantee that P is irreducible. Define $\overline{P}$ to be the matrix obtained from P by replacing the single entry $P_{i(w),t(w)}$ by $P_{i(w),t(w)} - \mu(w|i(w))$. Let q_m denote the probability that a path of length m does not contain w. Then

$$q_m = \mathbf{p}(\overline{P})^m \mathbf{1}.$$

But since P is irreducible, by Theorem 4.4.7 the spectral radius $\lambda = \lambda_{\overline{P}}$ of $\overline{P}$ is strictly less than the spectral radius of P, and the latter is 1. By Proposition 4.2.1 there is a constant $d > 0$ such that $(\overline{P}^m)_{IJ} \leqslant d\lambda^m$ for all I, J. Thus there is a constant $d' > 0$ such that

$$q_m = p(\overline{P})^m \mathbf{1} \leqslant d'\lambda^m$$

with $\lambda < 1$ as desired. $\qquad\square$

Theorem 9.4.9. *Let $X = X_G$ be an irreducible edge shift. Then the set of doubly transitive points of X is the complement of a null set with respect to any stationary Markov chain on G which assigns positive conditional probability to every edge of G.*

PROOF: For a block w in X_G, define the sets

$$M_k^+(w) = \{x \in X_G : w \text{ does } not \text{ appear in } x_{[k,\infty)}\}$$

and

$$M_k^-(w) = \{x \in X_G : w \text{ does } not \text{ appear in } x_{(-\infty,k]}\}.$$

We can express the set D of doubly transitive points in X as

$$(9\text{-}4\text{-}13) \qquad D = \left(\bigcup_{w \in \mathcal{B}(X_G)} \bigcup_{k \in \mathbb{Z}} M_k^+(w) \cup M_k^-(w)\right)^c.$$

By Proposition 9.4.6, it suffices to show that $M_0^+(w)$ and $M_0^-(w)$ are null sets for each path w. For each m, let $\mathcal{P}_m(w)$ denote the collection of paths v of length m which do not contain w as a subpath. Now, $M_0^+(w) \subseteq \bigcup_{v \in \mathcal{P}_m(w)} C_0(v)$. Thus to show that $M_0^+(w)$ is a null set it suffices to show that

$$\lim_{m \to \infty} \sum_{v \in \mathcal{P}_m(w)} \mu(C_0(v)) = 0.$$

But this is the content of Lemma 9.4.8. In a completely similar way, one shows that $M_0^-(w)$ is also a null set. Thus D is the complement of a null set. $\qquad\square$

Hence we may regard the set of doubly transitive points in an irreducible edge shift as a "universally" large set.

A *stationary process* on a shift space X is an assignment of probabilities $\mu(w)$ to the words w of X such that the probabilities of the symbols of X sum to 1 and for each word w in X, the following versions of (9–4–4) and (9–4–5) hold:

$$\mu(w) = \sum_{\{a:aw\in\mathcal{B}(X)\}} \mu(aw)$$

and

$$\mu(w) = \sum_{\{a:wa\in\mathcal{B}(X)\}} \mu(wa).$$

It follows that a stationary Markov chain on a graph G defines a stationary process on X_G.

What about sofic shifts? We can use the minimal right-resolving presentation $(G, \mathcal{L})$ of an irreducible sofic shift X to define a stationary process on X, and this gives us a way of measuring sizes of sets, such as cylinder sets and the set of doubly transitive points. First, endow G with a stationary Markov chain μ_G that assigns positive conditional probability to each edge. Then for each word w in X, express the pre-image of the cylinder set $C_0(w)$ as a finite disjoint union of cylinder sets in X_G

$$(\mathcal{L}_\infty)^{-1}(C_0(w)) = \bigcup_i C_0(\gamma^{(i)})$$

and define

(9–4–14) $$\mu(w) = \sum_i \mu_G(\gamma^{(i)}).$$

We leave it to the reader to verify that μ does indeed define a stationary process on X. Recall from Lemma 9.1.13 that images and pre-images, via $\mathcal{L}_\infty$, of doubly transitive points are doubly transitive. So by Theorem 9.4.9 the pre-image of the set of doubly transitive points in X can be viewed as the complement of a null set and therefore as a "large" set.

EXERCISES

9.4.1. Let A be an irreducible integral matrix with right Perron eigenvector $\mathbf{v}$. Define conditional probabilities on the edges of G_A by

$$\mu(e|i(e)) = \frac{v_{t(e)}}{v_{i(e)}\lambda_A}.$$

(a) Show that there is a unique stationary Markov chain with these conditional probabilities and find its unique stationary distribution.

(b) Show that for any cycle γ in G_A of length n

$$\frac{\mu(\gamma)}{\mu(i(\gamma))} = \frac{1}{\lambda_A^n}.$$

9.4.2. Let μ be a stationary Markov chain on G. Let f be any function on $\mathcal{V}(G)$. Show that

$$\sum_{e \in \mathcal{E}(G)} \mu(e)f(i(e)) = \sum_{e \in \mathcal{E}(G)} \mu(e)f(t(e)).$$

9.4.3. Show that for an irreducible sofic shift X, the assignment of probabilities to blocks given in equation (9–4–14) defines a stationary process on X.

9.4.4. Recall that a subset of a compact metric space is dense if its closure is the entire space. A set is *thick* if it contains a countable intersection of open dense sets; recall from the Baire Category Theorem (Theorem 6.1.24) that a thick set is dense. Show that for an irreducible shift X, the set of doubly transitive points of X is a thick set. This gives another sense in which almost invertible codes are one-to-one on most points.

Notes

The notion of degree of a code goes back to Hedlund [Hed5] who, along with colleagues at the Institute for Defense Analysis, developed for cryptographic purposes the basic properties of degree for endomorphisms of the full shift (i.e., factor codes from the full shift to itself). The notion was extended to endomorphisms of irreducible shifts of finite type and sofic shifts by Coven and Paul [CovP1,2,3]. Our treatment here follows that given in [KitMT].

Proposition 9.1.7, which shows how to recode a magic word to a magic symbol, is contained in [KitMT]. The Replacement Theorem (Theorem 9.2.5) is due to Ashley [Ash5] and was motivated by the work of Adler, Goodwyn, and Weiss [AdlGW], who established a weaker version of the Road Theorem. Theorem 9.2.8 is also due to Ashley.

Almost conjugacy was introduced by Adler and Marcus [AdlM] as a weakened version of conjugacy that could be completely classified by simple invariants. This concept has its roots in ergodic theory. Namely, Adler and Weiss [AdlW] constructed codes for the purpose of solving a classification problem in ergodic theory and then noticed that their codes satisfied stronger conditions which ultimately evolved into almost conjugacy. Theorem 9.3.2 was proved in [AdlM], but some of the basic ideas are in [AdlGW]. Lemma 9.3.4 is due to Kitchens, and Lemma 9.3.5 is taken from Marcus and Tuncel [MarT2]. Theorem 9.3.9 is in [AdlM].

Finally, we remark that the material in §9.4 is mostly standard probability theory; see [Bil], [KemS]. Theorem 9.4.9 is actually a special case of a more general result in ergodic theory; see §13.4.

Exercise 9.1.3 (in special cases) is in [Hed5], [CovP1], and [CovP3]; Exercise 9.1.4 is in [CovP3]; Exercise 9.1.5 is (partly) contained in [Nas2]; and Exercises 9.1.11 and 9.2.5 are due to J. Ashley.

CHAPTER 10

EMBEDDINGS AND FACTOR CODES

When can we embed one shift of finite type into another? When can we factor one shift of finite type onto another? The main results in this chapter, the Embedding Theorem and the Lower Entropy Factor Theorem, tell us the answers when the shifts have different entropies. For each theorem there is a simple necessary condition on periodic points, and this condition turns out to be sufficient as well. In addition, these periodic point conditions can be verified with relative ease.

We state and prove the Embedding Theorem in §10.1. The necessity of the periodic point condition here is easy. The sufficiency makes use of the fundamental idea of a marker set to construct sliding block codes. In §10.2 we prove the Masking Lemma, which shows how to represent embeddings in a very concrete form; we will use this in Chapter 11 to prove a striking application of symbolic dynamics to linear algebra. §10.3 contains the statement and proof of the Lower Entropy Factor Theorem, which is in a sense "dual" to the Embedding Theorem. The proof employs a marker construction similar to that of the Embedding Theorem. One consequence is an unequal entropy version of the Finite Equivalence Theorem.

§10.1. The Embedding Theorem

Suppose that X and Y are irreducible shifts of finite type. When is there an embedding from X into Y? If the embedding is also onto, then X and Y are conjugate. Chapter 7 treats this question in detail. Hence we restrict our attention here to *proper* embeddings, i.e., those that are not onto.

Two conditions quickly surface as necessary for there to be a proper embedding $\phi\colon X \to Y$.

The first condition is a consequence of the fact that any proper subshift of an irreducible shift of finite type has strictly smaller entropy (see Corollary 4.4.9). Namely, since Y is irreducible and $\phi(X)$ is a proper subshift of Y, we must have $h(X) = h(\phi(X)) < h(Y)$.

The second condition involves the action of an embedding on periodic points. Let $P(X)$ denote the set of all periodic points in X. Recall that $q_n(X)$ is the number with *least* period n. If $\phi\colon X \to Y$ is an embedding, then ϕ restricts to a one-to-one shift-commuting mapping from $P(X)$ into $P(Y)$. The existence of such a mapping is clearly equivalent to the condition $q_n(X) \leqslant q_n(Y)$ for all $n \geqslant 1$. We denote these equivalent conditions by $P(X) \hookrightarrow P(Y)$ and call this the *embedding periodic point condition*.

The following result, due to Krieger [Kri6], shows that these two simple necessary conditions are also sufficient for a proper embedding.

Theorem 10.1.1 (Embedding Theorem). *Let X and Y be irreducible shifts of finite type. Then there is a proper embedding of X into Y if and only if $h(X) < h(Y)$ and $P(X) \hookrightarrow P(Y)$.*

We will actually prove the following result.

Theorem 10.1.2. *Let X be a shift of finite type, and let Y be a mixing shift of finite type. Then there is a proper embedding of X into Y if and only if $h(X) < h(Y)$ and $P(X) \hookrightarrow P(Y)$.*

It is straightforward to reduce the Embedding Theorem to this result. After the proof of Theorem 10.1.2 we will state and prove a generalization, Corollary 10.1.9, where X is merely assumed to be a shift space.

One approach to constructing the embedding would be to represent both X and Y as edge shifts $X \cong X_G$ and $Y \cong X_H$ such that G embeds into H. Simple examples show that we cannot expect this to work for arbitrary edge shift representations; for instance, if H were the 1-state graph consisting of two self-loops, and G were any proper subgraph of $H^{[N]}$, $N \geqslant 2$, then $X = X_G, Y = X_H$ would satisfy the embedding conditions, but unless G has only one state, it would be too large to embed in H. One could try to modify G and H by constructing graphs G', H' such that $X_{G'} \cong X_G$, $X_{H'} \cong X_H$, and G' embeds in H'. At the beginning of §10.2 we show that this is always possible, but the proof of this uses the assumption that we already have an embedding of X_G into X_H.

Instead, we construct an explicit sliding block code $\phi\colon X_G \to X_H$ by stitching together two quite different encoding schemes, each corresponding to one of the two embedding conditions. These schemes use the fundamental notion of a *marker set*, which is a finite union F of cylinder sets in the domain X such that (a) F moves disjointly under the shift map for a "moderate" amount of time, and (b) a point whose orbit misses F for a "long" amount of time looks very much like a periodic point in X of "low" period. There is a set $\mathcal{C}$ of $(2m+1)$-blocks such that F consists of all points $x \in X$ for which $x_{[-m,m]} \in \mathcal{C}$. We can regard $\mathcal{C}$ as a set of "marker patterns" and use $\mathcal{C}$ to recognize "marker coordinates" of x, i.e., those coordinates i for which $x_{[i-m,i+m]} \in \mathcal{C}$ or, equivalently, for which $\sigma^i(x) \in F$.

Property (a) of F shows that successive marker coordinates for a point $x \in X$ must be spaced at least "moderately" far apart. The entropy embedding condition $h(X) < h(Y)$ then shows that the number of possible blocks between marker coordinates is smaller than the number of blocks in Y of the same length, so we can embed blocks between marker coordinates. However, were we to try this scheme for marker coordinates arbitrarily far apart, we would fail to produce a sliding block code since there would be no bound on how many symbols in a point we would need to observe before finding a marker pattern. This is where property (b) of F is used. When successive marker coordinates are separated by a "long" amount, the intervening block looks like a point of "low" period and we use an injection of $P(X)$ into $P(Y)$ to encode such blocks. Thus we use two different coding schemes, one for moderate intervals and one for long intervals. A remarkable feature of this construction is that these two schemes, based on quite different assumptions, can be stitched together into a seamless sliding block code.

This sketch of the proof shows that we are making use of the embedding periodic point condition only for "low" periods. However, we claim that since Y is mixing, the entropy condition $h(X) < h(Y)$ implies $q_n(X) \leqslant q_n(Y)$ for all sufficiently large n. To see this, use Proposition 4.1.15 and Corollary 4.5.13 to see that

$$(10\text{--}1\text{--}1) \qquad \limsup_{n\to\infty} \frac{\log q_n(X)}{n} \leqslant h(X) < h(Y) = \lim_{n\to\infty} \frac{\log q_n(Y)}{n}.$$

Thus the embedding periodic point condition $P(X) \hookrightarrow P(Y)$ supplies only a finite amount of information in addition to $h(X) < h(Y)$. At the end of this section, we will introduce a useful tool for checking this condition.

Our construction of the embedding follows the original construction due to Krieger [Kri6] with some modifications due to Boyle [Boy6]. It relies heavily on the assumption that Y is a shift of finite type. The two necessary conditions are not always sufficient when Y is merely assumed to be mixing sofic (see Exercise 10.1.13) and indeed the Embedding Problem for mixing sofic shifts remains unsolved.

We prepare for the proof of Theorem 10.1.2 by establishing some preliminary lemmas. The first allows us to make some "elbow room" in Y for later use in our marker codes.

Lemma 10.1.3. *Let X be a shift space, and let Y be a mixing shift of finite type such that $h(X) < h(Y)$ and $P(X) \hookrightarrow P(Y)$. Then there is a mixing shift of finite type W that is properly contained in Y such that $h(X) < h(W)$ and $P(X) \hookrightarrow P(W)$.*

PROOF: By recoding, we may assume that Y is an edge shift X_G. Since Y is mixing and $h(X) < h(Y)$, we have by (10–1–1):

$$(10\text{--}1\text{--}2) \qquad\qquad q_k(X) < q_k(Y) - k$$

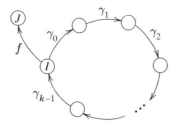

FIGURE 10.1.1.   A cycle with additional edge.

for all sufficiently large k. Fix such a k and let $y \in Y$ have least period k. Then $\gamma = y_{[0,k-1]}$ is a cycle in G which we may assume to be simple by passing to a higher edge graph (see Exercise 2.3.7). Since Y is mixing, we may also assume by shifting γ as necessary that the initial state I of γ has an outgoing edge $f \neq \gamma_0$. Since γ is simple, the edge f does not appear in γ (see Figure 10.1.1).

Let $Y^{(m)}$ denote the shift of finite type obtained from Y by forbidding the block γ^m. We show that for sufficiently large m the shift $W = Y^{(m)}$ has the required properties.

Put $A = A_G$. Let J be the terminal state of f. Since Y is mixing, we have

$$\lim_{n \to \infty} \frac{1}{n} \log (A^n)_{JI} = h(Y).$$

Hence we can fix n large enough so that

$$h(X) < \frac{1}{n+1} \log (A^n)_{JI}.$$

Next, assume that m is large enough that $m|\gamma| > n$. Since f does not appear in γ, every point of the form

$$\ldots f\, u_{-2}\, f\, u_{-1} . f\, u_0\, f\, u_1 \ldots,$$

where each u_i is a path in G from J to I of length n, does not contain γ^m, and so occurs in $Y^{(m)}$. Hence

$$h(X) < \frac{1}{n+1} \log (A^n)_{JI} \leqslant h(Y^{(m)}).$$

We next show that $Y^{(m)}$ is mixing. For suppose that u and v are blocks that occur in $Y^{(m)}$. By extending u to the right and v to the left as necessary, we may assume that the terminal edge of u and the initial edge of v do not lie on the cycle γ. Let w be any path from the terminal state of u to the initial state of v that does not contain γ as a subpath. Then uwv does not contain γ^m. Since u and v occur in $Y^{(m)}$, it follows that uwv occurs in $Y^{(m)}$, proving that $Y^{(m)}$ is irreducible. Since Y is mixing, it

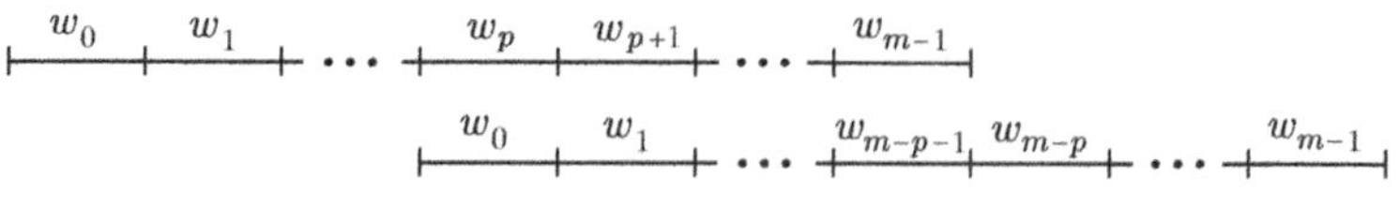

FIGURE 10.1.2. A p-periodic block.

has periodic points of all sufficiently large periods and in particular it has two periodic orbits disjoint from γ^{∞} whose lengths are relatively prime. These orbits are contained in $Y^{(m)}$ for large enough m, proving that $Y^{(m)}$ is mixing for all sufficiently large m (see Proposition 4.5.10(4)).

Hence there is an m_0 such that for $m \geqslant m_0$ we have $h(X) < h(Y^{(m)})$ and $Y^{(m)}$ is mixing. So, by (10–1–1), there is an i_0 such that

$$q_i(X) \leqslant q_i(Y^{(m_0)}) \qquad \text{for } i \geqslant i_0.$$

Since $Y^{(m)}$ increases with m, we see that, for all $m \geqslant m_0$,

(10–1–3) $q_i(X) \leqslant q_i(Y^{(m)}) \qquad \text{for } i \geqslant i_0.$

But for sufficiently large m, a point in Y with least period $< i_0$ that is *not* in the orbit of $y = \gamma^{\infty}$ must be in $Y^{(m)}$. Thus, using (10–1–2), for sufficiently large m we have

(10–1–4) $q_i(X) \leqslant \begin{cases} q_i(Y) = q_i(Y^{(m)}) & \text{for } i < i_0,\ i \neq k, \\ q_k(Y) - k = q_k(Y^{(m)}) & \text{for } i = k. \end{cases}$

Putting (10–1–3) and (10–1–4) together shows that $P(X) \hookrightarrow P(Y^{(m)})$ for sufficiently large m, completing the proof.   □

To discuss periodicity for blocks, we introduce the following terminology.

Definition 10.1.4. Let $w = w_0 w_1 \ldots w_{m-1}$ and $1 \leqslant p \leqslant m - 1$. Say that w is p-*periodic* if $w_0 w_1 \ldots w_{m-1-p} = w_p w_{p+1} \ldots w_{m-1}$.

If w is p-periodic, then it overlaps itself after a shift of p symbols to the right (see Figure 10.1.2). When p is small then w has a large self-overlap, while for large p the overlap is small. A block may be p-periodic for many choices of p.

Example 10.1.5. (a) Let $w = 11111$. Then w is p-periodic for $1 \leqslant p \leqslant 4$.
(b) Let $w = 11011$. Then w is both 3-periodic and 4-periodic.
(c) Let $w = 10000$. Then w is not p-periodic for any p.   □

Suppose that X is an edge shift and $w = w_0 w_1 \ldots w_{m-1}$ is a block in $\mathcal{B}_m(X)$ that is p-periodic. We would like to associate to w a unique point z of period p such that $z_{[0,m-1]} = w$. The obvious choice is

$$z = (w_0 w_1 \ldots w_{p-1})^{\infty}.$$

But this choice may not be unique if p is allowed to be greater that $m/2$. For instance, the block $w = 11011$ in Example 10.1.5(b) is both 3-periodic and 4-periodic, leading to the two periodic points $z = (110)^\infty$ and $z' = (1101)^\infty$ with $z_{[0,4]} = w = z'_{[0,4]}$. However, the following result shows that when the periods are assumed to be no greater than half the block length, then the resulting periodic point is unique. We use the notation $per(z)$ to denote the least period of a periodic point z.

Lemma 10.1.6. *Let X be an edge shift and $w \in \mathcal{B}_m(X)$. If w is p-periodic for some p with $1 \leqslant p \leqslant m/2$, then there is a unique point $z \in P(X)$ such that $per(z) \leqslant m/2$ and $z_{[0,m-1]} = w$.*

PROOF: We have already observed that $z = (w_0 w_1 \ldots w_{p-1})^\infty$ has period p and satisfies $z_{[0,m-1]} = w$. Suppose that $z' \in P(X)$ has least period $p' = per(z') \leqslant m/2$ and $z'_{[0,m-1]} = w$. Then by periodicity,

$$z_m = z_{m-p} = z'_{m-p} = z'_{m-p-p'},$$

and

$$z'_m = z'_{m-p'} = z_{m-p'} = z_{m-p'-p}.$$

Since $1 \leqslant p + p' \leqslant m$ and $z_{[0,m-1]} = w = z'_{[0,m-1]}$, it follows that

$$z_m = z'_{m-p-p'} = z_{m-p'-p} = z'_m.$$

Repeated application of this argument shows that $z_i = z'_i$ for $i \geqslant m$. A similar argument proves the same for $i < 0$, so that $z = z'$. □

Another formulation of this lemma is that if a point $z \in P(X)$ is known to have least period $\leqslant n$, then we can detect the value of $per(z)$ by observing any block of length $2n$ from z.

Our next lemma shows that a "locally periodic" block is periodic with small period. Suppose that $w \in \mathcal{B}_m(X)$, and that every subblock of w of some fixed length k is p-periodic, where p is allowed to vary with the subblock. We would like to conclude that w occurs in a point with period $\leqslant k$. In general this is not true. For instance, consider

$$w = 110110110110111011101,$$

which starts out looking like $(110)^\infty$ and ends up looking like $(1101)^\infty$. The reader can check that each subblock of w of length $k = 5$ is either 3-periodic, 4-periodic, or both. But w does not occur in any point with period $\leqslant 5$ (why?). However, if we assume the periods are less than half the observation length, we can obtain the desired periodic behavior. We state a version of this in the following result.

Lemma 10.1.7 (Periodic Lemma). *Let X be an edge shift and fix a block $w = w_0 \ldots w_{m-1} \in \mathcal{B}_m(X)$. Suppose that for some $N \geqslant 1$ every subblock of w of length $2N + 1$ is p-periodic for some $1 \leqslant p < N$ (where p is allowed to vary with the subblock). Then there is a unique $z \in P(X)$ such that $\mathrm{per}(z) < N$ and $z_{[0,m-1]} = w$.*

PROOF: For each subblock $w_{[k,k+2N]}$, let $z^{(k)}$ be the unique periodic point with $\mathrm{per}(z^{(k)}) < N$ such that

$$(10\text{--}1\text{--}5) \qquad\qquad z^{(k)}_{[k,k+2N]} = w_{[k,k+2N]}$$

guaranteed by Lemma 10.1.6. Then both $z^{(k)}$ and $z^{(k+1)}$ have period $< N$ and agree from $k + 1$ to $k + 2N$. The uniqueness part of Lemma 10.1.6 shows that $z^{(k)} = z^{(k+1)}$. Hence all the $z^{(k)}$ are equal; call the common point z. Then $\mathrm{per}(z) < N$ and (10–1–5) shows that $z_{[0,m-1]} = w$.   □

Next, we construct the marker set needed for the embedding construction.

Lemma 10.1.8 (Marker Lemma). *Let X be a shift space and $N \geqslant 1$. Then there is a marker set $F \subseteq X$ that is a finite union of cylinder sets such that*

(1) *the sets $\sigma^i(F)$ for $0 \leqslant i < N$ are disjoint, and*
(2) *if $\sigma^i(x) \notin F$ for $-N < i < N$, then $x_{[-N,N]}$ is p-periodic for some $p < N$.*

PROOF: In order to satisfy (1), we cannot have any periodic points in F of period $< N$. We will actually manufacture F out of $(2N + 1)$-blocks that are not p-periodic for every $p < N$.

Partition $\mathcal{B}_{2N+1}(X)$ into two classes $\mathcal{P}$ and $\mathcal{N}$: the class $\mathcal{P}$ consists of all $(2N + 1)$-blocks that are p-periodic for some $p < N$, and $\mathcal{N}$ is the complement. For $w \in \mathcal{B}_{2N+1}(X)$, let $[w]$ denote the cylinder set

$$[w] = C_{-N}(w) = \{x \in X : x_{[-N,N]} = w\}.$$

Order the blocks of $\mathcal{N}$ in some arbitrary fashion, say $w^{(1)}$, $w^{(2)}$, ..., $w^{(k)}$. With this ordering, we define sets $E^{(1)}$, $E^{(2)}$, ..., $E^{(k)}$ by putting

$$E^{(1)} = [w^{(1)}],$$

and

$$E^{(j)} = E^{(j-1)} \cup \left([w^{(j)}] \setminus \bigcup_{-N < i < N} \sigma^i(E^{(j-1)}) \right)$$

for $j = 2, 3, \ldots, k$. Let $F = E^{(k)}$, and notice that F is a finite union of cylinder sets.

Since all blocks $w^{(j)}$ in $\mathcal{N}$ are *not* p-periodic for any $p < N$, it follows that for each j the sets $\sigma^i([w^{(j)}])$, $0 \leqslant i < N$, are disjoint. From this, and the inductive definition of the $E^{(j)}$, it follows that the sets $\sigma^i(F)$ for $0 \leqslant i < N$ are also disjoint, verifying (1).

Observe that for $j = 1, 2, \ldots, k$

$$[w^{(j)}] \subseteq \bigcup_{-N < i < N} \sigma^i(E^{(j)})$$

so that

$$\bigcup_{j=1}^{k} [w^{(j)}] \subseteq \bigcup_{-N < i < N} \sigma^i(F).$$

Hence if $\sigma^i(x) \notin F$ for $-N < i < N$, then $x \notin \bigcup_{j=1}^{k} [w^{(j)}]$, and so $x_{[-N,N]} \notin \mathcal{N}$. Thus $x_{[-N,N]} \in \mathcal{P}$, so this block is p-periodic for some $p < N$. $\qquad\square$

This construction can be quite involved. Typically the number k of blocks in $\mathcal{N}$ is exponentially large in N, and at each stage $E^{(j)}$ may depend on $2N$ more coordinates than its predecessor $E^{(j-1)}$. Thus the cylinder sets in F will likely depend on exponentially many coordinates, so F can be exceedingly complicated.

We are now ready to prove Theorem 10.1.2.

PROOF OF THEOREM 10.1.2: By recoding, we may assume that X and Y are edge shifts and that $Y = X_G$ with G primitive. We may also assume that the W given by Lemma 10.1.3 has the form $W = X_H$ where H is a primitive subgraph of G that misses at least one vertex I_0 of G.

We define certain injections c_m and ψ used to construct the embedding $\phi \colon X \to Y$. We first define some auxiliary paths in G. Since G and H are primitive, there is an $L \geqslant 1$ satisfying the following two statements. For each vertex J in H there is a path $\gamma^-(J)$ in G from I_0 to J of length L that does not meet I_0 except at its initial state. Similarly, there is a path $\gamma^+(J)$ in G from J to I_0 of length L that does not meet I_0 except at its terminal state. We call these paths *transition paths* and L the *transition length* (see Figure 10.1.3).

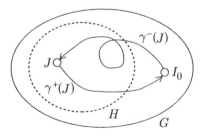

FIGURE 10.1.3. Transition paths.

Fix a vertex J_0 in H and let $\mathcal{C}_m$ denote the set of cycles in H of length m that begin and end at J_0. Since $h(X) < h(W) = h(X_H)$ and H is primitive, it follows that there is an N such that there is an injection

$$c_m \colon \mathcal{B}_m(X) \to \mathcal{C}_{m-2L}$$

for all $m \geqslant N$. The Marker Lemma applies to X and this N to produce a marker set $F \subseteq X$.

Since $P(X) \hookrightarrow P(X_H)$ there is a shift-commuting injection $\psi \colon P(X) \to P(X_H)$. This ψ automatically preserves least period.

For $x \in X$ let

$$\mathcal{M}(x) = \{i \in \mathbb{Z} : \sigma^i(x) \in F\}.$$

Call the integers in $\mathcal{M}(x)$ *marker coordinates* for x. These subdivide $\mathbb{Z}$ into *marker intervals*. It is convenient to use the notation $[i,j)$ for $\{i, i+1, \ldots, j-1\}$, and to call $j - i$ the *length* of $[i,j)$. Then finite marker intervals have the form $[i,j)$, where $i, j \in \mathcal{M}(x)$ while $k \notin \mathcal{M}(x)$ for $i < k < j$. There may also be infinite marker intervals: $(-\infty, j)$ is a marker interval if $j \in \mathcal{M}(x)$ but $k \notin \mathcal{M}(x)$ for $k < j$; also $[i, \infty)$ is a marker interval if $i \in \mathcal{M}(x)$ but $k \notin \mathcal{M}(x)$ for $k > i$; and $(-\infty, \infty)$ is a marker interval if $\mathcal{M}(x) = \varnothing$.

Each $x \in X$ is subdivided into pieces by its marker intervals. Our construction of the embedding ϕ will have the following crucial feature. For each marker interval T of x, the corresponding part $\phi(x)_T$ of the image point will be a path in G from I_0 to itself that does not meet I_0 except at its (finite) endpoints.

By part (1) of the Marker Lemma, all marker intervals have length at least N. We call a marker interval *moderate* if its length is $< 2N + 2L$, and *long* if it is infinite or has length $\geqslant 2N + 2L$. To define ϕ on x, we use the injections c_m on subblocks of x corresponding to moderate intervals, and the injection ψ on those corresponding to long intervals.

CASE 1: MODERATE INTERVALS. Let $[i,j)$ be a moderate interval of length $m = j - i$ so that $m < 2N + 2L$. Since $m \geqslant N$, the injection c_m is defined. We put

$$\phi(x)_{[i,j)} = \gamma^-(J_0) c_m(x_{[i,j)}) \gamma^+(J_0).$$

We next turn to long intervals. Let $T = [i,j)$ be a finite long interval, so $m = j - i \geqslant 2N + 2L$. Part (2) of the Marker Lemma shows that for $i \leqslant k < j - 2N$ the block $x_{[k,k+2N]}$ is p-periodic for some $p < N$. Then Lemma 10.1.7 shows that there is a unique point $z = z(T, x)$ of period $< N$ with $z_{[i,j)} = x_{[i,j)}$. The same argument applies to infinite marker intervals T to give a unique point $z = z(T, x)$ of period $< N$ with $z_T = x_T$. We shall use ψ on these periodic points $z(T, x)$ to define ϕ for long intervals.

CASE 2: FINITE LONG INTERVALS. Let $T = [i,j)$ be a long interval and $z = z(T, x)$ the point of period $< N$ with $z_T = x_T$. Put

$$\phi(x)_{[i,j)} = \gamma^-\big(i(\psi(z)_{i+L})\big) \psi(z)_{[i+L, j-L)} \gamma^+\big(i(\psi(z)_{j-L})\big).$$

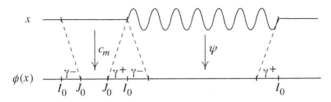

FIGURE 10.1.4. Mapping various parts of a point.

CASE 3: RIGHT-INFINITE LONG INTERVALS. Here $T = [i, \infty)$. Let $z = z(T, x)$ and put

$$\phi(x)_{[i,\infty)} = \gamma^-\big(i(\psi(z)_{i+L})\big)\psi(z)_{[i+L,\infty)}.$$

CASE 4: LEFT-INFINITE LONG INTERVALS. Here $T = (-\infty, j)$. Let $z = z(T, x)$ and put

$$\phi(x)_{(-\infty,j)} = \psi(z)_{(-\infty,j-L)}\gamma^+\big(i(\psi(z)_{j-L})\big).$$

CASE 5: BI-INFINITE LONG INTERVAL. Here $T = (-\infty, \infty)$ and x is already periodic with period $< N$. We put $\phi(x) = \psi(x)$.

Figure 10.1.4 depicts the action of ϕ on parts of a point corresponding to some of these cases.

We must now verify that this ϕ is a proper embedding of X into Y. First recall that the transition paths $\gamma^+(J)$ (resp. $\gamma^-(J)$) meet I_0 only at their terminal (resp. initial) state. The injections c_m and ψ map to paths in H, and so these paths never meet I_0. So for each marker interval T for x it follows that $\phi(x)_T$ is a path that meets I_0 only at its (finite) endpoints. Thus $\phi(x)$ is formed by concatenating (possibly infinite) paths that start and end at I_0 so that $\phi(x) \in X_G = Y$. It suffices to verify that ϕ is a sliding block code and is one-to-one.

To see that ϕ is a sliding-block code, we must show that ϕ commutes with the shift map and $\phi(x)_0$ is a function of $x_{[-M,M]}$ for some M. Clearly, ϕ commutes with the shift map, since it is defined only by marker patterns, not where they occur. Now choose m so that F is a union of cylinder sets corresponding to central $(2m + 1)$-blocks, and let $x \in X$. The coordinate 0 is contained in some marker interval T for x. To determine whether T is moderate or long, we need only examine x from $-2N - 2L - m$ to $2N + 2L + m$. If T is moderate, then $\phi(x)_0$ is determined by x_T, hence by $x_{[-2N-2L,2N+2L]}$. If T is long, then $z = z(T, x)$ has period $< N$, so is determined by $x_{[-2N,2N]}$. Furthermore $\phi(x)_T$, and in particular $\phi(x)_0$, is determined by z, hence by $x_{[-2N,2N]}$. In each case $\phi(x)_0$ is a function of $x_{[-2N-2L-m,2N+2L+m]}$. So ϕ is a sliding block code.

Finally, suppose that $\phi(x) = y$. We reconstruct x from y to prove that ϕ is one-to-one. First recall from above that if T is a marker interval for x, then $\phi(x)_T$ meets I_0 only at its (finite) endpoints. Hence

$$\mathcal{M}(x) = \{i \in \mathbb{Z} : i(y_i) = I_0\},$$

and we have recovered the marker coordinates of x from y.

If $T = [i, j)$ is a moderate interval (Case 1), then $x_{[i,j)} = c_{j-i}^{-1}(y_{[i+L,j-L)})$.

If T is a long interval, then we are in one of the Cases 2–5. In Case 2 (finite long), $T = [i, j)$ with $j - i \geqslant 2N + 2L$. Then $y_{[i+L,j-L)} = \psi(z)_{[i+L,j-L)}$, where $z = z(T, x)$. But since $\text{per}(z) < N$, we have that $\text{per}(\psi(z)) < N$, so Lemma 10.1.6 shows that $\psi(z)$ is determined by $\psi(z)_{[i+L,j-L)}$. Since ψ is one-to-one, we see that $x_T = x_{[i,j)} = z_{[i,j)}$ is also determined by $y_{[i+L,j-L)}$. In Case 3, $T = [i, \infty)$. Then $y_{[i+L,\infty)} = \psi(z)_{[i+L,\infty)}$, where $z = z(T, x)$. As in Case 2, the block $x_T = x_{[i,\infty)} = z_{[i,\infty)}$ is determined by $y_{[i+L,\infty)}$. Case 4 is treated as Case 3. In Case 5, $T = (-\infty, \infty)$ and ϕ agrees with ψ, so that x is determined by $y = \psi(x)$.

This completes the proof of Theorem 10.1.2 and therefore also the Embedding Theorem.   □

By approximating an arbitrary shift space by shifts of finite type, we can generalize Theorem 10.1.2 as follows. This result, rather than Theorem 10.1.1, is often called the Embedding Theorem.

Corollary 10.1.9. *Let X be a shift space and Y be a mixing shift of finite type. Then there is a proper embedding of X into Y if and only if $h(X) < h(Y)$ and $P(X) \hookrightarrow P(Y)$.*

PROOF: If there is an embedding of X into Y, then clearly $P(X) \hookrightarrow P(Y)$. If the embedding is proper then by Corollary 4.4.9, $h(X) < h(Y)$.

Conversely, assume that $h(X) < h(Y)$ and $P(X) \hookrightarrow P(Y)$. Let $X^{(m)}$ denote the shift of finite type consisting of all points for which every block of length m belongs to $\mathcal{B}_m(X)$. Then

$$(10\text{--}1\text{--}6) \qquad\qquad X = \bigcap_{m=1}^{\infty} X^{(m)}$$

and Proposition 4.4.6 shows that

$$h(X) = \lim_{m \to \infty} h(X^{(m)}).$$

Hence there is an m_0 for which $h(X^{(m)}) < h(Y)$ for $m \geqslant m_0$. By (10–1–1), we can therefore find n_0 such that $q_n(X^{(m_0)}) \leqslant q_n(Y)$ for $n \geqslant n_0$. Since the $X^{(m)}$ decrease, we have $q_n(X^{(m)}) \leqslant q_n(Y)$ for all $m \geqslant m_0$ and $n \geqslant n_0$. By (10–1–6) there is an $m \geqslant m_0$ such that $q_n(X^{(m)}) = q_n(X)$ for all $n < n_0$. Apply Theorem 10.1.2 to obtain a proper embedding $\phi \colon X^{(m)} \to Y$. The restriction of ϕ to X is the required proper embedding.   □

We remark that Corollary 10.1.9 does not hold if Y is assumed to be merely an irreducible shift of finite type (Exercise 10.1.17).

Finally, we discuss the problem of deciding when the embedding periodic point condition holds for irreducible shifts of finite type X, Y with $h(X) < h(Y)$. As mentioned earlier in this section, we need only check the condition for finitely many periods. But how can we tell in advance which periods we need to check?

To approach this problem, we first introduce some notation and terminology.

Recall from §6.4 that a list $\Lambda = \{\lambda_1, \lambda_2, \dots, \lambda_k\}$ of complex numbers is a collection of complex numbers where the order of listed elements is irrelevant, but repeated elements count. The union of two lists is formed by appending one list after the other, taking care to keep track of multiplicity, e.g., $\{3, 2, 1\} \cup \{2, 1, 1\} = \{3, 2, 2, 1, 1, 1\}$. For a list $\Lambda = \{\lambda_1, \lambda_2, \dots, \lambda_k\}$ define

$$\operatorname{tr}(\Lambda) = \sum_{j=1}^{k} \lambda_j \quad \text{and}$$
$$\Lambda^n = \{\lambda_1^n, \lambda_2^n, \dots, \lambda_k^n\}.$$

If $\Lambda = \operatorname{sp}^{\times}(A)$, then $\operatorname{tr}(\Lambda^n) = \operatorname{tr}(A^n)$ for all $n \geqslant 1$. From the Möbius Inversion Formula [NivZ], the number $q_n(X_A)$ of points in X_A with least period n is given by

$$(10\text{--}1\text{--}7) \qquad q_n(X_A) = \sum_{d \mid n} \mu\left(\frac{n}{d}\right) \operatorname{tr}(A^d)$$

where μ is the Möbius function defined by

$$\mu(m) = \begin{cases} (-1)^r & \text{if } m \text{ is the product of } r \text{ distinct primes,} \\ 0 & \text{if } m \text{ contains a square factor,} \\ 1 & \text{if } m = 1 \end{cases}$$

(see Exercise 6.3.1). This leads us to make the following definition.

Definition 10.1.10. Let $\Lambda = \{\lambda_1, \lambda_2, \dots, \lambda_k\}$. For $n \geqslant 1$ define the *n*th *net trace* of Λ to be

$$(10\text{--}1\text{--}8) \qquad \operatorname{tr}_n(\Lambda) = \sum_{d \mid n} \mu\left(\frac{n}{d}\right) \operatorname{tr}(\Lambda^d).$$

For instance,

$$\begin{aligned}
\mathrm{tr}_1(\Lambda) &= \mathrm{tr}(\Lambda), \\
\mathrm{tr}_2(\Lambda) &= \mathrm{tr}(\Lambda^2) - \mathrm{tr}(\Lambda), \\
\mathrm{tr}_3(\Lambda) &= \mathrm{tr}(\Lambda^3) - \mathrm{tr}(\Lambda), \\
\mathrm{tr}_4(\Lambda) &= \mathrm{tr}(\Lambda^4) - \mathrm{tr}(\Lambda^2), \\
\mathrm{tr}_5(\Lambda) &= \mathrm{tr}(\Lambda^5) - \mathrm{tr}(\Lambda), \\
\mathrm{tr}_6(\Lambda) &= \mathrm{tr}(\Lambda^6) - \mathrm{tr}(\Lambda^3) - \mathrm{tr}(\Lambda^2) + \mathrm{tr}(\Lambda).
\end{aligned}$$

Observe that the net trace of a list $\Lambda = \{\lambda_1, \lambda_2, \ldots, \lambda_k\}$ is the sum of the net traces of its individual elements, i.e.

$$(10\text{--}1\text{--}9) \qquad\qquad \mathrm{tr}_n(\Lambda) = \sum_{j=1}^{k} \mathrm{tr}_n(\lambda_j).$$

For a matrix A let $\mathrm{tr}_n(A)$ denote the nth net trace of $\mathrm{sp}^\times(A)$. Then comparing (10–1–7) and (10–1–8), we have

$$(10\text{--}1\text{--}10) \qquad\qquad q_n(X_A) = \mathrm{tr}_n(A),$$

and so we can reformulate the embedding periodic point condition as follows:

$$P(X_A) \hookrightarrow P(X_B) \iff \mathrm{tr}_n(A) \leqslant \mathrm{tr}_n(B) \text{ for all } n \geqslant 1.$$

In order to check this condition for a given pair of matrices, the following simple estimates are useful. First, observe that for any number λ and positive integer n,

$$|\lambda|^n - \sum_{m=1}^{\lfloor n/2 \rfloor} |\lambda|^m \leqslant |\mathrm{tr}_n(\lambda)| \leqslant \sum_{m=1}^{n} |\lambda|^m.$$

Using the formula for the sum of a geometric progression, it then follows that when $\lambda \neq 1$,

$$(10\text{--}1\text{--}11) \qquad |\lambda|^n - \frac{|\lambda|^{n/2+1} - |\lambda|}{|\lambda| - 1} \leqslant |\mathrm{tr}_n(\lambda)| \leqslant \frac{|\lambda|^{n+1} - |\lambda|}{|\lambda| - 1}.$$

It is straightforward to check that $\mathrm{tr}_n(1) = 0$ for $n > 1$. So, in the following estimates we will assume $n > 1$ and then we can ignore any eigenvalue equal to one.

Let A be a nonnegative integral matrix, and B be a primitive integral matrix with $\lambda_A < \lambda_B$. There is no harm in assuming that $\lambda_A \geqslant 1$ (and so

$\lambda_B > 1$) since otherwise X_A would be empty (see Exercise 10.1.6). Using (10–1–11) and (10–1–9) together with the triangle inequality we see that

$$\operatorname{tr}_n(B) = |\operatorname{tr}_n(B)| \geq \operatorname{tr}_n(\lambda_B) - \sum_{\substack{\mu \in \operatorname{sp}^\times(B) \\ \mu \neq \lambda_B}} |\operatorname{tr}_n(\mu)|$$

$$\geq \lambda_B^n - \frac{\lambda_B^{n/2+1} - \lambda_B}{\lambda_B - 1} - \sum_{\substack{\mu \in \operatorname{sp}^\times(B) \\ \mu \neq \lambda_B}} \frac{|\mu|^{n+1} - |\mu|}{|\mu| - 1}.$$

Similarly,

$$\operatorname{tr}_n(A) = |\operatorname{tr}_n(A)| \leq \sum_{\mu \in \operatorname{sp}^\times(A)} \frac{|\mu|^{n+1} - |\mu|}{|\mu| - 1}.$$

We conclude that a sufficient condition for $\operatorname{tr}_n(B) \geq \operatorname{tr}_n(A)$ is that

$$(10\text{–}1\text{–}12) \quad f(n) = \lambda_B^n - \frac{\lambda_B^{n/2+1} - \lambda_B}{\lambda_B - 1} - \sum_{\substack{\mu \in \operatorname{sp}^\times(B) \cup \operatorname{sp}^\times(A) \\ \mu \neq \lambda_B}} \frac{|\mu|^{n+1} - |\mu|}{|\mu| - 1}$$

be nonnegative. The contribution to $f(n)$ of the eigenvalues with modulus less than one is bounded by a constant. Thus

$$(10\text{–}1\text{–}13) \qquad\qquad f(n) \geq g(n) + K$$

where

(10–1–14)

$$g(n) = \lambda_B^n - \left(\frac{\lambda_B}{\lambda_B - 1}\right)(\lambda_B^{1/2})^n - \sum_{\substack{\mu \in \operatorname{sp}^\times(B) \cup \operatorname{sp}^\times(A) \\ \mu \neq \lambda_B, |\mu| > 1}} \left(\frac{|\mu|}{|\mu| - 1}\right)|\mu|^n,$$

and K is a constant. Now, $g(n)$ is of the form

$$(10\text{–}1\text{–}15) \qquad\qquad \lambda^n - \sum_{u \in U} a_u u^n$$

where $\lambda > 1$, the list U consists of positive numbers strictly less than λ, and each $a_u > 0$. Clearly any such function tends to infinity with n, and a calculus exercise shows that $g(n_0) \geq 0$ implies that g is monotonically increasing on the domain $n \geq n_0$ (Exercise 10.1.7). It follows that whenever $g(n_0) \geq \max\{0, -K\}$, then $f(n) \geq 0$ for all $n \geq n_0$.

Therefore, to check the embedding periodic point condition, we first find an n_0 such that $g(n_0) \geq \max\{0, -K\}$; one can obtain an explicit n_0 in terms of $\operatorname{sp}^\times(A)$ and $\operatorname{sp}^\times(B)$ (Exercise 10.1.8). So $\operatorname{tr}_n(B) \geq \operatorname{tr}_n(A)$ for all $n \geq n_0$. Then check to see if $\operatorname{tr}_n(B) \geq \operatorname{tr}_n(A)$ holds for all $n < n_0$. In practice, we will need to approximate the eigenvalues of A and B to finite precision; this will yield another expression of the form (10–1–15) to which we can apply the preceding argument.

Example 10.1.11. Let $A = \begin{bmatrix} 1 & 2 \\ 1 & 1 \end{bmatrix}$ and $B = \begin{bmatrix} 2 & 1 \\ 1 & 1 \end{bmatrix}$. We claim that X_A embeds in X_B.

First, we compute that $\chi_A(t) = t^2 - 2t - 1$ and $\chi_B(t) = t^2 - 3t + 1$. So $\mathrm{sp}^\times(A) = \{\mu_1, \mu_2\}$ where $\mu_1 = 1 + \sqrt{2}$, $\mu_2 = 1 - \sqrt{2}$, and $\mathrm{sp}^\times(B) = \{\lambda_1, \lambda_2\}$ where $\lambda_1 = (3 + \sqrt{5})/2$, $\lambda_2 = (3 - \sqrt{5})/2$. Using the approximations $\sqrt{2} \approx 1.414$ and $\sqrt{5} \approx 2.236$, we see that the eigenvalues of A are approximately $\{2.414, -.414\}$ and the eigenvalues of B are approximately $\{2.618, .382\}$, so $\lambda_A < \lambda_B$. It remains to check the embedding periodic point condition.

The first ten net traces of A are

$$2, 4, 12, 28, 80, 180, 476, 1120, 2772, 6640$$

and the first ten net traces of B are

$$3, 4, 15, 40, 120, 306, 840, 2160, 5760, 15000.$$

This looks promising, but to verify the net trace condition we need to use the general estimates above.

Using the approximations $2.414 < |\mu_1| < 2.415$, $.414 < |\mu_2| < .415$, $2.618 < |\lambda_1| < 2.619$, $.381 < |\lambda_2| < .382$, and $1.618 < |\lambda_1|^{1/2} < 1.619$, a straightforward manipulation shows that the expression $f(n)$ in (10–1–12) is bounded below by $g(n) + K$ where

$$g(n) = 2.618^n - \frac{2.619}{1.618}(1.619)^n - \frac{2.415}{1.414}(2.415)^n$$

and

$$K = \frac{2.618}{1.619} + \frac{2.414}{1.415} - \frac{.415}{.585} - \frac{.382}{.618} \approx 2.$$

Thus, according to the discussion above, the condition $\mathrm{tr}_n(B) \geqslant \mathrm{tr}_n(A)$ holds for all $n \geqslant n_0$ where n_0 is any number such that $g(n_0) \geqslant 0$. A computation reveals that $n_0 = 8$ will do, so $\mathrm{tr}_n(B) \geqslant \mathrm{tr}_n(A)$ for all $n \geqslant 8$. Since we have already checked the first ten net traces, we see that $P(X_A) \hookrightarrow P(X_B)$ as desired. $\square$

For the preceding example, Exercise 10.1.4 suggests an explicit embedding of X_A into X_B.

EXERCISES

10.1.1. Reduce the proof of the Embedding Theorem to the mixing case.

10.1.2. (a) Let X, Y, and Z be shifts of finite type with Z mixing. Suppose that $h(X) < h(Y)$, that there is a proper embedding of X into Y, and that $\zeta_Y = \zeta_Z$. Show that there is a proper embedding of X into Z.

(b) Let A, B and C be nonnegative integral matrices with B primitive. Suppose that $\lambda_A < \lambda_B$ and $\mathrm{sp}^\times(B) = \mathrm{sp}^\times(A) \cup \mathrm{sp}^\times(C)$. Show that X_A embeds in X_B.

(c) Let A, B, C, and D be nonnegative integral matrices with B primitive. Suppose that $\lambda_A < \lambda_B$, $\mathrm{sp}^\times(B) = \mathrm{sp}^\times(D) \cup \mathrm{sp}^\times(C)$, and X_A embeds in X_D. Show that X_A embeds in X_B.

10.1.3. For each of the following pairs of matrices, decide if X_A embeds (properly or improperly) in X_B.

(a)
$$A = \begin{bmatrix} 1 & 0 & 1 \\ 0 & 0 & 1 \\ 1 & 1 & 0 \end{bmatrix} \quad \text{and} \quad B = \begin{bmatrix} 2 \end{bmatrix}.$$

(b)
$$A = \begin{bmatrix} 1 & 1 & 0 \\ 0 & 1 & 1 \\ 1 & 0 & 1 \end{bmatrix} \quad \text{and} \quad B = \begin{bmatrix} 2 \end{bmatrix}.$$

(c)
$$A = \begin{bmatrix} 1 & 1 & 0 \\ 0 & 1 & 1 \\ 1 & 1 & 0 \end{bmatrix} \quad \text{and} \quad B = \begin{bmatrix} 2 \end{bmatrix}.$$

(d)
$$A = \begin{bmatrix} 2 \end{bmatrix} \quad \text{and} \quad B = \begin{bmatrix} 1 & 2 \\ 2 & 1 \end{bmatrix}.$$

(e)
$$A = \begin{bmatrix} 4 \end{bmatrix} \quad \text{and} \quad B = \begin{bmatrix} 4 & 1 \\ 1 & 0 \end{bmatrix}.$$

(f)
$$A = \begin{bmatrix} 4 \end{bmatrix} \quad \text{and} \quad B = \begin{bmatrix} 3 & 2 \\ 2 & 1 \end{bmatrix}.$$

[*Hint*: Use Exercise 10.1.2(a).]

(g)
$$A = \begin{bmatrix} 4 & 1 \\ 1 & 0 \end{bmatrix} \quad \text{and} \quad B = \begin{bmatrix} 3 & 2 \\ 2 & 1 \end{bmatrix}.$$

(h)
$$A = \begin{bmatrix} 8 \end{bmatrix} \quad \text{and} \quad B = \begin{bmatrix} 5 & 4 & 2 \\ 4 & 3 & 2 \\ 0 & 2 & 5 \end{bmatrix}.$$

[*Hint*: Use Exercise 10.1.2(c).]

(i)
$$A = \begin{bmatrix} 1 & 3 \\ 1 & 1 \end{bmatrix} \quad \text{and} \quad B = \begin{bmatrix} 3 & 1 \\ 1 & 1 \end{bmatrix}.$$

[*Hint*: Find approximations to the eigenvalues and use the method illustrated in Example 10.1.11.]

10.1.4. For the matrices A and B in Example 10.1.11 find an explicit embedding of X_A into X_B. [*Hint*: Perform state splittings on G_A and G_B to obtain graphs G'_A and G'_B such that G'_A embeds in G'_B.]

10.1.5. Show for mixing shifts of finite type X, Y that the following are equivalent.

(a) For all sufficiently large n, there is a proper embedding $X^n \to Y^n$.
(b) $h(X) < h(Y)$.

10.1.6. Show that if X_A is nonempty, then $\lambda_A \geqslant 1$. [*Hint:* Show that G_A contains a cycle.]

10.1.7. Let $g(n) = \lambda^n - \sum_{u \in U} a_u u^n$ where $\lambda > 1$, the list U consists of positive numbers strictly less than λ, and each $a_u > 0$. Show that whenever $g(n_0) \geqslant 0$, then g is monotonically increasing on the domain $n \geqslant n_0$.

10.1.8. Let A be a nonnegative integral matrix and B a primitive integral matrix with $\lambda_A < \lambda_B$. Find an explicit n_0, in terms of $\mathrm{sp}^\times(A)$ and $\mathrm{sp}^\times(B)$, such that the expression in (10–1–12) is nonnegative for all $n \geqslant n_0$ and hence give a finite procedure to determine when the embedding periodic point condition $P(X_A) \hookrightarrow P(X_B)$ holds. [*Hint:* See the discussion beginning on page 350.]

10.1.9. (a) Give examples to show that neither of the two conditions of the Embedding Theorem implies the other.

    (b) Show that for shifts of finite type the embedding periodic point condition $P(X) \hookrightarrow P(Y)$ implies $h(X) \leqslant h(Y)$.

    (c) Let

$$A = \begin{bmatrix} 4 \end{bmatrix} \text{ and } B = \begin{bmatrix} 3 & 1 \\ 1 & 3 \end{bmatrix}.$$

Verify that $q_n(X_A) < q_n(X_B)$ for $1 \leqslant n < \infty$, but $h(X_A) = h(X_B)$ (thus, even for mixing shifts of finite type, the strict inequalities, $q_n(X) < q_n(Y)$ ($1 \leqslant n < \infty$), do not imply $h(X) < h(Y)$).

10.1.10. (a) Give an example to show that for small m the shift of finite type $Y^{(m)}$ in the proof of Lemma 10.1.3 can fail to be mixing.

    (b) Find a sequence of arbitrarily long binary words $w^{(m)}$ such that for each m the shift of finite type obtained by deleting $w^{(m)}$ from the full 2-shift is reducible.

10.1.11. Prove the following stronger form of the Marker Lemma: Let X be a shift space and $r > N \geqslant 1$; then there is a set $F = F_{r,N} \subseteq X$ which is a finite union of cylinder sets such that

    (a) the sets $\sigma^i(F)$, $0 \leqslant i < N$ are disjoint, and

    (b) if $\sigma^i(x) \notin F$ for $-N < i < N$, then $x_{[-r,r]}$ is p-periodic for some $p < N$.

10.1.12. Let $m(N)$ be the minimum number m such that F (as constructed in the Marker Lemma) can be written as the union of cylinder sets of the form $C_{-m}(x_{[-m,m]})$. Show that $\limsup_{N \to \infty}(1/N) \log m(N) \leqslant 2h(X)$.

10.1.13. (a) Let Y be the even shift. Let X be an irreducible shift of finite type which is contained in Y and contains the fixed point 0^∞. Show that $X = \{0^\infty\}$.

    (b) Is the same true if, in part (a), 0^∞ is replaced by 1^∞?

    (c) Use part (a) to show that the entropy and periodic point conditions of the Embedding Theorem are not sufficient for mixing sofic shifts.

10.1.14. (a) Let X be a shift space. Show that for any fixed p, if there are arbitrarily long p-periodic blocks in $\mathcal{B}(X)$, then there is a periodic point in X with period p.

    (b) Use part (a) and the techniques of this section to prove Corollary 10.1.9 directly, not using the Embedding Theorem or Theorem 10.1.2.

10.1.15. Find an example to show that the Embedding Theorem fails if Y is not assumed to be irreducible but X is assumed irreducible.

*10.1.16. Find an example to show that the Embedding Theorem fails if X is not assumed to be irreducible but Y is assumed irreducible. Hint: try constructing an X with two irreducible components of period 2 connected by a transient path and an irreducible shift of finite type Y with period 2.

10.1.17. Find an example to show that Corollary 10.1.9 is false if Y is assumed to be merely irreducible. [*Hint*: Consider a mixing shift with no periodic points.]

§10.2. The Masking Lemma

If a graph G embeds into a graph H, then the edge shift X_G easily embeds into X_H by the corresponding 1-block code. However, as we indicated in §10.1, it may be possible to embed X_G into X_H but not to embed G into H; in fact, this is the typical case. Nevertheless, the following result shows that *up to conjugacy* every embedding of X_G into X_H is given by an embedding of graphs.

Proposition 10.2.1. *Suppose that X_G embeds into X_H. Then there are graphs G' and H' such that $X_{G'} \cong X_G$, $X_{H'} \cong X_H$, and G' is a subgraph of H'.*

PROOF: Let $\phi: X_G \to X_H$ be an embedding. Then $\phi(X_G)$ is a shift of finite type in X_H, so it must be N-step for some $N \geqslant 1$. By Theorem 2.3.2, $\phi(X_G)^{[N+1]}$ is an edge shift, say $X_{G'}$, whose symbols are edges in the graph $H' = H^{[N+1]}$, so G' is a subgraph of H'. Clearly $X_{G'} \cong X_G$ and $X_{H'} \cong X_H$. □

The Masking Lemma, stated below, strengthens this result by showing that we only need to modify H into a graph H', leaving G exactly as it is, and that G can be realized as a particular kind of subgraph of H'.

Definition 10.2.2. Let H be a graph with vertex set $\mathcal{V}$. For each subset $\mathcal{W}$ of $\mathcal{V}$ define the *induced subgraph of H from $\mathcal{W}$* to have vertex set $\mathcal{W}$ and edge set the collection of all edges in H that start and end in $\mathcal{W}$. An *induced subgraph* of H is one that is induced by some subset of $\mathcal{V}$.

If H has adjacency matrix B and G is an induced subgraph of H, then $A = A_G$ is a *principal submatrix* of B; i.e., A is obtained from B by deleting the jth row and jth column for a certain set of j's.

Lemma 10.2.3 (Masking Lemma). *Let G and H be graphs. Suppose that X_G embeds into X_H. Then there is a graph K such that $X_K \cong X_H$ and G is an induced subgraph of K.*

PROOF: Let G' and H' be the graphs constructed in Proposition 10.2.1. Then G' is a subgraph of H'. While G', as constructed, need not be an induced subgraph of H', by exercise 10.2.1(a) $(G')^{[2]}$ is an induced subgraph of $(H')^{[2]}$. So, by replacing G' by $(G')^{[2]}$ and H' by $(H')^{[2]}$, we may assume that G' is an induced subgraph of H'.

Since $A_G \approx A_{G'}$ there is a sequence of graphs

$$G' = G_0, G_1, \ldots, G_k = G$$

such that $A_{G_i} \approx A_{G_{i+1}}$ for $0 \leqslant i \leqslant k-1$. We will find a sequence of graphs

$$(10\text{-}2\text{-}1) \qquad\qquad H' = H_0,\, H_1,\, \ldots,\, H_k$$

such that each G_i is an induced subgraph of H_i and the elementary equivalence from A_{G_i} to $A_{G_{i+1}}$ extends to one from A_{H_i} to $A_{H_{i+1}}$. Once this is done, we set $K = H_k$. Then $A_K \approx A_{H'} \approx A_H$ and G is an induced subgraph of K. The following lemma constructs the graphs H_i in (10-2-1) one at a time.

Lemma 10.2.4. *Let G_1, G_2, and H_1 be graphs such that $A_{G_1} \approx A_{G_2}$ and G_1 is an induced subgraph of H_1. Then there is a graph H_2 such that $A_{H_2} \approx A_{H_1}$ and G_2 is an induced subgraph of H_2.*

PROOF: Put $A = A_{G_1}$ and $B = A_{G_2}$. Let $(R, S) : A \approx B$, so that $RS = A$ and $SR = B$. Recall from §7.2 the construction of the auxiliary graph $G_{R,S}$. This graph has vertex set the disjoint union of $\mathcal{V}(G_1)$ and $\mathcal{V}(G_2)$. It contains a copy of G_1, whose edges are called A-edges, and a copy of G_2, whose edges are called B-edges. For each $I \in \mathcal{V}(G_1)$ and $J \in \mathcal{V}(G_2)$ there are R_{IJ} edges from I to J, called R-edges, and S_{JI} edges from J to I, called S-edges.

We use H_1 to enlarge $G_{R,S}$, forming a new graph $\widetilde{G}$ as follows. Let $\mathcal{V} = \mathcal{V}(H_1) \setminus \mathcal{V}(G_1)$ and $\mathcal{E} = \mathcal{E}(H_1) \setminus \mathcal{E}(G_1)$. Add $\mathcal{V} \cup \mathcal{E}$ to the vertex set of $G_{R,S}$ to form the vertex set of $\widetilde{G}$. To avoid notational confusion, let I_e denote the added vertex corresponding to $e \in \mathcal{E}$. It is useful to think of vertices in $\mathcal{V}$ as being added to the "G_1-part" of $G_{R,S}$ and vertices I_e, for $e \in \mathcal{E}$, as being added to the "G_2-part" of $G_{R,S}$. For each $e \in \mathcal{E}$, add a new R-edge $\mathsf{r}(e)$ from the initial state of e to I_e, and a new S-edge $\mathsf{s}(e)$ from I_e to the terminal state of e. For each new R, S-path from I to I' that this creates, add to $\widetilde{G}$ a new A-edge from I to I'; similarly, for each new S, R-path from J to J' that this creates, add to $\widetilde{G}$ a new B-edge from J to J' (see Figure 10.2.1). This completes the construction of $\widetilde{G}$.

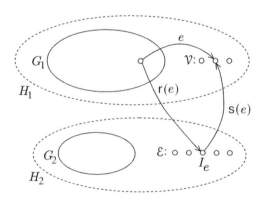

FIGURE 10.2.1. An augmented strong shift equivalence graph.

Each $e \in \mathcal{E}$ determines a new R, S-path $r(e)s(e)$ in $\widetilde{G}$ with the same initial and terminal states in H_1 as in $\widetilde{G}$, and these are the only additional R, S-paths. Hence the graph with vertices $\mathcal{V}(G_1) \cup \mathcal{V}$ and A-edges of $\widetilde{G}$ is isomorphic to H_1. Let H_2 be the graph with vertices $\mathcal{V}(G_2) \cup \{I_e : e \in \mathcal{E}\}$ and all B-edges of $\widetilde{G}$. Let $\widetilde{R}$ and $\widetilde{S}$ record the incidences of R-edges and S-edges in $\widetilde{G}$. Then $\widetilde{G}$ is the auxiliary graph $G_{\widetilde{R}, \widetilde{S}}$ corresponding to the elementary equivalence $(\widetilde{R}, \widetilde{S}): A_{H_1} \approx A_{H_2}$. Furthermore, G_2 is the subgraph of H_2 induced from the vertex subset $\mathcal{V}(G_2)$. This completes the proof of Lemma 10.2.4 and hence of the Masking Lemma. □

Example 10.2.5. Let H_1 and G_2 be the graphs shown in Figure 10.2.2 with G_1 the subgraph of H_1 indicated by dotted lines. Then the matrices

$$R = \begin{bmatrix} 1 \\ 1 \end{bmatrix} \quad \text{and} \quad S = \begin{bmatrix} 1 & 1 \end{bmatrix}$$

give an elementary equivalence $(R, S): A_{G_1} \approx A_{G_2}$. The auxiliary graph $G_{R,S}$ is shown in Figure 10.2.3.

The enlarged graph $\widetilde{G}$ constructed in the proof of Lemma 10.2.4 is depicted in Figure 10.2.4. Here H_2 has two additional vertices, corresponding to the edges e and f in H_1 not in G_1 (see Figure 10.2.2).

Note that H_2 has an induced subgraph isomorphic to G_2. Here

$$\widetilde{R} = \begin{bmatrix} 1 & 1 & 0 \\ 1 & 0 & 0 \\ 0 & 0 & 1 \end{bmatrix}, \qquad \widetilde{S} = \begin{bmatrix} 1 & 1 & 0 \\ 0 & 0 & 1 \\ 0 & 1 & 0 \end{bmatrix},$$

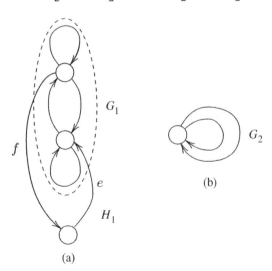

FIGURE 10.2.2. X_{G_2} embeds in X_{H_1}, but G_2 does not embed in H_1.

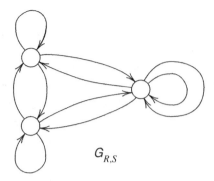

$$G_{R,S}$$

FIGURE 10.2.3.   Auxiliary graph for extending a strong shift equivalence.

and

$$A_{H_1} = \begin{bmatrix} 1 & 1 & 1 \\ 1 & 1 & 0 \\ 0 & 1 & 0 \end{bmatrix} = \widetilde{R}\widetilde{S}, \quad \widetilde{S}\widetilde{R} = \begin{bmatrix} 2 & 1 & 0 \\ 0 & 0 & 1 \\ 1 & 0 & 0 \end{bmatrix} = A_{H_2}.$$

Observe that A_{H_2} contains $[2] = A_{G_2}$ as a principal submatrix corresponding to the first diagonal entry.   □

Finally, we remark that the Masking Lemma is called a "Lemma" because Nasu [Nas5] invented it as a tool used to construct automorphisms (i.e., self-conjugacies) of shifts of finite type.

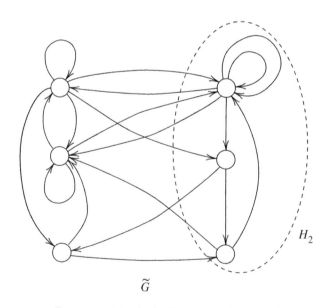

$$\widetilde{G}$$

$$H_2$$

FIGURE 10.2.4.   Embedding an auxiliary graph.

EXERCISES

10.2.1. (a) Show that if $X \subseteq Y$ are edge shifts, then $X^{[2]} = X_G$ and $Y^{[2]} = X_H$ are edge shifts where G is an induced subgraph of H.

(b) Show that if $X \subseteq Y$ are vertex shifts, then $X^{[2]} = X_G$ and $Y^{[2]} = X_H$ are edge shifts, but G need not be an induced subgraph of H.

10.2.2. Show that the full n-shift embeds in an edge shift X_H if and only if there exists $B \approx A_H$ such that B has a diagonal entry equal to n.

10.2.3. Let

$$A = \begin{bmatrix} 1 & 1 & 1 & 0 \\ 1 & 1 & 1 & 0 \\ 1 & 1 & 1 & 1 \\ 1 & 0 & 0 & 0 \end{bmatrix}.$$

Find a graph K such that $A_K \approx A$ and K contains a state with three self-loops.

10.2.4. For each of the following pairs of matrices A, B, find a graph K such that K contains G_A as an induced subgraph and $A_K \approx B$.

(a)

$$A = \begin{bmatrix} 1 & 1 \\ 1 & 0 \end{bmatrix}, \qquad B = \begin{bmatrix} 1 & 1 & 0 & 1 \\ 0 & 0 & 1 & 0 \\ 1 & 1 & 0 & 0 \\ 0 & 0 & 1 & 0 \end{bmatrix}.$$

(b)

$$A = \begin{bmatrix} 3 & 3 \\ 1 & 1 \end{bmatrix}, \qquad B = \begin{bmatrix} 0 & 1 \\ 1 & 4 \end{bmatrix}.$$

§10.3. Lower Entropy Factor Codes

When can one shift of finite type X be factored onto another Y? Clearly X must have as much entropy as Y, i.e., $h(X) \geqslant h(Y)$. In the equal entropy case $h(X) = h(Y)$ this problem is still not completely answered (see §12.2). However, in the lower entropy case $h(X) > h(Y)$ there is a simple necessary condition on periodic points that turns out to also be sufficient.

Suppose that $\phi \colon X \to Y$ is a factor code. If $x \in X$ has least period n, then $\phi(x) \in Y$ has least period that divides n. Thus a necessary condition for X to factor onto Y is that if X has a point of least period n, then Y has a periodic point whose least period divides n. We denote this by $P(X) \searrow P(Y)$ and call this the *factor periodic point condition*. It is easy to check that this condition is equivalent to the existence of a shift-commuting mapping from $P(X)$ into $P(Y)$.

The following result, due to Boyle [Boy1], shows that in the lower entropy case this necessary condition is also sufficient:

Theorem 10.3.1 (Lower Entropy Factor Theorem). *Let X and Y be irreducible shifts of finite type with $h(X) > h(Y)$. Then there is a factor code from X onto Y if and only if $P(X) \searrow P(Y)$.*

Before starting the proof, let us compare this with the Embedding Theorem. In the unequal entropy case, the factor periodic point condition plays a role analogous to the embedding periodic point condition. Both conditions can be verified by algorithms (Exercises 10.1.8 and 10.3.5). Sometimes it is extremely easy to decide whether $P(X) \searrow P(Y)$. For example, if Y contains a fixed point then $P(X) \searrow P(Y)$ is always true, while if X has a fixed point and Y does not, then $P(X) \searrow P(Y)$ must fail since there is no place for the fixed point to map to.

In the equal entropy case, the analogy between embeddings and factor codes breaks down. If X and Y are irreducible shifts of finite type and $h(X) = h(Y)$, then an embedding of X into Y is automatically a conjugacy. On the other hand, we have seen numerous examples of finite-to-one and hence entropy-preserving factor codes which are not conjugacies, for instance the finite-to-one codes in Chapter 8.

The proof of the Lower Entropy Factor Theorem is broken down into two steps. First we construct a shift of finite type W and a right-resolving factor code $\phi: W \to Y$ such that W embeds into X. This is done by "blowing up" the periods of finitely many periodic points in Y so that the conditions of the Embedding Theorem are met. The second step extends the factor code $\phi: W \to Y$ to $\tilde{\phi}: X \to Y$ using a marker construction similar to that used in the Embedding Theorem.

We leave it to the reader to reduce to the case where X and Y are mixing edge shifts.

If we could embed Y into X, the first step would be done by simply using $W = Y$. Otherwise, since $h(Y) < h(X)$, the only obstruction is the embedding periodic point condition. As we have already observed, the strict inequality $h(Y) < h(X)$ implies that there is an n_0 such that $q_n(Y) \leqslant q_n(X)$ for all $n \geqslant n_0$, so the only obstructions to embedding Y into X are a finite number of periodic points in Y of "low" period $< n_0$. We correct this situation by "blowing up" the periods of these points. The following lemma shows how to do this, one periodic orbit at a time. Recall that the orbit of a point z is the set $\{\sigma^n(z)\}$. If z is periodic, then its least period is also called the *length* of the orbit of z.

Lemma 10.3.2 (Blowing-up Lemma). *Let Y be a mixing edge shift, which does not consist of only a single point. Let z be a periodic point in Y with least period p, and $M \geqslant 1$. Then there is a mixing edge shift W and a right-resolving factor code $\phi: W \to Y$ such that*

(1) *the pre-image under ϕ of the orbit of z is an orbit of length Mp, and*

(2) *every periodic point that is not in the orbit of z has exactly one pre-image under ϕ.*

PROOF: Let $Y = \mathsf{X}_H$ and $z = \gamma^\infty$, where we can assume that γ is a simple cycle of length p (see Exercise 2.3.7). We form a new graph G by replacing

γ with a new simple cycle $\widetilde{\gamma}$ of length Mp which acts like γ coiled M times, taking care to modify edges incident to γ appropriately. A precise description of G is as follows.

Let $\gamma = e_0 e_1 \ldots e_{p-1}$ and $I_i = i(e_i)$ for $0 \leqslant i \leqslant p - 1$. Replace each I_i with M vertices $I_i^{(k)}$ for $0 \leqslant k \leqslant M - 1$, so that

$$\mathcal{V}(G) = \big(\mathcal{V}(H) \backslash \{I_i : 0 \leqslant i \leqslant p-1\}\big) \cup \{I_i^{(k)} : 0 \leqslant i \leqslant p-1, 0 \leqslant k \leqslant M-1\}.$$

The edges of G are of the following types.

(a) For each e_i in γ with $0 \leqslant i \leqslant p - 1$, there are edges $e_i^{(k)}$ from $I_i^{(k)}$ to $I_{i+1}^{(k)}$ for $0 \leqslant k \leqslant M - 1$ where $I_p^{(k)}$ is identified with $I_0^{(k+1)}$ and $I_0^{(M)}$ is identified with $I_0^{(0)}$.

(b) For each e in H not on γ that starts at one of the I_i but ends at a state J not in γ, there are M edges $e^{(k)}$ in $\mathcal{E}$ where $e^{(k)}$ starts at $I_i^{(k)}$ and ends at J.

(c) For each edge e in H not on γ that ends at an I_j but starts at a state J not in γ, there is one edge $e^{(0)}$ from J to $I_j^{(0)}$.

(d) For each edge e in H not on γ starting at an I_i and ending at an I_j, there are edges $e^{(k)}$ from $I_i^{(k)}$ to $I_j^{(0)}$ for $0 \leqslant k \leqslant M - 1$.

(e) For each edge e in H that is not incident to γ, there is an edge in G corresponding to e with the same initial and terminal states as in H.

This completes the definition of G. Figure 10.3.1 shows a simple case of this construction.

Let $W = X_G$. There is a natural graph homomorphism Φ from G to H defined by dropping parenthetic superscripts; by construction Φ is right-covering. By Proposition 8.2.2 we obtain a right-resolving factor code $\phi = \Phi_\infty : W \to Y$.

Clearly the pre-image of the orbit of γ^∞ is the orbit of $\widetilde{\gamma}^\infty$, where

$$\widetilde{\gamma} = e_0^{(0)} e_1^{(0)} \ldots e_{p-1}^{(0)} e_0^{(1)} \ldots e_{p-1}^{(1)} e_0^{(2)} \ldots e_{p-1}^{(M-2)} e_0^{(M-1)} \ldots e_{p-1}^{(M-1)};$$

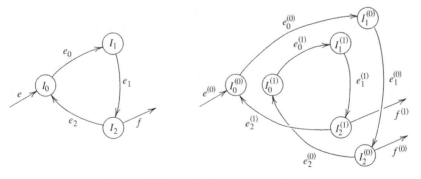

FIGURE 10.3.1. Blowing up a periodic orbit.

note that this orbit has length Mp. Let z be a periodic point in Y not in the orbit of γ^∞. If no edge in z is incident to γ, then by (e) above the pre-image of z contains exactly one point. If some edge of z is incident to γ, then there is an edge e in z that is not on γ but begins or ends at some I_j. In the former case, the pre-images of e are of type (b), hence all end at the same state. Thus e is a synchronizing word for Φ considered as a labeling of G. Since z is periodic, this word occurs infinitely often to the left. Since Φ is right-resolving, we see that z has a unique pre-image. A similar argument works in the case where the pre-images are of type (c) or (d).

Finally, since Y is mixing, condition (2) shows that G is aperiodic; since G is clearly irreducible, we conclude that W is mixing. $\qquad\square$

We next show how to extend the factor code $\phi\colon W \to Y$.

Lemma 10.3.3 (Extension Lemma). *Let $W \subseteq X$ be shifts of finite type and Y a mixing shift of finite type. Suppose that $\phi\colon W \to Y$ is a factor code and that $P(X) \searrow P(Y)$. Then ϕ can be extended to a factor code $\tilde{\phi}\colon X \to Y$.*

PROOF: The proof of the Extension Lemma is based on the same kinds of ideas as the proof of the Embedding Theorem. We use the Marker Lemma to construct a marker set F. As before, for any point $x \in X$, this F defines a set of marker coordinates. In the construction of embeddings, we used the marker coordinates to partition the set of integers into nonoverlapping half-open intervals, called marker intervals; then we encoded the blocks that fill in these intervals and checked that the image blocks strung together to define a point in Y. Here, we find it more convenient to encode blocks on enlarged intervals which overlap by one symbol. We will guarantee consistency of the encoding on the overlaps, and this will then give a well-defined image point in Y. As in the construction of embeddings, the marker intervals are never "short," and we encode blocks in one of two different ways, depending on the length of the marker interval. For marker intervals of "medium" length, we use surjections based on the entropy condition $h(X) > h(Y)$; for those of "large" length, we encode using the condition $P(X) \searrow P(Y)$ for points of low period.

There are some important differences between this proof and that of the Embedding Theorem. We must take care that our choices of surjections and transition paths work to extend ϕ and that the extension maps into (hence onto) Y. On the other hand, we do not need to worry about the map being one-to-one.

We may assume as usual that W, X, and Y are edge shifts, and that $\phi = \Phi_\infty$ is a 1-block code. Fix an (arbitrary) extension of Φ to the alphabet of X.

Write $Y = X_H$ where H is a primitive graph. Then there is a transition length N such that, for every pair e, f of edges in H and every $n > N$, there is a path of length n with initial edge e and terminal edge f. For $n = N + 1$ we denote this path by $\tau_{e,f}$.

Let F be the marker set produced by the Marker Lemma applied to X and N. As in our embedding construction, for each $x \in X$ the marker coordinates $\mathcal{M}(x) = \{i \in \mathbb{Z} : \sigma^i(x) \in F\}$ decompose $\mathbb{Z}$ into marker intervals: finite marker intervals $T = [i, j)$ where $i, j \in \mathcal{M}(x)$ and $k \notin \mathcal{M}(x)$ for $i < k < j$ and infinite marker intervals. For any finite marker interval $T = [i, j)$, we define the *closure* of T to be $T^* = [i, j]$. For infinite marker intervals, define $[i, \infty)^* = [i, \infty)$, $(-\infty, j)^* = (-\infty, j]$, and $(-\infty, \infty)^* = (-\infty, \infty)$. Note that the closures of marker intervals overlap on their boundaries.

By the marker construction, all marker intervals must have length at least N. We partition marker intervals into two types based on length. We use different thresholds than in the embedding construction, so also different terminology. A *medium* interval is a marker interval whose length is $\leqslant 2N$, while a *large* interval is a marker interval whose length is $> 2N$.

For each $n \geqslant 1$ the 1-block map Φ extends to $\Phi_n: \mathcal{B}_n(W) \to \mathcal{B}_n(Y)$. For $n \geqslant N + 1$ we extend Φ_n to $\mathcal{B}_n(X)$ so that Φ_n is consistent with Φ on the first and last symbols of a block, i.e., so that $\Phi_n(x_1 x_2 \ldots x_n)$ starts with $\Phi(x_1)$ and ends with $\Phi(x_n)$. This is possible by our choice of N.

Since $\phi: W \to Y$ is a factor code, the restriction $\phi|_{P(W)}$ of ϕ to $P(W)$ is a shift-commuting map onto $P(Y)$. By the factor periodic point condition, $P(X) \searrow P(Y)$, we can extend $\phi|_{P(W)}$ to a shift-commuting map $\psi: P(X) \to P(Y)$.

Lemma 10.1.7 and the Marker Lemma 10.1.8 together guarantee that for any $x \in X$ and any large marker interval T for x there is a periodic point $z = z(T, x) \in X$ of least period $< N$ such that $z_T = x_T$.

For a medium interval $T = [i, j)$ we define

$$\widetilde{\phi}(x)_{T^*} = \widetilde{\phi}(x)_{[i,j]} = \Phi_{j-i+1}(x_{[i,j]}),$$

which is a path in H.

For a finite large interval $T = [i, j)$, let $z = z(T, x)$. Define paths π^-, π^0, and π^+ by

$$\pi^0 = \psi(z)_{[i+N+1, j-N-1]},$$

$$\pi^- = \begin{cases} \Phi_{N+1}(x_{[i,i+N]}) & \text{if } x_{[i,j]} \in \mathcal{B}(W), \\ \tau_{\Phi(x_i), \psi(z)_{i+N}} & \text{if } x_{[i,j]} \notin \mathcal{B}(W), \end{cases}$$

$$\pi^+ = \begin{cases} \Phi_{N+1}(x_{[j-N,j]}) & \text{if } x_{[i,j]} \in \mathcal{B}(W), \\ \tau_{\psi(z)_{j-N}, \Phi(x_j)} & \text{if } x_{[i,j]} \notin \mathcal{B}(W), \end{cases}$$

where $\tau_{e,f}$ is defined as above. Set

$$\widetilde{\phi}(x)_{T^*} = \widetilde{\phi}(x)_{[i,j]} = \pi^- \pi^0 \pi^+.$$

We first verify that $\widetilde{\phi}(x)_{T^*}$ is a path in H. This is clear when $x_{[i,j]} \notin \mathcal{B}(W)$. When $x_{[i,j]} \in \mathcal{B}(W)$, the terminal edge of π^- is $\Phi(x_{i+N}) = \Phi(z_{i+N}) = \psi(z)_{i+N}$, the latter equality because the uniqueness part of Lemma 10.1.7 implies that $z \in W$ in this setting. The initial edge of π^0 is $\psi(z)_{i+N+1}$, so $\pi^-\pi^0$ is a path in H; similarly $\pi^0\pi^+$ is also. Analogous arguments apply to the three types of infinite marker intervals. So, $\widetilde{\phi}(x)_{T^*}$ is a path in H.

For each marker interval T, if e is the initial (resp. terminal) edge of x_{T^*}, then $\Phi(e)$ is the initial (resp. terminal) edge of $\widetilde{\phi}(x)_{T^*}$. Since marker intervals overlap in one coordinate, it follows that the blocks $\widetilde{\phi}(x)_{T^*}$ glue together to form a point in X_H. This completes the definition of $\widetilde{\phi}: X \to X_H = Y$.

The only thing left is to check that $\widetilde{\phi}$ extends ϕ. This follows from the observation that if x_{T^*} occurs in W, then $\widetilde{\phi}(x)_{T^*} = \phi(x)_{T^*}$. This is true by definition for medium intervals; for large intervals, it is clearly true on the transition portions $\pi^{\pm}$, and it is true on the intermediate part π^0 since $z = z(T, x) \in W$ as we observed above. This completes the proof of the lemma. $\qquad\square$

We remark that the shift of finite type assumption on W and X in Lemma 10.3.3 is not necessary (see Exercise 10.3.8).

PROOF OF THE LOWER ENTROPY FACTOR THEOREM: As usual, it is easy to reduce to the case where both X and Y are mixing edge shifts. Since $h(X) > h(Y)$, we have $q_n(X) > q_n(Y)$ for all sufficiently large n. If Y consists only of a single point, then the result is obvious. So, we can assume that Y does not consist of only a single point and we can repeatedly apply Lemma 10.3.2 to points in Y having low period; this shows that we can find a mixing edge shift W and a right-resolving factor code $\phi: W \to Y$ such that $h(X) > h(W)$ and $q_n(X) > q_n(W)$ for all n. By the Embedding Theorem, there is an embedding $\theta: W \to X$. Thus by Lemma 10.3.3 the factor code $\phi \circ \theta^{-1}: \theta(W) \to Y$ can be extended to a factor code $\widetilde{\phi}: X \to Y$. $\qquad\square$

In the case of irreducible sofic shifts, the entropy and periodic point conditions are not sufficient to obtain a lower entropy factor code (see Exercise 10.3.6 and [Boy1] for more information).

The first step in the proof of the Lower Entropy Factor Theorem can be used to produce finite equivalences and almost conjugacies where both legs are right-closing. We state this in Theorem 10.3.5 below, which requires the following definitions.

Definition 10.3.4. A *right-closing finite equivalence* is a finite equivalence in which both legs are right-closing. A *right-closing almost conjugacy* is an almost conjugacy in which both legs are right-closing.

Theorem 10.3.5. *Let X, Y be irreducible sofic shifts with minimal right-resolving presentations (G_X, L_X), (G_Y, L_Y).*

(1) *If $h(Y) < h(X)$, then there exist an irreducible sofic shift Z contained in X and a right-closing finite equivalence between Y and Z.*

(2) *If $h(Y) < h(X)$ and G_X and G_Y have the same period, then the finite equivalence can be chosen to be a right-closing almost conjugacy.*

PROOF: By passing to the minimal right-resolving presentations, we may assume that X and Y are irreducible edge shifts. As usual, we may also assume that they are mixing. As in the proof of the Lower Entropy Factor Theorem above, there are a mixing shift of finite type W, a right-resolving factor code $\phi \colon W \to Y$, and an embedding $\psi \colon W \to X$. Let Z be the (sofic) image of ψ. Since any embedding is a right-closing factor code onto its image, it follows that (W, ϕ, ψ) is a right-closing finite equivalence between Y and Z. This establishes (1). For (2), we show that (W, ϕ, ψ) is actually a right-closing almost conjugacy. For this, first recall from Theorem 9.1.11(2) that the degree of a finite-to-one code on an irreducible shift of finite type is the minimum number of pre-images of points in the range. Since ϕ is obtained from repeated use of Lemma 10.3.2 which produces factor codes that are one-to-one on all periodic points of sufficiently large least period, we conclude that ϕ is almost invertible. This, together with the fact that any embedding is almost invertible, shows that (W, ϕ, ψ) is indeed a right-closing almost conjugacy. $\square$

The period assumption in part (2) of the preceding result was never mentioned in the proof because it is vacuous when both X and Y are mixing: $(L_X)_\infty$ and $(L_Y)_\infty$ are almost invertible (by Proposition 9.1.6(2)), and so X_{G_X} and X_{G_Y} are mixing in this case.

The finite equivalence constructed in Theorem 10.3.5(1) is much stronger than required since the leg onto Z is a conjugacy. In Exercise 10.3.9 we outline an alternative construction in which neither leg is guaranteed to be a conjugacy, but which avoids the proof of the Embedding Theorem and therefore gives a simpler construction. An alternative construction for part 2 is suggested in Exercise 12.4.4.

Finally, we point out that Theorem 10.3.5 can be used to give a generalization of the Finite-State Coding Theorem 5.2.5.

Definition 10.3.6. Let X and Y be sofic shifts. A *finite-state (X, Y)-code* is a triple $(G, \mathcal{I}, \mathcal{O})$ where G is a graph with labelings $\mathcal{I}$ and $\mathcal{O}$, such that $(G, \mathcal{I})$ is right-resolving and presents Y and $(G, \mathcal{O})$ is right-closing and presents a subshift of X.

For there to be a finite-state (X, Y)-code $(G, \mathcal{I}, \mathcal{O})$, we must have that $h(Y) = h(X_G) \leqslant h(X)$. We treat here the unequal entropy case $h(Y) < h(X)$. The equal entropy case is more subtle, and is treated in Chapter 12.

By Proposition 5.1.11 a finite-state (X, Y)-code is, up to recoding, the same as a right-closing finite equivalence between Y and a subshift of X. Thus we can restate Theorem 10.3.5 as follows.

Theorem 10.3.7. *Let X, Y be irreducible sofic shifts with minimal right-resolving presentations (G_X, L_X), (G_Y, L_Y).*

(1) *If $h(Y) < h(X)$, then there is a finite-state (X, Y)-code.*

(2) *If $h(Y) < h(X)$ and G_X and G_Y have the same period, then there is a finite-state (X, Y)-code $(G, \mathfrak{I}, \mathfrak{O})$ such that $\mathfrak{I}_\infty$ and $\mathfrak{O}_\infty$ are almost invertible.*

When Y is an irreducible sofic shift, we may assume that the underlying graph of a finite-state (X, Y)-code is irreducible: simply pass to an irreducible component with maximal Perron value. Recall that when a graph G is irreducible, any right-resolving presentation $(G, \mathcal{L})$ of a full shift is automatically a road-coloring (see Exercise 5.1.4). Thus taking $Y = X_{[n]}$, we see that a finite-state $(X, X_{[n]})$-code yields a finite-state (X, n)-code as defined in §5.2. It then follows that Theorem 10.3.7(1) generalizes the Finite-State Coding Theorem 5.2.5 in the unequal entropy case. However, the constructions of finite-state (X, n)-codes implied by Theorem 10.3.7(1) or by the alternative construction outlined in Exercise 10.3.9 are usually much more complex than the state splitting construction given in Chapter 5.

EXERCISES

10.3.1. Reduce the proof of the Lower Entropy Factor Theorem to the mixing case.

10.3.2. In the notation of the Lower Entropy Factor Theorem, give an explicit definition for $\tilde{\phi}$ on infinite intervals.

10.3.3. For the following pairs of matrices A, B, decide if X_A factors onto X_B and if X_B factors onto X_A.

(a)
$$A = \begin{bmatrix} 0 & 1 & 1 \\ 1 & 0 & 1 \\ 1 & 1 & 0 \end{bmatrix} \text{ and } B = \begin{bmatrix} 1 & 1 \\ 1 & 0 \end{bmatrix}.$$

(b)
$$A = \begin{bmatrix} 0 & 1 & 0 & 0 & 0 \\ 1 & 0 & 1 & 0 & 0 \\ 0 & 0 & 0 & 1 & 0 \\ 0 & 0 & 0 & 0 & 1 \\ 1 & 0 & 0 & 0 & 0 \end{bmatrix} \text{ and } B = \begin{bmatrix} 0 & 1 & 0 & 0 \\ 0 & 0 & 1 & 0 \\ 1 & 0 & 0 & 1 \\ 1 & 0 & 0 & 0 \end{bmatrix}.$$

(c)
$$A = \begin{bmatrix} 0 & 1 & 0 & 0 \\ 0 & 0 & 1 & 0 \\ 8 & 0 & 0 & 1 \\ 1 & 0 & 0 & 0 \end{bmatrix} \text{ and } B = \begin{bmatrix} 0 & 1 & 0 \\ 1 & 0 & 1 \\ 1 & 0 & 0 \end{bmatrix}.$$

10.3.4. For mixing shifts of finite type X, Y with $h(X) > h(Y)$, show that X^n factors onto Y^n for sufficiently large n. Formulate a version of this for irreducible shifts of finite type.

10.3.5. (a) Show that if B is a primitive $n \times n$ matrix then $B^m > 0$ for all $m \geqslant 2^{n^2}$.

    (b) Explain why this gives a finite procedure to decide, given an arbitrary pair of primitive graphs G, H, whether $P(X_G) \searrow P(X_H)$.

    (c) Extend part (b) to irreducible graphs.

  *(d) Show that the estimate in part (a) can be improved dramatically: if B is a primitive $n \times n$ matrix then $B^m > 0$ for all $m \geqslant n^2 - n + 2$.

10.3.6. Show that the lower entropy factor theorem does not hold for mixing sofic shifts. Hint: try $X =$ the full 2-shift and $Y =$ the sofic shift defined by the labelled graph consisting of two cycles joined at a single vertex, one cycle of length 2, with both edges labelled by 0, and the other cycle of length 3, with all three edges labelled by 1.

10.3.7. Suppose that X and Y are mixing shifts of finite type, $h(X) < h(Y)$, and $P(X) \searrow P(Y)$. Show that there is a right-closing code from X into Y. [*Hint:* Start as in the proof of the Embedding Theorem, but use $h(X) < h(Y)$ to encode all but the low-period periodic points and $P(X) \searrow P(Y)$ to encode low-period periodic points; also, use Exercise 10.1.11 to guarantee, for any x and any finite or right-infinite long interval T, that x and the periodic point $z(T, x)$ agree on an interval that contains T and extends somewhat to the left of T.]

10.3.8. Show that Lemma 10.3.3 holds for arbitrary shift spaces $W \subseteq X$ and mixing shift of finite type Y.

10.3.9. Prove Theorem 10.3.5(1), without using the Embedding Theorem or Theorem 10.1.2 as follows. (As usual, assume $X = X_G, Y = X_H$, with G and H primitive.) For $N > 0$ let P_N denote the N shifts of the binary sequence $(10^{N-1})^\infty$; i.e., P_N consists of all binary sequences such that gaps between 1's are always of length $N - 1$. Let

$$W^{(N)} = \{(y, w) : y \in Y, w \in P_N\}.$$

Show that for some N, there exists a right-closing code from $W^{(N)}$ into X. (This yields a right-closing finite equivalence between Y and a sofic subshift of X.)

Notes

The Embedding Theorem, Corollary 10.1.9, and Theorem 10.3.5 are all due to Krieger [Kri6]. These results give various analogues of a central result in ergodic theory known as (Krieger's) Generator Theorem [Kri1]. Ergodic theory studies measure-preserving transformations on probability spaces (see §13.4). There is a notion of entropy for such transformations, and the Generator Theorem shows that any ergodic transformation T can, in some sense, be "embedded" in the full n-shift provided that the entropy of T is strictly less than the entropy $\log n$ of the full n-shift.

The Marker Lemma is a symbolic dynamics version of an important tool in ergodic theory, the Kakutani–Rokhlin tower construction (see [Kak], [Rok2], [Pet3]), which shows that if T is an ergodic transformation on a probability space X, then for any $N \geqslant 1$ and $\epsilon > 0$ there is a set F such that the sets $F, T(F), T^2(F), \ldots, T^{N-1}(F)$ are disjoint and have union with measure at least $1 - \epsilon$.

The Masking Lemma is due to Nasu, who used it as a tool for extending shift-commuting permutations of finite sets of periodic points to automorphisms (i.e., self-conjugacies) of shifts of finite type [Nas5, Lemma 3.18].

The Lower Entropy Factor Theorem, which is due to Boyle [Boy1], is an analogue of the Embedding Theorem. Boyle's paper contains other interesting results and examples pertaining to the embedding and lower entropy factor problems for irreducible shifts of finite type as well as irreducible sofic shifts.

Exercises 10.1.13 and 10.1.14 are in [Boy1]; Exercise 10.3.5(a) is in [CovP3]; Exercise 10.3.5(d) is in [Sen, Thm. 2.9]; Exercise 10.3.6 was manufactured from [Boy1]; Exercise 10.3.8 is in [Boy1]; and Exercise 10.3.9 is in [Kri6].

CHAPTER 11

REALIZATION

Invariants such as entropy, zeta function, and the dimension pair play an important role in studying shifts of finite type and sofic shifts. What values can these invariants take? Which numbers are entropies, which functions are zeta functions, which pairs are dimension pairs? Answers to these kinds of questions are called *realization theorems*.

In §11.1 we completely answer the entropy question. There is a simple algebraic description of the possible entropies of shifts of finite type and of sofic shifts. This amounts to characterizing the spectral radii of nonnegative integral matrices.

We focus on zeta functions in §11.2. Theorem 6.4.6 (see also Corollary 6.4.7) shows that the zeta function of an edge shift contains the same information as the nonzero spectrum of the adjacency matrix. Thus characterizing zeta functions of shifts of finite type is the same as characterizing the nonzero spectra of nonnegative integral matrices. We state a complete characterization of the nonzero spectra of primitive integral matrices due to Kim, Ormes and Roush [KimOR]. This result was conjectured by Boyle and Handelman [BoyH1], who proved several partial results and obtained a complete characterization of the nonzero spectra of primitive real matrices. The proofs of these results are too complicated to include here, but we illustrate some of the main ideas involved by treating some special cases such as when all eigenvalues are integers. A remarkable feature of the work of Boyle and Handelman is that a significant theorem in linear algebra is proven by using important tools from symbolic dynamics: the Embedding Theorem and the Masking Lemma from Chapter 10. At the end of §11.2, we state a complete characterization of zeta functions of mixing sofic shifts [BoyH1].

In §11.3 we consider the question of which invariant subgroups of dimension groups define dimension pairs in their own right.

§11.1. Realization of Entropies

Entropy is the most basic numerical invariant for shift spaces. For example, we saw in Chapter 8 that entropy is a complete invariant for finite equivalence. In Chapter 4 we computed the entropy of a shift of finite type to be $\log \lambda$ where λ is the spectral radius of a nonnegative integral matrix. Indeed, Theorem 4.4.4 shows that the set of such numbers is precisely the set of possible entropies of shifts of finite type, and similarly for sofic shifts by Theorem 4.3.3.

Can we characterize which numbers actually occur as entropies of shifts of finite type? Alternatively, can we completely describe the spectral radii of nonnegative integral matrices? Only countably many such numbers occur, but there are further restrictions that follow from integrality and the Perron–Frobenius Theorem. To describe these, we first give a quick review of some basic notions from algebra and number theory. The reader can consult [Her] or [Pol] for a full account.

Recall that $\mathbb{Z}[t]$ denotes the set of polynomials with integral coefficients in the variable t. From here on, by *polynomial* we mean an element of $\mathbb{Z}[t]$. Given a polynomial $f(t) = c_n t^n + c_{n-1} t^{n-1} + \cdots + c_1 t + c_0$ with $c_n \neq 0$, we call n the *degree* of $f(t)$ and c_n its *leading coefficient*. A polynomial $f(t)$ is *irreducible* if whenever $f(t) = g(t)h(t)$ with $g(t), h(t) \in \mathbb{Z}[t]$, then either $g(t) = \pm 1$ or $h(t) = \pm 1$. For example, $f(t) = t^2 - 2$ is irreducible (although were we to allow irrational coefficients, then $f(t)$ would factor as $(t - \sqrt{2})(t + \sqrt{2})$). A polynomial is *monic* if its leading coefficient is 1.

A complex number is *algebraic* if it is the root of a polynomial and it is an *algebraic integer* if it is the root of a monic polynomial. For example, $\sqrt{3}$ satisfies $t^2 - 3$, so $\sqrt{3}$ is an algebraic integer. On the other hand, the real number $\pi = 3.14159\ldots$ turns out not to be algebraic, although the proof of this took a long time to discover. For each algebraic integer λ there is a unique monic polynomial $f_\lambda(t)$, called the *minimal polynomial* of λ, which is the monic polynomial of smallest degree having root λ. It turns out that $f_\lambda(t)$ is irreducible and divides every polynomial having λ as a root, so that $f_\lambda(t)$ is the unique irreducible monic polynomial satisfied by λ. The roots of $f_\lambda(t)$ are the *algebraic conjugates* or simply *conjugates* of λ. The *degree* of λ, denoted by $\deg \lambda$, is the degree of its minimal polynomial $f_\lambda(t)$.

Example 11.1.1. (a) Let $\lambda = 2$. Then λ satisfies $t - 2$, so λ is an algebraic integer and clearly $f_\lambda(t) = t - 2$. Hence $\deg \lambda = 1$ and λ has no other algebraic conjugates. The reader should verify that if λ is an algebraic integer with degree 1, then λ must be an ordinary integer (sometimes called a rational integer). So, while every rational number is an algebraic number, the only rational numbers that are algebraic integers are the ordinary integers.

(b) Let $\lambda = (1 + \sqrt{5})/2$, the golden mean. Then λ satisfies $t^2 - t - 1$, and by the remark in part (a) λ cannot satisfy a polynomial of degree 1. Thus $f_\lambda(t) = t^2 - t - 1$ and so $\deg \lambda = 2$ and λ has algebraic conjugates λ and

$(1 - \sqrt{5})/2$.

(c) The polynomial $f(t) = t^3 - t - 1$ is irreducible and has roots $\lambda_1 \approx$ 1.3247 and $\lambda_2, \lambda_3 \approx -0.66236 \pm 0.56228i$, where $i = \sqrt{-1}$. Hence λ_1 is an algebraic integer of degree 3 whose other conjugates are λ_2 and λ_3.

(d) The polynomial $f(t) = t^3 + 3t^2 - 15t - 46$ is irreducible and has roots $\lambda_1 \approx 3.89167$, $\lambda_2 \approx -3.21417$, and $\lambda_3 \approx -3.67750$. Again, λ_1 is an algebraic integer of degree 3 whose other conjugates are λ_2 and λ_3. □

Suppose that A is a primitive integral matrix. Then its spectral radius $\lambda = \lambda_A$ satisfies the characteristic polynomial $\chi_A(t)$, so that λ is an algebraic integer. The Perron–Frobenius Theorem and Theorem 4.5.11 show that λ is a simple root of $\chi_A(t)$ and that all other roots of $\chi_A(t)$ have absolute value strictly smaller than λ. Since $f_\lambda(t)$ divides $\chi_A(t)$, it follows in particular that all the other algebraic conjugates λ_j of λ satisfy $|\lambda_j| < \lambda$. We incorporate these two conditions into the following definition.

Definition 11.1.2. A real number $\lambda \geqslant 1$ is a *Perron number* if it is an algebraic integer that strictly dominates all its other algebraic conjugates. The set of all Perron numbers is denoted by $\mathbb{P}$.

Example 11.1.3. Looking back at Example 11.1.1, we see that the following are Perron numbers: 2 from part (a), $(1 + \sqrt{5})/2$ from (b), λ_1 from (c), and λ_1 from (d). □

It is not hard to show that any algebraic integer $\lambda > 0$ that strictly dominates its algebraic conjugates must satisfy $\lambda \geqslant 1$ (Exercise 11.1.3), and so we could have replaced "$\lambda \geqslant 1$" by "$\lambda > 0$" in the definition of Perron number. Thus, our discussion above can be summarized by saying that if A is a primitive integral matrix, then $\lambda_A \in \mathbb{P}$. The converse, due to Lind [Lin3], [Lin4], is the main result of this section:

Theorem 11.1.4. *Let λ be a Perron number. Then there is a primitive integral matrix A such that $\lambda_A = \lambda$.*

Before proving this theorem, we will use it to characterize the spectral radii of general nonnegative integral matrices. Let A be a nonnegative integral matrix and let $A_1, A_2, \ldots, A_k$ denote its irreducible components. According to Lemma 4.4.3, the spectral radius of A is $\lambda_A = \max_{1 \leqslant j \leqslant k} \lambda_{A_j}$, so it is enough to characterize the spectral radii of irreducible integral matrices A. If $A = [0]$, then $\lambda_A = 0$. If $A \neq [0]$, then the period p of A is finite, and by the discussion at the beginning of §4.5, A^p is a block diagonal matrix, whose diagonal blocks are primitive. Hence, $\lambda_A^p = \lambda_{A^p}$ is Perron. This suggests the following definition.

Definition 11.1.5. A real number $\lambda \geqslant 1$ is a *weak Perron number* if it is an algebraic integer such that λ^p is a Perron number for some integer $p \geqslant 1$.

Our preceding analysis leads to the following result.

Theorem 11.1.6. *Let $\lambda > 0$. Then the following are equivalent.*

(1) λ *is the spectral radius of a nonnegative integral matrix.*

(2) λ *is the spectral radius of an irreducible integral matrix.*

(3) λ *is a weak Perron number.*

PROOF: We prove this assuming Theorem 11.1.4, which we prove later in this section.

(1) $\Rightarrow$ (3). This follows from the discussion above.

(3) $\Rightarrow$ (2). Since $\lambda^p \in \mathbb{P}$, there is a primitive integral matrix B such that $\lambda_B = \lambda^p$. Then λ is the spectral radius of the irreducible $p \times p$ block matrix

$$
\begin{bmatrix}
0 & Id & 0 & \dots & 0 \\
0 & 0 & Id & \dots & 0 \\
\vdots & \vdots & \vdots & \ddots & Id \\
B & 0 & 0 & \dots & 0
\end{bmatrix}.
$$

(2) $\Rightarrow$ (1). Obvious. $\square$

Example 11.1.7. (a) Let $\lambda = \sqrt{2}$. Note that λ is *not* Perron since its conjugate $-\sqrt{2}$ does not have strictly smaller absolute value. However, $\lambda^2 = 2$ is Perron. Here λ is the spectral radius of $\begin{bmatrix} 0 & 1 \\ 2 & 0 \end{bmatrix}$, which is irreducible but not primitive.

(b) Let $\lambda = (3 + \sqrt{5})/2$, which has conjugate $\mu = (3 - \sqrt{5})/2$. Since $|\mu| < \lambda$, we see that $\lambda \in \mathbb{P}$. By Theorem 11.1.4 there is a primitive integral matrix A with spectral radius λ. The reader should verify that $A = \begin{bmatrix} 2 & 1 \\ 1 & 1 \end{bmatrix}$ works.

(c) Let $\lambda = \lambda_1 \approx 1.3247$ from Example 11.1.1(c). Since $\lambda \in \mathbb{P}$, there must be a primitive integral matrix A with spectral radius λ. For example,

$$
A = \begin{bmatrix}
0 & 1 & 0 \\
0 & 0 & 1 \\
1 & 1 & 0
\end{bmatrix}
$$

is primitive and has $\chi_A(t) = t^3 - t - 1$, so that $\lambda = \lambda_A$.

(d) Let $\lambda = \lambda_1 \approx 3.89167$ be from Example 11.1.1(d). Since $\lambda \in \mathbb{P}$, there must be a primitive integral matrix A with $\lambda = \lambda_A$. Could A be 3×3? No, for then $\chi_A(t) = f_\lambda(t) = t^3 + 3t^2 - 15t - 46$, and so $\operatorname{tr} A = \lambda_1 + \lambda_2 + \lambda_3 = -3 < 0$, contradicting that A is nonnegative. So even though λ has degree 3, any primitive integral matrix with spectral radius λ must have size at least 4. The procedure we give below for constructing such an A produces a primitive integral 10×10 matrix A with $\lambda_A = \lambda$ [Lin3]. $\square$

Strict positivity of the second coefficient is not the only obstruction to finding an A whose size is $\deg \lambda$. Further restrictions are given in §11.2.

We now begin our preparations to prove Theorem 11.1.4. Suppose that $\lambda \in \mathbb{P}$ has conjugates $\lambda = \lambda_1, \lambda_2, \ldots, \lambda_d$, where $d = \deg \lambda$. Then the minimal polynomial $f_\lambda(t)$ of λ factors over $\mathbb{C}$ as

$$f_\lambda(t) = \prod_{j=1}^{d} (t - \lambda_j).$$

Next, we show how to manufacture from $f_\lambda(t)$ an integral matrix whose list of eigenvalues is exactly the set of conjugates of λ.

Definition 11.1.8. Let $f(t) = t^n + c_{n-1}t^{n-1} + \cdots + c_1 t + c_0 \in \mathbb{Z}[t]$. The *companion matrix* C_f of $f(t)$ is the $n \times n$ matrix

$$C_f = \begin{bmatrix} 0 & 1 & 0 & \ldots & 0 \\ 0 & 0 & 1 & \ldots & 0 \\ \vdots & \vdots & \vdots & \ddots & \vdots \\ 0 & 0 & 0 & \ldots & 1 \\ -c_0 & -c_1 & -c_2 & \ldots & -c_{n-1} \end{bmatrix}.$$

An easy calculation shows that the characteristic polynomial of C_f is $f(t)$, so that the list of eigenvalues of C_f is exactly the list of roots of $f(t)$ (counted with multiplicity). But of course C_f may have negative entries. For example, the polynomial $f(t) = t^3 + 3t^2 - 15t - 46$ in Example 11.1.1(d) has companion matrix

(11-1-1)
$$C_f = \begin{bmatrix} 0 & 1 & 0 \\ 0 & 0 & 1 \\ 46 & 15 & -3 \end{bmatrix}.$$

Although C_f is not a nonnegative matrix, as a linear transformation it still retains some "Perron geometry" and we will use this extensively in what follows. We make this property precise in the following definition.

Definition 11.1.9. A matrix B is *spectrally Perron* if there is a simple eigenvalue $\lambda \geq 1$ of B such that $\lambda > |\mu|$ for all other eigenvalues μ of B.

Observe that this largest (in modulus) eigenvalue coincides with the spectral radius λ_B. We use the terminology right (left) Perron eigenvector to refer to the corresponding right (left) eigenvectors, which are unique up to scale.

Observe also that if λ is a Perron number, then the companion matrix C_{f_λ} is a spectrally Perron integral matrix. In particular, the matrix C_f in (11-1-1) is spectrally Perron.

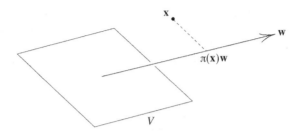

FIGURE 11.1.1.   Spectrally Perron geometry.

Let B be spectrally Perron, let $\mathbf{w}$ be a left Perron eigenvector for B, and let W be the space spanned by $\mathbf{w}$. According to the Jordan canonical form, there is a complementary B-invariant subspace V corresponding to the eigenvalues $|\mu| < \lambda$. Since $\mathbb{R}^d = V \oplus W$, we can project any $\mathbf{x} \in \mathbb{R}^d$ to W along V. Let $\pi(\mathbf{x})$ denote the multiple of $\mathbf{w}$ resulting from this operation, so that $\mathbf{x} = \pi(\mathbf{x})\mathbf{w} + \mathbf{v}$, where $\mathbf{v} \in V$ (see Figure 11.1.1). We call $\pi(\mathbf{x})$ the *dominant coordinate* of $\mathbf{x}$. Observe that all elements of V have dominant coordinate equal to zero.

The key idea of how to manufacture nonnegative matrices from spectrally Perron integral matrices while preserving spectral radius is contained in the following lemma.

Lemma 11.1.10. *Let B be a spectrally Perron integral $d \times d$ matrix. Suppose there are vectors $\mathbf{y}_1$, $\mathbf{y}_2$, $\ldots$, $\mathbf{y}_m \in \mathbb{Z}^d$ all with strictly positive dominant coordinate and nonnegative integers A_{ij} such that*

$$(11\text{--}1\text{--}2) \qquad \mathbf{y}_i B = \sum_{j=1}^{m} A_{ij}\mathbf{y}_j \qquad \text{for } 1 \leqslant i \leqslant m.$$

Then the $m \times m$ matrix $A = [A_{ij}]$ is a nonnegative integral matrix with spectral radius $\lambda_A = \lambda_B$.

PROOF: Let P be the $m \times d$ matrix defined by $\mathbf{e}_i P = \mathbf{y}_i$ for $1 \leqslant i \leqslant m$. Equation (11–1–2) shows that $AP = PB$. By Exercise 4.4.7, there is a nonnegative left eigenvector $\mathbf{u} \neq \mathbf{0}$ corresponding to λ_A, so that $\mathbf{u}A = \lambda_A \mathbf{u}$. Note that since each $\pi(\mathbf{y}_j) > 0$ and at least one coordinate of $\mathbf{u}$ is strictly positive, we have

$$\pi(\mathbf{u}P) = \pi\left(\sum_{j=1}^{m} u_j \mathbf{e}_j P\right) = \sum_{j=1}^{m} u_j \pi(\mathbf{y}_j) > 0,$$

so that $\mathbf{z} = \mathbf{u}P \neq \mathbf{0}$. Since $\mathbf{z}B = \mathbf{u}PB = \mathbf{u}AP = \lambda_A(\mathbf{u}P) = \lambda_A \mathbf{z}$, it follows that $\mathbf{z}$ is an eigenvector of B with eigenvalue λ_A. But the only eigenvectors of B having positive dominant coordinate lie in W, so that $\mathbf{z}$ is a multiple of $\mathbf{w}$. Hence $\lambda_A \mathbf{z} = \mathbf{z}B = \lambda_B \mathbf{z}$, so that $\lambda_A = \lambda_B$. $\qquad\square$

The rest of the proof of Theorem 11.1.4 consists of showing how to find the integral vectors $\mathbf{y}_j$ required to apply Lemma 11.1.10 and arranging that the resulting matrix A be primitive. To describe how to do this, we first introduce some notation and terminology.

Let $\lambda \in \mathbb{P}$ with minimal polynomial $f_\lambda(t)$, and let B be any spectrally Perron integral matrix with $\lambda_B = \lambda$ (for instance, B could be the companion matrix of $f_\lambda(t)$). Let d denote the size of B. As before, fix a left Perron eigenvector $\mathbf{w}$ for B. Let W be the span of $\mathbf{w}$, and let V be the complementary B-invariant subspace. Define $\pi_W \colon \mathbb{R}^d \to W$ to be projection to W along V, and $\pi_V \colon \mathbb{R}^d \to V$ as projection to V along W. Hence $\mathbf{x} = \pi_W(\mathbf{x}) + \pi_V(\mathbf{x})$ for all $\mathbf{x} \in \mathbb{R}^d$. Observe that

$$\pi_W(\mathbf{x}) = \pi(\mathbf{x})\mathbf{w}.$$

Note that since V and W are B-invariant subspaces, we have

$$\pi_W(\mathbf{x})B = \pi_W(\mathbf{x}B) \text{ and } \pi_V(\mathbf{x})B = \pi_V(\mathbf{x}B).$$

We will need to make some estimates using a norm. While any norm will do, we find it convenient to use the norm:

$$\|\mathbf{x}\| = \|\pi_W(\mathbf{x})\|_W + \|\pi_V(\mathbf{x})\|_V$$

where $\| \cdot \|_W$ and $\| \cdot \|_V$ are arbitrary norms on W and V. We assume that $\mathbf{w}$ is normalized so that $\|\mathbf{w}\| = 1$.

For $\theta > 0$ let

$$K_\theta = \{\mathbf{x} \in \mathbb{R}^d : \|\pi_V(\mathbf{x})\| < \theta\pi(\mathbf{x})\},$$

which is an open cone surrounding the ray through $\mathbf{w}$ (see Figure 11.1.2).

The following lemma shows that whenever $\mathbf{x}$ has strictly positive dominant coordinate, $\mathbf{x}B^n$ becomes closer and closer in direction to the left Perron eigenvector $\mathbf{w}$.

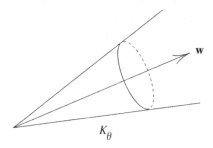

FIGURE 11.1.2. The open cone K_θ.

Lemma 11.1.11. *Let B be a spectrally Perron matrix. Fix $\theta > 0$. If $\mathbf{x}$ has strictly positive dominant coordinate, then $\mathbf{x}B^n$ is eventually contained in K_θ.*

PROOF: Since the spectral radius of B restricted to V is strictly less than λ_B, it follows from the Jordan canonical form that the growth rate of B on V is strictly less than λ_B; i.e., for $\mathbf{v} \in V$ we have $\|\mathbf{v}B^n\|/\lambda_B^n \to 0$ as $n \to \infty$ (Exercise 4.5.10). Since $\pi_V(\mathbf{x}) \in V$ it follows that there is an n_0 such that for $n \geqslant n_0$ we have

$$\|\pi_V(\mathbf{x}B^n)\| = \|\pi_V(\mathbf{x})B^n\| < \theta \lambda_B^n \pi(\mathbf{x}) = \theta\pi(\mathbf{x}B^n). \qquad \square$$

We will build our set of vectors $\mathbf{y}_1, \mathbf{y}_2, \ldots, \mathbf{y}_m \in \mathbb{Z}^d$, as in Lemma 11.1.10, by starting with *all* integral vectors with positive dominant coordinate within a certain distance from the origin: for $R > 0$ let

$$\Omega_R = \{\mathbf{z} \in \mathbb{Z}^d : \pi(\mathbf{z}) > 0 \quad \text{and} \quad \|\mathbf{z}\| \leqslant R\}.$$

By Lemma 11.1.11, given $\theta > 0$, for each $\mathbf{z} \in \Omega_R$ we know that $\mathbf{z}B^n \in K_\theta$ for all large enough n. The purpose of the following lemma is to show that *all* integral vectors in K_θ are nonnegative integral combinations of vectors in Ω_R provided we take R large enough. For large enough n, we will then use the collection $\{\mathbf{z}B^j : 0 \leqslant j \leqslant n-1, \mathbf{z} \in \Omega_R\}$ for our set of vectors $\mathbf{y}_1$, $\mathbf{y}_2, \ldots, \mathbf{y}_m \in \mathbb{Z}^d$ as in Lemma 11.1.10.

In the following lemma, we use the notation $\langle\Omega_R\rangle_+$ for the set of all nonnegative integral combinations of elements in Ω_R.

Lemma 11.1.12. *Let $\theta > 0$. For all sufficiently large R,*

$$K_\theta \cap \mathbb{Z}^d \subseteq \langle\Omega_R\rangle_+.$$

PROOF: First we claim that for large enough $r_0 > 0$ and $s_0 > 0$ every translate of the solid "disk"

$$E = \{\mathbf{y} \in \mathbb{R}^d : \|\pi_V(\mathbf{y})\| \leqslant r_0, 0 \leqslant \pi(\mathbf{y}) \leqslant s_0\}$$

contains at least one integer point. To see this, first recall from §6.1 (Exercise 6.1.2) that the metrics defined by any two norms on the same space are equivalent. Thus, for large enough r_0 and s_0, E contains an ordinary ball of Euclidean radius $\sqrt{d}$. Now, any such ball contains an integer point and so every translate of E contains an integer point.

Next put $r = 2r_0/\theta$ and $K_\theta(r) = \{\mathbf{x} \in K_\theta : \pi(\mathbf{x}) \leqslant r\}$. We claim that if $\mathbf{x} \in K_\theta \setminus K_\theta(r)$, then there is a translate of E containing $\mathbf{x}$ and contained in K_θ (see Figure 11.1.3).

If $\|\pi_V(\mathbf{x})\| \leqslant r_0$, then the translate $\mathbf{x} + E$ works because if $\mathbf{y} \in E$, then

$$\|\pi_V(\mathbf{x} + \mathbf{y})\| \leqslant \|\pi_V(\mathbf{x})\| + \|\pi_V(\mathbf{y})\| \leqslant r_0 + r_0$$
$$= \theta r < \theta\pi(\mathbf{x}) \leqslant \theta\pi(\mathbf{x} + \mathbf{y}).$$

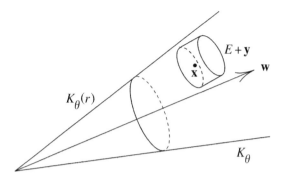

FIGURE 11.1.3. Translation of disk fitting in to a cone.

If $\|\pi_V(\mathbf{x})\| > r_0$, then the translate

$$\mathbf{x} - \frac{r_0}{\|\pi_V(\mathbf{x})\|}\pi_V(\mathbf{x}) + E$$

works because if $\mathbf{y} \in E$, then

$$\left\|\pi_V\left(\mathbf{x} - \frac{r_0}{\|\pi_V(\mathbf{x})\|}\pi_V(\mathbf{x}) + \mathbf{y}\right)\right\| \leqslant \left(1 - \frac{r_0}{\|\pi_V(\mathbf{x})\|}\right)\|\pi_V(\mathbf{x})\| + r_0$$

$$= \|\pi_V(\mathbf{x})\|$$

$$< \theta\pi(\mathbf{x})$$

$$\leqslant \theta\pi\left(\mathbf{x} - \frac{r_0}{\|\pi_V(\mathbf{x})\|}\pi_V(\mathbf{x}) + \mathbf{y}\right),$$

the last inequality following from $\pi(\pi_V(\mathbf{x})) = 0$.

Finally, let $r' = r + 2s_0$ and choose R large enough so that $K_\theta(r') \cap \mathbb{Z}^d \subset \Omega_R$ and $R > 2r_0 + 2\theta s_0 + 3s_0$. We show that $K_\theta \cap \mathbb{Z}^d \subset \langle \Omega_R \rangle_+$ by inductively proving that $K_\theta(r' + ns_0) \cap \mathbb{Z}^d \subset \langle \Omega_R \rangle_+$ for $n = 0, 1, 2 \ldots$.

For $n = 0$, this is true by our choice of R. Suppose inductively that $K_\theta(r' + ns_0) \cap \mathbb{Z}^d \subset \langle \Omega_R \rangle_+$. Let

$$\mathbf{z} \in \left[K_\theta(r' + (n+1)s_0) \setminus K_\theta(r' + ns_0)\right] \cap \mathbb{Z}^d.$$

Then

$$\mathbf{x} = \mathbf{z} - \frac{2s_0}{\pi(\mathbf{z})}\mathbf{z} \in K_\theta(r' + (n-1)s_0) \setminus K_\theta(r' + (n-2)s_0)$$

$$\subset K_\theta(r' + (n-1)s_0) \setminus K_\theta(r).$$

By the discussion above, some translate of E is contained in K_θ and contains both $\mathbf{x}$ and an integral point $\mathbf{y}$ (see Figure 11.1.4). Since

$$\pi(\mathbf{y}) \leqslant \pi(\mathbf{x}) + s_0 \leqslant r' + (n-1)s_0 + s_0 = r' + ns_0,$$

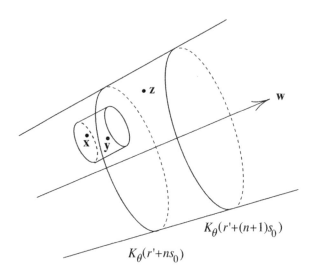

FIGURE 11.1.4. Inductive generation of lattice points.

we see that $\mathbf{y} \in K_\theta(r' + ns_0)$. Our inductive hypothesis shows that $\mathbf{y} \in \langle \Omega_R \rangle_+$. Since $\mathbf{z} = \mathbf{y} + (\mathbf{z} - \mathbf{y})$, it remains only to show that $\mathbf{z} - \mathbf{y} \in \langle \Omega_R \rangle_+$. To see this, first observe that

$$\pi(\mathbf{z} - \mathbf{y}) = \pi(\mathbf{z}) - \pi(\mathbf{y}) > (r' + ns_0) - (r' + ns_0) = 0,$$

so $\mathbf{z} - \mathbf{y}$ has strictly positive dominant coordinate. Also, since the diameter of E in the metric defined by our norm is at most $2r_0 + s_0$, we see that

$$\begin{aligned}
\|\mathbf{z} - \mathbf{y}\| &\leqslant \|\mathbf{z} - \mathbf{x}\| + \|\mathbf{x} - \mathbf{y}\| \\
&\leqslant 2s_0 \frac{\|\mathbf{z}\|}{\pi(\mathbf{z})} + 2r_0 + s_0 \\
&= 2s_0 \frac{\|\pi(\mathbf{z})\mathbf{w} + \pi_V(\mathbf{z})\|}{\pi(\mathbf{z})} + 2r_0 + s_0 \\
&\leqslant 2s_0(1 + \theta) + 2r_0 + s_0 \\
&< R.
\end{aligned}$$

Hence $\mathbf{z} - \mathbf{y} \in \Omega_R$ and $\mathbf{z} = \mathbf{y} + (\mathbf{z} - \mathbf{y})$ is in $\langle \Omega_R \rangle_+$. This proves that $K_\theta(r' + (n + 1)s_0) \cap \mathbb{Z}^d \subset \langle \Omega_R \rangle_+$, completing the inductive step and the proof.   □

PROOF OF THEOREM 11.1.4: Let $\lambda \in \mathbb{P}$, let B be a spectrally Perron integral matrix with $\lambda_B = \lambda$, and carry over the meanings of V, W, π_V, π_W, and π from our previous discussion. If $\lambda = 1$, use $A = [1]$. So we may assume that $\lambda > 1$. Fix $\theta > 0$ and choose R such that $K_\theta \cap \mathbb{Z}^d \subset \langle \Omega_R \rangle_+$ as in Lemma 11.1.12.

Since B is spectrally Perron, if $\mathbf{z}$ has strictly positive dominant coordinate, then $\mathbf{z}B^n \in K_\theta$ for large enough n by Lemma 11.1.11. In fact, since $\lambda > 1$ and $\pi(\mathbf{z}) > 0$, $\pi(\mathbf{z}B - \mathbf{z}) = (\lambda - 1)\pi(\mathbf{z}) > 0$, so again by Lemma 11.1.11, we have $\mathbf{z}B^n - \mathbf{z}B^{n-1} \in K_\theta$ for large enough n.

List the elements of Ω_R as $\mathbf{z}_1, \mathbf{z}_2, \dots, \mathbf{z}_k$. Choose n such that $\mathbf{z}_j B^n - \mathbf{z}_j B^{n-1} \in K_\theta$ for $1 \leqslant j \leqslant k$. Define $\mathbf{z}_{i,j} = \mathbf{z}_j B^i$, and consider the set of vectors $\{\mathbf{z}_{i,j} : 0 \leqslant i \leqslant n-1, 1 \leqslant j \leqslant k\}$. These vectors will play the role of the vectors $\mathbf{y}_1, \mathbf{y}_2, \dots, \mathbf{y}_m \in \mathbb{Z}^d$ in Lemma 11.1.10. Observe that each $\mathbf{z}_{i,j}$ has strictly positive dominant coordinate. We will show that each $\mathbf{z}_{p,q}B$ can be written as a nonnegative integral combination of the $\mathbf{z}_{i,j}$. When $0 \leqslant i \leqslant n-2$ this is easy:

$$\mathbf{z}_{i,j}B = \mathbf{z}_{i+1,j}.$$

For $i = n-1$, observe that since $\mathbf{z}_j B^n - \mathbf{z}_j B^{n-1} \in K_\theta \cap \mathbb{Z}^d \subset \langle \Omega_R \rangle_+$, there are nonnegative integers $C_{j,\ell}$ such that

$$\mathbf{z}_j B^n - \mathbf{z}_j B^{n-1} = \sum_{\ell=1}^{k} C_{j,\ell} \mathbf{z}_\ell.$$

We can then write

$$\mathbf{z}_{n-1,j}B = \mathbf{z}_j B^n = \mathbf{z}_j B^{n-1} + \sum_{\ell=1}^{k} C_{j,\ell} \mathbf{z}_\ell = \mathbf{z}_{n-1,j} + \sum_{\ell=1}^{k} C_{j,\ell} \mathbf{z}_\ell.$$

Let A be the matrix which expresses how, with these choices, each $\mathbf{z}_{p,q}B$ is written as a nonnegative integral combination of the $\mathbf{z}_{i,j}$, i.e., A is the matrix indexed by $\{(i,j) : 0 \leqslant i \leqslant n-1, 1 \leqslant j \leqslant k\}$ and defined by

$$A_{(i,j),(i+1,j)} = 1 \qquad \text{for } 0 \leqslant i \leqslant n-2,$$
$$A_{(n-1,j),(n-1,j)} = 1,$$
$$A_{(n-1,j),(0,\ell)} = C_{j,\ell},$$

with all other entries 0. According to Lemma 11.1.10, $\lambda_A = \lambda$. So if A were primitive, then we would be done.

But A need not even be irreducible. We will remedy this by replacing A with one of its irreducible components. To see how to do this, first observe that whenever an index (i,j) belongs to an essential irreducible component of A, then the index $(n-1,j)$, and hence also the self-loop at $(n-1,j)$, belongs to the same irreducible component. Thus, any irreducible component of A with maximal entropy is primitive. And by replacing A with such an irreducible component, we still have $\lambda_A = \lambda$. $\square$

We conclude this section by connecting what we have done thus far to the dimension groups introduced in Chapter 7. Recall from §7.5 that we associated to each $n \times n$ integral matrix A its dimension group Δ_A defined by

$$\Delta_A = \{\mathbf{v} \in \mathcal{R}_A : \mathbf{v}A^k \in \mathbb{Z}^n \quad \text{for all large enough } k\},$$

and an automorphism $\delta_A \colon \Delta_A \to \Delta_A$ defined by $\delta_A(\mathbf{v}) = \mathbf{v}A$. We will be interested in quotients of dimension pairs.

Definition 11.1.13. For $i = 1, 2$ let Δ_i be an abelian group and $\delta_i \colon \Delta_i \to \Delta_i$ an automorphism of Δ_i. Say that (Δ_2, δ_2) is a *quotient* of (Δ_1, δ_1) if there is an onto group homomorphism $\psi \colon \Delta_1 \to \Delta_2$ such that $\psi \circ \delta_1 = \delta_2 \circ \psi$. Such a ψ is called a *quotient mapping*.

The following relates quotients of dimension pairs to matrix equations.

Proposition 11.1.14. *Let A and B be integral matrices. Then (Δ_B, δ_B) is a quotient of (Δ_A, δ_A) if and only if there are integral matrices R and S and an $\ell \geqslant 0$ such that $AR = RB$ and $SR = B^\ell$.*

PROOF: The matrix equations are "half" of a shift equivalence from A to B with lag ℓ. The proof is a modification of the proof of Theorem 7.5.7, which we leave to the reader. □

In the following result, we will build on our proof of Theorem 11.1.4 to show that the dimension pair of a spectrally Perron integral matrix is a quotient of the dimension pair of a primitive integral matrix.

Theorem 11.1.15. *Let B be a spectrally Perron integral matrix. Then there is a primitive integral matrix A such that $\lambda_A = \lambda_B$ and (Δ_B, δ_B) is a quotient of (Δ_A, δ_A).*

PROOF: We modify the proof of Theorem 11.1.4 to construct the required primitive matrix A.

Let $\theta' > \theta$, so that $K_{\theta'} \supset K_\theta$. By Lemma 11.1.12 there is an $R > 0$ so that $K_{\theta'} \cap \mathbb{Z}^d \subset \langle \Omega_R \rangle_+$. As in the proof of Theorem 11.1.4, for sufficiently large n we have that

$$\mathbf{z}_j B^n - \mathbf{z}_j B^{n-1} \in K_\theta \cap \mathbb{Z}^d.$$

But this implies that, for sufficiently large n,

$$\mathbf{z}_j B^n - \mathbf{z}_j B^{n-1} - \sum_{\ell=1}^{k} \mathbf{z}_\ell \in K_{\theta'} \cap \mathbb{Z}^d \subset \langle \Omega_R \rangle_+.$$

Hence there are nonnegative integers $C_{j,\ell}$ such that

$$(11\text{--}1\text{--}3) \qquad \mathbf{z}_{n-1,j} B = \mathbf{z}_{n-1,j} + \sum_{\ell=1}^{k} C_{j,\ell} \mathbf{z}_\ell + \sum_{\ell=1}^{k} \mathbf{z}_\ell.$$

Let A be the matrix produced as in the proof of Theorem 11.1.4 using the expression (11–1–3). Since $A_{(n-1,j),(0,\ell)} = C_{j,\ell} + 1 > 0$, it follows that A is irreducible, hence primitive, since it has positive trace. We can also take the R in that proof large enough so that Ω_R generates $\mathbb{Z}^d$, since for large enough R, every standard basis vector is the difference of two elements of Ω_R. The $(nk) \times d$ matrix P defined by $\mathbf{e}_{ij}P = \mathbf{z}_{ij}$ satisfies $AP = PB$ as in the proof of Lemma 11.1.10. Since Ω_R is in $\mathbb{Z}^{nk}P$ and generates $\mathbb{Z}^d$, it follows that $\mathbb{Z}^{nk}P = \mathbb{Z}^d$. Hence for each standard basis vector $\mathbf{e}_j \in \mathbb{Z}^d$ there is a vector $\mathbf{v}_j \in \mathbb{Z}^{nk}$ with $\mathbf{v}_jP = \mathbf{e}_j$. Let S be the $d \times (nk)$ matrix whose jth row is $\mathbf{v}_j$. Then $SP = Id$. By Proposition 11.1.14 with $\ell = 0$, we see that (Δ_B, δ_B) is a quotient of (Δ_A, δ_A). $\square$

We remark that there is a converse to the preceding result; see Exercise 11.1.11.

Can Theorem 11.1.15 be strengthened to show that if B is a spectrally Perron integral matrix, then there is a primitive integral matrix such that (Δ_B, δ_B) is isomorphic to (Δ_A, δ_A), instead of merely a quotient? Equivalently, is there a primitive integral A such that $B \sim_{\mathbb{Z}} A$? The following example shows the answer is no.

Example 11.1.16. Let B be the companion matrix of $t^3 + 3t^2 - 15t - 46$. Then by Example 11.1.1(d), B is spectrally Perron. Suppose that A were a nonnegative integral matrix with $B \sim_{\mathbb{Z}} A$. Since the nonzero spectrum is an invariant of $\sim_{\mathbb{Z}}$ (see Corollary 7.4.12), it follows that A and B would then have the same nonzero spectrum, hence the same trace. But $\operatorname{tr} B = -3$ while $\operatorname{tr} A \geqslant 0$. Hence there can be no nonnegative integral (and hence no primitive integral) matrix A for which $(\Delta_A, \delta_A) \simeq (\Delta_B, \delta_B)$. $\square$

On the other hand, in Theorem 11.1.19 below, we will see that it is possible to realize dimension pairs of spectrally Perron matrices by matrices that satisfy the following weaker positivity property:

Definition 11.1.17. A matrix A (with possibly some entries negative) is called *eventually positive* if $A^N > 0$ for all sufficiently large N.

Recall from Theorem 4.5.8 that a nonnegative matrix A is primitive if and only if $A^N > 0$ for all large enough N. So primitive matrices are eventually positive.

According to Exercise 4.5.11 Perron–Frobenius theory for primitive matrices carries over to eventually positive matrices. In particular, eventually positive matrices are spectrally Perron, so it makes sense to speak of Perron eigenvectors for eventually positive matrices. Exercise 11.1.9 gives a criterion, in terms of eigenvalues and eigenvectors, for a matrix to be eventually positive.

Example 11.1.18. (a) Let $A = \begin{bmatrix} 2 & 1 \\ 1 & -1 \end{bmatrix}$. Computation reveals that A^3

has a negative entry but A^4, A^5, A^6, A^7 are all positive. Since any power $\geqslant 5$ of A can be written as the product of a power of A^4 and one of the matrices A^5, A^6, A^7, we see that $A^N > 0$ for all $N \geqslant 4$. Hence A is eventually positive.

(b) Let B be the companion matrix of $t^3 + 3t^2 - 15t - 46$. It turns out that B is eventually positive (see Exercise 11.1.9), but this requires large powers since $(B^{48})_{11} < 0$, as the reader can check.

(c) Let $A = \begin{bmatrix} -1 & -1 \\ -1 & -1 \end{bmatrix}$. Then $A^{2n} > 0$ for all $n \geqslant 1$, but $A^{2n+1} < 0$ for $n \geqslant 0$. Hence A is not eventually positive, even though infinitely many of its powers are positive. $\square$

Theorem 11.1.19. *Let B be an integral matrix. Then B is spectrally Perron if and only if there is an eventually positive integral matrix A with $A \sim_{\mathbb{Z}} B$.*

For example, the matrix B in Example 11.1.16 is not shift equivalent over $\mathbb{Z}$ to a primitive matrix, but according to Example 11.1.18(b) it *is* shift equivalent over $\mathbb{Z}$ to an eventually positive integral matrix, namely itself.

We will outline the proof of Theorem 11.1.19, referring the reader to Handelman [Han1], [Han2] for the complete proof.

As mentioned above, an eventually positive matrix is spectrally Perron. Since $\sim_{\mathbb{Z}}$ preserves nonzero spectrum, it follows that if $B \sim_{\mathbb{Z}} A$ with A eventually positive then B is spectrally Perron. The converse splits into two cases, depending on whether λ_B is irrational ([Han1]) or rational ([Han2]).

First suppose that λ_B is irrational. Let B be $d \times d$. Handelman [Han1] showed (and this is the key to the proof in this case) that there are (row) vectors $\mathbf{y}_1, \mathbf{y}_2, \ldots, \mathbf{y}_d \in \mathbb{Z}^d$ such that (1) integral combinations of the $\mathbf{y}_j$ exhaust all of $\mathbb{Z}^d$, (2) each $\mathbf{y}_j$ has strictly positive dominant coordinate, and (3) the left Perron eigenvector is a positive combination of the $\mathbf{y}_i$. Let P be the $d \times d$ matrix whose jth row is $\mathbf{y}_j$. Let S be the $d \times d$ matrix whose Ith row is the vector of coefficients which expresses $\mathbf{e}_I$ as an integral combination of the $\mathbf{y}_j$. Then $SP = Id$ and $P^{-1} = S$ is integral. Hence $A = PBP^{-1}$ is shift equivalent over $\mathbb{Z}$ to B.

It remains to show that A is eventually positive. This is equivalent to showing that for $1 \leqslant j \leqslant d$, $\mathbf{y}_j B^n$ is eventually a positive combination of the $\mathbf{y}_i$. By (2) and Lemma 11.1.11 for any $\theta > 0$, each $\mathbf{y}_j B^n$ is eventually in K_θ. It then remains to show that for sufficiently small θ, every element of K_θ is a positive combination of the $\mathbf{y}_i$. To see this, first observe that the set of positive combinations of the $\mathbf{y}_i$ is an open cone K (with polygonal cross-section). By (3), the left Perron eigenvector lies in the interior of K. Thus, for sufficiently small θ, K_θ is contained in K, and so every element of K_θ is a positive combination of the $\mathbf{y}_i$.

This proof shows that if λ_B is irrational, then B is actually similar over

$\mathbb{Z}$ (rather than merely shift equivalent over $\mathbb{Z}$) to an eventually positive matrix. This may fail, however, when λ_B is rational (hence an integer); the following shows that there is an additional condition imposed by similarity over $\mathbb{Z}$. Suppose that P is invertible over $\mathbb{Z}$ and $A = PBP^{-1}$ is eventually positive. Then A would have positive left and right Perron eigenvectors, say $\mathbf{w}$ and $\mathbf{v}$, which we can assume are integral and minimal (an integer vector is minimal if the greatest common divisor of its entries is 1). It follows that $\mathbf{w}P, P^{-1}\mathbf{v}$ are minimal left, right Perron eigenvectors for B. And since $\mathbf{w} \cdot \mathbf{v} \geqslant d$, we also have $(\mathbf{w}P) \cdot (P^{-1}\mathbf{v}) = \mathbf{w} \cdot \mathbf{v} \geqslant d$. So we conclude that B has minimal left, right Perron eigenvectors whose dot product is at least d. For a spectrally Perron matrix with integral spectral radius, a minimal left (respectively, right) Perron eigenvector is unique up to multiplication by ± 1, so the foregoing argument establishes the following result.

Lemma 11.1.20. *Let B be a spectrally Perron integral $d \times d$ matrix. Suppose that λ_B is a positive integer and let $\mathbf{w}$ and $\mathbf{v}$ be minimal integral left and right Perron eigenvectors of B. If B is similar over $\mathbb{Z}$ to an eventually positive matrix, then $|\mathbf{w} \cdot \mathbf{v}| \geqslant d$.*

Example 11.1.21. Let $B = \begin{bmatrix} 2 & 0 \\ 0 & 1 \end{bmatrix}$, which is spectrally Perron. However, $\mathbf{w} = \pm [\,1 \quad 0\,]$ and $\mathbf{v} = \pm \begin{bmatrix} 1 \\ 0 \end{bmatrix}$ are the minimal Perron eigenvectors, and $|\mathbf{w} \cdot \mathbf{v}| = 1 < 2$. Hence B cannot be similar over $\mathbb{Z}$ to an eventually positive matrix. $\qquad\square$

In [Han2], Handelman proves the converse of Lemma 11.1.20, so that, in the notation of the Lemma, if $|\mathbf{w} \cdot \mathbf{v}| \geqslant d$, then B is similar over $\mathbb{Z}$ to an eventually positive matrix. He also proves that if B is spectrally Perron, then there is a sequence of integral matrices

$$B \sim_{\mathbb{Z}} B_1 \sim_{\mathbb{Z}} B_2 \sim_{\mathbb{Z}} \cdots \sim_{\mathbb{Z}} B_k$$

where B_k has minimal integral Perron eigenvectors $\mathbf{w}, \mathbf{v}$ with $|\mathbf{w} \cdot \mathbf{v}| \geqslant d$. So, by the converse to Lemma 11.1.20, B is shift equivalent over $\mathbb{Z}$ to an eventually positive integral matrix as desired.

EXERCISES

11.1.1. Show that the characteristic polynomial of the companion matrix of a polynomial $f(t)$ is $f(t)$ itself.

11.1.2. Which of the following are Perron numbers?

(a) $\sqrt{5}$, (b) $\dfrac{1+\sqrt{5}}{2}$, (c) $\dfrac{1+\sqrt{3}}{2}$, (d) $\dfrac{\sqrt{5}}{2}$, (e) $\sqrt{1+\sqrt{2}}$, (f) $\dfrac{3-\sqrt{5}}{2}$.

11.1.3. Show that any algebraic integer $\lambda > 0$ that dominates its other algebraic conjugates must satisfy $\lambda \geqslant 1$.

11.1.4. Using the method of this section, find a primitive integral matrix whose spectral radius is the largest root of $t^3 + 3t^2 - 15t - 46$ from Example 11.1.1(d).

11.1.5. Let B be the companion matrix of $t^3 + 3t^2 - 15t - 46$. Let $\{\mathbf{y}_1, \ldots, \mathbf{y}_6\}$ be the set of six vectors $\pm\mathbf{e}_1$, $\pm\mathbf{e}_2$, $\pm\mathbf{e}_3$, and A a nonnegative integral matrix defined by equation (11–1–2). Explain why λ_A need not equal λ_B and thus why the hypothesis in Lemma 11.1.10 is necessary.

11.1.6. Let B be a spectrally Perron matrix. As in the text, let W be the space spanned by a left Perron eigenvector $\mathbf{w}$, let V be the B-invariant subspace corresponding to the other eigenvalues, and let π_W denote the projection onto W along V. Also, let $\mathbf{v}$ be a right eigenvector corresponding to the dominant eigenvalue.

(a) Show that V is the orthogonal complement of the span of $\mathbf{v}$.

(b) Show that
$$\pi_W(\mathbf{x}) = \left(\frac{\mathbf{x} \cdot \mathbf{v}}{\mathbf{w} \cdot \mathbf{v}} \right) \mathbf{w}.$$

11.1.7. Prove Proposition 11.1.14 and show that if A and B are eventually positive and $\lambda_A = \lambda_B$, then the matrices R and S in Proposition 11.1.14 can be chosen to be strictly positive. [*Hint*: Modify the proof of Theorem 7.5.7.]

11.1.8. Show that an irreducible matrix is primitive if and only if it is spectrally Perron. [*Hint*: See Exercise 4.5.13.]

11.1.9. (a) Show that a matrix is eventually positive if and only if it is spectrally Perron and has positive right and left Perron eigenvectors. [*Hint*: Verify that the conclusion of Theorem 4.5.12 holds for a matrix that is spectrally Perron and has positive right and left Perron eigenvectors.]

(b) Use this to show that the companion matrix C of the polynomial $t^3 + 3t^2 - 15t - 46$ is eventually positive.

11.1.10. Which of the following matrices are spectrally Perron? eventually positive?

$$\text{(a)} \begin{bmatrix} 0 & 1 \\ 1 & 0 \end{bmatrix}, \quad \text{(b)} \begin{bmatrix} 1 & 1 \\ 1 & 0 \end{bmatrix}, \quad \text{(c)} \begin{bmatrix} 1 & 1 \\ 1 & -1 \end{bmatrix},$$

$$\text{(d)} \begin{bmatrix} 2 & 1 \\ 1 & -1 \end{bmatrix}, \quad \text{(e)} \begin{bmatrix} 2 & -1 \\ -1 & -1 \end{bmatrix}.$$

11.1.11. Prove the following converse to Theorem 11.1.15: Let B be an integral matrix. If there is a primitive integral matrix A such that $\lambda_A = \lambda_B$ and (Δ_B, δ_B) is a quotient of (Δ_A, δ_A), then B is spectrally Perron.

11.1.12. (a) Show that the following are equivalent for a number λ.

    (i) λ is an algebraic integer

    (ii) λ^p is an algebraic integer for some positive integer p

    (iii) λ^p is an algebraic integer for all positive integers p

(b) Call a number λ *almost Perron* if it is an algebraic integer $\geqslant 1$ such that if μ is a conjugate of λ, then $|\mu| \leqslant \lambda$. For each integer $p \geqslant 1$, show that λ is almost Perron if and only if λ^p is almost Perron.

(c) Which of the numbers in Exercise 11.1.2 are almost Perron?

§11.2. Realization of Zeta Functions

In the previous section we characterized the set of entropies of shifts of finite type. What are their possible zeta functions? Recall from §6.4 that the zeta function of an edge shift X_A is equivalent to the nonzero spectrum $\mathrm{sp}^\times(A)$ of A, in the sense that each determines the other. Hence we are really asking to characterize the nonzero spectra of nonnegative integral matrices. As in §11.1, we shall focus on the primitive case. We first review some notation and terminology.

Recall from the end of §10.1 that a list $\Lambda = \{\lambda_1, \lambda_2, \ldots, \lambda_k\}$ is a collection of complex numbers where the order of listed elements is irrelevant but repeated elements count. Recall that for a list $\Lambda = \{\lambda_1, \lambda_2, \ldots, \lambda_k\}$ we defined

$$\mathrm{tr}(\Lambda) = \sum_{j=1}^{k} \lambda_j \quad \text{and}$$
$$\Lambda^n = \{\lambda_1^n, \lambda_2^n, \ldots, \lambda_k^n\}.$$

Let

$$f_\Lambda(t) = \prod_{j=1}^{k}(t - \lambda_j).$$

The main question we study in this section is the *Spectral Problem:* Characterize those lists Λ of nonzero complex numbers for which there is a primitive integral matrix A such that $\Lambda = \mathrm{sp}^\times(A)$.

We describe three necessary conditions (conditions 11.2.1, 11.2.2, 11.2.3 below) on Λ for it to be the nonzero spectrum of a primitive integral matrix A.

First observe that $\chi_A(t) = t^m f_\Lambda(t)$ where m is the multiplicity of the eigenvalue 0 for A. Since $\chi_A(t)$ is monic and has integer coefficients, the same holds for $f_\Lambda(t)$:

Condition 11.2.1 (Integrality Condition). $f_\Lambda(t)$ *is a monic polynomial (with integer coefficients).*

A monic polynomial having integer coefficients factors into irreducible monic polynomials with integer coefficients. It follows from this that the Integrality Condition is equivalent to requiring that every element λ in Λ be an algebraic integer and that all of its algebraic conjugates occur with the same multiplicity as λ.

By the Perron–Frobenius Theorem, if $\Lambda = \mathrm{sp}^\times(A)$, then λ_A appears in Λ, but just once. Furthermore, λ_A is strictly larger than all other entries in the list. Hence Λ must also satisfy the following condition:

Condition 11.2.2 (Perron Condition). *There is a positive entry in*
Λ*, occurring just once, that strictly dominates in absolute value all other*
entries. We denote this entry by λ_Λ.

Observe that Λ satisfies the Perron Condition if and only if the companion
matrix of $f_\Lambda(t)$ is spectrally Perron.

Recall from Definition 10.1.10 that the nth net trace of a list Λ is defined
to be

$$\mathrm{tr}_n(\Lambda) = \sum_{d|n} \mu\left(\frac{n}{d}\right) \mathrm{tr}(\Lambda^d)$$

where μ is the Möbius function. Recall also that if Λ is the nonzero spectrum of a nonnegative integral matrix A, then $\mathrm{tr}_n(\Lambda) = q_n(X_A) \geqslant 0$ for all
$n \geqslant 1$. This leads to the following condition.

Condition 11.2.3 (Net Trace Condition). $\mathrm{tr}_n(\Lambda) \geqslant 0$ *for all* $n \geqslant 1$.

The Integrality, Perron, and Net Trace Conditions are referred to collec-
tively as the *Spectral Conditions*. We have shown that they are necessary
for the Spectral Problem. The *Spectral Conjecture*, due to Boyle and Handelman [BoyH1], asserted that the spectral conditions are sufficient as well.
Kim, Ormes, and Roush [KimOR] proved the Spectral Conjecture.

Theorem 11.2.4. *Let* Λ *be a list of nonzero complex numbers satisfying*
the Integrality, Perron, and Net Trace Conditions. Then there is a primitive
integral matrix A *for which* $\Lambda = \mathrm{sp}^\times(A)$.

The proof of Theorem 11.2.4 is an amazing use of formal power series.
It is algorithmic in the sense that it is capable of producing a matrix that
satisfies the conclusion of the theorem. It does not make direct use of symbolic dynamics, but it was conjectured based on ideas in symbolic dynamics,
which provided tools to earlier prove several special cases. These tools were
used by Boyle and Handelman [BoyH1] to give a complete solution to the
analogous problem of characterizing the nonzero spectra of matrices with
nonnegative real entries, which represented a breakthrough in a classical
problem in linear algebra.

The proof of Theorem 11.2.4 is too involved to include here. Instead,
we will prove special cases using, in part, the Embedding Theorem and
Masking Lemma from chapter 10, illustrating the use of symbolic tools
used by Boyle and Handelman.

Example 11.2.5. (a) Let $\Lambda = \{2, 1\}$. This Λ satisfies the Spectral Con-
ditions (for the Net Trace Condition, observe that $\mathrm{tr}_n(\Lambda) = q_n(X_{[2]}) + q_n(X_{[1]})$). So by Theorem 11.2.4 there is a primitive integral A for which
$\Lambda = \mathrm{sp}^\times(A)$. Unfortunately, the natural choice $A = \begin{bmatrix} 2 & 0 \\ 0 & 1 \end{bmatrix}$ is not primi-

tive. In Example 11.2.10 below we will show that

$$A = \begin{bmatrix} 1 & 0 & 0 & 1 \\ 1 & 0 & 1 & 0 \\ 0 & 1 & 1 & 0 \\ 0 & 1 & 0 & 1 \end{bmatrix}$$

is primitive and has $\mathrm{sp}^\times(A) = \{2, 1\}$.

(b) Let $\Lambda = \{6, 1, 1, 1, 1, -5, -5\}$, which satisfies the Integrality and Perron Conditions. Then we compute $\mathrm{tr}_1(\Lambda) = 0$, $\mathrm{tr}_2(\Lambda) = 90$, but $\mathrm{tr}_3(\Lambda) = -30$. Hence the Net Trace Condition fails, so Λ cannot be the nonzero spectrum of a primitive integral matrix.

(c) Let $f(t) = t^3 - 4t^2 + 6t - 6$, and $\Lambda = \{\lambda_1, \lambda_2, \lambda_3\}$ be the list of the roots of $f(t)$, where $\lambda_1 \approx 2.57474$ and $\lambda_2 = \overline{\lambda}_3 \approx 0.71263 + 1.35000i$. Then Λ clearly satisfies the Integrality Condition and the Perron Condition holds because $|\lambda_2| = |\lambda_3| \approx 1.52655$. The first ten net traces are 4, 0, 6, 36, 120, 306, 756, 1848, 4878, and 12660. This looks promising and, in fact, simple estimates show that $\mathrm{tr}_n(\Lambda) \geqslant 0$ for all n (see Example 11.2.13). Thus Λ satisfies the Net Trace Condition. Kim, Ormes, and Roush applied their algorithm to this example, finding a primitive matrix of size 179 whose nonzero spectrum is Λ.   □

Before continuing, we discuss the spectral conditions.

First, we emphasize that the Net Trace Condition is necessary for Λ to be the nonzero spectrum of a nonnegative integral matrix (not just a primitive integral matrix). So the list Λ in Example 11.2.5(b) above cannot even be realized as the nonzero spectrum of a nonnegative integral matrix.

Next, observe that if A is primitive with $\lambda_A = \lambda$, then the size of A must be at least $\deg \lambda$ since $f_\lambda(t)$ divides $\chi_A(t)$. Hence a necessary condition for there to be a primitive integral matrix of size $\deg \lambda$ and spectral radius λ is that the list of conjugates of λ satisfy the Spectral Conditions. Recall that this condition was violated in Example 11.1.7(d), where $\mathrm{tr}_1(\Lambda) = -3 < 0$. But the higher net traces, $\mathrm{tr}_n(\Lambda)$ for $n \geqslant 2$, typically provide further obstructions.

Finally, recall that $\mathrm{tr}_n(A)$ denotes the nth net trace of the nonzero spectrum of a matrix A, and the embedding periodic point condition that is necessary for the Embedding Theorem can be reformulated as

$$P(X_A) \hookrightarrow P(X_B) \iff \mathrm{tr}_n(A) \leqslant \mathrm{tr}_n(B) \text{ for all } n \geqslant 1.$$

Before proceeding, we need some more notation. For a matrix B let $|B|$ denote the corresponding matrix of absolute values, so that $|B|_{IJ} = |B_{IJ}|$. Also, let B^+ and B^- denote the positive and negative parts of B, so that B^+ and B^- are the unique nonnegative matrices satisfying $B = B^+ - B^-$ and $|B| = B^+ + B^-$. For example, if $B = \begin{bmatrix} -1 & 2 \\ 3 & -4 \end{bmatrix}$, then $B^+ = \begin{bmatrix} 0 & 2 \\ 3 & 0 \end{bmatrix}$

and $B^- = \begin{bmatrix} 1 & 0 \\ 0 & 4 \end{bmatrix}$. Clearly the positive and negative parts of an integral matrix, as well as the absolute value, are also integral.

The following Proposition is a consequence of Theorem 11.2.4, but we will give a direct proof using the Embedding Theorem and Masking Lemma.

Proposition 11.2.6. *Let A be a primitive integral matrix and B be an integral matrix such that*

(1) $\lambda_{|B|} < \lambda_A$, *and*
(2) $\mathrm{tr}_n(|B|) \leqslant \mathrm{tr}_n(A)$ *for all $n \geqslant 1$.*

Then there is a nonnegative integral matrix C such that $\mathrm{sp}^\times(C) = \mathrm{sp}^\times(A) \cup \mathrm{sp}^\times(B)$. Moreover, if B^- is primitive, then C can be chosen primitive as well.

To see why this result follows from Theorem 11.2.4 we check that the spectral conditions follow from hypotheses (1) and (2) of the Proposition. The integrality condition is obvious. The Perron condition follows from (1) and the fact that $\lambda_B \leqslant \lambda_{|B|}$ for any integral matrix (Exercise 11.2.5). For the net trace condition, observe that

$$\begin{bmatrix} B^+ & B^- \\ B^- & B^+ \end{bmatrix} = \begin{bmatrix} Id & 0 \\ Id & Id \end{bmatrix} \begin{bmatrix} |B| & B^- \\ 0 & B \end{bmatrix} \begin{bmatrix} Id & 0 \\ -Id & Id \end{bmatrix},$$

so that

$$\mathrm{tr}_n(|B|) + \mathrm{tr}_n(B) = \mathrm{tr}_n \begin{bmatrix} |B| & B^- \\ 0 & B \end{bmatrix} = \mathrm{tr}_n \begin{bmatrix} B^+ & B^- \\ B^- & B^+ \end{bmatrix} \geqslant 0$$

and so by (2),

$$tr_n(A) + tr_n(B) \geqslant tr_n(|B|) + tr_n(B) \geqslant 0$$

for all $n \geqslant 1$.

Evidently, the primitivity of B^- is not needed to obtain a primitive C; it is an artifact of our direct proof.

PROOF OF PROPOSITION 11.2.6: The hypotheses (1) and (2) are exactly what are needed to invoke the Embedding Theorem (actually Theorem 10.1.2) to show that $X_{|B|}$ embeds into X_A. The Masking Lemma then applies to show that there is an $A_1 \approx A$ with $|B|$ as a principal submatrix. Hence, by replacing A with A_1, we can assume that A has the block form

$$A = \begin{bmatrix} U & V \\ W & |B| \end{bmatrix}.$$

Let

(11–2–1) $Q = \begin{bmatrix} Id & 0 & 0 \\ 0 & Id & 0 \\ 0 & Id & Id \end{bmatrix},$

where the first diagonal block has the size of U while the second and third diagonal blocks each have the size of $|B|$. Put

$$C = Q \begin{bmatrix} U & V & 0 \\ W & |B| & B^- \\ 0 & 0 & B \end{bmatrix} Q^{-1}$$

$$= \begin{bmatrix} Id & 0 & 0 \\ 0 & Id & 0 \\ 0 & Id & Id \end{bmatrix} \begin{bmatrix} U & V & 0 \\ W & |B| & B^- \\ 0 & 0 & B \end{bmatrix} \begin{bmatrix} Id & 0 & 0 \\ 0 & Id & 0 \\ 0 & -Id & Id \end{bmatrix}$$

$$= \begin{bmatrix} U & V & 0 \\ W & B^+ & B^- \\ W & B^- & B^+ \end{bmatrix}.$$

Thus C is a nonnegative integral matrix with $\mathrm{sp}^\times(C) = \mathrm{sp}^\times(A) \cup \mathrm{sp}^\times(B)$.

It remains to show that C is primitive provided that B^- is primitive. Now $\lambda_B \leqslant \lambda_{|B|}$ (Exercise 11.2.5), so $\lambda_B < \lambda_A$. Hence C is spectrally Perron. Thus, according to Exercise 11.1.8, it suffices to show that C is irreducible.

Let m denote the size of U and n the size of B. List the states of C as $I_1, \ldots, I_m, J_1, \ldots, J_n, K_1, \ldots, K_n$, corresponding to its block form. Observe that since B^- is primitive,

$$\begin{bmatrix} B^+ & B^- \\ B^- & B^+ \end{bmatrix}$$

is irreducible (why?). Hence all the J_j and K_k communicate with one another. Since A is primitive, for given I_i and J_j, there is a path in G_A from I_i to J_j and also one from J_j to I_i. These correspond to paths in G_C from I_i to either J_j or to K_j, and from J_j and K_j to I_i. Thus C is irreducible and hence primitive as claimed. $\qquad\square$

Example 11.2.7. (a) Let $A = \begin{bmatrix} 0 & 1 \\ 1 & 4 \end{bmatrix}$ and $B = [-4]$. Then

$$\mathrm{sp}^\times(A) = \{2 + \sqrt{5}, 2 - \sqrt{5}\}$$

and $\lambda_{|B|} = 4 < 2 + \sqrt{5} = \lambda_A$. Also, $\mathrm{tr}_n(|B|) \leqslant \mathrm{tr}_n(A)$ since $X_{|B|}$ clearly embeds into X_A. Here $B^+ = [0]$ and $B^- = [4]$, and the latter is primitive, so the proof of Proposition 11.2.6 gives

$$C = \begin{bmatrix} 0 & 1 & 0 \\ 1 & 0 & 4 \\ 1 & 4 & 0 \end{bmatrix}.$$

The reader should check that C is indeed primitive and that

$$\mathrm{sp}^\times(C) = \{2 + \sqrt{5}, 2 - \sqrt{5}, -4\} = \mathrm{sp}^\times(A) \cup \mathrm{sp}^\times(B).$$

(b) Again, let $A = \begin{bmatrix} 0 & 1 \\ 1 & 4 \end{bmatrix}$. We can apply the construction in Proposition 11.2.6 to any integral matrix B whose absolute value is strong shift equivalent to $[4]$. For example, let $F = \begin{bmatrix} 3 & 3 \\ 1 & 1 \end{bmatrix}$ and observe that

$$(11\text{--}2\text{--}2) \qquad [4] = [1 \ \ 1] \begin{bmatrix} 3 \\ 1 \end{bmatrix}, \qquad \begin{bmatrix} 3 \\ 1 \end{bmatrix} [1 \ \ 1] = \begin{bmatrix} 3 & 3 \\ 1 & 1 \end{bmatrix} = F$$

is an elementary equivalence from $[4]$ to F. Let

$$B = \begin{bmatrix} 3 & -3 \\ -1 & -1 \end{bmatrix},$$

so that $|B| = F$ and $\mathrm{sp}^\times(B) = \{1 + \sqrt{7}, 1 - \sqrt{7}\}$. The Masking Lemma applied to A and the elementary equivalence (11–2–2) gives

$$\begin{bmatrix} 0 & 1 \\ 1 & 4 \end{bmatrix} = \begin{bmatrix} 1 & 0 & 0 & 0 \\ 0 & 1 & 1 & 1 \end{bmatrix} \begin{bmatrix} 0 & 1 \\ 1 & 0 \\ 0 & 3 \\ 0 & 1 \end{bmatrix} \quad \text{and}$$

$$\begin{bmatrix} 0 & 1 \\ 1 & 0 \\ 0 & 3 \\ 0 & 1 \end{bmatrix} \begin{bmatrix} 1 & 0 & 0 & 0 \\ 0 & 1 & 1 & 1 \end{bmatrix} = \begin{bmatrix} 0 & 1 & 1 & 1 \\ 1 & 0 & 0 & 0 \\ 0 & 3 & 3 & 3 \\ 0 & 1 & 1 & 1 \end{bmatrix},$$

where the last matrix contains F as a principal submatrix in the lower right-hand corner. We then apply the construction in Proposition 11.2.6 to obtain

$$C = \begin{bmatrix} 0 & 1 & 1 & 1 & 0 & 0 \\ 1 & 0 & 0 & 0 & 0 & 0 \\ 0 & 3 & 3 & 0 & 0 & 3 \\ 0 & 1 & 0 & 0 & 1 & 1 \\ 0 & 3 & 0 & 3 & 3 & 0 \\ 0 & 1 & 1 & 1 & 0 & 0 \end{bmatrix}.$$

Again, the reader should verify that C is primitive and that

$$\mathrm{sp}^\times(C) = \{2 + \sqrt{5}, 2 - \sqrt{5}, 1 + \sqrt{7}, 1 - \sqrt{7}\} = \mathrm{sp}^\times(A) \cup \mathrm{sp}^\times(B). \qquad \square$$

Next we establish the special case of Theorem 11.2.4 where Λ contains only integers. Note that in this case the Integrality Condition for Λ is automatically satisfied.

Theorem 11.2.8. *Let Λ be a list of nonzero integers satisfying the Perron and Net Trace Conditions. Then Λ is the nonzero spectrum of a primitive integral matrix.*

PROOF: List the elements of Λ in decreasing order, so that

$$\lambda_1 > \lambda_2 \geqslant \lambda_3 \geqslant \ldots \geqslant \lambda_k,$$

so $\lambda_\Lambda = \lambda_1$. Write $\Lambda_j = \{\lambda_1, \lambda_2, \ldots, \lambda_j\}$. The following Proposition, whose proof we give later in this section, is a key ingredient.

Proposition 11.2.9. *Let* $\Lambda = \{\lambda_1, \lambda_2, \ldots, \lambda_k\}$ *be a list of nonzero integers in decreasing order that satisfies the Perron and Net Trace Conditions. Then*

$$\mathrm{tr}_n(\Lambda_j) \geqslant \mathrm{tr}_n(|\lambda_{j+1}|) \quad \textit{for all } n \geqslant 1 \textit{ and } 1 \leqslant j \leqslant k-1.$$

Assuming this Proposition, we complete the proof of Theorem 11.2.8.

We prove inductively for $j = 1, 2, \ldots, k$ that Λ_j is the nonzero spectrum of a primitive integral matrix.

Since $\lambda_\Lambda = \lambda_1 \geqslant 1$, we see that $\Lambda_1 = \{\lambda_1\} = \mathrm{sp}^\times([\lambda_1])$, so our claim is true when $j = 1$.

Suppose that $\Lambda_j = \mathrm{sp}^\times(A_j)$, where A_j is primitive integral. As in the proof of Proposition 11.2.6, it follows from Proposition 11.2.9 together with the Embedding Theorem and Masking Lemma that by replacing A_j with a matrix strong shift equivalent to it, we may assume A_j has the form

$$(11\text{--}2\text{--}3) \qquad\qquad A_j = \begin{bmatrix} U & V \\ W & |\lambda_{j+1}| \end{bmatrix}.$$

There are two cases, depending on the sign of λ_{j+1}.

CASE 1: $\lambda_{j+1} < 0$. Since $[\lambda_{j+1}]^- = [-\lambda_{j+1}]$ is primitive, the construction in Proposition 11.2.6 applies here. This results in the primitive matrix

$$A_{j+1} = Q \begin{bmatrix} U & V & 0 \\ W & -\lambda_{j+1} & -\lambda_{j+1} \\ 0 & 0 & \lambda_{j+1} \end{bmatrix} Q^{-1} = \begin{bmatrix} U & V & 0 \\ W & 0 & -\lambda_{j+1} \\ W & -\lambda_{j+1} & 0 \end{bmatrix}$$

(with Q as in (11–2–1)), which is clearly integral, and $\mathrm{sp}^\times(A_{j+1}) = \Lambda_{j+1}$.

CASE 2: $\lambda_{j+1} > 0$. A manipulation is required here to obtain primitivity. The graph $G = G_{A_j}$ has a state I with exactly λ_{j+1} self-loops, corresponding to the lower right entry in (11–2–3). We would like this state to have at least two distinct incoming edges which are not self-loops – equivalently, for the sum of the entries of V to be at least 2. If not, the following state splitting argument transforms G so that it satisfies this property.

If $\lambda_{j+1} \geqslant 2$, then we in-split state I by partitioning the incoming edges into two subsets, one being the self-loops at I and the other being the remaining incoming edges. Then the split state corresponding to the self-loops has λ_{j+1} self-loops to itself and at least two other incoming edges

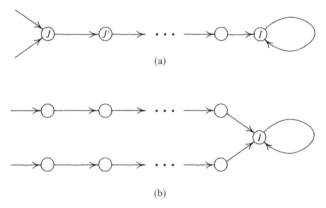

(a)

(b)

FIGURE 11.2.1.   Unzipping a path.

from another of the split states, as desired. So it remains only to consider
the case $\lambda_{j+1} = 1$.

We claim that we may assume that there is a state $J \neq I$ such that J
has at least two incoming edges. If this does not hold, then any nontrivial
in-splitting of state I will work: if $J \neq I$ is the terminal state of an outgoing
edge from I, then in the split graph J will have incoming edges from each
of the states split from I.

Now, there is a simple path from J to I, say of length k. Let J' be the
terminal state of the first edge of this path. In-split the graph at J in any
nontrivial way, creating a new graph H with a state J' with at least two
incoming edges, and such that there is a simple path of length $k-1$ from J'
to I (see Figure 11.2.1). Repeated application shows that we can "unzip"
the path, ultimately arriving at a graph in which I has exactly one self-loop
and at least two other incoming edges as desired.

This shows that in (11–2–3) we can write V as the sum of two nonnegative
integral matrices, $V = V_1 + V_2$, where $V_1 \neq 0$ and $V_2 \neq 0$. With Q as in
(11–2–1), put

$$(11\text{--}2\text{--}4) \quad A_{j+1} = Q \begin{bmatrix} U & V & V_1 \\ W & \lambda_{j+1} & 0 \\ 0 & 0 & \lambda_{j+1} \end{bmatrix} Q^{-1} = \begin{bmatrix} U & V_2 & V_1 \\ W & \lambda_{j+1} & 0 \\ W & 0 & \lambda_{j+1} \end{bmatrix}.$$

Then clearly $\mathrm{sp}^\times(A_{j+1}) = \Lambda_{j+1}$ and A_{j+1} is nonnegative and integral. An
argument similar to that in Proposition 11.2.6 shows that A_{j+1} is also
primitive (Exercise 11.2.2).   □

Example 11.2.10. Let $\Lambda = \{2, 1\}$. Then $\Lambda_1 = \{2\} = \mathrm{sp}^\times([2])$, so we
may put $A_1 = [2]$. The second higher block shift of $X_{[2]}$ is the edge shift
represented by the matrix

$$A'_1 = \begin{bmatrix} 1 & 1 \\ 1 & 1 \end{bmatrix},$$

which contains $[\lambda_2] = [1]$ as a principal submatrix (in the lower right-hand corner). The self-loop at state 2 has only one incoming edge other than itself. Thus if we "unzip" once by in-splitting state 1, we obtain

$$
A_1'' = \begin{bmatrix} 1 & 0 & 1 \\ 1 & 0 & 1 \\ 0 & 1 & 1 \end{bmatrix} = \begin{bmatrix} U & V \\ W & \lambda_2 \end{bmatrix},
$$

where the self-loop represented by the lower right hand corner has two incoming edges other than itself. Here $V = \begin{bmatrix} 1 \\ 1 \end{bmatrix} = \begin{bmatrix} 1 \\ 0 \end{bmatrix} + \begin{bmatrix} 0 \\ 1 \end{bmatrix} = V_1 + V_2$, and the method of Theorem 11.2.8 gives

$$
A_2 = \begin{bmatrix} U & V_2 & V_1 \\ W & \lambda_2 & 0 \\ W & 0 & \lambda_2 \end{bmatrix} = \begin{bmatrix} 1 & 0 & 0 & 1 \\ 1 & 0 & 1 & 0 \\ 0 & 1 & 1 & 0 \\ 0 & 1 & 0 & 1 \end{bmatrix}.
$$

The reader should verify that A_2 is indeed primitive and that $\mathrm{sp}^\times(A_2) = \{2, 1\}$. □

To complete the proof of Theorem 11.2.8, it remains only to prove Proposition 11.2.9. For this we need the following.

Lemma 11.2.11. *Let λ be an integer.*

(1) *If $\lambda \geqslant 1$, then $\mathrm{tr}_n(\lambda) \geqslant 0$ for all $n \geqslant 1$.*
(2) *If $\lambda \leqslant -1$, then $\mathrm{tr}_{2m}(\lambda) \geqslant 0$ for all $m \geqslant 1$.*

PROOF: If $\lambda \geqslant 1$ then $\mathrm{tr}_n(\lambda) = q_n(X_{[\lambda]}) \geqslant 0$ since this counts the number of points in $X_{[\lambda]}$ with least period n. Now assume that $\lambda \leqslant -2$. Since $|\mu(k)| \leqslant 1$ for all $k \geqslant 1$, it follows that

$$
\mathrm{tr}_{2m}(\lambda) \geqslant |\lambda|^{2m} - \sum_{j=1}^{2m-1} |\lambda|^j = |\lambda|^{2m} - \frac{|\lambda|^{2m} - |\lambda|}{|\lambda| - 1} \geqslant 0.
$$

We leave the remaining case $\lambda = -1$ as Exercise 11.2.7.

PROOF OF PROPOSITION 11.2.9: We will show that

(11–2–5) $\mathrm{tr}_n(\Lambda_{k-1}) \geqslant \mathrm{tr}_n(|\lambda_k|)$ for all $n \geqslant 1$.

It then follows that Λ_{k-1} satisfies the Net Trace Condition, and it clearly satisfies the Perron Condition, so another application of (11–2–5) shows that $\mathrm{tr}_n(\Lambda_{k-2}) \geqslant \mathrm{tr}_n(|\lambda_{k-1}|)$ for all $n \geqslant 1$. We then continue by backward induction to complete the proof.

First suppose $\lambda_k > 0$. Then the full λ_k-shift embeds into the full λ_1-shift, which in turn embeds into X_A, where A is the diagonal matrix with entries $\lambda_1, \lambda_2, \ldots, \lambda_{k-1}$ down the diagonal. Then $q_n(X_A) \geqslant q_n(X_{[\lambda_k]})$, which translates into (11–2–5).

Next suppose $\lambda_k < 0$. First assume that n is odd, so all divisors of n are odd and so $\mathrm{tr}_n(-\lambda_k) = -\mathrm{tr}_n(\lambda_k)$. Hence

$$\mathrm{tr}_n(\Lambda_{k-1}) - \mathrm{tr}_n(|\lambda_k|) = \mathrm{tr}_n(\Lambda_{k-1}) + \mathrm{tr}_n(\lambda_k) = \mathrm{tr}_n(\Lambda_k) \geqslant 0$$

by hypothesis, proving (11–2–5) for n odd.

When n is even, by Lemma 11.2.11,

$$\mathrm{tr}_n(\Lambda_{k-1}) = \mathrm{tr}_n(\lambda_1) + \sum_{j=2}^{k-1} \mathrm{tr}_n(\lambda_j) \geqslant \mathrm{tr}_n(\lambda_1) \geqslant \mathrm{tr}_n(|\lambda_k|),$$

the last inequality since $X_{[|\lambda_k|]}$ clearly embeds in $X_{[\lambda_1]}$. $\qquad\square$

How can we check the net trace condition for a given list Λ that satisfies the Integrality and Perron conditions? The estimates given at the end of §10.1 make this easy. In particular, we can find an explicit n_0 in terms of Λ such that $\mathrm{tr}_n(\Lambda) \geqslant 0$ for all $n \geqslant n_0$ (Exercise 11.2.6), and then check to see if $\mathrm{tr}_n(\Lambda) \geqslant 0$ for all $n < n_0$.

Example 11.2.12. Let $\Lambda = \{5, 1, 1, 1, -4, -4\}$, which satisfies the Integrality and Perron Conditions. The first six net traces are 0, 60, 0, 1080, 1080, 23820, which looks promising. We claim that in fact $\mathrm{tr}_n(\Lambda) \geqslant 0$ for all n. For this, first recall that $tr_n(1) = 0$ for $n > 1$. So, using (10–1–11) and (10–1–9), we have for $n > 1$ that

$$\begin{aligned}
\mathrm{tr}_n(\Lambda) &= \mathrm{tr}_n(5) + 2\,\mathrm{tr}_n(-4) \\
&\geqslant \mathrm{tr}_n(5) - 2|\,\mathrm{tr}_n(-4)| \\
&\geqslant 5^n - \frac{5^{n/2+1} - 5}{4} - 2\left(\frac{4^{n+1} - 4}{3}\right) \\
&= 5^n - (5/4)(\sqrt{5})^n - (8/3)4^n + (5/4 + 8/3).
\end{aligned}$$

Write this last expression as $f(n) = g(n) + (5/4 + 8/3)$ where $g(n) = 5^n - (5/4)(\sqrt{5})^n - (8/3)4^n$. As in the discussion given at the end of §10.1 (see Exercise 10.1.7), if we can find an n_0 such that $g(n_0) \geqslant 0$, then $g(n)$ is monotonically increasing on the domain $n \geqslant n_0$, so $f(n) \geqslant 0$ for $n \geqslant n_0$. A computation shows that $n_0 = 5$ works. So for $n \geqslant 5$ the nth net traces are all nonnegative. Since we have already checked that the first six net traces are nonnegative, it follows that Λ satisfies the Net Trace Condition. Hence by Theorem 11.2.8 we can conclude that Λ is the nonzero spectrum of a primitive integral matrix. $\qquad\square$

Example 11.2.13. For the list Λ in Example 11.2.5(c), recall that $\lambda_1 \approx$ 2.57474 and $|\lambda_2| = |\lambda_3| \approx 1.52655$, and Λ satisfies the Integrality and Perron Conditions. Using (10–1–11) and (10–1–9), we have

$$\begin{aligned}
\operatorname{tr}_n(\Lambda) &= \operatorname{tr}_n(\lambda_1) + \operatorname{tr}_n(\lambda_2) + \operatorname{tr}_n(\lambda_3) \\
&\geq \operatorname{tr}_n(\lambda_1) - |\operatorname{tr}_n(\lambda_2)| - |\operatorname{tr}_n(\lambda_3)| \\
&\geq \lambda_1^n - \frac{\lambda_1^{n/2+1} - \lambda_1}{\lambda_1 - 1} - 2\left(\frac{|\lambda_2|^{n+1} - |\lambda_2|}{|\lambda_2| - 1}\right) \\
&\geq 2.57^n - \frac{2.58^{n/2+1} - 2.57}{1.57} - 2\frac{1.53^{n+1} - 1.52}{.52} \\
&= g(n) + (2.57/1.57 + 3.04/.52)
\end{aligned}$$

where

$$g(n) = 2.57^n - \left(\frac{2.58}{1.57}\right)(2.58^{1/2})^n - \left(\frac{3.06}{.52}\right)(1.53^n).$$

A computation shows that the $g(5) \geq 0$, so we can conclude that for all $n \geq 5$ the nth net traces are nonnegative. Recall that in Example 11.2.5(c) we already checked that $\operatorname{tr}_n(\Lambda) \geq 0$ for all $n \leq 10$. So Λ satisfies the Net Trace Condition. □

Recall from §6.4 that the zeta function ζ_{σ_X} of a sofic shift X is a rational function p/q. Thus ζ_{σ_X} is determined by two lists of complex numbers: the roots of p and the roots of q (with multiplicity). While the zeta functions of mixing shifts of finite type have not yet been completely classified, it turns out that the zeta functions of mixing sofic shifts have been classified [BoyH1].

Theorem 11.2.14. *A function $\phi(t)$ is the zeta function of a mixing sofic shift if and only if*

$$\phi(t) = \frac{\prod_{\lambda \in \Lambda_1}(1 - t\lambda)}{\prod_{\lambda \in \Lambda_2}(1 - t\lambda)}$$

where Λ_1 and Λ_2 are lists of nonzero complex numbers such that

(1) Λ_1 *and* Λ_2 *each satisfy the Integrality Condition.*
(2) Λ_2 *satisfies the Perron Condition and* λ_{Λ_2} *strictly dominates all the elements of* Λ_1 *in absolute value.*
(3) $\operatorname{tr}_n(\Lambda_2) - \operatorname{tr}_n(\Lambda_1) \geq 0$ *for all* $n \geq 1$.

We leave it to the reader to verify the necessity of the conditions, and we refer the reader to [BoyH1] for a proof of sufficiency.

EXERCISES

11.2.1. Prove that a list Λ of nonzero complex numbers is the nonzero spectrum of a nonnegative integral matrix if and only if Λ can be written as the union $\Lambda = \bigcup_{k=1}^{m} \Lambda_k$ where for each k there is a positive integer p_k and a set $\widehat{\Lambda}_k$ such that

(a) $\Lambda_k = \{(\exp(2\pi\sqrt{-1}j/p_k))\lambda : 0 \leqslant j \leqslant p_k - 1, \lambda \in \widehat{\Lambda}_k\}$,

(b) $(\widehat{\Lambda}_k)^{p_k}$ satisfies the Spectral Conditions.

11.2.2. Complete the proof of Theorem 11.2.8 (Case 2) by showing that the matrix A_{j+1} in equation (11–2–4) is primitive.

11.2.3. Let G be an irreducible graph and $n \geqslant 2$ be a positive integer. Assume that there is a vertex which has n self loops. Let m be a positive integer. Show that there is an irreducible graph H such that A_H is strong shift equivalent to A_G and H has a vertex I which has n self loops and m incoming edges to I from vertices other than I.

11.2.4. Let Λ be a list of complex numbers.

(a) Show that the Net Trace Condition implies that $\text{tr}(\Lambda^n) \geqslant 0$ for all $n \geqslant 1$.

(b) Show that the converse is false even for lists that satisfy the Integrality and Perron Conditions. [*Hint*: Consider $\Lambda = \{2, i, i, -i, -i\}$.]

11.2.5. Show $\lambda_B \leqslant \lambda_{|B|}$ for any integral matrix B.

11.2.6. (a) Given a list of numbers $\Lambda = \{\lambda_1, \ldots, \lambda_k\}$ that satisfies the Perron condition, show that there is an n_0, depending on Λ, such that $\text{tr}_n(\Lambda) \geqslant 0$ for all $n \geqslant n_0$.

(b) For Λ as in part (a), find an explicit n_0, in terms of λ_Λ, k, and $\lambda_\Lambda - \max_{\{\lambda \in \Lambda,\ \lambda \neq \lambda_\Lambda\}} |\lambda_i|$, such that $\text{tr}_n(\Lambda) \geqslant 0$ for all $n \geqslant n_0$.

(c) Explain why this gives a finite procedure to decide if a list of k nonzero complex numbers that satisfies the Integrality and Perron Conditions also satisfies the Net Trace Conditions.

11.2.7. Show that $\text{tr}_1(-1) = -1$, $\text{tr}_2(-1) = 2$, and $\text{tr}_n(-1) = 0$ for all $n \geqslant 3$.

11.2.8. (a) Show that $\{2, 1\}$ cannot be realized as the nonzero spectrum of a 2×2 primitive integral matrix.

(b) What if $\{2, 1\}$ were replaced by $\{3, 1\}$?

11.2.9. Show that, for any integer λ and odd integer n,

$$\text{tr}_n(\lambda) \geqslant \lambda^n - \lambda^{n/2} \qquad \text{if } \lambda \geqslant 1,$$
$$\text{tr}_n(\lambda) \geqslant \lambda^n \qquad \text{if } \lambda \leqslant -1.$$

11.2.10. Let Λ_k be the list consisting of k once, $k - 2$ copies of 1, and two copies of $-(k - 1)$. Show that Λ_k satisfies the Net Trace Condition if and only if $k = 2, 3, 4$, or 5.

11.2.11. For each list below, find a primitive integral matrix whose nonzero spectrum is the given list.

(a) $\{3, 1, -2, -2\}$. [*Hint*: Start with the matrix $\begin{bmatrix} 2 & 1 \\ 1 & 2 \end{bmatrix}$.]

(b) $\{4, 2, -3, -3\}$. [*Hint*: start with the matrix $\begin{bmatrix} 3 & 1 \\ 1 & 3 \end{bmatrix}$.]

11.2.12. Verify the necessity of the conditions in Theorem 11.2.14.

§11.3. Pure Subgroups of Dimension Groups

Let A be an integral matrix with dimension pair (Δ_A, δ_A). Suppose that Δ is a δ_A-invariant subgroup of Δ_A. In this section we give a simple sufficient condition for $(\Delta, \delta_A|_\Delta)$ to be a dimension pair in its own right. This condition, which will be used in Chapter 12, is based on the following notion of pure subgroup.

Definition 11.3.1. Let Γ be a subgroup of $\mathbb{Q}^n$. A subgroup Δ of Γ is a *pure subgroup of Γ*, or is *pure in Γ*, if whenever $\mathbf{v} \in \Gamma$ and $q\mathbf{v} \in \Delta$ for a nonzero integer q, then $\mathbf{v} \in \Delta$.

We are mostly interested in the case where Γ is a dimension group and Δ is a subgroup of Γ. In this case, pure subgroups are sometimes called *closed subgroups* (for instance in [BoyMT]). Note that when Γ is a dimension group, it is a subgroup of $\mathbb{Q}^n$, and so the definition applies. There is a notion of pure subgroup for arbitrary abelian groups that reduces to the definition we have given for subgroups of $\mathbb{Q}^n$; see [Kap].

For a subset E of $\mathbb{Q}^n$ we let $\langle E \rangle_\mathbb{Q}$ denote the rational subspace of $\mathbb{Q}^n$ generated by E.

Proposition 11.3.2. *Let Γ be a subgroup of $\mathbb{Q}^n$ and Δ a subgroup of Γ. Then the following are equivalent.*

(1) Δ *is a pure subgroup of Γ.*
(2) $\Delta = \langle \Delta \rangle_\mathbb{Q} \cap \Gamma$.
(3) Δ *is the intersection of Γ with a rational subspace of $\mathbb{Q}^n$.*

Moreover, $\langle \Delta \rangle_\mathbb{Q} \cap \Gamma$ is the smallest pure subgroup of Γ containing Δ.

We leave the proof of this result as an exercise for the reader.

Example 11.3.3. (a) The only pure subgroups of $\mathbb{Z}$ are $\{0\}$ and $\mathbb{Z}$ itself, since the only rational subspaces of $\mathbb{Q}$ are $\{0\}$ and $\mathbb{Q}$.

(b) Pure subgroups Δ of $\mathbb{Z}^n$ are those that do not contain "holes" in the sense that if $\mathbf{z} \in \mathbb{Z}^n$ and $q\mathbf{z} \in \Delta$, then $\mathbf{z}$ must already be in Δ. For example, for each $\mathbf{v} \in \mathbb{Q}^n$ the subgroup $\{\mathbf{z} \in \mathbb{Z}^n : \mathbf{v} \cdot \mathbf{z} = 0\}$ is pure in $\mathbb{Z}^n$. In contrast, the subgroup $\Delta = \{[a, b] \in \mathbb{Z}^2 : a + b \text{ is even}\}$ is not pure in $\mathbb{Z}^2$, since $[2, 0] \in \Delta$ and $[2, 0] = 2 \cdot [1, 0]$ while $[1, 0] \notin \Delta$.

(c) Let
$$A = \begin{bmatrix} 2 & 1 & 0 \\ 1 & 1 & 1 \\ 0 & 1 & 2 \end{bmatrix}$$

and $\Gamma = \Delta_A$. Recall from Example 7.5.5 that $\Delta_A = \mathbb{Z}[1/3] \cdot [1, 1, 1] \oplus \mathbb{Z}[1/2] \cdot [1, 0, -1]$. Hence $\mathbb{Z}[1/3] \cdot [1, 1, 1]$ and $\mathbb{Z}[1/2] \cdot [1, 0, -1]$ are pure in Δ_A, being intersections of Δ_A with rational subspaces of $\mathbb{Q}^3$. There are many other pure subgroups, formed by intersecting Δ_A with other rational subspaces. On the other hand, proper subgroups of each of these pure subgroups cannot be pure. For example, $\mathbb{Z}[1/3] \cdot [2, 2, 2]$ is not pure in Δ_A.

(d) Let $A = \begin{bmatrix} 2 & 1 \\ 1 & 2 \end{bmatrix}$. By Exercise 7.5.2,

$$\Delta_A = \{[a, a + b] : a \in \mathbb{Z}[1/3], \ b \in \mathbb{Z}\}.$$

Hence $\{[a, a] : a \in \mathbb{Z}[1/3]\}$ and $\{[0, b] : b \in \mathbb{Z}\}$ are each pure subgroups of Δ_A. $\qquad\square$

Before proceeding, we need to review some ideas from abstract algebra.

Definition 11.3.4. A group is *finitely generated* if there is a finite subset of the group such that every group element is a sum of elements in the subset. An *integral basis* $\mathfrak{I}$ for an abelian group Δ is a finite subset of Δ such that every element of Δ can be expressed uniquely as a sum of elements of $\mathfrak{I}$. A finitely generated abelian group is *free* if it has an integral basis. A finitely generated abelian group Δ is *torsion-free* if whenever $x \in \Delta$, q is a nonzero integer and $qx = 0$, then $x = 0$.

It is not hard to see that any free abelian group must be torsion-free (Exercise 11.3.2). Thus, finite groups are not free. On the other hand, $\mathbb{Z}^n$ (with the usual additive group operation) is free since the set of standard basis vectors is an integral basis. In fact, up to isomorphism, these are the only free abelian groups. This is one of many fundamental facts about free abelian groups that we summarize in the following result.

Theorem 11.3.5.

(1) *An abelian group is free if and only if it is isomorphic to $\mathbb{Z}^n$ for some $n \geqslant 0$.*

(2) *Any subgroup of a free abelian group is free (in particular, any subgroup of $\mathbb{Z}^n$ is free).*

(3) *Any finitely generated torsion-free abelian group is free.*

(4) *If Γ is a free abelian group and Δ is a pure subgroup of Γ, then the quotient group Γ/Δ is free.*

(5) *If Γ is a free abelian group and Δ is a pure subgroup of Γ, then every integral basis of Δ can be enlarged to an integral basis of Γ.*

Proofs of statements (1), (2) and (3) are contained in many textbooks on abstract algebra, for example [BirM], [Kap]. Part (4) follows from (3) since the quotient of a free abelian group by a pure subgroup is finitely generated and torsion-free (Exercise 11.3.3(a)). For (5), one uses the integral basis for Γ/Δ, guaranteed by (4), to enlarge the integral basis of Δ to an integral basis of Γ (Exercise 11.3.3(b)).

The following result gives a simple sufficient condition for a δ_A-invariant subgroup of Δ_A, together with the restriction of δ_A, to be a dimension pair of an integral matrix.

Proposition 11.3.6. *Let A be an $n \times n$ integral matrix and Δ a pure δ_A-invariant subgroup of Δ_A. Then $(\Delta, \delta_A|_\Delta)$ is the dimension pair of an integral matrix. In fact, if C is the matrix of the restriction of A to $\langle \Delta \rangle_\mathbb{Q}$ with respect to an arbitrary integral basis of $\langle \Delta \rangle_\mathbb{Q} \cap \mathbb{Z}^n$, then $(\Delta, \delta_A|_\Delta) \simeq (\Delta_C, \delta_C)$.*

PROOF: Let $V = \langle \Delta \rangle_\mathbb{Q}$. Since Δ is δ_A-invariant, so is $V \cap \mathbb{Z}^n$. By Theorem 11.3.5(2), $V \cap \mathbb{Z}^n$ has an integral basis $\mathcal{J} = \{z_1, \ldots, z_m\}$. Let C be the matrix of $A|_V$ with respect to $\mathcal{J}$, i.e., C_{ij} is the coefficient of z_j in the unique expression for $z_i A$ as an integral combination of $\mathcal{J}$. In particular, C is an integral matrix.

Define $\phi: \mathbb{Q}^m \to V$ by $\phi(e_i) = z_i$ for $1 \leqslant i \leqslant m$. Then clearly ϕ is a linear (over $\mathbb{Q}$) isomorphism that intertwines C and $A|_V$ in the sense that $\phi(vC) = \phi(v)A$ for all $v \in \mathbb{Q}^m$. We complete the proof by showing that $\phi(\Delta_C) = \Delta$.

Since $\mathcal{J}$ is an integral basis for $V \cap \mathbb{Z}^n$,

$$(11\text{--}3\text{--}1) \qquad \phi(\mathbb{Z}^m) = V \cap \mathbb{Z}^n.$$

Since V is a subspace of the eventual range of A, it follows that C is nonsingular, so that

$$(11\text{--}3\text{--}2) \qquad \Delta_C = \{w \in \mathbb{Q}^m : wC^k \in \mathbb{Z}^m \quad \text{for some } k\}.$$

By (11–3–1) and (11–3–2), it follows that

$$\phi(\Delta_C) = \{v \in V : vA^k \in V \cap \mathbb{Z}^n \quad \text{for some } k\} = V \cap \Delta_A = \Delta,$$

the last equality holding since Δ is pure in Δ_A. $\qquad\square$

While the purity assumption gives a sufficient condition for a δ_A-invariant subgroup Δ of Δ_A to be a dimension group, it is certainly not necessary. For example, let $A = [\,1\,]$, so that $\Delta_A = \mathbb{Z}$ and δ_A is the identity. Then $\Delta = 2\mathbb{Z}$ is not pure in $\mathbb{Z}$, while $(\Delta, \delta_A|_\Delta)$ is isomorphic to (Δ_A, δ_A).

For $i = 1, 2$ let Δ_i be an abelian group and $\delta_i: \Delta_i \to \Delta_i$ an automorphism of Δ_i. In §11.1 we defined what it meant for (Δ_2, δ_2) to be a quotient of (Δ_1, δ_1). The following is a dual notion:

Definition 11.3.7. A pair (Δ_2, δ_2) is *embedded in* a pair (Δ_1, δ_1) if there is a δ_2-invariant *pure* subgroup Δ of Δ_1 such that $(\Delta_2, \delta_2) \simeq (\Delta, \delta_1|_\Delta)$.

Proposition 11.3.6 shows how to find many dimension pairs embedded in a given dimension pair (Δ_A, δ_A): simply take $\Delta = V \cap \Delta_A$ where V is any A-invariant rational subspace of V_A.

The following shows that there is a natural duality between quotients and embeddings.

Proposition 11.3.8. *Let A and B be integral matrices. Then (Δ_B, δ_B) is embedded in (Δ_A, δ_A) if and only if $(\Delta_{B^\mathsf{T}}, \delta_{B^\mathsf{T}})$ is a quotient of $(\Delta_{A^\mathsf{T}}, \delta_{A^\mathsf{T}})$.*

PROOF: Let n denote the size of A and m the size of B. First suppose that $(\Delta_{B^\mathsf{T}}, \delta_{B^\mathsf{T}})$ is a quotient of $(\Delta_{A^\mathsf{T}}, \delta_{A^\mathsf{T}})$. By Proposition 11.1.14, there are integral matrices R and S and a nonnegative integer ℓ such that

$$A^\mathsf{T} R = R B^\mathsf{T} \qquad \text{and} \qquad SR = (B^\mathsf{T})^\ell.$$

Taking transposes gives

$$(11\text{--}3\text{--}3) \qquad R^\mathsf{T} A = B R^\mathsf{T} \qquad \text{and} \qquad R^\mathsf{T} S^\mathsf{T} = B^\ell.$$

Define $\psi: \Delta_B \to \Delta_A$ by $\psi(\mathbf{w}) = \mathbf{w} R^\mathsf{T}$. Since R^T is integral and $R^\mathsf{T} A = B R^\mathsf{T}$, it follows that ψ is well-defined, it intertwines δ_B with δ_A, and $\Delta = \psi(\Delta_B)$ is a δ_A-invariant subgroup of Δ_A. Furthermore, since $R^\mathsf{T} S^\mathsf{T} = B^\ell$ and B is nonsingular on Δ_B, we see that ψ is one-to-one.

It remains to show that Δ is pure in Δ_A. For this we need to show that whenever $\mathbf{v} \in \Delta_A$ and $q\mathbf{v} \in \Delta$, then $\mathbf{v} \in \Delta$. To prove this, choose $\mathbf{w} \in \Delta_B$ such that $\psi(\mathbf{w}) = \mathbf{w} R^\mathsf{T} = q\mathbf{v}$. Then for large enough k we have

$$\left(\frac{\mathbf{w}}{q} R^\mathsf{T} \right) A^k = \left(\mathbf{v} \right) A^k \in \mathbb{Z}^n,$$

so that (11–3–3) gives

$$\left(\frac{\mathbf{w}}{q} \right) B^{k+\ell} = \left(\frac{\mathbf{w}}{q} \right) (R^\mathsf{T} A^k S^\mathsf{T}) \in \mathbb{Z}^m.$$

Thus $\mathbf{w}/q \in \Delta_B$, so that $\mathbf{v} = \psi(\mathbf{w}/q) \in \psi(\Delta_B) = \Delta$ as required.

Conversely, suppose that (Δ_B, δ_B) is embedded in (Δ_A, δ_A) with image a pure subgroup Δ. Let B' be the matrix of $A|_\Delta$ with respect to some integral basis $\mathcal{I}$ of $\langle \Delta \rangle_\mathbb{Q} \cap \mathbb{Z}^n$. By Proposition 11.3.6, $(\Delta_B, \delta_B) \simeq (\Delta_{B'}, \delta_{B'})$. Hence we may assume that $B = B'$. In particular, the size m of B is the number of elements in $\mathcal{I}$.

Let R be the $m \times n$ matrix whose rows are the elements of $\mathcal{I}$. Our assumption that $B = B'$ shows that $RA = BR$. Now $\langle \Delta \rangle_\mathbb{Q} \cap \mathbb{Z}^n$ is pure in $\mathbb{Z}^n$. According to Theorem 11.3.5(5), we can extend $\mathcal{I}$ to an integral basis $\widehat{\mathcal{I}}$ of $\mathbb{Z}^n$. Enlarge R to an $n \times n$ matrix $\widehat{R}$ by adding as new rows the elements in $\widehat{\mathcal{I}} \setminus \mathcal{I}$. Since each standard basis vector can be expressed as an integral combination of the rows of $\widehat{R}$, there is an integral $n \times n$ matrix $\widehat{S}$ such that $\widehat{S}\widehat{R} = Id$. Thus, $\widehat{S}$ and $\widehat{R}$ are inverses of one another, and so we also have $\widehat{R}\widehat{S} = Id$. Let S be the submatrix consisting of the first m columns of $\widehat{S}$. Then

$$Id = \widehat{R}\widehat{S} = \begin{bmatrix} R \\ * \end{bmatrix} [S \quad *],$$

where a $*$ in a matrix denotes irrelevant entries. It follows that $RS = Id$. Taking transposes of the equations $RA = BR$, $RS = Id$ and applying Proposition 11.1.14 shows that $(\Delta_{B^\mathsf{T}}, \delta_{B^\mathsf{T}})$ is a quotient of $(\Delta_{A^\mathsf{T}}, \delta_{A^\mathsf{T}})$. □

EXERCISES

11.3.1. Prove Proposition 11.3.2.

11.3.2. Show that any free abelian group must be torsion-free.

11.3.3. Assuming parts (1), (2) and (3) of Theorem 11.3.5,

    (a) Prove part (4) of Theorem 11.3.5. [*Hint*: Use part (3) of Theorem 11.3.5 and show that the quotient of a free abelian group by a pure subgroup is finitely generated and torsion-free.]

    (b) Prove part (5) of Theorem 11.3.5 by using the the integral basis for Γ/Δ, guaranteed by part (4), to enlarge the integral basis of Δ to form an integral basis of Γ.

    (c) Explain why we need the purity assumption in parts (4) and (5) of Theorem 11.3.5.

11.3.4. Find all pure invariant subgroups of the dimension group of the following matrices.

$$
\text{(a)} \quad \begin{bmatrix} 1 & 1 \\ 1 & 0 \end{bmatrix}, \qquad
\text{(b)} \quad \begin{bmatrix} 2 & 1 & 0 \\ 1 & 1 & 1 \\ 0 & 1 & 2 \end{bmatrix}, \qquad
\text{(c)} \quad \begin{bmatrix} 2 & 1 \\ 1 & 2 \end{bmatrix}.
$$

Notes

The characterization of entropies of shifts of finite type contained in Theorems 11.1.4 and 11.1.6 is due to Lind [Lin3], [Lin4]. While the proofs of these results do not make use of symbolic dynamical ideas, they do use the dynamics of linear transformations and so are very dynamical in spirit. Perrin [Perr] later observed that earlier work by Soittola on rational series (see [SaSo]) could be used to derive these results, and he provided a simpler proof of Soittola's main result. Theorem 11.1.19 is due to Handelman [Han1], [Han2], who was motivated by issues arising in the theory of C^*-algebras, rather than issues of symbolic dynamics. The condition that we have called spectrally Perron was called weak Perron in [Han1].

The arguments that we used to prove Proposition 11.2.6 and Theorem 11.2.8 are due to Boyle and Handelman [BoyH1, BouH2]. They used these to establish their Subtuple Theorem, which gave a mechanism for enlarging the spectrum of a primitive matrix and was a key component in their approach to the Spectral Conjecture.

An old (still unsolved) problem in linear algebra is the problem of characterizing the spectra of nonnegative real matrices. Using the techniques in §11.2, Boyle and Handelman [BoyH1] solved a variant of this problem: namely, they completely characterized the nonzero spectra of nonnegative real (as opposed to nonnegative integral) matrices. Their proof of Theorem 11.2.14 also applies to a more general class of dynamical systems, that of mixing finitely presented systems.

Propositions 11.3.6 and 11.3.8 are contained in [BoyMT]; Exercise 11.1.9 comes from Handelman [Han1]; and Exercise 11.2.3 is due to M. Boyle.

CHAPTER 12

EQUAL ENTROPY FACTORS

The Lower Entropy Factor Theorem of §10.3 completely solves the problem of when one irreducible shift of finite type factors onto another of strictly lower entropy. In contrast, the corresponding problem for equal entropy does not yet have a satisfactory solution. This chapter contains some necessary conditions for an equal-entropy factor code, and also some sufficient conditions. We will also completely characterize when one shift of finite type "eventually" factors onto another of the same entropy.

In §12.1 we state a necessary and sufficient condition for one irreducible shift of finite type to be a right-closing factor of another. While the proof of this is too complicated for inclusion here, we do prove an "eventual" version: for irreducible shifts of finite type X and Y with equal entropy we determine when Y^m is a right-closing factor of X^m for all sufficiently large m. In §12.2 we extend this to determine when X^m factors onto Y^m for all sufficiently large m (where the factor code need not be right-closing). This is analogous to Theorem 7.5.15, which showed that two irreducible edge shifts are eventually conjugate if and only if their associated matrices are shift equivalent.

At the end of §10.3 we proved a generalization of the Finite-State Coding Theorem from Chapter 5, where the sofic shifts had different entropies. This amounted to the construction of right-closing finite equivalences. In §12.3 we show that given two irreducible edge shifts with equal entropy $\log \lambda$, the existence of a right-closing finite equivalence forces an arithmetic condition on the entries of the corresponding Perron eigenvectors. This condition can be stated in terms of equivalence classes of ideals in the ring $\mathbb{Z}[1/\lambda]$. In §12.4 we show that this ideal class condition is also sufficient in the case of mixing shifts of finite type, and we give a complete characterization of when two irreducible sofic shifts are right-closing finitely equivalent.

§12.1. Right-Closing Factors

Let X be an irreducible shift of finite type with $h(X) \geqslant \log n$. In Theorem 5.5.8 we showed that X factors onto the full n-shift $X_{[n]}$. According to Proposition 8.1.12 when $h(X) = \log n$, the factor code from X to $X_{[n]}$ can be chosen to be right-closing: X is recoded to an edge shift X_G, then G is transformed by a sequence of out-splittings to a graph H with constant out-degree n, and the factor code is the composition of conjugacies induced by the out-splittings followed by the right-resolving factor code induced by a road-coloring of H.

For arbitrary irreducible shifts of finite type X, Y of the same entropy, if we try to factor X onto Y by a right-closing code, we run into an obstruction that can be formulated in terms of dimension pairs; for concreteness, let $X = X_A$ and $Y = X_B$ be edge shifts. We claim that if there is a right-closing factor code from X_A onto X_B, then (Δ_B, δ_B) must be a quotient of (Δ_A, δ_A).

This can best be explained using the interpretation of the dimension pair in terms of rays and beams (see §7.5). Recall that a ray in X_A is a subset of the form $\{y \in X_A : y_{(-\infty, m]} = x_{(-\infty, m]}\}$, and that a beam is a finite union of rays. There is a natural notion of equivalence of beams, and Δ_A is generated by equivalence classes of beams. The automorphism δ_A is induced by the action of σ_A on (equivalence classes of) beams.

First, let us see how a right-*resolving* factor code $\phi: X_A \to X_B$ induces a map $\psi: \Delta_A \to \Delta_B$. Recall from Proposition 8.2.2 that ϕ must be right-covering. It follows that ϕ maps beams in X_A to beams in X_B, and respects equivalence of beams. Also, every beam in X_B is the image of a beam in X_A. This gives an onto homomorphism $\psi: \Delta_A \to \Delta_B$ that intertwines δ_A with δ_B since ϕ intertwines σ_A with σ_B. Next, since the dimension pair is a conjugacy invariant and every right-closing factor code is a composition of a conjugacy and a right-resolving factor code (Proposition 5.1.11), it follows that a right-closing factor code $X_A \to X_B$ forces (Δ_B, δ_B) to be a quotient of (Δ_A, δ_A).

The existence of a quotient mapping $(\Delta_A, \delta_A) \to (\Delta_B, \delta_B)$ is not enough to guarantee the existence of a right-closing factor code $X_A \to X_B$, or even a sliding block code from X_A to X_B, since the periodic point condition $P(X) \searrow P(Y)$ may not be satisfied. However, Ashley [Ash6] showed that for mixing edge shifts these two conditions are both necessary and sufficient for a right-closing factor code.

Theorem 12.1.1. *Let X_A and X_B be mixing edge shifts with the same entropy. Then there is a right-closing factor code from X_A onto X_B if and only if $P(X_A) \searrow P(X_B)$ and (Δ_B, δ_B) is a quotient of (Δ_A, δ_A).*

The necessity of these conditions is a consequence of our discussion above. The sufficiency is much more difficult; we refer the reader to [Ash6]. This section concentrates on proving a version of Theorem 12.1.1 for a weakened notion of factor code.

Definition 12.1.2. A shift space Y is an *eventual factor* of a shift space X if Y^m is a factor of X^m for all sufficiently large m. Also, Y is a *right-closing eventual factor* of X if Y^m is a right-closing factor of X^m for all sufficiently large m. A *left-closing eventual factor* is defined similarly.

Example 12.1.7 below shows that Y may be an eventual factor of X without being a factor of X. Example 12.1.9 shows that it is possible for Y^m to be a factor of X^m for some m without Y being an eventual factor of X.

Since $p_n(X^k) = p_{nk}(X)$, the periodic point condition $P(X_A) \searrow P(X_B)$ is not needed for eventual factors. The remaining condition in Theorem 12.1.1, i.e., the condition that (Δ_B, δ_B) be a quotient of (Δ_A, δ_A), turns out to be both necessary and sufficient for a right-closing eventual factor. We state this result below in the more general context of eventually positive integral matrices. Recall from Exercise 4.5.11 that Perron–Frobenius theory for primitive matrices also applies to eventually positive matrices.

Definition 12.1.3. Let A and B be eventually positive integral matrices. Then B is an *eventual factor* of A if X_{B^m} is a factor of X_{A^m} for all sufficiently large m. Also, B is a *right-closing eventual factor* of A if X_{B^m} is a right-closing factor of X_{A^m} for all sufficiently large m. *Left-closing eventual factor* is defined similarly.

Theorem 12.1.4. *Let A and B be eventually positive integral matrices with $\lambda_A = \lambda_B$. Then B is a right-closing eventual factor of A if and only if (Δ_B, δ_B) is a quotient of (Δ_A, δ_A).*

PROOF: We first prove the necessity of the dimension pair quotient condition. So suppose that X_{B^m} is a right-closing factor of X_{A^m} for all sufficiently large m, say all $m \geqslant m_0$. Then $(\Delta_{B^m}, \delta_{B^m})$ is a quotient of $(\Delta_{A^m}, \delta_{A^m})$, so that by Proposition 11.1.14 there are integral matrices R_m and S_m and a nonnegative integer ℓ_m such that

$$A^m R_m = R_m B^m \qquad \text{and} \qquad S_m R_m = B^{\ell_m}$$

for all $m \geqslant m_0$. The reverse implication in Proposition 11.1.14 shows that we need only show that there is an $m \geqslant m_0$ such that $AR_m = R_mB$. Now a variation on the argument at the end of the proof of Theorem 7.5.15 shows that there is an $m \geqslant m_0$ such that if λ is a nonzero eigenvalue of A and $\mu \neq \lambda$ is a nonzero eigenvalue of B, then λ/μ is not an mth root of unity. By replacing R_m by $A^{dm}R_m$, where d is the size of A, we may also assume that the eventual kernel of A is contained in the kernel of R_m (recall that our matrices act on rational vectors by multiplication on the right and so, for instance, the kernel of R_m is the set of all rational vectors

v such that $\mathbf{v}R_m = 0$). The following lemma, whose proof is a modification of the argument in Lemma 7.5.16, then completes the proof of necessity. We leave the proof to the reader.

Lemma 12.1.5. *Let m be a positive integer and A and B be integral matrices such that the ratio of any nonzero eigenvalue λ of A to any nonzero eigenvalue $\mu \neq \lambda$ of B is not an mth root of unity. Suppose that R is an integral matrix such that $A^m R = RB^m$ and that the eventual kernel of A is contained in the kernel of R. Then $AR = RB$.*

Now for sufficiency. So suppose that (Δ_B, δ_B) is a quotient of (Δ_A, δ_A). Proposition 11.1.14 shows that there are integral matrices R and S and an integer $\ell \geqslant 0$ such that $AR = RB$ and $SR = B^\ell$. Since A and B are eventually positive and $\lambda_A = \lambda_B$, we can choose R and S to be strictly positive (Exercise 11.1.7). The following result details a state splitting construction used to construct the eventual factor codes. We will make use of this result in §12.4.

Theorem 12.1.6. *Let A and B be eventually positive integral matrices with $\lambda_A = \lambda_B$, and suppose that (Δ_B, δ_B) is a quotient of (Δ_A, δ_A). Let R and S be positive integral matrices and $\ell \geqslant 0$ be an integer such that $AR = RB$ and $SR = B^\ell$. Then for all sufficiently large m there is an out-splitting H of G_{A^m} for which each state $I \in \mathcal{V}(A)$ is split into states $[I, J, k]$, where $J \in \mathcal{V}(B)$ and $1 \leqslant k \leqslant R_{IJ}$, and the amalgamation matrix $\widetilde{R}$ defined by*

$$(12\text{–}1\text{–}1) \qquad \widetilde{R}_{[I,J,k],J'} = \begin{cases} 1 & \text{if } J = J', \\ 0 & \text{if } J \neq J' \end{cases}$$

satisfies

$$(12\text{–}1\text{–}2) \qquad A_H \widetilde{R} = \widetilde{R} B^m.$$

Theorem 12.1.6 then produces a right-closing factor code $X_{A^m} \to X_{B^m}$ by composing the conjugacy determined by the out-splitting H of G_{A^m} followed by the right-resolving factor code determined by (12–1–2) as in Proposition 8.2.6. This proves Theorem 12.1.4.

PROOF OF THEOREM 12.1.6. For $J \in \mathcal{V}(B)$ and $n \geqslant 1$, define the vector $\mathbf{x}^{(J,n)}$ by

$$\mathbf{x}^{(J,n)} = \mathbf{e}_J S A^n.$$

For $I \in \mathcal{V}(A)$ put

$$\mathbf{a}^{(I,n)} = \mathbf{e}_I (A^\ell - RS) A^n.$$

Note that $\mathbf{x}^{(J,n)} > 0$ for large enough n, but $\mathbf{a}^{(I,n)}$ may have some negative entries. Also observe that $\mathbf{a}^{(I,n)} \in \ker R$ since

$$\mathbf{a}^{(I,n)} R = \mathbf{e}_I (A^\ell - RS) A^n R = \mathbf{e}_I (A^\ell - RS) R B^n$$
$$= \mathbf{e}_I (A^\ell R - RSR) B^n = \mathbf{e}_I (A^\ell R - RB^\ell) B^n = \mathbf{0}.$$

Let $\lambda = \lambda_A = \lambda_B$. We next show that $\mathbf{x}^{(J,n)}$ grows like λ^n while $\mathbf{a}^{(I,n)}$ grows at a strictly smaller rate. More precisely, we show that

$$(12\text{--}1\text{--}3) \qquad\qquad \lim_{n\to\infty} \frac{1}{\lambda^n}\mathbf{x}^{(J,n)} > 0,$$

while

$$(12\text{--}1\text{--}4) \qquad\qquad \lim_{n\to\infty} \frac{1}{\lambda^n}\mathbf{a}^{(I,n)} = 0.$$

Inequality (12–1–3) follows from Perron–Frobenius Theory since $\mathbf{e}_J S > 0$ and A is eventually positive (see Theorem 4.5.12 and Exercise 4.5.11). For (12–1–4) we argue as follows. Let $V = \{\mathbf{u} \in \mathbb{R}^d : \mathbf{u}R = 0\}$, the real kernel of R. Observe that since $AR = RB$, V is A-invariant. Since R is positive and any left Perron eigenvector for A is positive, it follows that V cannot contain a left Perron eigenvector. Thus V is contained in the A-invariant subspace corresponding to eigenvalues whose absolute value is less than λ. Using the Jordan canonical form, this yields (12–1–4) (see Exercise 4.5.10).

Hence we can choose n large enough so that $A^n > 0$, $B^n > 0$, and, for each $I \in \mathcal{V}(A)$ and $J \in \mathcal{V}(B)$,

$$(12\text{--}1\text{--}5) \qquad\qquad |\mathbf{a}^{(I,n)}| \leqslant \mathbf{x}^{(J,n)}.$$

Put $m = n + \ell$. Define an out-splitting H of G_{A^m} as follows. Split each $I \in \mathcal{V}(A) = \mathcal{V}(G_{A^m})$ into $\sum_J R_{IJ}$ states, denoted by $[I, J, k]$, where $J \in \mathcal{V}(B)$ and $1 \leqslant k \leqslant R_{IJ}$, according to a partition $\mathcal{P}_I$ of $\mathcal{E}_I(G_{A^m})$ defined as follows. Fix an arbitrary $J_0 \in \mathcal{V}(B)$. For every $I \in \mathcal{V}(A)$, $J \in \mathcal{V}(B)$, and $1 \leqslant k \leqslant R_{IJ}$, let

$$\mathbf{y}^{[I,J,k]} = \begin{cases} \mathbf{x}^{(J,n)} & \text{if } (J,k) \neq (J_0,1), \\ \mathbf{x}^{(J_0,n)} + \mathbf{a}^{(I,n)} & \text{if } (J,k) = (J_0,1). \end{cases}$$

Let $\mathcal{P}_I$ partition $\mathcal{E}_I(G_{A^m})$ into sets indexed by $[I, J, k]$, where there are exactly $(\mathbf{y}^{[I,J,k]})_{I'}$ edges from I to I'.

To see that this make sense, we must verify that

(1) each $\mathbf{y}^{[I,J,k]} \geqslant 0$, and
(2) for each $I \in \mathcal{V}(A)$,

$$\sum_{J,k} \mathbf{y}^{[I,J,k]} = \mathbf{e}_I A^m.$$

Now (1) holds by our choice of n in (12–1–5). For (2), note that

$$\begin{aligned} \sum_{J,k} \mathbf{y}^{[I,J,k]} &= \sum_J R_{IJ}\mathbf{x}^{(J,n)} + \mathbf{a}^{(I,n)} \\ &= \mathbf{e}_I RSA^n + \mathbf{e}_I(A^{\ell+n} - RSA^n) \\ &= \mathbf{e}_I A^m. \end{aligned}$$

We will use the division and edge matrices associated to this splitting (see §2.4). The division matrix D is indexed by $\mathcal{V}(A) \times \mathcal{V}(H)$ and defined by

$$D_{I,[I',J',k']} = \begin{cases} 1 & \text{if } I = I', \\ 0 & \text{if } I \neq I'. \end{cases}$$

The edge matrix E is the $\mathcal{V}(H) \times \mathcal{V}(A)$ matrix whose $[I, J, k]$th row is $\mathbf{y}^{[I,J,k]}$. By Theorem 2.4.12 we see that

$$A^m = DE, \quad A_H = ED.$$

It remains to show that

$$A_H \widetilde{R} = \widetilde{R} B^m,$$

where $\widetilde{R}$ is defined by (12–1–1). Observe that $R = D\widetilde{R}$. Thus

$$(12–1–6) \qquad\qquad A_H \widetilde{R} = ED\widetilde{R} = ER.$$

We claim that

$$\mathbf{y}^{[I,J,k]} R = \mathbf{e}_J B^m.$$

This follows from $\mathbf{x}^{(J,n)} R = \mathbf{e}_J S A^n R = \mathbf{e}_J B^m$ and $\mathbf{a}^{(I,n)} \in \ker R$. From the definitions of $\widetilde{R}$ and E we then see that

$$\mathbf{e}_{[I,J,k]} \widetilde{R} B^m = \mathbf{e}_J B^m = \mathbf{y}^{[I,J,k]} R = \mathbf{e}_{[I,J,k]} ER,$$

and hence $\widetilde{R} B^m = ER$. Comparison with (12–1–6) shows that $A_H \widetilde{R} = \widetilde{R} B^m$. This completes the proof of Theorem 12.1.6, and hence of Theorem 12.1.4. $\qquad\square$

The proof of Theorem 12.1.1 uses Theorem 12.1.4 together with marker ideas from Chapter 10 and additional state splitting ideas.

Example 12.1.7. Let

$$A = \begin{bmatrix} 0 & 1 & 0 & 0 & 0 \\ 0 & 0 & 1 & 0 & 0 \\ 0 & 0 & 0 & 1 & 0 \\ 0 & 0 & 0 & 0 & 1 \\ 1 & 0 & 0 & 0 & 1 \end{bmatrix} \quad \text{and} \quad B = \begin{bmatrix} 0 & 1 & 0 \\ 0 & 0 & 1 \\ 1 & 1 & 0 \end{bmatrix}.$$

If we let

$$R = \begin{bmatrix} 1 & 0 & 0 \\ 0 & 1 & 0 \\ 0 & 0 & 1 \\ 1 & 1 & 0 \\ 0 & 1 & 1 \end{bmatrix} \quad \text{and} \quad S = \begin{bmatrix} 1 & 0 & 0 & 0 & 0 \\ 0 & 1 & 0 & 0 & 0 \\ 0 & 0 & 1 & 0 & 0 \end{bmatrix},$$

then
$$AR = RB \qquad \text{and} \qquad SR = Id.$$

Hence (Δ_B, δ_B) is a quotient of (Δ_A, δ_A). We also claim that $\lambda_A = \lambda_B$. For A and B are each companion matrices, so we can read off their characteristic polynomials from their bottom rows, and find that $\chi_A(t) = t^5 - t^4 - 1 = (t^3 - t - 1)(t^2 - t + 1)$ and $\chi_B(t) = t^3 - t - 1$. The roots of $t^2 - t + 1$ have modulus 1, and so $\lambda_A = \lambda_B$, as claimed. Then Theorem 12.1.4 shows that X_B is a right-closing eventual factor of X_A. However, $\operatorname{tr} A > 0$ while $\operatorname{tr} B = 0$. Hence there is no sliding block code from X_A to X_B since the fixed point of X_A has no place to be mapped. In particular, X_B is not a factor of X_A.   □

There is an algorithm to decide whether one dimension pair is a quotient of another [KimR4], but as with shift equivalence, this procedure is impractical to apply in many cases. It is therefore useful to have some easily checked necessary conditions for (Δ_B, δ_B) to be a quotient of (Δ_A, δ_A), which by Theorem 12.1.4 are also necessary conditions for X_B to be a right-closing eventual factor of X_A, and so a right-closing factor of X_A.

In order to state these conditions, we review some notation from Chapters 6 and 7. For a matrix A we denote its nonzero spectrum by $\operatorname{sp}^\times(A)$, its characteristic polynomial away from 0 by

$$\chi_A^\times(t) = \prod_{\lambda \in \operatorname{sp}^\times(A)} (t - \lambda),$$

and its Jordan form away from 0 by $J^\times(A)$. We say that $J^\times(B) \subseteq J^\times(A)$ if there is a one-to-one mapping θ from the list of Jordan blocks of B away from zero into the list of Jordan blocks of A away from zero such that $\theta(J_k(\mu)) = J_m(\mu)$, where $m \geqslant k$ (as usual, when we use the term "list" we mean that multiplicity is taken into account). Finally, for a polynomial $p(t) \in \mathbb{Z}[t]$ such that $p(0) = \pm 1$, recall the generalized Bowen–Franks group

$$BF_p(A) = \mathbb{Z}^n / \mathbb{Z}^n p(A).$$

It is straightforward to show that $J^\times(B) \subseteq J^\times(A)$ if and only if there is a one-to-one linear mapping $\psi \colon \mathcal{R}_B \to \mathcal{R}_A$ that intertwines A with B (Exercise 12.1.6). Note that if $J^\times(B) \subseteq J^\times(A)$, then $J^\times(B)$ is a principal submatrix of $J^\times(A)$, but the converse is not true (Exercise 12.1.7).

Proposition 12.1.8. *Let A and B be integral matrices such that (Δ_B, δ_B) is a quotient of (Δ_A, δ_A). Then*

(1) *$\chi_B^\times(t)$ divides $\chi_A^\times(t)$ (equivalently, $\operatorname{sp}^\times(B) \subseteq \operatorname{sp}^\times(A)$),*
(2) *$J^\times(B) \subseteq J^\times(A)$,*
(3) *$BF_p(B)$ is a quotient group of $BF_p(A)$ for every $p(t) \in \mathbb{Z}[t]$ for which $p(0) = \pm 1$.*

PROOF: Clearly (1) follows from (2). To prove (2), let R and S be integral matrices such that $AR = RB$ and $SR = B^\ell$ for some $\ell \geqslant 0$. Define $\psi \colon \mathcal{R}_{B^\mathsf{T}} \to \mathcal{R}_{A^\mathsf{T}}$ by $\psi(\mathbf{v}) = \mathbf{v}R^\mathsf{T}$. As in the proof of Theorem 7.4.6, the image of ψ is contained in $\mathcal{R}_{A^\mathsf{T}}$ and ψ is a linear transformation which intertwines $B^\mathsf{T}|_{\mathcal{R}_{B^\mathsf{T}}}$ with $A^\mathsf{T}|_{\mathcal{R}_{A^\mathsf{T}}}$. Since $SR = B^\ell$ and B^ℓ is nonsingular on $\mathcal{R}_{B^\mathsf{T}}$, it follows that ψ is one-to-one. Hence

$$J^\times(B^\mathsf{T}) = J(B^\mathsf{T}|_{\mathcal{R}_{B^\mathsf{T}}}) \subseteq J(A^\mathsf{T}|_{\mathcal{R}_{A^\mathsf{T}}}) = J^\times(A^\mathsf{T}).$$

Now from linear algebra (see Exercise 7.4.5) we know that $J^\times(A^\mathsf{T}) = J^\times(A)$ and similarly for B, which completes the proof of (2).

For (3) observe that the map $\mathbf{v} \mapsto \mathbf{v}R$ defines the required quotient map, as in the proof of Theorem 7.4.17. $\qquad\square$

This Proposition, together with Theorem 12.1.4, allows us to find many pairs X_A and X_B of edge shifts having the same entropy such that X_B is not a right-closing eventual factor of X_A. As we will see in the next section, for each example below, X_B is not even an eventual factor (hence not a factor) of X_A.

Example 12.1.9. Let $A = \begin{bmatrix} 2 & 1 \\ 1 & 2 \end{bmatrix}$ and $B = \begin{bmatrix} 1 & 2 \\ 2 & 1 \end{bmatrix}$. Then $\mathrm{sp}^\times(A) = \{3, 1\}$ while $\mathrm{sp}^\times(B) = \{3, -1\}$. By Proposition 12.1.8(1), neither of X_A, X_B is a right-closing eventual factor of the other. Note however that $A^2 = B^2$; in particular X_A^2 is conjugate to X_B^2. $\qquad\square$

Example 12.1.10. Let A and B be the matrices in Example 7.4.8. Recall that $J^\times(A) = J_1(5) \oplus J_2(1)$ and $J^\times(B) = J_1(5) \oplus J_1(1) \oplus J_1(1)$. So, $J^\times(B) \not\subseteq J^\times(A)$. By Proposition 12.1.8(2), neither of X_A, X_B is a right-closing eventual factor of the other. $\qquad\square$

Example 12.1.11. Let A and B be the matrices in Example 7.3.4. Recall that $BF(A) \simeq \mathbb{Z}_4$ while $BF(B) \simeq \mathbb{Z}_2 \oplus \mathbb{Z}_2$. Thus by Proposition 12.1.8(3), neither of X_A, X_B is a right-closing eventual factor of the other. $\qquad\square$

Using Proposition 11.3.8, we can prove a dual version of Theorem 12.1.4. (Recall that the definition of embedding of dimension pairs requires that the image be pure.)

Theorem 12.1.12. *Let A and B be eventually positive integral matrices with $\lambda_A = \lambda_B$. Then the following are equivalent.*

(1) *B is a left-closing eventual factor of A.*
(2) *$(\Delta_{B^\mathsf{T}}, \delta_{B^\mathsf{T}})$ is a quotient of $(\Delta_{A^\mathsf{T}}, \delta_{A^\mathsf{T}})$.*
(3) *(Δ_B, δ_B) is embedded in (Δ_A, δ_A).*

PROOF: Use Theorem 12.1.4 and Proposition 11.3.8. $\qquad\square$

We conclude by stating versions of Theorems 12.1.1 and 12.1.4 for irreducible edge shifts.

Theorem 12.1.13. *Let X_A and X_B be irreducible edge shifts with $h(X_A) = h(X_B)$ and let $p = \mathrm{per}(A)$. Let A' be an arbitrary irreducible component of A^p and B' an arbitrary irreducible component of B^p. Then*

(1) *X_B is a right-closing factor of X_A if and only if $P(X_A) \searrow P(X_B)$ and $(\Delta_{B'}, \delta_{B'})$ is a quotient of $(\Delta_{A'}, \delta_{A'})$.*

(2) *X_B is a right-closing eventual factor of X_A if and only if $\mathrm{per}(B)$ divides p and $(\Delta_{B'}, \delta_{B'})$ is a quotient of $(\Delta_{A'}, \delta_{A'})$.*

We leave the proof of this to the reader. We caution that one of the conditions for right-closing eventual factor codes stated in [BoyMT, 3.9] is not correct (see Exercise 12.3.7).

EXERCISES

12.1.1. Let A and B be the matrices in Example 12.1.7.

(a) Using Theorem 12.1.1 show that X_{B^m} is a right-closing factor of X_{A^m} for all $m \geqslant 2$.

(b) Using the proof of Theorem 12.1.6 as a rough guide, give a specific right-closing factor code $X_{A^3} \to X_{B^3}$.

12.1.2. Let

$$A = \begin{bmatrix} 1 & 1 & 1 & 3 \\ 0 & 0 & 2 & 1 \\ 0 & 2 & 0 & 2 \\ 1 & 0 & 0 & 0 \end{bmatrix}, \quad B = \begin{bmatrix} 0 & 2 & 1 \\ 2 & 0 & 2 \\ 1 & 1 & 0 \end{bmatrix},$$

$$R = \begin{bmatrix} 1 & 1 & 0 \\ 1 & 0 & 0 \\ 0 & 1 & 0 \\ 0 & 0 & 1 \end{bmatrix}, \quad S = \begin{bmatrix} 0 & 1 & 0 & 0 \\ 0 & 0 & 1 & 0 \\ 0 & 0 & 0 & 1 \end{bmatrix}.$$

(a) Show that $AR = RB$ and that $SR = \mathrm{Id}$ and so conclude that X_B is an eventual right-closing factor of X_A.

(b) Using the proof of Theorem 12.1.6 as a rough guide, give a specific right-closing factor code $X_{A^3} \to X_{B^3}$.

12.1.3. Complete the proof of the necessity of the dimension quotient condition in Theorem 12.1.4 by proving Lemma 12.1.5. [*Hint*: Modify the proof of Lemma 7.5.16.]

12.1.4. Let A be an irreducible integral matrix with $\lambda_A = n \in \mathbb{N}$. Give an explicit quotient mapping $(\Delta_A, \delta_A) \to (\Delta_{[n]}, \delta_{[n]})$.

12.1.5. Prove Theorem 12.1.13 assuming Theorems 12.1.1 and 12.1.4.

12.1.6. For a pair of matrices A, B, show that $J^\times(B) \subseteq J^\times(A)$ if and only if there is a one-to-one linear mapping $\psi \colon \mathcal{R}_B \to \mathcal{R}_A$ that intertwines A with B.

12.1.7. Give an example of a pair A, B of matrices such that $J^\times(B)$ is a principal submatrix of $J^\times(A)$, but for which $J^\times(B) \not\subseteq J^\times(A)$.

12.1.8. Complete the proof of Proposition 12.1.8(3).

§12.2. Eventual Factors of Equal Entropy

In this section, we characterize when one irreducible shift of finite type is an eventual factor of another having the same entropy. As usual, we first focus on mixing edge shifts.

Theorem 12.2.1. *Let X_A and X_B be mixing edge shifts with the same entropy. Then the following are equivalent.*

(1) *X_B is an eventual factor of X_A.*
(2) *There is a pure δ_A-invariant subgroup $\Delta \subseteq \Delta_A$ such that (Δ_B, δ_B) is a quotient of $(\Delta, \delta_A|_\Delta)$.*
(3) *There is an eventually positive integral matrix C such that B is a right-closing eventual factor of C and C is a left-closing eventual factor of A.*

PROOF: $(2) \Rightarrow (3)$. First apply Proposition 11.3.6 to realize $(\Delta, \delta_A|_\Delta)$ as a dimension pair (Δ_C, δ_C) of an integral matrix C. Next, apply Theorem 11.1.19 to show that we can choose C to be eventually positive. For this we need to know that C is spectrally Perron. To see this, first note that since $(\Delta_C, \delta_C) \simeq (\Delta, \delta_A|_\Delta)$, it follows that $\mathrm{sp}^\times(C) \subseteq \mathrm{sp}^\times(A)$. Also, (Δ_B, δ_B) is a quotient of (Δ_C, δ_C), so by Proposition 12.1.8(1), we see that $\lambda_A = \lambda_B$ is an eigenvalue of C. Hence C is spectrally Perron. We can then apply Theorems 12.1.4 and 12.1.12 to conclude (3).

$(3) \Rightarrow (1)$. Obvious.

$(1) \Rightarrow (2)$. We first handle the case where X_B is a factor (rather than an eventual factor) of X_A. Let n denote the size of A and m the size of B.

By Theorem 8.1.16, X_B is a finite-to-one factor of X_A. Then the Replacement Theorem 9.2.5 shows that there is an almost invertible factor code $\phi: X_A \to X_B$. By recoding as in Proposition 9.1.7, we may assume that $\phi = \Phi_\infty$ is a 1-block code with magic symbol $b \in \mathcal{E}(G_B)$. Let a be the unique Φ-pre-image of b.

Denote by $\mathcal{L}_b$ the set of paths π in G_B that begin with b. For each $\pi \in \mathcal{L}_b$ let $\mathbf{x}^\pi$ denote the vector indexed by $\mathcal{V}(A)$ that records the distribution of terminal states of the paths w in G_A such that $\Phi(w) = \pi$:

$$(\mathbf{x}^\pi)_I = \text{number of paths } w \text{ in } G_A \text{ ending at } I \text{ with } \Phi(w) = \pi.$$

Since b is a magic symbol, every w in G_A with $\Phi(w) = \pi$ must start with a. Since ϕ is finite-to-one, the labeled graph (G_A, Φ) has no graph diamonds, so that $\mathbf{x}^\pi$ is a 0–1 vector (see Figure 12.2.1(a)).

Similarly, let $\mathcal{R}_b$ denote the set of paths τ in G_B that end with b. For $\tau \in \mathcal{R}_b$ let $\mathbf{y}^\tau$ denote the vector recording the distribution of initial states of paths u in G_A with $\Phi(u) = \tau$. All such paths must end in a. As before, $\mathbf{y}^\tau$ is a 0–1 vector (see Figure 12.2.1(b)).

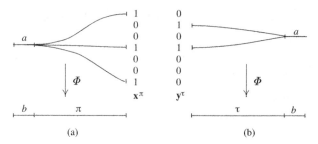

FIGURE 12.2.1.   Paths and distribution vectors.

View the vectors $\mathbf{x}^\pi$ as row vectors in $\mathbb{Z}^n$ and the $\mathbf{y}^\tau$ as column vectors. Let L be the subgroup of $\mathbb{Z}^n$ generated by the set $\{\mathbf{x}^\pi : \pi \in \mathcal{L}_b\}$. Put

$$\Delta = \langle L \rangle_{\mathbb{Q}} \cap \Delta_A.$$

Our proof that Δ satisfies condition (2) is based on two observations. The first is that for all $\pi \in \mathcal{L}_b$,

$$(12\text{--}2\text{--}1) \qquad \mathbf{x}^\pi A = \sum_{\{e \in \mathcal{E}(G_B):\ t(\pi) = i(e)\}} \mathbf{x}^{\pi e}.$$

To see this, note that if $I \in \mathcal{V}(A)$, then both $(\mathbf{x}^\pi A)_I$ and $\sum(\mathbf{x}^{\pi e})_I$ count the same thing, namely the number of paths in G_A that end at state I and map under Φ to a path in G_B of the form πe.

The second observation is that for all $\pi \in \mathcal{L}_b$ and $\tau \in \mathcal{R}_b$,

$$(12\text{--}2\text{--}2) \qquad \mathbf{x}^\pi \cdot \mathbf{y}^\tau = \begin{cases} 1 & \text{if } t(\pi) = i(\tau), \\ 0 & \text{if } t(\pi) \neq i(\tau). \end{cases}$$

To see this, first note that $\mathbf{x}^\pi \cdot \mathbf{y}^\tau$ counts the number of paths in G_A of the form wu such that $\Phi(w) = \pi$ and $\Phi(u) = \tau$. If $t(\pi) = i(\tau)$, then $\pi\tau \in \mathcal{B}(X_B)$, so there must be at least one such path wu. But since b is a magic symbol, such paths must begin and end with a. Since (G_A, Φ) has no graph diamonds, it follows that there is exactly one such path wu. If $t(\pi) \neq i(\tau)$, then $\pi\tau \notin \mathcal{B}(X_B)$, so there are no such paths wu. See Figure 12.2.1, where putting together (a) and (b) creates an allowed path in G_A if and only if $\pi\tau$ is allowed in G_B.

Since Δ is the intersection of a rational subspace with Δ_A, it follows that Δ is a pure subgroup of Δ_A. Observation (12–2–1) shows that Δ is δ_A-invariant. It remains to prove that (Δ_B, δ_B) is a quotient of $(\Delta, \delta_A|_\Delta)$.

We define a quotient mapping as follows. For each state $J \in \mathcal{V}(B)$, choose a path $\tau^J \in \mathcal{R}_b$ that begins with J. Define R to be the $n \times m$ matrix whose Jth column is $\mathbf{y}^{\tau^J}$. Clearly R is integral. Define $\phi \colon \Delta \to \mathbb{Q}^m$ by $\phi(\mathbf{v}) = \mathbf{v} R B^m$. We will show that ϕ defines a quotient mapping $(\Delta_A, \delta_A) \to (\Delta_B, \delta_B)$.

First, to see that ϕ intertwines $\delta_A|_\Delta$ with δ_B it is enough to show that

(12–2–3) $\qquad\qquad \mathbf{x}^\pi A R = \mathbf{x}^\pi R B \qquad$ for all $\pi \in \mathcal{L}_b$.

Now (12–2–2) shows that

(12–2–4) $\qquad\qquad \mathbf{x}^\pi R = \mathbf{e}_{t(\pi)} \qquad$ for all $\pi \in \mathcal{L}_b$.

Combining this with (12–2–1) shows that

$$\mathbf{x}^\pi A R = \sum_{\{e \in \mathcal{E}(G_B):\; t(\pi)=i(e)\}} \mathbf{x}^{\pi e} R = \sum_{\{e \in \mathcal{E}(G_B):\; t(\pi)=i(e)\}} \mathbf{e}_{t(e)},$$

while

$$\mathbf{x}^\pi R B = \mathbf{e}_{t(\pi)} B = \sum_{\{e \in \mathcal{E}(G_B):\; t(\pi)=i(e)\}} \mathbf{e}_{t(e)},$$

verifying (12–2–3).

Now, RB^m is integral and the image of the map $\mathbf{v} \mapsto \mathbf{v}RB^m$ is contained in the eventual range of B. By (12–2–3), it follows that $\phi(\Delta)$ is a δ_B-invariant subgroup of Δ_B.

We complete the proof (in the factor case) by showing that ϕ is onto. For this, observe from (12–2–4) and integrality of R that R applied to the subgroup L is exactly $\mathbb{Z}^m$. Suppose $\mathbf{v} \in \Delta_B$. Then $\mathbf{v}B^k \in \mathbb{Z}^m$ for some k, and thus $\mathbf{v}B^k = \mathbf{w}R$ for some $\mathbf{w} \in L$. This together with (12–2–3) yields $\mathbf{v}B^{k+m+n} = \phi(\mathbf{x})$, where $\mathbf{x} = \mathbf{w}A^n \in \Delta$. Thus $\mathbf{v}B^{k+m+n} \in \phi(\Delta)$, and so $\mathbf{v} \in \phi(\Delta)$ by δ_B-invariance.

We now turn to the case when X_B is only assumed to be an eventual factor of X_A. By what we have proved so far, there is an m_0 such that for all $m \geqslant m_0$ there is a pure δ_A^m-invariant subgroup Δ_m of Δ_{A^m} such that $(\Delta_{B^m}, \delta_{B^m})$ is a quotient of $(\Delta_m, \delta_{A^m}|_{\Delta_m})$. Recall that $\Delta_{A^m} = \Delta_A$, and so Δ_m may be regarded as a subgroup of Δ_A. An argument using the Jordan form (similar to that in Lemma 7.5.16) shows that if $m \geqslant m_0$ is chosen so that the ratio of any two distinct nonzero eigenvalues of A is not an mth root of unity, then Δ_m is actually δ_A-invariant. For such an m, use Proposition 11.3.6 to realize $(\Delta_m, \delta_A|_{\Delta_m})$ as the dimension pair of an integral matrix C. We can further assume that the ratio of any eigenvalue of C to a different eigenvalue of B is not an mth root of unity, so that Lemma 12.1.5 implies that (Δ_B, δ_B) is a quotient of $(\Delta_m, \delta_A|_{\Delta_m})$. $\qquad\square$

Note that the implication (1) $\Rightarrow$ (3) in this theorem gives a weak sense in which every entropy-preserving factor code from one mixing edge shift to another can be replaced by a composition of a left-closing factor code followed by a right-closing factor code. This contrasts with Exercise 8.1.13, which shows that not all entropy-preserving factor codes from one mixing shift of finite type to another can be decomposed into such factor codes in the usual sense.

There is an extension of Theorem 12.2.1 to irreducible shifts of finite type, whose proof we leave to the reader.

Proposition 12.2.2. *Let X_A and X_B be irreducible edge shifts with the same entropy. Let $p = \mathrm{per}(A)$. Let A' be an irreducible component of A^p and B' be an irreducible component of B^p. Then X_B is an eventual factor of X_A if and only if $\mathrm{per}(B)$ divides $\mathrm{per}(A)$ (in which case both $X_{A'}$ and $X_{B'}$ are mixing) and $X_{B'}$ is an eventual factor of $X_{A'}$.*

Although condition (2) of Theorem 12.2.1 gives an explicit necessary and sufficient condition, it is not clear whether it is decidable. However, there are some cases for which one can show decidability (see Exercise 12.2.6). Also there are some easily checkable necessary conditions, similar to Proposition 12.1.8.

Proposition 12.2.3. *Let X_A and X_B be irreducible edge shifts with the same entropy. If X_B is an eventual factor of X_A (in particular, if X_B is a factor of X_A), then*

(1) $\chi_B^\times(t)$ *divides* $\chi_A^\times(t)$ *(equivalently,* $\mathrm{sp}^\times(B) \subseteq \mathrm{sp}^\times(A)$*),*
(2) $J^\times(B) \subseteq J^\times(A)$*, and*
(3) $BF_p(B)$ *is a quotient group of* $BF_p(A)$ *for every* $p(t) \in \mathbb{Z}[t]$ *for which* $p(0) = \pm 1$.

PROOF: We deal with the mixing case, leaving the reduction to this case to the reader. Linear algebra shows that any matrix A is similar over $\mathbb{C}$ to A^T and, if A is integral, then A is equivalent via elementary operations over $\mathbb{Z}$ to A^T (see Exercises 7.4.5 and 7.4.9). Then using Proposition 12.1.8 together with Theorems 12.2.1, 12.1.4, and 12.1.12 gives all three conditions.   □

This proposition shows that everything we said about the nonexistence of right-closing eventual factors in Examples 12.1.9, 12.1.10, 12.1.11 can also be said for eventual factors. In particular, Example 12.1.9 gives a pair A, B of matrices such that X_B is not an eventual factor of X_A, but for which X_{B^2} is a factor of (indeed conjugate to) X_{A^2}.

Recall from Theorem 7.5.15 that A and B are shift equivalent if and only if X_A and X_B are eventually conjugate. This gives a "coding" interpretation of shift equivalence. We will give another "coding" interpretation by using the following concept.

Definition 12.2.4. Two shifts X and Y are *weakly conjugate* if each is a factor of the other.

For example, we know from Example 1.5.6 and Exercise 1.5.16 that the golden mean shift and the even shift are weakly conjugate, but they are not conjugate since one is a shift of finite type and the other is strictly sofic.

Theorem 12.2.5. *Let X_A and X_B be irreducible edge shifts. Then A and B are shift equivalent if and only if X_A and X_B are weakly conjugate. Thus for irreducible shifts of finite type eventual conjugacy is the same as weak conjugacy.*

PROOF: We treat the mixing case, leaving the irreducible case to the reader. If A and B are shift equivalent, then they satisfy the conditions of Theorem 12.1.1 for X_A and X_B to be factors (indeed, right-closing factors) of one another.

Conversely, first observe that X_A and X_B must have the same entropy. Applying Theorem 12.2.1 shows that there are pure subgroups $\Delta_1 \subseteq \Delta_A$ and $\Delta_2 \subseteq \Delta_B$ such that (Δ_B, δ_B) is a quotient of $(\Delta_1, \delta_A|_{\Delta_1})$ and (Δ_A, δ_A) is a quotient of $(\Delta_2, \delta_B|_{\Delta_2})$. But then

$$\dim\langle\Delta_A\rangle_\mathbb{Q} \geqslant \dim\langle\Delta_1\rangle_\mathbb{Q} \geqslant \dim\langle\Delta_B\rangle_\mathbb{Q} \geqslant \dim\langle\Delta_2\rangle_\mathbb{Q} \geqslant \dim\langle\Delta_A\rangle_\mathbb{Q}.$$

Hence the dimensions of all these subspaces must coincide. Since there is only one pure subgroup spanning each rational subspace, it follows that $\Delta_1 = \Delta_A$ and $\Delta_2 = \Delta_B$. Thus (Δ_B, δ_B) is a quotient of (Δ_A, δ_A) by an onto group homomorphism $\psi \colon \Delta_A \to \Delta_B$. By Proposition 7.5.6, ψ extends to an onto linear mapping $\mathcal{R}_A \to \mathcal{R}_B$. Since $\langle\Delta_A\rangle_\mathbb{Q} = \mathcal{R}_A$ and $\langle\Delta_B\rangle_\mathbb{Q} = \mathcal{R}_B$ have the same dimension, it follows that this mapping, and hence also ψ, is one-to-one. Thus $(\Delta_A, \delta_A) \simeq (\Delta_B, \delta_B)$, and so A and B are shift equivalent. $\qquad\square$

For mixing edge shifts, the dimension group condition in (2) of Theorem 12.2.1 together with the periodic point condition $P(X_A) \searrow P(X_B)$ subsume all known necessary conditions for the existence of an entropy-preserving factor code from X_A to X_B. It is not known whether these conditions are also sufficient.

We conclude this section by discussing extensions of our results to sofic shifts. There are differences. For example, there is a mixing sofic shift of entropy $\log n$ that does *not* factor onto the full n-shift [TroW] (see also [KarM]), in contrast with the situation for shifts of finite type (see Theorem 5.5.8). Furthermore, weak conjugacy and eventual conjugacy are not the same for mixing sofic shifts: we saw that the golden mean shift and the even shift are weakly conjugate, but they cannot be eventually conjugate since powers of the golden mean shift always have finite type while powers of the even shift are always strictly sofic. But much of what we have done for shifts of finite type can be carried over to sofic shifts. An essential tool for this is the *core matrix* introduced by Nasu [Nas3], which we discuss in §13.1. See also [WilS3], [WilS5] and [TroW].

EXERCISES

12.2.1. Complete the proof of (1) $\Rightarrow$ (2) in Theorem 12.2.1 by showing the following.

   (a) For any nonsingular matrix A and positive integer m, if the ratio of distinct eigenvalues of A is not an mth root of unity, then every A^m-invariant subspace is actually A-invariant.

   (b) Let A and B be matrices. There are arbitrarily large m such that the ratio of two distinct eigenvalues of A is not an mth root of unity, and the ratio of any eigenvalue of A to a distinct eigenvalue of B is also not an mth root of unity.

12.2.2. Let

$$A = \begin{bmatrix} 5 & 3 \\ 3 & 5 \end{bmatrix} \quad \text{and} \quad B = \begin{bmatrix} 6 & 8 \\ 1 & 4 \end{bmatrix}.$$

Show that X_A and X_B have the same entropy, zeta function and Jordan form away from zero, but that X_B is *not* an eventual factor of X_A. [*Hint*: Consider the generalized Bowen–Franks group BF_{1+t}.]

12.2.3. Show that the edge shifts X_A and X_B of Example 12.1.7 have the same entropy, but they have no common finite-type factor of the same entropy.

12.2.4. Prove Proposition 12.2.2.

12.2.5. Prove Proposition 12.2.3 and Theorem 12.2.5 in the periodic case.

12.2.6. Assume that A is primitive integral and has simple nonzero spectrum, i.e., that every nonzero eigenvalue has multiplicity one. Show that the eventual factor problem of determining when X_A eventually factors onto another mixing edge shift X_B is decidable. (Use the fact [KimR4] that the problem of determining when, for primitive integral matrices, one dimension pair is a quotient of another is decidable.)

§12.3.   Ideal Classes

Recall from §10.3 the notion of a finite-state (X,Y)-code, which generalized that of a finite-state (X,n)-code from §5.2. Given sofic shifts X and Y, a finite-state (X,Y)-code is a graph G together with two labelings $\mathfrak{I}$ and $\mathfrak{O}$ of G, such that the labeled graph $(G,\mathfrak{I})$ is right-resolving and presents Y and $(G,\mathfrak{O})$ is right-closing and presents a subshift of X. The existence of a finite-state (X,Y)-code clearly forces $h(Y) \leqslant h(X)$. We showed in Theorem 10.3.7 that when X and Y are irreducible and $h(Y) < h(X)$ there is always a finite-state (X,Y)-code. This section deals with the equal entropy case $h(Y) = h(X)$.

When X and Y are irreducible sofic shifts and $h(Y) = h(X)$, then for every finite-state (X,Y)-code $(G,\mathfrak{I},\mathfrak{O})$, $\mathfrak{O}_\infty$ (as well as $\mathfrak{I}_\infty$) must be onto. Hence a finite-state (X,Y)-code between irreducible sofic shifts of equal entropy is a finite equivalence with one leg right-resolving and the other leg right-closing. By Proposition 5.1.11, this is, up to recoding, the same as requiring both legs to be right-closing. Recall from §10.3 that such a finite equivalence is called a right-closing finite equivalence. Recall also that we treated right-resolving finite equivalences in §8.4.

Let X_A and X_B be irreducible edge shifts. We will see that a right-closing finite equivalence between X_A and X_B forces an arithmetic relation between the entries of the right Perron eigenvectors of A and B. Furthermore, this relation is essentially the only obstruction to such a finite equivalence, leading to a complete description in Theorem 12.3.7 of when two irreducible sofic shifts are right-closing finitely equivalent.

The arithmetic relation mentioned above is best formulated in terms of ideal classes, described as follows.

Let $\mathfrak{R}$ be a subring of the real numbers that contains 1. An *ideal* in $\mathfrak{R}$ is a subset of $\mathfrak{R}$ that is closed under linear combinations with coefficients in $\mathfrak{R}$. For a subset E of $\mathfrak{R}$ we let $\langle E \rangle_{\mathfrak{R}}$ denote the ideal in $\mathfrak{R}$ generated by E, i.e., the smallest ideal in $\mathfrak{R}$ containing E. Two ideals $\mathfrak{a}$ and $\mathfrak{b}$ in $\mathfrak{R}$ are said to be *equivalent*, denoted $\mathfrak{a} \sim \mathfrak{b}$, provided that $\mathfrak{b}$ is a nonzero scalar multiple of $\mathfrak{a}$; i.e., there is a real number $t \neq 0$ such that $\mathfrak{b} = t\mathfrak{a}$. Note that t is *not* required to be an element of $\mathfrak{R}$. The reader should check that $\sim$ is indeed an equivalence relation. Equivalence classes of ideals under $\sim$ are called *ideal classes*. The ideal class of an ideal $\mathfrak{a}$ is denoted by $[\![\mathfrak{a}]\!]_{\mathfrak{R}}$.

If $r \in \mathfrak{R}$, then $\langle r \rangle_{\mathfrak{R}} = r\mathfrak{R}$ consists of just the multiples of r; this is called the *principal ideal* generated by r. For instance, the ring $\mathfrak{R}$ itself is principal, since it has the form $\mathfrak{R} = \langle 1 \rangle_{\mathfrak{R}}$. We claim that the collection of all nonzero principal ideals is precisely the ideal class of the ideal $\mathfrak{R}$. For if $\mathfrak{a} = \langle r \rangle_{\mathfrak{R}}$ and $\mathfrak{b} = \langle s \rangle_{\mathfrak{R}}$ are nonzero principal ideals, then $\mathfrak{b} = (s/r)\mathfrak{a}$. Conversely, if $\mathfrak{a} = \langle r \rangle_{\mathfrak{R}}$ is principal and $\mathfrak{b} = t\mathfrak{a}$, then $\mathfrak{b} = \langle tr \rangle_{\mathfrak{R}}$.

We shall mainly use rings of the form $\mathfrak{R} = \mathbb{Z}[1/\lambda]$, the ring of polynomials in $1/\lambda$ with integer coefficients, where λ is a nonzero algebraic integer. Note that if $f_{\lambda}(t) = t^d + c_{d-1}t^{d-1} + \cdots + c_1 t + c_0$ is the minimal polynomial of λ, then

$$\lambda = -c_{d-1} - c_{d-2}\left(\frac{1}{\lambda}\right) - \cdots - c_1\left(\frac{1}{\lambda}\right)^{d-2} - c_0\left(\frac{1}{\lambda}\right)^{d-1}$$

expresses λ as a polynomial in $1/\lambda$, proving that $\mathbb{Z}[\lambda] \subseteq \mathbb{Z}[1/\lambda]$. For a subset $E \subseteq \mathbb{Z}[1/\lambda]$ we abbreviate $\langle E \rangle_{\mathbb{Z}[1/\lambda]}$ to $\langle E \rangle$, and for an ideal $\mathfrak{a}$ of $\mathbb{Z}[1/\lambda]$ we shorten $[\![\mathfrak{a}]\!]_{\mathbb{Z}[1/\lambda]}$ to $[\![\mathfrak{a}]\!]$. For a vector $\mathbf{v}$, we let $\langle \mathbf{v} \rangle$ denote the $\mathbb{Z}[1/\lambda]$-ideal generated by the entries of $\mathbf{v}$.

Let A be an irreducible integral matrix, and λ_A be its Perron eigenvalue. Using Gaussian elimination, we can find a right Perron eigenvector $\mathbf{v} = \mathbf{v}_A = [v_1, v_2, \ldots, v_n]^{\mathsf{T}}$ for A whose entries lie in the field of fractions of $\mathbb{Z}[\lambda]$. We can then clear denominators to assume that the v_j all lie in $\mathbb{Z}[\lambda] \subseteq \mathbb{Z}[1/\lambda]$. Define the *ideal class of A* to be the class $\mathfrak{I}(A)$ of the ideal generated by the entries of $\mathbf{v}_A$, i.e.,

$$\mathfrak{I}(A) = [\![\langle v_1, v_2, \ldots, v_n \rangle]\!] = [\![\langle \mathbf{v}_A \rangle]\!].$$

Note that any other right Perron eigenvector is a nonzero multiple of $\mathbf{v}$, so that the ideal generated by its entries is equivalent to $\langle v_1, v_2, \ldots, v_n \rangle$. This shows that $\mathfrak{I}(A)$ is well-defined.

Our first result shows that equality of ideal classes is necessary for a right-closing finite equivalence.

Theorem 12.3.1. *Let A and B be irreducible integral matrices. If X_A and X_B are right-closing finitely equivalent, then $\mathfrak{I}(A) = \mathfrak{I}(B)$.*

PROOF: Since finite equivalence preserves entropy, we see that $\lambda_A = \lambda_B$. Suppose that (W, ϕ_A, ϕ_B) is a finite equivalence with ϕ_A and ϕ_B right-closing. By Proposition 8.3.6 we may assume that W is an irreducible edge shift X_C. We can recode right-closing codes to right-resolving codes according to Proposition 5.1.11. Hence there are irreducible integral matrices A' and B', right-resolving factor codes $\psi_A = (\Psi_A)_\infty \colon X_{A'} \to X_A$ and $\psi_B = (\Psi_B)_\infty \colon X_{B'} \to X_B$, and conjugacies $X_C \to X_{A'}$ and $X_C \to X_{B'}$, related as follows.

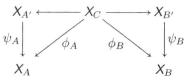

Now Proposition 8.2.2 shows that every right-resolving factor code from an irreducible edge shift is right-covering. Hence we can lift a right Perron eigenvector $\mathbf{v}_A$ for A to a right Perron eigenvector $\mathbf{v}_{A'}$ for A' by just copying entries using the associated map $\partial\Psi_A$ on states, i.e.,

$$(\mathbf{v}_{A'})_I = (\mathbf{v}_A)_{\partial\Psi_A(I)}.$$

This shows that A and A' have right Perron eigenvectors whose entries form the same set of numbers, so that $\mathfrak{I}(A') = \mathfrak{I}(A)$. Similarly, $\mathfrak{I}(B') = \mathfrak{I}(B)$.

Since $X_{A'} \cong X_{B'}$, it remains only to show that the ideal class of a matrix is invariant under elementary equivalence (and hence under conjugacy of edge shifts). For this, suppose that $(R, S) \colon \widetilde{A} \approx \widetilde{B}$, where $\widetilde{A}$ and $\widetilde{B}$ are irreducible integral matrices, so that

$$\widetilde{A} = RS \qquad \text{and} \qquad \widetilde{B} = SR.$$

Let $\lambda = \lambda_{\widetilde{A}} = \lambda_{\widetilde{B}}$, and $\mathbf{v}$ be a right Perron eigenvector for $\widetilde{A}$ with entries in $\mathbb{Z}[1/\lambda]$. Then $\mathbf{w} = S\mathbf{v}$ is a right Perron eigenvector for $\widetilde{B}$ and has entries in $\mathbb{Z}[1/\lambda]$. Since the entries in $\mathbf{w}$ are integral combinations of those in $\mathbf{v}$, clearly $\langle \mathbf{w} \rangle \subseteq \langle \mathbf{v} \rangle$. On the other hand,

$$\mathbf{v} = \frac{1}{\lambda}\widetilde{A}\mathbf{v} = \frac{1}{\lambda}RS\mathbf{v} = \frac{1}{\lambda}R\mathbf{w},$$

so that each entry in $\mathbf{v}$ is a linear combination of entries in $\mathbf{w}$ using coefficients in $\mathbb{Z}[1/\lambda]$. Thus $\langle \mathbf{v} \rangle \subseteq \langle \mathbf{w} \rangle$, so that $\langle \mathbf{v} \rangle = \langle \mathbf{w} \rangle$. Hence

$$\mathfrak{I}(\widetilde{B}) = [\![\langle \mathbf{w} \rangle]\!] = [\![\langle \mathbf{v} \rangle]\!] = \mathfrak{I}(\widetilde{A}).$$

Therefore the ideal classes of irreducible integral matrices that are strong shift equivalent coincide, completing the proof. $\square$

Note that the condition $\mathfrak{I}(A) = \mathfrak{I}(B)$ can be expressed arithmetically by saying that for suitably chosen $\mathbf{v}_A, \mathbf{v}_B$, each entry in $\mathbf{v}_A$ is a $\mathbb{Z}[1/\lambda]$-combination of entries in $\mathbf{v}_B$, and vice versa. Since $\mathbf{v}_A$ and $\mathbf{v}_B$ are not uniquely defined, the choice of $\mathbf{v}_B$ may depend on $\mathbf{v}_A$, and this is what makes the ideal condition, $\mathfrak{I}(A) = \mathfrak{I}(B)$, harder to check than you might expect. Nevertheless, it can be checked in many cases such as the following example.

Example 12.3.2. Let

$$A = \begin{bmatrix} 1 & 0 & 0 & 1 \\ 0 & 1 & 0 & 1 \\ 0 & 0 & 1 & 1 \\ 2 & 2 & 2 & 5 \end{bmatrix}.$$

We will show that A and A^{T} have different ideal classes, and so the corresponding edge shifts, $X_A, X_{A^{\mathsf{T}}}$, have the same entropy but are not right-closing finitely equivalent (and in particular are not conjugate).

As the reader can easily verify, A has an eigenvalue $\lambda = 3 + \sqrt{10}$ with right eigenvector $\mathbf{v} = (1, 1, 1, 2 + \sqrt{10})$ and left eigenvector $\mathbf{w} = (2, 2, 2, 2 + \sqrt{10})$. By the Perron–Frobenius Theorem, λ is the Perron eigenvalue and $\mathbf{v}, \mathbf{w}$ are right, left Perron eigenvectors. Observe that $1/\lambda = -3 + \sqrt{10}$, and so

$$\mathbb{Z}[1/\lambda] = \mathbb{Z}[-3 + \sqrt{10}] = \mathbb{Z}[\sqrt{10}] = \{a + b\sqrt{10} : a, b \in \mathbb{Z}\}.$$

Now,

$$\mathfrak{I}(A) = [\mathbf{v}] = [\langle 1 \rangle],$$

the equivalence class of principal ideals. Thus, it suffices to show that the ideal

$$\mathfrak{I} = \langle 2, 2 + \sqrt{10} \rangle = \langle 2, \sqrt{10} \rangle$$

is not principal.

For this, we first show that $\mathfrak{I} \neq \mathbb{Z}[\sqrt{10}]$, equivalently, that $1 \notin \mathfrak{I}$. Suppose to the contrary, that $1 \in \mathfrak{I}$. Then there are integers a, b, c, d such that

$$(a + b\sqrt{10})(2) + (c + d\sqrt{10})(\sqrt{10}) = 1,$$

or, equivalently,

$$(2a + 10d) + (2b + +c)\sqrt{10} = 1.$$

Since $\sqrt{10}$ is irrational, we conclude that $2a + 10d = 1$, a contradiction. So, $\mathfrak{I} \neq \mathbb{Z}[\sqrt{10}]$, as desired.

Next, we claim that 2 and $\sqrt{10}$ are *coprime* in $\mathbb{Z}[\sqrt{10}]$; this means that if $r \in \mathbb{Z}[\sqrt{10}]$ is a common divisor of 2 and $\sqrt{10}$ (i.e., $2/r$ and $\sqrt{10}/r$ both belong to $\mathbb{Z}[\sqrt{10}]$), then r is a *unit* in $\mathbb{Z}[\sqrt{10}]$ (i.e., invertible in $\mathbb{Z}[\sqrt{10}]$).

Assuming this for the moment, observe that if $\mathfrak{J}$ were a principal ideal, $\mathfrak{J} = \langle r \rangle$, then $r \in \mathbb{Z}[\sqrt{10}]$ would be a common divisor of 2 and $\sqrt{10}$, and so r would be a unit. But then we would have $\mathfrak{J} = \langle r \rangle = \mathbb{Z}[\sqrt{10}]$, contrary to what we have already proved above.

So it remains to show that 2 and $\sqrt{10}$ are coprime in $\mathbb{Z}[\sqrt{10}]$. For this, consider the function $N: \mathbb{Z}[\sqrt{10}] \to \mathbb{Z}$ defined by

$$N(a + b\sqrt{10}) = (a + b\sqrt{10})(a - b\sqrt{10}) = a^2 - 10b^2.$$

This function is multiplicative, i.e.,

$$N(rs) = N(r)N(s).$$

So if r divides r' in $\mathbb{Z}[\sqrt{10}]$, then $N(r)$ divides $N(r')$ in $\mathbb{Z}$. Note that if $r = a + b\sqrt{10}$ and $N(r) = \pm 1$, then $\pm(a - b\sqrt{10})$ is an inverse of r, and so r is a unit in $\mathbb{Z}[\sqrt{10}]$.

Now

$$N(2) = 4, \quad N(\sqrt{10}) = -10.$$

So if r is a common divisor of 2 and $\sqrt{10}$, then $N(r) = \pm 1$ or ± 2. We will show that $N(r) \neq \pm 2$.

Write $r = a + b\sqrt{10}$. If $N(r) = \pm 2$, then we have $a^2 - 10b^2 = \pm 2$. So, either 2 or -2 is a residue class of a perfect square modulo 10. But the only residue classes of perfect squares modulo 10 are $0, 1, 4, 5, 6, 9$. So, $N(r) \neq \pm 2$. Thus, $N(r) = \pm 1$, and so r is a unit in $\mathbb{Z}[\sqrt{10}]$. Hence 2 and $\sqrt{10}$ are coprime, as desired. $\qquad\square$

In a similar way, one can show that the familiar matrices $\begin{bmatrix} 4 & 1 \\ 1 & 0 \end{bmatrix}$ and $\begin{bmatrix} 3 & 2 \\ 2 & 1 \end{bmatrix}$ have different ideal classes (Exercise 12.3.5), and this gives another viewpoint on why the corresponding edge shifts are not conjugate.

In Example 12.3.4 below we exhibit mixing edge shifts X_A and X_C with the same entropy such that X_A is a factor of X_C, but X_A is *not* a right-closing eventual factor (in particular, not a right-closing factor) of X_C. For this, we isolate an idea from the proof of Theorem 12.3.1 in the following result.

Proposition 12.3.3. *Let A and B be irreducible integral matrices with $\lambda_A = \lambda_B$. If (Δ_B, δ_B) is a quotient of (Δ_A, δ_A), then $\mathfrak{J}(A) = \mathfrak{J}(B)$.*

PROOF: By Proposition 11.1.14, there are integral matrices R and S and an integer $\ell \geq 0$ such that $AR = RB$ and $SR = B^\ell$. Let $\mathbf{v}$ be a right Perron eigenvector for B. Then $\mathbf{u} = R\mathbf{v}$ is a right Perron eigenvector for A. Clearly $\langle \mathbf{u} \rangle \subseteq \langle \mathbf{v} \rangle$. Since

$$\mathbf{v} = \frac{1}{\lambda^\ell} B^\ell \mathbf{v} = \frac{1}{\lambda^\ell} SR\mathbf{v} = \frac{1}{\lambda^\ell} S\mathbf{u},$$

we see that $\langle \mathbf{v} \rangle \subseteq \langle \mathbf{u} \rangle$ as well. Hence $\mathfrak{J}(A) = \mathfrak{J}(B)$. $\qquad\square$

Example 12.3.4. Let A be the matrix in Example 12.3.2 and let $B = A^{\mathsf{T}}$. The corresponding edge shifts are almost conjugate since they are mixing and have the same entropy, so there is a mixing edge shift X_C such that X_B is a right-resolving factor of X_C and X_A is a left-resolving factor of X_C. Then by Theorem 12.1.4 and Proposition 12.3.3 $\mathfrak{I}(B) = \mathfrak{I}(C)$. But $\mathfrak{I}(A) \neq \mathfrak{I}(B)$. Hence $\mathfrak{I}(A) \neq \mathfrak{I}(C)$, so by Proposition 12.3.3 (Δ_A, δ_A) is not a quotient of (Δ_C, δ_C). Then Theorem 12.1.4 shows that X_A is not a right-closing eventual factor of X_C. $\qquad\square$

In the mixing case, the ideal class condition, together with entropy, turns out to be sufficient for right-closing finite equivalence.

Theorem 12.3.5. *Let A and B be primitive integral matrices. Then X_A and X_B are right-closing finitely equivalent if and only if $\lambda_A = \lambda_B$ and $\mathfrak{I}(A) = \mathfrak{I}(B)$. Moreover, the common extension can be chosen to be mixing.*

The proof of sufficiency, together with the "moreover," is given in §12.4. It depends on a realization result (Theorem 11.1.15), a result on eventual factors (Theorem 12.1.4), and some additional algebra and coding.

The irreducible case is slightly more complicated. But it can be reduced to the mixing case by the following result which we leave to the reader.

Proposition 12.3.6. *Let A and B be irreducible integral matrices. Let m be the least common multiple of their periods. Let A' and B' be irreducible components of A^m and B^m (so that A' and B' are primitive). Then X_A and X_B are right-closing finitely equivalent if and only if $X_{A'}$ and $X_{B'}$ are.*

This gives a slightly finer invariant than entropy and ideal class (see Exercise 12.3.6). And it shows us how to formulate the correct version of the ideal class condition for irreducible edge shifts. We state this in the generality of irreducible sofic shifts.

Theorem 12.3.7. *Let X and Y be irreducible sofic shifts, and G_X and G_Y be the underlying graphs of their minimal right-resolving presentations. Let m be the least common multiple of the periods of G_X and G_Y, and let A' and B' be irreducible components of $A_{G_X}^m$ and $A_{G_Y}^m$ (so A' and B' are primitive). Then X and Y are right-closing finitely equivalent if and only if $h(X) = h(Y)$ and $\mathfrak{I}(A') = \mathfrak{I}(B')$. Moreover, the common extension can be chosen to be irreducible with period m.*

PROOF: By Exercise 8.2.6(b), a right-closing finite equivalence (W, ϕ_X, ϕ_Y) between X and Y factors through such a finite equivalence between X_{G_X} and X_{G_Y}, as follows.

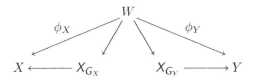

By Proposition 12.3.6 (only if), this induces a right-closing finite equivalence between $X_{A'}$ and $X_{B'}$, and then Theorem 12.3.1 completes the proof of necessity.

Conversely, suppose that $h(X) = h(Y)$ and $\mathfrak{I}(A') = \mathfrak{I}(B')$. Since A' and B' are primitive, we apply Theorem 12.3.5 to obtain a right-closing finite equivalence between $X_{A'}$ and $X_{B'}$. Then apply Proposition 12.3.6 (if) to get such a finite equivalence between X_{G_X} and X_{G_Y}, and thus also between X and Y. The mixing common extension for $X_{A'}$ and $X_{B'}$ from Theorem 12.3.5 translates to an irreducible common extension with period m for X and Y.   □

For a given Perron number λ, there are only finitely many ideal classes in the ring $\mathbb{Z}[1/\lambda]$ (see [BoyMT, Theorem 5.13]). This is deduced from the number-theoretic fact that there are only finitely many ideal classes in the ring $\mathbb{Z}[\lambda]$. It follows that the set of irreducible sofic shifts of a given entropy is partitioned into only finitely many equivalence classes under the relation of right-closing finite equivalence. In contrast, recall from §8.4 (see Exercise 8.4.3) that for right-resolving finite equivalence, typically there are infinitely many equivalence classes.

When λ is a positive integer then $\mathbb{Z}[1/\lambda]$ has only one ideal class (Exercise 12.3.1), and the same holds for $\lambda = (1 + \sqrt{5})/2$ [Sta]. So, in these cases, there is only one right-closing finite equivalence class. On the other hand, for $\lambda = \sqrt{10}$ there are at least two ideal classes, represented by the matrices in Example 12.3.2.

For a Perron number λ, it turns out that every ideal class in $\mathbb{Z}[1/\lambda]$ can be realized as the ideal class of a primitive integral matrix [BoyMT, Theorem 5.13]. Thus the number of equivalence classes of mixing edge shifts with entropy $\log \lambda$ under the relation of right-closing finite equivalence is exactly the number of ideal classes in $\mathbb{Z}[1/\lambda]$. However, the computation of this number is difficult in general.

We conclude by observing that to obtain a right-closing almost conjugacy (rather than finite equivalence), we only have to add a period condition to the entropy and ideal class conditions.

Theorem 12.3.8. *Let X and Y be irreducible sofic shifts, and G_X and G_Y be the underlying graphs of their minimal right-resolving presentations. Let $p = \mathrm{per}(G_X)$ and $q = \mathrm{per}(G_Y)$, and A', B' be irreducible components of $A_{G_X}^p$, $A_{G_Y}^q$. Then X and Y are right-closing almost conjugate if and only if $h(X) = h(Y)$, $p = q$, and $\mathfrak{I}(A') = \mathfrak{I}(B')$.*

PROOF: Necessity follows from Theorem 12.3.7 (only if), the Replacement Theorem 9.2.5(only if) and the fact that every right-closing almost conjugacy between X and Y must factor through such an almost conjugacy between X_{G_X} and X_{G_Y} (see Exercise 8.2.6(b)). For sufficiency, first pass from X, Y to the edge shifts X_{G_X}, X_{G_Y}. Then apply Theorem 12.3.7 (if) and the Replacement Theorem 9.2.5 (if). □

Note that the proof of this result depends on the Replacement Theorem and so makes essential use of the fact that the common extension in Theorem 12.3.5 can be chosen to be mixing.

EXERCISES

12.3.1. (a) Show that $\mathbb{Z}[1/\lambda]$ has exactly one ideal class if and only if it is a principal ideal domain; i.e., every ideal is principal.
(b) If λ is a positive integer, show that $\mathbb{Z}[1/\lambda]$ is a principal ideal domain.
(c) Explain why the ideal class condition was not needed in Chapter 5 when we constructed finite-state (X, n)-codes for irreducible sofic shifts X with entropy $\log n$.

12.3.2. Prove Proposition 12.3.6.

12.3.3. Let A be an irreducible integral matrix with period p, and let $A_1, \ldots, A_p$ denote the irreducible components of A^p. Show that $\mathfrak{I}(A_1) = \ldots = \mathfrak{I}(A_p)$.

12.3.4. Let

$$A = \begin{bmatrix} 0 & 1 & 0 & 0 & 0 \\ 0 & 0 & 1 & 0 & 0 \\ 0 & 0 & 0 & 1 & 0 \\ 0 & 0 & 0 & 0 & 1 \\ 1 & 0 & 0 & 0 & 1 \end{bmatrix} \quad \text{and} \quad B = \begin{bmatrix} 0 & 1 & 0 \\ 0 & 0 & 1 \\ 1 & 1 & 0 \end{bmatrix}$$

be the matrices from Example 12.1.7. Show that $\mathfrak{I}(A) = \mathfrak{I}(B)$, and thus X_A and X_B are right-closing finitely equivalent.

12.3.5. Let

$$A = \begin{bmatrix} 4 & 1 \\ 1 & 0 \end{bmatrix} \quad \text{and} \quad B = \begin{bmatrix} 3 & 2 \\ 2 & 1 \end{bmatrix}.$$

Show that $\mathfrak{I}(A) \neq \mathfrak{I}(B)$, and thus X_A and X_B are not right-closing finitely equivalent.

12.3.6. Let

$$A = \begin{bmatrix} 0 & 1 & 0 & 0 & 0 & 0 \\ 0 & 0 & 1 & 0 & 0 & 0 \\ 0 & 0 & 0 & 1 & 0 & 0 \\ 0 & 0 & 0 & 0 & 1 & 0 \\ 0 & 0 & 0 & 0 & 0 & 1 \\ 1 & 0 & 0 & 4 & 0 & 0 \end{bmatrix} \quad \text{and} \quad B = \begin{bmatrix} 0 & 0 & 1 & 1 & 0 & 0 \\ 0 & 0 & 1 & 0 & 0 & 0 \\ 0 & 0 & 0 & 0 & 1 & 1 \\ 0 & 0 & 0 & 0 & 1 & 0 \\ 1 & 1 & 0 & 0 & 0 & 0 \\ 1 & 0 & 0 & 0 & 0 & 0 \end{bmatrix}.$$

(a) Show that A and B are irreducible and both have period three.
(b) Show that X_A and X_B have the same entropy $\log((1 + \sqrt{5})/2)$. [*Hint:* Observe that A is the companion matrix of $t^6 - 4t^3 - 1$, and for B observe that the golden mean shift is a three-to-one factor of X_B.]
(c) Show that A and B have the same ideal class.
(d) Let A', B' be irreducible components of A^3, B^3. Show that $\mathfrak{I}(A') \neq \mathfrak{I}(B')$. Hence $X_{A'}$ and $X_{B'}$ are *not* right-closing finitely equivalent, so neither are X_A and X_B, even though A and B have the same ideal class and Perron eigenvalue. [*Hint:* See Exercise 12.3.5.]

12.3.7. Let X_A and X_B be irreducible edge shifts.

    (a) Show that if X_B is a right-closing eventual factor of X_A, then (Δ_B, δ_B) is a quotient of (Δ_A, δ_A) by a quotient mapping that takes Δ_A^+ onto Δ_B^+.

    (b) Show that the converse is false. [*Hint:* Consider the matrix A in Exercise 12.3.6.]

12.3.8. Let A and B be primitive integral matrices such that $\lambda_A = \lambda_B$ and $\chi_A^\times(t)$ and $\chi_B^\times(t)$ are irreducible. Show that $\Im(A) = \Im(B)$ if and only if A and B are shift equivalent.

12.3.9. Show that for any two nonzero ideals $\mathfrak{a}$ and $\mathfrak{b}$ of $\mathbb{Z}[1/\lambda]$, there is a number r such that $\mathfrak{a} \supseteq r\mathfrak{b}$.

12.3.10. A *module* over a commutative ring $\mathfrak{R}$ is the same as a vector space over a field except that the ring need not be a field (see [Her] for a precise definition). Show how any ideal in $\mathfrak{R}$ can be regarded as a module over $\mathfrak{R}$ and how to generalize the notion of ideal class to an equivalence relation on modules. Explain why this eliminates the need to assume that the entries of the right Perron eigenvector $\mathbf{v}_A$ (used in our definition of $\Im(A)$) has entries in $\mathbb{Z}[1/\lambda]$.

12.3.11. Consider the following properties for mixing edge shifts X_A, X_B:

    (a) Right-resolving finite equivalence

    (b) Right-closing finite equivalence

    (c) Finite equivalence

    Consider the corresponding properties of primitive integral matrices A, B:

    (a') $M_{G_A} = M_{G_B}$ (recall that M_G is the underlying graph of the minimal right-resolving factor of X_G)

    (b') $\lambda_A = \lambda_B$ and $\Im(A) = \Im(B)$

    (c') $\lambda_A = \lambda_B$

    Obviously, (a) $\Rightarrow$ (b) $\Rightarrow$ (c). From Theorems 8.4.5, 12.3.1, and 8.3.7, this yields that (a') $\Rightarrow$ (b') $\Rightarrow$ (c'). The implication (b') $\Rightarrow$ (c') is obvious. Prove the implication (a') $\Rightarrow$ (b') directly.

§12.4. Sufficiency of the Ideal Class Condition

This section completes the proof of Theorem 12.3.5, showing that equality of ideal classes is sufficient as well as necessary. Specifically, let A and B be primitive integral matrices satisfying $\lambda_A = \lambda_B$ and $\Im(A) = \Im(B)$. We will show that X_A and X_B are right-closing finitely equivalent, and moreover the common extension can be chosen to be mixing.

The proof is broken down into two steps.

STEP 1: Find a primitive integral matrix C such that $\lambda_C = \lambda_A = \lambda_B$ and both dimension pairs (Δ_A, δ_A) and (Δ_B, δ_B) are quotients of (Δ_C, δ_C).

STEP 2: Use C and the right-closing eventual factor codes induced by the quotient maps $(\Delta_C, \delta_C) \to (\Delta_A, \delta_A)$ and $(\Delta_C, \delta_C) \to (\Delta_B, \delta_B)$, as in Theorem 12.1.6, to construct a mixing shift of finite type X and right-closing factor codes $X \to X_A$ and $X \to X_B$.

We first handle Step 1. For a column vector $\mathbf{v} \in \mathbb{R}^n$ define $d_\mathbf{v} : \mathbb{R}^n \to \mathbb{R}$ by $d_\mathbf{v}(\mathbf{x}) = \mathbf{x} \cdot \mathbf{v}$. Let $\mathbf{v}$ be a vector whose entries lie in $\mathbb{Z}[1/\lambda]$. Recall our notation $\langle \mathbf{v} \rangle$ for the ideal in $\mathbb{Z}[1/\lambda]$ generated by the entries in $\mathbf{v}$.

Lemma 12.4.1. *Let A be an irreducible integral matrix with Perron eigenvalue λ, and let $\mathbf{v}$ be a right Perron eigenvector for A with entries in $\mathbb{Z}[1/\lambda]$. Then $d_{\mathbf{v}}(\Delta_A) = \langle \mathbf{v} \rangle$.*

PROOF: Observe that

$$(12\text{--}4\text{--}1) \qquad d_{\mathbf{v}}(\mathbf{x}A) = (\mathbf{x}A) \cdot \mathbf{v} = \mathbf{x} \cdot (A\mathbf{v}) = \lambda \mathbf{x} \cdot \mathbf{v} = \lambda d_{\mathbf{v}}(\mathbf{x}).$$

For $\mathbf{z} \in \mathbb{Z}^n$, $d_{\mathbf{v}}(\mathbf{z}) \in \langle \mathbf{v} \rangle$. This together with (12–4–1) shows that $d_{\mathbf{v}}(\Delta_A) \subseteq \langle \mathbf{v} \rangle$.

For each state $I \in \mathcal{V}(A)$, clearly $d_{\mathbf{v}}(\mathbf{e}_I) = v_I$, so by (12–4–1), $d_{\mathbf{v}}(\mathbf{e}_I A^k) = \lambda^k v_I$ for all $k \geqslant 0$. Since $\mathbf{e}_I A^k \in \Delta_A$ for large enough k, it follows that $\lambda^k v_I \in d_{\mathbf{v}}(\Delta_A)$, and so by (12–4–1) $v_I \in d_{\mathbf{v}}(\Delta_A)$. Thus all generators of $\langle \mathbf{v} \rangle$ are in $d_{\mathbf{v}}(\Delta_A)$, so $\langle \mathbf{v} \rangle \subseteq d_{\mathbf{v}}(\Delta_A)$. □

Let $\mathbf{v}_A$ and $\mathbf{v}_B$ be right Perron eigenvectors for A and B, respectively, with entries in $\mathbb{Z}[1/\lambda]$, where $\lambda = \lambda_A = \lambda_B$. Since $\mathfrak{I}(A) = \mathfrak{I}(B)$, we can arrange that $\langle \mathbf{v}_A \rangle = \langle \mathbf{v}_B \rangle$. Let d_A be the restriction of $d_{\mathbf{v}_A}$ to Δ_A, and define d_B similarly. By Lemma 12.4.1, $d_A(\Delta_A) = d_B(\Delta_B)$.

Whenever two maps have the same image, we can define a "fiber product" as we did for sliding block codes in §8.3. Thus we define the fiber product of d_A and d_B to be (Δ, ϕ_A, ϕ_B), where

$$\Delta = \{(\mathbf{x}, \mathbf{y}) \in \Delta_A \times \Delta_B : d_A(\mathbf{x}) = d_B(\mathbf{y})\},$$

and

$$\phi_A(\mathbf{x}, \mathbf{y}) = \mathbf{x}, \quad \phi_B(\mathbf{x}, \mathbf{y}) = \mathbf{y}.$$

The fiber product Δ inherits a group structure as a subgroup of $\Delta_A \times \Delta_B$. Since $d_A(\mathbf{x}A) = \lambda d_A(\mathbf{x})$ and $d_B(\mathbf{y}B) = \lambda d_B(\mathbf{y})$, it follows that Δ is $\delta_A \times \delta_B$-invariant. Let η be the restriction of $\delta_A \times \delta_B$ to Δ. Observe that both (Δ_A, δ_A) and (Δ_B, δ_B) are quotients of (Δ, η), using the maps ϕ_A and ϕ_B.

We next claim that Δ is pure in $\Delta_A \times \Delta_B$. For suppose that $(\mathbf{x}, \mathbf{y}) \in \Delta_A \times \Delta_B$ with $(n\mathbf{x}, n\mathbf{y}) \in \Delta$ for some integer $n \neq 0$. It follows from linearity of d_A and of d_B that $(\mathbf{x}, \mathbf{y}) \in \Delta$, verifying purity of Δ.

To complete Step 1, it remains to find a primitive integral matrix C such that $\lambda_C = \lambda_A = \lambda_B$ and (Δ, η) is a quotient of (Δ_C, δ_C). For this, we first show that there is an integral matrix D such that
  (a) $(\Delta_D, \delta_D) \simeq (\Delta, \eta)$,
  (b) $\lambda_D = \lambda_A = \lambda_B$, and
  (c) D is spectrally Perron.
Then we can apply Theorem 11.1.15 to produce the desired C.

To find D, first consider

$$E = \begin{bmatrix} A & 0 \\ 0 & B \end{bmatrix}.$$

Let n be the size of E. Observe that Δ_E is naturally identified with $\Delta_A \times \Delta_B$, so that Δ can be viewed as a pure δ_E-invariant subgroup of Δ_E. By Proposition 11.3.6, there is an integral matrix D, namely the matrix of $\delta_E|_{\langle \Delta \rangle_{\mathbb{Q}}}$ with respect to an integral basis for $\langle \Delta \rangle_{\mathbb{Q}} \cap \mathbb{Z}^n$, such that (a) holds. Observe that $\lambda = \lambda_A = \lambda_B$ is an eigenvalue of E with multiplicity two, and that λ strictly dominates the other eigenvalues of E. Let $\mathbf{w}_A$, $\mathbf{w}_B$ be left Perron eigenvectors for A, B. The real vector space W generated by $(\mathbf{w}_A, \mathbf{0})$ and $(\mathbf{0}, \mathbf{w}_B)$ spans the subspace of $\mathbb{R}^n$ corresponding to λ. Let V denote the real vector space spanned by Δ. Now λ_D is the spectral radius of $E|_V$. Hence for (b) and (c), it suffices to show that $V \cap W$ is a 1-dimensional space. By rescaling, we may assume that $d_A(\mathbf{w}_A) = d_B(\mathbf{w}_B)$. Then

$$V \cap W = \{t(\mathbf{w}_A, \mathbf{w}_B) : t \in \mathbb{R}\},$$

which is 1-dimensional as required.

We have therefore found D satisfying (a), (b), and (c). By Theorem 11.1.15, we can obtain from D a primitive integral matrix C such that $\lambda_C = \lambda_A = \lambda_B$ and (Δ_A, δ_A) and (Δ_B, δ_B) are both quotients of (Δ_C, δ_C). This completes Step 1.

We next turn to Step 2. We will construct a shift of finite type X and right-closing factor codes $\phi_A \colon X \to X_A$ and $\phi_B \colon X \to X_B$ by stitching together factor codes $X_{C^m} \to X_{A^m}$ and $X_{C^m} \to X_{B^m}$ obtained from Step 1 and Theorem 12.1.6. The construction is analogous to the marker construction for the Lower Entropy Factor Theorem in §10.3. Since entropies are equal here, we will need to introduce an auxiliary space to provide room for marker coordinates.

We construct only ϕ_B; the construction of ϕ_A is entirely similar. The shift X consists of pairs (x, y), where $x \in X_C$ and y is a binary sequence that obeys certain restrictions spelled out in Rules 1–4 below. The coordinates i for which $y_i = 1$ act as marker coordinates. We define ϕ_B on (x, y) by breaking up x into blocks determined by the marker coordinates in y, and using a map on each block that depends only on the length of the block. In order to obtain the right-closing property, we will need to use a "shifted" version of this construction.

We now give a detailed description of ϕ_B. We may assume that $C \neq [1]$, for otherwise $A = B = [1]$, and we are done.

For the construction we need a path $c = c_1 c_2 \ldots c_m$ in G_C that does not overlap itself, i.e., such that $c_j c_{j+1} \ldots c_m \neq c_1 c_2 \ldots c_{m-j+1}$ for $2 \leqslant j \leqslant m$. To find such a path c, first observe that since C is primitive and $\lambda_C > 1$, G_C must have a simple cycle γ and an edge e which meets γ but is not an edge of γ. Then for all $r \geqslant 1$, $c = \gamma^r e$ does not overlap itself. We can therefore choose M such that G_C has a path c of length M that does not overlap itself, and such that the construction in Theorem 12.1.6 works for all $m \geqslant M$ with respect to a positive integral solution, R, S, of the equations $CR = RB$ and $SR = B^\ell$.

Define the shift space X as the set of all pairs (x, y) of sequences with $x \in X_C$ and $y \in X_{[2]}$ subject to the following rules.

RULE 1. $y_{[i,i+2M-1]} \neq 0^{2M}$ for all i.

RULE 2. If $y_i = 1$, then $y_{[i+1,i+M-1]} = 0^{M-1}$.

RULE 3. If $x_{[i,i+M-1]} = c$, then $y_i = 1$.

RULE 4. If $y_i = 1$ and c does not appear in $x_{[i+1,i+3M-2]}$, then $y_{i+M} = 1$.

What do these Rules mean? Rules 1 and 2 say that whenever we see a marker coordinate i (i.e., an i for which $y_i = 1$), the next marker coordinate to the right appears at least M but no more than $2M$ places to the right. In particular, the sequences y satisfy the $(M-1, 2M-1)$ run-length limited constraint. Rule 3 says that if the block c begins in x at some coordinate i, then i is a marker coordinate. Finally, Rule 4 says that whenever i is a marker coordinate and the first appearance of c in $x_{[i+1,\infty)}$ does not begin before coordinate $i + 2M$, then $i + M$ is also a marker coordinate.

These rules show that X is defined using a finite set of forbidden blocks of length $3M$, so that X is a shift of finite type.

For $m \geqslant M$ let $[I, J, k]_m$ denote the element of the partition used in the out-splitting H of G_{C^m} in Theorem 12.1.6 (here we have included the dependence on m in the notation).

When out-splitting a graph, each edge e is replicated, one copy for each descendent of $t(e)$. In this case, an edge in H is specified by a choice of $(\pi, [I', J', k'])$, where π is a path of length m in G_C and $I' = t(\pi)$. By Propositions 8.2.6 and 8.2.2, the solution $\tilde{R}$ to the equation $A_H \tilde{R} = \tilde{R} B^m$ given in Theorem 12.1.6 defines a right-covering $X_H \to X_{B^m}$ by specifying, for each choice of I, J, k, and J', a bijection between

$$\{(\pi, [I', J', k']) : \pi \in [I, J, k]_m \quad \text{and} \quad I' = t(\pi)\}$$

and

$$\{\text{paths } \omega \text{ in } G_B \text{ of length } m \text{ such that } J = i(\omega) \text{ and } J' = t(\omega)\}.$$

We denote this bijection by ϕ_m, so that with the above notation,

$$(12\text{--}4\text{--}2) \qquad \omega = \phi_m(\pi, [I', J', k']).$$

Then these bijections obey the following property, whose proof is simply an interpretation of the right-covering property.

Lemma 12.4.2. *The state $[I, J, k]$ in H and the path ω in G_B determine the edge $(\pi, [I', J', k'])$ in H, according to (12–4–2).*

It is tempting to define $\phi_B \colon X \to X_B$ by encoding the block $x_{[i,j)}$ according to the bijection ϕ_{j-i} whenever i and j are successive marker coordinates. However, to guarantee that ϕ_B is right-closing, we need the gap size $j - i$ between i and j to be determined by information strictly to the left of these coordinates. To achieve this, we encode the block $x_{[i,j)}$ using ϕ_{j-i} whenever $i - 3M$ and $j - 3M$ are successive marker coordinates. To make this precise, let

$$F_y(i) = \min\{j > i : y_{j-3M} = 1\} \quad \text{and} \quad P_y(i) = \max\{j < i : y_{j-3M} = 1\}.$$

The following lemma, which is crucial for our construction, is an immediate consequence of Rules 1–4 above.

Lemma 12.4.3. *If $y_{i-3M} = 1$, then the next marker coordinate to the right of $i - 3M$ is uniquely determined by $x_{[i-3M,i)}$. More precisely, if (x,y), $(x',y') \in X$ with $y_{i-3M} = y'_{i-3M} = 1$ and $x_{[i-3M,i)} = x'_{[i-3M,i)}$, then $F_y(i) = F_{y'}(i)$.*

We are now ready to define ϕ_B. Let $(x,y) \in X$, and suppose that $i - 3M$ is a marker coordinate. Put $j = F_y(i)$, and let L_i denote the state $[I_i, J_i, k_i]$ determined by the unique I_i, J_i, k_i for which

$$x_{[i,j)} \in [I_i, J_i, k_i]_{j-i}.$$

Define ϕ_B by

$$\phi_B(x,y)_{[i,j)} = \phi_{j-i}(x_{[i,j)}, L_j).$$

Since $\phi_B(x,y)_{[i,j)}$ is a path in G_B from J_i to J_j, it follows that $\phi_B(x,y)$ is in X_B.

To show that ϕ_B is right-closing, we prove that whenever $y_{i-3M} = 1$, then $x_{(-\infty,i)}$ together with $z = \phi_B(x,y)$ determine $j = F_y(i)$ and $x_{[i,j)}$. It then follows that any left-infinite sequence of (x,y) together with the image sequence $\phi_B(x,y)$ in X_B inductively determine the entire sequence (x,y).

By Lemma 12.4.3, $x_{[i-3M,i)}$ determines the marker coordinate $j = F_y(i)$. By our definition of ϕ_B, we have

$$z_{[i,j)} = \phi_{j-i}(x_{[i,j)}, L_j).$$

By Lemma 12.4.2, L_i and $z_{[i,j)}$ determine $x_{[i,j)}$. So it remains to show that $x_{(-\infty,i)}$ and z determine L_i.

For this, let $n = P_y(i)$, and observe that by Lemma 12.4.2 the state L_n and $z_{[n,i)}$ determine L_i. But $x_{[n,i)}$ determines L_n since $x_{[n,i)}$ belongs to the partition element $[I_n, J_n, k_n]_{i-n}$. Thus $x_{[n,i)}$ and $z_{[n,i)}$ determine L_i. Hence $x_{(-\infty,i)}$ and z determine L_i, as desired. This completes the proof that ϕ_B is right-closing. A similar construction also works for ϕ_A, so long as we make sure M is large enough.

We will show that any irreducible component X' of X with maximal entropy is mixing and that both $\phi_A|_{X'}$ and $\phi_B|_{X'}$ are onto. This gives a right closing finite equivalence between X_A and X_B, with mixing common extension, as required.

We first show that $h(X) = \log \lambda$ (where $\lambda = \lambda_A = \lambda_B$).

For this, consider the map $\psi \colon X \to X_C$ defined by $\psi(x, y) = x$. It follows from Lemma 12.4.3 that ψ is right-closing and therefore finite-to-one. Let $x \in X_C$ be a point that contains the block c infinitely often to the left. We claim that x has exactly one pre-image under ψ. This follows from Rules 1–4 above and Lemma 12.4.3. In particular, the image of ψ contains the doubly transitive points of X_C, and since these points are dense we see that ψ is onto. Since ψ is finite-to-one, $h(X) = h(X_C) = \log \lambda$, as desired. Since ϕ_A and ϕ_B are right closing, and therefore finite-to-one, they are both onto.

Now, let X' be any irreducible component of X with maximal entropy. Then the restriction $\psi|_{X'}$ is still a factor code onto X_C, and each doubly transitive point of X_C has exactly one pre-image under $\psi|_{X'}$. Thus $\psi|_{X'}$ is an almost invertible factor code onto X_C. Since X' is irreducible and X_C is mixing, it then follows by (the easy half) of the Replacement Theorem 9.2.5 that X' is also mixing. Thus, the restrictions of ϕ_A and ϕ_B to X' do indeed yield a right closing finite equivalence with mixing common extension.

EXERCISES

12.4.1. Write down an explicit list of forbidden blocks that describes the shift of finite type X used in the proof in this section.

12.4.2. Let $\psi \colon X \to X_C$ be defined by $\psi(x, y) = x$ as in this section, and let $x \in X_C$. Show that if c does not appear infinitely often to the left in x, then x has exactly M pre-images under ψ.

12.4.3. Let X be the shift of finite type defined in this section. Is X mixing?

*12.4.4. Prove part 2 of Theorem 10.3.5 by using a construction similar to the construction of the shift X in this section, but without using the Embedding Theorem.

Notes

Theorem 12.1.1 is due to Ashley [Ash6], but is based in part on Theorem 12.1.4 due to Boyle, Marcus, and Trow [BoyMT]. This, in turn, was an outgrowth of an earlier result of Trow [Tro1]. Theorem 12.2.1 is due to Kitchens, Marcus, and Trow [KitMT], but some of the key ideas of that paper come from unpublished work of Kitchens. Parts (1) and (2) of Proposition 12.2.3, which were developed here as a corollary to Theorem 12.2.1, were actually established much earlier by Kitchens [Kit2], and part (1) is due independently to Nasu [Nas1]. Theorem 12.2.5 combines results that can be found in [Ash6] and [KitMT]. The ideal class as a complete invariant of right-closing finite equivalence is contained in [BoyMT]; this result was inspired by earlier work of Krieger [Kri5]. Example 12.3.2 is taken from S. Williams [WilS5].

Exercise 12.3.5 is taken from [Tro1]; Exercise 12.3.6 partly comes from J. Ashley; Exercise 12.3.8 is taken from [BoyMT].

CHAPTER 13

GUIDE TO ADVANCED TOPICS

Although we have seen many aspects of symbolic dynamics, there are still many more that we have not mentioned. This final chapter serves as a guide to the reader for some of the more advanced topics. Our treatment of each topic only sketches some of its most important features, and we have not included some important topics. For each topic we have tried to give sufficient references to research papers so that the reader may learn more. In many places we refer to papers for precise proofs and sometimes even for precise definitions. The survey paper of Boyle [Boy5] contains descriptions of some additional topics.

§13.1. More on Shifts of Finite Type and Sofic Shifts

The Core Matrix

Any shift of finite type X can be recoded to an edge shift X_G, and we can associate the matrix A_G to X. This matrix is not unique, but any two such matrices are shift equivalent, and in particular they must have the same Jordan form away from zero. This gives us a way of associating to X a particular Jordan form, or, equivalently, a particular similarity class of matrices. By Theorem 7.4.6, this similarity class is an invariant of conjugacy, and, by Proposition 12.2.3, it gives a constraint on finite-to-one factors between irreducible shifts of finite type.

Is there something similar for sofic shifts? For an irreducible sofic shift X we could consider the minimal right-resolving presentation $(G_X, \mathcal{L}_X)$ and use A_{G_X}. But there is also the minimal left-resolving presentation of X, and its adjacency matrix need not have the same Jordan form away from zero as A_{G_X} (see [Nas3]). Nevertheless, there is a canonical similarity class of matrices that one can associate to an irreducible sofic shift. For each finite-to-one irreducible presentation $\mathcal{G} = (G, \mathcal{L})$ of X there is a certain subspace $V_{\mathcal{G}}$ of $\mathbb{Q}^n$, where n denotes the size of A_G (see [Nas3] or [WilS3] for the precise definition). Any matrix which represents the restriction of

A_G to $V_{\mathcal{G}}$ is called a *core matrix* for X. It turns out that the similarity class of a core matrix depends only on X and not on the particular presentation. Moreover, this similarity class is an invariant of conjugacy. It also gives a constraint (similar to Proposition 12.2.3) on finite-to-one factors from one irreducible sofic shift to another [Nas3]. S. Williams [WilS3] showed that there is always an integral core matrix; i.e., there is an integral matrix in the similarity class of $A_G|_{V_{\mathcal{G}}}$, but that there need not be a nonnegative core matrix. By blending the core matrix with some ideas in §12.2, S. Williams [WilS5] and Trow and S. Williams [TroW] found further constraints on finite-to-one factors between irreducible sofic shifts.

CONSTRAINTS ON DEGREES OF FINITE-TO-ONE FACTOR CODES

In Chapter 9 we defined the degree of a finite-to-one factor code. Given that there is a finite-to-one factor code from one mixing shift of finite type X to another Y, the Replacement Theorem 9.2.5 shows that there is also a factor code with degree one. It is natural to ask whether, given an integer d, there must also be a factor code $X \to Y$ of degree d. It turns out that this is not always true, and that there are significant constraints on the possible degrees.

For simplicity, let us consider the special case of *endomorphisms* of an irreducible sofic shift (i.e., factor codes from the shift to itself). For the full n-shift, a theorem of Welch [Hed5] shows that if d is the degree of an endomorphism, then every prime that divides d must also divide n. In particular, the degrees of endomorphisms of the full 2-shift are exactly the powers of 2. Recall from §1.1 that there is a two-to-one endomorphism of the full 2-shift, and so its iterates realize all possible degrees.

This condition has been extended considerably. Boyle [Boy3] showed that if d is the degree of an endomorphism of an irreducible sofic shift X with entropy $\log \lambda$, then any prime which divides d must also divide the nonleading coefficients of the minimal polynomial $f_\lambda(t)$ of λ. Observe that when X is the full n-shift, $f_\lambda(t) = t - n$, so that Boyle's result implies Welch's. To obtain his result, Boyle showed that the degree d must be a unit (i.e., invertible) in the ring $\mathbb{Z}[1/\lambda]$, and then he proved the following purely algebraic result: for any monic polynomial $f(t)$ (with integer coefficients), a positive integer d is a unit in $\mathbb{Z}[1/\lambda]$ for each root λ of $f(t)$ if and only if every prime that divides d also divides the nonleading coefficients of $f(t)$.

For endomorphisms of an irreducible edge shift X_A, Trow [Tro3] went further. He showed that the degree d of an endomorphism of X_A must be a unit in $\mathbb{Z}[1/\mu]$ for every nonzero eigenvalue μ of A (in particular, for $\mu = \lambda_A$). Equivalently, every prime that divides d also divides the nonleading coefficients of the characteristic polynomial $\chi_A(t)$ of A. Trow and S. Williams [TroW] then generalized this result to irreducible sofic shifts, where the matrix A is replaced by the core matrix. Additional constraints on degrees of endomorphisms can be found in [Tro4] and [Boy3].

Boyle [Boy3] also established a constraint on the degrees of factor codes from one irreducible sofic shift X to another Y of the same entropy $\log \lambda$: the degree must be a product of two integers u and e, where u is a unit in $\mathbb{Z}[1/\lambda]$ and e belongs to a finite set E of positive integers, depending on X and Y. Trow [Tro3] showed that if $X = X_A$ and $Y = X_B$ are edge shifts, then whenever μ is a common eigenvalue of A and B with the same multiplicity, the integer u above must be a unit in $\mathbb{Z}[1/\mu]$. This result was extended to irreducible sofic shifts, again via the core matrix, by Trow and S. Williams [TroW].

The exact set of degrees that can occur for given irreducible sofic domain and range (even for given irreducible finite type domain and range) has not yet been determined completely. However, Boyle (cited in [Tro2]) showed that for an irreducible edge shift X_A, whenever d is an integer that satisfies Trow's constraint (i.e., that every prime that divides d also divides the nonleading coefficients of $\chi_A(t)$), then d can be realized as the degree of an endomorphism of some power X_{A^m}. In fact, d is the degree of a constant-to-one endomorphism of X_{A^m} for all m greater than or equal to the degree of the minimal polynomial of A.

Renewal Systems

A *renewal system* is the set of all (bi-infinite) sequences obtained by freely concatenating words, possibly of varying lengths, in a finite list. The list itself is called a *generating list*. Observe that a renewal system is an irreducible sofic shift: form a labeled graph by joining together, at one state, one cycle for each word of the generating list.

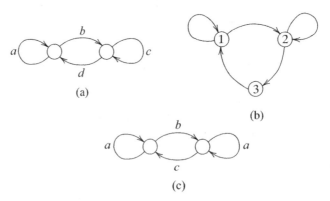

FIGURE 13.1.1.   Graphs for renewal systems.

Not every irreducible sofic shift is a renewal system. For example, let X be the edge shift X_G corresponding to the graph shown in Figure 13.1.1(a). Since $a^\infty, c^\infty \in X_G$, it follows that any generating list must contain a^m and c^n for some $m, n \geqslant 0$. But $a^m c^n$ is not an allowed word in X_G. Hence X_G

is not a renewal system. On the other hand, X_G is conjugate to a renewal system, namely the full 2-shift.

As another example, consider the vertex shift $\widehat{X}_H$ where H is shown in Figure 13.1.1(b). Now $\widehat{X}_H$ is not a renewal system for basically the same reason as X_G above. Nevertheless, it is conjugate to a renewal system: the 1-block code defined by the map $1 \mapsto 1$, $2 \mapsto 2$, and $3 \mapsto 2$ is a conjugacy from $\widehat{X}_H$ to the renewal system generated by the list $\{1, 22, 222\}$ [WilS4].

For one further example, let X be the irreducible sofic shift presented by the labeled graph shown in Figure 13.1.1(c). If X were a renewal system, then since $a^\infty.ba^\infty \in X$, the generating list would have to contain the word $a^n ba^m$ for some $n, m \geqslant 0$. Hence the list would have to contain the word $a^n ba^m a^n ba^m$, which is not allowed, giving a contradiction. In contrast to the preceding examples, S. Williams [WilS4] showed that X is not even conjugate to a renewal system. Note that X is strictly sofic.

This shows that there are irreducible sofic shifts that are not conjugate to renewal systems. R. Adler raised the following question:

Is every irreducible shift of finite type conjugate to a renewal system?

This question is still unanswered. However, Goldberger, Lind, and Smorodinsky [GolLS] showed that every number which occurs as the entropy of an irreducible shift of finite type (i.e., the roots of Perron numbers) is the entropy of a renewal system. Hence the answer to Adler's question is "yes" if we replace conjugacy by finite equivalence.

Almost Finite Type Shifts

Recall from Proposition 3.1.6 that a shift space has finite type if and only if it has a presentation $(G, \mathcal{L})$ such that $\mathcal{L}_\infty$ is a conjugacy. One way to weaken the finite type condition, still staying within the sofic setting, would be to require the existence of a presentation $(G, \mathcal{L})$ such that $\mathcal{L}_\infty$ is *bi-closing*, i.e., simultaneously right-closing and left-closing. Clearly, bi-closing is a weakening of conjugacy. An irreducible sofic shift which has a bi-closing presentation is said to have *almost finite type*. Of course, by passing to an irreducible component of maximal entropy, any almost finite type sofic shift has an irreducible bi-closing presentation. The class of almost finite type shifts was introduced in Marcus [Mar2], with the additional requirement that $\mathcal{L}_\infty$ be almost invertible. But Nasu [Nas3] observed that any irreducible sofic shift which has a bi-closing presentation automatically has an almost invertible bi-closing presentation, showing that the almost invertibility condition in the definition is redundant. A (possibly reducible) sofic shift that has a bi-closing presentation is called an *almost Markov shift* (see Boyle-Krieger [BoyK2]). Almost Markov shifts were introduced because they arise naturally in the conjugacy problem for almost finite type shifts.

Almost finite type shifts constitute a meaningful intermediate class in between irreducible shifts of finite type and irreducible sofic shifts for several reasons.

First, recall from §5.5 that the sliding block decoding theorem (Theorem 5.5.6) fails for irreducible sofic shifts in the case of equal entropy; i.e., there are irreducible sofic shifts X with entropy $h(X) = \log n$ such that no finite-state (X, n)-code has a sliding block decoder [KarM] (see also Trow and Williams [TroW]). On the other hand, for almost finite type sofic shifts, the sliding block decoding theorem does hold [KarM]. Thus almost finite type sofic shifts are a larger class of interesting shift spaces for which the sliding block decoding theorem applies.

Next, almost finite type shifts include constraints of practical interest. For instance, the minimal right-resolving presentations of the charge-constrained shifts of Example 1.2.7 are left-resolving (see Exercise 3.3.2); so these shifts have almost finite type. See Patel [Pat] and Immink [Imm2] for applications of these constraints.

Finally, recall from Exercise 8.2.6 that any right-closing factor code from an irreducible shift of finite type to an irreducible sofic shift X factors through $(\mathcal{L}_X)_\infty$, where $(G_X, \mathcal{L}_X)$ is the minimal right-resolving presentation of X. Here, the right-closing assumption is essential. But Boyle, Kitchens, and Marcus [BoyKM] showed that the almost finite type shifts are precisely those irreducible shifts X that do have a *unique minimum irreducible presentation*, i.e., an irreducible presentation $(G, \mathcal{L})$ such that every factor code $Z \to X$, with Z an irreducible shift of finite type, factors through $\mathcal{L}_\infty$. Such a presentation is unique up to conjugacy, i.e., conjugacy of the underlying edge shifts which intertwines the labelings. When it exists, the unique minimum irreducible presentation coincides with $(G_X, \mathcal{L}_X)$. Furthermore, Nasu [Nas3] (see also [BoyKM]) showed that an irreducible sofic shift has almost finite type if and only if its minimal right-resolving presentation is left-closing.

All of the characterizations of almost finite type shifts that we have given so far involve consideration of some presentation. S. Williams [WilS2] gave an intrinsic characterization in terms of intrinsically synchronizing words (recall from Exercise 3.3.4 that a block w is intrinsically synchronizing if whenever uw and wv are allowed blocks, then so is uwv). Williams also showed that almost finite type shifts are dramatically different from non-almost finite type sofic shifts in the following sense. For any irreducible sofic shift X that is not almost finite type, there are infinitely many irreducible presentations of X, no pair of which factor through a common presentation!

§13.2. Automorphisms of Shifts of Finite Type

In many areas of mathematics, objects are studied by means of their symmetries. This holds true in symbolic dynamics, where symmetries are expressed by automorphisms. An *automorphism* of a shift space X is a conjugacy from X to itself. The set of all automorphisms of a shift space X is a group under composition, so is naturally called the *automorphism group* of X, denoted $\mathrm{aut}(X)$. For a more thorough exposition than we give here, we refer the reader to Wagoner [Wag6].

Most of the work has focused on automorphism groups of mixing shifts of finite type. In this section, unless otherwise specified, our shift spaces will be mixing shifts of finite type. In fact, we will usually assume that our shifts are mixing edge shifts with positive entropy.

The goals are to understand $\mathrm{aut}(X)$ as a group (What kind of subgroups does it contain? How "big" is it?), and how it acts on the shift space X (Given σ-invariant subsets U, V, such as finite sets of periodic points, when is there an automorphism of X that maps U to V?). One hope is that the automorphism group would shed new light on the conjugacy problem for shifts of finite type. At the end of this section, we will discuss one result that does.

The symmetries of a graph G are the *graph automorphisms*, i.e., the graph isomorphisms of G to itself. The edge mapping Φ of a graph automorphism of G generates an automorphism Φ_∞ of X_G. These are precisely the automorphisms of X_G which are 1-block codes with 1-block inverses. But, assuming that the edge shift has positive entropy (which we do from here on), there are many more automorphisms. For instance, it is known that $\mathrm{aut}(X)$ contains an embedded copy of every finite group. For this, one first makes use of Cayley's Theorem from group theory that any finite group is a subgroup of a symmetric group on n letters S_n [Her]. Then one embeds S_n into $\mathrm{aut}(X)$ as follows. Let w be a block, and let D be a collection of blocks all of the same length such that for any $u, u' \in D$, the blocks wuw and $wu'w$ overlap only in the obvious way; i.e., either $wuw = wu'w$ or the terminal w of wuw matches the initial w of $wu'w$. Then any permutation π of D yields a well-defined automorphism of X by simply replacing wuw by $w\pi(u)w$ whenever wuw appears in a point of X. The block w is called a *marker*, and such an automorphism is often called a *marker automorphism* (the reader is warned that the definition of marker automorphism has not been consistent in the literature). For any mixing shift of finite type, one can find a block w and a collection of blocks D of any prescribed length as above, and so any symmetric group does indeed embed in $\mathrm{aut}(X)$ (see [BoyLR], [AdlKM]).

Marker automorphisms are central examples, because they are easy to construct and because compositions of marker automorphisms provide examples of interesting phenomena. A general question, which has not been satisfactorily answered, is to what extent do the marker automorphisms generate all automorphisms? Since marker automorphisms have finite-order (as elements of the group $\mathrm{aut}(X)$), a closely related question is to what extent do automorphisms of finite order generate all automorphisms? We will return to this question later in this section.

The fact that every symmetric group is a subgroup of aut(X) essentially goes back to Hedlund [Hed5], where this was proved for full shifts. That paper also contains an example of a pair of marker automorphisms ϕ, ψ of the full 2-shift, each with order two, but whose composition $\phi \circ \psi$ has infinite order. In fact, the subgroup generated by ϕ and ψ is free; i.e., there are no relations between ϕ and ψ in the group aut(X). Boyle, Lind, and Rudolph [BoyLR] proved that the automorphism group aut(X) of any mixing shift of finite type X contains the free group on two generators. In addition, they showed that aut(X) contains the direct sum of countably many copies of $\mathbb{Z}$, and also the direct sum of any countable collection of finite groups. Furthermore, they showed that aut(X) is not finitely generated as a group. In short, the automorphism group of a mixing shift of finite type (with positive entropy) is very big. On the other hand it is countable since there are only countably many sliding block codes, and it is known that it does not contain certain countable groups such as the group of rational numbers [BoyLR]. Automorphism groups have many other interesting properties as groups. For instance, the center of aut(X) (i.e., the set of elements of aut(X) that commute with all elements of aut(X)) consists only of the shift map and its powers [Rya2]. But there are some simple questions that remain unsolved: for example, are the automorphism groups of the full 2-shift and the full 3-shift isomorphic as groups?

Next, we consider the action of aut(X) on periodic points. Observe that any automorphism must permute, in a shift-commuting way, the set $Q_n(X)$ of points of least period n. It is natural to ask whether any such permutation of $Q_n(X)$ can be realized as the restriction of an automorphism. This was an open problem for a long time. The answer turns out to be no! Kim and Roush [KimR9] gave an example of a shift-commuting permutation of $Q_6(X_{[2]})$ which is not the restriction of an automorphism of $X_{[2]}$. And Kim, Roush, and Wagoner [KimRW1] found a mixing shift of finite type with exactly two fixed points x and y that cannot be interchanged by an automorphism. On the other hand, for every mixing shift of finite type X, there is a positive integer N such that for all $n \geqslant N$, any shift-commuting permutation of $Q_n(X)$ can be realized by an automorphism of X [BoyLR]. So the obstructions have more to do with periodic points of low period. Finally, we mention that Ashley [Ash6] showed that any shift-commuting mapping of a finite set of periodic points can be realized by an endomorphism of X, i.e., a factor code of X onto itself.

It is also natural to ask how the action of an automorphism on the points of one least period affects its action on the points of another least period. A key to addressing this question is a compatibility condition forced between two sequences of numbers, the sequence of sign numbers and the sequence of gyration numbers, which we now describe.

The sign numbers reflect how an automorphism permutes orbits. Precisely, for an automorphism ϕ of X, the *nth sign number* $s_n(\phi)$ of ϕ is the sign, ± 1, of the permutation induced by ϕ on the set of orbits in $Q_n(X)$.

The gyration numbers, on the other hand, reflect how an automorphism shifts points within periodic orbits. The precise definition is given as follows. For each n and each periodic orbit γ of least period n, select an arbitrary element x_γ of γ. Now $\phi(x_\gamma)$ and $x_{\phi(\gamma)}$ differ by a power of σ, say $k(\gamma)$, so that

$$\phi(x_\gamma) = \sigma^{k(\gamma)}(x_{\phi(\gamma)}).$$

Note that $k(\gamma)$ is defined only modulo n. Summing the $k(\gamma)$ over all γ of least period n, we get a number $g_n(\phi)$ modulo n, called the *nth gyration number* of ϕ. It turns out that this number is well-defined and independent of the choices of x_γ.

Boyle and Krieger discovered that these two sequences of numbers are related for certain kinds of automorphisms. In [BoyK1] they introduced the following *sign-gyration compatibility condition*, or SGCC for short. This condition states that for q odd,

$$g_{2^m q}(\phi) = \begin{cases} 0 & \text{if } \prod_{j=0}^{m-1} s_{2^j q}(\phi) = 1, \\ 2^{m-1} q & \text{if } \prod_{j=0}^{m-1} s_{2^j q}(\phi) = -1. \end{cases}$$

Note that a very special case of SGCC is that if $s_1(\phi) = 1$ (i.e., ϕ acts as an even permutation of the fixed points), then $g_2(\phi) = 0$. While the SGCC does not always hold, it does hold for any automorphism ϕ of a full shift that is a composition of finite-order automorphisms. This includes compositions of marker automorphisms of full shifts. So the action of such an automorphism on fixed points restricts the possible actions on periodic points of period two. The SGCC holds for other classes of automorphisms that we will mention later. The SGCC was one of the tools used in the examples of shift commuting permutations of $Q_n(X)$ mentioned above that cannot be realized by automorphisms.

One approach to understanding a group is to construct a homomorphism from the group into a simpler group, thereby dividing the complexity of the group into the image of the homomorphism, the kernel of the homomorphism, and how the image and kernel are put together. Such a homomorphism is often called a *representation*. One such representation, useful in the study of the automorphism group, is the dimension representation, which we now define. Recall, from Exercise 7.5.9 that any conjugacy from one edge shift X_A to another X_B maps beams to beams in a way that intertwines δ_A with δ_B. In this way, an automorphism ϕ of an edge shift defines a group automorphism $s_A(\phi)$ (in the sense that it respects the group structure) of the dimension group Δ_A. The map $s_A(\phi)$ also preserves Δ_A^+ and commutes with δ_A. We denote the set of all such automorphisms of Δ_A by $\mathrm{aut}(\Delta_A)$. The map $\phi \mapsto s_A(\phi)$ defines a homomorphism $s_A \colon \mathrm{aut}(X_A) \to \mathrm{aut}(\Delta_A)$ called the *dimension representation* of $\mathrm{aut}(X_A)$.

Now $\mathrm{aut}(\Delta_A)$ is usually much simpler than $\mathrm{aut}(X_A)$. For instance, if A has simple nonzero spectrum (i.e., each nonzero eigenvalue of A is simple), then $\mathrm{aut}(\Delta_A)$ is finitely generated and abelian. An exact description of the image of s_A is not known, except for some very special cases such as full shifts. It is known that s_A is not always onto: restrictions on elements of the image of s_A are forced by a certain combination of sign information and gyration information, related to the SGCC condition mentioned above [KimRW1].

Typically, the vast complexity of $\mathrm{aut}(X)$ is concentrated in the kernel of the dimension representation, elements of which are called *inert automorphisms*. A particularly nice, concrete class of inert automorphisms is the class of simple automorphisms introduced by Nasu [Nas5]. A *simple automorphism* is an automorphism which is conjugate to a code generated by a graph automorphism which fixes all vertices. Nasu used these automorphisms to improve earlier results on extending shift-commuting permutations of finite sets of periodic points.

Simple automorphisms are clearly of finite order, and so give a class of finite-order inert automorphisms. At one point it was conjectured that every inert automorphism is a composition of finite order automorphisms (known as the *Finite-Order Generation* or *FOG Conjecture*). If FOG were true, then the elements of finite order, together with a finitely generated abelian group, would generate all of $\mathrm{aut}(X)$, at least in the case of simple nonzero spectrum. FOG turns out to be false [KimRW2]. However, there is a version of it which is true in the "eventual" category: for every inert automorphism ϕ of X_A and all sufficiently large m, we can obtain ϕ as a composition of finite order automorphisms of X_{A^m} (see [Wag3]).

It is still not known if FOG holds in the special case of full shifts. In particular, the jury is still out on an earlier question of F. Rhodes: Is every automorphism of the full shift a composition of powers of the shift map and involutions (i.e., automorphisms with order two)?

There is a good deal that is known about inert automorphisms. For instance, Kim and Roush [KimR9] showed that SGCC must hold for any inert automorphism. Boyle and Fiebig [BoyF] completely classified the actions of compositions of finite order inert automorphisms on finite sets of periodic points; this involves many further constraints beyond SGCC.

It turns out that every inert automorphism of a shift of finite type Y contained in a shift of finite type X extends to an automorphism of X itself [KimR9]. But not all automorphisms of Y extend to X; for instance, we mentioned above that certain shift-commuting permutations of a finite set Y of periodic points do not extend. There are also examples where Y

has positive entropy. Boyle and Krieger [BoyK3] have reduced the problem of deciding when an automorphism ϕ of a subshift Y of a mixing shift of finite type X can be extended to an automorphism of X to a problem involving only periodic points: namely, given X and Y, they found a number $N = N(X, Y)$ such that any automorphism ϕ of Y can be extended to an automorphism of X if and only if the restriction of ϕ to the periodic points in Y of period at most N can be extended to an automorphism of X. In fact, their results apply to the more general problem of deciding when a conjugacy $\phi: Y \to Z$, with Y, Z being subshifts of a mixing shift of finite type X, extends to an automorphism of X.

Another tool, used in the study of the automorphism group, is a topological space which models elementary equivalences of matrices [Wag5]. The space, called RS, is actually a simplicial complex, i.e., a space built up out of vertices, edges, triangles, tetrahedra, and higher dimensional simplices. The vertices of RS are the square 0–1 matrices, the edges of RS are defined by elementary equivalences, and the triangles and higher dimensional simplices are defined by certain identities, involving the R and S matrices of an elementary equivalence. Strong shift equivalences are then viewed as paths in RS, and so the connected components of RS are naturally identified with the strong shift equivalence classes of 0–1 matrices. Automorphisms are then viewed as equivalence classes of closed paths in RS. In this way, as readers familiar with algebraic topology will recognize, for a 0–1 matrix A, the fundamental group of the connected component of RS (at the base point A) can be identified with $\text{aut}(X_A)$. If one allows nonnegative integral matrices instead of 0–1 matrices, one gets a simplicial complex whose fundamental group (at the base point A) can be identified with the quotient of $\text{aut}(X_A)$ modulo the group of simple automorphisms of X_A.

There is another simplicial complex, called S, whose fundamental group can be identified with $\text{aut}(\Delta_A)$. There is also a natural mapping $RS \to S$ such that the induced map on fundamental groups is identified with the dimension representation $s_A: \text{aut}(X_A) \to \text{aut}(\Delta_A)$. When viewed in this way, questions regarding the automorphism group can be treated in the spirit of algebraic topology. This perspective has played an important role in the construction of many of the counterexamples mentioned in this section.

Next, let us consider the conjugacy problem for shifts of finite type from the viewpoint of the automorphism group. The automorphism group is clearly an invariant of conjugacy; that is, if two shifts are conjugate, then their automorphism groups are isomorphic (as groups). It is natural to wonder how good an invariant of conjugacy it is. It cannot be a complete invariant of conjugacy. To see this, first observe that $\text{aut}(X_A)$ and $\text{aut}(X_{A^\top})$ are isomorphic, since any automorphism read backwards can be viewed as an automorphism of the transposed shift; on the other hand, X_A and $X_{A^\top}$ may fail to be conjugate (see Examples 7.4.19 and 12.3.2). It is not known if the only way for the automorphism groups of mixing edge shifts X_G and X_H to be isomorphic is for X_G to be conjugate to either X_H or $X_{H^\top}$. As we mentioned previously, it is not even known if the automorphism groups of the full 2-shift and the full 3-shift are isomorphic.

Nevertheless, the automorphism group does give some information about the conjugacy problem, at least in the reducible case. Kim and Roush [KimR11] found a pair of reducible matrices which are shift equivalent but not strong shift equivalent (later they showed how to find examples of irreducible matrices with the same property [KimR12]). Their example is as follows.

Let

$$U = \begin{bmatrix} 0 & 0 & 1 & 1 \\ 1 & 0 & 0 & 0 \\ 0 & 1 & 0 & 0 \\ 0 & 0 & 1 & 0 \end{bmatrix}, \quad A = \begin{bmatrix} U & 0 \\ Id & U \end{bmatrix},$$

$$B = \begin{bmatrix} U & 0 \\ U^n(U - Id) & U \end{bmatrix}, \quad \text{and} \quad C = \begin{bmatrix} Id & 0 \\ 0 & U^n(U - Id) \end{bmatrix}.$$

It is straightforward to check that C^{-1} is integral and $B = CAC^{-1}$. Hence A and B are shift equivalent over $\mathbb{Z}$. For $n \geqslant 23$ it turns out that $B \geqslant 0$, $R = C^{-1}B^n \geqslant 0$, and $S = CA \geqslant 0$. So the pair (R, S) defines a shift equivalence between A and B.

Now, any conjugacy ψ from X_A to X_B would define two automorphisms ϕ_1, ϕ_2 of X_U, one viewing U as a sink of A and B, the other viewing U as a source of A and B. Then ϕ_1 and ϕ_2 would induce corresponding elements s_1, s_2 of $\mathrm{aut}(\Delta_U)$. But, using constraints on the image of the dimension representation [KimRW1], Kim and Roush showed that at least one of s_1 or s_2 cannot be in the image of the dimension representation, and so ψ does not exist. Thus, A and B are not strong shift equivalent. We refer the reader to Wagoner [Wag6] for an exposition of the details.

It follows from Krieger [Kri8] that the automorphism group of a sofic shift embeds in the automorphism group of a shift of finite type. But beyond this, little is known about the automorphism group of sofic shifts.

Another approach to the study of automorphisms has been pioneered by Nasu [Nas6] under the rubric of *textile systems*.

§13.3. Symbolic Dynamics and Stationary Processes

Recall from §9.4 the notion of a stationary Markov chain on a graph. It is natural to wonder if, for a given graph G, there is a particular stationary Markov chain on G that in some sense distributes probabilities on paths as uniformly as possible. This would be the most "random" possible Markov chain, subject to the constraints imposed by G. If it were unique, then it would be a natural and canonical way of assigning probabilities.

There is a notion of entropy for stationary Markov chains, defined below, and this is used as a measure of randomness or uniformity of distribution. It turns out that when G is irreducible, there is a unique stationary Markov chain on G of maximal entropy, denoted μ_G, and so this is a canonical Markov chain associated to G.

Many results in symbolic dynamics were originally proved using μ_G. In this text we chose to avoid the use of μ_G simply because there are newer proofs that rely only on the basic combinatorial structure of G and X_G. Nevertheless, μ_G provides motivation and insight into symbolic dynamics problems, and, it illustrates a connection with ergodic theory, as we shall see in §13.4.

The entropy of a stationary Markov chain is naturally defined in the more general setting of stationary processes, introduced in §9.4. A stationary process over an alphabet $\mathcal{A}$ is an assignment μ of probabilities (nonnegative numbers) to blocks over $\mathcal{A}$ such that

$$(13\text{--}3\text{--}1) \qquad \sum_{a \in \mathcal{A}} \mu(a) = 1$$

and

$$(13\text{--}3\text{--}2) \qquad \mu(w) = \sum_{a \in \mathcal{A}} \mu(wa) = \sum_{a \in \mathcal{A}} \mu(aw)$$

for any block w over $\mathcal{A}$. Blocks over $\mathcal{A}$ are allowed to have zero probability. The *support* of a stationary process is defined to be the shift space over the alphabet $\mathcal{A}$ defined by forbidding the blocks with zero probability. A stationary process on a shift space X is one whose support is contained in X. In this section and §13.4, we assume that $\mathcal{A}$ is finite.

Let G be a graph. A *stationary process on* G is defined to be a stationary process on the set $\mathcal{A} = \mathcal{E}$ of edges of G whose support is contained in X_G. Equivalently, only allowed paths in G are permitted to have positive probability (but not all allowed paths in G need have positive probability). A stationary Markov chain on a graph G is a stationary process on G (this fact was key to Proposition 9.4.4).

To define the entropy of a stationary process, we first need to define a notion of entropy for probability vectors. Let $\mathbf{p} = (p_1, \dots, p_k)$ be a probability vector. The *entropy* of $\mathbf{p}$ is defined to be

$$h(\mathbf{p}) = -\sum_{i=1}^{k} p_i \log(p_i),$$

with $0 \log(0) = 0$ by convention. A simple calculation shows that for probability vectors of length k, we have

(13–3–3) $0 \leqslant h(\mathbf{p}) \leqslant \log k$,

with

$h(\mathbf{p}) = \log k$ if and only if $\mathbf{p} = (1/k, 1/k, \ldots, 1/k)$,

and

$h(\mathbf{p}) = 0$ if and only if $p_i = 1$ for some i.

In other words, entropy is maximized when the probabilities are equally divided, and entropy is minimized when all of the probability is concentrated in one component. For proofs of these facts, we refer the reader to Petersen [Pet3] or Walters [Wal].

Given a stationary process μ, for each n, assemble the probabilities of the paths of length n into a probability vector $\mathbf{p}^{(n)} = \mathbf{p}_\mu^{(n)}$. The *entropy of* μ is defined as

(13–3–4) $h(\mu) = \lim\limits_{n \to \infty} \dfrac{1}{n} h(\mathbf{p}_\mu^{(n)})$.

It is well known that the limit exists ([Pet3] or Walters [Wal]). In information theory, this notion of entropy is usually called "entropy rate" (see Cover and Thomas [CovT]).

For computational purposes, it is desirable to have a formula for the entropy of a stationary Markov chain μ directly in terms of its initial state probabilities and conditional probabilities. Indeed, the following is such a formula.

(13–3–5) $h(\mu) = - \sum\limits_{e \in \mathcal{E}(G)} \mu(i(e)) \mu(e|i(e)) \log(\mu(e|i(e)))$.

This result can be proved by a computation directly from the definition (13–3–4), but a more meaningful proof involves the notion of conditional entropy (see [Pet3] or Walters [Wal]).

Let G be an irreducible graph. We will exhibit a stationary process on G, in fact a stationary Markov chain, with maximal entropy among all stationary processes on G. Let $A = A_G$, $\lambda = \lambda_A$, and $\mathbf{w}, \mathbf{v}$ be left, right Perron eigenvectors for A, scaled so that $\mathbf{w} \cdot \mathbf{v} = 1$. By Exercise 4.5.14 (see also Theorem 4.5.12), there are roughly $(\sum_{I,J \in \mathcal{V}} w_J v_I) \lambda^n$ paths of length n. In fact, that result shows that for sufficiently large n there are at most $2(\sum_{I,J \in \mathcal{V}} w_J v_I) \lambda^n$ paths of length n. It follows from this and (13–3–3) that

$$h(\mathbf{p}^{(n)}) \leqslant \log\left(2 \sum\limits_{I,J \in \mathcal{V}} w_J v_I \lambda^n \right),$$

so that

$$h(\mu) \leqslant \log \lambda.$$

In other words, the entropy of every stationary process on G is bounded above by the entropy of X_G.

Since there are roughly λ^n paths of length n, a stationary process which assigns equal probability to paths of equal length would achieve maximal entropy $\log \lambda$. In general, there is no such stationary process. However, the following stationary Markov chain μ_G on G does assign equal probability to paths of equal length, up to constant factors independent of length.

For a state I, define the initial state probability

$$\mu_G(I) = w_I \cdot v_I,$$

and for an edge e, define the conditional probability

(13–3–6) $$\mu_G(e|i(e)) = \frac{v_{t(e)}}{\lambda v_{i(e)}}.$$

Since $\mathbf{w} \cdot \mathbf{v} = 1$ and $\mathbf{v}$ is a right Perron eigenvector, it follows that μ_G is indeed a stationary Markov chain on G. Using the initial state and conditional probabilities to define probabilities of paths, we see that for any path γ of length n from state I to state J,

(13–3–7) $$\mu_G(\gamma) = \frac{w_I v_J}{\lambda^n}.$$

From this and the definition (13–3–4), it follows that $h(\mu_G) = \log \lambda$. Alternatively, one can plug the initial state and conditional probabilities into (13–3–5) to establish the same result.

We conclude that μ_G achieves maximal entropy. This Markov chain was originally discovered by Shannon [Sha] (see also Parry [Par1]). Parry [Par1] showed that μ_G is the unique stationary Markov chain, in fact the unique stationary process, on G with maximal entropy. See Parry and Tuncel [ParT2] for a proof of this result using convexity. In summary, for an irreducible graph G, we have the following.

(1) Any stationary process μ on G satisfies $h(\mu) \leqslant \log \lambda$, and
(2) μ_G is the unique stationary process on G such that $h(\mu_G) = \log \lambda$.

For an irreducible sofic shift X with minimal right-resolving presentation $(G_X, \mathcal{L}_X)$, the stationary Markov chain μ_{G_X} defines a stationary process μ_X on X, as described at the end of §9.4, and μ_X is the unique stationary process on X of maximal entropy (see Coven and Paul [CovP2]).

The uniqueness of the stationary Markov chain of maximal entropy can be used to prove that an irreducible sofic shift contains no proper subshift of maximal entropy (Corollary 4.4.9). In fact, the original proof by Coven and Paul [CovP1] of this result was along these lines. In the following, we prove this in the special case of shifts of finite type: if X is an irreducible shift of finite type and $Y \subseteq X$ is a subshift with $h(Y) = h(X)$, then $X = Y$. By Proposition 4.4.6, we can approximate Y, in terms of entropy, from the outside by shifts of finite type, and so we may assume that Y is a shift of finite type. By recoding, we may assume that $X = X_G$ and $Y = X_H$ are both edge shifts, with G irreducible. By passing to an irreducible component of maximal entropy, we may assume that H is irreducible. Then both μ_G and μ_H are stationary Markov chains on G with maximal entropy. By uniqueness, we must have $\mu_G = \mu_H$, and so $G = H$ and therefore $X = Y$. This shows that indeed X contains no proper subshift of maximal entropy.

Finally, we briefly discuss the theory of equilibrium states. Recall that the entropy of a shift space X is the growth rate of the number of blocks in the shift. Given a continuous function $f \colon X \to \mathbb{R}$, one can also consider the growth rate of the total weights of blocks, where each block is weighted using f in a certain way (see [Wal, Chap. 9] for details). This growth rate is called the *pressure* of f, and denoted by $P_X(f)$. Topological entropy is a special case of pressure.

For a stationary process μ whose support is contained in X, let $E_\mu(f)$ denote the expected value of f with respect to μ. It turns out that the supremum, over the stationary measures μ, of the quantity $h(\mu) + E_\mu(f)$ equals $P_X(f)$. For an irreducible shift of finite type X and a suitably nice function f, there is a unique stationary measure called an *equilibrium state*, denoted $\mu_{X,f}$, which achieves the pressure. In particular, $\mu_{X_G,0} = \mu_G$ for an irreducible graph G. These results depend on a powerful generalization of the Perron–Frobenius Theorem, called Ruelle's Perron–Frobenius Theorem. This circle of ideas gives an interpretation of the classical thermodynamic formalism in a modern dynamical systems setting (see [Ruel], [Bow6], [Kea2], [Wal]).

§13.4. Symbolic Dynamics and Ergodic Theory

We show in this section how stationary Markov chains, and more generally stationary processes, on graphs can be viewed as measure-preserving transformations of probability spaces. This gives a connection between symbolic dynamics and ergodic theory. For a much more thorough introduction to ergodic theory, see Petersen [Pet3] and Walters [Wal]. For more on the relationship between symbolic dynamics and ergodic theory, see Denker, Grillenberger and Sigmund [DenGS], Parry and Tuncel [ParT2] and Parry [Par5].

We first describe measure-preserving transformations. This requires some basic measure theory (see, for instance, [Roy]), which we now briefly review. While our review is more or less self-contained (in the sense that all the definitions are given), it may be difficult to appreciate for those readers who have not had a course in measure theory or foundations of probability theory.

Let X be a set. A *sigma-algebra* $\mathcal{S}$ is a collection of subsets of X that is closed under countable unions and complements and also contains the set X itself. A *measurable space* $(X, \mathcal{S})$ is a set X together with a sigma-algebra $\mathcal{S}$ of subsets of X. A *measure* on a measurable space $(X, \mathcal{S})$ is a nonnegative function μ on $\mathcal{S}$ such that $\mu(\varnothing) = 0$ and for any countable collection of disjoint sets $E_1, E_2, \ldots$ in $\mathcal{S}$ we have

$$\mu\left(\bigcup_{i=1}^{\infty} E_i\right) = \sum_{i=1}^{\infty} \mu(E_i).$$

A *measure space* is a triple $(X, \mathcal{S}, \mu)$ where $(X, \mathcal{S})$ is a measurable space and μ is a measure on $(X, \mathcal{S})$.

A *probability measure* is a measure μ such that $\mu(X) = 1$. A *probability space* is a measure space $(X, \mathcal{S}, \mu)$ where μ is a probability measure. From here on, we consider only probability spaces.

It is convenient to assume that the probability space is *complete*, i.e., that $\mathcal{S}$ contains all subsets of sets in $\mathcal{S}$ that have measure zero. If a probability space is not complete, then one forms the *completion*, i.e., the probability space obtained by enlarging the sigma-algebra $\mathcal{S}$ to include all subsets of sets in $\mathcal{S}$ of measure zero (and declaring them to have measure zero).

A *homomorphism* from a probability space $(X, \mathcal{S}, \mu)$ to a probability space $(Y, \mathcal{T}, \nu)$ is a map $\phi\colon X \to Y$ which sends $\mathcal{S}$ to $\mathcal{T}$ and μ to ν in the sense that $\phi^{-1}(F) \in \mathcal{S}$ and $\mu(\phi^{-1}(F)) = \nu(F)$ for every $F \in \mathcal{T}$. An *isomorphism* of probability spaces is an invertible homomorphism, i.e., a homomorphism which is a bijection and whose inverse is also a homomorphism (fixed sets of measure zero in each space can be safely ignored).

Let $\mathcal{X} = (X, \mathcal{S}, \mu)$ be a measure space. A *measure-preserving transformation* of $\mathcal{X}$ is a homomorphism T of $\mathcal{X}$ to itself. An *isomorphism* (respectively, *homomorphism*) from one measure-preserving transformation $(\mathcal{X}, T)$ to another $(\mathcal{Y}, S)$ is an isomorphism ϕ (respectively, homomorphism) from $\mathcal{X}$ to $\mathcal{Y}$ which intertwines T and S, i.e., $\phi \circ T = S \circ \phi$. We then say that $(\mathcal{X}, T)$ is *isomorphic* to $(\mathcal{Y}, S)$ (respectively, $(\mathcal{Y}, S)$ is a *factor* of $(\mathcal{X}, T)$).

A fundamental goal of ergodic theory is to classify measure-preserving transformations up to isomorphism via explicit invariants. This is hopeless in general, but this goal has been achieved completely within certain interesting special classes of measure-preserving transformations (see Ornstein [Orn], Friedman and Ornstein [FriO], Halmos and Von Neumann [HalV]; see also [Pet3], [Wal]).

A measure-preserving transformation $((X, \mathcal{S}, \mu), T)$ is *ergodic* if whenever $E \in \mathcal{S}$ and $T^{-1}(E) \subseteq E$, then E has measure zero or one. It is a classical result of ergodic theory that every measure-preserving transformation can, in some sense, be decomposed into a family of ergodic measure-preserving transformations [Rok2]. These ergodic measure-preserving transformations can be viewed as analogues of the irreducible components of a graph, although often the ergodic decomposition involves an infinite (in fact, uncountable) family of ergodic measure-preserving transformations.

We can now see how a stationary process can be viewed as a measure-preserving transformation. For a shift space X, let $\mathcal{S}_X$ denote the smallest sigma-algebra which contains the cylinder sets. Note that $\mathcal{S}_X$ contains all finite unions of cylinder sets, and that the shift map σ preserves $\mathcal{S}_X$. By a probability measure on X we mean a probability measure on $(X, \mathcal{S}_X)$. Any stationary process μ with support contained in X assigns probabilities to blocks of X, and we can use these assignments to define measures on cylinder sets, in fact on finite unions of cylinder sets, as in §9.4. Standard measure theory then shows how to extend this assignment to all of $\mathcal{S}_X$ [Roy]. This yields a probability measure on X, which we call a *stationary measure*, also denoted μ. Since the process is stationary, it follows that the stationary measure μ satisfies $\mu(\sigma^{-1}(E)) = \mu(E)$ for all $E \in \mathcal{S}_X$. In other words, any stationary process with support in a shift space X defines a probability space $\mathcal{X} = (X, \mathcal{S}_X, \mu)$ such that $(\mathcal{X}, \sigma)$ is a measure-preserving transformation. Letting $\mathcal{X}_\mu$ denote the completion of $(X, \mathcal{S}_X, \mu)$, we then have the measure-preserving transformation $(\mathcal{X}_\mu, \sigma)$ (technically, $\mathcal{X}_\mu$ depends on the particular shift space X which contains the support of μ, but it turns out that all of these measure-preserving transformations are isomorphic for different choices of X). In particular, any stationary Markov chain μ on a graph G defines a stationary measure on X_G, and therefore defines a measure-preserving transformation $(\mathcal{X}_\mu, \sigma)$. In this case, the sets of measure zero coincide with the null sets as defined in 9.4.5.

Conversely, if $\mathcal{X}$ is the completion of a probability space $(X, \mathcal{S}_X, \mu)$ such that $(\mathcal{X}, \sigma)$ is a measure-preserving transformation, then $\mathcal{X}$ arises from a stationary process in this way [Bil]. So, we may view stationary processes as the measure-preserving transformations $(\mathcal{X}, T)$ with $T = \sigma$.

It is well-known ([Pet3], [Wal]) that the measure-preserving transformation defined by a stationary Markov chain μ on an irreducible graph G is ergodic if the support of μ is all of X_G (the latter condition is equivalent to requiring that the conditional probabilities of all edges of G be positive). Recall from Theorem 9.4.9 that the set of doubly transitive points of an irreducible edge shift X_G is the complement of a null set with respect to any stationary Markov chain whose support is all of X_G. This result extends to the more general class of ergodic stationary processes on G whose support is all of X_G.

As ergodic theory developed from the 1920's to the mid-1950's, the primary invariants of isomorphism for measure-preserving transformations were abstract invariants based on functional analysis. While these were successful in distinguishing some kinds of measure-preserving transformations, they were not very useful in distinguishing the measure-preserving transformations arising from stationary Markov chains. For instance, for a probability vector $(p_1, \ldots, p_n)$, the *Bernoulli $(p_1, \ldots, p_n)$-process*, a special type of stationary Markov chain on the full n-shift, is the stationary process which assigns probability $\mu(w_1 \ldots w_m) = \prod_{i=1}^{m} p_{w_i}$ to the block $w = w_1 \ldots w_m$. Up until 1957, it was not known if the measure-preserving transformation defined by the Bernoulli $(1/2, 1/2)$-process and the measure-preserving transformation defined by the Bernoulli $(1/3, 1/3, 1/3)$-process were isomorphic. This is where a notion of entropy for measure-preserving transformations enters the picture.

Entropy in this setting was developed by Kolmogorov [Kol] and Sinai [Sin1] (see [Pet3] or [Wal] for a definition). We denote the entropy of the measure-preserving transformation $(\mathcal{X}, T)$ by $h(\mathcal{X}, T)$. This entropy generalizes the notion of entropy for stationary processes, i.e., for a stationary process μ we have that $h(\mathcal{X}_\mu, \sigma) = h(\mu)$. Entropy turns out to be an invariant of isomorphism, and if $(\mathcal{Y}, S)$ is a factor of $(\mathcal{X}, T)$, then $h(\mathcal{Y}, S) \leqslant h(\mathcal{X}, T)$. The entropy of the measure-preserving transformation defined by the Bernoulli $(1/2, 1/2)$-process turns out to be $\log 2$, and the entropy of the measure-preserving transformation defined by the Bernoulli $(1/3, 1/3, 1/3)$-process turns out to be $\log 3$, proving that they are not isomorphic. Moreover, Ornstein [Orn] showed that entropy is a complete invariant of isomorphism for Bernoulli shifts. Friedman and Ornstein [FriO]) extended this to mixing Markov chains, i.e., stationary Markov chains on primitive graphs G with support all of X_G (see Shields [Shi] for an exposition of Ornstein's theory).

Suppose that G and H are irreducible graphs and $\phi \colon X_G \to X_H$ is a factor code. For a stationary measure μ on G, we define a stationary measure $\nu = \phi(\mu)$ on X_H by transporting μ to X_H the following way. For $A \in \mathcal{S}_{X_H}$, define

$$(13\text{--}4\text{--}1) \qquad\qquad \nu(A) = \mu(\phi^{-1}(A)).$$

Then ϕ defines a homomorphism $(\mathcal{X}_\mu, \sigma) \to (\mathcal{X}_\nu, \sigma)$, and so $h(\nu) \leqslant h(\mu)$.

Suppose that ϕ is actually a conjugacy. Then it defines an isomorphism, and so $h(\mu) = h(\nu)$. If $\mu = \mu_G$, then by uniqueness, we have $\nu = \mu_H$. Thus ϕ defines an isomorphism between $(\mathcal{X}_{\mu_G}, \sigma)$ and $(\mathcal{X}_{\mu_H}, \sigma)$. In fact, this holds whenever ϕ is merely an almost invertible factor code (see [AdlM]). This establishes the following result:

Let G, H be irreducible graphs, and let μ_G, μ_H be the stationary Markov chains of maximal entropy on G, H. If X_G, X_H are almost conjugate (in particular, if they are conjugate), then the measure-preserving transformations $(\mathcal{X}_{\mu_G}, \sigma)$, $(\mathcal{X}_{\mu_H}, \sigma)$ are isomorphic.

Hence conjugacies and almost conjugacies yield isomorphisms between measure-preserving transformations defined by stationary Markov chains of maximal entropy. In fact, the isomorphisms obtained in this way have some very desirable properties compared to the run-of-the-mill isomorphism. For instance, an isomorphism ϕ between stationary processes typically has an infinite window; i.e., to know $\phi(x)_0$, you typically need to know all of x, not just a central block $x_{[-n,n]}$. A conjugacy always has a finite window of uniform size. It turns out that an isomorphism obtained from an almost conjugacy, as well as its inverse, has finite expected coding length in the sense that to know $\phi(x)_0$, you need to know only a central block $x_{[-n(x),n(x)]}$, where the function $n(x)$ has finite expectation (see [AdlM]). In particular, by Theorem 9.3.2, whenever X_G and X_H are mixing edge shifts with the same entropy, the measure-preserving transformations (X_{μ_G}, σ) and (X_{μ_H}, σ) are isomorphic via an isomorphism with finite expected coding length.

Using these ideas, we now outline the original proof (see Coven and Paul [CovP1]) of the special case of Theorem 8.1.16 which asserts that an endomorphism ϕ of an irreducible edge shift X_G is finite-to-one.

For any stationary measure ν on X_G we can find a stationary measure μ on X_G such that $\nu = \phi(\mu)$ by first defining μ on sets in $\phi^{-1}(S_{X_G})$ by (13–4–1) and then extending μ to all of S_{X_G} (see Coven and Paul [CovP1] or Goodwyn [Goo] (Proposition 4)). This yields a homomorphism $(X_\mu, \sigma) \to (X_\nu, \sigma)$, and thus $h(\nu) \leqslant h(\mu)$. But if $\nu = \mu_G$, then, by uniqueness, $\mu = \mu_G$. Thus we see that ϕ defines a homomorphism $(X_{\mu_G}, \sigma) \to (X_{\mu_G}, \sigma)$. Now it follows from (13–3–7) that up to a uniform constant factor all cylinder sets of length n in X_G have μ_G-measure $1/\lambda^n$. Since ϕ defines a homomorphism, this yields a uniform upper bound on the number of pre-images of any block in X_G. This then gives a uniform upper bound on the number of pre-images of any point in X_G, as desired.

The notions of conjugacy, finite equivalence, almost conjugacy, embedding, factor code, and so on can all be generalized to the context of stationary measures, in particular stationary Markov chains. For instance, a conjugacy between two stationary measures is a map that is simultaneously a conjugacy of the underlying shift spaces and an isomorphism of the associated measure-preserving transformations. In particular, some of the results given in this text have been generalized to the context of stationary Markov chains. There is a substantial literature on this, and there are many open problems; see the expositions [ParT2] and [Par5] as well as the research papers [ParT1], [Par4], [ParS], [Sch1], [Kri7], [Tun1], [Tun2], [MarT1], [MarT2], [MarT3].

For an exposition of recent work in abstract ergodic theory see Rudolph [Rudo] and for expositions of smooth ergodic theory, see Mañé [Mane] and Katok and Hasselblatt [KatH].

§13.5. Sofic-like Shifts

Throughout most of this book, we have focused primarily on shifts of finite type and sofic shifts. We did this for several reasons. First, shifts of finite type and sofic shifts arise naturally in practical problems (Chapter 5) as well as models for geometric dynamical systems (Chapter 6). Second, the coding problems that we have considered in this book appear to be more tractable for shifts of finite type and sofic shifts than for shift spaces in general (although there are other classes of shift spaces for which the conjugacy problem has been solved; see substitution minimal shifts in §13.7). Third, sofic shifts can be described in a concrete and simple manner (via labeled graphs). Finally, sofic shifts have interesting, describable dynamical behavior (e.g., explicitly computable zeta functions).

But sofic shifts occupy only a small part of the universe of shift spaces. Recall, for instance, that there are uncountably many shift spaces, but only countably many sofic shifts.

In this section, we consider some classes of shifts which generalize sofic shifts. In §13.7 we will consider shifts which are very different from sofic shifts.

A *countable directed graph* (or *countable graph* for short) is a directed graph with countably many states and edges. For countable graphs, we have the usual notion of irreducibility (i.e., given any pair of states I, J, there is a path from I to J).

A *countable labeled directed graph* (*or countable labeled graph* for short) is a countable graph G together with a labeling $\mathcal{L}$ of the edges by a finite alphabet. Typically, the set X of bi-infinite sequences presented by $(G, \mathcal{L})$ is not closed and therefore not a shift space. But since X is shift-invariant, the closure $\overline{X}$ of X is a shift space, and we then say that $(G, \mathcal{L})$ is a *presentation* of $\overline{X}$. It turns out that every shift can be presented in this way.

If we require a little more of the presentation, then we get a proper class of shift spaces that generalizes the class of sofic shifts and has some interesting properties. A shift space that can be presented by an irreducible countable labeled graph is called a *coded system*. Of course, any coded system is irreducible. Coded systems were introduced by Blanchard and Hansel [BlaH1]. In that paper, they showed that any factor of a coded system is coded, thereby generalizing one of the key results for sofic shifts (Corollary 3.2.2). The following are equivalent characterizations of coded systems:

(1) X is a coded system.
(2) X has an irreducible right-resolving presentation (here "right-resolving" means the same as for finite presentations).
(3) X is the closure of the set of sequences obtained by freely concatenating the words in a (possibly infinite) list of words over a finite alphabet.

(4) X is the closure of the set of sequences obtained by freely concatenating the words in a (possibly infinite) uniquely decipherable set (see §8.1 for the definition of uniquely decipherable).

(5) X contains an increasing sequence of irreducible shifts of finite type whose union is dense in X.

For the equivalence of (1), (2), (3) and (4), see [BlaH1] and [FieF]; the equivalence of (1) and (5) is due to Krieger (cited in [BlaH2]).

It is natural to wonder which properties of irreducible sofic shifts extend to coded systems. Recall that every irreducible sofic shift has a periodic point, an intrinsically synchronizing word (see Exercise 3.3.4), finitely many futures of left-infinite sequences (see Exercise 3.2.8), finitely many follower sets of blocks, and a unique minimal right-resolving (finite) presentation. Clearly, the first property holds for all coded systems. Versions of the other properties hold for some nonsofic coded systems but not others. Specifically, consider the four properties:

(1) X is a sofic shift.

(2) X has countably many futures of left-infinite sequences.

(3) X has an intrinsically synchronizing word.

(4) X is a coded system.

For irreducible shifts (1) $\Rightarrow$ (2) $\Rightarrow$ (3) $\Rightarrow$ (4), and there are examples to show that none of these implications can be reversed (see [BlaH1], [FieF]).

A *synchronized system* is an irreducible shift which has an intrinsically synchronizing word [BlaH1]. This defines a special class of coded systems. The uniqueness of the minimal right-resolving presentation extends to synchronized systems in the following sense. For a synchronized system X, let $(G_X, \mathcal{L}_X)$ denote the labeled subgraph of the follower set presentation determined by states which are follower sets of intrinsically synchronizing words. Then there is a natural class $\mathcal{C}$ of (countable) presentations of X such that every $(G, \mathcal{L}) \in \mathcal{C}$ "collapses onto $(G_X, \mathcal{L}_X)$ by merging states" (in a sense analogous to Lemma 3.3.8 [FieF]). Also, various notions of right-closing, left-closing and almost finite type for synchronized systems were developed in [FieF].

D. Fiebig has obtained versions of some of the coding theorems in this text for coded systems. For instance, she established a version of the Finite Equivalence Theorem (Theorem 8.3.7) for coded systems: for coded systems the relation of having a common entropy-preserving extension is an equivalence relation within the class of coded systems, and entropy is a complete invariant [FieD2]. Also, she obtained partial results on the problems of classifying synchronized systems up to almost conjugacy, finite equivalence with right-closing legs, and other relations [FieD1], [FieD2].

On the other hand, many nice features of irreducible sofic shifts do not carry over to coded systems or even synchronized systems. For example, in contrast to Corollary 4.4.9, there are synchronized systems that have proper subshifts with full entropy [FieD2]. Also, many nice features of sliding block codes on irreducible sofic shifts do not carry over to coded systems or even synchronized systems. For instance, for sliding block codes between synchronized systems, entropy-preserving does not coincide with finite-to-one; in fact, equality of entropy is not sufficient for two coded systems to have a common finite-to-one extension [FieD2]. There are sliding block codes on coded systems for which each point has a finite, but not a uniformly bounded, number of pre-images, and there are sliding block codes for which there is a uniform upper bound on the number of pre-images of points, but no uniform upper bound on the number of pre-images of blocks, [FieD3]. There are several open problems along these lines (see [FieD1], [FieD2], [FieF]).

Finally, we discuss almost sofic shifts. A shift X is *almost sofic* if it can be approximated, in terms of entropy, arbitrarily well on the inside by sofic shifts or, more precisely, if there are sofic shifts $X_n \subseteq X$ such that $\lim_{n\to\infty} h(X_n) = h(X)$ (see [Pet2]). Clearly, every sofic shift is almost sofic. The almost sofic condition is natural because the Finite-State Coding Theorem (Theorem 5.2.5) holds for almost sofic shifts with excess entropy: in other words, if X is almost sofic and $h(X) > n$, then there is a finite-state (X, n)-code. Thus, whenever $h(X) > p/q$, there is a rate $p\!:\!q$ finite-state code from binary data into X. In fact, we claim that the Sliding Block Decoding Theorem (Theorem 5.5.6) extends to almost sofic shifts with excess entropy; this follows from the fact (Exercise 5.5.5) that every sofic shift can be approximated arbitrarily well on the inside by shifts of finite type.

Not every shift is almost sofic. For instance, recall that there are shifts with positive entropy and no periodic points at all. However, there are some interesting nonsofic shifts that are almost sofic. For instance, let F be any finite subset of $\mathbb{Z}^n$ such that $F = -F$; for instance, F could be the set of all vectors with entries ± 1. Now, let $(G, \mathcal{L})$ be the countable labeled graph with states $\mathbb{Z}^n$ and for each $x \in \mathbb{Z}^n$ and $f \in F$, an edge from x to $x + f$ labeled f. Let E be any subset of $\mathbb{Z}^n$, and let Γ_E be the subgraph of G determined by the states of E. Petersen [Pet2] showed that the shift on the alphabet F presented by $(\Gamma_E, \mathcal{L}|\Gamma_E)$ is almost sofic (actually Petersen showed much more than this). Note that these almost sofic shifts generalize the charge-constrained shifts (which are sofic) of Example 1.2.7 (set $n = 1$, $F = \{\pm 1\}$, and $E = [-c, c]$).

It is tempting to think that every coded system X is almost sofic: one could try to approximate X by the sofic shifts presented by the finite subgraphs of any irreducible right-resolving (countable) presentation of X. This does not work since there are examples of coded systems, even synchronized systems, which are not almost sofic (see [Pet2]). In particular, according to characterization (5) above of coded system, an irreducible shift X can contain a dense increasing union of irreducible shifts of finite type, whose entropies are bounded away from $h(X)$.

§13.6. Continuous Flows

In Chapter 6 we presented a brief glimpse of the role that symbolic dynamics plays in the study of dynamical systems. Recall that a dynamical system (M, ϕ) is a continuous map ϕ on a compact metric space M. Such a system is really a "discrete-time" dynamical system, where the iterates of ϕ on a point $x \in M$ keep track of the evolution of x at discrete ticks of a clock, but not at in-between times. A more realistic model of physical systems uses "continuous-time" to model evolution. The mathematical model for this is called a *continuous flow*, which is a family $\{\phi_t\}_{t\in\mathbb{R}} \colon X \to X$ of homeomorphisms ϕ_t from a compact metric space X to itself such that

(1) $\phi_t(x)$ is jointly continuous in t and x, and
(2) $\phi_s \circ \phi_t = \phi_{s+t}$ for all $s, t \in \mathbb{R}$.

Naturally, the *orbit* of a point $x \in X$ under a continuous flow $\{\phi_t\}$ is the set $\{\phi_t(x) \colon t \in \mathbb{R}\}$. Just as with discrete-time systems, continuous flows on noncompact spaces are also considered, but our interest here is mostly in compact spaces.

The set of solutions to a system of (sufficiently smooth) differential equations can be viewed as orbits of continuous flows [HirS], [Har]. One class of continuous flows of particular interest in the early days of symbolic dynamics was the class of geodesic flows, which we now briefly describe.

For surfaces, there are classical notions of tangent vector, curvature, and geodesic (see [Boo]). Geodesics are certain "distance-minimizing" curves. On a compact surface S of negative curvature, for each point x on S and each unit vector u tangent to S at x, there is a unique geodesic $\gamma_{x,u}$ which passes through x in the direction u. The geodesics are naturally parameterized by $\mathbb{R}$, the parameter being viewed as "time." Let $\gamma_{x,u}(t)$ denote the position of the geodesic $\gamma_{x,u}$ at time t.

The space X consisting of all such pairs (x, u) inherits a natural metric inherited from S, and with respect to this metric X is compact. The *geodesic flow* is the continuous flow $\{\phi_t\}_{t\in\mathbb{R}} \colon X \to X$ defined by $\phi_t(x, u) = (y, v)$, where $y = \gamma_{x,u}(t)$ and v is the unit tangent vector to $\gamma_{x,u}$ at y. See Figure 13.6.1.

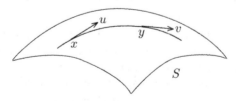

FIGURE 13.6.1.   Geodesic flow on a surface.

As we have described in the case of discrete-time, the idea of symbolic dynamics was to associate sequences of symbols with orbits of a continuous flow and to deduce properties of the orbits (and properties of the flow) from properties of the sequences. Hadamard is usually credited with the original idea, modeling geodesic flows in this way in a paper dating from 1898 [Had].

Recall from Chapter 6 that two prominent features of irreducible shifts of finite type and sofic shifts are density of periodic orbits and the existence of a point with dense forward orbit. About 1920, using the symbolic technique, Morse [Mor2] (see also [Mor1]) established these properties for certain geodesic flows. He also established the existence and density of "almost periodic" geodesics (see §13.7 for a definition of almost periodic). In the next few years, there was a flurry of activity in this subject, with many papers, notably Artin [Art], Nielsen [Nie1], [Nie2], Birkhoff [Bir2, pp. 238–248], and Koebe [Koe]. Of special interest were geodesic flows for surfaces of constant negative curvature. While much of this work focused on special noncompact surfaces, these authors also considered certain compact surfaces. Ergodic theory, born in the early 1930's, suggested new problems regarding the dynamics of geodesic flows; these were solved by Hedlund [Hed1], [Hed2], [Hed3] and Hopf [Hop1], [Hop2], among others.

Early expositions of symbolic dynamics for geodesic flows are contained in Morse [Mor3] and Gottschalk and Hedlund [GotH, Chapter 13]. Hedlund [Hed6] gives an interesting personal perspective. For a recent exposition of symbolic dynamics for geodesic flows, see Adler and Flatto [AdlF] and the references therein.

While shifts of finite type and sofic shifts did not formally appear on the scene until the 1960's and 1970's ([Sma], [Wei]), finite-type constraints were used to model geodesic flows much earlier in the work mentioned above (see also Morse and Hedlund [MorH1, pp. 822–824]).

The hyperbolic toral automorphisms, introduced in §6.5, are discrete-time dynamical systems that share many dynamical properties with geodesic flows, and so they are often viewed as discrete-time analogues of geodesic flows. Together, these two classes of systems motivated the more general class of hyperbolic dynamical systems, prominent in the modern theory of smooth dynamical systems (see Smale [Sma]).

Continuous flows give rise to discrete-time dynamical systems in two different ways. First, the *time–one map*, ϕ_1, of a continuous flow defines the discrete-time dynamical system (X, ϕ_1). Secondly, given a continuous flow $\{\phi_t\} \colon X \to X$ and a surface M "transverse" to the flow, define the "Poincare return-map" (or simply "return-map") $\phi \colon M \to M$, by setting $\phi(x) = \phi_t(x)$ where $t = t(x)$ is the smallest $t > 0$ such that $\phi_t(x) \in M$ (see [Rue2], [Nit], [ArrP]). Strictly speaking, (M, ϕ) need not define a discrete-time dynamical system because ϕ need not be continuous (in particular, there are often problems at the boundary of M). Nevertheless,

Poincare return-maps, rather than time–one maps, originally motivated the study of discrete-time dynamical systems.

Conversely, there is a way to manufacture a continuous flow out of a discrete-time dynamical system, described next.

Let (M, ϕ) be an (ordinary) dynamical system. Consider the product space $M \times \mathbb{R}$, and let $\sim$ denote the equivalence relation on $M \times \mathbb{R}$ generated by identifying $(x, s + 1)$ with $(\phi(x), s)$ for each $x \in M, s \in \mathbb{R}$. Let X denote the set of $\sim$-equivalence classes. There is a natural way to make X into a compact metric space (see Bowen and Walters [BowW]). Let $[(x, s)]$ denote the equivalence class of (x, s) and define for each $t \in \mathbb{R}$ the mapping $\phi_t : X \to X$ by

$$\phi_t([(x, s)]) = [x, s + t].$$

Then each $\phi_t : X \to X$ is a well-defined map, and the collection $\{\phi_t\} : X \to X$ constitutes a continuous flow called the *suspension flow over* (M, ϕ).

For a suspension flow, we visualize X as a rectangle with base M, side $[0, 1]$, and the top and bottom identified via $(x, 1) \sim (\phi(x), 0)$. The flow $\{\phi_t\}$ moves a point $(x, 0)$ vertically upwards until it reaches the top at $(x, 1)$, which is then identified with the point $(\phi(x), 0)$ at the bottom. It then continues moving up from $(\phi(x), 0)$ in the same way (see Figure 13.6.2). Sometimes the space X is called the mapping torus of (M, ϕ). Observe that in the special case that M is the circle and $\phi = Id$, then X is literally the two-dimensional torus.

An *equivalence* between two continuous flows, say $\{\phi_t\}_{t \in \mathbb{R}} : X \to X$ and $\{\psi_t\}_{t \in \mathbb{R}} : Y \to Y$, is a homeomorphism $\pi : X \to Y$ which maps orbits of $\{\phi_t\}$ to orbits of $\{\psi_t\}$ in an orientation preserving way; i.e., for all $x \in X$, $\pi(\phi_t(x)) = \psi_{f_x(t)}(\pi(x))$ and $f_x : \mathbb{R} \to \mathbb{R}$ is monotonically increasing (note in particular that the time parameter t need not be preserved exactly; i.e., f_x need not be the identity). Two flows are called *equivalent* if there is an equivalence from one to the other.

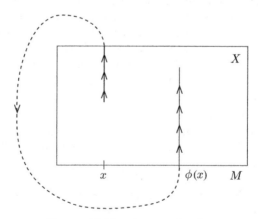

FIGURE 13.6.2.   A suspension flow.

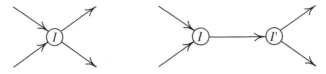

FIGURE 13.6.3. Graph expansion.

We say that two (discrete-time) dynamical systems (M, ϕ) and (M', ϕ') are *flow equivalent* if the corresponding suspension flows over (M, ϕ) and (M', ϕ') are equivalent. It is not hard to see that any conjugacy of dynamical systems yields a flow equivalence of the corresponding suspension flows, but generally flow equivalence is a much weaker relation.

We now consider flow equivalence for shifts of finite type. As usual, we may first recode to edge shifts. So, let $(M, \phi) = (X_G, \sigma)$ and $(M', \phi') = (X_H, \sigma)$. We say that the graphs G and H are flow equivalent whenever (X_G, σ) and (X_H, σ) are flow equivalent.

Parry and Sullivan [ParSu] showed that the relation of flow equivalence on graphs is generated by elementary equivalences together with one additional operation, which roughly corresponds to changing the time parameter along the suspension flow: namely, an *expansion* of a graph is a change in the local picture at one state I, where I is replaced by an edge from I to a new state I', and all outgoing edges from I are replaced by outgoing edges from I' (see Figure 13.6.3). Thus, graphs G and H are flow equivalent if and only if you can pass from G to H by a sequence of elementary equivalences, expansions, and inverses of expansions.

Using a matrix interpretation of expansion, Parry and Sullivan [ParSu] showed that the quantity $D(A_G) = \det(Id - A_G)$ is an invariant of flow equivalence. Then Bowen and Franks [BowF] showed that the particular Bowen–Franks group $BF(A_G) = BF_{1-t}(A_G)$ defined in §7.4, is also an invariant of flow equivalence. Actually, the Bowen–Franks group was first invented for use as an invariant of flow equivalence rather than conjugacy. Then Franks [Fran2] completely solved the flow equivalence problem for irreducible graphs. He showed that when G and H are irreducible graphs whose adjacency matrices have spectral radius greater than 1 (equivalently, X_G and X_H have positive entropy), then $D(A_G)$ and $BF(A_G)$ are a complete set of invariants of flow equivalence. Note that when the spectral radius of A_G is 1, then the flow consists of a single periodic orbit, and any two periodic orbits are clearly flow equivalent. It turns out that $BF(A_G)$ itself is almost a complete invariant: one can deduce the absolute value $|D(A_G)|$, but not always the sign of $D(A_G)$, from $BF(A_G)$.

Boyle [Boy4] solved versions of the finite-to-one factor problem for suspension flows over irreducible shifts of finite type and the almost conjugacy problem (called *almost flow equivalence*) for suspension flows over irreducible sofic shifts. In particular, he showed that for any pair X, Y of irreducible sofic shifts of positive entropy, X and Y are almost flow equivalent.

Huang [Hua1], [Hua2] has obtained results on flow equivalence for reducible graphs. Fujiwara and Osikawa [FujO] found invariants of flow equivalence for sofic shifts, but even irreducible sofic shifts have not yet been completely classified up to flow equivalence.

§13.7. Minimal Shifts

In this section we discuss minimal shifts, defined below. We will see that these shifts are very different from shifts of finite type and sofic shifts. The study of minimal shifts was a natural outgrowth of efforts in the early days of symbolic dynamics to prove the existence of (and to actually construct) geodesics on surfaces of negative curvature that are almost periodic, defined as follows.

For an (ordinary discrete-time) dynamical system (M, ϕ), a point $x \in M$ is called *almost periodic* if x returns close to its initial position (in space) with bounded gaps (in time); i.e., for each $\epsilon > 0$, there is some $N = N(\epsilon)$ such that the set $\{n \geqslant 0 : \rho(\phi^n(x), x) < \epsilon\}$ has gaps of size at most N. The orbit generated by such a point is also called almost periodic. Observe that if $M = X$ is a shift space and ϕ is the shift map, then an element $x \in X$ is almost periodic if and only if every allowed block in x appears in x with bounded gaps. In the early work of Birkhoff [Bir1], [Bir2] and Morse [Mor1], [Mor2], almost periodicity was called recurrence, but starting with [Hed4] recurrence came to mean something weaker (namely, that x returns arbitrarily close to itself, but not necessarily with bounded gaps).

There is an analogous notion of almost periodicity for continuous flows, and, via the geodesic flow, this gives a notion of almost periodicity for geodesics. Of course, periodic orbits are almost periodic, and so the real interest was in finding almost periodic geodesics which are not periodic.

The shift spaces that model almost periodic behavior are called minimal shifts (also called minimal sets): a shift space X is *minimal* if it contains no proper subshift. Any shift space contains at least one minimal shift, and two distinct minimal shifts cannot intersect [Bir1], [MorH1].

There are many equivalent characterizations of minimality, most notably that a shift space is minimal if and only if it is the closure of an almost periodic orbit [Bir1] (see also [Bir2], [MorH1], [GotH]). This is why minimal shifts model almost periodic behavior. In fact, every orbit in a minimal shift is almost periodic. Note that every almost periodic sequence is contained in exactly one minimal set, namely the closure of its orbit. Another equivalent condition for a shift to be minimal is that every orbit in X be dense.

The finite minimal shifts are simply the periodic orbits. Infinite minimal shifts are necessarily uncountable since they are compact perfect sets (i.e., each point is a limit point) [Rud]. While infinite minimal shifts are irreducible (since each, and hence at least one, forward orbit is dense), they are very different from irreducible sofic shifts. An infinite irreducible sofic shift contains orbits with quite varied behavior – lots of periodic orbits, many complicated proper minimal subshifts, as well as dense forward orbits. Infinite minimal shifts can have positive entropy ([Fur2], [HahK]) but since they can have no periodic points, the entropy formula in terms of periodic points from Theorem 4.3.6 fails completely. Properties of minimal shifts have been developed in many papers, for instance, [MorH1], [MorH2], [GotH], [CovK], [Got], [Hed4] [Pet1] (see [Ell] for an exposition on minimal dynamical systems).

Perhaps the most famous infinite minimal shift is the Morse shift [Mor2], which Morse devised specifically to exhibit a nonperiodic almost periodic geodesic. To define this shift, consider the following sequence of blocks. Let $B_0 = 0$, and for $n \geqslant 1$, define

$$B_{n+1} = B_n \overline{B}_n,$$

where $\overline{B}_n$ is the complement of B_n, obtained by interchanging the 1's and 0's in B_n. Since B_n is a prefix of B_{n+1} for each n, in the limit the blocks B_n define a right-infinite sequence x^+ called the *Morse sequence*, which is the (unique) sequence having each B_n as a prefix. The first few terms of the Morse sequence are

$$x^+ = 0110100110010110100101100110 1001 \ldots.$$

The Morse sequence is sometimes called the Morse–Thue sequence because it was actually discovered earlier in a different context by Thue [Thu] (1906). In fact, the Morse sequence can be found still earlier; see Prouhet [Pro] (1851).

Any right-infinite sequence x^+ determines a shift space in several equivalent ways: (1) the set of limit points in the full shift of $\{\sigma^n(x^-.x^+) : n \geqslant 0\}$ where x^- is any left-infinite sequence; (2) the shift space defined by declaring the forbidden blocks to be the blocks that do not appear in x^+; (3) the shift space defined by declaring the allowed blocks to be the blocks that do appear in x^+ (although not all blocks that appear in x^+ need occur in the shift space). The *Morse shift* X is the subshift determined in this way from the Morse sequence above. Morse [Mor2] showed that X is indeed an infinite minimal shift, equivalently that x is almost periodic but not periodic. Using this, he showed that for certain surfaces, almost periodic geodesics exist, and in fact are dense.

There is still another way to describe the Morse shift, and this suggests a method of constructing a broad family of minimal shifts. Let $C_0 = 01$ and $C_1 = 10$. It is not hard to see inductively that in the definition of the Morse sequence above, the block B_{n+1} is obtained from B_n by replacing each appearance of the symbol 0 with C_0 and each appearance of the symbol

1 with C_1. More generally, for a finite alphabet $\mathcal{A}$, let $\mathcal{C} = \{C_a : a \in \mathcal{A}\}$ be a list of blocks over $\mathcal{A}$, possibly of varying lengths. We assume that there is a symbol $a_0 \in \mathcal{A}$ such that C_{a_0} begins with a_0. We then obtain a sequence of blocks B_n by defining $B_0 = a_0$ and B_{n+1} to be the block obtained from B_n by replacing, for each symbol a, each appearance of a with C_a. Assuming that $\lim_{n\to\infty} |B_n| = \infty$, these blocks define a unique right-infinite sequence x^+ which has each B_n as a prefix. The *substitution shift* X generated by the list $\mathcal{C}$ is the shift space determined by x^+. Substitution shifts were formally defined (slightly differently) by Gottschalk [Got], but versions of this idea were implicit in earlier work. We will describe some properties of substitution shifts, but for a thorough exposition, see Queffelec [Que].

For a substitution $\mathcal{C}$ there is a naturally associated $|\mathcal{A}| \times |\mathcal{A}|$ matrix $A_{\mathcal{C}}$ whose (a, b) entry records the number of appearances of the symbol b in the block C_a. A substitution shift is minimal if and only if $A_{\mathcal{C}}$ is irreducible. Observe that when $A_{\mathcal{C}}$ is irreducible, the assumption above that $\lim_{n\to\infty} |B_n| = \infty$ is equivalent to the simple assumption that at least one of the blocks C_a has length at least 2. It turns out that when $A_{\mathcal{C}}$ is irreducible, the substitution shift X is independent of the special symbol a_0.

Substitution shifts that are minimal are naturally called *substitution minimal shifts* (or substitution minimal sets). Some classes of substitution minimal shifts have been completely classified up to conjugacy, for instance the shifts defined by substitutions on two symbols $0, 1$ where the blocks C_0 , C_1 have the same length (see Coven and Keane [CovK]). Substitution shifts have zero entropy (see Klein [Kle]). Mixing properties of substitution minimal shifts were investigated by Dekking and Keane [DekK]. Other properties of substitution minimal sets were established by Martin [Mart1], [Mart2].

Another way to generalize the Morse shift is as follows. For a binary block b, let $b^0 = b$, and let $b^1 = \bar{b}$, the complement of b. For binary blocks $b = b_1 \ldots b_n$ and $c = c_1 \ldots c_m$, define

$$b \times c = b^{c_1} \ldots b^{c_m}.$$

Given an arbitrary sequence of binary blocks $b^{(1)}, b^{(2)}, \ldots$, we get a new sequence of binary blocks defined by

$$B_1 = b^{(1)}, \quad B_{n+1} = B_n \times b^{(n+1)}.$$

If we assume that each block $b^{(n)}$ begins with the symbol 0, then B_n is a prefix of B_{n+1} for each n, and so the blocks B_n define a right-infinite sequence x^+, and therefore a shift space called a *generalized Morse shift* (see Keane [Kea1]). The Morse shift itself is the generalized Morse shift generated by the sequence of blocks $b^{(n)} = 01$ for all n. A generalized Morse shift is always minimal [Kea1].

Another interesting class of minimal shifts is the class of Toeplitz shifts. These shifts were motivated by a construction of Toeplitz [Toe] for exhibiting "almost periodic" functions on the real line. A bi-infinite sequence x is a *Toeplitz sequence* if the set of integers can be decomposed into arithmetic progressions such that x_i is constant on each arithmetic progression. Toeplitz sequences can be constructed with great flexibility. First fill in the coordinates of x only on some arithmetic progression, thinking of the remaining coordinates as "holes." Then fill in the holes by inductively setting x_i to have some constant value along an arithmetic progression within the holes. Continue this process until x is defined completely. A shift space X is a *Toeplitz shift* if it is the closure of the orbit of a Toeplitz sequence. Jacobs and Keane [JacK] introduced Toeplitz shifts and showed that every Toeplitz shift is minimal.

Toeplitz shifts are useful for exhibiting a wide variety of behaviors. For instance, Furstenberg's example [Fur2] of a minimal shift with positive entropy is actually a Toeplitz shift. See Markley and Paul [MarP1] and S. Williams [WilS1] for Toeplitz shifts with other specified behaviors.

There is some overlap among the classes of minimal shifts that we have already described. For instance, every infinite substitution minimal shift defined by a constant length substitution on $\{0, 1\}$ is either a generalized Morse shift or a Toeplitz shift [CovK].

Finally, we discuss the class of Sturmian shifts. This class of shifts was introduced in [MorH2] and constitutes one of the earliest general classes of shift spaces studied in symbolic dynamics. In the following, we define Sturmian shifts as shifts which model translations of the circle $\mathbb{T} = [0, 1)$ (see Example 6.2.2).

Let $\beta \in [0, 1)$ and consider the translation $\phi_\beta \colon [0, 1) \to [0, 1)$ defined by $\phi_\beta(x) = x + \beta \pmod 1$. Let $\mathcal{P}$ denote the partition of $[0, 1)$ given by $\mathcal{P} = \{[0, \beta), [\beta, 1)\}$. Then, in the usual symbolic spirit, we associate a binary sequence to each $t \in [0, 1)$ according to its itinerary relative to $\mathcal{P}$; that is, we associate to $t \in [0, 1)$, the bi-infinite sequence x defined by $x_i = 0$ if $\phi_\beta^i(t) \in [0, \beta)$ and $x_i = 1$ if $\phi_\beta^i(t) \in [\beta, 1)$. The set of such sequences x is not necessarily closed, but it is shift-invariant and so its closure is a shift space $X(\beta)$, called a *Sturmian shift*. This definition differs slightly from the original definition given in [MorH2], but it is equivalent. Hedlund showed that if β is irrational, then the Sturmian shift $X(\beta)$ is minimal [Hed4].

Minimal Sturmian shifts have some very interesting properties. For instance, they have "smallest" possible growth rate of blocks in the following sense. It is not hard to see that in an infinite shift space, the number of n-blocks which occur is at least $n + 1$. For a minimal Sturmian shift, the number of n-blocks which occur is exactly $n + 1$. Hence not only do these shifts have zero entropy, but they also have the smallest growth possible for infinite shift spaces. Coven and Hedlund [CovH] completely characterized the shift spaces with this smallest possible growth.

It turns out that Sturmian shifts are very different from substitution shifts. For instance, no Sturmian shift can be a factor of a constant length substitution [Kle].

In our discussion of minimal shifts we have completely left out their ergodic-theoretic properties. In fact, in many of the papers we have cited the major motivation and main results originate in ergodic theory.

§13.8. One-Sided Shifts

Up to this point, the reader may have the impression that in symbolic dynamics we study only spaces of *bi-infinite* sequences over *finite* alphabets of symbols. In this section, as well as in §13.9 and §13.10, we review some work that has been done within variations of this framework, allowing the index set of the sequences to be something other than $\mathbb{Z}$ and allowing the alphabet to be infinite.

Recall from §5.1 that a one-sided shift is defined as the set X^+ of all right-infinite sequences that appear in elements of an ordinary shift space X. We found this definition to be slightly more convenient than the obvious alternative definition: a set of right-infinite sequences defined by a list of forbidden blocks. By a compactness argument it is not hard to see that any one-sided shift X^+ is defined by a list of forbidden blocks, namely the list of all blocks which do not occur in X. So, our definition is (slightly) more general. While these definitions do not agree even for edge shifts, it is easy to see that they do agree for edge shifts defined by essential graphs.

A sliding block code ϕ with zero memory determines a map ϕ^+, also called a sliding block code, between the corresponding one-sided shifts. Hence any sliding block code on a shift space X can be shifted to define a sliding block code on the corresponding one-sided shift space X^+. For one-sided shifts, we can then define factor codes, embeddings, conjugacies, and automorphisms in the obvious way, i.e., sliding block codes which are, respectively, onto, one-to-one, invertible, and self-conjugacies.

A compactness argument can be used to show that a sliding block code (with zero memory) ϕ is onto if and only if the corresponding sliding block code ϕ^+ is onto (Exercise 8.1.14). On irreducible shifts of finite type and sofic shifts, a sliding block code (with zero memory) ϕ is finite-to-one if and only if ϕ^+ is finite-to-one, roughly because of the no-diamond property (Exercise 8.1.14).

Since memory is not allowed on the one-sided level, it is much harder for ϕ^+ to be invertible than for ϕ to be invertible. For a dramatic illustration of this, consider that the shift map σ is always one-to-one, but σ^+ is rarely one-to-one. Also, it turns out that for a 1-block conjugacy ϕ between irreducible edge shifts, ϕ^+ is a conjugacy if and only if ϕ is left-resolving. Finally, recall that the automorphism group $\mathrm{aut}(X_{[2]})$ of the full 2-shift is quite large (for instance, it contains a copy of every finite group, and it is not finitely generated), but it turns out that there are only two automorphisms of the one-sided full 2-shift: the identity map and the map which interchanges 0 and 1 (see [Hed5]).

Boyle, Franks, and Kitchens [BoyFK] have studied the automorphism group aut(X^+) of one-sided shifts of finite type. They found severe constraints on the finite subgroups contained in aut(X^+), and they completely characterized the finite subgroups contained in the automorphism group of a one-sided full shift. They also showed that any automorphism ϕ^+ of a one-sided shift of finite type is a composition of automorphisms of finite order.

In sharp contrast to the two-sided case, the conjugacy problem for one-sided shifts of finite type has been solved completely; we describe this as follows.

First recall that for an N-step shift of finite type X, the higher block shift $X^{[N+1]}$ is an edge shift X_G. This yields a one-sided conjugacy between X^+ and X_G^+. Thus we need only consider one-sided edge shifts.

Let G and T be graphs. We say that T is a *total amalgamation* of G if

(1) T is obtained from G via a sequence of out-amalgamations, and
(2) any out-amalgamation of T is graph isomorphic to T itself.

In terms of adjacency matrices, this means that

(1) A_T is obtained from A_G by iteratively deleting multiple copies of repeated columns and adding corresponding rows, and
(2) A_T has no repeated columns.

Clearly every graph has a total amalgamation. R. Williams [WilR2] showed that the total amalgamation T_G of a graph G is unique up to graph isomorphism, and that it completely determines the conjugacy class of X_G^+. That is, X_G^+ is conjugate to X_H^+ if and only if $T_G \cong T_H$.

This result gives a finite procedure for deciding conjugacy of a pair X_G^+, X_H^+ of one-sided edge shifts. Simply delete multiple columns and add corresponding rows of A_G until it has no repeated columns, yielding A_{T_G}, and do the same for A_H. Then compare A_{T_G} with A_{T_H} to see if they are similar via a permutation matrix. In contrast, recall that it is not known if the ordinary (two-sided) conjugacy problem for edge shifts is decidable, even for edge shifts defined by graphs with only two states.

For an irreducible graph G, the graph T_G is obtained by iteratively "collapsing" states in a way that yields a left-resolving conjugacy. Recall that the graph M_G, described in §8.4, is obtained by iteratively "collapsing" states in a way that yields a right-resolving factor code. Hence, we see that T_G is a partial step along the way to constructing $M_{G^\top}$.

Finally, we remark that the embedding problem for one-sided shifts of finite type and the conjugacy problem for one-sided sofic shifts have not been solved (even in the irreducible case), although Fujiwara [Fuj] obtained a partial result on the latter.

§13.9. Shifts with a Countable Alphabet

The full shift $\mathcal{A}^{\mathbb{Z}}$ over an arbitrary alphabet $\mathcal{A}$ is defined just as for finite alphabets, i.e.,

$$\mathcal{A}^{\mathbb{Z}} = \{x = (x_i) : x_i \in \mathcal{A} \quad \text{for all } i \in \mathbb{Z}\}.$$

We can use the same metric as in Chapter 6, namely

$$(13\text{-}9\text{-}1) \qquad\qquad \rho(x,y) = 2^{-k},$$

where k is the largest integer such that $x_{[-k,k]} = y_{[-k,k]}$ (with the conventions: $\rho(x,y) = 0$ if $x = y$ and $\rho(x,y) = 2$ if $x_0 \neq y_0$). With respect to this metric, convergence means the same as for finite alphabets: a sequence of points $x^{(n)} \in \mathcal{A}^{\mathbb{Z}}$ converges to a point $x \in \mathcal{A}^{\mathbb{Z}}$ if and only if for any k and all sufficiently large n (depending on k) the central blocks $x^{(n)}_{[-k,k]}$ and $x_{[-k,k]}$ agree. However, the full shift over an infinite alphabet is not compact, for if $a_1, a_2, \dots$ denotes an infinite collection of distinct symbols, then the sequence $a_1^{\infty}, a_2^{\infty}, \dots$ has no convergent subsequence.

We define an (infinite alphabet) *shift space* to be a closed (with respect to the metric ρ) shift-invariant subset X of the full shift. As with finite alphabets, this is equivalent to saying that X is determined by a list of forbidden blocks.

Sliding block codes on infinite alphabet shift spaces are defined in exactly the same way as for finite alphabets. However, the characterization of sliding block codes in terms of continuous maps (as in §6.2) is different. On a compact space, continuity coincides with uniform continuity: the "δ" can be chosen to depend only on the "ϵ," uniformly over the entire space. On a noncompact space, uniform continuity is a stronger condition. It is not hard to see that the sliding block codes on $\mathcal{A}^{\mathbb{Z}}$ are exactly the shift-commuting maps that are uniformly continuous with respect to the metric ρ.

Although an infinite alphabet shift space is not compact, it still makes sense to study the standard symbolic dynamics problems (conjugacy, embedding, factor code, finite equivalence) in this setting. But progress has been difficult to come by.

In contrast to finite alphabet shift spaces, the image of an infinite alphabet shift space via a sliding block code is not necessarily closed and therefore not necessarily a shift space. So we take as the definition of factor code a sliding block code $X \to Y$ whose image is dense in Y. However, some strange things happen with this definition. For instance, there are reasonable notions of entropy (such as the ones given later in this section) such that entropy can actually increase under factor codes.

In this section we focus on countably infinite alphabets. Recall from §13.5 that a countable graph is a directed graph with countably many states and edges. A *countable edge shift* is the set of all bi-infinite edge sequences of paths in a countable graph. Likewise a *countable vertex shift* is the set of all bi-infinite state sequences of paths in a countable graph. A countable graph G has an adjacency matrix A_G, defined just as for finite graphs, except that now it is countably-indexed and some of the entries may be infinite. Irreducibility for countable graphs, for nonnegative countably-indexed matrices, for countable edge shifts, and for countable vertex shifts is defined just as in the finite case.

One reason why edge shifts and vertex shifts have been considered in this setting is that they can be used to present coded systems (see §13.5). Another reason is that they can be used to model dynamical systems, via infinite Markov partitions. This includes certain maps of an interval (see Hofbauer [Hof1], [Hof2], [Hof3], [Hof4] and Newhouse [Newh]), "billiard" flows (see Kruger and Troubetskoy [KruT] and Bunimovitch, Sinai and Chernov [BunSC], [BunS]) and other mappings [Yur]. A third reason is that they constitute the natural analogue for symbolic dynamics of countable state Markov chains, a subject of great interest in probability theory (see Seneta [Sen]).

How should we define entropy for countable edge shifts X_G? We will suggest some possibilities below. We assume that the graph G is irreducible and *locally finite*, i.e., that each state has only finitely many incoming edges and outgoing edges (equivalently, the row and column sums of A_G are all finite). We remark that this can be viewed as a "local compactness" assumption.

We would like to define entropy as the growth rate of the number of paths in the graph, but this does not make sense since there may be infinitely many paths of length n. However, since G is locally finite, it does make sense to consider the growth rate of the number of paths that begin at a fixed arbitrary state I. Hence we define

$$(13\text{-}9\text{-}2) \qquad h^{(1)}(X_G) = \lim_{n \to \infty} \frac{1}{n} \log \sum_J (A_G^n)_{IJ}.$$

Since G is irreducible, this quantity is independent of the state I.

There are other reasonable definitions of entropy. For instance, we can consider the growth rate of the number of cycles, instead of the growth rate of paths, and define

$$(13\text{-}9\text{-}3) \qquad h^{(2)}(X_G) = \lim_{n \to \infty} \frac{1}{n} \log(A_G^n)_{II}.$$

Again, since G is irreducible, this definition does not depend on the choice of state I.

Another notion of entropy makes use of finite subgraphs. Put

$$h^{(3)}(X_G) = \sup_{H \subseteq G,\ H \text{ finite}} h(X_H).$$

Finally, for any metric ρ on X_G such that the shift map is uniformly continuous, there is a notion of entropy $h_\rho(X_G)$, depending on ρ. This is a special case of a notion of entropy for (noncompact) dynamical systems introduced by Bowen [Bow3]; see also [Wal].

How are these notions of entropy related? Clearly

$$h^{(2)}(X_G) \leqslant h^{(1)}(X_G).$$

If $h^{(3)}(X_G) < \infty$, then

$$h^{(2)}(X_G) = h^{(3)}(X_G)$$

(this is implicit in Gurevic [Gur1]).

It turns out that the entropies, $h^{(1)}$, $h^{(2)}$, and $h^{(3)}$ agree with entropies based on metrics. The entropy $h^{(1)}$ coincides with the entropy based on the metric ρ defined by (13–9–1), so that

$$h^{(1)}(X_G) = h_\rho(X_G)$$

(see Salama [Sal2] and Wagoner [Wag2]). Define a metric τ by identifying $\mathcal{A}$ with the set $\{1, 1/2, 1/3, \dots\}$ in an arbitrary way and putting

$$\tau(x,y) = \sum_{n=-\infty}^{\infty} \frac{|x_n - y_n|}{2^{|n|}}.$$

According to [Gur1] and [Sal2],

$$h^{(2)}(X_G) = h^{(3)}(X_G) = h_\tau(X_G).$$

Salama [Sal1] and Petersen [Pet2] give examples where entropy can increase via factors with respect to any of these definitions.

Partly motivated by these notions of entropy for countable edge shifts, Handel and Kitchens [HanK] have studied the relationship among various notions of entropy for dynamical systems on noncompact spaces.

In §13.3 we considered stationary processes. There we assumed that the alphabet is finite, but there is no harm in allowing the alphabet to be infinite. In particular, there are *countable-state stationary Markov chains*, defined just like ordinary (finite-state) stationary Markov chains except that there may be a countably infinite number of states.

States of such a Markov chain are naturally divided into three classes: transient, null recurrent, and positive recurrent, according to the probability of return and expected length of return to the state; when the Markov chain is irreducible, the class is independent of the state [Sen]. Vere-Jones [Ver1], [Ver2] (see also Seneta [Sen]) established an analogous classification of countably-indexed irreducible matrices, and, therefore, irreducible countable graphs. We describe this as follows.

In analogy with ordinary graphs, $R = 2^{h^{(2)}(X_G)}$ is viewed as the "Perron eigenvalue" (recall our convention that log uses base 2). Then, in analogy with (13–3–7), given a starting state I, the conditional probability of taking a path in G of length n is viewed as R^{-n}. Let $\ell_{II}(n)$ denote the number of "first-return" cycles at I, i.e., cycles at I which meet I only at the beginning and end. Then $\sum_n \ell_{II}(n)R^{-n}$ is viewed as the probability of return to state I, while $\sum_n n\ell_{II}(n)R^{-n}$ is viewed as the expected length of return to state I.

It turns out that $\sum_n \ell_{II}(n)R^{-n} \leqslant 1$. An irreducible countable graph G is called *transient* if $\sum_n \ell_{II}(n)R^{-n} < 1$ and *recurrent* if $\sum_n \ell_{II}(n)R^{-n} = 1$. A recurrent graph is called *null recurrent* if $\sum_n n\ell_{II}(n)R^{-n} = \infty$ and *positive recurrent* if $\sum_n n\ell_{II}(n)R^{-n} < \infty$. Thus the positive recurrent graphs are the ones for which each state almost surely returns to itself and the expected length of return is finite. Positive recurrent graphs turn out to be the graphs for which most of Perron–Frobenius theory carries over. Using this, Gurevic [Gur2] showed that for an irreducible, countable graph G with $h^{(3)}(X_G) < \infty$, the following conditions are equivalent:

(1) G is positive recurrent.
(2) There is a stationary process of maximal entropy on X_G.
(3) There is a unique stationary process of maximal entropy on X_G.
(4) There is a stationary Markov chain of maximal entropy on X_G.
(5) There is a unique stationary Markov chain of maximal entropy on X_G.

We refer the reader to Kitchens [Kit4] for a survey of these issues.

§13.10. Higher Dimensional Shifts

In this section we discuss higher dimensional shift spaces. For a more thorough introduction, we refer the reader to Schmidt [Sch2], [Sch3].

The *d-dimensional full A-shift* is defined to be $A^{\mathbb{Z}^d}$. Ordinarily, A is a finite alphabet, and here we restrict ourselves to this case. An element x of the full shift may be regarded as a function $x: \mathbb{Z}^d \to A$, or, more informally, as a "configuration" of alphabet choices at the sites of the integer lattice $\mathbb{Z}^d$.

For $x \in A^{\mathbb{Z}^d}$ and $F \subseteq \mathbb{Z}^d$, let x_F denote the restriction of x to F. The usual metric on the one-dimensional full shift naturally generalizes to a metric ρ on $A^{\mathbb{Z}^d}$ given by

(13–10–1) $$\rho(x, y) = 2^{-k},$$

where k is the largest integer such that $x_{[-k,k]^d} = y_{[-k,k]^d}$ (with the usual conventions when $x = y$ and $x_0 = y_0$). According to this definition, two points are close if they agree on a large cube $\{-k, \ldots, k\}^d$.

A *d-dimensional shift space* (or *d-dimensional shift*) is a closed (with respect to the metric ρ) translation-invariant subset of $\mathcal{A}^{\mathbb{Z}^d}$. Here "translation-invariance" means that $\sigma^{\mathbf{n}}(X) = X$ for all $\mathbf{n} \in \mathbb{Z}^d$, where $\sigma^{\mathbf{n}}$ is the translation in direction $\mathbf{n}$ defined by $(\sigma^{\mathbf{n}}(x))_{\mathbf{m}} = x_{\mathbf{m}+\mathbf{n}}$.

There is an equivalent definition of shift space based on lists of "forbidden patterns," as follows. A *shape* is a finite subset F of $\mathbb{Z}^d$. A *pattern* f on a shape F is a function $f \colon F \to \mathcal{A}$. Given a list $\mathcal{F}$ of patterns, put

$$X = X_{\mathcal{F}} = \{x \in \mathcal{A}^{\mathbb{Z}^d} : \sigma^{\mathbf{n}}(x)_F \notin \mathcal{F} \text{ for all } \mathbf{n} \in \mathbb{Z}^d \text{ and all shapes } F\}.$$

We say that a pattern f on a shape F *occurs* in a shift space X if there is an $x \in X$ such that $x_F = f$. Hence the analogue of the language of a shift space is the set of all occurring patterns.

A *d-dimensional shift of finite type* X is a subset of $\mathcal{A}^{\mathbb{Z}^d}$ defined by a finite list $\mathcal{F}$ of forbidden patterns. Just as in one dimension, a d-dimensional shift of finite type X can also be defined by specifying allowed patterns instead of forbidden patterns, and there is no loss of generality in requiring the shapes of the patterns to be the same. Thus we can specify a finite list $\mathcal{L}$ of patterns on a fixed shape F, and put

$$X = X_{\mathcal{L}^c} = \{x \in \mathcal{A}^{\mathbb{Z}^d} : \text{for all } \mathbf{n} \in \mathbb{Z}^d, \sigma^{\mathbf{n}}(x)_F \in \mathcal{L}\}.$$

In fact, there is no loss in generality in assuming that F is a d-dimensional cube $F = \{(n_1, \ldots, n_d) : 0 \leqslant n_i \leqslant k\}$.

Given a finite list $\mathcal{L}$ of patterns on a shape F, we say that a pattern f' on a shape F' is $\mathcal{L}$-*admissible* (in the shift of finite type $X = X_{\mathcal{L}^c}$) if each of its sub-patterns, whose shape is a translate of F, belongs to $\mathcal{L}$. Of course, any pattern which occurs in X is $\mathcal{L}$-admissible. But an $\mathcal{L}$-admissible pattern need not occur in X.

The analogue of vertex shift (or 1-step shift of finite type) in higher dimensions is defined by a collection of d transition matrices $A_1, \ldots, A_d$ all indexed by the same set of symbols $\mathcal{A} = \{1, \ldots, m\}$. We put

$$\Omega(A_1, \ldots, A_d) = \{x \in \{1, \ldots, m\}^{\mathbb{Z}^d} : A_i(x_{\mathbf{n}}, x_{\mathbf{n}+\mathbf{e}_i}) = 1 \quad \text{for all } \mathbf{n}, i\},$$

where $\mathbf{e}_i$ is as usual the ith standard basis vector and $A_i(a, b)$ denotes the (a, b)-entry of A_i. Such a shift space is called a *matrix subshift* [MarP2]. When $d = 2$, this amounts to a pair of transition matrices A_1 and A_2 with identical vertex sets. The matrix A_1 controls transitions in the horizontal direction and the matrix A_2 controls transitions in the vertical direction. Note that any matrix subshift is a shift of finite type and, in particular, can be specified by a list $\mathcal{L}$ of patterns on the unit cube $F = \{(a_1, \ldots, a_n) : a_i \in$

$\{0,1\}\}$; specifically, $\Omega(A_1, \ldots, A_n) = X_{\mathcal{L}^c}$ where $\mathcal{L}$ is the set of all patterns $f \colon F \to \{1, \ldots, m\}$ such that if $\mathbf{n}, \mathbf{n}+\mathbf{e}_i \in F$, then $A_i(f(\mathbf{n}), f(\mathbf{n}+\mathbf{e}_i)) = 1$. When we speak of admissible patterns for a matrix subshift, we mean $\mathcal{L}$-admissible patterns with this particular $\mathcal{L}$. Just as in one dimension, we can recode any shift of finite type to a matrix subshift.

Higher dimensional shifts of finite type are plagued with difficulties that we simply do not see in one dimension. One example of this is as follows. Recall from Exercise 2.2.10 that there is a simple method to determine if a one-dimensional edge shift, and therefore a one-dimensional shift of finite type, is nonempty. Recall also from Exercise 2.2.11 that there is an algorithm to tell, for a given finite list $\mathcal{L}$, whether a given block occurs in $X = X_{\mathcal{L}^c}$. The corresponding problems in higher dimensions, called the *nonemptiness problem* and the *extension problem*, turn out to be undecidable [Ber], [Rob]; see also [KitS3]. Even for two-dimensional matrix subshifts X, the decision problems are undecidable. On the other hand, there are some special cases where these problems are decidable. Markley and Paul [MarP2] defined a natural class of d-dimensional matrix subshifts by a condition analogous to the "overlapping property" of Theorem 2.1.8 for one-dimensional shifts of finite type. This class includes any two-dimensional matrix subshift such that A_1 commutes with A_2 and A_2^T. For this class, any admissible pattern on a cube must occur, and so the nonemptiness and extension problems are decidable. In dimensions two and three, Markley and Paul also gave a procedure to decide when a matrix subshift belongs to their class of shifts, and they explored notions of irreducibility and mixing in this setting.

A point x in an d-dimensional shift X is *periodic* if its orbit $\{\sigma^\mathbf{n}(x) : \mathbf{n} \in \mathbb{Z}^d\}$ is finite. Observe that this reduces to the usual notion of periodic point in one dimension. Now an ordinary (one-dimensional) nonempty shift of finite type is conjugate to an edge shift X_G, where G has at least one cycle. Hence a one-dimensional shift of finite type is nonempty if and only if it has a periodic point. This turns out to be false in higher dimensions ([Ber], [Rob]), and this fact is intimately related to the undecidability results mentioned above. While one can formulate a notion of zeta function for keeping track of numbers of periodic points, the zeta function is hard to compute, even for very special and explicit matrix subshifts, and it is not a rational function [Lin9].

In higher dimensions, the entropy of a shift is defined as the asymptotic growth rate of the number of occurring patterns in arbitrarily large cubes. In particular, for two-dimensional shifts it is defined by

$$h(X) = \lim_{n \to \infty} \frac{1}{n^2} \log |X_{[0,n-1] \times [0,n-1]}|,$$

where $X_{[0,n-1] \times [0,n-1]}$ denotes the set of patterns on the square

$$\{0, 1, \ldots, n-1\} \times \{0, 1, \ldots, n-1\}$$

that occur in X. We saw in Chapter 4 that it is easy to compute the entropy of a (one-dimensional) shift of finite type using linear algebra. But in higher dimensions, there is no analogous formula and, in fact, other than the group shifts mentioned below, the entropies of only a very few higher dimensional shifts of finite type have been computed explicitly [Sch2, Chap. 5], [Kas]. Even for the two-dimensional "golden mean" matrix subshift defined by the horizontal and vertical transition matrices,

$$A_1 = A_2 = \begin{bmatrix} 1 & 1 \\ 1 & 0 \end{bmatrix},$$

an explicit formula for the entropy is not known. However, there are good numerical approximations to the entropy of a matrix subshift (see [MarP3]).

Just as in one dimension, we have the notion of sliding block code for higher dimensional shifts. For finite alphabets $\mathcal{A}, \mathcal{B}$, a finite subset $F \subset \mathbb{Z}^d$ and a function $\Phi: \mathcal{A}^F \to \mathcal{B}$, the mapping $\phi = \Phi_\infty: \mathcal{A}^{\mathbb{Z}^d} \to \mathcal{B}^{\mathbb{Z}^d}$ defined by

$$\Phi_\infty(x)_\mathbf{n} = \Phi(x_{\mathbf{n}+F})$$

is called a sliding block code. By restriction we have the notion of sliding block code from one d-dimensional shift space to another. For d-dimensional shifts X and Y, the sliding block codes $\phi: X \to Y$ coincide exactly with the continuous translation-commuting maps from X to Y, i.e., the maps which are continuous with respect to the metric ρ from (13–10–1) and which satisfy $\phi \circ \sigma^\mathbf{n} = \sigma^\mathbf{n} \circ \phi$ for all $\mathbf{n} \in \mathbb{Z}^d$. Thus it makes sense to consider the various coding problems, in particular the conjugacy problem, in the higher dimensional setting. Entropy, periodic points, and notions of irreducibility, and mixing are all invariants of conjugacy. But usually these are hard to determine. For another conjugacy-invariant, see Geller and Propp [GelP].

While the study of d-dimensional shifts is still largely shrouded in mystery, there is a subclass which is somewhat tractable, namely d-dimensional shifts with group structure in the following sense. Let $\mathcal{A}$ be a (finite) group. Then the full d-dimensional shift over $\mathcal{A}$ is also a group with respect to the coordinate-wise group structure. A (higher-dimensional) *group shift* is a subshift of $\mathcal{A}^{\mathbb{Z}^d}$ which is also a subgroup (recall that one-dimensional group shifts were introduced at the end of §2.1; see also §1.6).

Kitchens and Schmidt [KitS2] (see also [KitS1]) showed that any group shift is actually a shift of finite type (in one dimension, a proof of this is outlined in Exercise 2.1.11). Concretely, they proved that if $\mathcal{A}$ is a finite group and $X \subset \mathcal{A}^{\mathbb{Z}^d}$ is a group shift, then there is a finite subset $F \subset \mathbb{Z}^d$ and a subgroup $H \subset \mathcal{A}^F$ such that

$$X = \{x \in \mathcal{A}^{\mathbb{Z}^d} : \sigma^\mathbf{n}(x)_F \in H \text{ for all } \mathbf{n} \in \mathbb{Z}^d \},$$

i.e., the allowed patterns on F which define X form a subgroup. Such a shift is called a *Markov subgroup*..

A basic and very important example of Markov subgroups due to Ledrappier [Led] uses $\mathcal{A} = \mathbb{Z}/2\mathbb{Z}$, $F = \{\mathbf{0}, \mathbf{e}_1, \mathbf{e}_2\}$, and $H \subset \mathcal{A}^F$ the subgroup of those elements whose entries sum to 0. This Markov subgroup is particularly easy to visualize: A configuration belongs to it if and only if whenever you lay down a translate of F in the integer lattice, the number of 1's that you see within F is even. Ledrappier proved that although this example is mixing with respect to Haar measure, it fails to have higher order mixing, and so provides a counterexample to a higher dimensional version of the notorious r-fold mixing problem for single measure-preserving transformations that remains open after more than seventy years [Rok3].

The special class of d-dimensional group shifts is much better understood than the class of d-dimensional shifts of finite type. For instance, there is a finite procedure to decide the extension problem for group shifts [KitS2] (see also [KitS1]). Of course, any group shift contains the point of all 0's and so is nonempty. The entropy of any group shift can be calculated explicitly, and any group shift properly contained in the 2-dimensional full shift over $\{0, 1\}$ has zero entropy [KitS3]. See [KitS4] and Lind, Schmidt, and Ward [LinSW] for related results. Kitchens and Schmidt [KitS2] showed how to use the algebraic structure to determine whether certain dynamical properties such as irreducibility and mixing hold for any given group shift. They also showed that the set of periodic points in a group shift is always dense [KitS2] (see also [KitS4]). However, even for this class of shifts, there is no known way of computing the zeta function.

Finally, we mention that the idea of substitution, as in §13.7, can be used to construct higher dimensional shifts of finite type related to tiling systems with interesting properties; see Mozes [Moz1], [Moz2] and Radin [Rad].

ADDENDUM

FOR THE SECOND EDITION

Since the original publication of this book in 1995 there has been an enormous amount of activity and progress in symbolic dynamics. Some outstanding problems have been solved, others have seen great advances, while still others have seen little or no headway. Furthermore, new areas of research suggested in part by the topics in this book have created profound connections with other mathematical disciplines that were never imagined when this book was written. In this Addendum we sketch some of these developments, focusing on those most closely related to the topics covered in this book. We can be neither comprehensive nor detailed within the space limits here, but we give references in a separate Addendum Bibliography of newer literature that should help the reader explore much further. Addendum citations marked with a * refer to the original Bibliography, while those without a * are in the Addendum Bibliography.

Other expository materials treat related aspects of what appears here. The article [CovN] by Coven and Nitecki gives an historical account of the origins of symbolic dynamics, including some primary sources. The book by Kitchens [Kit4*] contains a comprehensive analysis of one-sided and countable state shifts of finite type. A recent book by Almeida, Costa, Kyriakoglou, and Perrin [AlmCKP] emphasizes a more algebraic and language-theoretic approach to symbolic dynamics. Finally, Boyle's extensive survey [Boy3] paints a colorful panorama of current open problems in symbolic dynamics together with their background and history.

§A.1. Classification Problems

In Chapters 7–9 we discussed various notions for when two shifts of finite type are considered to be "the same" or equivalent. These include conjugacy (equivalently, strong shift equivalence), shift equivalence, finite equivalence, and almost conjugacy, to which we added flow equivalence in §13.6. The goal for each is to find a set of computable invariants that "classify" or decide equivalence.

For some notions this goal has been achieved. For example, by Theorem 7.5.8 shift equivalence is classified by the corresponding dimension triple, and Kim and Roush [KimR3*] showed that given two such triples there is an algorithm to decide whether they are isomorphic. Both finite equivalence and almost conjugacy can be decided using the simple invariants of entropy and period (see Theorems 8.3.7 and 9.3.2). However, for other equivalences, including conjugacy, this goal remains out of reach.

We remark that Boyle, Buzzi, and Gómez [BoyBG] have proven a version of Theorem 9.3.2 for a class of countable edge shifts that enjoy a strengthened version of positive recurrence mentioned at the end of §13.9. This is important since various smooth and piecewise smooth systems have symbolic representations using countable edge shifts from this class.

CONJUGACY

The most central classification problem is to decide whether two shifts of finite type are conjugate. Given matrices A and B over $\mathbb{Z}^+$, Theorem 7.2.7 shows that the corresponding shifts of finite type X_A and X_B are conjugate if and only if A and B are strong shift equivalent. However, there is currently no known way to decide this, even for 2×2 matrices. As a concrete example, it is not known whether

$$\begin{bmatrix} 1 & 4 \\ 3 & 1 \end{bmatrix} \quad \text{and} \quad \begin{bmatrix} 1 & 12 \\ 1 & 1 \end{bmatrix}$$

are strong shift equivalent (see Example 7.3.13).

Eilers and Kiming [EilK] performed large-scale computations on the conjugacy problem for the collection $\mathcal{U}$ of the 17,250 2×2 irreducible matrices with entry sum at most 25. Earlier work of Jensen resulted in a database of explicit conjugacies, decomposing $\mathcal{U}$ into 3522 "fine" equivalence classes from this database. In particular these classes are at least as fine as the strong shift equivalence classes. Two new algebraic invariants were introduced in [EilK], closely related to the ideal class invariant described in Theorem 12.3.1, but more easily computed. Both are invariants of the ideal class. These, together with 19 generalized Bowen–Franks group invariants (see §7.4), subdivided $\mathcal{U}$ into 2068 "coarse" equivalence classes (each of which is a disjoint union of fine classes). Since the ideal class and generalized Bowen–Franks groups are invariants of shift equivalence, the coarse classes are unions of shift equivalence classes. A total of 1471 coarse classes were also fine, and so for these the invariants used are complete. However, 410 coarse classes contained two fine classes, and others contained even more. Their work also resolved some previously open cases, for example showing that

$$\begin{bmatrix} 14 & 2 \\ 1 & 0 \end{bmatrix} \quad \text{and} \quad \begin{bmatrix} 13 & 5 \\ 3 & 1 \end{bmatrix}$$

are not shift equivalent, and hence not strong shift equivalent. Thus even for this restricted collection of small matrices our knowledge of strong shift equivalence is far from complete.

For many years it was hoped that strong shift equivalence was the same as shift equivalence, and this was known as the Shift Equivalence Problem or the Williams Conjecture (see §7.3). But this hope was dashed by Kim and Roush, first in the reducible case [KimR11*] and then in the primitive case [KimR12*]. Their arguments used the topological machinery introduced by Wagoner [Wag5*] and key advances in their joint paper [KimRW1*] with Wagoner. His later surveys [Wag1] and [Wag2] together with their references give a lively and accessible account of these developments together with further approaches to this problem. Among the open problems discussed in these surveys is whether each shift equivalence class decomposes into only finitely many strong shift equivalence classes. Surveying the wreckage, it appears that there is currently no working conjecture to classify shifts of finite type up to conjugacy, and it is not even clear whether strong shift equivalence is decidable.

According to Corollary 7.1.5, any conjugacy from an edge shift X_G to X_H is obtained by a sequence of state splittings and state amalgamations starting from G and ending at H. A special case would be to allow only in- and out-amalgamations. Given an edge shift X_G there is a graph H such that X_H is obtained from X_G only by amalgamations and H has the minimum number of vertices. To give an indication of the complexity of the conjugacy problem, Frongillo [Fro] showed that the problem of finding such an H from G is NP-hard.

It is possible that with constraints on the spectrum the Shift Equivalence Problem has a positive answer. One instance is the "Little Shift Equivalence Conjecture" [Boy3, Conj. 3.1], which claims that if A has a unique and simple nonzero integer eigenvalue n, then A is strong shift equivalent to $[n]$ (the eigenvalue condition guarantees that A is shift equivalent to $[n]$). For $n = 2$ this conjecture was confirmed by Ashley for all matrices up to size eight that have simple nonzero eigenvalue 2 and all row and column sums equal to 2, with the single exception (up to graph isomorphism) of the sum A of the permutation matrices corresponding to (12345678) and (1)(2)(374865) in cycle notation. Although this A is shift equivalent to [2] as noted above, it is still not known whether it is strong shift equivalent to [2]. More generally, there are essentially no general sufficient conditions known for conjugacy.

Both shift equivalence and strong shift equivalence between matrices are purely algebraic notions that make sense over any semi-ring R. If these notions agree over R we say that *the Shift Equivalence Conjecture holds over R*. For example, in his foundational paper [WilR2*], Williams showed this to be true over $R = \mathbb{Z}$. However, it is open for both $R = \mathbb{Q}^+$ and $R = \mathbb{Q}$ [Boy3, §5]. Kim and Roush proved the Little Shift Equivalence Conjecture over $\mathbb{Q}^+$ and over $\mathbb{R}$ [KimR6*]. Although these algebraic questions seem far removed from their dynamical origins, the invariants they suggest generalize to other symbolic dynamical systems. For instance, a shift of finite type commuting with the action of a finite group G can be presented by a matrix over the integral group ring of G. In this setting, Boyle and Sullivan [BoyS3] devised some K-theory invariants to resolve an old problem of Parry. For a comprehensive overview of this approach with an extensive bibliography, see the recent survey by Boyle and Schmieding [BoyS1].

The symbolic matrix approach to sofic shifts at the end of §7.2 shows how to express conjugacy between sofic shifts as strong shift equivalence of their defining matrices over a semi-ring R. As with shifts of finite type, there is no known algorithm for deciding conjugacy between sofic shifts.

Boyle and Krieger [BoyK2*] characterized eventual conjugacy of sofic shifts as shift equivalence over R of canonical matrices representing the shifts, analogous to Theorem 7.5.15 for shifts of finite type. Later, Kim and Roush [KimR5*] proved that shift equivalence over R is algorithmically decidable.

Thus for shifts of finite type and sofic shifts, eventual conjugacy is algebraic and decidable, while conjugacy is still shrouded in murky darkness.

Recently Jeandel [Jea4] has considered strong shift equivalence from a categorical perspective, enabling him to deploy novel machinery in a search for new invariants.

FLOW EQUIVALENCE

In §13.6 we introduced the notion of flow equivalence between shifts of finite type. Roughly speaking, flow equivalent shifts are those whose suspension flows, as depicted in Figure 13.6.2, are "time changes" of one another. As mentioned in §13.6, Franks showed that the Parry–Sullivan invariant and the Bowen–Franks group completely classify irreducible shifts of finite type of positive entropy up to flow equivalence. Boyle and Huang have extended this classification to cover the reducible case [Boy1, BoyH]. These developments were critical ingredients in the classification of Cuntz–Krieger algebras of real rank zero up to stable isomorphism, and more generally the spectacular classification of unital graph C^*-algebras up to stable isomorphism and also up to unital isomorphism by Eilers, Restorff, Ruiz, and Sørensen [EilRRS]. These developments illustrate the ongoing interplay between symbolic dynamics, including the countably infinite state case (see Exel and Laca [ExeL]), and C^*-algebras.

The *mapping class group* of a shift is the group of isotopy classes of orientation preserving self-homeomorphisms of its suspension flow. This group plays a role for flow equivalence analogous to that of the automorphism group of a shift for conjugacy. Boyle and Chuysurichay [BoyC] developed tools for answering questions about the mapping class group of a shift of finite type, and Schmieding and Yang [SchY] studied this group for low complexity shifts.

Silver and Williams [SilW1] discovered new invariants of knots using the symbolic dynamics of a shift of finite type derived from the fundamental group of the knot complement. These shifts also have an action by a finite group coming from the geometric setting, and they investigated the consequences of this additional structure in [SilW2]. Boyle, Carlsen, and Eilers [BoyCE1] reduced the classification up to flow equivalence of shifts of finite type with a finite group action to an algebraic question about symbolic matrices, and found an explicit and complete set of invariants when the acting group is $\mathbb{Z}/2\mathbb{Z}$.

The classification of sofic shifts up to flow equivalence is as yet unsolved, although recently this has been done for special classes of sofic shifts by Boyle, Carlsen, and Eilers [BoyCE2].

A drastically more general framework for studying such problems for general shift spaces has been developed by Matsumoto. For an outline of this approach and its connections with C^*-algebras, see [Boy3, §33]. For an account of this theory from a dynamical viewpoint, without the associated C^*-algebra motivations, see [Mat1].

TOPOLOGICAL ORBIT EQUIVALENCE

Two invertible dynamical systems (M, ϕ) and (N, ψ) are called *topologically orbit equivalent* if there is a homeomorphism $\theta \colon M \to N$ mapping every ϕ-orbit in M onto a ψ-orbit in N. Conjugate systems are clearly orbit equivalent. Although some useful invariants are known, the classification of shifts of finite type up to topological orbit equivalence is still unsolved (see [Boy3, Prob. 23.4(1)] and [Boy4, §9] for an account of what is known). But the implications of topological orbit equivalence for other systems have become much clearer since publication of the first edition.

Dye [Dye] introduced a measurable version of orbit equivalence in his pioneering work in the 1960s on approximately finite-dimensional operator algebras. In particular, he showed that every pair of ergodic measure-preserving transformations on nonatomic probability spaces are measurably orbit equivalent, and indeed ergodic actions even by different groups within a class he studied are all measurably orbit equivalent. Later, Ornstein and Weiss [OrnW] proved that this class includes all infinite amenable groups.

Topological orbit equivalence becomes interesting when the spaces are highly disconnected. If M is a Cantor set and $\phi \colon M \to M$ is a homeomorphism such that there are no proper closed ϕ-invariant subsets of M, then (M, ϕ) is called a *Cantor minimal system*. The minimal shifts described in §13.7 are important basic examples of these systems.

Remarkably, work of Giordano, Putnam, and Skau [GioPS] has led to a complete classification of Cantor minimal systems up to topological orbit equivalence in terms of an ordered group reminiscent of the dimension triple classification of shift equivalence described in §7.5 (both have their origins in operator algebras). To describe this invariant, begin with a Cantor minimal system (M, ϕ). Let $C(M, \mathbb{Z})$ be the set of continuous $\mathbb{Z}$-valued functions on M, which forms a group under pointwise addition. This group is ordered in the obvious way, and has a distinguished positive element 1. Let $B_\phi(M, \mathbb{Z})$ be the subgroup of all $f \in C(M, \mathbb{Z})$ such that $\int_M f \, d\mu = 0$ for all ϕ-invariant probability measures μ on M. For example, if $g \in C(M, \mathbb{Z})$ then $g - g \circ \phi \in B_\phi(M, \mathbb{Z})$. The quotient group $D_\phi = C(M, \mathbb{Z})/B_\phi(M, \mathbb{Z})$ inherits the order structure and distinguished element. Then Cantor minimal systems (M, ϕ) and (N, ψ) are topologically orbit equivalent if and only if there is a group isomorphism from D_ϕ to D_ψ preserving order and the distinguished element. Building on ideas of Vershik, a central tool in this study is representing minimal systems using a combinatorial structure called a Bratteli diagram. Recently these results have been extended to minimal actions of $\mathbb{Z}^d$ on Cantor sets [GioMPS].

For a very readable and elementary account of this theory and its motivating examples see Putnam's expository book [Put]. An extensive account of the role of dimension groups and Bratteli diagrams in dynamics is contained in the recent book by Durand and Perrin [DurP].

These ideas have led to the discovery of new kinds of abstract groups long sought by group theorists. Let (M, ϕ) be a Cantor minimal system. Its *topological full group* $[[\phi]]$ is the set of all homeomorphisms $\psi : M \to M$ such that $\psi(x) = \phi^{n(x)}(x)$ for some continuous function $n : M \to \mathbb{Z}$, which is a group under composition.

Topological full groups are countable groups enjoying many interesting properties. Matui [Mat3] showed that the commutator subgroup $[[\phi]]'$ of $[[\phi]]$ is simple (no nontrivial normal subgroups), and is finitely generated if and only if ϕ is conjugate to a shift. Grigorchuk and Medynets [GriM] conjectured that $[[\phi]]$, and hence $[[\phi]]'$, are amenable, and this was later confirmed by Juschenko and Monod [JusM].

Before this work there were no known examples of simple, infinite, finitely generated, and amenable groups (earlier methods used to construct finitely generated simple groups, dating back to Higman in the 1950s, inevitably led to nonamenable groups). As delightfully described in [Mat2 §4], deploying the Sturmian minimal shifts discussed in §13.7 and using results from [GioPS] leads to uncountably many nonisomorphic examples of such groups.

§A.2.   Factor Codes and Embeddings

As explained in §9.1, each finite-to-one factor code on an irreducible shift of finite type has a degree, d, which is the minimum number of pre-images and, equivalently, the "typical" number of pre-images of points. The case $d = 1$ thus corresponds to one-to-one almost everywhere factor codes.

One breakthrough is Trahtman's solution [Tra*] of the Road Problem (Theorem 5.1.3), which can be formulated as "every primitive graph with constant out-degree has a road coloring with factor degree $d = 1$." This purely graph-theoretic problem, which grew out of a coding problem in ergodic theory, was open for almost forty years. The solution, which built on earlier work of Friedman [Fri2*] and of Culik, Karhumäki, and Kari [CulKK*], is elegant and surprisingly short. The Road Problem is related to a class of synchronization problems, such as the still-unsolved Černý Conjecture 3.4.16 [Vol] on the length of the shortest synchronizing word for a deterministic automata.

The State Splitting Algorithm (or ACH algorithm) for constructing encoders with sliding block decoders (i.e., factor codes) was described in §5.4 and §5.5. The algorithm was widely used in practice in the 1980s and 1990s. Variants of the algorithm were used in the development of codes which became standards in the recording industry, including the (1,7) code [AdlHM*] for hard disk drives, the EFMPlus code for DVDs (see [Imm1] or [Imm2, §11.5.2]), and the code used in the first generation of Linear Tape Open [AshJMS]. Today this algorithm is used mainly as a proof of concept in new coding schemes, such as radio frequency identification [BarRYY], weakly constrained codes [EliMS], and parity preserving codes [RotS].

The structure of finite-to-one factor codes was developed in §8.1 and §9.1. Boyle [Boy7*] made further progress on decomposing such codes, motivated by work of Trow [Tro6*]. In particular, Boyle showed that if $\mathcal{B}$ denotes the class of all finite-to-one factor codes from irreducible shifts of finite type to sofic shifts, then each map in $\mathcal{B}$ can be written as a composition of maps in $\mathcal{B}$ in only finitely many essentially distinct ways, and that there is a procedure for finding all such compositions.

Allahbakhshi and Quas [AllQ] introduced a notion of degree, called *class degree*, for infinite-to-one factor codes, analogous to the notion of degree for finite-to-one factor codes. The class degree is the minimum number of so-called transition classes analogous to the fibers of a finite-to-one factor code. Allahbakhshi, Hong, and Jung [AllHJ] established several dynamical properties of class degree for infinite-to-one factor codes. Yoo showed that any infinite-to-one factor code can be decomposed into a finite-to-one factor code and a factor code of class degree one [Yoo].

Class degree was motivated by problems involving invariant measures on shifts of finite type and sofic shifts. Suppose that X is a shift of finite type, that $\phi: X \to Y$ is a factor code, and that ν is an ergodic invariant measure on Y. Then the number of ergodic invariant measures on X that map to ν and maximize entropy among all such measures is bounded above by the class degree. More recent work on this subject can be found in Allahbakhshi, Antonioli, and Yoo [AllAY]. As an application of the result in [Yoo] mentioned above, he showed that in the special case that ν is Markov there is a unique invariant measure of relative maximal entropy that projects to ν. This answered a question raised by Boyle and Petersen [BoyP](Problem 3.16).

Thomsen [Tho] generalized both the Embedding Theorem 10.1.1 and the Lower Entropy Factor Theorem 10.3.1 to special classes of sofic shifts. The essential new idea is that of an irreducible component for an arbitrary shift, generalizing §4.4 but using a recursive definition involving synchronizing words and periodic points. This structure clarifies constraints on codes, especially between sofic shifts, and explains why these theorems cannot hold for all mixing shifts. Krieger [Kri] gives a complete necessary, sufficient, and decidable set of conditions for the Lower Entropy Factor Theorem for all mixing sofic shifts.

The problem of deciding the existence of a finite-to-one (equivalently, entropy-preserving) factor code from one mixing shift of finite type to another remains open. In particular, there appears to have been no progress on the question of whether the necessary dimension group and periodic point conditions mentioned near the end of §12.2 are sufficient.

§A.3. Symbolic Models for Smooth Systems

Markov partitions, introduced in §6.5, continue to be an essential tool in modeling smooth dynamical systems (see the survey of symbolic dynamics and Markov partitions by Adler [Adl]). The main examples discussed in §6.5, namely hyperbolic toral automorphisms and Smale's horseshoe, have finitely many "rectangles" and so are modeled by a shift of finite type with a finite alphabet.

As described in §13.9, there is an analogue of finite type shifts using a countably infinite alphabet and a countable edge shift. In a remarkable advance, Sarig [Sar1] showed that an arbitrary $C^{1+\epsilon}$ surface diffeomorphism with positive topological entropy has Markov partitions with countably many rectangles, and hence can be modeled by a countable edge shift. The generality of this result, which assumes nothing other than positive entropy, was completely unexpected. Sarig uses this to prove two outstanding conjectures for $C^{1+\epsilon}$ surface diffeomorphisms, one due to A. Katok [Kat, Prob. 7.4], concerning the growth rate of the number of periodic points and the other due to Buzzi [Buz, Conj. 1], concerning the number of measures of maximal entropy. See also [LimS].

In a related development, Boyle and Downarowicz [BoyD] discovered that the problem of finding shifts which factor onto a given dynamical system is surprisingly deep, and their results have provided researchers in smooth dynamics with new tools (see [Dow] for an expository account).

Symbolic dynamics models for smooth dynamical systems grew out of symbolic codings for the geodesic flows on surfaces of negative curvature beginning with Hadamard [Had*]. This continues to be a very active area of research. For an overview see the survey by S. Katok and Ugarcovici [KatU] and its extensive bibliography; related material is contained in [AbrK] and [Frie].

For surfaces with zero curvature, or *flat surfaces*, the geodesic flow has a different dynamical character. Consider the unit square with opposite sides identified, creating a torus with two circles which we label A and B. Then the unit speed flow in an irrational direction produces a "cutting sequence" of successive intersections with A or B, and it turns out that the resulting bi-infinite symbolic sequence in $\{A, B\}^{\mathbb{Z}}$ completely specifies the starting point on $A \cup B$ and the direction. Indeed, this is an equivalent description of the Sturmian minimal shifts described in §13.7. This analysis has been extended to the regular octagon with opposite sides identified by Smillie and Ulcigrai [SmiU], and to other regular polygons by Davis [Dav]. The study of flows on flat surfaces and their connections with billiard flows, interval exchange transformations, and moduli space is very active; for an informal and lucid account see the survey by Zorich [Zor].

§A.4. Realization

In Chapter 11 we discussed the possible values that invariants like entropy and the zeta function can take. Here we describe further work in this direction.

ENTROPY, TILINGS, AND PERRON NUMBERS

Recall from §11.1 that a real number $\lambda \geqslant 1$ is called a *Perron number* if it is an algebraic integer such that $\lambda > |\mu|$ for all other algebraic conjugates μ of λ, and is called a *weak Perron number* if some power of it is a Perron number. Theorems 11.1.4 and 11.1.6 show that the logarithms of Perron numbers are precisely the entropies of mixing shifts of finite type, while those of weak Perron numbers are the entropies of general shifts of finite type (and also of sofic shifts).

The idea behind the constructive part of these theorems was adapted by Thurston and Kenyon in their study of the possible expansion constants of self-similar tilings of the line and the plane. For a lucid general account of tilings and symbolic dynamics, including terminology used here, see the survey by Robinson [Rob].

A tiling of $\mathbb{R}$ is a covering of $\mathbb{R}$ by closed intervals with only finitely many distinct lengths and disjoint interiors. It is called self-similar if there is an expansion constant λ such that the image of each tile under multiplication by λ is a union of tiles. Thurston [Thu1] showed that λ is an expansion constant for a self-similar tiling of $\mathbb{R}$ if and only if it is a Perron number.

We consider the plane as the set $\mathbb{C}$ of complex numbers. Here a tile is a compact connected set whose interior is dense in the set. A tiling of $\mathbb{C}$ is a covering by tiles which are translates of only finitely many shapes and whose interiors are disjoint. Just as for $\mathbb{R}$, a tiling of $\mathbb{C}$ is called self-similar if there is an expansion constant $\lambda \in \mathbb{C}$ such that λ times each tile is a union of tiles. In order to describe such constants, Thurston called $\lambda \in \mathbb{C}$ a *complex Perron number* if it is an algebraic integer such that $|\lambda| > |\mu|$ for every algebraic conjugate μ of λ, with the exception of the complex conjugate of λ when

λ is not real. He showed [Thu1] that the expansion constant for a self-similar tiling must be a complex Perron number. Kenyon [Ken] developed techniques to construct self-similar tilings from complex Perron numbers, but unfortunately his proof that this always works turned out to have a gap (see [KenS]). It remains unknown whether every complex Perron number is the expansion constant for a self-similar tiling of the plane. Kenyon and Solomyak [KenS] investigated which expanding affine maps of $\mathbb{R}^n$ could be the analogues of expansion constants for self-similar tilings, and obtained necessary conditions which they conjectured are also sufficient.

In his last paper, Thurston [Thu2] studied maps of $[0,1]$ to itself for which the union of the forward orbits of all critical points forms a finite set. This set divides $[0,1]$ into intervals, each of which is mapped to a union of these intervals, a condition reminiscent of a Markov partition. He showed that $\log \lambda$ is the entropy of such a map if and only if λ is a weak Perron number, and investigated questions arising from his large-scale computer experiments on the distribution of Perron numbers and their conjugates. Some of these questions were answered by Calegari and Huang [CalH], who found in particular that the number of Perron numbers of fixed degree n and size at most T is given by $D_n T^{n(n+1)/2}\big(1 + O(1/T)\big)$ as $T \to \infty$, where D_n is a complicated but explicitly computable constant.

Yazdi [Yaz] defined the *Perron–Frobenius degree* $\deg_{PF} \lambda$ of a Perron number λ to be the dimension of the smallest primitive matrix with spectral radius λ. Using a geometric idea originating from Example 11.1.1(d), he gave lower bounds on $\deg_{PF} \lambda$ in terms of the second largest algebraic conjugate of λ, and constructed cubic λ with arbitrarily large $\deg_{PF} \lambda$. However, computing $\deg_{PF} \lambda$ appears to be quite difficult. For instance, all that is currently known for the cubic λ from this example is that $4 \leqslant \deg_{PF} \lambda \leqslant 10$ (see [Lin3*] for the upper bound).

In Exercise 11.1.12 we defined a number $\lambda \geqslant 1$ to be *almost Perron* provided that it is an algebraic integer such that $|\mu| \leqslant \lambda$ for every algebraic conjugate μ of λ. Thurston [Thu2, Prop. 5.1] showed that a number is almost Perron if and only if it is weak Perron. This equivalence was mentioned earlier in [Lin4*], and more recently reproved by Brunotte [Bru]. Boyd [Boyd] showed that if λ is weak Perron, then its minimal irreducible polynomial over $\mathbb{Q}$ has the form $g(x^n)$, and hence the set of conjugates of λ is invariant under rotation by all nth roots of unity. While investigating examples of cellular automata with irrational behavior, Korec [Kor] introduced a class of algebraic numbers, and subsequently Schinzel [Schi] showed that this class coincides with the weak Perron numbers.

SPECTRAL REALIZATIONS AND CONJECTURES

In §11.2 we discussed the problem of describing the possible zeta functions of mixing shifts of finite type. This is equivalent to characterizing those lists Λ of nonzero complex numbers for which there is a primitive integral matrix A with $\text{sp}^\times(A) = \Lambda$.

We saw that there are three simple and verifiable necessary conditions on Λ, namely the Integrality Condition, the Perron Condition, and the Net Trace Condition. The Spectral Conjecture, which was open for twenty years, asserted that these conditions are sufficient. As described in Theorem 11.2.4, Kim, Ormes, and Roush [KimOR*] solved this problem in the affirmative with an astonishing proof using formal power series. Their proof is algorithmic, and provides a procedure for finding the required primitive matrix. In Example 11.2.5(c) we gave an explicit cubic polynomial whose roots Λ satisfy all three necessary conditions, and in [KimOR*] they applied their algorithm to find a primitive integral matrix of size 179 whose nonzero eigenvalues are just the three cubic roots. Matrices of this size are manageable by using a more compact presentation using polynomials instead of integers (see [BoyL2] for a systematic treatment of this approach and its uses).

The question of describing the spectra of nonnegative real matrices has a long history. By focusing on just the nonzero spectrum, one can hope to reach more definitive answers. Let S be a subring of $\mathbb{R}$ containing 1. We can then ask to characterize the nonzero spectra of primitive matrices with entries in S. As in the case $S = \mathbb{Z}$, there are three necessary conditions: the same Perron condition, the condition that $\prod_{\lambda \in \Lambda}(t - \lambda) \in S[t]$, and if $S \neq \mathbb{Z}$ the trace condition that $\text{tr}(\Lambda^k) \geqslant 0$ and if $\text{tr}(\Lambda^k) > 0$ then $\text{tr}(\Lambda^{nk}) > 0$ for all $n \geqslant 1$. The Spectral Conjecture for S claims that these conditions are also sufficient.

In groundbreaking work, Boyle and Handelman [BoyH1*] showed this conjecture to be true when $S = \mathbb{R}$, as well as various other cases, using ideas from symbolic dynamics that were completely different from those used in [KimOR*] for $S = \mathbb{Z}$. This conjecture for general S still remains open.

A companion problem is the Generalized Spectral Conjecture for S, which claims that if A is a matrix over S whose nonzero spectrum satisfies the necessary conditions for the Spectral Conjecture, then A is strong shift equivalent over S to a primitive matrix. This is known to be true for $S = \mathbb{Z}$ and $S = \mathbb{Q}$ when the spectrum has only rational elements, but remains open in general.

§A.5. Automorphism Groups of Shifts

In §13.2 we discussed properties of the group of automorphisms of a shift space, focusing on mixing shifts of finite type where the automorphism group plays an important role in classification problems. Despite some progress, there are many aspects of these groups that remain mysterious.

For instance, it is not known whether $\text{aut}(X_{[2]})$ and $\text{aut}(X_{[3]})$ are isomorphic as groups. The only useful tool we know of for showing the automorphism groups of two shifts of finite type are not isomorphic is Ryan's theorem [Rya2*]. This states that if X_A is an irreducible shift of finite type, and we denote the shift map by σ_A instead of σ_{X_A}, then the center of $\mathrm{aut}(X_A)$ consists of only the powers of σ_A. One consequence of this is that $\mathrm{aut}(X_{[4]})$ and $\mathrm{aut}(X_{[2]})$ are not isomorphic. If they were, their centers would be preserved and so $\sigma_{[4]}$ would be mapped to either $\sigma_{[2]}$ or its inverse. But $\sigma_{[4]}$ has a square root in $\mathrm{aut}(X_{[4]})$ since $X_{[2]}^2 \cong X_{[4]}$, while $\sigma_{[2]}$ has no square root in $\mathrm{aut}(X_{[2]})$ by [Hed5*, Thm. 18.1].

The main techniques for studying the automorphism group come from considering its action on related structures such as the periodic points and the dimension triple. Recently, Frisch, Schlank, and Tamuz [FriST] introduced another such structure for full shifts called a Furstenberg boundary. Starting with the subset of points whose pasts are eventually constant, they construct a compact space on which the automorphism group acts with strong focusing properties. Using earlier results from the dynamics of such actions, they show that every normal amenable subgroup of $\mathrm{aut}(X_{[n]})$ is contained in the group consisting of powers of the shift. This gives a new proof of Ryan's theorem in the case of full shifts, since if ϕ is in the center of $\mathrm{aut}(X_{[n]})$, then the subgroup generated by ϕ and $\sigma_{[n]}$ is normal and also amenable since it is commutative, and hence ϕ is a power of the shift.

RANGE OF THE DIMENSION REPRESENTATION

Let X_A be a mixing shift of finite type and $(\Delta_A, \Delta_A^+, \delta_A)$ be its dimension triple (see §7.5). As described in §13.2 there is a dimension representation s_A from $\mathrm{aut}(X_A)$ to $\mathrm{aut}(\Delta_A)$, where $\mathrm{aut}(\Delta_A)$ denotes the group of order-preserving automorphisms of Δ_A that commute with δ_A. An obstruction coming from the sign-gyration compatibility condition shows that s_A is not always onto. A basic open problem is to determine the exact range of s_A.

It is known that a solution to this problem would have two important consequences: (1) the generalization of the classification of irreducible shifts of finite type (as yet unsolved) to the reducible case and (2) the solution of the extension problem, which asks whether an automorphism of a shift within X_A can be extended to an automorphism on all of X_A. The latter problem has been essentially solved for finite unions of periodic points, but even there it is quite subtle (see [Boy3, §4] for details).

VIRTUAL FOG CONJECTURE

Let $\mathrm{aut}_0(X_A)$ denote the kernel of the dimension representation s_A, which is a normal subgroup of $\mathrm{aut}(X_A)$ that contains most of the mystery of $\mathrm{aut}(X_A)$. Elements of $\mathrm{aut}_0(X_A)$ are called *inert*. We do not know its commutator subgroup, nor its finite-index subgroups.

Let $\mathrm{fog}(X_A)$ be the subgroup of $\mathrm{aut}_0(X_A)$ generated by its elements of finite order. For a long time it was conjectured that $\mathrm{fog}(X_A) = \mathrm{aut}_0(X_A)$, but this turned out to be false [KimRW2*]. The Virtual FOG Conjecture is that if X_A is a mixing shift of finite type, then $\mathrm{aut}_0(X_A)/\mathrm{fog}(X_A)$ is a finite group. A simple version of this due to Rhodes, still unsolved, is whether every automorphism of the full 2-shift is a composition of shift-commuting involutions and a power of the shift.

PROPERTIES OF AUTOMORPHISMS OF SHIFTS OF FINITE TYPE

Recall from Exercise 6.3.13 that we defined a dynamical system (M, ϕ) to be *expansive* if there is a constant $c > 0$ such that if $x, y \in M$ and $\rho_M(\phi^n(x), \phi^n(y)) < c$ for all $n \in \mathbb{Z}$, then $x = y$. Expansiveness is an important and multifaceted notion in dynamics.

Let X_A be a mixing shift of finite type. Is there a procedure to decide whether a given $\phi \in \mathrm{aut}(X_A)$ is expansive? If ϕ is expansive, must it be conjugate to a shift of finite type?

D. Fiebig gave a simple example of a reducible shift of finite type with an expansive automorphism that is not of finite type, and Boyle [Boy2] gave some examples of strictly sofic shifts that cannot commute with mixing shifts of finite type. Nasu [Nas] gave the first general result addressing this question, showing that if an expansive automorphism has zero memory or zero anticipation, then it has finite type. A context in which to view this problem is the "expansive subdynamics" of the $\mathbb{Z}^2$-action generated by the shift and the automorphism (see §A.6 below).

Although automorphisms are given by seemingly simple sliding block codes, there are very few exact calculations of their entropy. Lind [Lin7*] used marker constructions to show that for each irreducible shift of finite type the set of entropies of its automorphisms is dense in $[0, \infty)$. In these constructions entropy is generated by information being transmitted using compositions of involutions, much like buckets of water are passed by a bucket brigade with each person switching their bucket from one side to the other. This leads to values of entropy such as $\log \frac{9}{2}$, and in general logarithms of a class of algebraic numbers. Must the entropy of an automorphism be the logarithm of an algebraic number? Can we characterize the set of such entropies? To what extent, if any, does this set vary with the shift of finite type?

SHIFTS WITH LOW COMPLEXITY

The richness of the automorphism group of a shift of finite type is closely related to the complexity of the underlying space. By considering shifts whose complexity is more constrained one can hope to obtain a much better understanding of their automorphism groups.

Let X be a shift space with shift map σ and $\Sigma \subseteq \mathrm{aut}(X)$ be the subgroup generated by σ. In 1971 Coven [Cov1*] showed that if X is a substitution minimal shift contained in the 2-shift created with constant length substitutions, then $\mathrm{aut}(X)/\Sigma$ is either trivial or of order two. For example, the automorphism group of the Morse minimal shift from §13.7 is generated by Σ and the involution interchanging 0 and 1. Host and Parreau [HosP] extended this to more general substitution minimal shifts X for which they prove that $\mathrm{aut}(X)/\Sigma$ is a finite group. This was further extended by Salo and Törmä [SalT] to certain minimal substitutions with nonconstant length.

Some work on automorphism groups of shifts with low complexity grew out of attempts to solve a long-standing conjecture about two-dimensional patterns. An old result of Morse and Hedlund [MorH2*] says that if x is an element in a shift space with the property that for some n the number of n-blocks occurring in x is less than or equal to n, then x must be a periodic point. Nivat's Conjecture, a two-dimensional version of this, claims that if $x \in \mathcal{A}^{\mathbb{Z}^2}$ has the property that there are some n and k for which the number of $n \times k$ patterns in x is less than or equal to nk, then x has a nonzero period, i.e., there is a nonzero $\mathbf{p} \in \mathbb{Z}^2$ such that $x_{\mathbf{n}+\mathbf{p}} = x_{\mathbf{n}}$ for all $\mathbf{n} \in \mathbb{Z}^2$. Direct analogues of this conjecture in dimensions three and higher are known to fail badly.

Sander and Tijdeman [SanT] proved that Nivat's Conjecture is true for $n \times 2$ rectangles, followed by Cyr and Kra's proof [CyrK4] for $n \times 3$ rectangles. Epifanio, Koskas, and Mignosi [EpiKM] proved the conjecture provided that the bound nk is replaced by $nk/144$, which was strengthened by Quas and Zamboni [QuaZ], who assumed a bound of only $nk/16$.

A new attack on Nivat's Conjecture was initiated by Cyr and Kra [CyrK1] by considering the $\mathbb{Z}^2$-shift on the orbit closure of x in $\mathcal{A}^{\mathbb{Z}^2}$, and using the ideas and results from the expansive subdynamics for such actions [BoyL1] (see §A.6). In particular, with this new approach they showed that Nivat's Conjecture is valid if the bound nk is replaced by $nk/2$, a significant advance. The conjecture itself remains open.

Using completely different ideas coming from algebraic geometry, Kari and Szabados [KarS] very recently showed that if $x \in \mathcal{A}^{\mathbb{Z}^2}$ has no nonzero periods, then for all but finitely many pairs of n and k the number of $n \times k$ patterns in x is at least $nk + 1$, so that in a sense Nivat's Conjecture is "almost true." Using these methods, Kari and Moutot [KarM] have recently proved that Nivat's Conjecture is true if x is contained in a minimal $\mathbb{Z}^2$-shift.

Cyr and Kra's approach to Nivat's Conjecture led them to a more general study of automorphism groups of shifts with various flavors of low complexity. They began with an $x \in \mathcal{A}^{\mathbb{Z}^2}$ that satisfies the assumption in Nivat's Conjecture for some n and k. Let Y be the orbit closure of x in $\mathcal{A}^{\mathbb{Z}^2}$. Form the set $X \subseteq \mathcal{A}^{\mathbb{Z} \times \{0,1,\dots,k-1\}}$ of restrictions of all points in Y to coordinates in $\mathbb{Z} \times \{0, 1, \dots, k-1\}$, which is both closed and invariant under the horizontal shift. We can think of X as a (one-dimensional) shift with alphabet $\mathcal{A}^k$, and from the assumption on x this shift will have a quantifiable "low complexity." Furthermore, the vertical shift can be interpreted as an automorphism of this low-complexity shift. More generally, if X is an arbitrary shift, we can quantify its complexity by introducing $b_X(n) := |\mathcal{B}_n(X)|$. They showed in [CyrK2] that if X is a transitive shift with $b_X(n) = O(n)$, then $\mathrm{aut}(X)/\Sigma$ has bounded exponent, i.e., there is an N such that every element has order a divisor of N. With the weaker assumption that $b_X(n) = o(n^2)$ they prove in [CyrK3] that every element of $\mathrm{aut}(X)/\Sigma$ has finite order. An edifying example by Salo [Sal] provides a Toeplitz shift with $b_X(n) = O(n^{1.757})$ and for which $\mathrm{aut}(X)/\Sigma$ is infinite.

This research direction is undergoing vigorous development. The papers [CyrK2] and [DonDMP] obtain further results when $b_X(n)$ grows linearly, each including the Sturmian minimal shifts from §13.7 but also handling situations with different kinds of additional assumptions. Much of this is generalized in [CyrK5], and [CyrK6] contains a more detailed analysis when $b_X(n) = o(n^2)$. If X is an irreducible shift with zero entropy, then Cyr, Franks, Kra, and Petite [CyrFKP] showed that certain groups such as $SL(n, \mathbb{Z})$ for $n \geqslant 3$ and the Baumslag–Solitar groups $BS(1, n)$ for $n \geqslant 2$ cannot be isomorphic to a subgroup of $\mathrm{aut}(X)$. Their proof involves the study of the group-theoretic notion of distortion within the automorphism group.

In general, the relationship between the complexity of a shift and the structure of its automorphism group is still not well understood, but enough results are available to indicate that this will continue to be a fruitful and interesting area.

§A.6. Higher Dimensional Shifts

Among the most important and exciting developments in the past two decades has been the extension of symbolic dynamics to the higher dimensional setting. As described in §13.10, bi-infinite sequences of symbols from an alphabet $\mathcal{A}$ are replaced by arrays of such symbols indexed by $\mathbb{Z}^d$. A $\mathbb{Z}^d$-*shift space* (or simply $\mathbb{Z}^d$-*shift*) is a closed subset of $\mathcal{A}^{\mathbb{Z}^d}$ that is invariant under the $\mathbb{Z}^d$-shift. There are then straightforward extensions to the $\mathbb{Z}^d$ setting of dynamical notions such as finite type, sofic, sliding block code, conjugacy, factor code, embedding, mixing, periodic points, and topological entropy. Their study is often referred to as $\mathbb{Z}^d$-symbolic dynamics. This book is largely about $\mathbb{Z}$-symbolic dynamics.

Some basic results for $d = 1$ extend to the general $\mathbb{Z}^d$ setting. For instance, topological conjugacies coincide with bijective sliding block codes, both topological entropy and having finite type are conjugacy invariants [RadS, Thm. 2], and entropy cannot increase under factor codes and cannot decrease under embeddings.

However, $\mathbb{Z}^d$-symbolic dynamics when $d \geqslant 2$ is far more than a routine generalization of the 1-dimensional case. A fundamental reason is that while $\mathbb{Z}$-shifts of finite type can be understood using graphs and linear algebra, when $d \geqslant 2$ there are $\mathbb{Z}^d$-shifts of finite type that can carry out computations and indeed can encode a universal Turing machine. This explains why there are undecidable problems such as determining whether a $\mathbb{Z}^d$-shift of finite type is empty, and means that any general theory of $\mathbb{Z}^d$-symbolic dynamics will necessarily involve the theory of computation. Thus both the mathematical tools used and the kinds of results possible differ drastically between the cases $d = 1$ and $d \geqslant 2$. See [Lin] for an introductory survey of $\mathbb{Z}^d$-symbolic dynamics.

EXAMPLES

Attempts to understand simple, concrete, yet puzzling examples of $\mathbb{Z}^d$-shifts of finite type have stimulated much of the work in this area. Some examples originated in statistical physics as discrete models, while others arose from constrained data storage in d-dimensional arrays (e.g., higher dimensional generalizations of run-length limited shifts that were discussed in §2.5 and Chapter 5). Here we describe a few key examples of $\mathbb{Z}^2$-shifts of finite type and their historical importance.

Ledrappier's example. Consider $\mathcal{A} = \{0, 1\}$ as the group $\mathbb{Z}/2\mathbb{Z}$. Then $\mathcal{A}^{\mathbb{Z}^2}$ is a compact abelian group under coordinate-wise addition. Let

$$X = \left\{ x \in \mathcal{A}^{\mathbb{Z}^2} : x_{\mathbf{n}} + x_{\mathbf{n}+\mathbf{e}_1} + x_{\mathbf{n}+\mathbf{e}_2} = 0 \quad \text{for all } \mathbf{n} \in \mathbb{Z}^2 \right\},$$

which is a closed, shift-invariant subgroup. As mentioned in §13.10, Ledrappier [Led*] constructed this example to give a negative answer to the two-dimensional version of a long-standing and still open problem in ergodic theory. Its main feature is that, although there are no correlations between widely separated pairs of coordinates, there are correlations between special but widely separated triples of coordinates. As for entropy, it is an easy exercise to show that $h(X) = 0$.

This example started a far-reaching and important research direction exploring the rich interplay between algebra and dynamics, which we sketch in §A.9. Deeply buried within Ledrappier's example are algebraic structures analogous to eigenspaces for toral automorphisms which explain essentially all of its dynamical behavior [EinL, Exam. 6.8].

Three-colored chessboard. Let $\mathcal{A} = \{0, 1, 2\}$ and $X \subseteq \mathcal{A}^{\mathbb{Z}^2}$ be the set of points such that the symbols assigned to pairs of coordinates either horizontally or vertically adjacent are different. In the 1960's physicists studied X as a 2-dimensional discrete model of ice, and were keenly interested in the value of $h(X)$ to validate the accuracy of their model. In a landmark paper, Lieb [Lie] proved that $h(X) = \log(4/3)^{3/2}$, one of the first exactly solved models in statistical physics. No similar simple expression is known if the number of symbols is at least four. For an extensive account of exactly solved models see Baxter's book [Bax1].

Two-dimensional golden mean shift. Let $X \subseteq \{0, 1\}^{\mathbb{Z}^2}$ be the set of points for which 1's cannot occur next to each other either horizontally or vertically. This is a two-dimensional version of the golden mean shift of Example 1.2.3.

In the statistical physics literature X is called the hard square model of a lattice gas. Attempts to compute $h(X)$, or at least find accurate approximations, go back to the 1970's, but no simple expression has been found. Markley and Paul [MarP3*] tackled this and similar examples by treating an allowed rectangular array as a succession of horizontal slices with transitions governed by a transfer matrix, providing them a way to get a crude estimate of $h(X)$. Refinements of this approach allowed Friedland, Lundow, and Markström [FriLM] to rigorously compute $h(X)$ to fifteen decimal places. Recent work by Chan and Rechnitzer gives improved convincing lower [ChaR1] and upper [ChaR2] bounds on $h(X)$. Baxter [Bax2] used variations of the transfer matrix method, and making reasonable assumptions on the speed of convergence gave a value of $h(X)$ he believes is accurate to 43 decimal places, but this is not rigorous. Pavlov [Pav1] took a different approach, showing that successive differences of entropies of horizontal strips of increasing height converge exponentially fast to $h(X)$. However, this method does not give explicit constants for the rate of convergence, and so in its current form it cannot be used for rigorous numerical estimates.

Wang tiles. Start with a finite collection $\mathcal{W}$ of unit squares with colored edges. Let $X_{\mathcal{W}}$ denote the set of all tilings of $\mathbb{R}^2$ by translates of tiles in $\mathcal{W}$ with vertices in $\mathbb{Z}^2$ such that abutting edges have the same color. Then $X_{\mathcal{W}}$ is a $\mathbb{Z}^2$-shift of finite type with alphabet $\mathcal{W}$ called a *Wang tile shift*. Much as an arbitrary one-dimensional shift of finite type can be recoded to be 1-step, every $\mathbb{Z}^2$-shift of finite type is conjugate to a Wang tile shift.

In the early 1960's Hao Wang was studying problems in logic related to automated theorem proving. He noticed that these could be recast into more easily understood and visually appealing problems about tilings (see [Wan1] and his personal account [Wan2]). Immediately, connections with the theory of computation became apparent: there is no algorithm that decides on input $\mathcal{W}$ whether $X_{\mathcal{W}} = \varnothing$, nor one that decides whether $X_{\mathcal{W}}$ contains a periodic point. These results are typical of what has been termed the "swamp of undecidability" for $\mathbb{Z}^d$-symbolic dynamics [Lin].

Wang observed that if every nonempty $X_{\mathcal{W}}$ were to contain periodic points, certain decidability questions could be answered positively. But this hope was dashed, first by his student Berger [Ber*] and in more streamlined form by Robinson [Rob*], who constructed tile sets for which every allowed tiling has a hierarchical structure that prevents periodicity.

The study of Wang tile shifts continues to undergo vigorous development. For instance, Chen, Hu, Lai, and Lin [CheHLL] recently showed using an extensive computer enumeration of millions of cases that if only three colors are allowed, then every nonempty Wang tile shift contains periodic points (that this fails when there are five colors was already known to Berger).

Domino tilings. Let $\mathcal{D}$ be the set of Wang tiles displayed below using two

colors indicated by solid and broken lines. Figure A.6.1 shows a partial Wang tiling of $\mathbb{R}^2$ by $\mathcal{D}$ and the result of erasing the common broken lines, creating a tiling by 2×1 and 1×2 dominoes.

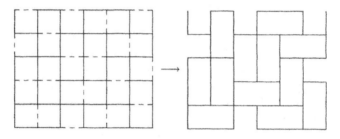

FIGURE A.6.1.   Wang tiling converted to domino tiling.

This Wang tile shift $X_{\mathcal{D}}$ is called the *domino shift* or *dimer model*, and has been intensively studied for decades (see [BurP], [CohEP], [Sch1], and especially [Ein] where its mixing properties are investigated). In statistical physics it models solids composed of dimers, or matched pairs of adjacent lattice points. In 1961 Kasteleyn computed the number of domino tilings of a rectangle, and his formula suggests that the entropy of the domino shift is given by

$$h(X_{\mathcal{D}}) = \frac{1}{4} \int_0^1 \int_0^1 \log(4 - 2\cos 2\pi s - 2\cos 2\pi t) \, ds \, dt.$$

Burton and Pemantle [BurP] have given a rigorous proof of this formula, and related entropy to the growth of the number of spanning forests in the Cayley graph of $\mathbb{Z}^2$. This quantity will make another appearance in §A.9 as the entropy of another system, although the connection remains mysterious.

ENTROPY AND MIXING

The previous examples indicate that computing the entropy of higher dimensional shifts of finite type, either exactly or approximately, poses formidable difficulties. Adding to this sense of despair is the unavoidable swamp of undecidability, which means that it is hopeless to attempt a general theory that will apply to every individual case.

But in 2010, Hochman and Meyerovitch [HocM] employed a sort of mathematical jujutsu. Instead of yielding to the roadblocks thrown up by the theory of computation, they used this theory to give computational characterizations of behaviors of classes of shifts of finite type. Their paper and another by Hochman [Hoc1] have strongly influenced how researchers view $\mathbb{Z}^d$-symbolic dynamics. These papers provide a roadmap for attacking various problems: find computation-theoretic obstructions, which are usually apparent, and then show by constructions that there are no others.

Let us see how this roadmap works for entropy. Let C be a fixed cube in $\mathbb{Z}^d$ and $\mathcal{P} = \{f_j \colon C \to \mathcal{A}\}$ be a set of "allowed" patterns on C. These data determine the shift of finite type

$$X_{\mathcal{P}} = \left\{ x \in \mathcal{A}^{\mathbb{Z}^d} : \sigma^{\mathbf{n}}(x)_C \in \mathcal{P} \quad \text{for all } \mathbf{n} \in \mathbb{Z}^d \right\},$$

and every $\mathbb{Z}^d$-shift of finite type can be put into this form.

Let $Q_n = \{0, 1, \ldots, n-1\}^d$, and $\mathcal{B}_n(X_{\mathcal{P}}) = \{x_{Q_n} : x \in X_{\mathcal{P}}\}$ be the collection of Q_n-blocks appearing in $X_{\mathcal{P}}$. In §13.10 we defined the entropy of $X_{\mathcal{P}}$ to be

$$h(X_{\mathcal{P}}) = \lim_{n \to \infty} \frac{1}{n^d} \log |\mathcal{B}_n(X_{\mathcal{P}})|.$$

However, we cannot directly compute $\mathcal{B}_n(X_{\mathcal{P}})$ since there is no way in general to decide whether a pattern on Q_n can be extended to a point in $X_{\mathcal{P}}$.

To deal with this situation, let $\widetilde{\mathcal{B}}_n(\mathcal{P}) \subseteq \mathcal{A}^{Q_n}$ be the set of locally admissible patterns, i.e., patterns whose restriction to every translate of C within Q_n is in $\mathcal{P}$. Clearly $\mathcal{B}_n(X_{\mathcal{P}}) \subseteq \widetilde{\mathcal{B}}_n(\mathcal{P})$, and also $|\widetilde{\mathcal{B}}_n(\mathcal{P})|$ is readily computed. In 1978, Ruelle [Rue, Thm. 3.4] showed that

$$h(X_{\mathcal{P}}) = \lim_{n \to \infty} \frac{1}{n^d} \log |\widetilde{\mathcal{B}}_n(\mathcal{P})|,$$

but in a thermodynamic setting that obscured its combinatorial meaning. Later and independently, Friedland [Fri] proved this using a complicated argument that was subsequently simplified in [HocM, Thm. 3.1]. Each term in the limit is at least $h(X_{\mathcal{P}})$ and log is a computable function. We conclude that there is a Turing machine which computes a sequence $\{r_n\}$ of rational numbers such that $r_n \geqslant h(X_{\mathcal{P}})$ and $r_n \to h(X_{\mathcal{P}})$. This condition on $h(X_{\mathcal{P}})$ is called *right recursively enumerable*, and is the "apparent" computational obstruction for entropy.

Conversely, suppose that $d \geqslant 2$ and $h \geqslant 0$ is right recursively enumerable. Hochman and Meyerovitch showed how to engineer a $\mathbb{Z}^d$-shift of finite type with entropy h, so that this is the only obstruction for entropy. Their construction starts with a highly rigid zero entropy scaffolding on which they successively model the Turing machine approximations to h. This work provides for $d \geqslant 2$ a complete computation-theoretic description of the possible entropies of $\mathbb{Z}^d$-shifts of finite type quite different from the algebraic one for $d = 1$ given in Theorem 11.1.6.

Does the set of entropies of $\mathbb{Z}^d$-shifts of finite type change if we impose a mixing condition? From §11.1 we know this happens when $d = 1$, since restricting to the set of mixing shifts of finite type shrinks the set of entropies from logs of weak Perron numbers to logs of Perron numbers.

For $d \geqslant 2$ a number of mixing conditions have been explored. Let $X \subseteq \mathcal{A}^{\mathbb{Z}^d}$ be a shift of finite type and $F \subseteq \mathbb{Z}^d$. We say that a pattern $p\colon F \to \mathcal{A}$ is *admissible* if $p = x_F$ for some $x \in X$. We call X *strongly irreducible* if there is an $M \geqslant 1$ such that whenever $p\colon F \to \mathcal{A}$ and $q\colon G \to \mathcal{A}$ are admissible patterns on (possibly infinite) subsets and the distance between F and G is at least M, then the pattern on $F \cup G$ using p and q is also admissible.

Strong irreducibility is a powerful assumption with many consequences. These include the decidability of the extension problem [HocM, Cor. 3.5] and the computability of entropy [HocM, Thm. 1.3] (a real number r is *computable* if there is a sequence $\{r_n\}$ of rationals computed by a Turing machine such that $|r - r_n| < 1/n$). Computable numbers are always right recursively enumerable, but not conversely, so assuming strong irreducibility strictly shrinks the set of entropies. However, there is no known characterization of the entropies of strongly irreducible $\mathbb{Z}^d$-shifts of finite type.

If the shapes F and G in the definition of strong irreducibility are confined to be finite d-dimensional rectangles, the resulting weaker condition is called *block gluing*. Pavlov and Schraudner [PavS] showed that with this condition the entropy is computable by a Turing machine which on input n computes the rational r_n in at most $O(e^{n^2})$ steps. There are computable numbers that do not satisfy this property, and hence cannot be the entropy of a block gluing (nor strongly irreducible) $\mathbb{Z}^d$-shift of finite type. They also gave quantitative sufficient conditions for a computable number to be such an entropy.

As the size of the rectangles increases, we might want to require more separation between them, yielding a condition weaker than block gluing. Let $f\colon \mathbb{N} \to \mathbb{N}$ be nondecreasing and assume that $f(n) = o(n)$. A $\mathbb{Z}^d$-shift of finite type is called *f-block gluing* if the separation between the rectangles F and G in block gluing is required to be at least $f(\max\{\operatorname{diam}(F), \operatorname{diam}(G)\})$. Gangloff and Hellouin de Menibus [GanH] proved that if $\sum_{n=1}^{\infty} f(n)/n^2$ converges computably, then the entropy of every f-block gluing $\mathbb{Z}^d$-shift of finite type is computable. They were unable to decide whether this threshold is sharp, specifically whether if the series diverges then the possible entropies are the right recursively enumerable numbers.

EFFECTIVENESS IN SYMBOLIC DYNAMICS

Entropy is but one example of the rich interplay between symbolic dynamics and the theory of computation. This interplay has not only changed how we view $\mathbb{Z}^d$-symbolic dynamics, but also who views it, with theoretical computer scientists becoming seriously involved. Jeandel [Jea1] describes the computational aspects of these dynamics, and Hochman [Hoc4] supplies a comprehensive survey.

One computational notion that has become increasingly important is effectiveness. Recall that a cylinder set in $\{0,1\}^{\mathbb{N}}$ is specified by a pattern of 0's and 1's on a finite subset of $\mathbb{N}$. We say that a closed set $X \subseteq \{0,1\}^{\mathbb{N}}$ is *effective* if its complement is the union of cylinder sets that can be enumerated by a Turing machine.

Similarly we say that a $\mathbb{Z}^d$-shift X is effective if there is a Turing machine which enumerates a list of forbidden patterns that defines X. Using a computable bijection between $\mathbb{Z}^d$ and $\mathbb{N}$ and a recoding of symbols, we can view X as an effective subset of $\{0,1\}^{\mathbb{N}}$. Clearly a $\mathbb{Z}^d$-shift of finite type is effective, but even for $d = 1$ there are many effective shifts that do not have finite type or are even sofic. For instance, an S-gap shift from Example 1.2.6 is effective if and only if the list of forbidden gaps can be enumerated by a Turing machine. Thus the prime gap shift is effective. But there are many S-gap shifts that are not effective, since there are only countably many Turing machines but uncountably many S-gap shifts. We remark that S-gap shifts have been studied extensively by Climenhaga and Thompson [ClmT1], Dastjerdi and Jangjoo [DasJ*], and Baker and Ghenciu [BakG]).

Two sets X and Y are said to be Medvedev equivalent if there exist partial computable functionals from X into Y and vice versa. Simpson [Sim] showed that every Medvedev equivalence class can be realized by a $\mathbb{Z}^2$-shift of finite type. Roughly speaking, this means that $\mathbb{Z}^2$-shifts of finite type encompass all possible degrees of computational complexity.

There is a more dynamical result along these lines due to Aubrun and Sablik [AubS] and, independently, to Durand, Romaschenko, and Shen [DurRS]. This result says that every effective $\mathbb{Z}^d$-shift X can be "simulated" by a $\mathbb{Z}^{d+1}$-shift of finite type Y, meaning that X is a d-dimensional slice of a factor of Y. More concretely, they show that if X is an effective $\mathbb{Z}$-shift, then the $\mathbb{Z}^2$-shift consisting of points whose rows are in X and whose columns are constant is a factor of a $\mathbb{Z}^2$-shift of finite type. This shows that any dynamical behavior present in an effective $\mathbb{Z}$-shift, no matter how pathological, can be simulated by a $\mathbb{Z}^2$-shift of finite type. This line of research originated from work of Hochman [Hoc1], who showed that it is possible to simulate any effective $\mathbb{Z}^d$-shift by using a $\mathbb{Z}^{d+2}$-shift of finite type.

CONJUGACY, FACTOR CODES, AND EMBEDDINGS

In Chapter 7 we saw how to decompose conjugacies between $\mathbb{Z}$-shifts of finite type into splitting codes and amalgamation codes. Extensions of this idea to the $\mathbb{Z}^d$ setting have been given by Johnson and Madden [JohM1] and by Schraudner [Schr]. Their work leads to a notion of strong shift equivalence that characterizes conjugacy.

However, when $d \geqslant 2$ the conjugacy problem for $\mathbb{Z}^d$-shifts of finite type must be undecidable, since any decision procedure would in particular solve the nonemptiness problem, which is known to be undecidable. Another aspect has been explored by Jeandel and Vanier [JeaV1], who characterized the computational complexity of deciding conjugacy, factoring, and embedding. Surprisingly, they show that it is harder to decide whether one $\mathbb{Z}^d$-shift of finite type factors onto another than whether they are conjugate.

Recall that factor codes are surjective sliding block codes. Meester and Steif [MeeS] developed relationships in the $\mathbb{Z}^d$ setting among entropy-preserving, finite-to-one, and almost invertible factor codes. Their work included a $\mathbb{Z}^d$ version of Theorem 8.1.16, giving several equivalent conditions for a factor code to be finite-to-one.

According to Theorem 5.5.8, every $\mathbb{Z}$-shift of finite type with entropy at least $\log N$ factors onto the full N-shift. Johnson and Madden [JohM2] asked whether this holds in the $\mathbb{Z}^d$ setting, and gave some partial positive answers that were later improved by Desai [Des2]. But Boyle and Schraudner [BoyS2] constructed counterexamples when the entropy equals $\log N$. Later, Boyle, Pavlov, and Schraudner [BoyPS] gave counterexamples where the entropy of the shift strictly exceeds $\log N$. On the other hand, in the latter case they showed if the shift is also assumed to have block gluing, then there is such a factor code. These results were extended to factoring onto general $\mathbb{Z}^d$-shifts of finite type by Briceño, McGoff, and Pavlov [BriMP].

Lightwood [Lig1, Lig2] proved versions of the Embedding Theorem 10.1.1 for $\mathbb{Z}^2$-shifts of finite type, but with some restrictions. Quas and Trow [QuaT] showed that every $\mathbb{Z}^2$-shift of finite type properly contains $\mathbb{Z}^2$-shifts of finite type with entropy arbitrarily close to that of the ambient shift. Desai [Des1] improved this by proving that for any $\mathbb{Z}^d$-shift of finite type X, the entropies of subshifts of X of finite type are dense in $[0, h(X)]$.

McGoff and Pavlov [McgP] have taken a different approach to these problems. They imposed a probabilistic structure on the collection of all $\mathbb{Z}^d$-shifts of finite type and investigated the probability that one such shift is conjugate to, factors onto, and embeds into another, allowing a sensible notion of "typical" behavior.

Hochman [Hoc2] made a detailed study of the automorphism groups of $\mathbb{Z}^d$-shifts of finite type for $d \geqslant 2$, and showed that their behavior differs, often dramatically, from that described in §13.2 for $\mathbb{Z}$-shifts of finite type.

PERIODIC POINTS

Periodic points for $\mathbb{Z}$-shifts of finite type serve as important invariants, are used to define the zeta function, and have a growth rate that equals entropy.

A point in a $\mathbb{Z}^d$-shift is *periodic* if its orbit is finite. When $d \geqslant 2$ periodic points play a less central role. For example, there are $\mathbb{Z}^2$-shifts of finite type with positive entropy that have no periodic points at all. In fact, one of the most frustrating open problems is whether every strongly irreducible $\mathbb{Z}^d$-shift of finite type has periodic points. This is easily seen to be true when $d = 1$, and an argument by Lightwood [Lig2] shows that it is also true for $d = 2$. But it remains open for all $d \geqslant 3$. This problem is of particular interest since if a strongly irreducible shift of finite type contains one periodic point, then it contains a dense set of periodic points [CecC2].

A point in a $\mathbb{Z}^d$-shift has *square period* n if it is fixed by all shifts in $n\mathbb{Z}^d$. Jeandel and Vanier [JeaV2] investigated the sets of positive integers n that can arise as the square periods for given $\mathbb{Z}^d$-shifts of finite type. They characterized such sets as exactly those that are computable by a nondeterministic Turing machine in polynomial time.

Sofic Shifts

A $\mathbb{Z}^d$-shift is *sofic* if it is a factor of a $\mathbb{Z}^d$-shift of finite type. Using essentially the same argument that the entropy of a $\mathbb{Z}^d$-shift of finite type is right recursively enumerable, one can show that the entropy of an effective $\mathbb{Z}^d$-shift also has this property. Since the collection of sofic shifts lies between those of finite type and effective shifts, all three collections have the same set of entropies. In particular, for every sofic shift there is a shift of finite type with the same entropy.

A major open problem, due to Weiss, is whether every sofic $\mathbb{Z}^d$-shift is a factor of a $\mathbb{Z}^d$-shift of finite type with the same entropy. This is true when $d = 1$ by using the minimal right resolving presentation from §3.3. Desai [Des1] showed that if one is allowed to use slightly higher entropy then such a factor code is always possible. But Weiss's problem remains open for all $d \geqslant 2$.

Another approach to understanding sofic $\mathbb{Z}^d$-shifts is to mimic the follower set characterization from Theorem 3.2.10. Motivation for this comes from work by French, Ormes, and Pavlov [FreOP], who studied follower sets in $\mathbb{Z}$-shift spaces. In particular, they strengthened Theorem 3.2.10 (which characterized sofic shifts as those with only finitely many follower sets) by showing that if X is a $\mathbb{Z}$-shift and if the number of follower sets of all words of length n in X is no more than $\log(n + 1)$, then X is a sofic shift. They conjecture that the bound $\log(n + 1)$ can be replaced by n.

Now let X be a $\mathbb{Z}^d$-shift. For a finite set $F \subset \mathbb{Z}^d$ and pattern f on F, its *extender set* is the collection of all patterns on $\mathbb{Z}^d \setminus F$ that when glued together with f give a point in X. Ormes and Pavlov [OrmP] have shown that if for some n the number of extender sets for patterns on the cube of side length n is at most n, then X must be sofic. This is of interest even in the case $d = 1$ where it gives a bound different from that discussed in the previous paragraph. However, the general question of characterizing when a $\mathbb{Z}^d$-shift is sofic, in terms of extender sets, does not currently have an answer.

Expansive Subdynamics

Sometimes the points in a $\mathbb{Z}^d$-shift are determined by their coordinates near a lower dimensional subspace of $\mathbb{R}^d$. Consider Ledrappier's example $X \subseteq (\mathbb{Z}/2\mathbb{Z})^{\mathbb{Z}^2}$ which we discussed earlier. Let L be a line in $\mathbb{R}^2$ through the origin, and L_r be the strip in $\mathbb{R}^2$ of points within distance r of L. Suppose first that L has slope 1. An easy geometric argument shows that there is an $s > 0$ such that if $r > 2$ and x and y are points in X that agree on $L_r \cap \mathbb{Z}^2$, then they also agree on $L_{r+s} \cap \mathbb{Z}^2$, hence on $L_{r+2s} \cap \mathbb{Z}^2$, and so on, and hence $x = y$. Hence using only the shifts within $L_r \cap \mathbb{Z}^2$ we have an analogue of the expansiveness we already defined for dynamical systems. This argument works except when the line is horizontal, vertical, or has slope -1, and for these cases no strip, no matter how wide, will determine a point.

A *hyperplane* H in $\mathbb{R}^d$ is a $(d-1)$-dimensional subspace. For $r > 0$ let H_r be the set of points within distance r of H. Let X be a $\mathbb{Z}^d$-shift. We say that H is *expansive for* X if there is an $r > 0$ such that if x and y are in X and $x_{\mathbf{n}} = y_{\mathbf{n}}$ for all $\mathbf{n} \in H_r \cap \mathbb{Z}^d$, then $x = y$; if no such r exists then H is *nonexpansive for* X. Let $\mathcal{E}(X)$ denote the set of expansive hyperplanes for X and $\mathcal{N}(X)$ the nonexpansive ones.

Let S_{d-1} be the unit sphere in $\mathbb{R}^d$. Each hyperplane corresponds to a pair of antipodal unit vectors in its orthogonal complement. Thus, sets of hyperplanes correspond to subsets of S_{d-1} that are invariant under the antipodal map, sets that we call *projective*. We can then use the standard topology on S_{d-1} to define a compact topology on the space of all hyperplanes. An argument that an expansive hyperplane codes nearby hyperplanes shows that $\mathcal{E}(X)$ is open in this topology, and hence $\mathcal{N}(X)$ is compact.

A fundamental fact [BoyL1, Thm. 3.7] is that if X is infinite then $\mathcal{N}(X)$ must be nonempty: nontrivial shifts always have some nonexpansive behavior. There are examples in which every nonexpansive hyperplane has trivial intersection with $\mathbb{Z}^d$, illustrating the need to use all hyperplanes, not just those spanned by the lattice points they contain, in order to "see" nonexpansiveness.

Nonexpansiveness along lower dimensional subspaces has a similar definition, and it turns out that a subspace is nonexpansive if and only if it is contained in a nonexpansive hyperplane. These ideas, called *expansive subdynamics*, can be formulated beyond the symbolic setting as well, and apply to d commuting homeomorphisms of a compact metric space.

A key use of expansive subdynamics is in studying how dynamical behavior along subspaces changes as we vary the subspace. The connected components of the open set $\mathcal{E}(X)$ are called the *expansive components* of X. The guiding philosophy in [BoyL1] is that within an expansive component dynamical properties are either constant (e.g., having a gluing property analogous to being Markov) or vary nicely (e.g., directional entropy or zeta function). However, when passing from one expansive component to another through $\mathcal{N}(X)$, behavior typically changes abruptly, analogous to a phase transition. For instance, in Ledrappier's example the directional entropy introduced by Milnor [Mil] is given by a simple formula in each expansive component, but the formula is different for distinct components [BoyL1, Exam. 6.4].

Which compact projective sets can arise as $\mathcal{N}(X)$ for a $\mathbb{Z}^d$-shift X? When $d = 2$ a construction in [BoyL1] shows that, with the possible exception of singleton sets corresponding to lines of irrational slope, every projective compact set can be realized. It took over a decade before Hochman [Hoc3] resolved this remaining case in the affirmative, using an elaborate construction. Currently there is no complete characterization of the nonexpansive sets when $d \geqslant 3$, although [BoyL1, Thm. 4.4] shows that every compact projective set in S_{d-1} which properly contains an isolated point does arise as a nonexpansive set for a $\mathbb{Z}^d$-shift.

Suppose we restrict consideration to $\mathbb{Z}^d$-shifts of finite type. There are only countably many such shifts, drastically reducing the possible nonexpansive sets from the general shift case. In his recent thesis, Zinoviadis [Zin] showed that when $d = 2$ the nonexpansive sets for shifts of finite type are exactly the compact projective sets that are effectively closed. This means that if we take S_1 to be $[0, 1]$ after identifying endpoints, then there is a Turing machine that outputs a sequence $\{a_n\}$ of rational numbers and another sequence $\{r_n\}$ of rational radii such that the complement of the set is the union of the intervals $(a_n - r_n, a_n + r_n)$. Again there is no similar complete answer when $d \geqslant 3$.

Barge and Olimb [BarO] adapted the expansive dynamics approach to investigate tiling dynamical systems. It has also been used by Sablik [Sab] to study the directional behavior of cellular automata maps.

§A.7. Equilibrium States

Why does water freeze? How does a slight change in an external parameter like temperature result in such a drastic transformation of a physical system? Statistical physics attempts to understand such phenomena using simple models that involve $\mathbb{Z}^d$-shifts. This is a vast and rapidly growing subject; even the classical Ising model described below, based on the $\mathbb{Z}^d$-full 2-shift, is the focus of hundreds of research papers every year. For lucid self-contained surveys of the mathematical aspects of statistical physics, see the expository article by Cipra [Cip] and the book chapter by Georgii, Häggström, and Maes [GeoHM].

Topological Pressure

In §13.3 we showed that if X_G is an irreducible edge shift, then there is a unique probability measure μ_G that maximizes measure-theoretic entropy over all shift-invariant measures on X_G, indicating its special nature. By modifying this set-up in ways suggested by statistical physics, we can find other measures with special properties, and furthermore we can extend these ideas to $\mathbb{Z}^d$-shifts. This leads to the notion of topological pressure, which generalizes topological entropy.

Whereas the topological entropy of a $\mathbb{Z}^d$-shift space X is the growth rate of the number of finite configurations in X, the topological pressure is a weighted version of this growth rate, where each block is weighted by a real-valued continuous function $f\colon X \to \mathbb{R}$, in a way made precise below. The generalization of a measure of maximal entropy to this setting is known as an equilibrium state. Equilibrium states are central to the interface between ergodic theory and symbolic dynamics. In this section, we consider questions concerning uniqueness of equilibrium states, efficient computation of topological pressure, and the systems of conditional probabilities induced by an equilibrium state. The latter touches on the relationship among equilibrium states, Gibbs states, and Markov random fields.

To define topological pressure, fix a continuous function $f\colon X \to \mathbb{R}$ on a $\mathbb{Z}^d$-shift X (also called a *potential*). Set $Q_n = \{0, \dots, n-1\}^d$, and put $S_n f(x) = \sum_{\mathbf{k} \in Q_n} f(\sigma^{\mathbf{k}}(x))$. Let

$$p_n(X, f) := \sum_{w \in B_{Q_n}(X)} \sup_{\{x \in X : x_{Q_n} = w\}} \exp(S_n f(x)).$$

Then the *topological pressure of f* is defined as

$$P_X(f) := \lim_{n \to \infty} \frac{1}{n^d} \ln p_n(X, f),$$

where the limit exists by [Rue, Thm. 3.4]. The topological pressure of the zero function is a scaled version of topological entropy, namely $P_X(0) = (\ln 2)h(X)$, where the scale factor is an artifact of our convention of using log to the base 2 in our definition of entropy.

Pressure has a natural interpretation in statistical mechanics, which concerns systems involving a large number of particles, situated throughout space, with each particle in one condition or another. Treating the particle positions as sites on the d-dimensional lattice $\mathbb{Z}^d$ and the possible conditions as symbols in a finite alphabet, a configuration of the physical system can be represented as an element of a $\mathbb{Z}^d$-shift space X. A continuous function $f\colon X \to \mathbb{R}$ is interpreted as assigning a potential energy to any given configuration, and topological pressure has a related thermodynamic interpretation. The normalization by $1/n^d$ in the definition of $P_X(f)$ makes it natural to interpret topological pressure as a measure of pressure per unit volume, which in thermodynamics is called the Gibbs free energy per site.

There are several classical statistical mechanical models of interest. In the case of the *Ising model* of ferromagnetism, X is the full shift over two symbols $\{\pm 1\}$, which indicate the "spin" of a particle at a lattice site, and

$f(x) = -\sum_{i=1}^{d} x_{\mathbf{e}_i} x_{\mathbf{0}}$, where the $\mathbf{e}_i$ are the standard basis vectors, giving the strength of interaction between neighboring particles. In the case of the *hard square model*, X is the $\mathbb{Z}^d$ golden mean shift over the alphabet $\{0, 1\}$ indicating the presence (1) or absence (0) of a particle at a lattice site, and the function $f(x) = x_{\mathbf{0}}$ simply specifies whether or not a site is occupied. The *Potts model* is a generalization of the Ising model with a larger alphabet.

These models actually involve a scalar multiple βf of f, where β reflects a physical parameter. For the Ising and Potts models, β is the inverse temperature k/T where k is a constant, while for the hard square model β represents the propensity of a site to be occupied. In these models, of particular interest is the *topological pressure function* $\beta \mapsto P_X(\beta f)$. Phase transitions are then indicated by abrupt changes in this function.

THE VARIATIONAL PRINCIPLE AND EQUILIBRIUM STATES

For a shift-invariant Borel probability measure μ on a $\mathbb{Z}^d$-shift X, there is a notion of measure-theoretic entropy $h_\mu(X, \sigma)$ (which was discussed briefly in §13.4 in the case $d = 1$). Let $f \colon X \to \mathbb{R}$ be continuous. The celebrated variational principle [Kel, §4.4] says that

$$P_X(f) = \sup_\mu \left\{ h_\mu(X, \sigma) + \int_X f \, d\mu \right\},$$

where the supremum is over all shift-invariant Borel probability measures on X. A semicontinuity argument shows that this supremum is always attained, and any measure which does so is called an *equilibrium state* for f. There is a tension between two competing quantities in this result, the randomness of the system as measured by entropy and the expected value of the external function f, and these are in balance at equilibrium states. When $f = 0$ an equilibrium state is simply a measure of maximal entropy.

Topological pressure can be defined for any dynamical system and real-valued continuous function and there is a corresponding variational principle, along with the notion of equilibrium state, in this generality (although there are examples showing that the supremum need not always be attained). This framework has had an enormous impact on the understanding of smooth invariant measures for smooth dynamical systems. Much of the recent work on equilibrium states in the literature has focused on smooth systems. We will not discuss this here in any detail but instead refer to the surveys of Sarig [Sar2], on thermodynamical formalism for countable state Markov shifts and applications to surface diffeomorphisms, and Chazottes and Keller [ChaK], on pressure and equilibrium states for both symbolic and smooth dynamical systems.

Uniqueness of Equilibrium States

Krieger [Kri2*] showed that for a locally constant function f on a mixing $\mathbb{Z}$-shift of finite type, the equilibrium state, denoted μ_f, is unique, and he gave an explicit formula for μ_f generalizing the formulas (13-3-6) and (13-3-7) based on the Perron–Frobenius Theorem. R. Bowen extended this to functions f on mixing $\mathbb{Z}$-shifts of finite type satisfying exponentially decaying variations [Bow6*, Thm. 1.22], using an infinite-dimensional version of the Perron–Frobenius Theorem due to Ruelle. Bowen later generalized this to expansive dynamical systems that satisfy (1) a condition known as *specification*, which allows arbitrarily many partial orbits of any lengths to be shadowed with bounded gaps by a single orbit, and (2) a weaker condition on the decay of variations of f that has come to be known as the *Bowen property* [BowR].

More recently, Climenhaga and Thompson [ClmT2] and Pavlov [Pav2] have returned to Bowen's work and established very general results on uniqueness of equilibrium states for $\mathbb{Z}$-shift spaces that satisfy weakened versions of specification and for functions that satisfy weakened versions of the Bowen property. In [ClmT2] specification is weakened by requiring that it hold only for a "large set" of partial orbits. In [Pav2] specification is weakened by allowing the gaps to be unbounded but not too large, in particular sublogarithmic, and Pavlov showed that sublogarithmic is optimal.

In the case $d = 1$, uniqueness of the equilibrium state seems to fail mainly for complicated shift spaces X or poorly behaved functions f. But in the case $d = 2$ uniqueness can fail even on the full shift for very simple functions. This happens for certain values of the parameter β in the Ising model, Potts model, and hard square model. For the Ising and Potts models, there is an explicit critical value β_c, and for $\beta \neq \beta_c$ there is a unique equilibrium state for βf if and only if $\beta < \beta_c$. This phenomenon is known as a *phase transition*. For the Potts model with q letters, the value of β_c is known to be $(1/2)\ln(1 + \sqrt{q})$ (in particular, $(1/2)\ln(1 + \sqrt{2})$ for the Ising model); this was proven only recently by Beffara and Duminil-Copin [BefD]. For the critical value $\beta = \beta_c$, uniqueness is known to hold for the Ising model and certain Potts models. The situation is much less clear for the hard square model [GeoHM, Thm. 3.3]. Burton and Steif [BurS] showed that the measure of maximal entropy can fail to be unique even for strongly irreducible shifts of finite type in dimension $d = 2$.

Recall the unique equilibrium state μ_f for a locally constant function f on a mixing $\mathbb{Z}$-shift of finite type. Bremont [Bre] (see also Chazottes, Gambaudo, and Ugalde [ChaGU]) showed that the zero temperature limit $\lim_{\beta \to \infty} \mu_{\beta f}$ exists and described the limit. On the other hand, Chazottes and Hochman [ChaH] (see also van Enter and Ruszel [EntR]) showed that the limit need not exist for continuous functions with exponentially decaying variations on mixing $\mathbb{Z}$-shifts of finite type, nor for locally constant functions on higher dimensional shifts of finite type such as the the $\mathbb{Z}^d$-full 2-shift for $d \geqslant 3$.

COMPUTATION OF TOPOLOGICAL PRESSURE

As discussed earlier, for $d \geqslant 2$ the computation of topological entropy for $\mathbb{Z}^d$-shifts of finite type is notoriously difficult. Computation of topological pressure can only be harder. However, excellent estimates have been obtained for the topological pressure of certain functions on certain shifts of finite type, for instance by Friedland and Peled [FriP].

In some cases one can use equilibrium states to approximate the topological pressure as follows.

If μ is a shift-invariant measure on X, there is an associated information function I_μ on X given by the conditional entropy of a site given its past and defined μ-almost everywhere. Furthermore, $h_\mu(X, \sigma) = \int_X I_\mu \, d\mu$. Thus

$$P_X(f) = \int_X \left(I_\mu + f \right) d\mu.$$

Now let X be a $\mathbb{Z}^d$-shift of finite type satisfying *topological strong spatial mixing* [Bri1], a strengthening of strong irreducibility that holds for a wide variety of shifts of finite type including the full shift, the golden mean shift, and the k-colored chessboard over $\mathbb{Z}^2$ for $k \geqslant 5$, where the colors of nearest neighbors must be different.

Let f be a locally constant function, and suppose that ν is an arbitrary invariant measure. If the standard approximations to I_μ converge uniformly on the topological support of ν, then the topological pressure can be expressed as the integral of the *same* integrand but with respect to ν instead of μ [Bri1]:

$$P_X(f) = \int \left(I_\mu + f \right) d\nu.$$

If ν is an atomic periodic point measure, then we need only compute the values of I_μ and f at finitely many points (in some cases, just one point). Unfortunately, even for very simple f, the function I_μ can be very complicated and it may be very difficult to compute exactly even at a single point. However, in the case $d = 2$, if the uniform convergence of the approximations is also exponential, then it is possible to efficiently approximate the value of I_μ at a specific point, and therefore $P_X(f)$, in the sense that there is a Turing machine that on input n returns an approximation that is guaranteed to be accurate to within $1/n$ in time polynomial in n. This gives a rigorous sense for which, in some cases, topological pressure can be efficiently approximated.

This scheme works for all parameter values of the Ising model, all non-critical values of the Potts model, a wide range of parameter values for the hard square model [AdaBMP], and a range of coloring models (for the

Ising model this is actually of little interest since there is a closed form expression for the topological pressure due to Onsager [Ons]). The essential ideas for this go back to the work of Weitz [Weit] and Gamarnik and Katz [GamK] in approximation algorithms, and of J. Chayes, L. Chayes, and Schonmann [ChaCS] in statistical mechanics, and have been developed in the symbolic dynamics context by Marcus and Pavlov [MarP] and Briceño [Bri1, Bri2].

EQUILIBRIUM STATES, GIBBS STATES, AND MARKOV RANDOM FIELDS

One might expect a measure of maximal entropy for a $\mathbb{Z}^d$-shift space to have equal measure on all cylinders of a given finite shape $\Lambda \subseteq \mathbb{Z}^d$ of lattice sites, but this is not true even for mixing $\mathbb{Z}$-shifts of finite type. However, for a $\mathbb{Z}^d$-shift of finite type X with measure of maximal entropy μ and for μ-almost every $x \in X$, the conditional probability on compatible configurations u on Λ conditioned on x_{Λ^c} is uniform; here, compatible means that the configuration, $u \vee x_{\Lambda^c}$, which agrees with u on Λ and agrees with x_{Λ^c} on Λ^c, belongs to X. Such a measure is called a *uniform Gibbs state*. In short, every measure of maximal entropy on a shift of finite type is a uniform Gibbs state.

There is a generalization of this result for equilibrium states. Recall that the entropy $h(\mathbf{p})$ of a probability measure $\mathbf{p}$ on a finite set of size n is uniquely maximized by the uniform measure $p_i \equiv 1/n$. Similarly, the sum $h(\mathbf{p}) + \mathbf{p} \cdot \mathbf{u}$ for a real vector $\mathbf{u}$ is uniquely maximized by the Gibbs–Boltzmann distribution $p_i = Z^{-1} \exp(-u_i)$, where Z is a normalization constant. A *Gibbs state* with respect to a continuous function f is a measure for which the conditional probabilities of configurations compatible with x_{Λ^c} are given by a Gibbs–Boltzmann distribution, where the vector $\mathbf{u}$ is expressed in terms of f and x_{Λ^c} in a natural way [Kel, Chap. 5]. When f is identically zero, we have $\mathbf{u} = \mathbf{0}$, and the definition reduces to that of a uniform Gibbs state.

On a $\mathbb{Z}^d$-shift of finite type, every equilibrium state is a Gibbs state. This result is known as the Lanford–Ruelle Theorem [Rue, Thm. 4.2]. The Lanford–Ruelle Theorem can fail on shift spaces that are not of finite type. However, Meyerovitch [Mey] showed that every equilibrium state for a sufficiently regular potential satisfies a condition that he called the topological Gibbs condition. In general, this is weaker than being Gibbs, although sometimes equivalent, such as on shifts of finite type. The topological Gibbs condition is motivated by a definition of Gibbs states introduced by Capocaccia [Cap], which is equivalent to the one described above in terms of conditional probabilities, but applies to dynamical systems more general than shift spaces.

The Lanford–Ruelle Theorem has a partial converse which pre-dated it, due to Dobrushin [Dob]. Namely, for a sufficiently regular function f on a shift space that satisfies a strong mixing condition (the so-called D-

condition, defined in [Rue, §4.1]), any shift-invariant Gibbs state for f is an equilibrium state for f. The Lanford–Ruelle Theorem taken together with this partial converse is known as the Dobrushin–Lanford–Ruelle Theorem and says that for a $\mathbb{Z}^d$-shift of finite type which satisfies the D-condition, equilibrium states and shift-invariant Gibbs states coincide. Relative versions of this theorem in random environments have been established by Seppäläinen [Sep] and Barbieri, Gómez, Marcus, and Taati [BarGMT].

A Gibbs state defined by a continuous function f on a shift space can be alternatively described by a so-called *interaction* defined on configurations on finite sets; the relation between f and its corresponding interaction is given in [Rue, §3.2]. An interaction is *nearest neighbor* if it is nonzero only on configurations on single sites and sets consisting of two adjacent sites. The interactions for the Ising, Potts, and hard square models are all nearest neighbor. A Gibbs state for a nearest neighbor interaction satisfies the following conditional independence condition: given a finite subset $\Lambda \subseteq \mathbb{Z}^d$, we have that x_Λ and $x_{\Lambda^c \setminus \partial \Lambda}$ are independent conditioned on $x_{\partial \Lambda}$. A measure which satisfies this property is called a *Markov random field*. A nearest neighbor Gibbs state is thus always a Markov random field. The classical Hammersley–Clifford Theorem [HamC], [Gri, §7.2] asserts that the converse is true provided that the shift space X has a safe symbol, i.e., a symbol s such that for all $x \in X$ replacing a coordinate of x by s results in a point still in X. The same holds for shift-invariant Markov random fields and Gibbs states defined by a shift-invariant nearest neighbor interaction.

Chandgotia and Meyerovitch [ChaM] developed a formalism to show that a version of the Hammersley–Clifford theorem holds for certain special shifts of finite type which do not have a safe symbol; the most prominent example of this type is the two-dimensional three-colored chessboard. However, they also found examples where the theorem fails dramatically. Roughly speaking, they parameterized specifications of shift-invariant Markov random fields as a vector space M_X and Gibbs states given by shift-invariant nearest neighbor interactions as a subspace G_X. They then found a nearest neighbor shift of finite type X such that the dimension of G_X is finite, but the dimension of M_X is uncountable! On the other hand Chandgotia [Cha] showed that a version of the Hammersley–Clifford Theorem holds under a weaker condition (weaker than safe symbol), namely a notion of folding akin to that introduced by Nowakowski and Winkler [NowW].

Factors of Gibbs states

Given a factor code π on an irreducible edge shift X_G, and a Markov chain μ on G, the image measure $\pi_* \mu$ is called a *hidden Markov chain*; see Boyle and Petersen [BoyP] for a comprehensive treatment of hidden Markov chains from the symbolic dynamics perspective. See also Verbitskiy [Verb] for a discussion of the thermodynamics of hidden Markov chains.

It is not hard to show that a Markov chain is a Gibbs state for a very simple potential. While a hidden Markov chain need not be (and typically is not) Markov, there are cases when it is Gibbs for some potential; see Chazottes and Ugalde [ChaU1]. This raises the question of when the image of a Gibbs state with respect to a sufficiently "regular" potential is also a Gibbs state with respect to some potential. In other words when is the Gibbs property preserved by factor codes? This question has been studied by Chazottes and Ugalde [ChaU2] and Pollicott and Kempton [PolK] on full shifts. More recent work on this subject suggests that a kind of "fiber-wise mixing" condition on the factor code should characterize preservation of the Gibbs property. There has been a flurry of activity on this question, including Hong [Hon], Kempton [Kem], Piraino [Pir1, Pir2], and Yayama [Yay]. We remark that in these works, the notion of Gibbs state considered is that of a Bowen-Gibbs state, which is different from the notion that we have described here, although the notions coincide for "sufficiently regular" potentials.

§A.8. Symbolic Dynamics over Countable Groups

So far we have studied arrays of symbols indexed by the groups $\mathbb{Z}$ or $\mathbb{Z}^d$. But the basic ideas easily carry over to general countable groups. The resulting systems lead to a multitude of fascinating problems with deep connections to other areas of mathematics, many relating the properties of the group with the dynamics of their shifts.

We can sketch here only some of the major developments in this rapidly expanding area. The most important of these is the revolution launched in 2008 by Lewis Bowen with his discovery of profoundly new ways for thinking about dynamical notions.

GROUPS AND THEIR SHIFTS

Let Γ be a countably infinite discrete group. Denote typical elements of Γ by s, t, u, and the group operation by multiplication with identity 1_Γ. For a finite alphabet $\mathcal{A}$, the set $\mathcal{A}^\Gamma$ of arrays $x = (x_s)_{s \in \Gamma}$ is a compact metric space. An element $t \in \Gamma$ acts as a shift on $x \in \mathcal{A}^\Gamma$ by defining $(tx)_s = x_{t^{-1}s}$. The use of t^{-1} in this definition is needed so that the associative law $(t_t t_2)x = t_1(t_2 x)$ holds; when $\Gamma = \mathbb{Z}$ this results in the right shift. A Γ-*shift* is a closed subset of $\mathcal{A}^\Gamma$ that is invariant under Γ.

A *pattern* is an element of $\mathcal{A}^F$, where F is a finite subset of Γ. Using an obvious extension of terminology from the case $\Gamma = \mathbb{Z}^d$, a Γ-shift is determined by a collection of forbidden patterns. It is a Γ-*shift of finite type* if this collection can be chosen to be finite.

Suppose that $F \subseteq \Gamma$ is finite and $\Phi : \mathcal{A}^F \to \mathcal{A}$. These define a *sliding block code* $\phi = \Phi_\infty : \mathcal{A}^\Gamma \to \mathcal{A}^\Gamma$ given by $\phi(x)_s = \Phi\big((s^{-1}x)_F\big)$. Then ϕ is a continuous map that commutes with the Γ-action, and all such maps arise this way.

We briefly recall a few fundamental classes of groups together with some useful notation. A group Γ is *finitely generated* if there is a finite set $S \subseteq \Gamma$ whose elements generate Γ. Associated with (Γ, S) is its *Cayley graph*, whose vertex set is Γ and for each $s \in \Gamma$ and $a \in S$ contains an edge labeled a from s to sa. The *word metric* d_S on Γ is the shortest distance in the Cayley graph. For example, for each $r \geqslant 1$ the Cayley graph of the *free group* $\mathbb{F}_r$ of rank r using the standard generators is a tree.

A group is *residually finite* if it contains a decreasing sequence of finite-index normal subgroups with trivial intersection.

The definition of entropy for $\mathbb{Z}^d$-shifts makes crucial use of sets like cubes whose boundaries are relatively small. Generalizing this, a group Γ is *amenable* if it contains a sequence $F_1 \subseteq F_2 \subseteq F_3 \subseteq \ldots$ of finite sets (called a *Følner sequence*) such that $|sF_j \cap F_j|/|F_j| \to 0$ as $j \to \infty$ for every $s \in \Gamma$. Amenability is a fundamental notion whose ubiquity results from having many equivalent definitions.

SURJUNCTIVITY

When is a set F finite? One definition is the existence of a one-to-one correspondence between F and an "external" model $\{1, 2, \ldots, n\}$. But an alternative definition, pioneered by Dedekind, is that every one-to-one map of F into itself is also onto.

We have already seen dynamical analogues of Dedekind finiteness. Suppose that the sliding block code $\phi: \mathcal{A}^{\mathbb{Z}} \to \mathcal{A}^{\mathbb{Z}}$ is one-to-one. We claim that ϕ must also be onto. There are (at least) two ways to see this.

The periodic point argument observes that for every $n \geqslant 1$ the finite set of points with period n is mapped by ϕ into itself, and hence onto itself since ϕ is one-to-one. Thus the image of ϕ is compact and contains the dense set of all periodic points, and so ϕ is onto. The entropy argument uses that $\phi(\mathcal{A}^{\mathbb{Z}})$ is a subshift of $\mathcal{A}^{\mathbb{Z}}$ having the same entropy as $\mathcal{A}^{\mathbb{Z}}$ since ϕ is one-to-one. But by Corollary 4.4.9 the only such shift is $\mathcal{A}^{\mathbb{Z}}$ itself, so that ϕ is onto.

In 1973 Gottschalk [Got] coined the term *surjunctive* to describe groups Γ for which every embedding of $\mathcal{A}^{\Gamma}$ into itself is also onto. He briefly pointed out that several types of groups are surjunctive, and raised the problem of whether *every* group is surjunctive. For instance, a version of the periodic point argument shows that every residually finite group (such as $\mathbb{F}_r$) is surjunctive, and an analogue of the entropy argument can be used to show that all amenable groups are also surjunctive.

Despite the seeming simplicity of this problem, it has resisted many efforts, and remains unsolved. As we will see, an affirmative answer, although widely viewed as unlikely, would have profound consequences.

Sofic Groups

Amenable groups are strikingly different from free groups, and yet both are surjunctive. Nearly twenty years after Gottschalk's work, Gromov [Gro] showed that they share the common property of having finite approximations by permutation groups, and that this property alone is sufficient to establish surjunctivity. He called groups with this property "initially subamenable." Shortly afterwards, Weiss [Weis] distilled the symbolic part of Gromov's ideas, and also coined the name "sofic" for such groups. Be aware that the meaning of "sofic" for groups has no relation to its meaning for shifts, except that both express some form of finiteness.

Gromov's lengthy paper contains far more, and is an exploration of surjunctivity phenomena as a theme in quite different mathematical areas. A good example, cited in [Gro, Thm. 1.B′], is that if a complex polynomial map from $\mathbb{C}^n$ to itself is one-to-one, then it must also be onto.

The finite approximations used to define soficity can be briefly described as follows. Denote the permutation group of a finite set K by $\mathrm{sym}(K)$. A *sofic approximation* to a group Γ is a sequence $\Sigma = \{\pi_n \colon \Gamma \to \mathrm{sym}(K_n)\}$ of maps from Γ to finite permutation groups that are asymptotically homomorphic, i.e., for all s and t in Γ

$$\frac{|\{k \in K_n : \pi_n(st)k = \pi_n(s)\pi_n(t)k\}|}{|K_n|} \to 1 \quad \text{as } n \to \infty,$$

and are asymptotically free, i.e., for every $s \neq 1_\Gamma$,

$$\frac{|\{k \in K_n : \pi_n(s)k \neq k\}|}{|K_n|} \to 1 \quad \text{as } n \to \infty.$$

A group is *sofic* if it has a sofic approximation. For example, if Γ is residually finite with decreasing sequence $\{\Gamma_n\}$ of finite-index normal subgroups, then we can take $K_n = \Gamma/\Gamma_n$ and π_n to be the action of Γ on Γ/Γ_n by left translation. Here the π_n are honest homomorphisms, but in general they are not.

The Gromov–Weiss result is that sofic groups are surjunctive. If all groups were sofic, then Gottschalk's surjunctivity problem would have an affirmative answer, with important consequences. For instance, using operator algebra techniques, Kaplansky proved that if Γ is an arbitrary countable group, and if $f, g \in \mathbb{C}[\Gamma]$ such that $fg = 1$, then $gf = 1$. He conjectured that this remains true when $\mathbb{C}$ is replaced by an arbitrary field, a problem known today as the Kaplansky Direct Finiteness Conjecture. This conjecture can be verified if Γ is assumed surjunctive, and hence would be solved if every group were sofic. Pestov [Pes] has given a lucid survey of this and the related Connes Embedding Conjecture, together with connections to logic. The book by Capraro and Lupini [CapL] provides a more extensive account from the point of view of operator algebras. It is widely expected that non-sofic groups exist, but as of this writing no examples are known.

SOFIC ENTROPY

We defined entropy for $\mathbb{Z}$-shifts as the growth rate of the number of n-blocks as $n \to \infty$. This can be easily recast as the growth rate of the number of partial orbits of length n up to a fixed tolerance. Suppose that Γ is an amenable group with Følner sequence $\{F_n\}$. Then we can similarly define the entropy of a Γ-shift using the growth rate of partial orbits over the F_n. This was part of a general program to extend dynamics from a single transformation to actions of amenable groups, culminating in 1987 with the exhaustive account by Ornstein and Weiss [OrnW].

Near the end of their paper Ornstein and Weiss included a simple but devastating example. Let $\mathbb{F}_2$ be the free group with generators a and b. Consider the map

$$\phi \colon (\mathbb{Z}/2\mathbb{Z})^{\mathbb{F}_2} \to (\mathbb{Z}/2\mathbb{Z} \times \mathbb{Z}/2\mathbb{Z})^{\mathbb{F}_2}$$

defined by $(\phi x)_s = (x_s + x_{sa}, x_s + x_{sb})$. Both spaces are compact groups under coordinate-wise addition, and ϕ is a group homomorphism. It is easy to verify that ϕ is onto and has kernel consisting of the two constant points. At that time it was a strongly held belief that any reasonable entropy theory must assign entropy $\log n$ to the full n-shift, and also that it can only decrease under taking factors. Yet here the full 4-shift is a two-to-one factor of the full 2-shift! This single example made attempts to extend entropy beyond amenable groups appear hopeless.

But in 2008 Lewis Bowen upended this belief with a series of discoveries establishing a coherent entropy theory for sofic groups [Bow1, Bow2]. His profound insight was to shift from "internal" ways to define entropy, such as using partial orbits or partitions within the space, to "external" finite models of the action based on sofic approximations. This greater generality requires considerable technical machinery, and is often difficult to apply in concrete situations. For instance, it is a nontrivial problem to compute the entropy of the trivial identity action of a sofic group on a finite set. Bowen [Bow3] gave an introductory account of the main ideas in his address to the 2018 ICM. He has informed us that he drew inspiration from our book for arguments that he adapted to the sofic case. The book by Kerr and Li [KerL] contains an extensive account of this new viewpoint, combined with a masterful presentation of its implementation starting from the basics.

Sofic entropy has two novel aspects. In some cases there are no finite external models for a given sofic approximation, and for these sofic entropy equals $-\infty$. Also, sofic entropy might depend on the choice of sofic approximation. For amenable groups all sofic approximations give the same value. Certain degenerate examples are known, in which one sofic approximation yields a value of $-\infty$ while another gives a positive value. More significantly, Airey, Bowen, and Lin [AirBL] recently used probabilistic methods to establish the existence of different sofic approximations to a specific group and a shift of finite type over this group whose sofic entropy has distinct positive values for the two approximations.

This raises a number of intriguing questions. Does every nonamenable group admit two sofic approximations that give different values of sofic entropy for some action? Are there checkable sufficient conditions under which all sofic approximations give the same value? More concretely, what are the possible values of sofic entropy for $\mathbb{F}_r$-shifts of finite type?

Further Topics

We conclude this section by mentioning a few of the many interesting developments and problems about shifts over countable groups.

Entropies for Γ-shifts of finite type. Earlier we saw that the set $\mathcal{P}$ of entropies of $\mathbb{Z}$-shifts of finite type consists of the logarithms of weak Perron numbers, while for $\mathbb{Z}^d$ with $d \geqslant 2$ the corresponding description is the set $\mathcal{R}$ of all nonnegative right recursively enumerable numbers. What happens for general sofic groups?

Barbieri [Barb] recently developed techniques to transfer the answer for $\mathbb{Z}^2$ to some other groups. In particular, he showed that if $\Gamma = \Gamma_1 \supseteq \Gamma_2 \supseteq \cdots \supseteq \Gamma_r \supseteq \Gamma_{r+1} = \{1_\Gamma\}$ for some $r \geqslant 2$, with Γ_{j+1} normal in Γ_j and $\Gamma_j/\Gamma_{j+1} \cong \mathbb{Z}$, then the set of entropies of Γ-shifts of finite type equals $\mathcal{R}$. He also asked whether for a given finitely generated amenable group the set of such entropies must be either $\mathcal{P}$ or $\mathcal{R}$. Indeed, an intriguing problem is whether this might hold for all sofic groups.

Free shifts. A Γ-shift X is *free* (or *strongly aperiodic*) if whenever $sx = x$ for $s \in \Gamma$ and $x \in X$, then $s = 1_\Gamma$. Since $\mathbb{Z}$-shifts of finite type always contain periodic points they are never free. But there are several ways to construct free $\mathbb{Z}$-shifts, for example the Morse minimal shift. Certain Wang tile spaces provide examples of free $\mathbb{Z}^2$-shifts that even have finite type. Which groups admit free shifts, and what are their properties?

Remarkably, Gao, Jackson, and Seward [GaoJS1] showed that *every* countable group admits free shifts. They later expanded their ideas into a lengthy monograph [GaoJS2]. By using the Lovász Local Lemma from probability, Aubrun, Barbieri, and Thomassé [AubBT] gave a conceptually simpler proof of this result, and also investigated properties of free shifts. Bernshteyn [Ber] amplified this use of the Lovász Lemma to construct free Γ-shifts with alphabet $\mathcal{A}$ that are "large" in various senses, including having entropy arbitrarily close to $\log |\mathcal{A}|$ when Γ is sofic.

Suppose we ask instead: which groups admit a free shift of finite type? As mentioned above, $\mathbb{Z}$ does not while $\mathbb{Z}^2$ does. Piantadosi [Pia] proved that free nonabelian groups cannot have free shifts of finite type. The "large-scale" geometry of the group seems to play an important role here. Recall the Cayley graph of a group Γ having a finite set S of generators. Using the word metric d_S, if we remove from the Cayley graph a large ball of radius r around 1_Γ, the number of infinite connected components of the remainder stabilizes as $r \to \infty$ to a value called the number of *ends* of Γ. For instance, $\mathbb{Z}$ has two ends, $\mathbb{Z}^2$ has one end, and $\mathbb{F}_2$ has infinitely many ends. An old result due to Freudenthal and Hopf is that an infinite finitely generated group has either one, two, or infinitely many ends. Cohen [Coh] proved that a finitely generated group with at least two ends cannot have free shifts of finite type. For groups with one end the situation is more complicated. Jeandel [Jea2] has shown that polycyclic groups with one end, such as the discrete Heisenberg group, do admit free shifts of finite type. However, Jeandel has also shown in [Jea3] that finitely presented groups admitting free shifts of finite type must have a solvable word problem. There are examples of finitely presented groups with unsolvable word problem and also having only one end, and these cannot admit free shifts of finite type.

Domino problems. Wang tiles are sometimes called dominoes, and deciding whether or not a given set of Wang tiles can tile the plane is called the domino problem for $\mathbb{Z}^2$. Since every $\mathbb{Z}^2$-shift of finite type is conjugate to a Wang shift, this domino problem is just the nonemptiness problem for $\mathbb{Z}^2$-shifts of finite type. More generally, the *domino problem for* Γ asks whether there is an algorithm which, for each finite set of forbidden patterns, decides whether the resulting Γ-shift of finite type is empty or not.

Which groups have a decidable domino problem? Many groups like $\mathbb{Z}^2$ do not. Ballier and Stein [BalS] observed that earlier work on related problems shows that all free groups have a decidable domino problem. Indeed, they conjectured that a group has a decidable domino problem if and only if it contains a free subgroup with finite index, and gave some evidence for this. Their conjecture remains unsolved.

Garden of Eden theorems. The criterion developed in §8.1 for a sliding block code to be onto can be formulated over general groups. A *bubble* for a sliding block code $\phi: \mathcal{A}^\Gamma \to \mathcal{A}^\Gamma$ is a pair of distinct points differing in only finitely many coordinates having the same image under ϕ. A glance at Figure 8.1.3 shows why "bubble" is more descriptive than the term "point diamond" used there. Theorem 8.1.16 and Corollary 4.4.9 imply that a sliding block code $\phi: \mathcal{A}^\mathbb{Z} \to \mathcal{A}^\mathbb{Z}$ fails to be onto if and only if it has bubbles.

A *d-dimensional cellular automaton* is the same as a sliding block code from $\mathcal{A}^{\mathbb{Z}^d}$ to itself, and hence can be iterated. These were introduced by von Neumann to provide a theoretical model for self-reproducing machines. Their study was propelled by the immense popularity in the 1970's of Conway's Game of Life, a particular 2-dimensional cellular automaton with simple rules but whose iterations have complicated enough behavior to allow encoding of a universal Turing machine.

If a cellular automaton is not onto, a configuration outside of its image is called a *Garden of Eden*, a biblical allusion coined by Tukey for an idyllic state to which we can never return. The criterion mentioned above was extended to $\mathbb{Z}^d$ to show that a d-dimensional cellular automaton has Gardens of Eden if and only if it has bubbles, by Moore (necessity) and Myhill (sufficiency). Ceccherini-Silberstein, Machì, and Scarabotti [CecMS] extended this further to all amenable groups. Gromov [Gro] gave an alternative proof for amenable groups as part of his study of the surjunctivity problem.

There are groups for which this criterion fails, however. Let $\mathbb{F}_2$ be the free group generated by a and b. We will define a sliding block code $\phi \colon \{0,1\}^{\mathbb{F}_2} \to \{0,1\}^{\mathbb{F}_2}$ using "majority rule." For $x \in \{0,1\}^{\mathbb{F}_2}$ and $s \in \mathbb{F}_2$, look at the coordinates of x at sa, sa^{-1}, sb, and sb^{-1}. If three or four of them are 0, put $(\phi x)_s = 0$; if three or four of them are 1, put $(\phi x)_s = 1$; in case of a tie, put $(\phi x)_s = x_s$. The point of all 0's and a point with a single 1 clearly form a bubble for ϕ, but it is easy to show that ϕ is still onto [CecC1, §5.10].

Indeed, very recently Bartholdi [Bart] showed that for every nonamenable group there are onto sliding block codes with bubbles (as in the example above) and also sliding block codes that are not onto and yet have no bubbles. Together with [CecMS], this shows that amenability of a group is characterized by the equivalence of having no bubbles and being onto for sliding block codes.

The elegant book by Ceccherini-Silberstein and Coornaert [CecC1] provides a self-contained account of these ideas and much more.

§A.9. Symbolic Representations of Algebraic Actions

In §6.5 we described how to obtain a symbolic representation of an invertible dynamical system using partitions. In Example 6.5.10 we exhibited a Markov partition for a toral automorphism A of $\mathbb{T}^2$, where $\mathbb{T}$ denotes $\mathbb{R}/\mathbb{Z}$, together with its corresponding shift of finite type X. By Proposition 6.5.8 there is a factor map $\phi \colon (X, \sigma) \to (\mathbb{T}^2, A)$ that turns out to be one-to-one except over the union of the projections to $\mathbb{T}^2$ of the two one-dimensional eigenspaces of A. In the terminology of Definition 9.3.8, the map ϕ is almost invertible, and hence "most" points in $\mathbb{T}^2$ have a unique symbolic representation. Such representations are extremely useful, allowing us to study dynamical properties of the system represented using the much simpler and combinatorial structures of shifts of finite type or other shift spaces.

All hyperbolic toral automorphisms have Markov partitions. Adler and Marcus [AdlM*] combined this fact with Theorem 9.3.2 to show that two hyperbolic toral automorphisms (possibly on tori of different dimensions) are almost conjugate if and only if they have the same entropy.

The geometric approach originally used to obtain Markov partitions for hyperbolic toral automorphisms ignored the underlying group structure of $\mathbb{T}^n$, enabling it to be greatly extended to smooth maps of manifolds having hyperbolic behavior. But could the group structure also be used, perhaps allowing generalizations of symbolic representations to other algebraic settings, including actions of groups more general than $\mathbb{Z}$?

The first hint came in Vershik's 1992 paper [Vers], in which he obtained a Markov partition for the toral automorphism $A = \begin{bmatrix} 0 & 1 \\ 1 & 1 \end{bmatrix}$ by purely algebraic methods (this matrix is obtained from the one used in Example 6.5.10 by simply swapping coordinates). We will recast Vershik's result using an algebraic framework capable of great generality.

Let us begin with the group $T = \mathbb{T}^{\mathbb{Z}}$ and its shift σ. Operations on elements $t = (t_n) \in T$ are more conveniently expressed by identifying t with a formal bi-infinite series $t(z) = \sum t_n z^n$. For instance, $(\sigma t)(z) = z^{-1} t(z)$.

Let $f(z) = z^2 - z - 1$, the characteristic polynomial of A, and define $f^*(z) = f(z^{-1})$. For $t \in T$ the product $t(z) f^*(z)$ is well-defined, and we put

$$T_f = \{ t \in T : t(z) f^*(z) = 0 \}$$
$$= \{ t \in T : t_{n+2} - t_{n+1} - t_n = 0 \quad \text{for all } n \in \mathbb{Z} \},$$

and denote the restriction of σ to T_f by α_f. It is easy to check that the map $\psi : T_f \to \mathbb{T}^2$ given by $\psi(t) = \begin{bmatrix} t_0 \\ t_1 \end{bmatrix}$ is a group isomorphism, and also

$$\psi(\sigma(t)) = \begin{bmatrix} t_1 \\ t_2 \end{bmatrix} = \begin{bmatrix} t_1 \\ t_0 + t_1 \end{bmatrix} = A \begin{bmatrix} t_0 \\ t_1 \end{bmatrix}.$$

Thus (T_f, α_f) is just an alternative way of describing the toral automorphism A on $\mathbb{T}^2$.

Let $q : \mathbb{R} \to \mathbb{R}/\mathbb{Z} = \mathbb{T}$ be the quotient map, and choose the cross-section $q^{-1} : \mathbb{T} \to (-1, 0]$. We can apply q^{-1} to $t \in T_f$ coordinate-wise. Observe that $q((q^{-1}t)(z) f^*(z)) = t(z) f^*(z) = 0$, and so $(q^{-1}t)(z) f^*(z)$ must have integer coordinates. A simple estimate shows that these can be only 0 or 1, and thus $(q^{-1}t)(z) f^*(z) \in \{0, 1\}^{\mathbb{Z}}$. To return to t from this symbolic representation, we would like to multiply by $1/f^*(z)$ and apply q. We can make sense of this as follows.

Let $f(z) = (z - \lambda)(z - \mu)$, where $\lambda = (1 + \sqrt{5})/2$ and $\mu = (1 - \sqrt{5})/2$. Using partial fractions and geometric series, we obtain that

$$w^{\Delta}(z) := \frac{1}{f^*(z)} = \frac{1/\sqrt{5}}{z^{-1} - \lambda} - \frac{1/\sqrt{5}}{z^{-1} - \mu}$$
$$= -\frac{1}{\lambda \sqrt{5}} \left(1 + \lambda^{-1} z^{-1} + \lambda^{-2} z^{-2} + \dots \right) - \frac{z}{\sqrt{5}} \left(1 + \mu z + \mu^2 z^2 + \dots \right).$$

Clearly $w^{\Delta} \in \ell^1(\mathbb{Z}, \mathbb{R})$. Hence if $x \in \{0, 1\}^{\mathbb{Z}}$, then $x(z) w^{\Delta}(z)$ is well-defined since each coordinate is a convergent series. Let $t^{\Delta} = q(w^{\Delta})$.

Then $t^\Delta \in T_f$ since $t^\Delta(z)f^*(z) = q(w^\Delta(z)f^*(z)) = q(1) = 0$. Hence we can define the continuous shift-commuting map $\phi\colon \{0,1\}^{\mathbb{Z}} \to T_f$ by $\phi(x) = q(x(z)w^\Delta(z)) = x(z)t^\Delta(z)$. Furthermore, if $t \in T_f$ then

$$\phi\big((q^{-1}t)(z)f^*(z)\big) = q\big((q^{-1}t)(z)f^*(z)f^*(z)^{-1}\big) = t(z),$$

so that ϕ is onto. In other words, if we "lift" a point $t \in T_f$ to have real coordinates and convolve with f^* to obtain a point with integer coordinates, the result is mapped back to t by applying ϕ.

However, ϕ has bubbles. For instance, if $x_{[-2,0]} = 011$, form the point x' by replacing this block with 100. Since $x'(z) - x(z) = f^*(z)$, it follows that $\phi(x) = \phi(x')$, and so (x, x') is a bubble. We can repeat this to remove bubbles with larger strings of consecutive 1's, for example $01111 \to 10011 \to 10100$. Let $X \subseteq \{0,1\}^{\mathbb{Z}}$ be the golden mean shift obtained by forbidding the block 11. The replacement procedure above will eventually convert any point $x \in \{0,1\}^{\mathbb{Z}}$ having infinitely many 0's in each direction into a point $x' \in X$ with $\phi(x') = \phi(x)$. It turns out that with a bit more work one can show that the restriction of ϕ to X is onto, finite-to-one, and one-to-one on the doubly transitive points, and hence an almost invertible symbolic representation of (T_f, α_f), and thus also of $(\mathbb{T}^2, A)$.

This method can be adapted to general $f(z) \in \mathbb{Z}[z^{\pm 1}]$ provided that $f(z)$ does not vanish on $\mathbb{S} = \{z \in \mathbb{C} : |z| = 1\}$. For then $1/f^*(e^{2\pi i\theta})$ is a smooth function of $\theta \in [0,1]$, so its Fourier series $\sum c_n e^{2\pi i n\theta}$ has coefficients decaying rapidly to 0 as $|n| \to \infty$, and so we can use $w^\Delta(z) = \sum c_n z^n$ as before. Symbolic representations of hyperbolic automorphisms of $\mathbb{T}^2$ were obtained algebraically by Siderov and Vershik [SidV], and for those of arbitrary dimensional tori by Kenyon and Vershik [KenV]. Bubble removal, which amounts to finding efficient digit expansions using an algebraic number as base, is more complicated, and results in symbolic representations by shifts that are in general only sofic rather than of finite type.

However, if f does vanish on $\mathbb{S}$ then this method cannot even get off the ground since there is no suitable way to interpret $1/f^*(z)$. Indeed any attempt to find a symbolic representation using a mixing shift of finite type X is doomed to fail, since any continuous shift-commuting map from such an X to T_f must map all of X to a single point [Sch2, Cor. 5.3]. There are some one-sided analogues of t^Δ in this case, leading to subtler kinds of symbolic systems, but their properties are not well understood [Sch2, Sch3].

Our framework for describing Vershik's example can be extended verbatim to arbitrary countable groups Γ. Let $\mathbb{Z}[\Gamma]$ denote the integral group ring of Γ, consisting of all finite formal sums $f = \sum_{s \in \Gamma} f(s)s$, where $f(s) \in \mathbb{Z}$ and $f(s) = 0$ for all but finitely many $s \in \Gamma$. Put $f^* = \sum f(s)s^{-1}$. Similarly, let $T = \mathbb{T}^\Gamma$, and identify $t \in T$ with a formal series $t = \sum_{s \in \Gamma} t(s)s$. Then the product $t \cdot f^*$ is well-defined, and we put $T_f = \{t \in T : t \cdot f^* = 0\}$.

Also, Γ acts on T by left multiplication via $s't = \sum_s t(s)s's$, and since T_f is defined using right multiplication it is preserved by this action. The result is a homomorphism α_f from Γ to the group aut(T_f) of continuous automorphisms of T_f. The pair (T_f, α_f) is called the *principal algebraic Γ-action defined by f*. More generally, an *algebraic Γ-action* is a pair (Y, α), where Y is a compact abelian group and $\alpha\colon \Gamma \to \mathrm{aut}(Y)$ is a homomorphism.

When $\Gamma = \mathbb{Z}^d$ the powerful tools of commutative algebra combined with Pontryagin duality can be applied. The book [Sch3*] contains an extensive account of a dynamics $\leftrightarrow$ algebra "dictionary" for algebraic $\mathbb{Z}^d$-actions, translating dynamical properties into equivalent algebraic properties. To cite just one example, the proof that a mixing $\mathbb{Z}^2$-action generated by two commuting toral automorphism is also mixing of all orders depends on a deep result about additive relations in fields [Sch3*, Prop. 2.7.8].

There is a natural identification of $\mathbb{Z}[\mathbb{Z}^d]$ with the Laurent polynomial ring $\mathbb{Z}[z_1^{\pm 1}, \ldots, z_d^{\pm 1}]$. If $f \in \mathbb{Z}[\mathbb{Z}^d]$ does not vanish on $\mathbb{S}^d$, we can again use the Fourier coefficients of $1/f^*$ to define $w^\Delta \in \ell^1(\mathbb{Z}^d, \mathbb{R})$ and its projection to $x^\Delta \in T_f$. Such points are called *homoclinic*, and are basic tools for algebraic actions [LinS1]. We can use x^Δ to define a symbolic cover as before, but bubble elimination is much more complicated. In general, Einsiedler and Schmidt [EinS] were only able to obtain a $\mathbb{Z}^d$-shift of equal entropy. But in certain cases, such as $f = 3 - z_1 - z_2$, they showed via percolation arguments that this process yields a $\mathbb{Z}^2$-shift of finite type that is almost conjugate to (T_f, α_f).

As soon as we pass to noncommutative Γ, many of the tools used for $\mathbb{Z}^d$ no longer work. The simplest case is the discrete Heisenberg group, generated by a, b, and c together with relations $ac = ca$, $bc = cb$, and $ba = abc$. The recent survey [LinS2] explains what is known about algebraic actions of this group, and describes many open problems.

For one more example of symbolic representations, let $\Gamma = \mathbb{F}_2$, the free group with generators a and b, and let $f = 3 - a - b$. In [LinS3] it is shown that the map $\phi\colon \{0, 1, 2\}^{\mathbb{F}_2} \to T_f$ using $1/f^*$ as described above is a symbolic representation. Furthermore, if $\{0, 1, 2\}^{\mathbb{F}_2}$ is equipped with product measure with each symbol having measure $1/3$, then ϕ is a measure-preserving isomorphism to T_f equipped with Haar measure. This principal algebraic action (T_f, α_f) is the first example of a measure-preserving action of $\mathbb{F}_2$ whose Bernoullicity is not apparent.

The study of algebraic actions is extensive and expanding, with deep connections to operator algebras. The paper by Li and Thom [LiT] and the book by Kerr and Li [KerL] are excellent sources of more information.

The story of computing entropy for principal algebraic actions is particularly compelling. For $f(z) = \prod_{j=1}^n (z - \lambda_j) \in \mathbb{Z}[\mathbb{Z}]$, Yuzvinskii [Yuz] proved in 1967 that $h(\alpha_f) = \sum_j \log^+ |\lambda_j|$, where $\log^+ u = \max\{0, \log u\}$. For $0 \neq f(z_1, \ldots, z_d) \in \mathbb{Z}[\mathbb{Z}^d]$, Lind, Schmidt, and Ward [LinSW*] showed that

$$h(\alpha_f) = \int_0^1 \cdots \int_0^1 \log|f(e^{2\pi i u_1}, \ldots, e^{2\pi i u_d})| \, du_1 \ldots du_d,$$

a quantity called the *logarithmic Mahler measure* of f, and well-known to number theorists. This formula reduces to Yuzvinskii's result when $d = 1$ via Jensen's formula [Mah]. When $f(z_1, z_2) = 4 - z_1 - z_2 - z_1^{-1} - z_2^{-1}$, then $h(\alpha_f)$ is four times the entropy of the domino shift mentioned previously, although this is mysterious since there is no known structural connection between these two dynamical systems.

From this explicit formula one might conclude that it is possible to characterize the set of possible entropies of algebraic $\mathbb{Z}^d$-actions, as we have already seen is the case with $\mathbb{Z}^d$-shifts of finite type. There are just two possible answers, either a countable subset of $[0, \infty]$ or all of $[0, \infty]$, but we don't know which is correct! According to [LinSW*, Thm. 4.6], deciding this is equivalent to a problem of Lehmer from 1932, which in our language amounts to knowing whether, given an arbitrary $\epsilon > 0$, there is an $f \in \mathbb{Z}[\mathbb{Z}]$ with $0 < h(\alpha_f) < \epsilon$. The smallest known value, discovered by Lehmer himself, is for $f(z) = z^{10} + z^9 - z^7 - z^6 - z^5 - z^4 - z^3 + z + 1$. Here $h(\alpha_f) = \log \lambda$, where $\lambda \approx 1.17628$ is the only root of f outside the unit circle. There is a huge literature on this problem, including extensive computations [MosRW] showing that Lehmer's polynomial gives the smallest positive value among all polynomials of degree at most 54.

If Γ is not commutative there is no obvious candidate to replace Mahler measure. But in 2006, Deninger [Den] noticed that Mahler measure is a special case of a notion from operator theory called the Fuglede–Kadison determinant. Using this, he was able to conjecture a general formula.

To motivate the definition for this notion, first consider an $n \times n$ complex matrix C with eigenvalues $\lambda_1, \ldots, \lambda_n$. Then C^*C is self-adjoint with eigenvalues $|\lambda_1|^2, \ldots, |\lambda_n|^2$. The spectral measure of the operator C^*C provided by functional analysis is then $\mu_{C^*C} = \frac{1}{n} \sum_{j=1}^n \delta_{|\lambda_j|^2}$, where δ_r denotes the point mass at r. Hence we can compute the (normalized) absolute value of the determinant of C by

$$|\det C|^{1/n} = \prod_{j=1}^n |\lambda_j|^{1/n} = \exp\left[\frac{1}{2} \int_0^\infty \ln t \, d\mu_{C^*C}(t)\right].$$

Now let $f \in \mathbb{Z}[\Gamma]$. There is an associated bounded operator C_f on the Hilbert space $\ell^2(\Gamma)$ given by convolution by f. Then $C_f^*C_f$ is a self-adjoint operator, so by standard functional analysis it has a spectral measure $\mu_{C_f^*C_f}$ on $[0, \infty)$. The *Fuglede–Kadison determinant* of C_f is

$$\det C_f = \exp\left[\frac{1}{2} \int_0^\infty \ln t \, d\mu_{C_f^*C_f}(t)\right].$$

If $0 \neq f \in \mathbb{Z}[\mathbb{Z}^d]$, it is easy to check that $h(\alpha_f) = \log \det C_f$. Deninger suggested this might apply to all amenable groups, and proved it for a restricted class. After a succession of intermediate results by several authors, Li [Li] and Li and Thom [LiT] proved Deninger's conjecture (and more) for all amenable groups. One consequence is that $h(\alpha_f) = h(\alpha_{f^*})$, which is highly nontrivial since there is no direct dynamical connection between α_f and α_{f^*}.

The invention of sofic entropy opened the door for further generalization, culminating in a tour de force of analysis by Hayes [Hay]. He proved that if Γ is a sofic group, and if $f \in \mathbb{Z}[\Gamma]$ has the property that multiplication by f on $\mathbb{Z}[\Gamma]$ is injective, then the sofic entropy of α_f is independent of sofic approximation and $h(\alpha_f) = \log \det C_f$. It was previously known that for those f not satisfying this assumption the entropy is infinite, and so this gives a complete answer in the most general case.

We have barely scratched the surface of the rich interplay between algebraic actions, duality theory, algebraic geometry, and operator algebras. Even principal algebraic actions provide an endless supply of interesting examples whose dynamical properties are yet to be fully explored.

BIBLIOGRAPHY

[Adl1] R. Adler, *The torus and the disk*, IBM J. of Research and Development **31** (1987), 224–234.

[Adl2] R. Adler, *Symbolic dynamics and Markov partitions*, Bull AMS **35** (1998), 1–56.

[AdlCH] R. L. Adler, D. Coppersmith, and M. Hassner, *Algorithms for sliding block codes – an application of symbolic dynamics to information theory*, IEEE Trans. Inform. Theory **29** (1983), 5–22.

[AdlF] R. Adler and L. Flatto, *Geodesic flows, interval maps, and symbolic dynamics*, Bull. Amer. Math. Soc. **25** (1991), 229–334.

[AdlFKM] R. L. Adler, J. Friedman, B. Kitchens, and B. H. Marcus, *State splitting for variable-length graphs*, IEEE Trans. Inform. Theory **32** (1986), 108–113.

[AdlGW] R. L. Adler, L. W. Goodwyn, and B. Weiss, *Equivalence of topological Markov shifts*, Israel J. Math. **27** (1977), 48–63.

[AdlHM] R. L. Adler, M. Hassner, and J. Moussouris, *Method and apparatus for generating a noiseless sliding block code for a* $(1, 7)$ *channel with rate 2/3*, US Patent 4,413,251 (1982).

[AdlKM] R. Adler, B. Kitchens, and B. Marcus, *Finite group actions on subshifts of finite type*, Ergod. Th. & Dynam. Sys. **5** (1985), 1–25.

[AdlKoM] R. Adler, A. Konheim, and M. McAndrew, *Topological entropy*, Trans. Amer. Math. Soc. **114** (1965), 309–319.

[AdlM] R. Adler and B. Marcus, *Topological Entropy and Equivalence of Dynamical Systems*, Mem. Amer. Math. Soc. **219**, 1979.

[AdlW] R. Adler and B. Weiss, *Similarity of Automorphisms of the Torus*, Mem. Amer. Math. Soc. **98**, 1970.

[AhoHU] A. V. Aho, J. E. Hopcroft, and J. D. Ullman, *The Design and Analysis of Computer Algorithms*, Addison-Wesley, Reading, MA, 1974.

[AigZ] M. Aigner and G. Ziegler, *Proofs from the Book*, Springer, 6th ed., 2018.

[ArrP] D. Arrowsmith and L. Place, *An Introduction to Dynamical Systems*, Cambridge Univ. Press, Cambridge, 1990.

[Art] E. Artin, *Ein mechanisches System mit quasiergodishen Baknen*, Abhandlungen aus dem Mathematishen Seminar der Hamburgishen Universitat 3 (1924), 170–175.

[ArtM] M. Artin and B. Mazur, *On periodic points*, Annals of Math. **81** (1965), 82–99.

[Ash1] J. J. Ashley, *On the Perron–Frobenius eigenvector for non-negative integral matrices whose largest eigenvalue is integral*, Linear Algebra Appl. **94** (1987), 103–108.

[Ash2] J. J. Ashley, *Performance bounds in constrained sequence coding*, Ph.D. Thesis, University of California, Santa Cruz (1987).

[Ash3] J. J. Ashley, *A linear bound for sliding block decoder window size*, IEEE Trans. Inform. Theory **34** (1988), 389–399.

[Ash4] J. J. Ashley, *Marker automorphisms of the one-sided d-shift*, Ergod. Th. & Dynam. Sys. **10** (1990), 247–262.

[Ash5] J. J. Ashley, *Bounded-to-1 factors of an aperiodic shift of finite type are 1-to-1 almost everywhere factors also*, Ergod. Th. & Dynam. Sys. **10** (1990), 615–626.

[Ash6] J. J. Ashley, *Resolving factor maps for shifts of finite type with equal entropy*, Ergod. Th. & Dynam. Sys. **11** (1991), 219–240.

[Ash7] J. J. Ashley, *An extension theorem for closing maps of shifts of finite type*, Trans. Amer. Math. Soc. **336** (1993), 389–420.

[AshB] J. J. Ashley and M.-P. Béal, *A note on the method of poles for code construction*, IEEE Trans. Inform. Theory **40** (1994), 512-517.

[AshM] J. J. Ashley and B. H. Marcus, *Canonical encoders for sliding block decoders*, SIAM J. Discrete Math **8** (1995), 555-605.

[AshMR1] J. J. Ashley, B. H. Marcus, and R. M. Roth, *Construction of encoders with small decoding look-ahead for input-constrained channels*, IEEE Trans. Inform. Theory **41** (1995), 55–76.

[AshMR2] J. J. Ashley, B. H. Marcus, and R. M. Roth, *On the decoding delay of encoders for input-constrained channels*, IEEE Trans Info Theory **42** (1996), 1948–1956.

[AshMT] J. J. Ashley, B. H. Marcus, and S. Tuncel, *The classification of one-sided Markov chains*, Ergod. Th. & Dynam. Sys. **17** (1997), 269–295.

[Bak1] K. Baker, *Strong shift equivalence of non-negative integer 2×2 matrices*, Ergod. Th. & Dynam. Sys. **3** (1983), 501–508.

[Bak2] K. Baker, *Strong shift equivalence and shear adjacency of nonnegative square integer matrices*, Linear Algebra Appl. **93** (1987), 131–147.

[Bea1] M.-P. Béal, *Codes circulaires, automates locaux et entropie*, Theor. Computer Sci. **57** (1988), 283–302.

[Bea2] M.-P. Béal, *The method of poles: a coding method for constrained channels*, IEEE Trans. Inform. Theory **IT-36** (1990), 763–772.

[Bea3] M.-P. Béal, *Codage Symbolique*, Masson, Paris, 1993.

[Bed] T. Bedford, *Generating special Markov partitions for hyperbolic toral automorphisms using fractals*, Ergod. Th. & Dynam. Sys. **6** (1986), 325–333.

[BelH] M. Belongie and C. Heegard, *Run length limited codes with coding gains from variable length graphs*, preprint.

[Berg] K. Berg, *On the conjugacy problem for K-systems*, PhD Thesis, University of Minnesota, 1967.

[Ber] R. Berger, *The Undecidability of the Domino Problem*, Mem. Amer. Math. Soc. **66**, 1966.

[BerP] J. Berstel and D. Perrin, *Theory of Codes*, Academic Press, New York, 1985.

[Bil] P. Billingsley, *Probability and Measure*, Wiley, New York, 1979.

[Bir1] G. D. Birkhoff, *Quelques theorems sur le mouvement des systems dynamiques*, Bull. Soc. Math. de France **40** (1912), 305–323.

[Bir2] G. D. Birkhoff, *Dynamical systems*, Amer. Math. Soc. Colloquium Publ. **9** (1927).

[BirM] G. Birkhoff and S. Maclane, *Algebra*, Macmillan, 2nd ed., New York, 1979.

[Bla] R. E. Blahut, *Digital Transmission of Information*, Addison-Wesley, Reading, Massachusetts, 1990.

[BlaH1] F. Blanchard and G. Hansel, *Systemes codés*, Theor. Computer Sci. **44** (1986), 17–49.

[BlaH2] F. Blanchard and G. Hansel, *Sofic constant-to-one extensions of subshifts of finite type*, Proc. Amer. Math. Soc. **112** (1991), 259–265.

[BloC] L. S. Block and W. S. Coppel, *Dynamics in One Dimension*, Springer–Verlag Lecture Notes in Mathematics **1513**, New York, 1992.

[BonM] J. Bondy and U. Murty, *Graph Theory with Applications*, Macmillan, London, 1976.

[Boo] W. M. Boothby, *An Introduction to Differentiable Manifolds and Riemannian Geometry*, Academic Press, Orlando, 1986.

[Bow1] R. Bowen, *Markov partitions for Axiom A diffeomorphisms*, Amer. J. Math. **92** (1970), 725–747.

[Bow2] R. Bowen, *Markov partitions and minimal sets for Axiom A diffeomorphisms*, Amer. J. Math. **92** (1970), 907–918.

[Bow3] R. Bowen, *Entropy for group endomorphisms and homogeneous spaces*, Trans. Amer. Math. Soc. **153** (1971), 401–414; Errata: 181(1973), 509-510.

[**Bow4**] R. Bowen, *Symbolic dynamics for hyperbolic flows*, Amer. J. Math. **95** (1973), 429-460.

[**Bow5**] R. Bowen, *Smooth partitions of Anosov diffeomorphisms are weak Bernoulli*, Israel J. Math. **21** (1975), 95–100.

[**Bow6**] R. Bowen, *Equilibrium States and the Ergodic Theory of Anosov Diffeomorphisms*, Springer Lecture Notes in Math. **470**, New York, 1975.

[**Bow7**] R. Bowen, *On Axiom A Diffeomorphisms*, AMS–CBMS Reg. Conf. **35**, Providence, 1978.

[**Bow8**] R. Bowen, *Markov partitions are not smooth*, Proc. Amer. Math. Soc. **71** (1978), 130–132.

[**BowF**] R. Bowen and J. Franks, *Homology for zero-dimensional basic sets*, Annals of Math. **106** (1977), 73–92.

[**BowL**] R. Bowen and O. E. Lanford, *Zeta functions of restrictions of the shift transformation*, Proc. Symp. Pure Math. A.M.S. **14** (1970), 43–50.

[**BowW**] R. Bowen and P. Walters, *Expansive one-parameter flows*, J. Diff. Eqn. **12** (1972), 180-193.

[**Boy1**] M. Boyle, *Lower entropy factors of sofic systems*, Ergod. Th. & Dynam. Sys. **3** (1983), 541–557.

[**Boy2**] M. Boyle, *Shift equivalence and the Jordan form away from zero*, Ergod. Th. & Dynam. Sys. **4** (1984), 367–379.

[**Boy3**] M. Boyle, *Constraints on the degree of a sofic homomorphism and the induced multiplication of measures on unstable sets*, Israel J. Math **53** (1986), 52–68.

[**Boy4**] M. Boyle, *Almost flow equivalence for hyperbolic basic sets*, Topology **31** (1992), 857–864.

[**Boy5**] M. Boyle, *Symbolic dynamics and matrices*, in *Combinatorial and Graph Theoretic Problems in Linear Algebra*, IMA Volumes in Math. and its Appl. (ed. R. Brualdi, et. al.) **50**, 1993, pp. 1–38.

[**Boy6**] M. Boyle, Unpublished lecture notes.

[**Boy7**] M. Boyle, *Factoring factor maps*, J. London Math. Soc. **57** (1998), 491-502.

[**BoyF**] M. Boyle and U.-R. Fiebig, *The action of inert finite order automorphisms on finite subsystems of the shift*, Ergod. Th. & Dynam. Sys. **11** (1991), 413–425.

[**BoyFK**] M. Boyle, J. Franks, and B. Kitchens, *Automorphisms of one-sided subshifts of finite type*, Ergod. Th. & Dynam. Sys. **10** (1990), 421–449.

[**BoyH1**] M. Boyle and D. Handelman, *The spectra of nonnegative matrices via symbolic dynamics*, Annals of Math. **133** (1991), 249–316.

[**BoyH2**] M. Boyle and D. Handelman, *Algebraic shift equivalence and primitive matrices*, Trans. Amer. Math. Soc. **336** (1993), 121–149.

[**BoyKM**] M. Boyle, B. Kitchens, and B. H. Marcus, *A note on minimal covers for sofic systems*, Proc. Amer. Math. Soc. **95** (1985), 403–411.

[**BoyK1**] M. Boyle and W. Krieger, *Periodic points and automorphisms of the shift*, Trans. Amer. Math. Soc. **302** (1987), 125–149.

[**BoyK2**] M. Boyle and W. Krieger, *Almost Markov and shift equivalent sofic systems*, Proceedings of Maryland Special Year in Dynamics 1986–87, Springer-Verlag Lecture Notes in Math **1342** (1988), 33–93.

[**BoyK3**] M. Boyle and W. Krieger, *Automorphisms and subsystems of the shift*, J. Reine Angew. Math. **437** (1993), 13–28.

[**BoyLR**] M. Boyle, D. Lind, and D. Rudolph, *The automorphism group of a shift of finite type*, Trans. Amer. Math. Soc. **306** (1988), 71–114.

[**BoyMT**] M. Boyle, B. H. Marcus, and P. Trow, *Resolving Maps and the Dimension Group for Shifts of Finite Type*, Mem. Amer. Math. Soc. **377**, 1987.

[**BunS**] L. A. Bunimovitch and Y. G. Sinai, *Markov partitions for dispersed billiards*, Comm. Math. Phys. **73** (1980), 247–280; erratum: 107(1986), 357–358.

[BunSC] L. A. Bunimovitch, Y. G. Sinai, and N. Chernov, *Markov partitions for two-dimensional billiards*, Uspekhi Mat. Nauk **45, no. 3** (1990), 97–134; Russian Math. Surveys, 45 (no.3) (1990), 105-152.

[Caw] E. Cawley, *Smooth Markov partitions and toral automorphisms*, Ergod. Th. & Dynam. Sys. **11** (1991), 633–651.

[Cer] J. Cerny, *Poznamka K homogenym s Konoecnymi automati*, Mat.–fyz. Cas. SAV. **14** (1964), 208–215.

[ColE] P. Collet and J.-P. Eckmann, *Iterated Maps of the Interval as Dynamical Systems*, Birkhauser, Basel, 1981.

[CorC] A. Cordon and M. Crochemore, *Partitioning a graph in* $O(|A| \log_2 |V|)$, Theor. Computer Sci. **19** (1982), 85–98.

[CorG] D. Corneil and C. Gottlieb, *An efficient algorithm for graph isomorphism*, J. Assoc. Comput. Mach. **17** (1970), 51–64.

[Cov1] E. Coven, *Endomorphisms of substitution minimal sets*, Z. Wahrscheinlichkeitstheorie **20** (1971), 129–133.

[Cov2] E. Coven, *Topological entropy of block maps*, Proc. Amer. Math. Soc. **78** (1980), 590–594.

[CovH] E. Coven and G. A. Hedlund, *Sequences with minimal block growth*, Math. Systems Theory **7** (1973), 138–153.

[CovK] E. M. Coven and M. S. Keane, *The structure of substitution minimal sets*, Trans. Amer. Math. Soc. **162** (1971), 89–102.

[CovP1] E. Coven and M. Paul, *Endomorphisms of irreducible shifts of finite type*, Math. Systems Theory **8** (1974), 167–175.

[CovP2] E. Coven and M. Paul, *Sofic systems*, Israel J. Math. **20** (1975), 165–177.

[CovP3] E. Coven and M. Paul, *Finite procedures for sofic systems*, Monats. Math. **83** (1977), 265–278.

[CovT] T. Cover and J. Thomas, *Elements of Information Theory*, Wiley, New York, 1991.

[CulKK] Karel Culik II, Juhani Karhumäki, and Jarkko Kari, *A note on synchronized automata and road coloring problem*, Internat. J. Found. Comput. Sci **13** (2002), 459–471.

[Cun] J. Cuntz, *A class of C*-algebras and topological Markov chains II: reducible chains and the Ext-functor for C*-algebras*, Inventiones Math. **63** (1981), 25–40.

[CunK1] J. Cuntz and W. Krieger, *Topological Markov chains with dicyclic dimension groups*, J. fur Reine und Angew. Math. **320** (1980), 44–51.

[CunK2] J. Cuntz and W. Krieger, *A class of C*-Algebras and topological Markov chains*, Inventiones Math. **56** (1980), 251–268.

[DasJ] D. A. Dastjerdi and S. Jangjoo, *Dynamics and topology of S-gap shifts*, Topology and its Applications **159** (2012), 2654–2661.

[DekK] M. Dekking and M. Keane, *Mixing properties of substitutions*, Z. Wahrscheinlichkeitstheorie **42** (1978), 23–33.

[DemGJ] J. Demongeot, E. Goler, and M. Jchisente, *Dynamical Systems and Cellular Automata*, Academic Press, Orlando, 1985.

[DenGS] M. Denker, C. Grillenberger, and K. Sigmund, *Ergodic Theory on Compact Spaces*, Springer-Verlag Lecture Notes in Math. **527**, New York, 1976.

[Dev] R. Devaney, *An Introduction to Chaotic Dynamical Systems*, Addison-Wesley, Reading, MA, 1987.

[Dug] J. Dugundji, *Topology*, Allyn and Bacon, Boston, 1970.

[Edg] G. A. Edgar, *Measure, Topology, and Fractal Geometry*, Springer, New York, 1990.

[Edw] Harold Edwards, *Riemann's Zeta-Function*, Academic Press, New York, 1974.

[**Eff**] E. Effros, *Dimensions and C*-algebras*, AMS– CBMS Reg. Conf. **46**, Providence, 1981.

[**Ell**] R. Ellis, *Lectures on Topological Dynamics*, Benjamin, New York, 1969.

[**Eve1**] S. Even, *On information lossless automata of finite order*, IEEE Trans. Elect. Comput. **14** (1965), 561–569.

[**Eve2**] S. Even, *Graph Algorithms*, Computer Science Press, Potomac, Maryland, 1979.

[**FarTW**] D. Farmer, T. Toffoli, and S. Wolfram, *Cellular Automata, Proceedings of Interdisciplinary Workshop (Los Alamos, 1983)*, North–Holland, Eindhoven, 1984.

[**FieD1**] D. Fiebig, *Common closing extensions and finitary regular isomorphism for synchronized systems*, in *Symbolic Dynamics and Its Applications*, Contemporary Mathematics **135** (ed. P. Walters), Providence, 1992, pp. 125–138.

[**FieD2**] D. Fiebig, *Common extensions and hyperbolic factor maps for coded systems*, Ergod. Th. & Dynam. Sys. **15** (1995), 517–534.

[**FieD3**] D. Fiebig, Personal communication.

[**FieF**] D. Fiebig and U.-R. Fiebig, *Covers for coded systems*, in *Symbolic Dynamics and Its Applications*, Contemporary Mathematics **135** (ed. P. Walters), Providence, 1992, pp. 139–180.

[**Fin**] D. Finkbeiner, *An Introduction to Vector Spaces and Linear Transformations*, Freeman, San Francisco, 1966.

[**Fis1**] R. Fischer, *Sofic systems and graphs*, Monats. Math. **80** (1975), 179–186.

[**Fis2**] R. Fischer, *Graphs and symbolic dynamics*, Colloq. Math. Soc. János Bólyai: Topics in Information Theory **16** (1975), 229–243.

[**For1**] G. D. Forney, Jr., *Convolutional codes I: Algebraic structure*, IEEE Trans. Inform. Theory **IT-16** (1970), 720–738.

[**For2**] G. D. Forney, Jr., *Algebraic structure of convolutional codes, and algebraic system theory*, in *Mathematical System Theory*, (ed. by A. C. Antoulas), Springer-Verlag, 1991, pp. 527–558.

[**ForT**] G. D. Forney, Jr. and M. Trott, *The dynamics of linear codes over groups: state spaces, trellis diagrams and canonical encoders*, IEEE Trans. Inform. Theory **39** (1993), 1491–1513.

[**Fra1**] P. A. Franaszek, *Sequence-state coding for digital transmission*, Bell Sys. Tech. J. **47** (1968), 143–155.

[**Fra2**] P. A. Franaszek, *On synchronous variable length coding for discrete noiseless channels*, Inform. Control **15** (1969), 155–164.

[**Fra3**] P. A. Franaszek, *Sequence-state methods for run-length-limited coding*, IBM J. of Research and Development **14** (1970), 376–383.

[**Fra4**] P. A. Franaszek, *Run-length-limited variable length coding with error propagation limitation*, US Patent 3,689,899 (1972).

[**Fra5**] P. A. Franaszek, *On future-dependent block coding for input-restricted channels*, IBM J. of Research and Development **23** (1979), 75–81.

[**Fra6**] P. A. Franaszek, *Synchronous bounded delay coding for input restricted channels*, IBM J. of Research and Development **24** (1980), 43–48.

[**Fra7**] P. A. Franaszek, *A general method for channel coding*, IBM J. of Research and Development **24** (1980), 638–641.

[**Fra8**] P. A. Franaszek, *Construction of bounded delay codes for discrete noiseless channels,*, IBM J. of Research and Development **26** (1982), 506–514.

[**Fra9**] P. A. Franaszek, *Coding for constrained channels: a comparison of two approaches*, IBM J. of Research and Development **33** (1989), 602–607.

[**FraT**] P. A. Franaszek and J. A. Thomas, *On the optimization of constrained channel codes*, IEEE ISIT (1993).

[**Fran1**] J. Franks, *Homology and Dynamical Systems*, AMS–CBMS Reg. Conf. **49**, Providence, 1982.

[**Fran2**] J. Franks, *Flow equivalence of subshifts of finite type.*, Ergod. Th. & Dynam. Sys. **4** (1984), 53–66.

[**FreWD**] C. French, J. Wolf, and G. Dixon, *Signaling with special run length constraints for a digital recording channel*, IEEE-Magnetics **24** (1988), 2092–2097.

[**Frie**] D. Fried, *Finitely presented dynamical systems*, Ergod. Th. & Dynam. Sys. **7** (1987), 489–507.

[**Fri1**] J. Friedman, *A note on state splitting*, Proc. Amer. Math. Soc. **92** (1984), 206–208.

[**Fri2**] J. Friedman, *On the road coloring problem*, Proc. Amer. Math. Soc. **110** (1990), 1133–1135.

[**FriO**] N. Friedman and D. Ornstein, *On isomorphism of weak Bernoulli transformations*, Advances in Math. **5** (1970), 365–394.

[**Fro**] G. Frobenius, *Über Matrizen aus nicht negativen Elementen*, Sitz. der Preuss. Akad. der Wiss., Berlin (1912), 456–477.

[**Fuj**] M. Fujiwara, *Conjugacy for one-sided sofic systems*, in *Advanced Series in Dynamical Systems, vol.2, Dynamical Systems and Singular Phenomena* (ed. by G. Ikegami), World Scientific, Singapore, 1986, pp. 189–202.

[**FujO**] M. Fujiwara and M. Osikawa, *Sofic systems and flow equivalence*, Mathematical Reports of the College of General Education, Kyushu University **16, no. 1** (1987).

[**Fur1**] H. Furstenberg, *Stationary Processes and Prediction Theory*, Princeton Univ. Press, Princeton, 1960.

[**Fur2**] H. Furstenberg, *Disjointness in ergodic theory, minimal sets, and a problem in Diophantine approximation*, Math. Systems Theory **1** (1967), 1–49.

[**Gan**] F. R. Gantmacher, *Matrix Theory, Volume II*, Chelsea Publishing Company, New York, 1960.

[**GelP**] W. Geller and J. Propp, *The fundamental group for a $\mathbb{Z}^2$-shift*, Ergod. Th. & Dynam. Sys. **15** (1995), 1091-1118.

[**Gil**] R. Gilman, *Classes of linear automata*, Ergod. Th. & Dynam. Sys. **7** (1987), 105–118.

[**GolLS**] J. Goldberger, D. Lind, and M. Smorodinsky, *The entropies of renewal systems*, Israel J. Math. **33** (1991), 1–23.

[**Goo**] L. W. Goodwyn, *Comparing topological entropy with measure-theoretic entropy*, Amer. J. Math. **94** (1972), 366–388.

[**Got**] W. H. Gottschalk, *Substitution minimal sets*, Trans. Amer. Math. Soc. **109** (1963), 467–491.

[**GotH**] W. H. Gottschalk and G. A. Hedlund, *Topological Dynamics*, Amer. Math. Soc. Colloquium Publ. **36** (1955).

[**Gra**] R. Gray, *Sliding-block source coding*, IEEE Trans. Inform. Theory **21** (1975), 357–368.

[**Gre**] M. Greenberg, *Lectures on Algebraic Topology*, Benjamin, Reading, MA, 1968.

[**Gur1**] B. M. Gurevic, *Topological entropy of enumerable Markov chains*, Soviet Math. Dokl. **4,10** (1969), 911–915.

[**Gur2**] B. M. Gurevic, *Shift entropy and Markov measures in the path space of a denumerable graph*, Soviet Math. Dokl. **3,11** (1970), 744–747.

[**Had**] J. Hadamard, *Les surfaces a courbures opposées et leurs lignes geodesiques*, Journal de Mathematiques Pures et Appliqué **4** (1898), 27–73.

[**HahK**] F. Hahn and Y. Katznelson, *On the entropy of uniquely ergodic transformations*, Trans. Amer. Math. Soc. **126** (1967), 335–360.

[**HalV**] P. Halmos and J. Von Neumann, *Operator methods in classical mechanics*, Annals of Math. **43** (1942), 332–350.

[HanK] M. Handel and B. Kitchens, *Metrics and entropy for non-compact spaces*, Israel J. Math. **91** (1995), 253-271.

[Han1] D. Handelman, *Positive matrices and dimension groups affiliated to C*-Algebras and topological Markov chains*, J. Operator Th. **6** (1981), 55–74.

[Han2] D. Handelman, *Eventually positive matrices with rational eigenvectors*, Ergod. Th. & Dynam. Sys. **7** (1987), 193–196.

[Har] P. Hartmann, *Ordinary Differential Equations*, Wiley, New York, 1964.

[HarW] G. H. Hardy and E. M. Wright, *An Introduction to the Theory of Numbers* (Fifth Ed.), Oxford Univ. Press, Oxford, 1979.

[Has] M. Hassner, *A nonprobabilistic source and channel coding theory*, Ph.D. Thesis, University of California at Los Angeles (1980).

[Hed1] G. A. Hedlund, *On the metrical transitivity of the geodesics on a surface of constant negative curvature*, Proc. Nat. Acad. of Sci. **20** (1934), 87–140.

[Hed2] G. A. Hedlund, *On the metrical transitivity of the geodesics on a closed surface of constant negative curvature*, Annals of Math. **35** (1934), 787–808.

[Hed3] G. A. Hedlund, *The dynamics of geodesic flows*, Bull. Amer. Math. Soc. **45** (1939), 241–260.

[Hed4] G. A. Hedlund, *Sturmian minimal sets*, Amer. J. Math. **66** (1944), 605–620.

[Hed5] G. A. Hedlund, *Endomorphisms and automorphisms of the shift dynamical system*, Math. Systems Theory **3** (1969), 320–375.

[Hed6] G. A. Hedlund, *What is symbolic dynamics?*, Unpublished manuscript.

[HeeMS] C. D. Heegard, B. H. Marcus, and P. H. Siegel, *Variable-length state splitting with applications to average runlength-constrained (ARC) codes*, IEEE Trans. Inform. Theory **37** (1991), 759–777.

[Her] I. Herstein, *Topics in Algebra* 2nd ed., Wiley, New York, 1975.

[Hil] G. W. Hill, *Researches in the lunar theory*, Amer. J. Math. **1** (1878), 5–26, 129–147, 245–260.

[HirS] M. Hirsch and S. Smale, *Differential Equations, Dynamical Systems and Linear Algebra*, Academic Press, New York, 1974.

[Hof1] F. Hofbauer, *β-shifts have unique maximal measure*, Monats. Math. **85** (1978), 189–198.

[Hof2] F. Hofbauer, *On intrinsic ergodicity of piecewise monotonic transformations with positive entropy*, Israel J. Math **34** (1979), 213–236.

[Hof3] F. Hofbauer, *On intrinsic ergodicity of piecewise monotonic transformations with positive entropy II*, Israel J. Math **38** (1981), 107–115.

[Hof4] F. Hofbauer, *The structure of piecewise monotonic transformations*, Ergod. Th. & Dynam. Sys. **1** (1981), 159-178.

[Hof5] F. Hofbauer, *Piecewise invertible dynamical systems*, Prob. Th. and Rel. Fields **72** (1986), 359–386.

[Hol1] H. D. L. Hollmann, *A block decodable $(d, k) = (1, 8)$ runlength-limited rate 8/12 code*, IEEE Trans. Inform. Theory **40** (1994), 1292-1296.

[Hol2] H. D. L. Hollmann, *On the construction of bounded-delay encodable codes for constrained systems*, IEEE Trans. Inform. Theory **41** (1995), 1354 – 1378.

[Hop1] E. Hopf, *Fuchsian groups and ergodic theory*, Trans. Amer. Math. Soc. **39** (1936), 299-314.

[Hop2] E. Hopf, *Statitstik der geodatischen Linien Mannigfaltigkeiten negativer Krummung*, S.-B. Saechsischer Akad. Wiss. Math.-Nat. Kl. **91** (1939), 261-304.

[HopU] J. E. Hopcroft and J. D. Ullman, *Introduction to Automata Theory, Languages, and Computation*, Addison-Wesley, Reading, 1979.

[Hua1] D. Huang, *Flow equivalence of reducible shifts of finite type*, Ergod. Th. & Dynam. Sys. **14** (1994), 695–720.

[Hua2] D. Huang, *Flow equivalence of reducible shifts of finite type and Cuntz-Krieger algebras*, J. Reine Angew. Math **462** (1995), 185–217.

[Huf1] D. A. Huffman, *The synthesis of sequential switching circuits*, J. Franklin Inst. **257** (1954), 161–190 & 275–303.

[Huf2] D. A. Huffman, *Canonical forms for information lossless finite-state machines*, IRE Trans. Circuit Theory **6** (1959, Special Supplement), 41–59.

[HurKC] L. Hurd, J. Kari, and K. Culik, *The topological entropy of cellular automata is uncomputable*, Ergod. Th. & Dynam. Sys. **12** (1992), 255–265.

[Imm1] K. A. S. Immink, *Runlength-limited sequences*, Proc. IEEE **78** (1990), 1745–1759.

[Imm2] K. A. S. Immink, *Coding Techniques for Digital Recorders*, Prentice-Hall, New York, 1991.

[Imm3] K. A. S. Immink, *Block-decodable runlength-limited codes via look-ahead technique*, Phillips J. Res. **46** (1992), 293–310.

[Imm4] K. A. S. Immink, *Constructions of almost block decodable runlength-limited codes*, IEEE Transactions on Information Theory **41** (1995), 284–287.

[JacK] K. Jacobs and M. Keane, *0-1 sequences of Toeplitz type*, Z. Wahrscheinlichkeitstheorie **13** (1969), 123–131.

[Jon] N. Jonoska, *Sofic shifts with synchronizing presentations*, Theor. Computer Sci. **158** (1996), 81–115.

[Kak] S. Kakutani, *Induced measure-preserving transformations*, Proc. Japan Acad. **19** (1943), 635–641.

[Kam] H. Kamabe, *Minimum scope for sliding block decoder mappings*, IEEE Trans. Inform. Theory **35** (1989), 1335–1340.

[Kap] I. Kaplansky, *Infinite Abelian Groups*, Univ. of Michigan Press, Ann Arbor, 1954.

[KarM] R. Karabed and B. H. Marcus, *Sliding-block coding for input-restricted channels*, IEEE Trans. Inform.Theory **34** (1988), 2–26.

[Kas] P. W. Kastelyn, *The statistics of dimers on a lattice*, Physica A **27** (1961), 1209–1225.

[KatH] A. Katok and B. Hasselblatt, *Introduction to the Modern Theory of Dynamical Systems*, Cambridge Univ. Press, Cambridge, 1995.

[Kea1] M. S. Keane, *Generalized Morse sequences*, Z. Wahrscheinlichkeitstheorie **10** (1968), 335–353.

[Kea2] M. S. Keane, *Ergodic theory and subshifts of finite type*, in *Ergodic Theory, Symbolic Dynamics and Hyperbolic Spaces (ed. T. Bedford, et. al.)*, Oxford Univ. Press, Oxford, 1991, pp. 35-70.

[KeaS1] M. Keane and M. Smorodinsky, *Bernoulli schemes of the same entropy are finitarily isomorphic*, Annals of Math. **109** (1979), 397–406.

[KeaS2] M. Keane and M. Smorodinsky, *Finitary isomorphisms of irreducible Markov shifts*, Israel J. Math **34** (1979), 281–286.

[KemS] J. Kemeny and J. Snell, *Finite Markov Chains*, Van Nostrand, Princeton, 1960.

[KhaN1] Z. A. Khayrallah and D. Neuhoff, *Subshift models and finite-state codes for input-constrained noiseless channels: a tutorial*, EE Tech Report, Univ. Delaware, 90-9-1, 1990.

[KhaN2] Z. A. Khayrallah and D. Neuhoff, *On the window size of decoders for input constrained channels*, Proc. Internat. Symp. Information Theory and Its Applic. (1990), 1-2.5–1-2.7.

[KimOR] Ki Hang Kim, Nicholas A. Ormes, and Fred W. Roush, *The spectra of nonnegative integer matrices via formal power series*, J. Amer. Math. Soc. **13** (2000), 773–806.

[KimR1] K. H. Kim and F. W. Roush, *Some results on decidability of shift equivalence*, J. Combinatorics, Info. Sys. Sci. **4** (1979), 123–146.

[KimR2] K. H. Kim and F. W. Roush, *On strong shift equivalence over a Boolean semiring*, Ergod. Th. & Dynam. Sys **6** (1986), 81–97.

[KimR3] K. H. Kim and F. W. Roush, *Decidability of shift equivalence*, Proceedings of Maryland Special Year in Dynamics 1986–87, Springer-Verlag Lecture Notes in Math. **1342** (1988), 374–424.

[KimR4] K. H. Kim and F. W. Roush, *Decidability of epimorphisms of dimension groups and certain modules*, Ergod. Th. & Dynam. Sys. **9** (1989), 493–497.

[KimR5] K. H. Kim and F. W. Roush, *An algorithm for sofic shift equivalence*, Ergod. Th. & Dynam. Sys. **10** (1990), 381–393.

[KimR6] K. H. Kim and F. W. Roush, *Full shifts over $\mathbb{R}_+$ and invariant tetrahedra*, P.U.M.A. Ser. B **1** (1990), 251–256.

[KimR7] K. H. Kim and F. W. Roush, *Path components of matrices and shift equivalence over $\mathbb{Q}_+$*, Linear Algebra Appl. **145** (1991), 177–186.

[KimR8] K. H. Kim and F. W. Roush, *Solution of two conjectures in symbolic dynamics*, Proc. Amer. Math. Soc. **112** (1991), 1163–1168.

[KimR9] K. H. Kim and F. W. Roush, *On the structure of inert automorphisms of subshifts*, P.U.M.A. Ser. B **2** (1991), 3-22.

[KimR10] K. H. Kim and F. W. Roush, *Strong shift equivalence of boolean and positive rational matrices*, Linear Algebra Appl. **161** (1992), 153–164.

[KimR11] K. H. Kim and F. W. Roush, *Williams' conjecture is false for reducible subshifts*, J. Amer. Math. Soc. **5** (1992), 213–215.

[KimR12] K. H. Kim and F. W. Roush, *The Williams' Conjecture is false for irreducible subshifts*, Annals Math. **149** (1999), 545–558.

[KimRW1] K. H. Kim, F. W. Roush, and J. B. Wagoner, *Automorphisms of the dimension group and gyration numbers of automorphisms of a shift*, J. Amer. Math. Soc. **5** (1992), 191–211.

[KimRW2] K. H. Kim, F. W. Roush, and J. B. Wagoner, *Characterization of inert actions on periodic points I and II*, Forum Math **12** (2000), 565-602, 671-712.

[Kit1] B. Kitchens, *Continuity properties of factor maps in ergodic theory*, Ph.D. Thesis, University of North Carolina, Chapel Hill (1981).

[Kit2] B. Kitchens, *An invariant for continuous factors of Markov shifts*, Proc. Amer. Math. Soc. **83** (1981), 825–828.

[Kit3] B. Kitchens, *Expansive dynamics on zero-dimensional groups*, Ergod. Th. & Dynam. Sys. **7** (1987), 249–261.

[Kit4] Bruce Kitchens, *Symbolic Dynamics: One-sided, Two-sided and Countable State Markov Shifts*, Springer Universitext, Berlin, 1998.

[KitMT] B. Kitchens, B. Marcus and P. Trow, *Eventual factor maps and compositions of closing maps*, Ergod. Th. & Dynam. Sys. **11** (1991), 85–113.

[KitS1] B. Kitchens and K. Schmidt, *Periodic points, decidability and Markov subgroups*, Dynamical Systems, Proceedings of the special year (J. C. Alexander, ed.), Springer Lecture Notes in Math. **1342** (1988), 440–454.

[KitS2] B. Kitchens and K. Schmidt, *Automorphisms of compact groups*, Ergod. Th. & Dynam. Sys. **9** (1989), 691–735.

[KitS3] B. Kitchens and K. Schmidt, *Markov subgroups of $(\mathbb{Z}/2\mathbb{Z})^{\mathbb{Z}^2}$*, in *Symbolic Dynamics and Its Applications*, Contemporary Mathematics **135** (ed. P. Walters), Providence, 1992, pp. 265–283.

[KitS4] B. Kitchens and K. Schmidt, *Mixing sets and relative entropies for higher-dimensional Markov shifts*, Ergod. Th. & Dynam. Sys. **13** (1993), 705–735.

[Kle] B. G. Klein, *Homomorphisms of symbolic dynamical systems*, Math. Systems Theory **6** (1972), 107–122.

[Koe] P. Koebe, *Riemannian Manigfaltigkesten und nichteuclidische Raumformen IV*, Sitz. der Preuss. Akad. der Wiss. (1929), 414–457.

[Koh1] Z. Kohavi, *Minimization of incompletely specified switching circuits*, Research Report of Polytechnic Institute of Brooklyn, New York (1962).

[Koh2] Z. Kohavi, *Switching and Finite Automata Theory*, Second Ed., Tata McGraw-Hill, New Delhi, 1978.

[Kol] A. N. Kolmogorov, *New metric invariants of transitive dynamical systems and automorphisms of Lebesgue spaces*, Dokl. Akad. Nauk SSSR **119** (1958), 861–864.

[Kri1] W. Krieger, *On entropy and generators of measure preserving transformations*, Trans. Amer. Math. Soc. **149** (1970), 453–464, erratum: **168** (1972), 519.

[Kri2] W. Krieger, *On the uniqueness of the equilibrium state*, Math. Systems Theory **8** (1974), 97–104.

[Kri3] W. Krieger, *On a dimension for a class of homeomorphism groups*, Math. Ann. **252** (1980), 87–95.

[Kri4] W. Krieger, *On dimension functions and topological Markov chains*, Inventiones Math. **56** (1980), 239–250.

[Kri5] W. Krieger, *On the finitary coding of topological Markov chains*, Available at www.numdam.org/item/RCP25_1981__29__67_0/, R.C.P. 25, Université Louis Pasteur, Institute de Recherche Mathematiques Avancee, Strasbourg **29** (1981), 67-92.

[Kri6] W. Krieger, *On the subsystems of topological Markov chains*, Ergod. Th. & Dynam. Sys. **2** (1982), 195–202.

[Kri7] W. Krieger, *On the finitary isomorphisms of Markov shifts that have finite expected coding time,*, Z. Wahrscheinlichkeitstheorie **65** (1983), 323–328.

[Kri8] W. Krieger, *On sofic systems I*, Israel J. Math. **48** (1984), 305–330.

[Kri9] W. Krieger, *On sofic systems II*, Israel J. Math. **60** (1987), 167–176.

[KruT] T. Kruger and S. Troubetzkoy, *Markov partitions and shadowing for non-uniformly hyperbolic systems with singularities*, Ergod. Th. & Dynam. Sys. **12** (1992), 487–508.

[Kur] A. A. Kurmit, *Information–Lossless Automata of Finite Order*, Wiley (translated from Russian), New York, 1974.

[Led] François Ledrappier, *Un champ markovien peut être d'entropie nulle et mélangeant*, C. R. Acad. Sci. Paris Sér. A-B **287** (1978), A561–A563.

[LemC] A. Lempel and M. Cohn, *Look-ahead coding for input-restricted channels*, IEEE Trans. Inform. Theory **28** (1982), 933–937.

[LinC] S. Lin and D. J. Costello, Jr., *Error Control Coding, Fundamentals and Applications*, Prentice-Hall, Englewood Cliffs, 1983.

[Lin1] D. Lind, *Ergodic group automorphisms and specification*, Springer Lecture Notes in Math. **729** (1978), 93–104.

[Lin2] D. Lind, *Dynamical properties of quasihyperbolic toral automorphisms*, Ergod. Th. & Dynam. Sys. **2** (1982), 49–68.

[Lin3] D. Lind, *Entropies and factorizations of topological Markov shifts*, Bull. Amer. Math. Soc. **9** (1983), 219–222.

[Lin4] D. Lind, *The entropies of topological Markov shifts and a related class of algebraic integers*, Ergod. Th. & Dynam. Sys. **4** (1984), 283–300.

[Lin5] D. Lind, *Applications of ergodic theory and sofic systems to cellular automata*, Physica D **10D** (1984), 36–44.

[Lin6] D. Lind, *The spectra of topological Markov shifts*, Ergod. Th. & Dynam. Sys. **6** (1986), 571–582.

[Lin7] D. Lind, *Entropies of automorphisms of topological Markov shifts*, Proc. Amer. Math. Soc. **99** (1987), 589–595.

[**Lin8**] D. Lind, *Perturbations of shifts of finite type*, SIAM J. Discrete Math. **2** (1989), 350–365.

[**Lin9**] D. Lind, *A zeta function for $\mathbb{Z}^d$-actions*, London Math. Soc. Lecture Note Ser. (Proceedings of the Warwick Symposium on $\mathbb{Z}^d$-Actions, 1993-1994) **228** (1996), 433–450.

[**LinSW**] D. Lind, K. Schmidt, and T. Ward, *Mahler measure and entropy for commuting automorphisms of compact groups*, Inventiones Math. **101** (1990), 593–629.

[**LoeM**] H. A. Loeliger and T. Mittelholzer, *Convolutional codes over groups*, IEEE Trans Info Theory **42** (1996), 1660-1686.

[**Mane**] R. Mañé, *Ergodic Theory and Differentiable Dynamics*, Springer, New York, 1987.

[**Man**] A. Manning, *Axiom A diffeomorphisms have rational zeta functions*, Bull. London Math. Soc. **3** (1971), 215–220.

[**Mar1**] B. Marcus, *Factors and extensions of full shifts*, Monats. Math. **88** (1979), 239–247.

[**Mar2**] B. H. Marcus, *Sofic systems and encoding data*, IEEE Trans. Inform. Theory **31** (1985), 366–377.

[**MarR**] B. H. Marcus and R. M. Roth, *Bounds on the number of states in encoder graphs for input-constrained channels*, IEEE Trans. Inform. Theory **IT-37** (1991), 742–758.

[**MarRS**] B. H. Marcus, R. M. Roth, and P. H. Siegel, *Constrained systems and coding for recording*, chapter in *Handbook on Coding Theory* (ed. by R. Brualdi, C. Huffman and V. Pless), Elsevier (1998), 1635-1764.

[**MarS**] B. Marcus and P. H. Siegel, *Constrained codes for partial response channels*, Proc. Beijing International Workshop on Information Theory (1988), DI-1.1–1.4.

[**MarSW**] B. H. Marcus, P. H. Siegel, and J. K. Wolf, *Finite-state modulation codes for data storage*, IEEE J. Selected Areas in Communications **10** (1992), 5–37.

[**MarT1**] B. Marcus and S. Tuncel, *Entropy at a weight-per-symbol and embeddings of Markov chains*, Inventiones Math. **102** (1990), 235–266.

[**MarT2**] B. Marcus and S. Tuncel, *The weight-per-symbol polytope and scaffolds of invariants associated with Markov chains*, Ergod. Th. & Dynam. Sys. **11** (1991), 129–180.

[**MarT3**] B. Marcus and S. Tuncel, *Matrices of polynomials, positivity, and finite equivalence of Markov chains*, J. Amer. Math. Soc. **6** (1993), 131–147.

[**MarP1**] N. Markley and M. Paul, *Almost automorphic symbolic minimal sets without unique ergodicity*, Israel J. Math. **34** (1979), 259–272.

[**MarP2**] N. Markley and M. Paul, *Matrix subshifts for $\mathbb{Z}^\nu$ symbolic dynamics*, Proc. London Math. Soc. **43** (1981), 251–272.

[**MarP3**] N. Markley and M. Paul, *Maximal measures and entropy for $\mathbb{Z}^\nu$ subshifts of finite type*, in *Classical Mechanics and Dynamical Systems* (ed. R. Devaney and Z. Nitecki), Dekker Notes **70**, 135–157.

[**Mart1**] J. Martin, *Substitution minimal flows*, Amer. J. Math. **93** (1971), 503–526.

[**Mart2**] J. Martin, *Minimal flows arising from substitutions of non-constant length*, Math. Systems Theory **7** (1973), 72–82.

[**McE**] R. McElieece, *The Theory of Information and Coding*, Addison-Wesley, Reading, 1977.

[**MilT**] G. Miles and R. K. Thomas, *The breakdown of automorphisms of compact topological groups*, Studies in Probability and Ergodic Theory, Advances in Mathematics Supplementary Studies **2** (1978), 207–218.

[**Mil**] A. Miller, *Transmission*, U. S. Patent 3,108,261 (1963).

[**Min**] H. Minc, *Nonnegative Matrices*, Wiley, New York, 1988.

[Mon] John Monforte, *The digital reproduction of sound*, Scientific American (Dec., 1984), 78–84.

[Moo] E. F. Moore, *Gedanken-experiments on sequential machines*, Automata Studies, Annals of Math. Studies, Princeton Univ. Press, Princeton, 1956, pp. 129–153.

[Mor1] M. Morse, *A one-to-one representation of geodesics on a surface of negative curvature*, Amer. J. Math. **43** (1921), 33–51.

[Mor2] M. Morse, *Recurrent geodesics on a surface of negative curvature*, Trans. Amer. Math. Soc. **22** (1921), 84–100.

[Mor3] M. Morse, *Symbolic Dynamics, Lectures of 1937–1938, Notes by Rufus Oldenburger*, University Microfilms, 300 N. Zeeb Road, Ann Arbor, MI 48106.

[MorH1] M. Morse and G. A. Hedlund, *Symbolic dynamics*, Amer. J. Math. **60** (1938), 815–866.

[MorH2] M. Morse and G. A. Hedlund, *Symbolic dynamics II. Sturmian trajectories*, Amer. J. Math. **62** (1940), 1–42.

[MorH3] M. Morse and G. A. Hedlund, *Unending chess, symbolic dynamics and a problem in semigroups*, Duke Math. J. **11** (1944), 1–7.

[Moz1] S. Mozes, *Tilings, substitutions and the dynamical systems generated by them*, J. d'Analyse Math. **53** (1989), 139–186.

[Moz2] S. Mozes, *A zero entropy, mixing of all orders tiling system*, in *Symbolic Dynamics and Its Applications*, Contemporary Mathematics **135** (ed. P. Walters), Providence, 1992, pp. 319–326.

[Nas1] M. Nasu, *Uniformly finite-to-one and onto extensions of homomorphisms between strongly connected graphs*, Discrete Math. **39** (1982), 171–197.

[Nas2] M. Nasu, *Constant-to-one and onto global maps of homomorphisms between strongly connected graphs*, Ergod. Th. & Dynam. Sys. **3** (1983), 387–411.

[Nas3] M. Nasu, *An invariant for bounded-to-one factor maps between transitive sofic subshifts*, Ergod. Th. & Dynam. Sys. **5** (1985), 89–105.

[Nas4] M. Nasu, *Topological conjugacy for sofic systems*, Ergod. Th. & Dynam. Sys. **6** (1986), 265–280.

[Nas5] M. Nasu, *Topological conjugacy for sofic systems and extensions of automorphisms of finite subsystems of topological Markov shifts*, Proceedings of Maryland Special Year in Dynamics 1986-87, Springer-Verlag Lecture Notes in Math. **1342** (1988), 564–607.

[Nas6] M. Nasu, *Textile Automorphisms*, Mem. Amer. Math. Soc. **546**, 1995.

[Newh] S. Newhouse, *On some results of Hofbauer on maps of the interval*, in *Dynamical Systems and Related Topics* (ed. by K. Shiraiwa), World Scientific, Singapore, 1992, pp. 407–421.

[New] M. Newman, *Integral Matrices*, Academic Press, New York, 1972.

[Nie1] J. Nielsen, *Zur Topologie der geschlossenen zweiseitgen Flachen*, Mathematikerkongresses i Kopenhavn (1925), 263–358.

[Nie2] J. Nielsen, *Untersuchungen zur Topologie der geschlossenen zweiseitigen Flachen*, Acta Mathematica **50** (1927), 189–358.

[Nit] Z. Nitecki, *Differentiable Dynamics*, M.I.T. Press, Cambridge, MA, 1971.

[NivZ] I. Niven and H. S. Zuckerman, *An Introduction to the Theory of Numbers*, Wiley, New York, 1980.

[Obr] G. L. O'Brien, *The road coloring problem*, Israel J. Math. **39** (1981), 145–154.

[Orn] D. Ornstein, *Bernoulli shifts with the same entropy are isomorphic*, Advances in Math. **4** (1970), 337–352.

[Par1] W. Parry, *Intrinsic Markov chains*, Trans. Amer. Math. Soc. **112** (1964), 55–66.

[Par2] W. Parry, *Symbolic dynamics and transformations of the unit interval*, Trans. Amer. Math. Soc. **122** (1966), 368–378.

[**Par3**] W. Parry, *A finitary classification of topological Markov chains and sofic systems*, Bull. London Math. Soc. **9** (1977), 86–92.

[**Par4**] W. Parry, *Finitary isomorphisms with finite expected code-lengths*, Bull. London Math. Soc. **11** (1979), 170–176.

[**Par5**] W. Parry, *Notes on coding problems for finite state processes*, Bull. London Math. Soc. **23** (1991), 1–33.

[**ParP**] W. Parry and M. Pollicott, *Zeta Functions and the Periodic Orbit Structure of Hyperbolic Dynamical Systems*, Astérisque (Soc. Math. de France) **187–188** (1990).

[**ParS**] W. Parry and K. Schmidt, *Natural coefficients and invariants for Markov shifts*, Inventiones Math. 76 (1984), 15–32.

[**ParSu**] W. Parry and D. Sullivan, *A topological invariant for flows on one-dimensional spaces*, Topology **14** (1975), 297–299.

[**ParT1**] W. Parry and S. Tuncel, *On the classification of Markov chains by finite equivalence*, Ergod. Th. & Dynam. Sys. **1** (1981), 303–335.

[**ParT2**] W. Parry and S. Tuncel, *Classification Problems in Ergodic Theory*, LMS Lecture Note Series **67**, Cambridge Univ. Press, Cambridge, 1982.

[**ParW**] W. Parry and R. F. Williams, *Block coding and a zeta function for finite Markov chains*, Proc. London Math. Soc. **35** (1977), 483–495.

[**Pat**] A. M. Patel, *Zero-modulation encoding in magnetic recording*, IBM J. of Research and Development **19** (1975), 366–378.

[**PerRS**] M. Perles, M. O. Rabin, and E. Shamir, *The theory of definite automata*, IEEE. Trans. Electron. Computers **12** (1963), 233–243.

[**Perr**] D. Perrin, *On positive matrices*, Theoret. Computer Sci. **94** (1992), 357–366.

[**PerS**] D. Perrin and M-P Schutzenberger, *Synchronizing prefix codes and automata and the road coloring problem*, in *Symbolic Dynamics and Its Applications*, Contemporary Mathematics **135** (ed. P. Walters), 1992, pp. 295–318.

[**Per**] O. Perron, *Zur Theorie der Matrizen*, Math. Ann. **64** (1907), 248–263.

[**Pet1**] K. E. Petersen, *A topologically strongly mixing symbolic minimal set*, Trans. Amer. Math. Soc. **148** (1970), 603–612.

[**Pet2**] K. Petersen, *Chains, entropy, coding*, Ergod. Th. & Dynam. Sys. **6** (1986), 415–448.

[**Pet3**] K. Petersen, *Ergodic Theory*, Cambridge Univ. Press, Cambridge, 1989.

[**Pin**] J. E. Pin, *On two combinatorial problems arising from automata theory*, Annals of Discrete Math. **17** (1983), 535–548.

[**Pir**] P. Piret, *Convolutional Codes*, MIT Press, Cambridge, MA, 1988.

[**Poi**] H. Poincare, *Sur les courbes definés par les equationes differentieles*, J. Math. Pures et Appliqués **4** (1885), 167–244.

[**Pol**] H. Pollard, *The Theory of Algebraic Numbers*, MAA Carus Math. Monographs **9**, New York, 1961.

[**ProM**] M. H. Protter and C. B. Morrey, *A First Course in Real Analysis*, Undergraduate Texts in Mathematics, Springer-Verlag, New York, 1977.

[**Pro**] E. Prouhet, *Memoire sur quelques relations entre les puissance des nombres*, CR Acad. Sci., Paris **33** (1851), 31.

[**Que**] M. Queffélec, *Substitution Dynamical Systems–Spectral Analysis*, Springer–Verlag Lecture Notes in Math. **1294**, New York, 1987.

[**RabS**] M. Rabin and D. Scott, *Finite automata and their decision procedures*, IBM J. of Research and Development **3** (1959), 114–125.

[**Rad**] C. Radin, *Global order from local sources*, Bull. Amer. Math. Soc. **25** (1991), 335–364.

[**Rat**] M. Ratner, *Invariant measure with respect to an Anosov flow on a 3-dimensional manifold*, Soviet Math. Doklady **10** (1969), 586–588.

[Res] A. Restivo, *Finitely generated sofic systems*, Theor. Computer Sci. **65** (1989), 265–270.

[Rob] R. M. Robinson, *Undecidability and nonperiodicity for tilings of the plane*, Inventiones Math. **12** (1971), 177–209.

[Rok1] V. Rokhlin, *A "general" measure-preserving transformation is not mixing*, Dokl. Akad. Nauk SSSR **60** (1948), 349–351.

[Rok2] V. Rokhlin, *Selected topics in metric theory of dynamical systems*, Uspekhi Mat. Nauk **4** (1949), 57–128; English translation in Amer. Math. Soc. Transl. (Ser. 2) **49** (1966), 171–240.

[Rok3] V. A. Rokhlin, *On endomorphisms of compact commutative groups*, Izvestiya Akad. Nauk SSSR Ser. Mat. (Russian) **13** (1949), 329–340.

[Roy] H. L. Royden, *Real Analysis*, Macmillan, New York, 1968.

[Rud] W. Rudin, *Principles of Mathematical Analysis* 3rd ed., McGraw-Hill, New York, 1976.

[Rudo] D. Rudolph, *Fundamentals of Measurable Dynamics*, Oxford Univ. Press, New York, 1990.

[Rue1] D. Ruelle, *Statistical Mechanics: Rigorous Results*, Benjamin, New York, 1969.

[Rue2] D. Ruelle, *Elements of Differentiable Dynamics and Bifurcation Theory*, Academic Press, Boston, 1989.

[Rue3] D. Ruelle, *Dynamical Zeta Functions for Piecewise Monotone Maps of the Interval*, CRM Monograph Series, AMS, Providence, 1994.

[Rya1] J. Ryan, *The shift and commutativity*, Math. Systems Theory **6** (1972), 82–85.

[Rya2] J. Ryan, *The shift and commutativity II*, Math. Systems Theory **8** (1974), 249–250.

[Sal1] I. A. Salama, Ph.D. Thesis, University of North Carolina, Chapel Hill (1984).

[Sal2] I. A. Salama, *Topological entropy and recurrence of countable chains*, Pacific J. Math. **134** (1988), 325–341; erratum **140** (1989), 397–398.

[Sal3] I. A. Salama, *On the recurrence of countable topological Markov chains*, in *Symbolic Dynamics and Its Applications*, Contemporary Mathematics **135** (ed. P. Walters), Providence, 1992, pp. 349–360.

[SaSo] A. Salomaa and M. Soittola, *Automata-Theoretic Aspects of Formal Power Series*, Springer-Verlag, New York, 1978.

[Sch1] K. Schmidt, *Invariants for finitary isomorphisms with finite expected code lengths*, Inventiones Math. **76** (1984), 33–40.

[Sch2] K. Schmidt, *Algebraic Ideas in Ergodic Theory*, AMS–CBMS Reg. Conf. **76**, Providence, 1990.

[Sch3] K. Schmidt, *Dynamical Systems of Algebraic Origin*, Birkhauser, New York, 1995.

[Sen] E. Seneta, *Non-negative Matrices and Markov Chains* (Second Ed.), Springer, New York, 1980.

[Ser] C. Series, *Symbolic dynamics for geodesic flows*, Acta Math. **146** (1981), 103–128.

[Sha] C. Shannon, *A mathematical theory of communication*, Bell Sys. Tech. J. **27** (1948), 379–423, 623–656.

[Shi] P. Shields, *The Theory of Bernoulli Shifts*, Univ. of Chicago Press, Chicago, 1973.

[Sim] G. Simmons, *Introduction to Topology and Modern Analysis*, McGraw-Hill, New York, 1963.

[Sin1] Y. G. Sinai, *The notion of entropy of a dynamical system*, Dokl. Akad. Nauk SSSR **125** (1959), 768–771.

[Sin2] Y. G. Sinai, *Markov partitions and C-diffeomorphisms*, Funct. Anal. & Appl. **2** (1968), 64–89.

[Sma] S. Smale, *Differentiable dynamical systems*, Bull. Amer. Math. Soc. **73** (1967), 747–817.

[Sta] H. Stark, *An Introduction to Number Theory*, M.I.T. Press, Cambridge, Mass., 1978.

[SteT] I. Stewart and D. Tall, *Algebraic Number Theory*, Chapman and Hall, London, 1979.

[SweC] N. Swenson and J. Cioffi, *A simplified design approach for run-length limited sliding block codes*, preprint (based upon Ph.D. Dissertation by N. Swenson, Stanford Univ.) (1991).

[Thu] A. Thue, *Uber unendliche Zeichreihen*, Videnskbssolskabets Skrifter, I Mat.-nat. Kl., Christiana (1906).

[Toe] Toeplitz, *Beispeile zur theorie der fastperiodischen Funktionen*, Math. Ann. **98** (1928), 281–295.

[Tra] A. N. Trahtman, *The road coloring problem*, Israel J. Math. **172** (2009), 51–60.

[Tro1] P. Trow, *Resolving maps which commute with a power of the shift*, Ergod. Th. & Dynam. Sys. **6** (1986), 281–293.

[Tro2] P. Trow, *Degrees of constant-to-one factor maps*, Proc. Amer. Math. Soc. **103** (1988), 184–188.

[Tro3] P. Trow, *Degrees of finite-to-one factor maps*, Israel J. Math. **71** (1990), 229–238.

[Tro4] P. Trow, *Divisibility constraints on degrees of factor maps*, Proc. Amer. Math. Soc. **113** (1991), 755–760.

[Tro5] P. Trow, *Constant-to-one factor maps and dimension groups*, in *Symbolic Dynamics and Its Applications*, Contemporary Mathematics **135** (ed. P. Walters), Providence, 1992, pp. 377–390.

[Tro6] P. Trow, *Decompositions of finite-to-one factor maps*, Israel J. Math. **91** (1995), 129–155.

[TroW] P. Trow and S. Williams, *Core dimension group constraints for factors of sofic shifts*, Ergod. Th. & Dynam. Sys. **13** (1993), 213–224.

[Tun1] S. Tuncel, *Conditional pressure and coding*, Israel J. Math. **39** (1981), 101–112.

[Tun2] S. Tuncel, *A dimension, dimension modules and Markov chains*, Proc. London Math. Soc. **46** (1983), 100–116.

[Ver1] D. Vere-Jones, *Geometric ergodicity in denumerable Markov chains*, Quart. J. Math. Oxford (2) **13** (1962), 7–28.

[Ver2] D. Vere-Jones, *Ergodic properties of nonnegative matrices – I*, Pacific J. Math. **22** (1967), 361–386.

[Wag1] J. B. Wagoner, *Markov partitions and K_2*, Pub. Math. IHES **65** (1987), 91–129.

[Wag2] J. B. Wagoner, *Topological Markov chains, C^*-algebras and K_2*, Advances in Math. **71** (1988), 133–185.

[Wag3] J. B. Wagoner, *Eventual finite generation for the kernel the dimension group representation*, Trans. Amer. Math. Soc. **317** (1990), 331–350.

[Wag4] J. B. Wagoner, *Higher-dimensional shift equivalence and strong shift equivalence are the same over the integers*, Proc. Amer. Math. Soc. **109** (1990), 527–536.

[Wag5] J. B. Wagoner, *Triangle identities and symmetries of a subshift of finite type*, Pacific J. Math. **44** (1990), 181–205.

[Wag6] J. B. Wagoner, *Classification of subshifts of finite type revisited*, in *Symbolic Dynamics and Its Applications*, Contemporary Mathematics **135** (ed. P. Walters), Providence, 1992, pp. 423–444.

[Wal] P. Walters, *An Introduction to Ergodic Theory*, Springer Graduate Texts in Math **79**, New York, 1982.

[WeaW] A. D. Weathers and J. K. Wolf, *A new rate 2/3 sliding block code for the* (1, 7) *runlength constraint with the minimal number of encoder states,,* IEEE Trans. Inform. Theory **37** (1991), 908–913.

[Weig] T. Weigandt, *Magneto-optic recording using a* (2, 18, 2) *run-length-limited code,* MS Thesis, MIT (1991).

[Wei] B. Weiss, *Subshifts of finite type and sofic systems,* Monats. Math. **77** (1973), 462–474.

[Wil1] J. C. Willems, *System theoretic models for the analysis of physical systems,* Richerche di Automatica **10** (1979), 71–106.

[Wil2] J. C. Willems, *Models for dynamics,* Dynamics Reported **2** (1989), 171–269.

[WilR1] R. F. Williams, *Classification of 1-dimensional attractors,* Proc. Sym. Pure Math. A.M.S. **14** (1970), 341–361.

[WilR2] R. F. Williams, *Classification of subshifts of finite type,* Annals of Math. 98 (1973), 120–153; erratum, Annals of Math. 99 (1974), 380–381.

[WilR3] R. F. Williams, *Shift equivalence of matrices in* $GL(2, Z)$, in *Symbolic Dynamics and Its Applications,* Contemporary Mathematics **135** (ed. P. Walters), Providence, 1992, pp. 445–451.

[WilS1] S. Williams, *Toeplitz minimal flows which are not uniquely ergodic,* Z. Wahrscheinlichkeitstheorie **67** (1984), 95–107.

[WilS2] S. Williams, *Covers of non-almost-finite-type systems,* Proc. Amer. Math. Soc. **104** (1988), 245–252.

[WilS3] S. Williams, *A sofic system which is not spectrally of finite type,* Ergod. Th. & Dynam. Sys. **8** (1988), 483–490.

[WilS4] S. Williams, *Notes on renewal systems,* Proc. Amer. Math. Soc. **110** (1990), 851–853.

[WilS5] S. Williams, *Lattice invariants for sofic shifts,* Ergod. Th. & Dynam. Sys. **11** (1991), 787–801.

[Wil] R. Wilson, *Introduction to Graph Theory* (2nd ed.), Academic Press, New York, 1979.

[Yur] M. Yuri, *Multidimensional maps with infinite invariant measures and countable state sofic shifts,* Indagationes Mathematicae **6** (1995), 355-383.

ADDENDUM BIBLIOGRAPHY

*Citations in the Addendum marked with a * refer to
items in the original Bibliography*

[AbrK] Adam Abrams and Svetlana Katok, *Adler and Flatto revisited: cross-sections for geodesic flow on compact surfaces of constant negative curvature*, Studia Math. **246** (2019), 167–202.

[AdaBMP] Stefan Adams, Raimundo Briceño, Brian Marcus, and Ronnie Pavlov, *Representation and poly-time approximation for pressure of $\mathbb{Z}^2$ lattice models in the non-uniqueness region*, J. Stat. Phys. **162** (2016), 1031–1067.

[Adl] Roy L. Adler, *Symbolic dynamics and Markov partitions*, Bull. Amer. Math. Soc. (N.S.) **35** (1998), 1–56.

[AirBL] Dylan Airey, Lewis Bowen, and Frank Lin, *A topological dynamical system with two different positive sofic entropies*, arXiv:1911.08272.

[AllAY] Mahsa Allahbakhshi, John Antonioli, and Jisang Yoo, *Relative equilibrium states and class degree*, Ergodic Theor. Dyn. Syst. **39** (2019), 865–888.

[AllHJ] Mahsa Allahbakhshi, Soonjo Hong, and Uijin Jung, *Structure of transition classes for factor codes on shifts of finite type*, Ergodic Theor. Dyn. Syst. **35** (2015), 2353–2370.

[AllQ] Mahsa Allahbakhshi and Anthony Quas, *Class degree and relative maximal entropy*, Trans. Amer. Math. Soc **365** (2013), 1347–1368.

[AlmCKP] Jorge Almeida, Alfredo Costa, Revekka Kyriakoglou, and Dominique Perrin, *Profinite Semigroups and Symbolic Dynamics*, Springer Lecture Notes in Mathematics (to appear).

[AshJMS] J. Ashley, G. Jaquette, B. Marcus, and P. Seeger, *Runlength limited encoding/decoding with robust resync*, United States Patent 5,696,649, 1999.

[AubBT] Nathalie Aubrun, Sebastián Barbieri, and Stéphan Thomassé, *Realization of aperiodic subshifts and uniform densities in groups*, Groups Geom. Dyn. **13** (2019), 107–129.

[AubS] Nathalie Aubrun and Mathieu Sablik, *Simulation of effective subshifts by two-dimensional subshifts of finite type*, Acta Appl. Math. **126** (2013), 35–63.

[BakG] S. Baker and A.E. Ghenciu, *Dynamical properties of S-gap shifts and other shift spaces*, J. Math. Anal. Appl. **430** (2015), 633–647.

[BalS] Alexis Ballier and Maya Stein, *The domino problem on groups of polynomial growth*, Groups Geom. Dyn. **12** (2018), 93–105.

[Barb] Sebastián Barbieri, *On the entropies of subshifts of finite type on countable amenable groups*, arXiv:1905.10015.

[BarGMT] S. Barbieri, R. Gómez, B. Marcus, and S. Taati, *Equivalence of relative Gibbs and relative equilibrium measures for actions of countable amenable groups*, Nonlinearity **33** (2020), 2409–2454.

[Bart] Laurent Bartholdi, *Amenability of groups is characterized by Myhill's theorem*, J. Eur. Math. Soc. **21** (2019), 3191–3197.

[BarO] Marcy Barge and Carl Olimb, *Asymptotic structure in substitution tiling spaces*, Ergodic Theor. Dyn. Syst. **34** (2014), 55–94.

[BarRYY] A. Barbero, E. Rosnes, G. Yang, and O. Ytrehus, *Near-field passive RFID communication: channel model and code design*, IEEE Trans. Commun. **62** (2014), 1716–1726.

[Bax1] R. J. Baxter, *Exactly Solved Models in Statistical Physics*, Academic Press, London, 1982.

[Bax2] R. J. Baxter, *Planar lattice gases with nearest-neighbor exclusion*, Ann. Comb. **3** (1999), 191–203.

[**BefD**] Vincent Beffara and Hugo Duminil-Copin, *The self-dual point of the two-dimensional random-cluster model is critical for* $q \geqslant 1$, Probab. Theory Relat. Fields **153** (2012), 511–542.

[**Ber**] Anton Bernshteyn, *Building large free subshifts using the Local Lemma*, Groups Geom. Dyn. **13** (2019), 1417–1436.

[**Bow1**] Lewis Bowen, *A measure-conjugacy invariant for free group actions*, Ann. of Math. (2) **171** (2010), 1387–1400.

[**Bow2**] Lewis Bowen, *Measure conjugacy invariants for actions of countable sofic groups*, J. Amer. Math. Soc. **23** (2010), 217–245.

[**Bow3**] Lewis Bowen, *A brief introduction to sofic entropy theory*, Proceedings of the International Congress of Mathematicians—Rio de Janeiro 2018. Vol. III. Invited lectures, World Scientific Publications, Hackensack, NJ, 2018, pp. 1847–1866.

[**BowR**] Rufus Bowen, *Some systems with unique equilibrium states*, Math. Systems Theory **8** (1974/75), 193–202.

[**Boyd**] David Boyd, *Irreducible polynomials with many roots of maximum modulus*, Acta Arith. **68** (1994), 85–88.

[**Boy1**] Mike Boyle, *Flow equivalence of shifts of finite type via positive factorizations*, Pacific J. Math. **204** (2002), 273–317.

[**Boy2**] Mike Boyle, *Some sofic shifts cannot commute with nonwandering shifts of finite type*, Illinois J. Math. **48** (2004), 1267–1277.

[**Boy3**] Mike Boyle, *Open problems in symbolic dynamics*, Geometric and Probabilistic Structures in Dynamics, Contem. Math. **469**, American Mathematical Society, 2008, pp. 69–118. Updates at www.math.umd.edu/~mboyle/open.

[**Boy4**] Mike Boyle, *The work of Kim and Roush in symbolic dynamics*, Acta Appl. Math. **126** (2013), 17–27.

[**BoyBG**] Mike Boyle, Jerome Buzzi, and Ricardo Gómez, *Almost isomorphism for countable state Markov shifts*, J. Reine Angew. Math. **592** (2006), 23–47.

[**BoyC**] Mike Boyle and Sompong Chuysurichay, *The mapping class group of a shift of finite type*, J. Mod. Dynam. **13** (2018), 115–145.

[**BoyCE1**] Mike Boyle, Toke Meier Carlsen, and Søren Eilers, *Flow equivalence of sofic shifts*, Israel J. Math. **225** (2018), 111–146.

[**BoyCE2**] Mike Boyle, Toke Meier Carlsen, and Søren Eilers, *Flow equivalence of G-shifts*, arXiv:1512.05238v2 (2019).

[**BoyD**] Mike Boyle and Tomasz Downarowicz, *The entropy theory of symbolic extensions*, Invent. Math. **156** (2004), 119–161.

[**BoyH**] Mike Boyle and Danrun Huang, *Poset block equivalence of integral matrices*, Trans. Amer. Math. Soc. **355** (2003), 3861–3886.

[**BoyL1**] Mike Boyle and Douglas Lind, *Expansive subdynamics*, Trans. Amer. Math. Soc. **349** (1997), 55–102.

[**BoyL2**] Mike Boyle and Douglas Lind, *Small polynomial matrix presentations of nonnegative matrices*, Linear Algebra Appl. **355** (2002), 49–70.

[**BoyP**] Mike Boyle and Karl Petersen, *Hidden Markov processes in the context of symbolic dynamics*, Entropy of Hidden Markov Processes and Connections to Dynamical Systems, LMS Lecture Note Ser. **385**, Cambridge University Press, Cambridge, 2011, pp. 5–71.

[**BoyPS**] Mike Boyle, Ronnie Pavlov, and Michael Schraudner, *Multidimensional sofic shifts without separation and their factors*, Trans. Amer. Math. Soc. **362** (2010), 4617–4653.

[**BoyS1**] Mike Boyle and Scott Schmieding, *Symbolic dynamics and the stable algebra of matrices*, arXiv:2006.01051.

[**BoyS2**] Mike Boyle and Michael Schraudner, $\mathbb{Z}^d$ *shifts of finite type without equal entropy full shift factors*, J. Difference Equ. Appl. **15** (2009), 47–52.

[**BoyS3**] Mike Boyle and Michael C. Sullivan, *Equivariant flow equivalence for shifts of finite type, by matrix equivalence over group rings*, Prof. London. Math. Soc. (3) **91** (2005), 184–214.

[**Bre**] Julien Bremont, *Gibbs measures at temperature zero*, Nonlinearity **16** (2003), 419–426.

[**Bri1**] Raimundo Briceño, *The topological strong spatial mixing property and new conditions for pressure approximation*, Ergodic Theor. Dyn. Syst. **38** (2018), 1658–1696.

[**Bri2**] Raimundo Briceño, *An SMB approach for pressure representation in amenable virtually orderable groups*, J. Anal. Math., to appear.

[**BriMP**] Raimundo Briceño, Kevin McGoff, and Ronnie Pavlov, *Factoring onto $\mathbb{Z}^d$ subshifts with the finite extension property*, Proc. Amer. Math. Soc. **146** (2018), 5129–5240.

[**Bru**] Horst Brunotte, *Algebraic properties of weak Perron numbers*, Tatra Mt. Math. Publ. **56** (2013), 27–33.

[**BurP**] Robert Burton and Robin Pemantle, *Local characteristics, entropy and limit theorems for spanning trees and domino tilings via transfer-impedances*, Ann. Prob. **21** (1993), 1329–1371.

[**BurS**] R. Burton and J. Steif, *Nonuniqueness of measures of maximal entropy for subshifts of finite type*, Ergodic Theor. Dyn. Syst. **14** (1994), 213–235.

[**Buz**] Jérôme Buzzi, *Maximal entropy measures for piecewise affine surface diffeomorphisms*, Ergodic Theor. Dyn. Syst. **29** (2009), 1723–1763.

[**CalH**] Frank Calegari and Zili Huang, *Counting Perron numbers by absolute value*, J. London Math. Soc. (2) **96** (2017), 181–200.

[**Cap**] D. Capocaccia, *A definition of Gibbs state for a compact set with $\mathbb{Z}^\nu$ action*, Commun. Math. Phys. **48** (1976), 85–88.

[**CapL**] Valerio Capraro and Martino Lupini, *Introduction to Sofic and Hyperlinear Groups and Connes' Embedding Conjecture*, Lecture Notes in Math. **2136**, Springer, Heidelberg, 2015.

[**CecC1**] Tullio Ceccherini-Silberstein and Michel Coornaert, *Cellular Automata and Groups*, Springer, New York, 2012.

[**CecC2**] Tullio Ceccherini-Silberstein and Michel Coornaert, *On the density of periodic configurations in strongly irreducible subshifts*, Nonlinearity **25** (2012), 2119–2131.

[**CecMS**] T. G. Ceccherini-Silberstein, A. Machì, and F. Scarabotti, *Amenable groups and cellular automata*, Ann. Inst. Fourier (Grenoble) **49** (1999), 673–685.

[**Cha**] Nishant Chandgotia, *Generalisation of the Hammersley–Clifford theorem on bipartite graphs*, Trans. Amer. Math. Soc. **369** (2017), 7107–7137.

[**ChaM**] Nishant Chandgotia and Tom Meyerovitch, *Markov random fields, Markov cocycles and the 3-colored chessboard*, Israel J. Math. **215** (2016), 909–964.

[**ChaCS**] J. T. Chayes, L. Chayes, and R. H. Schonmann, *Exponential decay of connectivities in the two-dimensional Ising model*, J. Stat. Phys. **49** (1987), 433–445.

[**ChaGU**] J.-R. Chazottes, J.-M. Gambaudo, and E. Ugalde, *Zero-temperature limit of one-dimensional Gibbs states via renormalization: the case of locally constant potentials*, Ergodic Theor. Dyn. Syst. **31** (2011), 1109–1161.

[**ChaH**] J.-R. Chazottes and M. Hochman, *On the zero-temperature limit of Gibbs states*, Commun. Math. Phys. **297** (2010), 265–281.

[**ChaK**] J.-R. Chazottes and G. Keller, *Pressure and Equilibrium States in Ergodic Theory*, Mathematics of Complexity and Dynamical Systems. Vols. 1–3, Springer, New York, 2012, 1422–1437.

[ChaR1] Yao-Ban Chan and Andrew Rechnitzer, *Accurate lower bounds on 2-D constraint capacities from corner transfer matrices*, IEEE Trans. Inform. Theory **60** (2014), 3845–3858.

[ChaR2] Yao-Ban Chan and Andrew Rechnitzer, *Upper bounds on the growth rates of independent sets in two dimensions via corner transfer matrices*, Linear Algebra Appl. **555** (2018), 139–156.

[ChaU1] J.-R. Chazottes and E. Ugalde, *Projection of Markov measures may be Gibbsian*, J. Stat. Phys. **111** (2003), 1245–1272.

[ChaU2] J.-R. Chazottes and E. Ugalde, *On the preservation of Gibbsianness under symbol amalgamation*, Entropy of Hidden Markov Processes and Connections to Dynamical Systems, LMS Lecture Note Ser. **385**, Cambridge University Press, Cambridge, 2011, pp. 72–97.

[CheHLL] Hung-Hsun Chen, Wen-Guei Hu, De-Jan Lai, and Song-Sun Lin, *Nonemptiness problems of Wang tiles with three colors*, Theoretical Comput Sci. **547** (2014), 34–45.

[Cip] Barry Cipra, *An introduction to the Ising model*, Amer. Math. Monthly **94** (1987), 937–959.

[ClmT1] Vaughn Climenhaga and Daniel J. Thompson, *Intrinsic ergodicity beyond specification: β-shifts, S-gap shifts, and their factors*, Israel J. Math. **192** (2012), 785–817.

[ClmT2] Vaughn Climenhaga and Daniel J. Thompson, *Equilibrium states beyond specification and the Bowen property*, J. Lond. Math. Soc. (2) **87** (2013), 401–427.

[Coh] David Bruce Cohen, *The large scale geometry of strongly aperiodic subshifts of finite type*, Adv. Math. **308** (2017), 599–626.

[CohEP] C. Cohen, N. Elkies, and J. Propp, *Local statistics for random domino tilings of the Aztec diamond*, Duke Math. J. **85** (1996), 117–166.

[CovN] Ethan M. Coven and Zbigniew Nitecki, *On the genesis of symbolic dynamics as we know it*, Colloq. Math. **110** (2008), 227–242.

[CyrFKP] Van Cyr, John Franks, Bryna Kra, and Samuel Petite, *Distortion and the automorphism group of a shift*, J. Mod. Dyn. **13** (2018), 147–161.

[CyrK1] Van Cyr and Bryna Kra, *Nonexpansive $\mathbb{Z}^2$-subdynamics and Nivat's Conjecture*, Trans. Amer. Math. Soc. **367** (2015), 6487–6537.

[CyrK2] Van Cyr and Bryna Kra, *The automorphism group of a shift of linear growth: beyond transitivity*, Forum Math. Sigma **3** (2015), e5, 27pp.

[CyrK3] Van Cyr and Bryna Kra, *The automorphism group of a shift with subquadratic growth*, Proc. Amer. Math. Soc. **144** (2016), 513–621.

[CyrK4] Van Cyr and Bryna Kra, *Complexity of short rectangles and periodicity*, J. Eur. Comb. **52** (2016), 146–173.

[CyrK5] Van Cyr and Bryna Kra, *The automorphism group of a minimal shift of stretched exponential growth*, J. Mod. Dyn. **10** (2016), 483–495.

[CyrK6] Van Cyr and Bryna Kra, *The automorphism group of a shift of slow growth is amenable*, Ergodic Theor. Dyn. Syst. (to appear).

[Dav] Diana Davis, *Cutting sequences on translation surfaces*, New York J. Math. **20** (2014), 399–429.

[Den] Christopher Deninger, *Fuglede-Kadison determinants and entropy for actions of discrete amenable groups*, J. Amer. Math. Soc. **19** (2006), 737–758.

[Des1] Angela Desai, *Subsystem entropy for $\mathbb{Z}^d$-sofic shifts*, Indag. Mathem. **17** (2006), 353–359.

[Des2] Angela Desai, *A class of $\mathbb{Z}^d$ shifts of finite type which factors onto lower entropy full shifts*, Proc. Amer. Math. Soc. **137** (2009), 2613–2621.

[Dob] R. Dobrushin, *Description of a random eld by means of conditional probabilities and conditions for its regularity*, Teor. Verojatnost. i Primenen **13** (1968), 201–229.

[DonDMP] Sebastian Donoso, Fabien Durand, Alejandro Maass, and Samuel Petite, *On automorphism groups of Toeplitz subshifts*, Discrete Anal. (2017), Paper no. 11, 19 pp.

[Dow] Tomasz Downarowicz, *Entropy in Dynamical Systems*, New Mathematical Monographs **18**, Cambridge University Press, Cambridge, 2011.

[DurP] Fabien Durand and Dominique Perrin, *Dimension Groups and Dynamical Systems*, Cambridge University Press (to appear).

[DurRS] Bruno Durand, Andrei Romaschenko, and Alexander Shen, *Fixed-point tile sets and their applications*, J. Comput. System Sci. **78** (2012), 731–764.

[Dye] H. Dye, *On groups of measure preserving transformations I*, Amer. J. Math. **81** (1959), 119–159; *II* **85** (1963), 551–576.

[EilK] Søren Eilers and Ian Kiming, *On some new invariants for shift equivalence for shifts of finite type*, J. Number Theory **132** (2012), 502–510.

[EilRRS] Søren Eilers, Gunnar Restorff, Efren Ruiz, and Adam P. W. Sørensen, *Geometric classification of graph C^*-algebras over finite graphs*, Canad. J. Math. **70** (2018), 294–353.

[Ein] Manfred Einsiedler, *Fundamental cocycles and tiling spaces*, Ergodic Theor. Dyn. Syst. **21** (2001), 777–800.

[EinL] Manfred Einsiedler and Douglas Lind, *Algebraic $\mathbb{Z}^d$-actions of entropy rank one*, Trans. Amer. Math. Soc. **356** (2004), 1799–1831.

[EinS] M. Einsiedler and K. Schmidt, *Markov partitions and homoclinic points of algebraic $\mathbb{Z}^d$-actions*, Proc. Steklov Inst. Math. **1997** (216), 259–279.

[EliMS] O. Elishco, T. Meyerovitch, and M. Schwartz, *Semiconstrained systems*, IEEE Trans. Inform. Theory **62** (2016), 1688–1702.

[EntR] A. C. D. van Enter and W. M. Ruszel, *Chaotic temperature dependence at zero temperature*, J. Stat. Phys. **2007** (127), 567–573.

[EpiKM] Chiara Epifanio, Michel Koskas, and Filippo Mignosi, *On a conjecture in bidimensional words*, Theoret. Comput. Sci. **299** (2003), 123–150.

[ExeL] Ruy Exel and Marcelo Laca, *The K-theory of Cuntz-Krieger algebras for infinite matrices*, J. Reine Angew. Math. **512** (1999), 119–172.

[FreOP] Thomas French, Nic Ormes, and Ronnie Pavlov, *Subshifts with slowly growing numbers of follower sets*, Contemp. Math. **678** (2016), 192–203.

[Frie] David Fried, *Ideal tilings and symbolic dynamics for negatively curved surfaces*, Ergodic Theor. Dyn. Syst. **31** (2011), 1697–1726.

[Fri] Shmuel Friedland, *On the entropy of $\mathbb{Z}^d$ subshifts of finite type*, Linear Algebra Appl. **252** (1997), 199–220.

[FriLM] Shmuel Friedland, Per Håkan Lundlow, and Klas Markström, *The 1-vertex transfer matrix and accurate estimation of channel capacity*, IEEE Trans. Inform. Theory **56** (2010), 3692–3699.

[FriP] Shmuel Friedland and Uri Peled, *The pressure, densities and first-order phase transitions associated with multidimensional SOFT*, Trends in Mathematics, BirkhuserSpringer, Basel, 2011, pp. 179–220.

[FriST] Joshua Frisch, Tomer Schlank, and Omer Tamuz, *Normal amenable subgroups of the automorphism group of the full shift*, Ergodic Theor. Dyn. Syst. **39** (2019), 1290–1298.

[Fro] Rafael Frongillo, *Optimal state amalgamation is NP-hard*, Ergodic Theor. Dyn. Syst. **39** (2019), 1857–1869.

[GamK] David Gamarnik and Dmitriy Katz, *Sequential cavity method for computing free energy and surface pressure*, J. Stat. Phys. **137** (2009), 205–232.

[GanH] Silvère Gangloff and Benjamin Hellouin de Menibus, *Effect of quantified irreducibility on the computability of subshift entropy*, Discrete Contin. Dyn. Syst. **39** (2019), 1975–2000.

[GaoJS1] Su Gao, Steve Jackson, and Brandon Seward, *A coloring property for countable groups*, Math. Proc. Camb. Philos. Soc. **147** (2009), 579–592.

[GaoJS2] Su Gao, Steve Jackson, and Brandon Seward, *Group Colorings and Bernoulli Subflows*, Mem. Amer. Math. Soc. **241**, American Mathematical Society, 2016.

[GeoHM] Hans-Otto Georgii, Olle Häggström, and Christian Maes, *The random geometry of equilibrium phases*, Phase Transit. Crit. Phenom. **18**, Academic Press, San Diego, 2001, pp. 1–142.

[GioMPS] Thierry Giordano, Hiroki Matui, Ian Putnam, and Christian Skau, *Orbit equivalence for Cantor minimal $\mathbb{Z}^d$-systems*, Invent. Math. **179** (2010), 119–158.

[GioPS] Thierry Giordano, Ian Putnam, and Christian Skau, *Topological orbit equivalence and C^*-crossed products*, J. Reine Angew. Math. **469** (1995), 51–111.

[Got] Walter Gottschalk, *Some general dynamical notions*, Recent Advances in Topological Dynamics, Lecture Notes in Math. **138**, Springer, Berlin, pp. 120–125.

[Gri] Geoffrey Grimmett, *Probability on Graphs*, 2nd Edition, Cambridge University Press, Cambridge, 2018.

[GriM] Rostislav I. Grigorchuk and Konstantin Medynets, *On algebraic properties of topological full groups*, arXiv:1105.0719v4.

[Gro] M. Gromov, *Endomorphisms of symbolic algebraic varieties*, J. Eur. Math. Soc. **1** (1999), 109–197.

[HamC] J. M. Hammersley and P. Clifford, *Markov fields on finite graphs and lattices*, 1968, www.statslab.cam.ac.uk/~grg/books/hammfest/hamm-cliff.pdf.

[Hay] Ben Hayes, *Fuglede-Kadison determinants and sofic entropy*, Geom. Funct. Anal. **26** (2016), 520–606.

[Hoc1] Michael Hochman, *On the dynamics and recursive properties of multidimensional symbolic systems*, Invent. Math. **176** (2009), 131–167.

[Hoc2] Michael Hochman, *On the automorphism groups of multidimensional shifts of finite type*, Ergodic Theor. Dyn. Syst. **30** (2010), 809–840.

[Hoc3] Michael Hochman, *Non-expansive directions for $\mathbb{Z}^2$ actions*, Ergodic Theor. Dyn. Syst. **31** (2011), 91–112.

[Hoc4] Michael Hochman, *Multidimensional shifts of finite type and sofic shifts*, Combinatorics, Words and Symbolic Dynamics (Valérie Berthé and Michel Rigo, eds.), Cambridge University Press, Cambridge, 2016, pp. 296–358.

[HocM] Michael Hochman and Tom Meyerovitch, *A characterization of the entropies of multidimensional shifts of finite type*, Ann. Math. **171** (2010), 2011–2038.

[Hon] S. Hong, *Loss of Gibbs property in one-dimensional mixing shifts of finite type*, Qual. Theory of Dynam. Sys. **19** (2020), 21pp.

[HosP] B. Host and F. Parreau, *Homomorphismes entre systèmes dynamiques définis par substitutions*, Ergodic Theor. Dyn. Syst. **9** (1989), 469–477.

[Imm1] K. A. S. Immink, *EFMPlus, 8–16 modulation code*, United States Patent Number 5,696,505.

[Imm2] K. A. S. Immink, *Codes for Mass Data Storage Systems*, 2nd Edition, Shannon Foundation Publishers, Eindhoven, The Netherlands, 2004.

[Jea1] Emmanuel Jeandel, *Computability in symbolic dynamics*, Pursuit of the Universal (Lecture Notes in Computer Science), Springer, 2016, pp. 124–131.

[Jea2] Emmanuel Jeandel, *Aperiodic subshifts on nilpotent and polycyclic groups*, arXiv:1510.02360v2.

[Jea3] Emmanuel Jeandel, *Aperiodic subshifts of finite type on groups*, arXiv:1501.06831.

[Jea4] Emmanuel Jeandel, *Strong shift equivalence as a categorical notion*, Preprint (2020), 23 pp.

[JeaV1] Emmanuel Jeandel and Pascal Vanier, *Hardness of conjugacy, embedding and factorization of multidimensional subshifts*, J. Comput. System Sci. **81** (2015), 1648–1664.

[JeaV2] Emmanuel Jeandel and Pascal Vanier, *Characterizations of periods of multi-dimensional shifts*, Ergodic Theor. Dyn. Syst. **35** (2015), 431–460.

[JohM1] Aimee Johnson and Kathleen Madden, *The decomposition theorem for two dimensional shifts of finite type*, Proc. Amer. Math. Soc. **127** (1999), 1533–1543.

[JohM2] Aimee Johnson and Kathleen Madden, *Factoring higher dimensional shifts of finite type onto the full shift*, Ergodic Theor. Dyn. Syst. **25** (2005), 811–822.

[JusM] Kate Juschenko and Nicolas Monod, *Cantor systems, piecewise translations and simple amenable groups*, Ann. of Math. (2) **178** (2013), 775–787.

[Kar] Jarkko Kari, *Low-complexity tilings of the plane*, Lecture Notes in Comput. Sci., Springer, Cham, 2019.

[KarM] Jarkko Kari and Etienne Moutot, *Nivat's conjecture and pattern complexity in algebraic subshifts*, Theoret. Comput. Sci. **777** (2019), 379–386.

[KarS] Jarkko Kari and Michal Szabados, *An algebraic geometric approach to Nivat's conjecture*, Inf. and Comp. **271** (2020), 104481.

[Kat] Anatole Katok, *Fifty years of entropy in dynamics: 1958–2007*, J. Mod. Dyn. **1** (2007), 545–596.

[KatU] Svetlana Katok and Ilie Ugarcovici, *Symbolic dynamics for the modular surface and beyond*, Bull. Amer. Math. Soc. (N.S.) **44** (2007), 87–132.

[Kel] Gerhard Keller, *Equilibrium States in Ergodic Theory*, London Mathematical Society Student Texts, Cambridge University Press, 1998.

[Kem] T. Kempton, *Factors of Gibbs measures for subshifts of finite type*, Bull. London Math. Soc. **43** (2011), 751–764.

[Ken] Richard Kenyon, *The construction of self-similar tilings*, Geom. Funct. Anal. **6** (1996), 471–488.

[KenS] Richard Kenyon and Boris Solomyak, *On the characterization of expansion maps for self-affine tilings*, Discrete Comput. Geom. **43** (2010), 577–593.

[KenV] Richard Kenyon and Anatoly Vershik, *Arithmetic construction of sofic partitions of hyperbolic toral automorphisms*, Ergodic Theor. Dyn. Syst. **18** (1998), 357–372.

[KerL] David Kerr and Hanfeng Li, *Ergodic Theory: Independence and Dichotomies*, Springer, 2016.

[Kor] Ivan Korec, *Irrational speeds of configuration growth in generalized Pascal triangles*, Theoret. Computer Sci. **112** (1993), 399–412.

[Kri] Wolfgang Krieger, *On images of sofic systems*, arXiv:1101.1750v2.

[Li] Hanfeng Li, *Compact group automorphisms, addition formulas and Fuglede-Kadison determinants*, Ann. of Math. (2) **176** (2012), 303–347.

[LiT] Hanfeng Li and Andreas Thom, *Entropy, determinants, and L^2-torsion*, J. Amer. Math. Soc. **27** (2014), 239–292.

[Lie] E. H. Lieb, *Exact solution of the problem of the entropy of two-dimensional ice*, Phys. Rev. Lett. **18** (1967), 692–694.

[Lig1] Sam Lightwood, *Morphisms form nonperiodic $\mathbb{Z}^2$ subshifts I*, Ergodic Theor. Dyn. Syst. **23** (2003), 587–609.

[Lig2] Sam Lightwood, *Morphisms form nonperiodic $\mathbb{Z}^2$ subshifts II*, Ergodic Theor. Dyn. Syst. **24** (2004), 1227–1260.

[LimS] Yuri Lima and Omri Sarig, *Symbolic dynamics for three dimensional flows with positive topological entropy*, J. Eur. Math. Soc. **21** (2019), 199–256.

[Lin] Douglas Lind, *Multi-dimensional symbolic dynamics*, Symbolic Dynamics and its Applications, Proc. Sympos. Applied Mathematics, American Mathematical Society, 2004, pp. 61–79.

[**LinS1**] Douglas Lind and Klaus Schmidt, *Homoclinic points of algebraic $\mathbb{Z}^d$-actions*, J. Amer. Math. Soc. **12** (1999), 953–980.

[**LinS2**] Douglas Lind and Klaus Schmidt, *A survey of algebraic actions of the discrete Heisenberg group*, Russ. Math. Surv. **70** (2015), 657–714.

[**LinS3**] Douglas Lind and Klaus Schmidt, *A Bernoulli algebraic action of a free group*, arXiv:1905.09966v1.

[**Mah**] K. Mahler, *An application of Jensen's formula to polynomials*, Mathematika **7** (1960), 98–100.

[**MarP**] Brian Marcus and Ronnie Pavlov, *An integral representation for topological pressure in terms of conditional probabilities*, Israel J. Math. **207** (2015), 395–433.

[**Mat1**] K. Matsumoto, *Presentations of subshifts and their topological conjugacy invariants*, Documenta Math. **4** (1999), 285–340.

[**Mat2**] Nicolás Matte Bon, *Subshifts with slow complexity and simple groups with the Liouville property*, Geom. Frunct. Anal. **24** (2014), 1637–1659.

[**Mat3**] Hiroki Matui, *Some remarks on topological full groups of Cantor minimal systems*, Int. J. of Math. **17** (2006), 231–251.

[**McgP**] Kevin McGoff and Ronnie Pavlov, *Factor maps and embeddings for random $\mathbb{Z}^d$ shifts of finite type*, Israel J. Math. **230** (2019), 239–273.

[**MeeS**] Ronald Meester and Jeffrey Steif, *Higher dimensional subshifts of finite type, factor maps and measures of maximal entropy*, Pacific J. Math. **200** (2001), 497–510.

[**Mey**] T. Meyerovitch, *Gibbs and equilibrium measures for some families of subshifts*, Ergodic Theor. Dyn. Syst. **33** (2013), 934–953.

[**Mil**] John Milnor, *On the entropy geometry of cellular automata*, Complex Systems **2** (1988), 357–386.

[**MosRW**] Michael J. Mossinghoff, Georges Rhin, and Qiang Wu, *Minimal Mahler measures*, Experiment. Math. **17** (2008), 451–458.

[**Nas**] M. Nasu, *Textile systems and one-sided resolving automorphisms and endomorphisms of the shift*, Ergodic Theor. Dyn. Syst. **28** (2008), 167–209.

[**NowW**] Richard Nowakowski and Peter Winkler, *Vertex-to-vertex pursuit in a graph*, Discrete Math. **43** (1983), 235–239.

[**Ons**] L. Onsager, *Crystal statistics. I. A two-dimensional model with an order-disorder transition*, Phys. Rev. (2) **65** (1944), 117–149.

[**OrmP**] Nic Ormes and Ronnie Pavlov, *Extender sets and multidimensional subshifts*, Ergodic Theor. Dyn. Syst. **36** (2016), 908–923.

[**OrnW**] Donald Ornstein and Benjamin Weiss, *Entropy and isomorphism theorems for actions of amenable groups*, J. Anal. Math. **48** (1987), 1–141.

[**Pav1**] Ronnie Pavlov, *Approximating the hard square entropy constant with probabilistic methods*, Ann. Prob. **40** (2012), 2362–2399.

[**Pav2**] Ronnie Pavlov, *On non-uniform specification and uniqueness of the equilibrium state in expansive systems*, Nonlinearity **32** (2019), 2441–2460.

[**PavS**] Ronnie Pavlov and Michael Schraudner, *Entropies realizable by block gluing shifts of finite type*, J. Anal. Math. **126** (2015), 113–174.

[**Pes**] Vladimir G. Pestov, *Hyperlinear and sofic groups: a brief guide*, Bull. Symbolic Logic **14** (2008), 449–480.

[**Pia**] Steven T. Piantadosi, *Symbolic dynamics on free groups*, Discrete Contin. Dyn. Syst. **20** (2008), 725–738.

[**Pir1**] M. Piraino, *Projections of Gibbs states for Hölder potenials*, J. Stat. Phys. **170** (2018), 952–961.

[**Pir2**] M. Piraino, *Single site factors of Gibbs measures*, Nonlinearity **33** (2020), 742–761.

[PolK] M. Pollicott and T. Kempton, *Factors of Gibbs measures for full shifts*, Entropy of Hidden Markov Processes and Connections to Dynamical Systems, LMS Lecture Note Ser. **385**, Cambridge University Press, Cambridge, 2011, pp. 246–257.

[Put] Ian Putnam, *Cantor Minimal Systems*, University Lecture Series **70**, American Mathematical Society, 2018.

[QuaT] Anthony Quas and Paul Trow, *Subshifts of multidimensional shifts of finite type*, Ergodic Theor. Dyn. Syst. **20** (2000), 859–874.

[QuaZ] Anthony Quas and Luca Zamboni, *Periodicity and local complexity*, Theoret. Comput. Sci. **319** (2004), 229–240.

[RadS] Charles Radin and Lorenzo Sadun, *Isomorphism of hierarchical structures*, Ergodic Theor. Dyn. Syst. **21** (2001), 1239–1248.

[Rob] E. Arthur Robinson, Jr., *Symbolic dynamics and tilings in $\mathbb{R}^d$*, Symbolic Dynamics and its Applications, Proceedings of Symposia in Applied Mathematics, American Mathematical Society, 2004, pp. 81–119.

[RotS] R. Roth and P. Siegel, *On parity-preserving constrained coding*, Proc. Internat. Sympos. Information Theory (2018), IEEE, Piscataway, New Jersey, 1804–1808.

[Rue] D. Ruelle, *Thermodynamic Formalism*, 2nd Edition, (Update of 1st Edition from 1978), Cambridge University Press, Cambridge, 2004.

[Sab] Mathieu Sablik, *Directional dynamics for cellular automata: A sensitivity to initial condition approach*, Theoret. Comput. Sci. **400** (2008), 1–18.

[Sal] Ville Salo, *Toeplitz subshift whose automorphism group is not finitely generated*, Colloq. Math. **146** (2017), 53–76.

[SalT] Ville Salo and Ilkka Törmä, *Block maps between primitive uniform and Pisot substitutions*, Ergodic Theor. Dyn. Syst. **35** (2015), 2292–2310.

[SanT] J. W. Sander and R. Tijdeman, *The rectangle complexity of functions on two-dimensional lattices*, Theoret. Comput. Sci. **270** (2002), 857–863.

[Sar1] Omri Sarig, *Symbolic dynamics for surface diffeomorphisms with positive entropy*, J. Amer. Math. Soc. **26** (2013), 341–426.

[Sar2] Omri Sarig, *Thermodynamic formalism for countable Markov shifts*, Proc. Sympos. Pure Math. **89**, American Mathematical Society, Providence, R.I., (2015), 81–117.

[Schi] Andrzej Schinzel, *A class of algebraic numbers*, Tatra Mt. Math. Publ. **11** (1997), 35–42.

[Sch1] Klaus Schmidt, *Tilings, fundamental cocycles and fundamental groups of symbolic $\mathbb{Z}^d$-actions*, Ergodic Theor. Dyn. Syst. **18** (1998), 1473–1525.

[Sch2] Klaus Schmidt, *Quotients of $l^\infty(\mathbb{Z},\mathbb{Z})$ and symbolic covers of toral automorphisms*, Representation Theory, Dynamical Systems, and Asymptotic Combinatorics, Amer. Math. Soc. Transl. Ser. 2, **217**, 2006, pp. 223–246.

[Sch3] Klaus Schmidt, *Representations of toral automorphisms*, Topology Appl. **205** (2016), 88–116.

[Schr] Michael Schraudner, *A matrix formalism for conjugacies of higher-dimensional shifts of finite type*, Colloq. Math. **110** (2008), 493–515.

[SchY] Scott Schmieding and Kitty Yang, *The mapping class group of a minimal shift*, arXiv:1810.08847.

[Sep] Timo Seppäläinen, *Entropy, limit theorems, and variational principles for disordered lattice systems*, Commun. Math. Phys. **171** (1995), 233–277.

[SidV] N. Sidorov and A. Vershik, *Bijective arithmetic codings of hyperbolic automorphisms of the 2-torus, and binary quadratic forms*, J. Dynam. Control Systems **4** (1998), 365–399.

[SilW1] Daniel S. Silver and Susan G. Williams, *Knot invariants from symbolic dynamical systems*, Trans. Amer. Math. Soc. **351** (1999), 3243–3265.

[SilW2] Daniel S. Silver and Susan G. Williams, *An invariant of finite group actions on shifts of finite type*, Ergodic Theor. Dyn. Syst. **25** (2005), 1985–1996.

[Sim] Stephen G. Simpson, *Medvedev degrees of two-dimensional subshifts of finite type*, Ergodic Theor. Dyn. Syst. **34** (2014), 679–688.

[SmiU] John Smillie and Corinna Ulcigrai, *Beyond Sturmian sequences: coding linear trajectories in the regular octagon*, Proc. Lond. Math. Soc. (3) **102** (2011), 291–340.

[Tho] Klaus Thomsen, *On the structure of a sofic shift space*, Trans. Amer. Math. Soc. **356** (2004), 3557–3619.

[Thu1] William Thurston, *Groups, Tilings, and Finite State Automata*, AMS Colloquium Lecture Notes, Boulder, CO, 1989.

[Thu2] William P. Thurston, *Entropy in dimension one*, Frontiers in Complex Dynamics, Princeton Math. Ser. **51**, Princeton Univ. Press, Princeton, NJ, 2014, pp. 339–384.

[Verb] Evgeny Verbitskiy, *Thermodynamics of Hidden Markov Processes*, Entropy of Hidden Markov Processes and Connections to Dynamical Systems, LMS Lecture Note Ser. **385**, Cambridge Univ. Press, Cambridge, 2011, pp. 258–272.

[Vers] A. M. Vershik, *Arithmetic isomorphism of hyperbolic automorphisms of a torus and of sofic shifts*, Funktsional. Anal. i Prilozhen. **26** (1992), 22–27.

[Vol] Mikhail Volkov, *Synchronizing automata and the Černý Conjecture*, Proc. 2nd Intl. Conf. Language and Automata Theory and Applications, Springer-Verlag Lecture Notes in Computer Science **5196**, 2008, 11–27.

[Wag1] J. B. Wagoner, *Strong shift equivalence theory and the shift equivalence problem*, Bull. Amer. Math. Soc. (N.S.) **36** (1999), 271–296.

[Wag2] J. B. Wagoner, *Strong shift equivalence theory*, Symbolic Dynamics and its Applications, Proc. Sympos. Applied Mathematics, American Mathematical Society, 2004, pp. 121–154.

[Wan1] Hao Wang, *Proving theorems by pattern recognition II*, Bell Syst. Tech. J. **40** (1961), 1–41.

[Wan2] Hao Wang, *Notes on a class of tiling problems*, Fund. Math. **82** (1974), 295–305.

[Weis] Benjamin Weiss, *Sofic groups and dynamical systems*, Sankhya Ser. A **62** (2000), 350–359.

[Weit] Dror Weitz, *Counting independent sets up to the tree threshold*, Ann. Sympos. Theory of Computing (STOC) **38** (2006), 140–149.

[Yay] Y. Yayama, *On factors of Gibbs measures for almost additive potentials*, Ergodic Theor. Dyn. Syst. **36** (2016), 276–309.

[Yaz] Mehdi Yazdi, *Lower bound for the Perron-Frobenius degrees of Perron numbers*, Ergodic Theor. Dyn. Syst. (to appear).

[Yoo] Jisang Yoo, *Decomposition of infinite-to-one factor codes and uniqueness of relative equilibrium states*, J. Modern Dynamics **13** (2018), 271–284.

[Yuz] S. A. Yuzvinskii, *Calculation of the entropy of a group-endomorphism (Russian)*, Sibirsk. Mat. Z. **8** (1967), 230–239, Engl. transl. Sib. Math. J. **8**, 172–178 (1968).

[Zin] Charalampos Zinoviadis, *Hierarchy and expansiveness in two-dimensional subshifts of finite type*, Ph. D. dissertation (2016), arXiv:1603.05464.

[Zor] Anton Zorich, *Flat surfaces*, Frontiers in Number Theory, Physics, and Geometry I, Springer, Berlin, 2006, pp. 437–583.

NOTATION INDEX

INDEX

Page numbers where an entry is defined are underlined, those where the entry occurs in a theorem, proposition, etc. are in **bold italic**, and those where the entry occurs in an exercise are in *regular italic*.

A

adjacency matrix, 35
  symbolic, 65
adjugate matrix, 113
algebraic conjugate, 370
algebraic integer, 370
  degree of, 370
algebraic number, 370
almost conjugacy, 323
almost finite type shift, 433
almost flow equivalence, 455
almost invertible factor code, 314
almost Markov shift, 433
almost periodic point, 456
almost sofic shift, 451
alphabet, 1
amalgamation code, 218
amalgamation matrix, 278
approximate eigenvector, 151
Approximate Eigenvector Algorithm, *154*, *156*
automorphism, *22*, 435
  inert, 438
automorphism group, 435

B

Baire Category Theorem, **181**
basis
  integral, 398
beam, 257
Bernoulli process, 447
bi-closing code, 313
bi-resolving code, 313
binary sequence, 2
block, 2
  central, 3
  concatenation of, 3
  empty, 2
  forbidden, 5
  length of, 2
  occurring in a point, 5
  period of, 342

presentation of, 66
progressive overlap of, 12
block map, 15
  increasing window size of, 16
Bolzano–Weierstrass Theorem, **180**
bounded set, 180
Bowen–Franks group, 248
  generalized, 250
bubble, 269, 507

C

Cantor diagonal argument, 20
Cantor minimal system, 475
characteristic polynomial, 100
charge constrained shift, 7, *11*
circle rotation, 185
class degree, 477
Classification Theorem, **230**
closed set, 178
closed subgroup, 397
closure of set, 181
co-prime, 419
code
  amalgamation, 55
  block, 23
  convolutional, 24
  finite-state (X,Y), 365
  finite-to-one, 266
  sliding block, *see* sliding block code
  splitting, 55
coded system, 449
communicating classes, 119
compactness, 177
companion matrix, 373
complete splitting, 58
concatenation, 3
conjugacy, 18
  elementary, 232
  eventual, 260
  weak, 414
conjugate, 370
constant-to-one code, 313
context free shift, 7
context-free language, 68
context-free shift
  follower sets in, 73
  not sofic, 68
continuity, 177
continuous flow, 452